Oscar Zariski: Collected Papers
Volume I
Foundations of Algebraic Geometry and Resolution of Singularities

Mathematicians of Our Time Series

Gian-Carlo Rota, Editor

1. Carles Loewner: Theory of Continuous Groups
   Notes by Harley Flanders and Murray H. Protter

2. Oscar Zariski: Collected Papers
   Volume I
   Foundations of Algebraic Geometry and Resolution of Singularities
   Edited by H. Hironaka and D. Mumford

Oscar Zariski

Collected Papers

Volume I
Foundations of Algebraic Geometry
and Resolution of Singularities

Edited by
H. Hironaka and D. Mumford

The MIT Press
Cambridge, Massachusetts, and London, England

This book was designed by The MIT Press Design Department.
It was printed on Finch Textbook Offset
by The Maple Press Company
and bound by The Maple Press Company
in the United States of America.

Library of Congress Cataloging in Publication Data

Zariski, Oscar, 1899-
Collected papers.

(Mathematicians of our time, v. 2)
CONTENTS: v. 1. Foundations of algebraic geometry and resolution of singularities.
Bibliography: p.
1. Mathematics—Collected works. I. Series.
QA3.Z37 510'.8 73-171558
ISBN 0-262-08049-4

Oscar Zariski

Contents

(Bracketed numbers are from the Bibliography)

Part II
Resolution of Singularities

Reprints of Papers

Preface

The series "Mathematicians of Our Time" embraces, at least in principle, the works of living mathematicians; therefore, the term "collected works," as applied to this series stands of necessity for an open-ended entity, because the author—contrary to the old cliché that "mathematics is a young man's game"—may still be actively engaged in research and therefore continue to produce papers while the "collected" works is being printed. Thus, in my case, the bibliography of papers that appears in the first volume does not include two papers, one in course of publication and another in preparation. (These two papers form, together with the last paper [89] of the list printed in the first volume, a sequence of three papers under the common title "General theory of saturation and saturated local rings.") At any rate, the present plans envisage the publication of four volumes. These will include all my published works, with the following exceptions (the numbers in brackets refer to the bibliography as printed in this volume):

1. Books [6,25,72,75].
2. Lecture notes [87].
3. Expository articles in fields to which I have made no original contribution myself [1,3,5,11]. All these articles deal with the foundations of set theory, and I wrote them in my early postgraduate years in Rome at the urging and with the encouragement of my teacher F. Enriques, whose primary interest at that time was in the philosophy and history of science and who was editor of a series of books entitled "Per la Storia e la Filosofia delle Matematiche." As the reader can see, the book listed under [6] was published in that series.

The editorial preparation and the writing of introductions to each volume is entrusted to the capable hands of younger men, who are experts in the field of algebraic geometry and who at one time or another have been either my students at Harvard or have been closely associated with me in some capacity at Harvard or elsewhere. Thus, the editors of the first volume, H. Hironaka and D. Mumford, are truly leaders in the field of algebraic geometry and have studied at Harvard.

While all the papers printed in these collected works belong, without exception, to algebraic geometry, the reader will undoubtedly notice that beginning with the year 1937 the nature of my work underwent a radical change. It became strongly *algebraic* in character, both as to methods used and as to the very formulation of the problem studied (these problems, nevertheless, always have had, and never ceased to have in my mind, their origin and motivation in algebraic *geometry*). A few words on how this change came about may be of some interest to the reader. When I was nearing the age of 40, the circumstances that led me to this radical change of direction in my research (a change that marked the beginning of what was destined to become my chief contribution to algebraic geometry) were in part personal in character, but chiefly they had to do with the objective situation that prevailed in algebraic geometry in the 1930s.

In my early studies as a student at the University of Kiev in the Ukraine, I was interested in algebra and also in number theory (by tradition, the latter subject is strongly cultivated in Russia). When I became a student of the University of Rome in 1921, algebraic geometry reigned supreme in that university. I had the great fortune of finding there on the faculty three great mathematicians, whose very names now symbolize and are identified with classical algebraic geometry: G. Castelnuovo, F. Enriques, and F. Severi. Since even within the classical framework of algebraic geometry the algebraic background was clearly in evidence, it was inevitable that I should be attracted to that field. For a long time, and in fact for almost ten years *after* I left Rome in 1927 for a position at the Johns Hopkins University in Baltimore, I felt quite happy with the kind of "synthetic" (an adjective dear to my Italian teachers) geometric proofs that constituted the very life stream of classical algebraic geometry (Italian style). However, even during my Roman period, my algebraic tendencies were showing and were clearly perceived by Castelnuovo, who once told me: "You are here with us but are not one of us." This was said not in reproach but good-naturedly, for Castelnuovo himself told me time and time again that the methods of the Italian geometric school had done all they could do, had reached a dead end, and were inadequate for further progress in the field of algebraic geometry. It was with this perception of my algebraic inclination that Castelnuovo suggested to me a problem for my doctoral dissertation, which was closely related to Galois theory (see [2 and 12]).

Both Castelnuovo and Severi always spoke to me in the highest possible terms of S. Lefschetz's work in algebraic geometry, based on topology; they both were of the opinion that topological methods would play an increasingly important role in the development of algebraic geometry. Their views, very amply justified by future developments, have strongly influenced my own work for some time. This explains the topological trend in my work during the period 1929 to 1937 (see [15,16,17,20,22,27,28,29]). During that period I made frequent trips from Baltimore to Princeton to talk to and consult with Lefschetz, and I owe a great deal to him for his inspiring guidance and encouragement.

The breakdown (or the breakthrough, depending on how one looks at it) came when I wrote my Ergebnisse monograph *Algebraic Surfaces* [25]. At that time (1935) modern algebra had already come to life (through the work of Emmy Noether and the important treatise of B. L. van der Waerden), but while it was being applied to some aspects of the foundations of algebraic geometry by van der Waerden, in his series of papers "*Zur algebraischen Geometrie,*" the deeper aspects of *birational* algebraic geometry (such as the problem of reduction of singularities, the properties of fundamental loci and exceptional varieties of birational transormation, questions pertaining to complete linear systems and complete "continuous" systems of curves on surfaces, and so forth) were largely, or even entirely, virgin territory as far as algebraic exploration was concerned. In my Ergebnisse monograph I tried my best to present the underlying ideas of the ingenious geometric methods and proofs with which the Italian geometers were handling these deeper aspects of the whole theory of surfaces, and in all probability I succeeded, but at a price. The price was my own personal loss of the geometric paradise in which I was living so happily theretofore. I began to feel distinctly unhappy about the rigor of the original proofs I was trying to sketch (without losing in the least my admiration for the imaginative geometric spirit that permeated these proofs); I became convinced that the whole structure must be done over again by purely algebraic methods. After spending a couple of years just studying modern algebra, I had to begin somewhere, and it was not by accident that I began with the problem of local uniformization and reduction of singularities. At that time there appeared the Ergebnisse monograph *Idealtheorie* of W. Krull, emphasizing valuation theory and the concept of integral dependence and integral closure. Krull said somewhere in his monograph that the *general* concept of valuation (including, therefore, nondiscrete valuations and valuations of rank > 1) was not likely to have applications in algebraic geometry. On the contrary, after some trial tests (such as the valuation-theoretic analysis of the notion of infinitely near base points; see title [35]), I felt that this concept could be extremely useful for the analysis of singularities and for the problem of reduction of singularities. At the same time I noticed some promising connections between integral closure and complete linear systems; a systematic study of these connections later led me to the notions of normal varieties and normalization. However, I also concluded that this program could be successful only provided that much of the preparatory work be done for ground fields that are not algebraically closed. I restricted myself to characteristic zero: for a short time, the quantum jump to $p \neq 0$ was beyond the range of either my intellectual curiosity or my newly acquired skills in algebra; but it did not take me too long to make that jump; see for instance [48,49,50] published in 1943 to 1947.

I carried out this initial program of work primarily in the four papers [37,39,40,41] published in 1939 and 1940. From then on, for more than 30 years, my work ranged over a wide variety of topics in algebraic geometry. It is not my intention here, nor is it the purpose of this preface, to brief the reader on the nature of these topics and the results obtained or the manner in which my papers can be grouped together in various categories, according to the principal topics treated. This is the task of the editors of the various volumes. I will say only a few words about the present (first) volume.

The papers collected in this volume are divided in two groups: (1) *foundations,* meaning primarily properties of normal varieties, linear systems, birational transformations, and so on, and (2) *local uniformization and resolution of singularities.* These two subdivisions correspond precisely to the twofold aim I set to myself in my first concerted attack on algebraic geometry by purely algebraic methods—an undertaking and a state of mind about which I have already said a few words earlier. As a matter of fact, of the four main papers that I mentioned earlier as being the chief fruit of my first huddle with modern algebra and its applications to algebraic geometry, exactly two [37 and 40] belong to "foundations," while the other two [39, 41] belong to the category "resolution of singularities and local uniformization."

In 1950 I gave a lecture at the International Congress of Mathematicians at Harvard; the title of that lecture was "The fundamental ideas of abstract algebraic geometry" [60]. This is a good illustration of how relative in nature is what we call "abstract" at a given time. Certainly that lecture was very "abstract" for *that* time when compared with the reality of the Italian geometric school. Because it dealt only with projective varieties, that lecture, viewed at the present time, however, after the great generalization of the subject due to Grothendieck, appears to be a very, very concrete brand of mathematics. There is no doubt that the concept of "schemes" due to Grothendieck was a sound and inevitable generalization of the older concept of "variety" and that this generalization has introduced a new dimension into the conceptual content of algebraic geometry. What is more important is that this generalization has met what seems to me to be the true test of any generalization, that is, its effectiveness of solving, or throwing new light on, old problems by generalizing the terms of the problem (for example: the Riemann-Roch theorem for varieties of any dimension; the problem of the completeness of the characteristic linear series of a complete algebraic system of curves on a surface, both in characteristic zero and especially in characteristic $p \neq 0$; the computation of the fundamental group of an algebraic curve in characteristic $p \neq 0$).

But a mathematical theory cannot thrive indefinitely on greater and greater generality. A proper balance must ultimately be maintained between the generality and the concreteness of the structure studied, and usually this balance is restored after a period in which it was temporarily (and understandably) lost. There are signs at the present moment of the pendulum swinging back from "schemes," "motives," and so on toward concrete but difficult unsolved questions concerning the old pedestrian concept of a projective variety (and even of algebraic surfaces). There is no lack of such problems. It suffices to mention such questions as (1) criteria of rationality of higher varieties; (2) the study of cycles of codimension > 1 on any given variety; (3) even for divisors D on a variety there is the question of the behavior of the numerical function of n: dim $|nD|$ and finally (4) problems, such as reduction of singularities or the behavior of the zeta function, which are still unsolved when the ground field is of characteristic $p \neq 0$ (and is respectively algebraically closed or a finite field). These are new tasks that face the younger generation; I wholeheartedly wish that generation good speed and success.

Oscar Zariski

Cambridge, Massachusetts
January 1972

Bibliography of Oscar Zariski

(Entries preceded by an asterisk are reprinted in this volume.)

[1] *I fondamenti della teoria degli insiemi di Cantor,* Period. Mat., serie 4, vol. 4 (1924) pp. 408-437.

[2] *Sulle equazioni algebriche contenenti linearmente un parametro e risolubili per radicali,* Atti Accad. Naz. Lincei Rend. Cl. Sci. Fis. Mat. Natur., serie V, vol. 33 (1925) pp. 80-82.

[3] *Gli sviluppi più recenti della teoria degli insiemi e il principio di Zermelo,* Period. Mat., serie 4, vol. 5 (1925) pp. 57-80.

[4] *Sur le développement d'une fonction algébroide dans un domaine contenant plusieurs points critiques,* C. R. Acad. Sci., Paris, vol. 180 (1925) pp. 1153-1156.

[5] *Il principio de Zermelo e la funzione transfinita di Hilbert,* Rend. Sem. Mat. Roma, serie 2, vol. 2 (1925) pp. 24-26.

[6] *R. Dedekind, Essenza e Significato dei Numeri. Continuità e Numeri Irrazionali, Traduzione dal tedesco e note storico-critiche di Oscar Zariski* ("Per la Storia e la Filosofia delle Matematiche" series), Stock, Rome, 1926, 306 pp. The notes fill pp. 155-300.

[7] *Sugli sviluppi in serie delle funzioni algebroidi in campi contenenti più punti critici,* Atti Accad. Naz. Lincei Mem. Cl. Sci. Fis. Mat. Natur., serie VI, vol. 1 (1926) pp. 481-495.

[8] *Sull'impossibilità di risolvere parametricamente per radicali un'equazione algebrica* $f(x,y) = 0$ *di genere* $p > 6$ *a moduli generali,* Atti Accad. Naz. Lincei Rend. Cl. Sci. Fis. Mat. Natur., serie VI, vol. 3 (1926) pp. 660-666.

[9] *Sulla rappresentazione conforme dell'area limitata da una lemniscata sopra un cerchio,* Atti Accad. Naz. Lincei Rend. Cl. Sci. Fis. Mat. Natur., serie VI, vol. 4 (1926) pp. 22-25.

[10] *Sullo sviluppo di una funzione algebrica in un cerchio contenente più punti critici,* Atti Accad. Naz. Lincei Rend. Cl. Sci. Fis. Mat. Natur., serie VI, vol. 4 (1926), pp. 109-112.

[11] *El principio de la continuidad en su desarrolo histórico,* Rev. Mat. Hisp.-Amer., serie 2, vol. 1 (1926) pp. 161-166, 193-200, 233-240, 257-260.

[12] *Sopra una classe di equazioni algebriche contenenti linearmente un parametro e risolubili per radicali,* Rend. Circolo Mat. Palermo, vol. 50 (1926) pp. 196-218.

[13] *On a theorem of Severi,* Amer. J. Math., vol. 50 (1928) pp. 87-92.

[14] *On hyperelliptic θ-functions with rational characteristics,* Amer. J. Math., vol. 50 (1928) pp. 315-344.

[15] *Sopra il teorema d'esistenza per le funzioni algebriche di due variabili,* Atti Congr. Interna. Mat. 2, Bologna, vol. 4 (1928) pp. 133-138.

[16] *On the problem of existence of algebraic functions of two variables possessing a given branch curve,* Amer. J. Math., vol. 51 (1929) pp. 305-328.

[17] *On the linear connection index of the algebraic surfaces $z^n = f(x,y)$,* Proc. Nat. Acad. Sci. U.S.A., vol. 15 (1929) pp. 494-501.

[18] *On the moduli of algebraic functions possessing a given monodromie group,* Amer. J. Math., vol. 52 (1930) pp. 150-170.

[19] *On the irregularity of cyclic multiple planes,* Ann. of Math., vol. 32 Math., vol. 53 (1931) pp. 309-318.

[20] *On the irregularity of cyclic multiple planes,* Amer. J. Math., vol. 32 (1931) pp. 485-511.

[21] *On quadrangular 3-webs of straight lines in space,* Abh. Math. Sem. Univ. Hamburg, vol. 9 (1932) pp. 79-83.

[22] *On the topology of algebroid singularities,* Amer. J. Math., vol. 54 (1932) pp. 453-465.

[23] *On a theorem of Eddington,* Amer. J. Math., vol. 54 (1932) pp. 466-470.

[24] *Parametric representation of an algebraic variety,* Symposium on Algebraic Geometry, Princeton University, 1934-1935, mimeographed lectures, Princeton, 1935, pp. 1-10.

[25] *Algebraic Surfaces,* Ergebnisse der Mathematik, vol. 3, no. 5, Springer-Verlag, Berlin, 1935, 198 pp; second supplemented edition, with appendices by S. S. Abyankar, J. Lipman, and D. Mumford, Ergebnisse der Mathematik, vol. 61, Springer-Verlag, Berlin-Heidelberg-New York, 1971, 270 pp.

[26] (with S. F. Barber) *Reducible exceptional curves of the first kind,* Amer. J. Math., vol. 57 (1935) pp. 119-141.

[27] *A topological proof of the Riemann-Roch theorem on an algebraic curve,* Amer. J. Math., vol. 58 (1936) pp. 1-14.

[28] *On the Poincaré group of rational plane curves,* Amer. J. Math., vol. 58 (1936) pp. 607-619.

[29] *A theorem on the Poincaré group of an algebraic hyper surface,* Ann. of Math., vol. 38 (1937) pp. 131-141.

[30] *Generalized weight properties of the resultant of* $n + 1$ *polynomials in* n *indeterminants,* Trans. Amer. Math. Soc., vol. 41 (1937) pp. 249-265.

[31] *The topological discriminant group of a Riemann surface of genus* p, Amer. J. Math., vol. 59 (1937) pp. 335-358.

[32] *A remark concerning the parametric representation of an algebraic variety,* Amer. J. Math., vol. 59 (1937) pp. 363-364.

[33] (In Russian) *Linear and continuous systems of curves on an algebriac surface,* Progress of Mathematical Sciences, Moscow, vol. 3 (1937).

*[34] *Some results in the arithmetic theory of algebraic functions of several variables,* Proc. Nat. Acad. Sci. U.S.A., vol. 23 (1937) pp. 410-414.

*[35] *Polynomial ideals defined by infinitely near base points,* Amer. J. Math., vol. 60 (1938) pp. 151-204.

*[36] (with O. F. G. Schilling) *On the linearity of pencils of curves on algebraic surfaces,* Amer. J. Math., vol. 60 (1938) pp. 320-324.

*[37] *Some results in the arithmetic theory of algebraic varieties,* Amer. J. Math., vol. 61 (1939) pp. 249-294.

*[38] (with H. T. Muhly) *The resolution of singularities of an algebraic curve,* Amer. J. Math., vol. 61 (1939) pp. 107-114.

*[39] *The reduction of the singularities of an algebraic surface,* Ann. of Math., vol. 40 (1939) pp. 639-689.

*[40] *Algebraic varieties over ground fields of characteristic zero,* Amer. J. Math., vol. 62 (1940) pp. 187-221.

*[41] *Local uniformization on algebraic varieties,* Ann. of Math., vol. 41 (1940) pp. 852-896.

* [42] *Pencils on an algebraic variety and a new proof of a theorem of Bertini,* Trans. Amer. Math. Soc., vol. 50 (1941) pp. 48-70.

* [43] *Normal varieties and birational correspondences,* Bull. Amer. Math. Soc., vol. 48 (1942) pp. 402-413.

* [44] *A simplified proof for the resolution of singularities of an algebraic surface,* Ann. of Math., vol. 43 (1942) pp. 583-593.

* [45] *Foundations of a general theory of birational correspondences,* Trans. Amer. Math. Soc., vol. 53 (1943) pp. 490-542.

* [46] *The compactness of the Riemann manifold of an abstract field of algebraic functions,* Bull. Amer. Math. Soc., vol. 45 (1944) pp. 683-691.

* [47] *Reduction of the singularities of algebraic three dimensional varieties,* Ann. of Math., vol. 45 (1944) pp. 472-542.

* [48] *The theorem of Bertini on the variable singular points of a linear system of varieties,* Trans. Amer. Math. Soc., vol. 56 (1944) pp. 130-140.

[49] *Generalized semi-local rings,* Summa Brasiliensis Mathematicae, vol. 1, fasc. 8 (1946) pp. 169-195.

* [50] *The concept of a simple point of an abstract algebraic variety,* Trans. Amer. Math. Soc., vol. 62 (1947) pp. 1-52.

[51] *A new proof of Hilbert's Nullstellensatz,* Bull. Amer. Math. Soc., vol. 53 (1947) pp. 362-368.

[52] *Analytical irreducibility of normal varieties,* Ann. of Math., vol. 49 (1948) pp. 352-361.

[53] *A simple analytical proof of a fundamental property of birational transformations,* Proc. Nat. Acad. Sci. U.S.A., vol. 35 (1949) pp. 62-66.

[54] *A fundamental lemma from the theory of holomorphic functions on an algebraic variety,* Ann. Mat. Pura Appl., s. 4, vol. 29 (1949) pp. 187-198.

[55] *Quelques questions concernant la théorie des fonctions holomorphes sur une variété algébrique,* Colloque d'Algèbre et Théorie des Nombres, Paris, 1949, pp. 129-134.

[56] *Postulation et genre arithmétique,* Colloque d'Algèbre et Théorie des Nombres, Paris, 1949, pp. 115-116.

[57] *Hilbert's characteristic function and the arithmetic genus of an algebraic variety,* Trans. Amer. Math. Soc., vol. 69 (1950) pp. 78-88.

[58] *Theory and applications of holomorphic functions on algebraic varieties over arbitrary ground fields,* Mem. Amer. Math. Soc., no. 5 (1951) pp. 1-90.

[59] *Sur la normalité analytique des variétés normales,* Ann. Inst. Fourier (Grenoble), vol. 2 (1950) pp. 161-164.

[60] *The fundamental ideas of abstract algebraic geometry,* Proc. Internat. Cong. Math., Cambridge, 1950, pp. 77-89.

[61] *Complete linear systems on normal varieties and a generalization of a lemma of Enriques-Severi,* Ann. of Math., vol. 55 (1952) pp. 552-592.

* [62] *Le problème de la réduction des singularités d'une variété algébrique,* Bull. Sci. Mathématiques, vol. 78 (January-February 1954) pp. 1-10.

[63] *Interprétations algébro-géométriques du quatorzième problème de Hilbert,* Bull. Sci. Math., vol. 78 (July-August 1954) pp. 1-14.

[64] *Applicazioni geometriche della teoria delle valutazioni,* Rend. Mat. e Appl., vol. 13, fasc. 1-2, Roma (1954) pp. 1-38.

[65] (with S. Abhyankar) *Splitting of valuations in extensions of local domains,* Proc. Nat. Acad. Sci. U.S.A., vol. 41 (1955) pp. 84-90.
84-90.

[66] *The connectedness theorem for birational transformations,* Algebraic Geometry and Topology (Symposium in honor of S. Lefschetz), Princeton University Press, 1955, pp. 182-188.

[67] *Algebraic sheaf theory* (Scientific report on the second Summer Institute), Bull. Amer. Math. Soc., vol. 62 (1956) pp. 117-141.

[68] (with I. S. Cohen) *A fundamental inequaltiy in the theory of extensions of valuations,* Illinois J. Math., vol. 1 (1957) pp. 1-8.

[69] *Introduction to the problem of minimal models in the theory of algebraic surfaces,* Publ. Math. Soc. Japan, no. 4 (1958) pp. 1-89.

[70] *The problem of minimal models in the theory of algebraic surfaces,* Amer. J. Math., vol. 80 (1958) pp. 146-184.

[71] *On Castelnouvo's criterion of rationality* $p_a = P_2 = 0$ *of an algebraic surface,* Illinois J. Math., vol. 2 (1958) pp. 303-315.

[72] (with Pierre Samuel and cooperation of I. S. Cohen) *Commutative Algebra,* vol. I. D. Van Nostrand Company, Princeton, N.J., 1958.

[73] *On the purity of the branch locus of algebraic functions,* Proc. Nat. Acad. Sci. U.S.A., vol. 44 (1958) pp. 791-796.

[74] *Proof that any birational class of non-singular surfaces satisfies the descending chain condition.* Mem. Fac. Sci., Kyoto Univ., s. A, vol. 32, Mathematics, no. 1 (1959) pp. 21-31.

[75] (with Pierre Samuel) *Commutative Algebra,* vol. II. D. Van Nostrand Company, Princeton, N.J., 1960.

[76] (with Peter Falb) *On differentials in function fields,* Amer. J. Math., vol. 83 (1961) pp. 542-556.

[77] *On the superabundance of the complete linear systems* $|nD|$ *(n-large) for an arbitrary divisor D on an algebraic surface.* Atti del Convegno Internazionale di Geometria Algebrica tenuto a Torino, Maggio 1961, pp. 105-120.

* [78] *La risoluzione delle singolarità delle superficie algebriche immerse.* Nota I e II. Atti Accad. Naz. Lincei Rend., Cl. Sci. Fis. Mat. Natur., serie VIII, vol. 31, fasc. 3-4 (Settembre-Ottobre 1961) pp. 97-102; e fasc. 5 (Novembre 1961) pp. 177-180.

[79] *The theorem of Riemann-Roch for high multiples of an effective divisor on an algebraic surface.* Ann. Math., vol. 76 (1962) pp. 560-615.

[80] *Equisingular points on algebraic varieties.* Seminari dell'Istituto Nazionale di Alta Matematica, 1962-1963, Edizioni Cremonese, Roma, 1964, pp. 164-177.

[81] *Studies in equisingularity I. Equivalent singularities of plane algebroid curves.* Amer. J. Math., vol. 87 (1965) pp. 507-536.

[82] *Studies in equisingularity II. Equisingularity in co-dimension 1 (and characteristic zero).* Amer. J. Math., vol. 87 (1965) pp. 972-1006.

[83] *Characterization of plane algebroid curves whose module of differentials has maximum torsion.* Proc. Nat. Acad. Sci. U.S.A., vol. 56, (1966) pp. 781-786.

* [84] *Exceptional singularities of an algebroid surface and their reduction.* Accad. Naz. Lincei Rend., Cl. Sci. Fis. Mat. Natur. serie VIII, vol. 43, fasc. 3-4 (Settembre-Ottobre 1967) pp. 135-146.

[85] *Studies in equisingularity III. Saturation of local rings and equisingularity.* Amer. J. Math., vol. 90 (1968) pp. 961-1023.

[86] *Contributions to the problem of equisingularity.* Centro Internazionale Matematico Estivo (C.I.M.E.) Questions on Algebraic varieties. III ciclo, Varenna, 7-17 Settembre 1969, Edizioni Cremonese, Roma, 1970, pp. 261-343.

[87] *An Introduction to the Theory of Algebraic Surfaces,* Lecture Notes in Mathematics, No. 87, Springer-Verlag, Berlin, 1969.

[88] *Some open questions in the theory of singularities.* Bull. Amer. Math. Soc., vol. 77 (1971) pp. 481-491.

[89] *General theory of saturation and of saturated local rings.* I. *Saturation of complete local domains of dimension one having arbitrary coefficient fields (of characteristic zero).* Amer. J. Math., vol. 93 (1971) pp. 573-648.

Part I

Foundations of Algebraic Geometry

Introduction by D. Mumford

In the period 1937-1947, Oscar Zariski completely reoriented his research and began to introduce ideas from abstract algebra into algebraic geometry. Along with B. L. van der Waerden and André Weil, he undertook to completely rewrite the foundations of algebraic geometry without making any use of topological or analytical methods. There were two motivations for this: first, it became clear to Zariski, particularly after writing his Ergebnissebericht *Algebraic Surfaces* [25]† that many of the classical Italian "proofs" were not merely controversial but were really incomplete and imprecise at certain points. Second, it had become clear that it was both logical and useful to develop an "abstract" theory of algebraic geometry valid over an arbitrary ground field.

Two of the three main algebraic concepts that Zariski introduced into geometry and exploited most successfully were that of the integral closure of a ring and that of a valuation ring.‡ He introduces integral closure in [37], announced in [34], and applied it immediately to the resolution of singularities of curves and surfaces: [38], [39]. He introduces valuation rings in [35] to give an ideal-theoretic treatment of the Italian concept of the base conditions imposed by infinitely near points. The results of [35] were applied by Zariski and Schilling in [36] to give an algebraic proof of the theorem that an irrational pencil on an algebraic surface X can have base points only at the singularities of X. Zariski returned to the problem of base conditions and gave a new simpler treatment in Appendixes 3-5 to Volume II of *Commutative Algebra* [75]. Moreover, his results were extended to arbitrary rational singular points by Lipman.[2]

In [43] and [45], Zariski studied the effect of birational maps on normal singularities. In particular the theorem on p. 522 of [45] is the one which has been known ever since as "Zariski's Main Theorem." It states that if $f\colon X \to Y$ is a birational, regular map of varieties and if the local ring $o_{y,Y}$ of some point $y \in Y$ is integrally closed, then either $f^{-1}(y)$ is one point and f^{-1} is regular at y, or all components of $f^{-1}(y)$ have dimension $\geqslant 1$. This theorem has been generalized a great deal in recent years. Grothendieck[1] partly with Deligne, Chapter 4, Sections 8.12 and 18.12, prove the following

†Bracketed numbers refer to the Bibliography found at the front of this volume. This list includes all of Zariski's published work.

‡A third is that of the completion of a local ring which he exploited successfully in papers on formal holomorphic functions. (See Introduction by M. Artin in Volume II.)

theorem which they call Zariski's Main Theorem:

If $f: X \to Y$ is a quasi-finite morphism of schemes (that is, $f^{-1}(y)$ is finite for all $y \in Y$) and Y is quasi-compact, then f can be factored: $X \overset{g}{\to} Z \overset{h}{\to} Y$, where g is an open immersion and h is finite; that is, $\forall$ open affines Spec(A) in Y, h^{-1}(Spec(A)) is an open affine Spec(B) in Y, and B is finite and integral over A. If one applies this to the case where Y is normal (that is, $o_{y,Y}$ integrally closed) and f is birational, it follows that in that case f itself must be an open immersion, which is essentially the original form of Zariski's Main Theorem.

Following the methods of Zariski's original proof, Peskine[6] has proved a local version of this generalization:

If $A \subset B$ are 2 rings, with B finitely generated over A, and $P \subset B$ is a prime ideal which is "isolated" (that is, P is maximal and minimal among prime ideals P' such that $P \cap A = P' \cap A$) then $\exists\ s \in B - P$ which is integral over A, such that $B_s = (A')_s$, where A' = integral closure of A in B.

Finally, in papers [40] and [50], Zariski explored the algebraic significance of the geometric idea of a nonsingular, or simple point of a variety. In particular, in [50], the concept of a regular local ring is used. He discovered the basic fact that over an imperfect ground field k of characteristic p, there are really two different concepts of simple point: *regular* in the sense of having a regular local ring, and *smooth* in the sense that the usual Jacobian criterion is satisfied. (See definitions A and B in the Introduction to [50]; we are following Grothendieck's terminology.) Zariski shows that:

x is a smooth point $\Leftrightarrow$ corresponding points x' over algebraically closed $k' \supset k$ are smooth
$\Leftrightarrow$ corresponding points x' over algebraically closed $k' \supset k$ are regular
$\Rightarrow$ x is a regular point

but that the last arrow cannot be reversed in general when k is imperfect. The main result of the paper is that the set of regular points is open (the corresponding question for smooth is trivial). For more general schemes, Nagata[5] discovered that sadly this was not necessarily true. Samuel[7] and Nagata[4] gave other criteria for the set of regular points to be open, and Grothendieck[1] with his concept of an "excellent ring" has extended Nagata's work and formalized it, Chapter 0, Sections 22, 23; Chapter 4, Section 7. See also Tate[8] for the effect of regular but nonsmooth points on the genus of a curve. One of the motivations for Zariski's work on simple points is his study of the two Bertini theorems in [42] and [48], which provide an excellent illustration of the distinction between the concepts of regular and smooth. Bertini's theorems state that if $\{D(\lambda)\}_{\lambda \epsilon P}$ is a linear system of divisors on a variety X in characteristic zero, then (a) almost all of the $D(\lambda)$ have singularities only at the singularities of X and the base points of the linear system, and (b) either almost all the $D(\lambda)$ are irreducible, or else the linear system is composite with a pencil. Zariski proves (b) in [42], but in the middle of his proof (Lemma 5) uses characteristic zero. He returned to the question later in his *Introduction to the problem of minimal models in the theory of algebraic surfaces* [69, Section I.6], where he proves that if characteristic $= p$, either almost all $D(\lambda) = p^n \cdot E(\lambda)$, where $E(\lambda)$ is irreducible, or else the linear system is composite with a pencil.† Another proof can be found in Matsusaka.[3] It is Theorem (a) however that is most interesting because it is definitely *false* in characteristic p. Zariski shows in [48] that, in fact, the generic member of the pencil D^*, as a variety over the ground field $k(P)$, is regular except at the base points of the system and the singularities of X; but only if D^* is also smooth over $k(P)$ will almost all the $D(\lambda)$ have the same property.

† If I may be allowed to interject a personal note, as a student of Zariski's I once attended a seminar on this book where each week Zariski would select, just before the hour, one of us to give that week's lecture. I happened to be the one called upon the week that we proved Bertini's theorem and, being a novice who was struggling to see any geometry at all behind the algebra, I started by saying that "for simplicity" we might as well restrict ourselves to characteristic zero. The response was a laugh but a firm request to keep the characteristic p case in mind.

References

1. A. Grothendieck, *Eléments de la géometric algébrique,* Inst. Hautes Études Sci., Pub. Math., no. 4, 8, etc.

2. J. Lipman, *Rational singularities, with applications to algebraic surfaces and unique factorization,* Inst. Hautes Etudes Sci., Pub. Math., no. 36 (1970) p. 195.

3. T. Matsusaka, *The Theorem of Bertini on linear systems,* Mem. Coll. Sci. Kyoto, vol. 26 (1951) p. 51.

4. M. Nagata, *A jacobian criterion of simple points,* Ill. J. Math., vol. 1 (1957) p. 427.

5. M. Nagata, *On the closedness of singular loci,* Publ. Math. de l'I.H.E.S., no. 1 (1959) p. 29.

6. C. Peskine, *Une généralisation du "Main Theorem" de Zariski,* Bull. Sci. Mathématiques vol. 90 (1966) p. 119.

7. P. Samuel, *Simple subvarieties of analytic varieties,* Proc. Nat. Acad. Sci. U.S.A. vol. 41 (1955) p. 647.

8. J. Tate, *Genus change in inseparable extensions of function fields,* Proc. Amer. Math. Soc. vol. 3 (1952) p. 400.

Reprints of Papers

Reprinted from the Proceedings of the NATIONAL ACADEMY OF SCIENCES,
Vol. 23, No. 7, pp. 410–414. July, 1937.

SOME RESULTS IN THE ARITHMETIC THEORY OF ALGEBRAIC FUNCTIONS OF SEVERAL VARIABLES

BY OSCAR ZARISKI

DEPARTMENT OF MATHEMATICS, JOHNS HOPKINS UNIVERSITY

Communicated June 12, 1937

1. *Introduction.*—In this Note we give an outline of results arrived at in the course of an investigation dealing with the arithmetic theory of algebraic functions of several variables. The full treatment will be given in a forthcoming paper. We are dealing exclusively with the theory of zero-dimensional ideals. These ideals represent "points," or "places" in the natural and intuitive sense of the word.* We also point out that in the theory of these ideals we have, in a manner of speaking, the maximum deviation from the feature which dominates the Dedekind-Weber theory: the possibility of factoring ideals into power products of prime ideals.†

2. *Ramified Ideals.*—Let Σ be a field of algebraic functions of r variables $x_1, x_2, \ldots, x_r$. We denote by $\mathfrak{o}$ the ring of integral functions in Σ and by $P = K[x_1, x_2, \ldots, x_r]$ the ring of polynomials in $x_1, \ldots, x_r$. The field K of constants is assumed algebraically closed and of characteristic zero.

The dimension of a prime ideal $\mathfrak{p}$ in $\mathfrak{o}$ is, by definition, the dimension (degree of transcendentality) of the quotient field of the ring of residual classes $\mathfrak{o}/\mathfrak{p}$. Since K is algebraically closed, it follows that if, in particular, $\mathfrak{p}$ is zero-dimensional, then $\omega \equiv c \pmod{\mathfrak{p}}$, $c < K$, for any ω in $\mathfrak{o}$. We shall say that ω *has the value* c, $\omega = c$, *at the "place"* $\mathfrak{p}$. Zero-dimensional prime ideals in $\mathfrak{o}$ are divisorless and are characterized by this property. If then $\mathfrak{q}_1\mathfrak{q}_2$ are primary ideals belonging to two distinct prime ideals, they are relatively prime and therefore their intersection $[\mathfrak{q}_1\mathfrak{q}_2]$ coincides with the product $\mathfrak{q}_1\mathfrak{q}_2$.

Let $\mathfrak{X} = (x_1-a_1, x_2-a_2, \ldots, x_r-a_r)$ be a prime zero-dimensional ideal in P. The ideal $\mathfrak{X}$, considered as an ideal in $\mathfrak{o}$ (with the same basis elements $x_\alpha - a_\alpha$) decomposes into primary components:

$$\mathfrak{X} = [\mathfrak{q}_1, \mathfrak{q}_2, \ldots, \mathfrak{q}_\sigma] = \mathfrak{q}_1\mathfrak{q}_2 \ldots \mathfrak{q}_\sigma, \qquad (1)$$

belonging to distinct prime zero-dimensional ideals, $\mathfrak{p}_1, \mathfrak{p}_2, \ldots, \mathfrak{p}_\sigma$. The ideals $\mathfrak{p}_i$, $i = 1, 2, \ldots \sigma$, are the only places at which $x_1 = a_1$, $x_2 = a_2, \ldots, x_r = a_r$. Let ρ_i be the exponent of $\mathfrak{q}_i$, i.e., the smallest integer such that $\mathfrak{p}_i^{\rho_i} \equiv 0\ (\mathfrak{q}_i)$. $\mathfrak{X}$ is a *ramified ideal,* if some ρ_i is > 1, i.e., if some $\mathfrak{q}_i$ is different from $\mathfrak{p}_i$.

We consider any function ω in $\mathfrak{o}$ and we form the polynomial $F(x_\alpha, t) = N(t - \omega)$. It is not difficult to prove the following: *Every solution* $x_\alpha = a_\alpha$, $\omega = c$ *of the equation* $F(x_\alpha; \omega) = 0$ *consists of values which the functions* $x_1, \ldots, x_r$, ω *assume at some place* $\mathfrak{p}$. Let then $\omega \equiv c_i(\mathfrak{p}_i)$, so that c_i is a root of the equation $F(a_\alpha, t) = 0$. *If* ρ_i *is the exponent of* $\mathfrak{q}_i$, *then* c_i *is a root of* $F(a_\alpha;\ t)$, *of multiplicity* $s_i \geqq \rho_i$. From this follows the corollary: *If* $\mathfrak{X}$ *is a ramified ideal, then, no matter what function* ω *we consider in* $\mathfrak{o}$ *the polynomial in* t, $N(t - \omega)$, *has a multiple root when the* x_α's *are replaced by the* a_α's.

On the other hand, let $\mathfrak{X}$ be non-ramified. Then it can be proved that $\sigma = n$, and that $\mathfrak{X}$ *factors into exactly* n *distinct prime factors:*

$$\mathfrak{X} = \mathfrak{p}_1\mathfrak{p}_2 \ldots \mathfrak{p}_n,$$

where n is the degree of Σ with respect to $K(x_\alpha)$. We can find a function ω which assumes n distinct values $c_1, c_2, \ldots, c_n$ at the places $\mathfrak{p}_1, \mathfrak{p}_2, \ldots, \mathfrak{p}_n$. The corresponding polynomial $F(a_\alpha, t)$ has then n distinct roots c_i, hence all simple.

Hence $\mathfrak{X}$ is ramified if and only if the discriminant of the polynomial $N(t - \omega)$ vanishes at $x_\alpha = a_\alpha$, ω being *any* function in $\mathfrak{o}$. We conclude with the

Theorem. *Let* $\omega_1 = 1, \omega_2 = \omega, \ldots, \omega_n = \omega^{n-1}$ *and let* $\mathfrak{T}$ *be the ideal in* P *generated by the determinants* $|S(\omega_i\omega_j)|$ *of the traces, as* ω *varies in* $\mathfrak{o}$. *A prime ideal* $\mathfrak{X}$ *in* P *is ramified, if and only if* $\mathfrak{T} \equiv 0\ (\mathfrak{X})$.

Remark. The inequality $s_i \geqq \rho_i$ is only a rough estimate, and examples can be given in which $s_i > \rho_i$. One may look for a relation between s_i and the length λ_i of the primary component $\mathfrak{q}_i$. But also here we cannot expect to have always $s_i = \lambda_i$, because $s_1 + s_2 + \ldots + s_\sigma = n$, and $\lambda_1 + \lambda_2 + \ldots + \lambda_\sigma$ is the length of the ideal $\mathfrak{X}$, i.e., the rank of $\mathfrak{o}/\mathfrak{X}$, and this rank is not always equal to n.‡

3. *Relative and Absolute Ramified Ideals.*—If $\rho_i > 1$, we say that $\mathfrak{p}_i$ is *relative ramified* (relative to the independent variables x_i.)

For relative unramified ideals the following can be proved: *If* $\mathfrak{p}$ *is unramified with respect to the variables* x_α, *and if* $x_\alpha \equiv a_\alpha(\mathfrak{p})$, *then given any element* ω *in* $\mathfrak{o}$, *there exists, for any positive integer* m, *a congruence of the form*

$$\omega \equiv \varphi_{m-1}\,(x_1 - a_1, \ldots, x_r - a_r)\ (\mathfrak{p}^m), \qquad (2)$$

where φ_{m-1} *is a polynomial of degree* $m - 1$ *in* $x_1 - a_1, \ldots, x_r - a_r$; *this polynomial is uniquely determined.*

It is clear that for a given ω, $\varphi_m = \varphi_{m-1} + \psi_m$, where ψ_m is homogeneous of degree m in $x_1 - a_1, \ldots, x_r - a_r$. We have then, from (2), *a formal power series for any function* ω *at the place* $\mathfrak{p}$.

We now give the characterization of relative ramified ideals.

Theorem. *Let* $\mathfrak{z}_\alpha$ *be the ideal generated by all the derivatives* $\frac{\partial F}{\partial \omega}$ *as* ω *varies in* $\mathfrak{o}$. *If* $\mathfrak{p}$ *is relative ramified, then* $\mathfrak{z}_\alpha \equiv 0\ (\mathfrak{p})$, *and conversely.*

In the proof of this theorem it was necessary to consider the Galois extension of Σ and some properties of the Galois group, locally at the place $\mathfrak{p}$. The above theorem is equivalent to the following statement: *$\mathfrak{p}$ is relative ramified, if the value* c *which a function in* $\mathfrak{o}$ *assumes at* $\mathfrak{p}$ *is always a multiple root of the polynomial* $F(a_\alpha; t)$ $[= N(t - \omega)_{x_\alpha = a_\alpha}]$.

Dealing with relative ramified ideals, we encounter in the case $r > 1$ a new fact. Suppose that we pass from the variables $x_1, \ldots, x_r$ to new independent variables $y_1, \ldots, y_r$, such that the ring of integral functions remains the same. An ideal ramified with respect to the x's may very well be unramified with respect to the y's. However, if $r > 1$, *there may exist ideals* $\mathfrak{p}$ *which remain relative ramified for any choice of the independent variables.* We call such ideals *absolutely ramified.*

Let $\mathfrak{Z}$ be the h.c.d. of all the ideals $\mathfrak{z}_x$, as the independent variables are altered. Obviously, $\mathfrak{p}$ *is an absolutely ramified ideal if and only if* $\mathfrak{Z} \equiv 0(\mathfrak{p})$. Concerning $\mathfrak{Z}$, we prove that it is *an ideal of dimension at most* $r - 2$. Thus the locus of absolutely ramified places is at most of dimension two less than the dimension of the field. These places play the same rôle as the isolated singularities of an algebraic variety.

The absolutely ramified ideals admit the following intrinsic characterization:

An ideal $\mathfrak{p}$ *is absolutely ramified if and only if the ring of residual classes* $\mathfrak{p}/\mathfrak{p}^2$ *is of rank* $> r$.

4. *The Normal Variety for the Ring* $\mathfrak{o}$.—We choose a set of elements $y_1, y_2, \ldots, y_m$ in $\mathfrak{o}$ such that every other element in $\mathfrak{o}$ can be expressed as a polynomial in $y_1, y_2, \ldots, y_m$. We can take, for instance, for such a set the independent variables x_α together with the elements of an integral base of $\mathfrak{o}$ over the ring of polynomials $K[x_\alpha]$. The functions y_α define in the affine $S_m(y_1, y_2, \ldots, y_m)$ an algebraic r-dimensional variety V_r. The following properties of V_r can be established:

(1) There is a (1, 1) correspondence between the places $\mathfrak{p}$ in $\mathfrak{o}$ and the points of V_r;

(2) The singular points of V_r are those and only those which correspond to the absolutely ramified places of $\mathfrak{o}$; as a consequence, the locus of singularities of V_r is at most of dimension $r - 2$;**

(3) If m is sufficiently high, the rank of the ring $\mathfrak{p}/\mathfrak{p}^m$ is equal to $s_0\binom{m}{2} + s_1\binom{m}{1} + s_2$, where s_0, s_1, s_2 are integers. The point of V_r which corresponds to $\mathfrak{p}$ is s_0-fold for V_r.

We call V_r the *normal variety* for the ring $\mathfrak{o}$, since V_r obviously depends on $\mathfrak{o}$ itself and not on the choice of the independent variables. Two normal varieties for the ring $\mathfrak{o}$ are, by (2), in birational correspondence, free from fundamental points (at finite distance) on either variety.

5. *Properties of the Conductor.*—The importance of the notion of the conductor, both in question of ramification and of singularities, is well

known from the arithmetic theory of functions of one variable. Some interesting applications of the conductor in the theory of functions of several variables have been given by Schmeidler. I shall now give some properties of the conductor in relation to zero-dimensional ideals.

Let S be the ring of polynomials in $x_1, \ldots, x_r$ and in some primitive element ω in $\mathfrak{o}$. The conductor shall be denoted by $\mathfrak{A}_\omega$. It is known (Schmeidler[4]) that $\mathfrak{A}_\omega$ is an unmixed ideal of dimension $r - 1$. Two questions can be formulated: (1) characterize the zero-dimensional ideals $\mathfrak{X}$ in P which are divisors of the norm $N(\mathfrak{A}_\omega)$; (2) characterize the zero-dimensional ideals $\mathfrak{p}$ in $\mathfrak{o}$ which are divisors of $\mathfrak{A}_\omega$. The answers are given by the following theorems:

(1) $N(\mathfrak{A}_\omega) \not\equiv 0\ (\mathfrak{X})$, *if and only if every element* ξ *in* $\mathfrak{o}$ *satisfies a congruence of the form:*

$$\xi \equiv c_0 + c_1\omega + \ldots + c_{n-1}\omega^{n-1}(\mathfrak{X}),\ c_i < K,$$

i.e., if $1, \omega, \ldots, \omega^{n-1}$ is a linear base of the ring $\mathfrak{o}/\mathfrak{X}$. Hence the rank of $\mathfrak{o}/\mathfrak{X}$ is then exactly n. Another question is this: *How can one characterize from the point of view of the ideal theory in* $\mathfrak{o}$ *those ideals* $\mathfrak{X}$ *in* P *for which the rank of* $\mathfrak{o}/\mathfrak{X}$ *is* $=$ n?

(2) $\mathfrak{A}_\omega \not\equiv 0(\mathfrak{p})$, *if and only if* $\mathfrak{p} = (\mathfrak{X}, \omega - c)$, where $\mathfrak{X} = (x_1 - a_1, \ldots, x_r - a_r)$, $x_\alpha \equiv a_\alpha\ (\mathfrak{p})$, $\omega \equiv c(\mathfrak{p})$.

This last theorem implies that if $\mathfrak{A}_\omega \not\equiv 0(\mathfrak{p})$, then $\mathfrak{p}$ possesses a base of at most $r + 1$ elements $x_1 - a_1, \ldots, x_r - a_r, \omega - c$, and from this follows easily: Rank of $\mathfrak{p}/\mathfrak{p}^2 \leqq r + 1$. It is possible to prove also the converse, namely: *If* $\mathfrak{p}/\mathfrak{p}^2$ *is of rank* $\leqq$ r + 1, *then for a proper choice of the independent variables* x_i *and of the primitive element* ω, *we will have* $\mathfrak{A}_\omega \not\equiv 0(\mathfrak{p})$.

If then T *denotes the h.c.d. of all the conductors of rings of polynomials of* r + 1 *elements of* $\mathfrak{o}$, *then the zero-dimensional prime divisors of* T *are those and only those ideals* $\mathfrak{p}$ *for which the rank of* $\mathfrak{p}/\mathfrak{p}^2$ *is* $>$ r + 1.

* For the various definitions of an abstract point of a field of algebraic functions see van der Waerden,[7] Jung,[1] Schmeidler[4,5] (point = divisor of the first kind) and Ostrowski[3] (point = an arbitrary valuation of the field).

† On the other extreme are the ideals of maximum dimension $r - 1$ (r—the number of independent variables). It is well known that for these ideals the factorization property can be restored by using the notion of "quasigleichheit," due to van der Waerden.[6]

‡ It has been pointed out to me by Ore that the multiplicities s_i are the degrees of the irreducible factors of $F(x_a;\ t)$ (mod $\mathfrak{X}^m$), when m is sufficiently high. That this factorization (mod $\mathfrak{X}^m$) is unique (mod $\mathfrak{X}^{m-\delta}$), δ—independent of m, can be easily proved for fields of algebraic numbers and also for fields of functions of one variable (see Ore[2]). For $r > 1$ the proof is somewhat different.

** In particular, if $r = 1$, V_1 is a curve free from singularities (in the affine space S_m). In view of the existence of an integral independent base, we have in this case $m = n + 1$, where n is the degree of Σ with respect to $K(x)$. We obtain in exactly the same manner a curve free from singularities in a projective space, if we use two homogeneous variables x_1, x_2 and consider homogeneous ideals in $\mathfrak{o}$. For properties of such ideals, see A. Weil.[8]

[1] Jung, H., *Algebraische Flächen*, Hannover (1925).

[2] Ore, O., "Über den Zusammenhang zwischen den definierenden Gleichungen und der Idealtheorie in algebraischen Körpern." I, *Math. Ann.*, **96,** 313–352 (1926).

[3] Ostrowski, A., "Untersuchungen zur arithmetischen Theorie der Körper," *Math. Zeit.*, **39,** 392–404 (1934).

[4] Schmeidler, W., "Grundlagen einer Theorie der algebraischen Funktionen mehrerer Veränderlicher," *Math. Zeit.*, **28,** 116–141 (1928).

[5] Schmeidler, W., "Über Verzweigungspunkte bei Körpern von algebraischen Funktionen mehrerer Veränderlicher," *Jour. reine angew. Math.*, **167,** 251 (1931).

[6] van der Waerden, B. L., "Zur Productzerlegung der Ideale in ganz abgeschlossenen Ringen," *Math. Ann.*, **101,** 293–308 (1929).

[7] van der Waerden, B. L., "Zur Idealtheorie der ganz abgeschlossenen Ringe," *Math. Ann.*, **101,** 309–311 (1929).

[8] Weil, A., "Arithmétique et géometrie sur les varietés algébriques," *Actualités scientifiques et industrielles*, n. **206**.

POLYNOMIAL IDEALS DEFINED BY INFINITELY NEAR BASE POINTS.*

By OSCAR ZARISKI.

Introduction. The linear systems of curves, in the plane or on an algebraic surface, which are theoretically of importance, are the *complete systems.* For complete linear systems the defining linear conditions are *base conditions,* by which the curves of the system are constrained to pass with assigned multiplicities through an assigned set of base points. The set of base points may consist in part of proper points and in part of points infinitely near and in the successive neighborhoods of the proper points ([10], p. 27). However vague this geometric terminology may sound, it is nevertheless true that the facts involved have a precise algebraic meaning, and satisfactory definitions are available in terms of analytical branches and of intersection multiplicities of such branches. But it is equally true that although a well rounded geometric theory can and has been developed along these lines ([1], pp. 327-399), the arithmetic content of the notion of infinitely near points still remains somewhat obscure. It is the main purpose of the present investigation to develop an arithmetic theory parallel to the geometric theory of infinitely near points (in the plane or on a surface without singularities). By this we mean primarily a systematic study of those polynomials ideals in $\mathfrak{k}[x, y]$, or formal power series ideals (in two indeterminates), which adequately describe linear conditions having the character of base conditions. We call these ideals *complete ideals* (II, 12) by analogy with the terminology used in the theory of linear systems. We always suppose that the underlying field $\mathfrak{k}$ is algebraically closed and of characteristic zero, but the theory could be extended with a few modifications to fields of any characteristic. At any rate, the hypothesis that the characteristic is zero is not used in the first four sections of Part I. For possible generalizations to spaces of higher dimension than 2 it would be important to consider also fields which are not algebraically closed.

The class of complete ideals enjoys several striking properties, which respond, however, to a high geometric expectation. *This class is closed under all standard operations on ideals* (except addition) : *the intersection, the product and the quotient of two complete ideals is a complete ideal* (II, 12). Moreover, *a complete ideal has a unique factorization into simple complete ideals* (I, 7), an ideal being simple if it is not the product of ideals different from the unit ideal.

* Received November 8, 1937.

We define complete ideals in terms of *valuation ideals.* By valuation ideals in the polynomial ring $\mathfrak{k}[x, y]$ we mean the contracted ideals of the ideals of any valuation ring (belonging to some valuation of the field $\mathfrak{k}(x, y)$) which contains x and y. Valuation ideals are complete ideals, and, in particular, the simple complete ideals are valuation ideals. Most of our work is a study of valuation ideals (briefly: v-ideals) whether from an axiomatic point of view (Part I) or from the point of view of formal power series (Part II). The study of the behaviour of valuation ideals under quadratic transformations (I, 4 and 5) leads to reduction theorems (Theorems 4. 4 and 5. 3) which form the basis of many inductive proofs. *A simple v-ideal of kind* $k+1$ (I, 6) *represents the arithmetic analogue of the notion of a point infinitely near and in the k-th neighborhood of a proper point.*

The treatment deals explicitly only with polynomial rings and rings of holomorphic functions. It is clear, however, that the results carry over automatically to algebraic surfaces without singularities, since any set of base conditions at a simple point P of an algebraic surface is described by a complete ideal in the ring of holomorphic functions of the uniformizing parameters at P.

We make one more remark. In many instances the proofs do not depend on the fact that we have only two indeterminates. On the other hand, in many points a generalization to any number of variables faces new difficulties. In spaces of higher dimension the base conditions may be of a more complicated type: besides base conditions at an isolated base point, it is possible to have infinitely near base curves, or infinitely near base surfaces, etc. At an isolated base point we may also have such base conditions as are given by infinitesimal base curves (the case of an assigned tangent plane at a base point is the simplest example). The algebro-geometric theory has as yet no firm grasp on these eventualities. A generalization of the present treatment to any number of variables would therefore represent not merely an arithmetization of a known chapter of classical algebraic geometry.

CONTENTS.

Part I.

PART II.

PART I.

1. Valuation ideals in rings of polynomials in two indeterminates. Let $\mathfrak{k}$ be an algebraically closed field of characteristic zero and Σ—a pure transcendental extension of $\mathfrak{k}$ of dimension (degree of transcendentality) two. We consider a valuation B of Σ, i. e. an homorphic mapping of the multiplicative group Σ (the element 0 excluded) upon an ordered abelian group Γ: $a \rightarrow v(a)$ $=$ *value of* a, $a \in \Sigma$, $a \neq 0$, $v(a) \in \Gamma$, satisfying the valuation axioms:

$$(1)\ v(a \cdot b) = v(a) + v(b); \qquad (2)\ v(a+b) \geqq \min.(v(a), v(b)); \qquad (3)\ v(a^*) \neq 0,$$

for some a^* in Σ ([2], p. 101; [7]).[1] We assume, moreover, that the elements of the underlying field $\mathfrak{k}$, other than 0, have value zero in the given valuation B.

Let $\mathfrak{B}$ be the valuation ring of B (the set of all elements of Σ whose value is $\geqq 0$). Any ideal $\mathfrak{a}$ in $\mathfrak{B}$ has the following self-evident property: $a \equiv 0(\mathfrak{a})$, $v(b) \geqq v(a)$ implies $b \equiv 0(\mathfrak{a})$. Conversely, any subset of $\mathfrak{B}$ with this property constitutes, together with 0, an ideal. Consequently, given any two ideals $\mathfrak{a}$, $\mathfrak{a}'$ in $\mathfrak{B}$, either $\mathfrak{a} \equiv 0(\mathfrak{a}')$ or $\mathfrak{a}' \equiv 0(\mathfrak{a})$. We say that $\mathfrak{a}$ *precedes* $\mathfrak{a}'$ (and that $\mathfrak{a}'$ *follows* $\mathfrak{a}$) if $\mathfrak{a}' \equiv 0(\mathfrak{a})$, but $\mathfrak{a}' \neq \mathfrak{a}$. The ideals in $\mathfrak{B}$ form then an ordered set. The unit ideal $\mathfrak{B}$ is the first element of this set. The immediate successor of $\mathfrak{B}$ is the ideal $\mathfrak{P}$ consisting of all elements whose value is positive.

Let $\mathfrak{O}$ be a domain of integrity in Σ, *contained in the valuation ring* $\mathfrak{B}$. An ideal $\mathfrak{A}$ in $\mathfrak{O}$ shall be called a *valuation ideal*, or briefly, a *v-ideal, belonging to* or *for the valuation B*, if $\mathfrak{A}$ is the contracted ideal of an ideal $\mathfrak{a}$ in $\mathfrak{B}$, i. e.

[1] Note the following consequence of the above axioms: *if* $v(a) < v(b)$, *then* $v(a+b) = v(a)$. Proof: We have $v(a) = v(a \cdot 1) = v(a) + v(1)$, whence $v(1) = 0$. Also $v(-1) + v(-1) = v((-1)^2) = v(1) = 0$, consequently $v(-1) = 0$, since Γ is an ordered group. Hence $v(-b) = v(b)$, and, by axiom (2), $v(a-b) \geqq \min.(v(a), v(b))$. Replacing in this relation a by $a+b$ we find $v(a) \geqq \min.(v(a+b), v(b))$, and our assertion follows.

if $\mathfrak{A} = [\mathfrak{O}, \mathfrak{a}]$. If the reference to the specific valuation B is omitted and if we speak of $\mathfrak{A}$ as a v-ideal, we shall mean then that $\mathfrak{A}$ is a valuation ideal for some valuation B of Σ such that the valuation ring of B contains $\mathfrak{O}$. We have thus defined in $\mathfrak{O}$ a special class of ideals, the class of valuation ideals. We agree to include the zero ideal in this class. Also the valuation ideals in $\mathfrak{O}$, *belonging to a given valuation* B, enjoy the property: $a \equiv 0(\mathfrak{A})$, $v(b) \geqq v(a)$ implies $b \equiv 0(\mathfrak{A})$, where $\mathfrak{A}$ is a v-ideal and a, b are elements of $\mathfrak{O}$. Conversely, any subset of $\mathfrak{O}$ with this property constitutes, together with the zero element of Σ, a valuation ideal belonging to B. The valuation ideals in $\mathfrak{O}$, belonging to a given valuation B, form an ordered set: $\mathfrak{A}$ precedes $\mathfrak{A}'$ if $\mathfrak{A}' \equiv 0(\mathfrak{A})$ and $\mathfrak{A}' \neq \mathfrak{A}$. Since a v-ideal $\mathfrak{A}$ in $\mathfrak{O}$, for the valuation B, is the contracted ideal of an ideal $\mathfrak{a}$ in $\mathfrak{B}$, $\mathfrak{A}$ is also the contracted ideal of its extended ideal $\mathfrak{B}\mathfrak{A}$ in $\mathfrak{B}$. There is thus a $(1, 1)$ correspondence between the v-ideals in $\mathfrak{O}$ belonging to the valuation B and their extended ideals in the valuation ring $\mathfrak{B}$. These extended ideals form in general a proper subset of the set of all ideals of $\mathfrak{B}$.

Let, in particular, $\mathfrak{O}$ be the ring of polynomials in x, y, where we assume that x and y are generating elements of Σ ($\Sigma = k(x, y)$) and elements of the valuation ring.[2] From the fact that every ideal in $\mathfrak{O}$ possesses a finite base and from the valuation axiom (2), it follows that any ideal $\mathfrak{M}$ in $\mathfrak{O}$ contains elements of smallest possible value in B. If α is this minimum, $\alpha = \min\{v(a)\}$, $a \varepsilon \mathfrak{M}$, we shall write $\alpha = v(\mathfrak{M})$ and we shall regard α as *the evaluation of the ideal* $\mathfrak{M}$. Two ideals $\mathfrak{M}$, $\mathfrak{M}'$ will be said to be *equivalent*, in symbols $\mathfrak{M} \sim \mathfrak{M}'$, if $v(\mathfrak{M}) = v(\mathfrak{M}')$. The class $\{\mathfrak{M}\}$ of all ideals equivalent to $\mathfrak{M}$ contains one and only one v-ideal for B, namely the ideal $\mathfrak{A}$ consisting of all elements whose value is $\geqq v(\mathfrak{M})$. Clearly $\mathfrak{A}$ is a divisor of any ideal of the class. In the ordered set of v-ideals of $\mathfrak{O}$, belonging to the valuation B, $\mathfrak{A}$ precedes $\mathfrak{A}'$ if $v(\mathfrak{A}) < v(\mathfrak{A}')$.

In the sequel we shall be dealing with a fixed valuation B and it will be understood that when we speak of a v-ideal in $\mathfrak{O}$ we mean a v-ideal "belonging to the given valuation B."

The valuation ring $\mathfrak{B}$ contains the divisorless ideal $\mathfrak{P}$, consisting of all elements whose value is > 0. We shall denote by $\mathfrak{p}$ the contracted ideal of $\mathfrak{P}$ in $\mathfrak{O}$; $\mathfrak{p}$ is obviously a prime ideal in $\mathfrak{O}$. By the *dimension* r *of the valuation* B is meant the dimension of the ideal $\mathfrak{P}$, i. e. the degree of transcendentality of the field of residual classes $\mathfrak{B}/\mathfrak{P}$ over $\mathfrak{k}$.[3] Evidently, r is either 0 or 1.

---

[2] Note that if x is any element in Σ, then either x or $1/x$ belongs to $\mathfrak{B}$, since $v(1/x) = -v(x)$.

[3] By hypothesis, 0 is the only element of $\mathfrak{k}$ which belongs to $\mathfrak{P}$. Hence $\mathfrak{B}/\mathfrak{P}$ contains a subfield isomorphic to $\mathfrak{k}$. We identify this subfield with $\mathfrak{k}$.

THEOREM 1. *In the ordered set of v-ideals in* $\mathfrak{O}$ *every v-ideal* $\mathfrak{A}$ *has an immediate successor* $\mathfrak{A}'$. *If the valuation B is of dimension zero, then* $\mathfrak{A}'$ *is a maximal subideal of* $\mathfrak{A}$, *and the ring* $\mathfrak{A}/\mathfrak{A}'$ *is isomorphic to the underlying field* $\mathfrak{k}$.

Proof. By the valuation axiom (3), there exists an element a^* such that $v(a^*) > 0$. Let $a^* = f(x,y)/g(x,y)$, $f \varepsilon \mathfrak{O}$, $g \varepsilon \mathfrak{O}$. Since $v(f) = v(a^*) + v(g)$ and $v(g) \geqq 0$, also $v(f) > 0$. If then $\alpha = v(\mathfrak{A})$, there exist in $\mathfrak{A}$ elements whose value is greater than α, for instance, the elements of $\mathfrak{A}f$. The totality of all polynomials whose value is greater than α constitutes, together with the element 0, a v-ideal $\mathfrak{A}'$, contained in $\mathfrak{A}$, and clearly there exist no v-ideals which follow $\mathfrak{A}$ (proper multiples of $\mathfrak{A}$) and precede $\mathfrak{A}'$ (proper divisors of $\mathfrak{A}'$).

Assume that B is of dimension 0, whence $\mathfrak{B}/\mathfrak{P}$ is an algebraic extension of the underlying field $\mathfrak{k}$. Since $\mathfrak{k}$ is algebraically closed, it follows that $\mathfrak{B}/\mathfrak{P}$ is isomorphic to $\mathfrak{k}$. Let f be an element of $\mathfrak{A}$ of smallest possible value, $v(f) = v(\mathfrak{A})$, and let ϕ be any other element in $\mathfrak{A}$. Since $v(\phi) \geqq v(f)$, we have $v(\phi/f) \geqq 0$, i. e. ϕ/f belongs to $\mathfrak{B}$. Hence $\phi/f \equiv c(\mathfrak{P})$, where c is in $\mathfrak{k}$, $v(\phi/f - c) = v[(\phi - cf)/f] > 0$, i. e. $v(\phi - cf) > v(f)$, and consequently $\phi - cf \equiv 0(\mathfrak{A}')$. This shows that $\mathfrak{A}/\mathfrak{A}' \simeq \mathfrak{k}$ and also that $\mathfrak{A}'$ is a maximal subideal of $\mathfrak{A}$, q. e. d.

If B is of dimension 1, it defines an homomorphism of the field Σ upon the field $\mathfrak{B}/\mathfrak{P}$ of algebraic functions of one variable, i. e. B is a "divisor" of Σ. If $\mathfrak{O}$ contains elements which *mod* $\mathfrak{p}$ are transcendental with respect to $\mathfrak{k}$, i. e. if $\mathfrak{p}$ is a 1-dimensional ideal in $\mathfrak{O}$, we are dealing with a *divisor of the first kind* with respect to $\mathfrak{O}$. The ideal $\mathfrak{p}$ is then a principal ideal, say $\mathfrak{p} = (f)$, where f is an irreducible polynomial, and the v-ideals in $\mathfrak{O}$ are the ideals $\mathfrak{p}^n = (f^n)$, $n = 0, 1, 2, \cdots$. If, however, $\mathfrak{p}$ is 0-dimensional, we are dealing with a *divisor of the second kind* with respect to $\mathfrak{O}$. The v-ideals in $\mathfrak{O}$ are in this case certain primary ideals belonging to $\mathfrak{p}$. We need not consider separately this case, because it reduces to the case of 0-dimensional valuations. In fact, we may consider an arbitrary valuation B_1 of the field $\Sigma_1 = \mathfrak{B}/\mathfrak{P}$ (a point of the Riemann surface of the field Σ_1). The given valuation B of Σ followed up by the valuation B_1 of Σ_1 defines an homomorphism of Σ upon the underlying field $\mathfrak{k}$ (together with symbol ∞), hence a 0-dimensional valuation B' of Σ. The v-ideals in $\mathfrak{O}$ belonging to B will be among the v-ideals belonging to B'.

From now on we shall only consider 0-dimensional valuations. If B is 0-dimensional, then the v-ideal $\mathfrak{p} = [\mathfrak{P}, \mathfrak{O}]$ is prime and 0-dimensional, since it is the immediate successor of the unit ideal $\mathfrak{O}$ and since, by Theorem 1, $\mathfrak{O}/\mathfrak{p} \simeq \mathfrak{k}$. Replacing, if necessary, x and y by $x - c$, $y - d$, where $x \equiv c(\mathfrak{p})$, $y \equiv d(\mathfrak{p})$, we may assume that $x \equiv 0(\mathfrak{p})$, $y \equiv 0(\mathfrak{p})$, whence $\mathfrak{p} = (x, y)$.

Starting with $\mathfrak{O} = \mathfrak{q}_0$ and with its successor $\mathfrak{p} = \mathfrak{q}_1$, we form the simple sequence of v-ideals

$$(1) \qquad \mathfrak{q}_0, \mathfrak{q}_1, \mathfrak{q}_2, \cdots, \mathfrak{q}_i, \cdots$$

where $\mathfrak{q}_{i+1}$ is the immediate successor of $\mathfrak{q}_i$. *Each* $\mathfrak{q}_{i+1}$ *is a maximal subideal of its predecessor* $\mathfrak{q}_i$. We shall call a *Jordan sequence* any sequence of ideals having this last mentioned property. It is clear that all the ideals $\mathfrak{q}_i$, $i \geqq 1$, in the Jordan sequence (1), are primary ideals belonging to $\mathfrak{p}$ $(= \mathfrak{q}_1)$. In fact: (1) $\mathfrak{q}_i \equiv 0(\mathfrak{p})$; (2) $ab \equiv 0(\mathfrak{q})$, $a \not\equiv 0(\mathfrak{q})$ imply that $v(a) + v(b) > v(a)$, whence $v(b) > 0$ and $b \equiv 0(\mathfrak{p})$; (3) $v(\mathfrak{q}_1) = v(\mathfrak{p}) < v(\mathfrak{p}^2) < v(\mathfrak{p}^3) < \cdots < v(\mathfrak{p}^i)$, whence $v(\mathfrak{p}^i) \geqq v(\mathfrak{q}_i)$ and consequently $\mathfrak{p}^i \equiv 0(\mathfrak{q}_i)$.

Two cases are possible: (a) either the intersection of all the ideals $\mathfrak{q}_i$ is the zero-ideal; (b) or this intersection is a certain ideal $\mathfrak{p}_1 \neq (0)$. In the first case the Jordan sequence (1) contains *all* the v-ideals of $\mathfrak{O}$ belonging to the given valuation B. We investigate now the second case.

We first prove that $\mathfrak{p}_1$ *is a prime ideal.* In fact, let $ab \equiv 0(\mathfrak{p}_1)$, $a \not\equiv 0(\mathfrak{p}_1)$, and let $\mathfrak{q}_r$ be the last ideal in the sequence $\{\mathfrak{q}_i\}$ which contains a. Then $v(a) = v(\mathfrak{q}_r)$, whence $v(a) \leqq \rho v(\mathfrak{p})$, where ρ is the exponent of the ideal $\mathfrak{q}_r$. No power $\mathfrak{p}^n$ of $\mathfrak{p}$ can belong to all the ideals $\mathfrak{q}_i$, since $\mathfrak{p}^n/\mathfrak{p}$ is of finite rank with respect to $\mathfrak{k}$. Hence, for any integer $n > 0$ there exists an integer α_n such that $v(\mathfrak{q}_{\alpha_n}) > v(\mathfrak{p}^n)$. Since $ab \equiv 0(\mathfrak{q}_i)$ for any value of i, it follows that $v(a) + v(b) > nv(\mathfrak{p})$, n arbitrary. In view of the inequality $v(a) \leqq \rho v(\mathfrak{p})$, we deduce that also $v(b) > nv(\mathfrak{p})$, n—an arbitrary integer. In particular, if $\mathfrak{q}_m$ is any ideal in the sequence $\{\mathfrak{q}_i\}$ and if ρ_m is its exponent, we will have $v(\mathfrak{q}_m) \leqq \rho_m v(\mathfrak{p}) < v(b)$, whence $b \equiv 0(\mathfrak{q}_m)$, for any m. It follows that $b \equiv 0(\mathfrak{p}_1)$, i. e. $\mathfrak{p}_1$ is a prime ideal.

The ideal $\mathfrak{p}_1$ is one-dimensional, say (f), where f is an irreducible polynomial. The inequality $v(b) > nv(\mathfrak{p})$, n arbitrary, holds for any element b of $\mathfrak{p}_1$, and this shows that the value group of B is non-archimedean (B is a "special" valuation, of rank 2. See [2], p. 113). It is not difficult to see that *all the v-ideals in* $\mathfrak{O}$ *for B are of the form* $\mathfrak{p}_1{}^m\mathfrak{q}_n$, $m, n = 0, 1, 2 \cdots$. In fact, let F_1 and F_2 be any two polynomials in $\mathfrak{O}$ and let $F_1 = f^{m_1}G_1$, $F_2 = f^{m_2}G_2$, where G_1 and G_2 are not divisible by f. If $m_1 > m_2$, then $v(F_1) > v(F_2)$, since $v(f) > v(G_2)$. If $m_1 = m_2$, then $v(F_1) > v(F_2)$ or $v(F_1) = v(F_2)$ according as $v(G_1) > v(G_2)$ or $v(G_1) = v(G_2)$. Hence the set of all polynomials whose value is not less than the value of a given polynomial f^mG

$(G \not\equiv 0(f))$ coincides with the ideal $\mathfrak{p}_1^m \mathfrak{q}_n$, where $v(\mathfrak{q}_n) = v(G)$. This form of the v-ideals brings out clearly the well-known decomposition of the valuation B into two valuations of rank 1 and the nature of the value group, as consisting in this case of pairs of integers. B decomposes into two valuations, B' and $\bar{B}$. B' is the one-dimensional valuation defined by the prime 1-dimensional ideal $\mathfrak{p}_1$, and its valuation ring $\mathfrak{B}'$ is the set of rational functions $\frac{F(x,y)}{G(x,y)}$, $G \not\equiv 0(f)$. B' maps Σ upon the field $\bar{\Sigma} = \mathfrak{k}(\bar{x}, \bar{y})$ of algebraic functions of one variable, where $\bar{\Sigma}$ is the quotient field of the ring $\mathfrak{O}/\mathfrak{p}_1$. $\bar{B}$ is a valuation of $\bar{\Sigma}$, and the sequence $\{\mathfrak{q}_i\}$ (with x, y replaced by $\bar{x}, \bar{y}$) is the sequence of the valuation ideals in the ring $\bar{\mathfrak{O}} = \mathfrak{k}[\bar{x}, \bar{y}]$ which belong to $\bar{B}$.[4]

2. A characteristic property of Jordan sequence of valuation ideals. Given a Jordan sequence of ideals in $\mathfrak{O}$: $\mathfrak{q}_0, \mathfrak{q}_1, \mathfrak{q}_2, \cdots$, where $\mathfrak{q}_0 = \mathfrak{O}$, $\mathfrak{q}_1 = \mathfrak{p} = (x, y)$, we ask under what conditions will there exist a valuation of Σ for which the given sequence $\{\mathfrak{q}_i\}$ is the sequence of (zero-dimensional) v-ideals. In other words: *under what conditions does the given sequence $\{\mathfrak{q}_i\}$ belong to a valuation of Σ?*

THEOREM 2.1. *A necessary and sufficient condition in order that a Jordan sequence $\{\mathfrak{q}_i\}$ of 0-dimensional ideals in $\mathfrak{O}$ belong to a valuation of the field Σ, is that the quotient $\mathfrak{q}_i$: (a) belong to the sequence, for any i and for any elements a in $\mathfrak{O}$.*

THEOREM 2.2. *A necessary and sufficient condition in order that a Jordan sequence $\{\mathfrak{q}_i\}$ of 0-dimensional ideals in $\mathfrak{O}$ belong to a valuation of the field Σ, is that the congruences*

$$(2) \qquad \mathfrak{q}_i \mathfrak{q}_{j+1} : \mathfrak{q}_j \equiv 0(\mathfrak{q}_{i+1})$$

hold true for any pair $\mathfrak{q}_i, \mathfrak{q}_j$ of ideals of the sequence.

The characterization given in Theorem 2.2 has the advantage of involving only operations within the given sequence $\{\mathfrak{q}_i\}$. We prove both theorems simultaneously.

(1) *The conditions are necessary.* As to Theorem 2.1, let $v(a) = \alpha$, $\mathfrak{q}_i : (a) = \mathfrak{q}'$, and let b, c be any two elements of $\mathfrak{O}$ such that $b \equiv 0(\mathfrak{q}')$, $v(c) \geqq v(b)$. Since $ba \equiv 0(\mathfrak{q}_i)$, we have $v(c) + v(a) \geqq v(b) + v(a) \geqq v(\mathfrak{q}_i)$, whence $ca \equiv 0(\mathfrak{q}_i)$, and consequently $c \equiv 0(\mathfrak{q}')$. Hence $\mathfrak{q}'$ enjoys the property:

[4] It is possible to have $\mathfrak{p}_1 = (0)$ also in the case of a valuation of rank 2. This happens when the component B' (divisor of Σ) of the valuation B, of rank 2, is a divisor of the second kind with respect to $\mathfrak{O}$. (In geometric terms: the divisor B' is an exceptional curve which has been transformed into a point of the plane (x, y)).

$b \equiv 0(\mathfrak{q}')$, $v(c) \geqq v(b)$, implies $c \equiv 0(\mathfrak{q}')$, and is therefore a v-ideal belonging to the given valuation B. Since $\mathfrak{q}_i \equiv 0(\mathfrak{q}')$, necessarily $\mathfrak{q}' = \mathfrak{q}_j$, for some $j \leqq i$.

The necessity of the condition of Theorem 2.2 is proved in a similar manner. Let b be an element of $\mathfrak{q}_i\mathfrak{q}_{j+1} : \mathfrak{q}_j$, whence $b\mathfrak{q}_j \equiv 0(\mathfrak{q}_i\mathfrak{q}_{j+1})$. Since $v(\mathfrak{q}_i\mathfrak{q}_{j+1}) = v(\mathfrak{q}_i) + v(\mathfrak{q}_{j+1})$, we have $v(b) + v(\mathfrak{q}_j) \geqq v(\mathfrak{q}_i) + v(\mathfrak{q}_{j+1})$, and since $v(\mathfrak{q}_j) < v(\mathfrak{q}_{j+1})$, it follows, $v(b) > v(\mathfrak{q}_i)$, i. e. $b \equiv 0(\mathfrak{q}_{i+1})$, q. e. d.

(2) *The conditions are sufficient.* We introduce the following notations: if ξ, η are elements of $\mathfrak{O}$, we write $\xi \leqq \eta$ (or $\eta \geqq \xi$), if the congruence $\xi \equiv 0(\mathfrak{q}_i)$ always implies $\eta \equiv 0(\mathfrak{q}_i)$; we write $\xi < \eta$ (or $\eta > \xi$) if there exists in the sequence $\{\mathfrak{q}_i\}$ an ideal $\mathfrak{q}_m$ such that $\eta \equiv 0(\mathfrak{q}_m)$, $\xi \not\equiv 0(\mathfrak{q}_m)$. We now prove the following lemma, assuming that the condition of Theorem 2.1 *or* that of Theorem 2.2 is satisfied.

LEMMA. *If ξ, η, ζ are elements of $\mathfrak{O}$ and if $\xi\eta \equiv 0(\mathfrak{q}_i)$, then $\eta \leqq \zeta$ implies $\xi\zeta \equiv 0(\mathfrak{q}_i)$ and $\eta < \zeta$ implies $\xi\zeta \equiv 0(\mathfrak{q}_{i+1})$.*

In other words: *$\eta \leqq \zeta$ implies $\xi\eta \leqq \xi\zeta$; $\eta < \zeta$ implies that either $\xi\eta < \xi\zeta$ or that $\xi\eta$ and $\xi\zeta$ belong to all the ideals $\mathfrak{q}_i$ of the sequence.*

Assume the condition of Theorem 2.1, and let $\mathfrak{q}_i : (\xi) = \mathfrak{q}_h$. Since $\eta \equiv 0(\mathfrak{q}_h)$, it follows that if $\eta \leqq \zeta$, then also $\zeta \equiv 0(\mathfrak{q}_h)$, whence $\xi\zeta \equiv 0(\mathfrak{q}_i)$, and this proves the first part of the lemma. Let $\mathfrak{q}_{i+1} : (\xi) = \mathfrak{q}_s$, $s \geqq h$, and let ϕ and ψ be any pair of elements of $\mathfrak{q}_h$. We have $\phi\xi \equiv 0(\mathfrak{q}_i)$, $\psi\xi \equiv 0(\mathfrak{q}_i)$, whence there exist elements c, d in the underlying field $\mathfrak{k}$ such that $c\phi\xi + d\psi\xi \equiv 0(\mathfrak{q}_{i+1})$, since $\mathfrak{q}_i/\mathfrak{q}_{i+1}$ is of rank 1 with respect to $\mathfrak{k}$. Hence $c\phi + d\psi \equiv 0(\mathfrak{q}_s)$, for any two elements ϕ, ψ in $\mathfrak{q}_h$ and for appropriate elements c, d in $\mathfrak{k}$, i. e. $\mathfrak{q}_h/\mathfrak{q}_s$ is at most of rank 1 with respect to $\mathfrak{k}$, and s is either h or $h+1$. Since $\eta \equiv 0(\mathfrak{q}_h)$, it follows that if $\eta < \zeta$, then $\zeta \equiv 0(\mathfrak{q}_{h+1}) = 0(\mathfrak{q}_s)$, whence $\xi\zeta = 0(\mathfrak{q}_{i+1})$, q. e. d.

Assume the condition of Theorem 2.2. If η belongs to all the ideals $\mathfrak{q}_i$ of the sequence, the same will be true for ζ, and both parts of the lemma are trivial. Assume that there exists a *last* ideal $\mathfrak{q}_h$ which contains η: $\eta \equiv 0(\mathfrak{q}_h)$, $\eta \not\equiv 0(\mathfrak{q}_{h+1})$. If $\eta \leqq \zeta$, then also $\zeta \equiv 0(\mathfrak{q}_h)$, whence $\xi\eta\zeta \equiv 0(\mathfrak{q}_i\mathfrak{q}_h)$. Assume, if possible, $\xi\zeta \not\equiv 0(\mathfrak{q}_i)$. There will then exist an ideal $\mathfrak{q}_j$ in our sequence, $j < i$, such that $\xi\zeta \equiv 0(\mathfrak{q}_j)$, $\xi\zeta \not\equiv 0(\mathfrak{q}_{j+1})$. Since $j + 1 \leqq i$, $\mathfrak{q}_i \equiv 0(\mathfrak{q}_{j+1})$, the congruence $\xi\eta\zeta \equiv 0(\mathfrak{q}_i\mathfrak{q}_h)$ implies $\xi\eta\zeta \equiv 0(\mathfrak{q}_{j+1}\mathfrak{q}_h)$. Now $\mathfrak{q}_{j+1}$ is a maximal subideal of $\mathfrak{q}_j$ and $\xi\zeta$ is in $\mathfrak{q}_j$ but not in $\mathfrak{q}_{j+1}$; hence $\mathfrak{q}_j = (\xi\zeta, \mathfrak{q}_{j+1})$, and consequently, $\eta\mathfrak{q}_j = (\eta\xi\zeta, \eta\mathfrak{q}_{j+1}) \equiv 0(\mathfrak{q}_{j+1}\mathfrak{q}_h)$, since $\eta \equiv 0(\mathfrak{q}_h)$. It follows that $\eta \equiv 0(\mathfrak{q}_{j+1}\mathfrak{q}_h : \mathfrak{q}_j) \equiv 0(\mathfrak{q}_{h+1})$, in contradiction with our hypothesis $\eta \not\equiv 0(\mathfrak{q}_{h+1})$. This proves the first part of the lemma.

Let now $\eta < \zeta$, and hence $\zeta \equiv 0(\mathfrak{q}_{h+1})$. We have then $\xi\eta\zeta \equiv 0(\mathfrak{q}_i\mathfrak{q}_{h+1})$

and also by the first part of the lemma, just proved, $\xi\zeta \equiv 0(\mathfrak{q}_i)$. Since $\mathfrak{q}_{h+1}$ is a maximal subideal of $\mathfrak{q}_h$ and since η is in $\mathfrak{q}_h$ but not in $\mathfrak{q}_{h+1}$, we have $\mathfrak{q}_h = (\eta, \mathfrak{q}_{h+1})$, and consequently $\xi\zeta\mathfrak{q}_h = (\xi\zeta\eta, \xi\zeta\mathfrak{q}_{h+1}) \equiv 0(\mathfrak{q}_i\mathfrak{q}_{h+1})$. Hence $\xi\zeta \equiv 0(\mathfrak{q}_i\mathfrak{q}_{h+1} : \mathfrak{q}_h) \equiv 0(\mathfrak{q}_{i+1})$, q. e. d.

The rest of the proof is based solely on the above Lemma. Let $\mathfrak{p}_1$ be the intersection of the ideals $\mathfrak{q}_i$ of our sequence. *We prove that $\mathfrak{p}_1$ is a prime ideal.* We first observe, that x and y cannot both belong to $\mathfrak{q}_2$, since $\mathfrak{q}_1 = \mathfrak{p} = (x,y)$. Let, for instance, $x \equiv 0(\mathfrak{q}_1)$, $x \not\equiv 0(\mathfrak{q}_2)$, whence $y \geqq x$. We deduce from our lemma, that if a given power of x, say x^m, belongs to an ideal $\mathfrak{q}_i$ of the sequence, then also all the power products x^ky^l, $k + l \geqq m$, belong to $\mathfrak{q}_i$, i. e. $\mathfrak{p}^m \equiv 0(\mathfrak{q}_i)$. Since $\mathfrak{p}^m/\mathfrak{p}$ is of finite rank with respect to $\mathfrak{k}$, it follows that *no power of x can belong to all the ideals $\mathfrak{q}_i$, i. e.* to $\mathfrak{p}_1$.

We next observe that *all the ideals $\mathfrak{q}_i$, $i > 1$, are primary ideals belonging to* $\mathfrak{p}$ $(=\mathfrak{q}_1)$. In fact: (1) $\mathfrak{q}_i \equiv 0(\mathfrak{p})$. (2) Let $ab \equiv 0(\mathfrak{q}_i)$, $a \not\equiv 0(\mathfrak{q}_i)$, whence $a < ab$. It is not possible to have $b \leqq 1$, because this would imply, by our lemma, $ab \leqq a$, in contradiction with $ab > a$. Hence $1 < b$ and since 1 is in $\mathfrak{q}_0$, b must be in $\mathfrak{q}_1$, i. e. in $\mathfrak{p}$. (3) We have $1 < x$, hence, by our Lemma, $x < x^2$ (since no power of x belongs to $\mathfrak{p}_1$), whence, again by the Lemma, $x^2 < x^3$, and generally, $x < x^2 < x^3 < \cdots < x^i$. Consequently $x^2 \equiv 0(\mathfrak{q}_2)$, $x^3 \equiv 0(\mathfrak{q}_3), \cdots, x^i \equiv 0(\mathfrak{q}_i)$. The congruence $x^i \equiv 0(\mathfrak{q}_i)$ implies, as we have just shown, the congruence $\mathfrak{p}^i \equiv 0(\mathfrak{q}_i)$.

To prove that $\mathfrak{p}_1$ is prime, let $\xi\eta \equiv 0(\mathfrak{p}_1)$ and assume, if possible, that $\xi \not\equiv 0(\mathfrak{p}_1)$, $\eta \not\equiv 0(\mathfrak{p}_1)$. Let $\mathfrak{q}_r$ be the last ideal of the sequence $\{\mathfrak{q}_i\}$ which contains ξ and let similarly $\mathfrak{q}_s$ be the last ideal containing η. If m and n are the exponents of the primary ideals $\mathfrak{q}_r$ and $\mathfrak{q}_s$ respectively, we have $x^m \equiv 0(\mathfrak{q}_r)$, $x^n \equiv 0(\mathfrak{q}_s)$, whence $x^m \geqq \xi$, $x^n \geqq \eta$, and consequently, by our lemma, $x^{m+n} \geqq \xi\eta$, i. e. $x^{m+n} \equiv 0(\mathfrak{p}_1)$, and this is impossible. Hence $\mathfrak{p}_1$ is prime, necessarily either the zero ideal or one-dimensional, since $\mathfrak{p}_1 \equiv 0(\mathfrak{p})$.

We now consider the well ordered descending set S of the ideals $\mathfrak{q}_{mn} = \mathfrak{p}_1{}^m\mathfrak{q}_n$, where clearly $\mathfrak{q}_{mn} \equiv 0(\mathfrak{q}_{m_1n_1})$ if $m > m_1$ or if $m = m_1$ and $n > n_1$. Here $\mathfrak{q}_{0n} = \mathfrak{q}_n$ and if $\mathfrak{p}_1 = (0)$ it is understood that S coincides with the sequence $\{\mathfrak{q}_n\}$. In either case, the intersection of all the ideals $\mathfrak{q}_{mn}$ of the set is the zero ideal. Hence, given any element ξ in $\mathfrak{O}$, there will exist a *first* ideal in the set S, say $\mathfrak{q}_{ij}$, which does not contain ξ. Let η be another element in $\mathfrak{O}$ and let $\mathfrak{q}_{i'j'}$ be the first ideal in the set which does not contain η. We complete and modify our notations $\xi \leqq \eta$, $\xi < \eta$ introduced above, as follows: we write $\xi \not> \eta$ if $\mathfrak{q}_{i'j'} \equiv 0(\mathfrak{q}_{ij})$, and $\xi < \eta$ if $\mathfrak{q}_{i'j'} \equiv 0(\mathfrak{q}_{ij})$ and $\mathfrak{q}_{i'j'} \neq \mathfrak{q}_{ij}$. It is obvious that if $\xi \not> \eta$ and $\eta \not> \zeta$, then $\xi \not> \zeta$, and if $\xi < \eta$ and $\eta \not> \zeta$, or if $\xi \not> \eta$ and $\eta < \zeta$, then $\xi < \zeta$. From our Lemma and from the fact that $\mathfrak{p}_1$ is a principal ideal, it follows in a straight-forward manner that the relations $\xi \not> \eta$, $\xi < \eta$ imply, for any element ζ in $\mathfrak{O}$, the relations $\xi\zeta \not> \eta\zeta$, $\xi\zeta < \eta\zeta$ respectively. It is also evident that if $\xi \not> \eta$, $\xi \not> \eta'$, then $\xi \not> \eta \pm \eta'$, and if $\xi < \eta$ and $\xi < \eta'$ then $\xi < \eta \pm \eta'$.

Let $\mathfrak{B}$ be the set of all elements in the field Σ which can be put in the form η/ξ, $\eta, \xi \in \mathfrak{O}$, $\xi \not> \eta$. *We prove that* $\mathfrak{B}$ *is a valuation ring.*

First, $\mathfrak{B}$ *is a ring.* In fact, let $\eta/\xi \in \mathfrak{B}$, $\eta_1/\xi_1 \in \mathfrak{B}$. Since $\xi \not> \eta$ and $\xi_1 \not> \eta_1$, we have $\xi\xi_1 \not> \eta\xi_1$ and $\xi\xi_1 \not> \xi\eta_1$, whence $\xi\xi_1 \not> \eta\xi_1 \pm \xi\eta_1$, i. e.

$$\frac{\eta}{\xi} \pm \frac{\eta_1}{\xi_1} = \frac{\eta\xi_1 \pm \xi\eta_1}{\xi\xi_1} \in \mathfrak{B}.$$

Also, since $\xi\xi_1 \not> \xi\eta_1 \not> \eta\eta_1$, the product $\eta\eta_1/\xi\xi_1$ belongs to $\mathfrak{B}$.

To prove that $\mathfrak{B}$ is a valuation ring, it is sufficient to show that given any two elements a, b in $\mathfrak{B}$, then either ab^{-1} or $a^{-1}b$ is in $\mathfrak{B}$ ([2], p. 102). In other words, we have to show that given any two elements ξ, η in $\mathfrak{O}$, either ξ/η or η/ξ must belong to $\mathfrak{B}$. But this is obvious, since one of the two relations $\xi \not> \eta$, $\eta \not> \xi$ must hold true.

Finally, the valuation abstractly defined by the valuation ring $\mathfrak{B}$ is not the trivial one, in which every element has value zero. In other words, the ring $\mathfrak{B}$ does not contain all the elements of the field Σ. In fact, take two elements ξ, η in $\mathfrak{O}$ such that definitely, $\xi < \eta$. We assert that ξ/η does not belong to $\mathfrak{B}$. Assuming the contrary, we must be able to put ξ/η in the form η_1/ξ_1, where $\xi_1 \not> \eta_1$ and $\xi\xi_1 = \eta\eta_1$. By our lemma, $\xi_1 \not> \eta_1$ implies $\eta\xi_1 \not> \eta\eta_1$, while $\xi < \eta$ implies $\xi\xi_1 < \eta\xi_1$, i. e. $\eta\eta_1 < \eta\xi_1$, giving two contradictory relations.

It remains to show that the ideals $\mathfrak{q}_{mn}$ of our well ordered set S are the v-ideals belonging to the valuation B defined by the valuation ring $\mathfrak{B}$. Consider any ideal $\mathfrak{q}_{mn}$ and let ξ, η be elements in $\mathfrak{O}$ such that $\xi \equiv 0(\mathfrak{q}_{mn})$, $v(\eta) \geqq v(\xi)$. Since $v(\eta/\xi) \geqq 0$, we must have $\eta/\xi = \eta_1/\xi_1$, where $\xi_1 \not> \eta_1$. Hence $\xi\xi_1 \not> \xi\eta_1$, i. e. $\xi\xi_1 \not> \eta\xi_1$, and this implies $\xi \not> \eta$. Hence $\eta \equiv 0(\mathfrak{q}_{mn})$, and thus $\mathfrak{q}_{mn}$ enjoys the property: $\xi \equiv 0(\mathfrak{q}_{mn}), v(\eta) \geqq v(\xi)$ implies $\eta \equiv 0(\mathfrak{q}_{mn})$. Hence $\mathfrak{q}_{mn}$ is a v-ideal belonging to the valuation B. That the set $S = \{\mathfrak{q}_{mn}\}$ contains all the v-ideals for B, is implied by the fact that each $\mathfrak{q}_{mn}$ $(n \neq 0)$ is a maximal subideal of its immediate predecessor $\mathfrak{q}_{m,n-1}$, and that $\mathfrak{q}_{m0}$ is the intersection of the ideals which precede it in S.

3. Further properties of v-ideals. We consider the sequence $\{\mathfrak{q}_i\}$ of 0-dimensional v-ideals in $\mathfrak{O}$ belonging to a fixed valuation B of Σ, where $\mathfrak{q}_1 = \mathfrak{p} = (x, y)$ and $\mathfrak{q}_0 = \mathfrak{O}$. We may assume $x \not\equiv 0(\mathfrak{q}_2)$. Since $\mathfrak{q}_1/\mathfrak{q}_2$ is of rank 1 with respect to $\mathfrak{k}$, there exists elements c, d in $\mathfrak{k}$ such that $cx + dy \equiv 0(\mathfrak{q}_2)$, $d \neq 0$. We then replace y by $cx + dy$ and thus we may assume $y \equiv 0(\mathfrak{q}_2)$, whence $\mathfrak{q}_2 = (y, x^2)$.

Let $\mathfrak{p}^h$ be the highest power of $\mathfrak{p}$ which divides a given v-ideal $\mathfrak{q}_i$; $\mathfrak{q}_i \equiv 0(\mathfrak{p}^h)$, $\mathfrak{q}_i \not\equiv 0(\mathfrak{p}^{h+1})$. Every polynomial f in $\mathfrak{q}_i$ is then of the form $f = f_h + f_{h+1}, \cdots$, where f_i is homogeneous of degree i in x and y. We shall call f_h the *subform* of f (for particular polynomials in $\mathfrak{q}_i$, f_h may be identically zero). As f varies in $\mathfrak{q}_i$, its subform f_h generates a linear system of forms, of a certain dimension $r \geqq 0$ (a $\mathfrak{k}$-module of rank $r+1$). Denote this system by $\Omega(\mathfrak{q}_i)$. We define in a similar manner the symbol $\Omega(\mathfrak{A})$ for any ideal $\mathfrak{A}$ in $\mathfrak{O}$.

THEOREM 3. *Let* $\mathfrak{q}_i \equiv 0(\mathfrak{p}^h)$, $\mathfrak{q}_i \not\equiv 0(\mathfrak{p}^{h+1})$ *and let* $\mathfrak{A} = [\mathfrak{q}_i, \mathfrak{p}^k]$, $k \geqq h$. *If* $\Omega(\mathfrak{A})$ *is of dimension* r, *then* $\Omega(\mathfrak{A})$ *coincides with the system of forms (of degree* k*) which are divisible by* y^{k-r}, *and, moreover, there exists a v-ideal* $\mathfrak{q}_j$ *in the sequence* $\{\mathfrak{q}_n\}$ *such that* $\mathfrak{A} = \mathfrak{p}^r\mathfrak{q}_j$.

Proof. Since $\mathfrak{q}_i$ contains polynomials ϕ whose subforms are exactly of degree h, it also contains polynomials (such as $x^{k-h}\phi$) whose subforms are of degree k. Hence $\mathfrak{A} \not\equiv 0(\mathfrak{p}^{k+1})$, and $\Omega(\mathfrak{A})$ consists of forms of degree k. Let $f = f_k + f_{k+1} + \cdots$ be a polynomial belonging to $\mathfrak{A}$ such that $f_k \neq 0$, and let $f_k = y^\rho \psi_{k-\rho}$, $\psi_{k-\rho} \not\equiv 0(y)$. Since y^ρ and $\psi_{k-\rho}$ are relatively prime, every form g, of degree $m \geqq k-1$, can be expressed as a linear combination $A\psi_{k-\rho} + By^\rho$, where A and B are forms of degree $m-k$ and $m-\rho$ respectively. It follows that given any integer $n \geqq 0$, it is possible to find two polynomials $P^{(n)}(x,y)$, $Q^{(n)}(x,y)$ of the form

$$(3) \qquad \begin{aligned} P^{(n)}(x,y) &= y^\rho + A_{\rho+1}(x,y) + \cdots + A_{\rho+n}(x,y), \\ Q^{(n)}(x,y) &= \psi_{k-\rho} + B_{k-\rho+1}(x,y) + \cdots + B_{k-\rho+n}(x,y), \end{aligned}$$

where A_i, B_i are forms of degree i, in such a manner as to have

$$(4) \qquad f \equiv P^{(n)}Q^{(n)}(\mathfrak{p}^{k+n+1}).$$

In fact, we have for the unknown form A_i, B_i the equations

$$A_{\rho+i}\psi_{k-\rho} + B_{k-\rho+i}y^\rho = f_{k+i} - \sum_{j=1}^{i-1} A_{\rho+j}B_{k-\rho+i-j}, \qquad (i = 1, 2, \cdots, n),$$

and these equations can be solved successively for $A_{\rho+1}$, $B_{k-\rho+1}$; $A_{\rho+2}$, $B_{k-\rho+2}$; etc. We take n sufficiently high, so as to have $\mathfrak{p}^{k+n+1} \equiv 0(\mathfrak{A})$.[5] For such a value of n we will have

$$(5) \qquad P^{(n)}Q^{(n)} \equiv 0(\mathfrak{A}).$$

[5] Since $\mathfrak{A} = [\mathfrak{q}_i, \mathfrak{p}^k]$ and $\mathfrak{q}_i$ is a primary ideal belonging to $\mathfrak{p}$, also $\mathfrak{A}$ is a primary ideal belonging to $\mathfrak{p}$.

This implies $P^{(n)}Q^{(n)} \equiv 0(\mathfrak{q}_i)$. Now the subform $\psi_{k-\rho}$ of $Q^{(n)}$ is not divisible by y and, by our choice of the variables x, y, we have $v(y) > v(x)$. Hence $v(Q^{(n)}) = v(x^{k-\rho})$, and consequently, if $\phi_{k-\rho}$ is *any* form of degree $k - \rho$ in x, y, we have $v(\phi_{k-\rho}) \geqq v(Q^n)$. Consequently, $v(P^{(n)}\phi_{k-\rho}) \geqq v(P^{(n)}Q^{(n)})$, and, since $P^{(n)}Q^{(n)} \equiv 0(\mathfrak{q}_i)$, also $P^{(n)}\phi_{k-\rho} \equiv 0(\mathfrak{q}_i)$. This congruence holds true for any form $\phi_{k-\rho}$ of degree $k - \rho$, whence $P^{(n)}\mathfrak{p}^{k-\rho} \equiv 0(\mathfrak{q}_i)$. Since the subform $y^\rho\psi_{k-\rho}$ of $P^n\phi_{k-\rho}$ is of degree k, we have $P^{(n)}\phi_{k-\rho} \equiv 0(\mathfrak{p}^k)$, consequently, since $\mathfrak{A} = [\mathfrak{q}_i, \mathfrak{p}^k]$,

$$P^{(n)}\phi_{k-\rho} \equiv 0(\mathfrak{A}) \tag{6}$$

and

$$P^{(n)}\mathfrak{p}^{k-\rho} \equiv 0(\mathfrak{A}). \tag{6'}$$

We have then, in view of (6), that if $\mathfrak{A}$ contains a polynomial f whose subform f_k is divisible by y^ρ but not by $y^{\rho+1}$, $\mathfrak{A}$ also contains polynomials whose subforms are of degree k and are arbitrarily assigned forms divisible by y^ρ. This shows that $\Omega(\mathfrak{A})$ consists of all the forms which are divisible by a certain power of y, say y^ρ, where necessarily $\rho = k - r$, if r is the dimension of $\Omega(\mathfrak{A})$. This proves the first part of the theorem.

Let $\mathfrak{A}' = \mathfrak{A} : \mathfrak{p}^r$ and let $\mathfrak{A}' \sim \mathfrak{q}_j$, i. e. let $\mathfrak{q}_j$ be the v-ideal such that $v(\mathfrak{A}') = v(\mathfrak{q}_j)$. There exists such an ideal $\mathfrak{q}_j$ *in the sequence* $\{\mathfrak{q}_n\}$, since $\mathfrak{A}$, a primary 0-dimensional ideal, cannot belong to all the ideals $\mathfrak{q}_n$ (whose intersection $\mathfrak{p}_1$ is at least one-dimensional) and since $\mathfrak{A} \equiv 0(\mathfrak{A}')$. We have $\mathfrak{A}'\mathfrak{p}^r \equiv 0(\mathfrak{A}) \equiv 0(\mathfrak{q}_i)$, whence $\mathfrak{q}_j\mathfrak{p}^r \equiv 0(\mathfrak{q}_i)$, since $v(\mathfrak{q}_j\mathfrak{p}^r) = v(\mathfrak{A}'\mathfrak{p}^r) \geqq v(\mathfrak{q}_i)$. *We assert that* $\mathfrak{q}_j\mathfrak{p}^r$ *is also contained in the ideal* $\mathfrak{p}^k$, i. e. $\mathfrak{q}_j$ is contained in $\mathfrak{p}^{k-r}$. In fact, assume the contrary. There will then exist in $\mathfrak{q}_j$ a polynomial $F = F_\sigma + F_{\sigma+1} + \cdots$, whose subform F_σ is of degree $\sigma < k - r$, i. e. $\sigma < \rho$. The polynomial $x^{k-\sigma}F$ belongs to $\mathfrak{q}_i$, since $\mathfrak{q}_j\mathfrak{p}^r \equiv 0(\mathfrak{q}_i)$ and $k - \sigma > r$. It also belongs to $\mathfrak{p}^k$. Hence $x^{k-\sigma}F \equiv 0(\mathfrak{A})$, and this is impossible, since the subform $x^{k-\sigma}F_\sigma$ of $x^{k-\sigma}F$ is of degree k and is at most divisible by y^σ, $\sigma < \rho$. As a result, we have $\mathfrak{q}_j\mathfrak{p}^r \equiv 0(\mathfrak{q}_i)$ *and* $\mathfrak{q}_j\mathfrak{p}^r \equiv 0(\mathfrak{p}^k)$, whence

$$\mathfrak{q}_j\mathfrak{p}^r \equiv 0(\mathfrak{A}). \tag{7}$$

On the other hand, let f be any polynomial belonging to $\mathfrak{A}$, and let us first assume that its subform f_k is not divisible by $y^{\rho+1}$, $f_k = y^\rho\psi_{k-\rho}$, $\psi_{k-\rho} \not\equiv 0(y)$. As above, we determine the polynomials $P^{(n)}$ and $Q^{(n)}$, given by (3), so as to satisfy (4), and we again choose n sufficiently high so that $\mathfrak{p}^{k+n+1} \equiv 0(\mathfrak{A})$. Then the congruence (5) holds true and consequently also (6′), whence $P^{(n)} \equiv 0(\mathfrak{A}') \equiv 0(\mathfrak{q}_j)$, since $k - \rho = r$. Moreover, if n is sufficiently high, we will also have $\mathfrak{p}^{n+k+1} \equiv 0(\mathfrak{q}_j\mathfrak{p}^r)$. For such a value of n we deduce immediately from (4) that $f \equiv 0(\mathfrak{q}_j\mathfrak{p}^r)$, since we have just seen that $P^{(n)}$ belongs to $\mathfrak{q}_j$ and since $Q^{(n)} \equiv 0(\mathfrak{p}^r)$. Thus, we have shown that every polynomial f in $\mathfrak{A}$ belongs to $\mathfrak{q}_j\mathfrak{p}^r$, provided the subform of f is not divisible by $y^{\rho+1}$. If the subform f_k of f is divisible by $y^{\rho+1}$, we consider in $\mathfrak{A}$ a polynomial $\bar{f} = \bar{f}_k + \cdots$, such that $\bar{f}_k \not\equiv 0(y^{\rho+1})$. By the preceding result, we have $\bar{f} \equiv 0(\mathfrak{q}_j\mathfrak{p}^r)$ and also $\bar{f} + f \equiv 0(\mathfrak{q}_j\mathfrak{p}^r)$, whence again $f \equiv 0(\mathfrak{q}_j\mathfrak{p}^r)$. It is therefore proved that

$$\mathfrak{A} \equiv 0(\mathfrak{q}_j\mathfrak{p}^r),$$

and comparing with (7), we deduce

$$\mathfrak{A} = \mathfrak{q}_j\mathfrak{p}^r,$$

and this proves our theorem.

The following consequences can be drawn from Theorem 3:

COROLLARY 3.1. *$\mathfrak{A}$ cannot admit a factor $\mathfrak{p}^\sigma$ with $\sigma > r$ $(r = k - \rho)$,* i. e. *if $\mathfrak{A} = \mathfrak{p}^\sigma\mathfrak{A}_1$ is a product representation of $\mathfrak{A}$, then $\sigma \leqq r$.* In fact, the subforms of $\mathfrak{A}$ form then a system $\Omega(\mathfrak{A})$ of dimension $\geqq \sigma$.

COROLLARY 3.2 (special case $k = h$). *If $\mathfrak{q}_i \equiv 0(\mathfrak{p}^h)$, $\mathfrak{q}_i \not\equiv 0(\mathfrak{p}^{h+1})$ and if $\Omega(\mathfrak{q}_i)$ is of dimension r, then $\mathfrak{q}_i = \mathfrak{p}^r\mathfrak{q}_j$, where $\mathfrak{q}_j$ is an ideal in the sequence $\{\mathfrak{q}_n\}$, and $\mathfrak{q}_i$ does not admit as a factor a higher power of $\mathfrak{p}$ than $\mathfrak{p}^r$. In particular, if $r = 0$, then $\mathfrak{q}_i$ does not admit factors $\mathfrak{p}$ and every element of $\Omega(\mathfrak{q}_i)$ coincides, to within a constant factor in $\mathfrak{k}$, with y^h.*

COROLLARY 3.3. *If $\mathfrak{q}_i \equiv 0(\mathfrak{p}^h)$, $\mathfrak{q}_i \not\equiv 0(\mathfrak{p}^{h+1})$ and if $\mathfrak{p}^k\mathfrak{q}_i \sim \mathfrak{q}_m$* $(k \geqq 0)$, *then*

$$[\mathfrak{q}_m, \mathfrak{p}^{h+k}] = \mathfrak{p}^k\mathfrak{q}_i.$$

In fact, let $\mathfrak{A} = [\mathfrak{q}_m, \mathfrak{p}^{h+k}]$. We have $\mathfrak{q}_m \not\equiv 0(\mathfrak{p}^{h+k+1})$, hence, by Theorem 3, $\mathfrak{A} = \mathfrak{p}^r\mathfrak{q}_j$, where $\mathfrak{q}_j$ is some v-ideal of our sequence $\{\mathfrak{q}_n\}$ and r is the dimension of $\Omega(\mathfrak{A})$. Since $\mathfrak{p}^k\mathfrak{q}_i \sim \mathfrak{q}_m$, we have $\mathfrak{p}^k\mathfrak{q}_i \equiv 0(\mathfrak{q}_m)$ and also $\mathfrak{p}^k\mathfrak{q}_i \equiv 0(\mathfrak{p}^{h+k})$, since $\mathfrak{q}_i \equiv 0(\mathfrak{p}^h)$. Consequently $\mathfrak{p}^k\mathfrak{q}_i \equiv 0(\mathfrak{A}) \equiv 0(\mathfrak{p}^r\mathfrak{q}_j)$. Since the dimension of $\Omega(\mathfrak{p}^k\mathfrak{q}_i)$ is at least k and since $\Omega(\mathfrak{p}^k\mathfrak{q}_i)$ is *a* subset of $\Omega(\mathfrak{A})$ (in view of the assumption $\mathfrak{q}_i \not\equiv 0(\mathfrak{p}^{h+1})$), it follows $k \leqq r$. Now, $\mathfrak{q}_m \sim \mathfrak{p}^k\mathfrak{q}_i$,

$$v(\mathfrak{p}^k\mathfrak{q}_i) = v(\mathfrak{q}_m) \leqq v(\mathfrak{A}) = v(\mathfrak{p}^r\mathfrak{q}_j), \text{ i. e. } v(\mathfrak{q}_i) \leqq v(\mathfrak{p}^{r-k}\mathfrak{q}_j),$$

whence $\mathfrak{p}^{r-k}\mathfrak{q}_j \equiv 0(\mathfrak{q}_i)$ and $\mathfrak{p}^r\mathfrak{q}_j \equiv 0(\mathfrak{p}^k\mathfrak{q}_i)$. Since we also have $\mathfrak{p}^k\mathfrak{q}_i \equiv 0(\mathfrak{p}^r\mathfrak{q}_j)$, it follows that $\mathfrak{A} = \mathfrak{p}^r\mathfrak{q}_j = \mathfrak{p}^k\mathfrak{q}_i$, *q. e. d.*

4. v-ideals and quadratic transformations. We consider the quadratic transformation T:

$$x' = x,\ y' = y/x; \qquad x = x',\ y = x'y',$$

having at $x = y = 0$ a fundamental point, and we denote by $\mathfrak{O}'$ the ring of polynomials in x', y'. $\mathfrak{O}$ is a subring of $\mathfrak{O}'$, and moreover $\mathfrak{O}'$ is contained in the valuation ring of the given valuation B, since, we have assumed $v(y) > v(x)$, whence $v(y') > 0$. Let $\{\mathfrak{q}'_j\}$ be the sequence of 0-dimensional v-ideals in $\mathfrak{O}'$, where $\mathfrak{q}'_1 = \mathfrak{p}' = (x', y')$. We wish to study the connection between the v-ideals $\mathfrak{q}_i$ in $\mathfrak{O}$ and the v-ideals $\mathfrak{q}'_j$ in $\mathfrak{O}'$. Note that $\mathfrak{q}_1 = \mathfrak{p} = (x, y)$, $\mathfrak{q}_2 = (y, x^2)$, whence $\mathfrak{O}'\mathfrak{p} = (x', x'y') = (x')$, where $(x') = \mathfrak{O}'x'$, and $\mathfrak{O}'\mathfrak{q}_2 = (x'y', x'^2) = x'\mathfrak{p}'$.

THEOREM 4. 1. *The extended ideal of an ideal $\mathfrak{q}$ in the sequence $\{\mathfrak{q}_i\}$ is of the form $x'^h\mathfrak{q}'$, where $\mathfrak{q}'$ is an ideal in the sequence $\{\mathfrak{q}'_j\}$ and where $\mathfrak{q} \equiv 0(\mathfrak{p}^h)$, $\mathfrak{q} \not\equiv 0(\mathfrak{p}^{h+1})$. Moreover, $\mathfrak{q}$ is the contracted ideal of $x'^h\mathfrak{q}'$.*

Proof. If $f(x, y) = f_\sigma(x, y) + f_{\sigma+1}(x, y) + \cdots$ is a polynomial in x, y and $f_\sigma(x, y)$ is its subform, then

$$f(x, y) = f(x', x'y) = x'^\sigma f_\sigma(1, y') + x'^{\sigma+1} f_{\sigma+1}(1, y') + \cdots,$$

whence, considering f as an element of $\mathfrak{O}'$, we have $f \equiv 0(x'^\sigma)$, $f \not\equiv 0(x'^{\sigma+1})$. By hypothesis, the subform of any polynomial belonging to $\mathfrak{q}$ is of degree $\geqq h$ and $\mathfrak{q}$ contains a polynomial whose subform is exactly of degree h. Hence $\mathfrak{O}'\mathfrak{q} = x'^h\mathfrak{q}'$, where $\mathfrak{q}' \not\equiv 0(x')$.

We show that $\mathfrak{q}'$ is a v-ideal. We have

(8) $$v(\mathfrak{q}') = v(\mathfrak{q}) - hv(x).$$

Let ω be an element of $\mathfrak{O}'$ such that $v(\omega) \geqq v(\mathfrak{q}')$, and let

$$\omega = F(x', y') = F(x, y/x) = \frac{G(x, y)}{x^\sigma},$$

where F and G are polynomials and $G(x, y) \equiv 0(\mathfrak{p}^\sigma)$ (since clearly the subform of G is of degree $\geqq \sigma$). Since $v(\omega) \geqq v(\mathfrak{q}')$ it follows, by (8), $(h - \sigma)v(x) + v(G) \geqq v(\mathfrak{q})$. Let k be a non-negative integer such that

$$h_1 = h - \sigma + k \geqq 0.$$

We will have then

(9) $$v(x^{h_1}G) \geqq v(x^k\mathfrak{q}) = v(\mathfrak{p}^k\mathfrak{q}).$$

Let $\mathfrak{p}^k\mathfrak{q} \sim \mathfrak{q}_m$, where $\mathfrak{q}_m$ belongs to the sequence $\{\mathfrak{q}_i\}$ of v-ideals in $\mathfrak{O}$. The inequality (9) implies $x^{h_1}G \equiv 0(\mathfrak{q}_m)$, and since $x^{h_1}G \equiv 0(\mathfrak{p}^{h_1+\sigma}) \equiv 0(\mathfrak{p}^{h+k})$, we have

$$x^{h_1}G \equiv 0([\mathfrak{q}_m, \mathfrak{p}^{h+k}]).$$

By Corollary 3. 3 of the preceding section it follows that

$$x^{h_1}G \equiv 0(\mathfrak{q}\mathfrak{p}^k),$$

whence $x^{h_1}G$ is of the form

$$x^{h_1}G = \Sigma A_i(x, y)B_i(x, y),$$

where $A_i \equiv 0(\mathfrak{q})$, $B_i \equiv 0(\mathfrak{p}^k)$. But then B_i/x^k is a polynomial in x', y', and putting $B'_i = B_i/x^k$ we have

$$x^{h-\sigma}G = x^{h_1}G/x^k = \Sigma\, A_i B'_i \equiv 0(\mathfrak{O}'\mathfrak{q}) \equiv 0(x'^h\mathfrak{q}'),$$

whence $\omega = G/x^\sigma \equiv 0(\mathfrak{q}')$. We have thus proved that the ideal $\mathfrak{q}'$ enjoys the property which characterizes the v-ideals: if $v(\omega) \geqq v(\mathfrak{q}')$, then $\omega \equiv 0(\mathfrak{q}')$.

It remains to prove that $\mathfrak{q}'$ belongs to the sequence $\{\mathfrak{q}'_j\}$. But this is obvious, since $\mathfrak{q}'$ is necessarily a 0-dimensional ideal[6] or the unit ideal. That the contracted ideal $\bar{\mathfrak{q}}$ of $x'^h\mathfrak{q}'$ coincides with $\mathfrak{q}$ follows immediately from the fact that $v(\bar{\mathfrak{q}}) \geqq v(x'^h\mathfrak{q}') = v(\mathfrak{q})$, whence $\bar{\mathfrak{q}} \equiv 0(\mathfrak{q})$, while on the other hand we must have, of course, $\mathfrak{q} \equiv 0(\bar{\mathfrak{q}})$. The theorem is thus proved.

The next theorem is in a sense the converse of the preceding theorem.

THEOREM 4.2. *For any ideal $\mathfrak{q}'$ in the sequence $\{\mathfrak{q}'_j\}$ there exists an integer h such that $x'^h\mathfrak{q}'$ is the extended ideal of an ideal $\mathfrak{q}$ belonging to the sequence $\{\mathfrak{q}_i\}$.*

Proof. If $\phi_1(x', y'), \phi_2(x', y'), \cdots, \phi_k(x', y')$ is a base of $\mathfrak{q}'$ and if we write these polynomials in the form of quotients: $\phi_i(x', y') = \psi_i(x, y)/x^m$, with a common denominator x^m, where $\psi_1, \psi_2, \cdots, \psi_k$ are polynomials in x, y, then we see that $x'^m\mathfrak{q}'$ is the extended ideal of the ideal $(\psi_1, \psi_2, \cdots, \psi_k)$ in $\mathfrak{O}$. Thus, there exist integers m such that $x'^m\mathfrak{q}'$ is an extended ideal of an ideal in $\mathfrak{O}$. This will be true for all sufficiently high integers m, since if $\mathfrak{O}'\bar{\mathfrak{q}} = x'^m\mathfrak{q}'$, then $\mathfrak{O}'\mathfrak{p}\bar{\mathfrak{q}} = x'^{m+1}\mathfrak{q}'$. Let h be the smallest possible value of m, and let $\mathfrak{q}$ be the contracted ideal of $x'^h\mathfrak{q}'$:

$$\mathfrak{q} = [\mathfrak{O}, x'^h\mathfrak{q}'], \qquad \mathfrak{O}'\mathfrak{q} = x'^h\mathfrak{q}'. \tag{10}$$

Since the contracted ideal of (x'^h) is $\mathfrak{p}^h$, it follows $\mathfrak{q} \equiv 0(\mathfrak{p}^h)$. Moreover, since the primary ideal $\mathfrak{q}'$ belongs to the prime ideal $\mathfrak{p}' = (x', y')$, we have $x'^n \equiv 0(x'^h\mathfrak{q}')$, if n is sufficiently high. Passing to the contracted ideals we find $\mathfrak{p}^n \equiv 0(\mathfrak{q})$. Hence there exists an ideal $\mathfrak{q}_n$ in the sequence $\{\mathfrak{q}_i\}$ of v-ideals in $\mathfrak{O}$ such that $v(\mathfrak{q}_n) = v(\mathfrak{q})$. Let $\mathfrak{p}^\sigma$ be the highest power of $\mathfrak{p}$ which divides

[6] If n is a sufficiently high integer, then $x^{n+h} \equiv 0(\mathfrak{q})$, whence $x'^n \equiv 0(\mathfrak{q}')$. Let $f(x, y)$ be a polynomial in $\mathfrak{q}$ whose subform is $f_h(x, y)$, $f_h \neq 0$. If $f = f_h + f_{h+1} + \cdots$, then the polynomial $f_h(1, y') + x' f_{h+1}(1, y') + \cdots$ belongs to $\mathfrak{q}'$. It follows that also $[f_h(1, y')]^n$ belongs to $\mathfrak{q}'$, hence $(x'^n, [f_h(1, y')]^n) \equiv 0(\mathfrak{q}')$, and consequently $\mathfrak{q}'$ is 0-dimensional, or is the unit ideal, since $f_h(1, y') \neq 0$.

$\mathfrak{q}_n$, $\mathfrak{q}_n \equiv 0(\mathfrak{p}^\sigma)$, $\mathfrak{q}_n \not\equiv 0(\mathfrak{p}^{\sigma+1})$. We have $\mathfrak{q} \equiv 0(\mathfrak{p}^h)$ and $\mathfrak{q} \not\equiv 0(\mathfrak{p}^{h+1})$, consequently $\sigma \leqq h$, because $\mathfrak{q} \equiv 0(\mathfrak{q}_n)$. Let $\mathfrak{O}'\mathfrak{q}_n = x'^\sigma \mathfrak{q}'_j$, where $\mathfrak{q}'_j$, by the preceding theorem, is a v-ideal in $\mathfrak{O}'$. Since $\mathfrak{q}_n \sim \mathfrak{q}$, we have $v(x'^\sigma \mathfrak{q}'_j) = v(x'^h \mathfrak{q}')$. If $\sigma = h$, then $v(\mathfrak{q}'_j) = v(\mathfrak{q}')$, whence $\mathfrak{q}'_j = \mathfrak{q}'$, since both are v-ideals. In this case $\mathfrak{q}_n$ and $\mathfrak{q}'$ must coincide, since both are contracted ideals of $x'^h\mathfrak{q}'$, and the theorem is proved.

Assume $\sigma < h$. We have

$$\mathfrak{q} \equiv 0([\mathfrak{q}_n, \mathfrak{p}^h]). \tag{11}$$

If $f(x, y)$ is any polynomial belonging to $[\mathfrak{q}_n, \mathfrak{p}^h]$, then $f = x'^h f'(x'y')$ and $f \equiv 0(x'^\sigma \mathfrak{q}'_j)$. Since $v(x'^\sigma \mathfrak{q}'_j) = v(x'^h \mathfrak{q}')$, it follows that $f'(x', y') \equiv 0(\mathfrak{q}')$. Hence $f \equiv 0(x'^h \mathfrak{q}')$ and consequently, by (10), $f \equiv 0(\mathfrak{q})$. We have therefore $[\mathfrak{q}_n, \mathfrak{p}^h] \equiv 0(\mathfrak{q})$, and, by (11), it follows

$$\mathfrak{q} = [\mathfrak{q}_n, \mathfrak{p}^h]. \tag{12}$$

We have $\mathfrak{q}_n \equiv 0(\mathfrak{p}^\sigma)$, $\mathfrak{q}_n \not\equiv 0(\mathfrak{p}^{\sigma+1})$ and $h > \sigma$. We can then apply Theorem 3 and we obtain $\mathfrak{q} = \mathfrak{p}^r \mathfrak{q}_a$, where $\mathfrak{q}_a$ is again a v-ideal in $\mathfrak{O}$. Here r is the dimension of $\Omega(\mathfrak{q})$ and consequently $r > 0$, because $\mathfrak{p}^{h-\sigma}\mathfrak{q}_n \equiv 0(\mathfrak{q})$, whence $r \geqq h - \sigma > 0$. We have then

$$\mathfrak{O}'\mathfrak{q} = \mathfrak{O}'\mathfrak{p}^r \mathfrak{q}_a = x'^r \mathfrak{O}' \mathfrak{q}_a = x'^h \mathfrak{q}',$$

whence $\mathfrak{O}'\mathfrak{q}_a = x'^{h-r}\mathfrak{q}'$. This contradicts our hypothesis that h is the smallest integer such that $x'^h\mathfrak{q}'$ is an extended ideal. Hence $\sigma < h$ is impossible, and the theorem is proved.

If $\mathfrak{A}$ is any ideal in $\mathfrak{O}$ and if $\mathfrak{O}'\mathfrak{A} = x'^h \mathfrak{A}'$, $\mathfrak{A}' \not\equiv 0(x')$, we shall call $\mathfrak{A}'$ the *transformed ideal* of $\mathfrak{A}$ (under the quadratic transformation T), in symbols: $\mathfrak{A}' = T(\mathfrak{A})$. By Theorems 4. 1 and 4. 2, the transform of any v-ideal $\mathfrak{q}$ in $\mathfrak{O}$ is a v-ideal $\mathfrak{q}'$ in $\mathfrak{O}'$, and every v-ideal $\mathfrak{q}'$ in $\mathfrak{O}'$ is the transform of at least one v-ideal $\mathfrak{q}$ in $\mathfrak{O}$, everything referred to a fixed valuation. There may be more than one v-ideal in $\mathfrak{O}$ whose transform is $\mathfrak{q}'$. To find them, we again consider the smallest integer h such that $x'^h\mathfrak{q}'$ is an extended ideal. It follows from the preceding proof that if $\mathfrak{q}$ is the contracted ideal of $x'^h\mathfrak{q}'$, then $T(\mathfrak{q}) = \mathfrak{q}'$, and from Theorem 4. 1 it follows, that any other v-ideal $\bar{\mathfrak{q}}$ in $\mathfrak{O}$ such that $T(\bar{\mathfrak{q}}) = \mathfrak{q}'$ must be the contracted ideal of $x'^\sigma\mathfrak{q}'$, where σ is some integer greater than h. For a given integer σ greater than h, the contracted ideal of $x'^\sigma\mathfrak{q}'$ may or may not be a v-ideal in $\mathfrak{O}$, but let us at any rate examine this contracted ideal. Let us denote it by $\mathfrak{A}_\sigma$. Since $x'^\sigma\mathfrak{q}'$ is the extended ideal of $\mathfrak{A}_\sigma$ and *also* of $\mathfrak{p}^{\sigma-h}\mathfrak{q}$, it follows

$$\mathfrak{p}^{\sigma-h}\mathfrak{q} \equiv 0(\mathfrak{A}_\sigma). \tag{13}$$

For the same reason we have $v(\mathfrak{A}_\sigma) = v(\mathfrak{p}^{\sigma-h}\mathfrak{q})$. Let $\mathfrak{q}_m$ be the v-ideal equivalent to both $\mathfrak{A}_\sigma$ and $\mathfrak{p}^{\sigma-h}\mathfrak{q}$, $\mathfrak{q}_m \sim \mathfrak{A} \sim \mathfrak{p}^{\sigma-h}\mathfrak{q}$. By Theorem 3, Corollary 3. 3, we have $\mathfrak{p}^{\sigma-h}\mathfrak{q} = [\mathfrak{q}_m, \mathfrak{p}^\sigma]$. Now, $\mathfrak{A}_\sigma \sim \mathfrak{q}_m$ implies $\mathfrak{A}_\sigma \equiv 0(\mathfrak{q}_m)$ and since $\mathfrak{O}'\mathfrak{A}_\sigma = x'^\sigma\mathfrak{q}'$ we also have $\mathfrak{A}_\sigma \equiv 0(\mathfrak{p}^\sigma)$. Consequently $\mathfrak{A}_\sigma \equiv 0(\mathfrak{p}^{\sigma-h}\mathfrak{q})$, and hence, by (13), $\mathfrak{A}_\sigma = \mathfrak{p}^{\sigma-h}\mathfrak{q}$. We therefore can state the following theorem.

THEOREM 4. 3. *If $\mathfrak{q}'$ is a v-ideal in $\mathfrak{O}'$, belonging to the sequence $\{\mathfrak{q}'_j\}$, and if h is the smallest integer such that $x'^h\mathfrak{q}'$ is an extended ideal of an ideal in $\mathfrak{O}$, then the contracted ideal $\mathfrak{q}$ of $x'^h\mathfrak{q}'$ is a v-ideal in $\mathfrak{O}$, a member of the sequence $\{\mathfrak{q}_i\}$, and $T(\mathfrak{q}) = \mathfrak{q}'$. If σ is any integer $\geqq h$, the contracted ideal of $x'^\sigma\mathfrak{q}'$ is $\mathfrak{p}^{\sigma-h}\mathfrak{q}$, but need not be a v-ideal. In particular, the v-ideals $\mathfrak{q}_i$ whose transform is the given v-ideal $\mathfrak{q}'$ are all of the form $\mathfrak{p}^{\sigma-h}\mathfrak{q}$, $\sigma \geqq h$.*

We shall regard $\mathfrak{q}$ as the transform of $\mathfrak{q}'$ by T^{-1}: $\mathfrak{q} = T^{-1}(\mathfrak{q}')$. By the definition of $\mathfrak{q}$, the system $\Omega(\mathfrak{q})$ of the subforms of $\mathfrak{q}$ must be of dimension $r = 0$, i. e. every form in $\Omega(\mathfrak{q})$ differs from y^h by a factor c, $c \varepsilon \mathfrak{k}$. In fact, if it were $r > 0$, then $\mathfrak{q}$ could be put in the form $\mathfrak{q} = \mathfrak{p}^r\bar{\mathfrak{q}}$, where $\bar{\mathfrak{q}}$ is also a v-ideal (Corollary 3. 2), and we would have $\mathfrak{O}'\bar{\mathfrak{q}} = x'^{h-r}\mathfrak{q}'$, contrary to our assumption that h is the smallest integer such that $x'^h\mathfrak{q}'$ is an extended ideal. Conversely, if $\mathfrak{q}$ is a v-ideal in $\mathfrak{O}$, belonging to the sequence $\{\mathfrak{q}_i\}$, and if $\Omega(\mathfrak{q})$ is of dimension zero, then $\mathfrak{q}$ is so related to its transform $\mathfrak{q}' = T(\mathfrak{q})$, that $\mathfrak{q} = T^{-1}\mathfrak{q}'$, i. e. if $\mathfrak{O}'\mathfrak{q} = x'^h\mathfrak{q}'$, then h is the smallest integer such that $x'^h\mathfrak{q}'$ is an extended ideal. In fact, in the contrary case, $\mathfrak{q}$ could be put in the form $\mathfrak{p}^r\bar{\mathfrak{q}}$, $r > 0$, (by Theorem 4. 3), contrary to the hypothesis that $\Omega(\mathfrak{q})$ is of dimension zero. Thus, there is a one to one correspondence between the ideals $\mathfrak{q}'$ of the sequence $\{\mathfrak{q}'_j\}$ and those ideals $\mathfrak{q}$ of the sequence $\{\mathfrak{q}_i\}$ whose system $\Omega(\mathfrak{q})$ of subforms is of dimension zero: to each $\mathfrak{q}'$ there corresponds a unique ideal $\mathfrak{q} = T^{-1}(\mathfrak{q}')$, and $\mathfrak{q}' = T(\mathfrak{q})$.

Let $\mathfrak{q}'_\alpha$, $\mathfrak{q}'_\beta$ be two distinct v-ideals in $\mathfrak{O}'$, and let $\mathfrak{q}_i = T^{-1}(\mathfrak{q}'_\alpha)$, $\mathfrak{q}_j = T^{-1}(\mathfrak{q}'_\beta)$, be their transforms in $\mathfrak{O}$. Suppose that $i < j$, whence $\mathfrak{q}_j \equiv 0(\mathfrak{q}_i)$. *We assert that in such a case also $\alpha < \beta$.* In fact, let $\mathfrak{q}_i \equiv 0(\mathfrak{p}^h)$, $\mathfrak{q}_i \not\equiv 0(\mathfrak{p}^{h+1})$ and let $\mathfrak{q}_j \equiv 0(\mathfrak{p}^\sigma)$, $\mathfrak{q}_j \not\equiv 0(\mathfrak{p}^{\sigma+1})$, whence $\mathfrak{O}'\mathfrak{q}_i = x'^h\mathfrak{q}'_\alpha$, $\mathfrak{O}'\mathfrak{q}_j = x'^\sigma\mathfrak{q}'_\beta$, and clearly, $\sigma \geqq h$. Evidently $\mathfrak{p}^{\sigma-h}\mathfrak{q}_i \equiv 0(x'^\sigma\mathfrak{q}'_\alpha)$. Supposing that $\alpha > \beta$, whence $\mathfrak{q}'_\alpha \equiv 0(\mathfrak{q}'_\beta)$, we would have $x'^\sigma\mathfrak{q}'_\alpha \equiv 0(x'^\sigma\mathfrak{q}'_\beta)$, and passing to the contracted ideals in $\mathfrak{O}$, we would get by Theorem 4. 3 $\mathfrak{p}^{\sigma-h}\mathfrak{q}_i \equiv 0(\mathfrak{q}_j)$. The equality $\sigma = h$ is excluded, because $\mathfrak{q}_j \equiv 0(\mathfrak{q}_i)$ and $\mathfrak{q}_j \neq \mathfrak{q}_i$. Hence $\sigma > h$, but then the congruence $\mathfrak{p}^{\sigma-h}\mathfrak{q}_i \equiv 0(\mathfrak{q}_j)$ is in contradiction with the fact that the $\Omega(\mathfrak{q}_j)$ (consisting of form of degree σ) is of dimension zero. Hence our assumption $\alpha > \beta$ leads to a contradiction, and consequently it is proved that $i < j$ implies $\alpha < \beta$. If then $T^{-1}(\mathfrak{q}'_j) = \mathfrak{q}_{\alpha_j}$, the indices α_j form an ascending sequence, $\alpha_0 < \alpha_1 < \alpha_2 < \cdots$. It is immediately verified that $\alpha_0 = 0$, $\alpha_1 = 2$. Hence $\alpha_j > j$, if $j > 0$. Let now $\mathfrak{q}_s$ be any v-ideal in the sequence $\{\mathfrak{q}_i\}$ and let $T(\mathfrak{q}_s) = \mathfrak{q}'_\sigma$. By Theorem 4.3 we have $\mathfrak{q}_s = \mathfrak{p}^\rho \mathfrak{q}_{\alpha_\sigma}, \rho \geqq 0$, whence $\mathfrak{q}_s \equiv 0(q_{\alpha\sigma})$, and $s \geqq \alpha_\sigma$, i. e. $s > \sigma$. Observing that the length of the ideal $\mathfrak{q}_s$ is equal to s, we can reassume the preceding results in the following theorem:

THEOREM 4.4. *If* $\mathfrak{q}_{\alpha_j} = T^{-1}(\mathfrak{q}'_j)$, *then* $\alpha_j > j$, *if* $j \geqq 1$. *Moreover, if* $\mathfrak{q}_s$ *is any ideal in the sequence* $\{\mathfrak{q}_i\}$ *and if* $\mathfrak{q}'_\sigma = T(\mathfrak{q}_s)$, *then* $s > \sigma$, *i. e. length of* $\mathfrak{q}_s >$ *length of* $\mathfrak{q}_\sigma$.

5. Simple and composite *v*-ideals. We say that an ideal $\mathfrak{A}$ in $\mathfrak{O}$ is *simple*, if $\mathfrak{A}$ cannot be represented as the product of two ideals, both different from the unit ideal, i. e. if $\mathfrak{A} = \mathfrak{B}\mathfrak{C}$, $\mathfrak{B} \neq (1)$ implies $\mathfrak{C} = (1)$. An ideal is *composite* if it is not simple.

THEOREM 5.1. *A composite v-ideal can be represented as a product of v-ideals different from the unit ideal.*

Proof. Let $\mathfrak{A}$ be a v-ideal in $\mathfrak{O}$, belonging to some valuation B, and let $\mathfrak{A} = \mathfrak{B}\mathfrak{C}$, $\mathfrak{B} \neq (1)$, $\mathfrak{C} \neq (1)$. Let $\mathfrak{B}_1$, $\mathfrak{C}_1$ be the v-ideals belonging to B such that $\mathfrak{B} \sim \mathfrak{B}_1$, $\mathfrak{C} \sim \mathfrak{C}_1$. Since $\mathfrak{B} \equiv 0(\mathfrak{B}_1)$, $\mathfrak{C} \equiv 0(\mathfrak{C}_1)$, we have $\mathfrak{A} \equiv 0(\mathfrak{B}_1\mathfrak{C}_1)$. On the other hand, $v(\mathfrak{A}) = v(\mathfrak{B}) + v(\mathfrak{C}) = v(\mathfrak{B}_1) + v(\mathfrak{C}_1) = v(\mathfrak{B}_1\mathfrak{C}_1)$, whence $\mathfrak{B}_1\mathfrak{C}_1 \equiv 0(\mathfrak{A})$, since $\mathfrak{A}$ is a v-ideal. We conclude that $\mathfrak{A} = \mathfrak{B}_1\mathfrak{C}_1$, and it remains to prove that $\mathfrak{B}_1 \neq (1)$ and $\mathfrak{C}_1 \neq (1)$. Assume the contrary, and let, for instance $\mathfrak{B}_1 = (1)$. Then $\mathfrak{A} = \mathfrak{C}_1$, whence $\mathfrak{C} \equiv 0(\mathfrak{A})$, and consequently $\mathfrak{C} = \mathfrak{A}$ since $\mathfrak{A} = \mathfrak{B}\mathfrak{C} \equiv 0(\mathfrak{C})$. We have then $\mathfrak{A} = \mathfrak{B}\mathfrak{A}$, and this implies ([2], p. 36), $\mathfrak{B} = (1)$, contrary to hypothesis.

Consider the given valuation B and the corresponding sequences $\{\mathfrak{q}_i\}$, $\{\mathfrak{q}'_j\}$ of 0-dimensional v-ideals in the polynomial rings $\mathfrak{O} = \mathfrak{k}[x,y]$, $\mathfrak{O}' = \mathfrak{k}[x',y']$ respectively. Let, as before, $T^{-1}(\mathfrak{q}'_j) = \mathfrak{q}_{\alpha_j}$. It is clear that all the simple v-ideals of the sequence $\{\mathfrak{q}_i\}$, except the ideal $\mathfrak{p} = \mathfrak{q}_1$, belong to the sequence $\{\mathfrak{q}_{\alpha_j}\}$, since any ideal $\mathfrak{q}_i$ $(i \neq 1)$, not in the sequence $\{\mathfrak{q}_{\alpha_j}\}$, is, by Theorem 4.3, either of the form $\mathfrak{p}^\rho\mathfrak{q}_{\alpha_j}$, $\rho > 0$, $\mathfrak{q}_{\alpha_j} \neq (1)$, or of the form $\mathfrak{p}^\rho$, $\rho > 1$. Now suppose that $\mathfrak{q}_{\alpha_j}$, for a given j, is a composite ideal. Then we can write, by Theorem 5.1, $\mathfrak{q}_{\alpha_j} = \mathfrak{q}_s\mathfrak{q}_t$, where $\mathfrak{q}_s, \mathfrak{q}_t$ are in the sequence $\{\mathfrak{q}_i\}$. Hence $\mathfrak{q}'_j = T(\mathfrak{q}_{\alpha_j}) = \mathfrak{q}'_\sigma\mathfrak{q}'_\tau$, where $\mathfrak{q}'_\sigma = T(\mathfrak{q}_s)$ and $\mathfrak{q}'_\tau = T(\mathfrak{q}_t)$. We have seen above that $\Omega(\mathfrak{q}_{\alpha_j})$ is of dimension zero; therefore neither $\mathfrak{q}_s$ nor $\mathfrak{q}_t$ can be a power of $\mathfrak{p}$. It follows that $\mathfrak{q}'_\sigma \neq (1)$ and $\mathfrak{q}'_\tau \neq (1)$, i. e. $\mathfrak{q}'_j$ is composite. We conclude then that if $\mathfrak{q}'_j$ is a simple v-ideal in $\mathfrak{O}'$, then its transform $\mathfrak{q}_{\alpha_j} = T^{-1}(\mathfrak{q}'_j)$ is also a simple v-ideal. Much more difficult is to prove the converse:

THEOREM 5.2. *The transform* $T(\mathfrak{q}_i)$ *of a simple v-ideal* $\mathfrak{q}_i$ *in* $\mathfrak{O}$ *is a simple v-ideal (in* $\mathfrak{O}'$*).*

The proof of this theorem will be given in Part II of the paper (Corollary 11.2), where we shall characterize the simple v-ideals from the point of view of formal power series. Here we shall use this theorem without proof.

Let $\mathfrak{P}_1, \mathfrak{P}_2, \cdots, \mathfrak{P}_i, \cdots$ be the sequence of simple v-ideals, different from (1), as they occur in the sequence $\{\mathfrak{q}_i\}$:

$$(14) \qquad \mathfrak{P}_1 \supset \mathfrak{P}_2 \supset \cdots, \qquad \mathfrak{P}_1 = \mathfrak{p} = (x, y), \ \mathfrak{P}_2 = \mathfrak{q}_2 = (y, x^2).$$

Let similarly $\mathfrak{P}'_1, \mathfrak{P}'_2, \cdots$ be the sequence of simple v-ideals, different from (1), as they occur in the sequence $\{\mathfrak{q}'_\nu\}$. By the preceding results, especially by Theorem 5.2, there is (1,1) correspondence between the ideals $\mathfrak{P}_i$, $i > 1$, and the ideals $\mathfrak{P}'_a$, where to $\mathfrak{P}_i$ corresponds $T(\mathfrak{P}_i) = \mathfrak{P}'_a$, and $\mathfrak{P}_i = T^{-1}(\mathfrak{P}'_a)$. Moreover, by Theorem 4.4, if $T(\mathfrak{P}_i) = \mathfrak{P}'_\alpha$, and $T(\mathfrak{P}_j) = \mathfrak{P}'_\beta$, then $i < j$ implies $\alpha < \beta$. Since $T(\mathfrak{P}_2) = \mathfrak{P}'_1$, we conclude with the following theorem:

THEOREM 5.3. $T(\mathfrak{P}_i) = \mathfrak{P}'_{i-1}$, *i. e. the transform of the simple v-ideal* $\mathfrak{P}_i$ *by the quadratic transformation* T *is the simple v-ideal* $\mathfrak{P}'_{i-1}$.

As a consequence of this theorem, it follows incidentally that the sequence $\{\mathfrak{q}_n\}$ contains infinitely many simple v-ideals. In fact, if we assume that any sequence $\{\mathfrak{q}_n\}$ of v-ideals in any polynomial ring contains always at least $k > 0$ simple v-ideals (it always contains at least one, namely the prime ideal $\mathfrak{q}_1 = \mathfrak{p} = (x, y)$), and if we apply this assumption to the sequence $\{\mathfrak{q}'_\nu\}$ of v-ideals in the polynomial ring $\mathfrak{O}'$, we deduce immediately, by Theorem 5.3, that any sequence $\{\mathfrak{q}_n\}$ contains at least $k + 1$ simple v-ideals.

6. Properties of simple *v*-ideals. Heretofore we have been dealing with a fixed valuation B and with the v-ideals in $\mathfrak{O}$ belonging to B. Now, a v-ideal belonging to B may also occur as a v-ideal for many other valuations. Consider, in particular, the i-th simple v-ideal $\mathfrak{P}_i$ for B, and let $\bar{B}$ be another valuation for which $\mathfrak{P}_i$ is a v-ideal. We assert that $\mathfrak{P}_i$ *is also the i-th simple v-ideal for* B. The assertion is trivial for $i = 1$, because $\mathfrak{P}_1 = \mathfrak{p} = (x, y)$. We may then proceed by induction, assuming that our assertion is true for $i - 1$. Let $\{\mathfrak{q}_n\}$ and $\{\bar{\mathfrak{q}}_n\}$ be the sequences of v-ideals in $\mathfrak{O}$ for the valuations B and $\bar{B}$ respectively. Since $\mathfrak{P}_i$ occurs in both sequences, and since the ideals $\mathfrak{q}_n$ and $\bar{\mathfrak{q}}_n$ are primary ideals belonging to the prime 0-dimensional ideals $\mathfrak{q}_1$ and $\bar{\mathfrak{q}}_1$ respectively, it follows that $\mathfrak{q}_1 = \bar{\mathfrak{q}}_1 = \mathfrak{p} = (x, y)$. Furthermore, since $\mathfrak{P}_i$ is simple, its system $\Omega(\mathfrak{P}_i)$ of subforms is of dimension 0. If $cx + dy$ is the base of $\Omega(\mathfrak{P}_i)$, we must have $v(cx + dy) > v(\mathfrak{p})$ in B and also

$v(cx+dy) > v(\mathfrak{p})$ in $\bar{B}$. We may therefore assume that $v(y) > v(x)$ in both valuations B and $\bar{B}$. We apply the quadratic transformation $T: x' = x,\ y' = y/x$ to both sequences $\{\mathfrak{q}_n\}$ and $\{\bar{\mathfrak{q}}_n\}$. The sequences $\{\mathfrak{q}'_\nu\}$ and $\{\bar{\mathfrak{q}}'_\nu\}$ of v-ideals in $\mathfrak{O}' = \mathfrak{k}[x', y']$ belonging to the valuations B and $\bar{B}$ respectively, will consist of primary ideals belonging to the prime ideal $\mathfrak{p}' = (x', y')$. The transform $\mathfrak{P}'_{i-1}$ of $\mathfrak{P}_i$ belongs to both sequences and is the $(i-1)$-th simple v-ideal in the sequence $\{\mathfrak{q}'_\nu\}$. Hence, by our induction, $\mathfrak{P}'_{i-1}$ is also the $(i-1)$-th simple ideal in the sequence $\{\bar{\mathfrak{q}}'_\nu\}$. As a consequence, $\mathfrak{P}_i$ must be the i-th simple ideal in the sequence $\{\bar{\mathfrak{q}}_n\}$, and this proves our assertion.

Thus, given a simple v-ideal $\mathfrak{P} = \mathfrak{P}_i$ in $\mathfrak{O}$, there is uniquely determined an integer i, such that $\mathfrak{P}$ is the i-th simple v-ideal in the sequence of simple v-ideals of any valuation for which $\mathfrak{P}$ is a valuation ideal. We shall say that $\mathfrak{P}$ is *a simple v-ideal of kind i*, by analogy with the terminology of the geometric theory of infinitely near points in the plane, where a point $O^{(i)}$, infinitely near the point $O^{(1)} \equiv (0,0)$, is said to be of kind i, if it is in the $(i-1)$-th neighborhood of $O^{(1)}$. The identity of the two concepts will appear from the formal power series considerations of Part II. However, already at this stage, the analogy appears from the fact, that while it takes $i-1$ successive quadratic transformations to transform a point $O^{(i)}$ of kind i into a proper point (a point of kind 1), it takes as well $i-1$ successive quadratic transformations to transform a simple v-ideal $\mathfrak{P}_i$ of kind i into a simple v-ideal of kind 1, i. e. into a prime 0-dimensional ideal.

THEOREM 6.1. *If $\{\mathfrak{P}_\alpha\}$ and $\{\bar{\mathfrak{P}}_\alpha\}$ are the sequences of simple v-ideals in $\mathfrak{O}$ belonging to valuations B and $\bar{B}$ respectively and if, for a given i, we have $\mathfrak{P}_i = \bar{\mathfrak{P}}_i$, then also $\mathfrak{P}_\alpha = \bar{\mathfrak{P}}_\alpha$ for any $\alpha < i$. In other words, the $i-1$ simple v-ideals which precede a given simple v-ideal $\mathfrak{P}_i$ of kind i in a given valuation B for which $\mathfrak{P}_i$ is a v-ideal, are uniquely determined by $\mathfrak{P}_i$, being independent of the valuation B.*

Proof by induction. The theorem is trivial for $i = 1$. Assume that the theorem is true for simple v-ideals of kind $i-1$, and apply the quadratic transformation T. We will have then in $\mathfrak{O}'$: $\mathfrak{P}'_{i-1} = \bar{\mathfrak{P}}'_{i-1}$, where $\mathfrak{P}'_{\alpha-1} = T(\mathfrak{P}_\alpha)$ and $\bar{\mathfrak{P}}'_{\alpha-1} = T(\bar{\mathfrak{P}}_\alpha)$. Hence, by our induction, $\mathfrak{P}'_\alpha = \bar{\mathfrak{P}}'_\alpha$, for $\alpha = 1, 2, \cdots, i-2$, whence $\mathfrak{P}_\alpha = \bar{\mathfrak{P}}_\alpha$ for $\alpha = 2, \cdots, i-1$, because $\mathfrak{P}_\alpha$ as well as $\bar{\mathfrak{P}}_\alpha$ is the contracted ideal of the ideal $x'^h\mathfrak{P}'_{\alpha-1}$ $(= x'^h\bar{\mathfrak{P}}'_{\alpha-1})$, where h is the smallest integer such that $x'^h\mathfrak{P}'_{\alpha-1}$ is an extended ideal of an ideal in $\mathfrak{O}$. Moreover, $\mathfrak{P}_1 = \mathfrak{P}'_1$, since $\mathfrak{P}_1$ and $\mathfrak{P}'_1$ are the prime ideals belonging to $\mathfrak{P}_i$ and $\mathfrak{P}'_i$ respectively q. e. d.

A much stronger theorem can be proved:

THEOREM 6.2. *Under the hypothesis $\mathfrak{P}_i = \bar{\mathfrak{P}}_i$ of Theorem 6.1, the set of v-ideals for B which precede $\mathfrak{P}_i$ (divisors of $\mathfrak{P}_i$) coincides with the set of v-ideals for $\bar{B}$ which precede $\bar{\mathfrak{P}}_i$.*

Proof by induction with respect to i. For $i = 1$ the theorem is trivial. Assume that the theorem is true for simple v-ideals of kind $i - 1$. Let $\mathfrak{q}_k$ be a v-ideal for B such that $\mathfrak{P}_i \equiv 0(\mathfrak{q}_k)$. If $\mathfrak{q}_k$ is simple, then it is also a v-ideal for $\bar{B}$, by the preceding theorem. If $\mathfrak{q}_k$ is composite, we know by Theorem 5.1 that it can be factored into simple v-ideals belonging to B. Since $\mathfrak{q}_k$ is a proper divisor of $\mathfrak{P}_i$, only factor $\mathfrak{P}_j$, $j < i$, can occur. Let

$$\mathfrak{q}_k = \mathfrak{P}_1^{\alpha_1}\mathfrak{P}_2^{\alpha_2} \cdots \mathfrak{P}_{i-1}^{\alpha_{i-1}}, \qquad \mathfrak{P}_1 = \mathfrak{p} = (x, y).$$

We consider separately two cases: (1) $\alpha_1 = 0$, (2) $\alpha_1 > 0$.

(1) First case: $\alpha_1 = 0$. We apply our quadratic transformation T. We find then

$$T(\mathfrak{q}_k) = \mathfrak{q}'_\sigma = \mathfrak{P}'^{\alpha_2}_1 \cdots \mathfrak{P}'^{\alpha_{i-1}}_{i-2}. \tag{15}$$

Since the factor $\mathfrak{P}_1 (= \mathfrak{p})$ does not occur in the factorization of $\mathfrak{q}_k$, the system $\Omega(\mathfrak{q}_k)$ of subforms of $\mathfrak{q}_k$ is of dimension zero. Hence $\mathfrak{q}_k = T^{-1}(\mathfrak{q}'_\sigma)$. Since also $\mathfrak{P}_i = T^{-1}(\mathfrak{P}'_{i-1})$ and since $\mathfrak{P}_i \equiv 0(\mathfrak{q}_k)$, we have, by Theorem 4.4, $\mathfrak{P}'_{i-1} \equiv 0(\mathfrak{q}'_\sigma)$. Now $\mathfrak{P}'_{i-1}$ is a v-ideal in $\mathfrak{O}'$ for both B and $\bar{B}$. Hence, by our induction, $\mathfrak{q}'_\sigma$ is also a v-ideal for $\bar{B}$. But then also $\mathfrak{q}_k$ must be a v-ideal in $\mathfrak{O}$ for $\bar{B}$, since $\mathfrak{q}_k = T^{-1}(\mathfrak{q}'_\sigma)$.

(2) Second case: $\alpha_1 > 0$. We now use an induction with respect to k, i.e. we assume it has been already proved that all the v-ideals $\mathfrak{q}_j$ for B, $j < k$, are also v-ideals for $\bar{B}$. Since in the factorization (15) the factor $\mathfrak{p}$ occurs to the power α_1, it follows that the system $\Omega(\mathfrak{q}_k)$ of subforms of $\mathfrak{q}_k$ is of dimension α_1. Hence, by Theorem 3 (Corollary 3.2) we can write $\mathfrak{q}_k = \mathfrak{p}^{\alpha_1}\mathfrak{q}_t$, where $\mathfrak{q}_t$ is a v-ideal for B. Let $\mathfrak{p}^{\alpha_1-1}\mathfrak{q}_t \sim \mathfrak{q}_s$. We have $\mathfrak{q}_k \equiv 0(\mathfrak{p}\mathfrak{q}_s)$; on the other hand, since $v(\mathfrak{p}^{\alpha_1-1}\mathfrak{q}_t) = v(\mathfrak{q}_s)$, it follows that $v(\mathfrak{p}\mathfrak{q}_s) = v(\mathfrak{p}^{\alpha_1}\mathfrak{q}_t) = v(\mathfrak{q}_k)$, whence $\mathfrak{p}\mathfrak{q}_s \equiv 0(\mathfrak{q}_k)$. Consequently $\mathfrak{q}_k = \mathfrak{p}\mathfrak{q}_s$. We now consider the k-th v-ideal $\bar{\mathfrak{q}}_k$ for $\bar{B}$. We must suppose that the factor $\mathfrak{P}_1$ occurs also in the factorization of $\bar{\mathfrak{q}}_k$, since otherwise $\bar{\mathfrak{q}}_k$ would also be a v-ideal for B, by the preceding case $\alpha_1 = 0$, whence necessarily $\bar{\mathfrak{q}}_k = \mathfrak{q}_k$, since k is the length of both ideals $\mathfrak{q}_k$, $\bar{\mathfrak{q}}_k$. Let then $\bar{\mathfrak{q}}_k = \mathfrak{p}\bar{\mathfrak{q}}_\sigma$. By our induction, $\bar{\mathfrak{q}}_\sigma$ is a v-ideal for B, since $\sigma < k$; i.e. $\bar{\mathfrak{q}}_\sigma = \mathfrak{q}_\sigma$. Hence $\mathfrak{q}_k = \mathfrak{p}\mathfrak{q}_s$, $\bar{\mathfrak{q}}_k = \mathfrak{p}\mathfrak{q}_\sigma$. Since of the two ideals $\mathfrak{q}_s$, $\mathfrak{q}_\sigma$ one is a divisor of the other, the same is true of the ideals $\mathfrak{q}_k$ and $\bar{\mathfrak{q}}_k$. But these two ideals have the same length k, consequently $\mathfrak{q}_k = \bar{\mathfrak{q}}_k$, q. e. d.

Remark. The following example shows that a composite v-ideal does not determine the v-ideals preceding it. Let B be the valuation defined by the branch $y = x^{3/2}$ and let $\bar{B}$ be the valuation determined by the branch $y = x^{2/3}$. We have then $\mathfrak{q}_1 = \mathfrak{p} = (x, y)$, $\mathfrak{q}_2 = (y, x^2)$, $\mathfrak{q}_3 = \mathfrak{p}^2$, while $\bar{\mathfrak{q}}_1 = \mathfrak{p} = \mathfrak{q}_1$, $\bar{\mathfrak{q}}_2 = (x, y^2) \neq \mathfrak{q}_2$, $\bar{\mathfrak{q}}_3 = \mathfrak{p}^2 = \mathfrak{q}_3$.

7. A factorization theorem for *v*-ideals. We know from the preceding sections that any v-ideal $\mathfrak{A}$, belonging to a given valuation B, can be factored into simple v-ideals belonging to the same valuation B. The question arises as to the *unicity of this factorization.* The unicity of the factorization may be *a priori* intended in more than one way. In the first place, we *may fix some valuation* B to which $\mathfrak{A}$ belongs, and we may ask whether the factorization of $\mathfrak{A}$ into simple v-ideals *belonging to* B is unique. We may go a step further and ask whether the factorization of $\mathfrak{A}$, if unique for a given valuation B, *is independent of* B. Finally, we may formulate the unicity of factorization of $\mathfrak{A}$ in its strongest possible form and assert, that $\mathfrak{A}$ *can be factored in a unique manner into simple v-ideals,* where we allow *a priori* that the simple v-factors may belong to different valuations. It is this strongest form of the unicity theorem which we proceed to prove. It will, of course, follow from this theorem, that the simple v-factors are v-ideals for any valuation for which $\mathfrak{A}$ is a v-ideal.

We prove, however, a stronger theorem, from which the unique factorization of v-ideals into simple v-ideals will follow:

THEOREM 7.1. *Let* $\mathfrak{A}_1, \mathfrak{A}_2, \cdots, \mathfrak{A}_k$ *and* $\bar{\mathfrak{A}}_1, \bar{\mathfrak{A}}_2, \cdots, \bar{\mathfrak{A}}_s$ *be two sets of simple v-ideals belonging to valuations* $B_1, B_2, \cdots, B_k$ *and* $\bar{B}_1, \bar{B}_2, \cdots, \bar{B}_s$ *respectively. If*

$$\mathfrak{A}_1^{\alpha_1}\mathfrak{A}_2^{\alpha_2}\cdots\mathfrak{A}_k^{\alpha_k} = \bar{\mathfrak{A}}_1^{\bar{\alpha}_1}\bar{\mathfrak{A}}_2^{\bar{\alpha}_2}\cdots\bar{\mathfrak{A}}_s^{\bar{\alpha}_s}, \tag{16}$$

where the α_i*'s and* $\bar{\alpha}_j$*'s are positive integers, then necessarily* $k = s$ *and, for a proper arrangement of the indices,* $\mathfrak{A}_i = \bar{\mathfrak{A}}_i$, $\alpha_i = \bar{\alpha}_i$.

This theorem is stronger, because a product of v-ideals, in particular the power product $\Pi\mathfrak{A}_i^{\alpha_i}$, is not necessarily a v-ideal.[7]

Proof. Let the simple v-ideals $\mathfrak{A}_i$ and $\bar{\mathfrak{A}}_j$ be of kind h_i and $\bar{h}_j$ respectively[8]; and let $m = \max.(h_1, \cdots, h_k, \bar{h}_1, \cdots, \bar{h}_s)$. The theorem is trivial

[7] For instance, let $\mathfrak{A}_1 = (y, x^2)$, $\mathfrak{A}_2 = (x, y^2)$. Both $\mathfrak{A}_1$, $\mathfrak{A}_2$ are v-ideals, but $\mathfrak{A}_1\mathfrak{A}_2 = (xy, \mathfrak{p}^3)$ is not a v-ideal, because the system $\Omega(\mathfrak{A}_1\mathfrak{A}_2)$ of subforms is of dimension zero and its base xy is not the power of a linear form (see Theorem 3).

[8] We assume that all the ideals $\mathfrak{A}_i$, $\bar{\mathfrak{A}}_j$ are zero-dimensional (necessarily primary). The case of one-dimensional simple v-ideals is trivial, because any such ideal is prime,

in the case $m = 1$. In fact, in this case the ideals $\mathfrak{A}_i$ and $\bar{\mathfrak{A}}_j$ are prime zero-dimensional ideals, and the power products on both sides of (16) coincide with $[\mathfrak{A}_1^{\alpha_1}, \mathfrak{A}_2^{\alpha_2}, \cdots, \mathfrak{A}_k^{\alpha_k}]$ and $[\bar{\mathfrak{A}}_1^{\bar{\alpha}_1}, \bar{\mathfrak{A}}_2^{\bar{\alpha}_2}, \cdots, \bar{\mathfrak{A}}_s^{\bar{\alpha}_s}]$ respectively. The theorem follows in this case from the unicity of the decomposition of a zero-dimensional ideal into primary components. We may then prove our theorem by induction with respect to m.

The partial power products on both sides of (16) consisting of factors which belong to one and the same zero-dimensional ideal, must be equal to each other. Hence it is sufficient to prove the theorem for the case in which the ideals $\mathfrak{A}_i$, $\bar{\mathfrak{A}}_j$ all belong to one and the same prime ideal, say to $\mathfrak{p} = (x, y)$. The ideal $\mathfrak{p}$ consists then, for each of the given valuations B_i, $\bar{B}_j$, of all the polynomials whose value is > 0. We may assume that $v(x) = v(\mathfrak{p})$ in any of the valuations B_i, $\bar{B}_j$. Let us now apply the quadratic transformations $T: x' = x$, $y' = y/x$, and let $T(\mathfrak{A}_i) = \mathfrak{A}'_i$, $T(\bar{\mathfrak{A}}_j) = \bar{\mathfrak{A}}'_j$. If we have, in the valuation B_i, $v\left(\frac{y - c_i x}{x}\right) > 0$, then $\mathfrak{A}'_i$ will be a primary ideal in $\mathfrak{O}' = \mathfrak{k}[x', y']$ belonging to the ideal $(x', y' - c_i)$. A similar remark holds for $\bar{\mathfrak{A}}'_j$. At any rate, $\mathfrak{A}'_i$ will be a simple v-ideal of kind $h_i - 1$ and $\bar{\mathfrak{A}}'_j$ will be a simple v-ideal of kind $\bar{h}_j - 1$. We must remember, however, that the transform of a simple v-ideal of kind 1, i. e. of $\mathfrak{p}$, is the unit ideal $\mathfrak{O}'$. If then $\mathfrak{A}_1 = \bar{\mathfrak{A}}_1 = \mathfrak{p}$, where we allow now that one or both of the exponents α_1, $\bar{\alpha}_1$ may be zero, operating by T on (16) we get

$$\mathfrak{A}'^{\alpha_2}_2 \mathfrak{A}'^{\alpha_3}_3 \cdots \mathfrak{A}'^{\alpha_k}_k = \bar{\mathfrak{A}}'^{\bar{\alpha}_2}_2 \bar{\mathfrak{A}}'^{\bar{\alpha}_3}_3 \cdots \bar{\mathfrak{A}}'^{\bar{\alpha}_s}_s.$$

Since $\max.(h_i - 1, \bar{h}_j - 1) = m - 1$, we have by our induction, $k = s$, $\alpha_i = \bar{\alpha}_i$, $\mathfrak{A}'_i = \bar{\mathfrak{A}}'_i$ $(i > 1)$, and (16) becomes

$$\mathfrak{p}^{\alpha_1} \mathfrak{A}_2^{\alpha_2} \cdots \mathfrak{A}_k^{\alpha_k} = \mathfrak{p}^{\bar{\alpha}_1} \mathfrak{A}_2^{\alpha_2} \cdots \mathfrak{A}_k^{\alpha_k}.$$

Now α_1 is the dimension of the system $\Omega(\mathfrak{p}^{\alpha_1} \mathfrak{A}_2^{\alpha_2} \cdots \mathfrak{A}_k^{\alpha_k})$ and similarly $\bar{\alpha}_1$ is the dimension of $\Omega(\mathfrak{p}^{\bar{\alpha}_1} \mathfrak{A}_2^{\alpha_2} \cdots \mathfrak{A}_k^{\alpha_k})$. Hence $\alpha_1 = \bar{\alpha}_1$ and the theorem is proved.

PART II.

8. Algebraic and transcendental valuations. In this part of the paper we shall use the apparatus of formal power series in order to derive further

---

and is therefore a principal ideal (f), where f is an irreducible polynomial. The one-dimensional factors and their exponents on both sides of (16) must be the same and may be deleted.

properties of valuation ideals in the ring of polynomials. The use of formal power series is clearly indicated by the fact that a zero-dimensional valuation of the field Σ of rational functions of x and y is essentially a local property of the field. It is known that any zero-dimensional valuation B of rank 1, in which x and y have positive values, can be obtained by the following construction: We put

$$x = A(t) = \alpha_1 t^{a_1} + \alpha_2 t^{a_2} + \cdots, \qquad y = B(t) = \beta_1 t^{b_1} + \beta_2 t^{b_2} + \cdots,$$

where the coefficients α and β belong to the underlying field $\mathfrak{k}$ and where the exponents a_i, b_i of each power series $A(t)$ and $Q(t)$ form a monotonic increasing sequence of positive real numbers. By substitution of these power series every element r of Σ takes a definite form

$$r = \gamma_1 t^{c_1} + \gamma_2 t^{c_2} \cdots \quad (\gamma_1 \neq 0,\ c_1 \gtreqless 0),$$

and the valuation B is obtained by putting $v(r) = c_1$. We may eliminate formally t between $x = A(t)$ any $y = B(t)$ and we may thus define the valuation B by putting $y = P(x) = \delta_1 x^{d_1} + \delta_2 x^{d_2} + \cdots$, where the exponents are again increasing positive real numbers.

Now we may effect the substitution $x = A(t)$, $y = B(t)$ not only in any rational function r of x, y but also in any formal power series $\xi = \sum_{i,j} a_{ij} x^i y^j$, $i, j \geqq 0$, i, j-integers, and in any quotient of such formal power series. In this manner the valuation B defines a valuation B^* of the field Σ^* of meromorphic functions of x, y, and the valuation ring of B^* contains the ring $\mathfrak{O}^*$ of holomorphic functions ξ.

The special case in which the exponents a_i, b_i of the power series $A(t)$, $B(t)$ are integers, is the only one which is of interest in the classical theory of algebraic and analytic functions. The corresponding valuations B may be called *algebroid* or *analytic*, while non-analytic valuations may be referred to as *transcendental valuations*. For an algebroid valuation the power series $P(x)$ is an ordinary Puiseux series, i. e. the exponents $d_1, d_2, \cdots$ are rational numbers with fixed denominator:

$$d_i = m_i/n, \qquad (n, m_1, m_2, \cdots) = 1.$$

If the branch $y = P(x)$ is *algebraic*, i. e. belongs to an algebraic curve $f(x, y) = 0$, f-irreducible, then the valuation B is *algebraic* and is effectively of rank 2, being composed of the prime divisor defined by the prime ideal (f) and of a valuation of the field of rational functions on the curve $f(x, y) = 0$. In all cases, if B is algebroid, the induced valuation B^* of the field of meromorphic functions is of rank 2. If, namely, we denote by $P_0 = P, P_1, \cdots, P_{n-1}$

the n determinations of the power series $P(x)$, corresponding to the n determinations of $x^{1/n}$, then $\zeta = \prod_{i=0}^{n-1} (y - P_i)$ is an holomorphic function of x, y, i. e. an element of $\mathfrak{O}^*$ and is indecomposable in $\mathfrak{O}^*$. Given any element ξ of $\mathfrak{O}^*$, we have a unique decomposition $\xi = \zeta^\rho \xi_1$, $\xi_1 \not\equiv 0(\zeta)$. The substitution $y = P(x)$ does not annihilate ξ_1 and hence we find for ξ_1 a definite representation $\xi_1 = \gamma_1 x^{c_1} + \gamma_2 x^{c_2} + \cdots$, $\gamma_1 \neq 0$. We define B^* by putting $v(\xi) = (\rho, c_1)$.

It is evident that, conversely, any valuation B of Σ of rank 2 is algebraic, and if the values of x and y are positive, B can be defined by putting y equal to a Puiseux series in x, *provided that the divisor of which B is composed is of the first kind with respect to $\mathfrak{O}$.*

In the sequel we will have no occasion to use transcendental valuations. The results of Part I enable us, in fact, to prove the following theorem:

THEOREM 8. 1. *Every valuation ideal in $\mathfrak{O}$ belongs to an algebroid (and even to an algebraic) valuation of Σ.*

Proof.[9] It is sufficient to prove this assertion for 0-dimensional (primary) v-ideals, because v-ideals possessing a 1-dimensional component can belong only to algebraic valuations. Let $\{\mathfrak{q}_i\}$ be the sequence of zero-dimensional v-ideals in $\mathfrak{O}$ belonging to a valuation B. We wish then to prove that given any ideal in the sequence, say $\mathfrak{q}_n$, there exists an algebraic valuation for which $\mathfrak{q}_n$ is a valuation ideal. The nature of our proof requires that a stronger assertion be established. We propose to prove that *there exists an algebraic valuation $\bar{B}$ such that in the Jordan sequence $\{\bar{\mathfrak{q}}_i\}$ of the 0-dimensional v-ideals belonging to $\bar{B}$, the first n ideals $\bar{\mathfrak{q}}_1, \bar{\mathfrak{q}}_2, \cdots, \bar{\mathfrak{q}}_n$ coincide with $\mathfrak{q}_1, \mathfrak{q}_2, \cdots, \mathfrak{q}_n$.* This we prove by induction with respect to n, assuming then that this assertion has been already established for $n - 1$, *for any choice of the generators x, y of Σ and for any valuation of Σ whose valuation ring contains $\mathfrak{k}[x, y]$.* Assuming, as usual, that x and y have positive values in B, we use the quadratic transformation $T: x' = x$, $y' = y/x$, getting the ring $\mathfrak{O} = \mathfrak{k}[x', y']$ and the sequence of v-ideals $\{\mathfrak{q}'_i\}$ in $\mathfrak{O}'$ belonging to B. By our induction, there exists an algebraic valuation $\bar{B}$ of Σ whose valuation ring contains $\mathfrak{O}'$ and such that the ideals $\mathfrak{q}'_1, \mathfrak{q}'_2, \cdots, \mathfrak{q}'_{n-1}$ are v-ideals belonging to $\bar{B}$. Let $\{\bar{\mathfrak{q}}_i\}$ be the sequence of v-ideals in $\mathfrak{O}$ belonging to $\bar{B}$. We now use the results of section 4. The $n - 1$ ideals $T^{-1}\mathfrak{q}'_i$, $i = 1, 2, \cdots, n - 1$ must be members of both sequences $\{\mathfrak{q}_i\}$ and $\{\bar{\mathfrak{q}}_i\}$; here $T^{-1}\mathfrak{q}'_i$ is the contracted ideal of $x'^{h_i}\mathfrak{q}'_i$, where h_i is defined as the smallest integer such that $x'^{h_i}\mathfrak{q}'_i$ is an extended ideal of an ideal in $\mathfrak{O}$. Let $T^{-1}\mathfrak{q}'_i = \mathfrak{q}_{\alpha_i} = \bar{\mathfrak{q}}_{\alpha_i}$, $i = 1, 2, \cdots, n - 1$ (the indices α_i are the same, since

[9] Note that the proof makes no use of the Theorem 5. 2.

the index j of $\mathfrak{q}_j$ is its length). We also know that $\alpha_1 < \alpha_2 \cdots < \alpha_{n-1}$ and that $\alpha_{n-1} \geqq n$. Moreover, if $i \leqq \alpha_{n-1}$, then $\mathfrak{q}_i$ is necessarily of the form $\mathfrak{p}^\rho \mathfrak{q}_{\alpha_\sigma}$, $\sigma \leqq n-1$, and also $\bar{\mathfrak{q}}_i$ is of the form $\mathfrak{p}^{\bar{\rho}} \bar{\mathfrak{q}}_{\alpha_{\bar{\sigma}}}$, $\bar{\sigma} \leqq n-1$. *We assert that* $\mathfrak{q}_i = \bar{\mathfrak{q}}_i$, $i = 1, 2, \cdots, \alpha_{n-1}$. Suppose that we know already that $\mathfrak{q}_j = \bar{\mathfrak{q}}_j$, for all $j < i \leqq \alpha_{n-1}$. If $\mathfrak{q}_i$ (or $\bar{\mathfrak{q}}_i$) coincides with one of the ideals $\mathfrak{q}_{\alpha_j}$, $j \leqq \alpha_{n-1}$, there is nothing to prove: we will have $\mathfrak{q}_i = \mathfrak{q}_{\alpha_\sigma} = \bar{\mathfrak{q}}_{\alpha_\sigma} = \bar{\mathfrak{q}}_i$. In the contrary case, we have $\mathfrak{q}_i = \mathfrak{p}^\rho \mathfrak{q}_{\alpha_\sigma}$, $\rho > 0$. It is not difficult to see that $\mathfrak{q}_i$ is then also of the form: $\mathfrak{q}_i = \mathfrak{p}\mathfrak{q}_j$, where necessarily $j < i$. In fact, let $\mathfrak{p}^{\rho-1}\mathfrak{q}_{\alpha_\sigma} \sim \mathfrak{q}_j$ (the equivalence being intended in the sense of the valuation B and $\mathfrak{q}_j$ being a v-ideal for B). Then $\mathfrak{p}^{\rho-1}\mathfrak{q}_{\alpha_\sigma} \equiv 0(\mathfrak{q}_j)$, whence $\mathfrak{q}_i \equiv 0(\mathfrak{p}\mathfrak{q}_j)$. On the other hand it is clear that $\mathfrak{q}_i \sim \mathfrak{p}\mathfrak{q}_j$, whence $\mathfrak{p}\mathfrak{q}_j \equiv 0(\mathfrak{q}_i)$. Hence $\mathfrak{q}_i = \mathfrak{p}\mathfrak{q}_j$. In a similar manner we find for $\bar{\mathfrak{q}}_i$ a representation of the form $\bar{\mathfrak{q}}_i = \mathfrak{p}\bar{\mathfrak{q}}_\mu$. Since $\mu < i$, we have $\bar{\mathfrak{q}}_\mu = \mathfrak{q}_\mu$, whence of the two ideals $\mathfrak{q}_j$ and $\bar{\mathfrak{q}}_\mu$ one is a divisor of the other ($\mathfrak{q}_j \equiv 0(\bar{\mathfrak{q}}_\mu)$ or $\bar{\mathfrak{q}}_\mu \equiv 0(\mathfrak{q}_j)$ according as $j \geqq \mu$ or $\mu \geqq j$). As a consequence it is also true that of the two ideals $\mathfrak{q}_i$ and $\bar{\mathfrak{q}}_i$, one is a divisor of the other. Now both $\mathfrak{q}_i$ and $\bar{\mathfrak{q}}_i$ have the same length i. Consequently $\mathfrak{q}_i = \bar{\mathfrak{q}}_i$, and this proves our assertion. Since the equality $\mathfrak{q}_i = \bar{\mathfrak{q}}_i$ holds for $i = 1, 2, \cdots, \alpha_{n-1}$ and since $\alpha_{n-1} \geqq n$, it follows that $\mathfrak{q}_1, \mathfrak{q}_2, \cdots, \mathfrak{q}_n$ belong as v-ideals to the algebraic valuation $\bar{B}$, and this proves our theorem.

Using Theorem 8.1 and the results of sections 5, 6, it is possible to give a very simple proof of the well-known fact that *every valuation B of Σ is the limit of algebraic valuations.* Let $\{\mathfrak{q}_i\}$ be the sequence of 0-dimensional v-ideals belonging to B. We may assume that the intersection of the ideals $\mathfrak{q}_i$ is the 0-ideal, since otherwise B itself is algebraic. In the proof of the Theorem 8.1 it has been shown that for any value of k there exists an algebraic valuation B_k of Σ, for which the ideals $\mathfrak{q}_1, \mathfrak{q}_2, \cdots, \mathfrak{q}_k$ are valuation ideals. Let r/s be any element of the valuation ring $\mathfrak{B}$ of B, $r, s \in \mathfrak{O}$. There will then exist an integer n such that $r \equiv 0(\mathfrak{q}_n)$, $s \not\equiv 0(\mathfrak{q}_{n+1})$. This integer will depend only on s. For all values of k such that $k \geqq n+1$, the ideals $\mathfrak{q}_n$ and $\mathfrak{q}_{n+1}$ will also be v-ideals for B_k, and hence for all such values of k r/s will also belong to the valuation ring $\mathfrak{B}_k$ of B_k. As a consequence, we have $\lim_{k\to\infty} (\mathfrak{B}_k \cap \mathfrak{B}) = \mathfrak{B}$, i. e. B can be regarded as the limit of the valuation B_k.

Using the characteristic property of the Jordan sequence $\{\mathfrak{q}_i\}$ of v-ideals established in section 2, and the properties of simple ideals given in sections 5 and 6, we may go a step further and gain an insight into the manner in which transcendental valuations are constructed. Given a simple v-ideal $\mathfrak{P}_{k+1}$ we know that it determines uniquely the sequence of simple v-ideals $\mathfrak{P}_1, \mathfrak{P}_2, \cdots, \mathfrak{P}_k$ which precede $\mathfrak{P}_{k+1}$ in any valuation to which $\mathfrak{P}_{k+1}$ belongs. We ask now the following question: given $\mathfrak{P}_k$ (and hence given the entire sequence $\mathfrak{P}_1, \mathfrak{P}_2, \cdots, \mathfrak{P}_k$)

in how many ways is it possible to choose $\mathfrak{P}_{k+1}$? To answer this question, we apply $k-1$ successive quadratic transformations, getting a ring $\bar{\mathfrak{O}}$ of polynomials of variables X, Y, in which to $\mathfrak{P}_k$ there corresponds a simple v-ideal $\bar{\mathfrak{P}}_1$ of kind one, i. e. $\bar{\mathfrak{P}}_1$ is prime and 0-dimensional, say $\bar{\mathfrak{P}}_1 = (X, Y)$. Any maximal subideal of $\bar{\mathfrak{P}}_1$ is of the form $(aX + bY, \bar{\mathfrak{P}}_1^2)$, where a, b are in $\mathfrak{k}$ and are not both zero. It is obvious that any such maximal subideal of $\bar{\mathfrak{P}}_1$ belongs to some valuation (for instance, to any valuation defined by putting $X = bt + \cdots, Y = -at + \cdots$) and is moreover a simple v-ideal. These maximal subideals are in (1, 1) correspondence with the ratio $z = a/b$, i. e. with the places of the purely transcendental field $\mathfrak{k}(z)$. Going back to our original ring $\mathfrak{O}$, we see that the set of simple v-ideals $\mathfrak{P}_{k+1}$ of kind $k+1$ such that $\mathfrak{P}_1, \mathfrak{P}_2, \cdots, \mathfrak{P}_k, \mathfrak{P}_{k+1}$ belong to one and the same valuation, is in (1, 1) correspondence with the set consisting of the elements of the underlying field $\mathfrak{k}$ and of the symbol ∞. Starting with $\mathfrak{P}_1$ we can then construct, in infinitely many ways, an infinite sequence $\mathfrak{P}_1, \mathfrak{P}_2, \cdots, \mathfrak{P}_k, \cdots$ of simple v-ideals, such that, for any k, the ideals $\mathfrak{P}_1, \mathfrak{P}_2, \cdots, \mathfrak{P}_k$ belong to some valuation B_k, which we may suppose to be algebraic. We assert that *the infinite sequence* $\{\mathfrak{P}_k\}$ *defines a valuation* B *of* Σ. Obviously then $B = \operatorname{Lim} B_k$. To see this, we first observe, that by Theorem 6. 2, the infinite sequence $\{\mathfrak{P}_k\}$ determines uniquely an infinite Jordan sequence $\{\mathfrak{q}_j\}$ which contains the sequence $\{\mathfrak{P}_k\}$ and which has the property that the elements of the sequence which precede a given $\mathfrak{P}_i$ are v-ideals for all the valuations B_k, $k \geqq i$. It follows immediately that the congruence $\mathfrak{q}_{m+1}\mathfrak{q}_n : \mathfrak{q}_m \equiv 0(\mathfrak{q}_{n+1})$ holds true for any two ideals $\mathfrak{q}_m$, $\mathfrak{q}_n$ of the sequence $\{\mathfrak{q}_j\}$, since the ideals which occur in this congruence are v-ideals belonging to B_k, when k is sufficiently large. As a consequence the sequence effectively defines a valuation of Σ, *q. e. d.*

9. Valuation ideals in the ring of holomorphic functions.[10] It has been pointed out in the preceding section that any 0-dimensional valuation B of the field $\Sigma = \mathfrak{k}(x, y)$ in which x and y have positive values, defines a valuation B^* of the field Σ^* of meromorphic functions of x, y, whose valuation ring contains the ring $\mathfrak{O}^* = \mathfrak{k}\{x, y\}$ of holomorphic functions of x, y. We have then also valuation ideals in $\mathfrak{O}^*$ belonging to B^*. It is clear that the prime ideal defined by B^* in $\mathfrak{O}^*$ is the 0-dimensional ideal $\mathfrak{p}^* = (x, y)$, i. e. the extended ideal of the ideal $\mathfrak{p} = (x, y)$ in $\mathfrak{O}$. Let $\{\mathfrak{q}_i\}$ be the sequence of 0-dimensional v-ideals in $\mathfrak{O}$ belonging to B, and let $\mathfrak{q}^*_i = \mathfrak{O}^*\mathfrak{q}_i$ be the extended

[10] Results of this and of the following sections will be later applied toward the proof of Theorem 5. 2. The properties of simple v-ideals derived in Part I and based on Theorem 5. 2 are therefore not to be used until a proof of this theorem has been given (Corollary 11. 2). We may, however, use Theorem 8. 1 (see footnote on p. 175).

ideal of $\mathfrak{q}_i$ in $\mathfrak{O}^*$. It is clear that $v(q^*_{i+1}) > v(\mathfrak{q}^*_i)$ in B^*, since $v(\mathfrak{q}^*_i) = v(\mathfrak{q}_i)$. Let $f_1, f_2, \cdots, f_k$ be a base of $\mathfrak{q}_i$ and let f be an element of $\mathfrak{q}_i$ not in $\mathfrak{q}_{i+1}$. Since $\mathfrak{q}_i/\mathfrak{q}_{i+1} \cong \mathfrak{k}$, we have $f_i \equiv c_i f(\mathfrak{q}_{i+1})$, $c_i \varepsilon \mathfrak{k}$. If then $\xi = \xi_1 f_1 + \cdots + \xi_k f_k$ is any element of $\mathfrak{q}^*_i$ ($\xi_j \varepsilon \mathfrak{O}^*$), we have $\xi \equiv f \cdot (c_1\xi_1 + \cdots + c_k\xi_k)(\mathfrak{q}^*_{i+1})$. Now $c_1\xi_1 + \cdots + c_k\xi_k \equiv c(\mathfrak{p}^*)$, where $c \varepsilon \mathfrak{k}$, and

$$(f)\mathfrak{p}^* \equiv 0(\mathfrak{q}^*_i\mathfrak{p}^*) \equiv 0(\mathfrak{O}^*\mathfrak{q}_i\mathfrak{p}) \equiv 0(\mathfrak{q}^*_{i+1}),$$

since $\mathfrak{q}_i\mathfrak{p} \equiv 0(\mathfrak{q}_{i+1})$. Hence $\xi \equiv cf(\mathfrak{q}^*_{i+1})$, and this shows that $\mathfrak{q}^*_i/\mathfrak{q}^*_{i+1} \simeq \mathfrak{k}$, i. e. $\mathfrak{q}^*_{i+1}$ is a maximal subideal of $\mathfrak{q}^*_i$. Hence the sequence $\{\mathfrak{q}^*_i\}$ is a Jordan sequence in $\mathfrak{O}^*$, and since $v(\mathfrak{q}^*_{i+1}) > v(\mathfrak{q}^*_i)$, the sequence $\{\mathfrak{q}^*_i\}$ is the sequence of v-ideals in $\mathfrak{O}^*$ belonging to B^*, i. e. *the zero-dimensional v-ideals in $\mathfrak{O}^*$ belonging to B^* are the extended ideals of the zero-dimensional v-ideals in $\mathfrak{O}$ belonging to B.* It is of course evident that $\mathfrak{q}_i$ is the contracted ideal of $\mathfrak{q}^*_i$.

The theorems of sections 3, 4,[11] the notions of simple and composite ideals introduced in section 5 and Theorem 5. 1 carry over without modification to ideals in $\mathfrak{O}^*$. In order to see this, a perusal of the proof is not at all necessary. It is sufficient to take into account quite generally the nature of the relationship between the polynomial ideals in $\mathfrak{O}$ and the power series ideals in $\mathfrak{O}^*$. There is a $(1, 1)$ correspondence between primary 0-dimensional ideal in $\mathfrak{O}$ belonging to the prime ideal $\mathfrak{p} = (x, y)$ and the 0-dimensional ideals in $\mathfrak{O}^*$. The correspondence is such that to an ideal $\mathfrak{q}$ in $\mathfrak{O}$ there corresponds its extended ideal $\mathfrak{q}^*$ in $\mathfrak{O}^*$ and $\mathfrak{q}$ is the contracted ideal of $\mathfrak{q}^*$. In fact, let $\mathfrak{q}^*$ be any 0-dimensional ideal in $\mathfrak{O}^*$. If ρ is the exponent of $\mathfrak{q}^*$, the $\mathfrak{q}^*$ possesses a base consisting of the power products x^iy^j, $i + j = \rho$, and of a set of polynomials F_a of degree $< \rho$. Hence $\mathfrak{q}^* = (F_a, x^iy^j)$ and therefore $\mathfrak{q}^*$ is the extended ideal of a zero-dimensional primary polynomial ideal $\mathfrak{q}$ belonging to $\mathfrak{p}$ $(= (x, y))$. Let $\mathfrak{q}^* = (f_1, f_2, \cdots, f_k)$, where $f_1, f_2, \cdots, f_k$ is a base of $\mathfrak{q}$, and let $\mathfrak{q}'$ be the contracted ideal in $\mathfrak{O}$ of $\mathfrak{q}^*$. If F is any polynomial in $\mathfrak{q}'$, $F = \sum_{i=1}^{k} \xi_i(x, y)f_i$, $\xi_i \varepsilon \mathfrak{O}^*$, and if we denote by $A_i^{(n)}$ the partial sum of the terms of degree $\leqq n$ in the power series $\xi_i(x, y)$, then the polynomial $F - \Sigma A_i^{(n)} f_i$ does contain terms of degree $< n + 1$, whence $F \equiv \Sigma A_i^{(n)} f_i(\mathfrak{p}^{n+1})$. Now, if n is sufficiently high, then $\mathfrak{p}^{n+1} \equiv 0(\mathfrak{q})$, whence $F \equiv 0(\mathfrak{q})$. Hence $\mathfrak{q}' \equiv 0(\mathfrak{q})$, and since $\mathfrak{q}'$ is the contracted ideal of $\mathfrak{q}^*$, it follows $\mathfrak{q}' = \mathfrak{q}$. Thus every zero-dimensional ideal $\mathfrak{q}^*$ in $\mathfrak{O}^*$ is the extended ideal of one and only

[11] In the case of formal power series the quadratic transformation $x' = x$, $y' = y/x$ leads from the ring $k\{x, y\}$ of formal power series in x and y to the larger ring of $k\{x', y'\}$ of formal power series in x' and y'.

one primary ideal $\mathfrak{q}$ in $\mathfrak{O}^*$, and $\mathfrak{q}$ is the contracted ideal of $\mathfrak{q}^*$. In particular, $\mathfrak{q}^*$ is simple or composite, according as $\mathfrak{q}$ is simple or composite.

10. The notion of a general element of an ideal of formal power series. The ring $\mathfrak{O}^*$ of holomorphic functions of x and y contains only one prime 0-dimensional prime ideal, namely the ideal $\mathfrak{p}^* = (x, y)$, and every 0-dimensional ideal $\mathfrak{A}$ in $\mathfrak{O}^*$ is necessarily primary and belongs to $\mathfrak{p}^*$. Since any ideal in $\mathfrak{O}^*$ has a finite base, given a 0-dimensional ideal $\mathfrak{A}$, it belongs to a finite exponent ρ, i. e. $\mathfrak{p}^{*\rho} \equiv 0(\mathfrak{A})$. In other words, $\mathfrak{A}$ contains all the formal power series which contain terms of lowest degree $\geqq \rho$. Since $\mathfrak{A}$ is at any rate a linear $\mathfrak{k}$-module, it follows that the condition in order that an element $\xi = \sum_{i,j \geqq 0} a_{ij}, x^i y^j$ of $\mathfrak{O}^*$ belong to $\mathfrak{A}$ is expressed by linear homogeneous relations between the coefficients a_{ij}, $i + j \leqq \rho$. With every linear relation

$$\sum_{i,j<\rho} c_{ij} a_{ij} = 0, \qquad c_{ij} \varepsilon \mathfrak{k}, \tag{17}$$

satisfied by all the elements ξ of $\mathfrak{A}$, we associate the function

$$E = \sum c_{ij} x^{-i} y^{-j}. \tag{17'}$$

The set of all the functions E obtained in this manner is called the "inverse system" $\mathfrak{A}^{-1}$ of the ideal $\mathfrak{A}$ (Macaulay,[5,6] Lasker[3]). If $E_1, E_2, \cdots, E_r$ belong to $\mathfrak{A}^{-1}$, then also $\alpha_1 E_1 + \cdots + \alpha_r E_r$ is in $\mathfrak{A}^{-1}$, $\alpha_1, \cdots, \alpha_r \varepsilon \mathfrak{k}$. From the fact that if $\xi = \Sigma a_{ij} x^i y^j$ belongs to the ideal $\mathfrak{A}$, also

$$x\xi = \Sigma a_{ij} x^{i+1} y^j \quad \text{and} \quad y\xi = \Sigma a_{ij} x^i y^{j+1}$$

belong to $\mathfrak{A}$, it follows immediately that if $E = c_{ij} x^{-i} y^{-j}$ is an element of the inverse system, then also the following relations are true for any element $\xi = \Sigma a_{ij} x^i y^j$ in $\mathfrak{A}$:

$$\sum c_{i+1,j} a_{ij} = 0, \qquad \sum c_{i,j+1} a_{ij} = 0.$$

This shows that the inverse system $\mathfrak{A}^{-1}$ *becomes an* $\mathfrak{O}^*$*-module, provided that we define multiplication as follows:*

$$x^\alpha y^\beta \sum_{i,j \geqq 0} c_{ij} x^{-i} y^{-j} = \sum_{\substack{i \geqq \alpha \\ j \geqq \beta}} c_{ij} x^{-i+\alpha} y^{-j+\beta}, \qquad \alpha, \beta \geqq 0.$$

In fact, with this definition of multiplication by power products $x^\alpha y^\beta$, the above equations $\Sigma c_{i+1,j}a_{ij} = 0$, $\Sigma c_{i,j+1}a_{ij} = 0$ signify that if E is in $\mathfrak{A}^{-1}$, also xE and yE are in $\mathfrak{A}^{-1}$. Multiplication of E by any formal power series in $\mathfrak{O}^*$ is then to be defined formally by the requirement of the distributive law of multiplication. It is then not difficult to see that $\mathfrak{A}^{-1}$ *consists of those and only those functions E which have the property that $\xi E = 0$ for any element ξ in $\mathfrak{A}$.* It is evident that if $\mathfrak{A}$ and $\mathfrak{B}$ are ideals in $\mathfrak{O}^*$, then $\mathfrak{A}^{-1} = \mathfrak{B}^{-1}$ implies $\mathfrak{A} = \mathfrak{B}$. Moreover, it is not difficult to show that given any set $\mathfrak{C}^{-1}$ of functions E which is an $\mathfrak{O}^*$-module and for which there exists at least one element $\xi \neq 0$ in $\mathfrak{O}^*$ such that $\xi E = 0$ for any E in $\mathfrak{C}^{-1}$, then the set of all elements ξ satisfying this condition forms an ideal $\mathfrak{A}$, and $\mathfrak{C}^{-1}$ is the inverse system of $\mathfrak{A}$.

We now come to the definition of a concept which shall be very useful in the sequel. Let $\tau = t_{ij}x^iy^j$ be a formal power series whose coefficients t_{ij} belong to an extension field $\mathfrak{K}$ of $\mathfrak{k}$. We shall say that τ is a *variable element of an ideal* $\mathfrak{A}$, if the coefficients t_{ij} do not all belong already to $\mathfrak{k}$ (so that at least one of the t_{ij}'s is transcendental with respect to $\mathfrak{k}$) and if they satisfy (in $\mathfrak{K}$) all the linear relations (17) (with a_{ij} replaced by t_{ij}), which are satisfied by the coefficients of all the actual power series belonging to $\mathfrak{A}$. In other words, we require that $\tau E = 0$ for any function E in $\mathfrak{A}^{-1}$. We shall say that a variable element τ of $\mathfrak{A}$ is a *general element of* $\mathfrak{A}$, if the t_{ij}'s do not satisfy any other algebraic relation, algebraically independent of the above linear relations. Finally, we shall say that a variable element τ of $\mathfrak{A}$ is a *quasi-general* element of $\mathfrak{A}$, if the t_{ij}'s do not satisfy *linear relations* other than those which are satisfied by the coefficients of a general element of $\mathfrak{A}$.

In a similar manner we define the notions of a variable element, of a general element and of a quasi-general element of the inverse system $\mathfrak{A}^{-1}$. The condition that a function $E = \Sigma c_{ij}x^{-i}y^{-j}$ belong to $\mathfrak{A}^{-1}$ is expressed by a certain set (δ) of homogeneous linear equations between the c_{ij}'s. These equations are obtained by expressing the fact that $E\xi_i = 0$, $i = 1, 2, \cdots, k$, where $\xi_1, \xi_2, \cdots, \xi_k$ is a base of the ideal $\mathfrak{A}$. A function $E = \Sigma e_{ij}x^{-i}y^{-j}$, where the e_{ij} are elements of an extension field $\mathfrak{k}(e_{ij})$ of $\mathfrak{k}$, shall be called a *variable element of* $\mathfrak{A}^{-1}$, if the coefficients e_{ij} satisfy all the relations of the set (δ), i. e. if $E\xi = 0$ for any ξ in $\mathfrak{A}$. If the coefficients e_{ij} do not satisfy algebraic (*or* linear) relations algebraically independent of the linear relations (δ), then E shall be called a *general* (*or quasi-general*) element of $\mathfrak{A}^{-1}$.

Let $E = \Sigma e_{ij}x^{-i}y^{-j}$ and $\tau = \Sigma t_{ij}x^iy^j$ be variable elements of $\mathfrak{A}^{-1}$ and of $\mathfrak{A}$ respectively, where we assume that the e_{ij} and t_{ij} are elements of one and the same field $\mathfrak{K}$.

We have $\xi E = 0$ for any element $\xi = \Sigma a_{ij}x^iy^j$ of $\mathfrak{A}$. Since the a_{ij}'s are arbitrary elements of underlying field $\mathfrak{k}$ subject to the only condition of satisfying a finite set of linear relations $\Sigma c_{ij}a_{ij} = 0$, where $\Sigma c_{ij}x^{-i}y^{-j}$ is an element of $\mathfrak{A}^{-1}$, it follows that the relation $\xi E = 0$ remains true if we regard the a_{ij} as indeterminates connected by the linear relations $\Sigma c_{ij}a_{ij} = 0$. But then the relation $\xi E = 0$ holds also after the specialization $a_{ij} \rightarrow t_{ij}$, since the t_{ij} also satisfy all the relations $\Sigma c_{ij}t_{ij} = 0$. *We conclude that* $\tau E = 0$. The following assertion is now easily derived:

If E *is a quasi-general element of* $\mathfrak{A}^{-1}$, *then* $\tau E = 0$ *implies that* τ *is a variable element of* $\mathfrak{A}$, *and if* τ *is a quasi-general element of* $\mathfrak{A}$ *then* $\tau E = 0$ *implies that* E *is a variable element of* $\mathfrak{A}^{-1}$. Proof straightforward.

Let $\mathfrak{A}', \mathfrak{A}''$ be two (distinct or coincident) ideals in $\mathfrak{O}^*$ and let $\tau' = \Sigma t'_{ij}x^iy^j$, $\tau'' = \Sigma t''_{ij}x^iy^j$ be variable elements of $\mathfrak{A}'$ and $\mathfrak{A}''$ respectively. Let $\mathfrak{K}' = \mathfrak{k}(t'_{ij})$ and $\mathfrak{K}'' = \mathfrak{k}(t''_{ij})$. We consider a pure transcendental base $\{u\}$ of $\mathfrak{K}''$ over $\mathfrak{k}$ and we adjoin the elements of this base to the field $\mathfrak{K}'$, *adjunction to be regarded as a pure transcendental extension of* $\mathfrak{K}'$. We obtain in this manner the field $\mathfrak{K}'(\{u\})$ having $\mathfrak{k}(\{u\})$ as subfield, and we then adjoin to $\mathfrak{K}'(\{u\})$ all the elements of $\mathfrak{K}''$ which are algebraic with respect to $\mathfrak{k}(\{u\})$. In the field $\mathfrak{K} = \mathfrak{k}(t'_{ij}, t''_{ij})$ obtained in this manner, any algebraic relation between the t'_{ij} and the t''_{ij} must be a consequence of algebraic relations between the t'_{ij}'s alone in $\mathfrak{K}'$ and the t''_{ij} alone in $\mathfrak{K}''$. Now that the t'_{ij} and the t''_{ij} have been properly imbedded in a common field, we form the product $\tau'\tau'' = \tau = \Sigma t_{ij}x^iy^j$, $t_{ij} \varepsilon \mathfrak{K}$, and we refer to τ as *the direct product of the variable elements* τ', τ'' *of* $\mathfrak{A}'$ *and* $\mathfrak{A}''$.

THEOREM 10. 1. *The direct product* $\tau = \tau'\tau''$ *of variable elements of two ideals* $\mathfrak{A}', \mathfrak{A}''$ *is a variable element of the product* $\mathfrak{A}'\mathfrak{A}''$ *of the two ideals. If* τ', τ'' *are general or quasi-general elements of* $\mathfrak{A}'$ *and* $\mathfrak{A}''$ *respectively, then* τ *is a quasi-general (not necessarily general) element of* $\mathfrak{A}'\mathfrak{A}''$.

Proof. Let $\mathfrak{A}'\mathfrak{A}'' = \mathfrak{B}$. It is well known (and is of immediate verification) that $\mathfrak{B}^{-1} = \mathfrak{A}'^{-1} : \mathfrak{A}''$, i. e. $\mathfrak{B}^{-1}$ consists of all functions E such that $E\mathfrak{A}'' \varepsilon \mathfrak{A}'^{-1}$. Let then $E_0 = \Sigma c_{ij}x^{-i}y^{-j}$ be an element in $\mathfrak{B}^{-1}$ and let $\tau''E_0 = \Sigma \sigma_{ij}x^{-i}y^{-j}$, $\sigma_{ij} \varepsilon \mathfrak{K}$. *We assert that* $\tau''E_0$ *is a variable element of* $\mathfrak{A}'^{-1}$. In fact, consider any element $\xi'' = \Sigma a''_{ij}x^iy^j$ of $\mathfrak{A}''$. Since $E_0\mathfrak{A}'' \varepsilon \mathfrak{A}'^{-1}$, the product $\xi''E_0 = \Sigma \sigma_{ij}^{(0)}x^{-i}y^{-j}$ is an element of the inverse system $\mathfrak{A}'^{-1}$. Hence the coefficients $\sigma_{ij}^{(0)}$ must satisfy the linear relations (δ'') which are satisfied by the coefficients of the general elements of $\mathfrak{A}'^{-1}$. The coefficients $\sigma_{ij}^{(0)}$ being linear forms in the coefficients a''_{ij} of ξ'', we get by substitution into the equations (δ'') a set of linear homogeneous relations between the a''_{ij}. Since these relations hold true for the coefficients a''_{ij} of any element ξ'' of $\mathfrak{A}''$, it follows that they are also satisfied by the coefficients t''_{ij} of the variable element τ'' of $\mathfrak{A}''$. But then it follows that the coefficients σ_{ij} of $\tau'' E_0$ must satisfy the linear equations (δ''), whence $\tau'' E_0$ is a variable element of $\mathfrak{A}'^{-1}$. Since τ' is a variable element of $\mathfrak{A}'$, it follows by a previously proved result, that $\tau'\tau'' E_0 = 0$, i. e. $\tau E_0 = 0$. Since the relation $\tau E_0 = 0$ holds for any element E_0 of $\mathfrak{B}^{-1}$, it follows that τ is a variable element of $\mathfrak{B}$, and this proves the first part of the theorem.

Now suppose that τ' and τ'' are general (or quasi-general) elements of the ideals $\mathfrak{A}'$ and $\mathfrak{A}''$ respectively. To prove that in this case τ is a quasi-general element of $\mathfrak{B}$ we have to show that if $\Sigma c_{ij} t_{ij} = 0$ is any linear homogeneous relation between the t_{ij}, then $E = \Sigma c_{ij} x^{-i} y^{-j}$ belongs to the inverse system $\mathfrak{B}^{-1}$. Now the t_{ij} are bilinear forms in the two sets of coefficients t'_{ij} and t''_{ij}. Substituting these bilinear forms we get $\Sigma c_{ij} t_{ij} = H(t'_{ij}, t''_{ij})$, where H is a bilinear form in the t'_{ij} and the t''_{ij}. The relation $H(t'_{ij}, t''_{ij}) = 0$ must be a consequence of the *linear relations* between the t'_{ij} and the t''_{ij} separately, since τ is the direct product of τ' and τ''. Now τ' is a quasi-general element of $\mathfrak{A}'$. As a consequence, any *linear* relation between the t'_{ij} arises from a function E' in $\mathfrak{A}'^{-1}$ and is therefore not destroyed if each t'_{ij} is replaced by $t'_{i-1,j}$ or by $t'_{i,j-1}$ (multiplication of E' by x or by y respectively). Hence the relation $H(t'_{ij}, t''_{ij}) = 0$ implies the relations $H(t'_{i-1,j}, t''_{ij}) = 0$ and $H(t'_{i,j-1}, t''_{ij}) = 0$. It is clear that these two relations correspond to the relations $\Sigma c_{ij} t_{i-1,j} = 0$ and $\Sigma c_{ij} t_{i,j-1} = 0$, which therefore must be true relations between the t_{ij}. As a consequence the functions $E = c_{ij} x^{-i} y^{-j}$ corresponding to the various linear relations $\Sigma c_{ij} t_{ij} = 0$ between the t_{ij}, form an $\mathfrak{O}^*$-module. From this it follows immediately that $E\tau = 0$, for any E in this module. Now $E\tau = E\tau'\tau'' = 0$ implies that $E\tau'$ is a variable element of $\mathfrak{A}''^{-1}$, since τ'' is a quasi-general element of $\mathfrak{A}''$. The condition of belonging to $\mathfrak{A}''^{-1}$ is expressed by a certain set (δ'') of linear equations, and thus the coefficients of $E\tau'$ must satisfy these equations. Since τ' is a quasi-general element of $\mathfrak{A}'$, these linear equations (δ'') must be satisfied by $E\xi'$, ξ'—an arbitrary element of $\mathfrak{A}'$, and hence $E\mathfrak{A}' \varepsilon \mathfrak{A}''^{-1}$. As a consequence $E \varepsilon \mathfrak{A}''^{-1} : \mathfrak{A}' = \mathfrak{B}^{-1}$, which proves our theorem.[12]

[12] That τ need not be the general element of $\mathfrak{A}'\mathfrak{A}''$ is shown by the following example. Let $\mathfrak{p} = (x, y)$, $\mathfrak{q} = (x^2, y^2)$, so that $\mathfrak{p}\mathfrak{q} = \mathfrak{p}^3$. The general element τ' of $\mathfrak{p}$ is $t'_{10}x + t'_{01}y + t'_{20}x^2 + \cdots$, where the t'_{ij}'s are indeterminates. Similarly

$$\tau'' = t''_{20}x^2 + t''_{02}y^2 + t''_{30}x^3 + \cdots$$

is the general element of $\mathfrak{q}$. Now

$$\tau = \tau'\tau'' = t_{30}x^3 + t_{21}x^2 + t_{12}xy^2 + t_{03}y^3 + \cdots,$$

where the t_{ij} satisfy *one non-linear* relation $t_{30}t_{03} = t_{21}t_{12}$.

The preceding theorem implies that if an ideal $\mathfrak{B}$ is composite, then a suitable quasi-general element τ of $\mathfrak{B}$ (namely the direct product of the general elements of the factors of $\mathfrak{B}$) is reducible in the algebraic closure of the field of the coefficients t_{ij} of τ. We are interested in the question of the extent to which the above considerations can be inverted. *What can be said about an ideal $\mathfrak{B}$, if its general element τ is reducible in the algebraic closure of the coefficients of τ?* That the ideal $\mathfrak{B}$ need not in general be composite, is illustrated by the example $\mathfrak{B} = (x^2, y^2)$. The ideal $\mathfrak{B}$ is simple, but its general element

$$\tau = t_{20}x^2 + t_{02}y^2 + t_{30}x^3 + \cdots \quad (t_{00} = t_{10} = t_{01} = t_{11} = 0)$$

is reducible, since

$$\tau = (\sqrt{t_{20}}x + \sqrt{-t_{02}}y + \cdots)(\sqrt{t_{20}}x - \sqrt{-t_{02}}y + \cdots).$$

However, in the special case of valuation ideals we can prove the following theorem:

THEOREM 10. 2. *If $\mathfrak{A}$ is a valuation ideal and if the general element t of $\mathfrak{A}$ is reducible, $t = t_1t_2 \cdots t_k$ (in the algebraic closure of its coefficients), then $\mathfrak{A}$ is composite; t is the direct product of its irreducible factors t_i, each irreducible factor t_i is the general element of a valuation ideal $\mathfrak{A}_i$, and $\mathfrak{A} = \mathfrak{A}_1\mathfrak{A}_2 \cdots \mathfrak{A}_k$.*

Proof. Let $t = \Sigma t_{ij}x^iy^j$ and let $t = t_1t_2 \cdots t_k$ be the factorization of t into irreducible factors $t_i = t_i(x, y)$, belonging to the ring of formal power series of x, y with coefficients in the algebraic closure of $\mathfrak{k}(\cdots t_{ij} \cdots)$. We assume of course that no t_i is a unit, i. e. that t_i does not contain a term independent of x and y. The valuation B to which $\mathfrak{A}$ belongs can be assumed to be algebraic. Substituting into t and into t_i the Puiseux expansion $y = P(x)$ which determines the valuation B, we are able to attach to these elements definite values $v(t)$, $v(t_i)$. It is clear that $v(t) = v(\mathfrak{A})$. Let $v(t_i) = \alpha_i$, whence $v(t) = \alpha_1 + \alpha_2 + \cdots + \alpha_k$. There exist elements in $\mathfrak{k}\{x, y\}$ whose value in B equals α_i: they can be obtained by specializing the coefficients of the formal power series t_i in such a manner as not to annihilate the coefficient of the leading term x^{α_i} of $t_i(x, P(x))$. As a consequence there exists a v-ideal $\mathfrak{A}_i$ for B such that $v(\mathfrak{A}_i) = \alpha_i$. Since $\alpha_i > 0$, no $\mathfrak{A}_i$ is the unit ideal. Since $v(t_i) = v(\mathfrak{A}_i)$, it follows that t_i is a variable element of $\mathfrak{A}_i$, and hence $t = \Pi t_i$

is a variable element of the ideal $\mathfrak{A}_1\mathfrak{A}_2\cdots\mathfrak{A}_k$. But since t is a general element of $\mathfrak{A}$, necessarily $\mathfrak{A}\equiv 0(\mathfrak{A}_1\mathfrak{A}_2\cdots\mathfrak{A}_k)$. On the other hand we have

$$v(\mathfrak{A}) = v(t) = \Sigma v(t_i) = \Sigma v(\mathfrak{A}_i) = v(\Pi\mathfrak{A}_i).$$

Hence $\Pi\mathfrak{A}_i\equiv 0(\mathfrak{A})$, and consequently $\mathfrak{A}=\mathfrak{A}_1\mathfrak{A}_2\cdots\mathfrak{A}_k$.

Let τ_i be the general element of $\mathfrak{A}_i$ and let us consider the *direct* product $\tau=\tau_1\tau_2\cdots\tau_k$. By Theorem 10.1, τ is a variable element of $\mathfrak{A}$. On the other hand, any algebraic relation between the coefficients of the power series τ leads also to a true relation between the coefficients of t, since t is obtained from τ by the specialization $\tau_i\rightarrow t_i$. Consequently τ *must be a general element of* $\mathfrak{A}$. It can be then identified with t, and it is thus seen that in the original factorization $t=\Pi t_i$, each t_i is a general element of $\mathfrak{A}_i$ and that the product is a direct product (unicity of factorization of t into irreducible factors). The theorem is proved.

COROLLARY. *The general element of a simple v-ideal is absolutely irreducible* (i. e. irreducible in $\mathfrak{K}\{x,y\}$, $\mathfrak{K}$ being any extension field of $\mathfrak{k}$).

11. The characterization of simple v-ideals. Let

$$(18)\qquad y=y_1=\sum_{i=1}^{k} c_i x^{\alpha_i/\nu}+\sum_{j=0}^{\infty} t_j x^{(\alpha_{k+1}+j)/\nu},\qquad 0<\alpha_1<\alpha_2<\cdots$$

be a Puiseux series, *in which we assume that* $(\nu,\alpha_1,\alpha_2,\cdots,\alpha_g)=1$, $g\leqq k+1$, *and that the first k coefficients $c_1,\cdots,c_k$ are in the underlying field $\mathfrak{k}$, while the remaining coefficients are indeterminates.*

THEOREM 11.1. *Given a formal power series $\xi(x,y)=\Sigma a_{ij}x^iy^j$ with indeterminate coefficients a_{ij}, there exists a set of linear forms $F_m(a_{ij})$, $G_n(a_{ij})$ which have the following properties:* (1) *the relations $F_m(a_{ij}^{(0)})=0$ $(a_{ij}^{(0)}\,\varepsilon\,\mathfrak{k})$ give necessary conditions that the equation $\xi_0(x,y)=\Sigma a_{ij}^{(0)}x^iy^j=0$ admit a uniformization of type* (18), *the t_j's being replaced by special values $t_j^{(0)}$ in $\mathfrak{k}$;* (2) *the relations $F_m(a_{ij}^{(0)})=0$ and the inequalities $G_n(a_{ij}^{(0)})\neq 0$ give necessary and sufficient condition in order that $\xi_0(x,y)=0$ admit the above uniformization and that $\xi_0(x,y)$ be an irreducible element of $\mathfrak{k}\{x,y\}$. The set of all elements ξ in $\mathfrak{k}\{x,y\}$ whose coefficients a_{ij} satisfy the relations $F_m=0$, is a simple ideal.*

Proof. The theorem is true for $g=0$, in which case $\nu=1$. In fact, let

$\bar{y}=\sum_{i=1}^{k} c_i x^{\alpha_i}$ ($\bar{y}=0$, if $k=0$). If the equation $\xi(x,y)=0$ is uniformizable by the expansion $y=y_1$, then we must have $\xi(x,y_1)=0$, and this shows that $\xi(x,\bar{y})$ is divisible by $x^{\alpha_{k+1}}$. This condition is expressed by a certain set of linear homogeneous equations $F_m(a_{ij})=0$ between the coefficients of $\xi(x,y)$. Should moreover $\xi(x,y)$ be irreducible, then we must have

$$\xi(x,y)=(y-y_1)\eta(x,y),$$

where $\eta(x,y)$ is a unit. As a consequence, the coefficient a_{01} must be different from 0. Conversely, if $F_m(a_{ij})=0$ and $a_{01}\neq 0$, then $\xi(x,y)$ is divisible by $y-y_1$ and is irreducible in $\mathfrak{k}\{x,y\}$.

We assume that the theorem is true for $g-1$. Let $\nu=\nu_1 n$, $\alpha_1=\nu_1\alpha'_1$, $(n,\alpha'_1)=1$. We put

(19) $$x=\bar{x}^n,\qquad y=\bar{x}^{\alpha'_1}(c_1+\bar{y}).$$

If $\xi(x,y)=\Sigma a_{ij}x^iy^j$ can be uniformized by the expansion $y=y_1$, then ξ must be divisible by the product $\prod_{i=1}^{\nu}(y-y_1^{(i)})$, where $y_1=y_1^{(1)},\cdots,y_1^{(\nu)}$ are the conjugates of y_1. From this it follows immediately that ξ cannot contain terms x^iy^j in which $\nu i+\alpha_1 j<\nu\alpha_1$, whence

(20) $$a_{ij}=0,\text{ for all } i,j,\text{ such that } \nu i+\alpha_1 j<\nu\alpha_1.$$

Substituting (19) we find, in view of (20),

(20′) $$\xi(x,y)=\bar{x}^{\nu\alpha'_1}\bar{\xi}(\bar{x},\bar{y})=\bar{x}^{\nu\alpha'_1}\Sigma\bar{a}_{ij}\bar{x}^i\bar{y}^j,$$

while the equation of the branch (18) becomes

(21) $$\bar{y}=\bar{y}_1=\sum_{i=2}^{k}c_i\bar{x}^{(\alpha_i-\alpha_1)/\nu_1}+\sum_{j=0}^{\infty}t_j\bar{x}^{(\alpha_{k+1}+j-\alpha_1)/\nu_1}.$$

The equation $\bar{\xi}(\bar{x},\bar{y})=0$ admits a uniformization by means of an expansion $\bar{y}=\bar{y}_1$ of type (21). Since $\alpha_2>\alpha_1$, there can be no constant term in $\bar{\xi}(\bar{x},\bar{y})$. This constant term arises from the terms $a_{ij}x^iy^j$ of $\xi(x,y)$ in which $\nu i+\alpha_1 j=\nu\alpha_1$, i. e. $ni+\alpha'_1 j=n\alpha'_1\nu_1$, and is therefore equal to

$$a_{\nu_1\alpha'_1,0}+a_{(\nu_1-1)\alpha'_1,n}c_1^{\,n}+\cdots+a_{0,n\nu_1}c_1^{\,n\nu_1}.$$

Consequently

(22) $$a_{\nu_1\alpha'_1,0}+a_{(\nu_1-1)\alpha'_1,n}c_1^{\,n}+\cdots+a_{0,n\nu_1}c_1^{\,n\nu_1}=0.$$

It is clear that $(\alpha_2-\alpha_1,\alpha_3-\alpha_1,\cdots,\alpha_g-\alpha_1,\nu_1)=1$, hence we are in the case $g-1$. By our induction, we have a set of linear forms $\bar{F}(\bar{a}_{ij})$ and $\bar{G}(\bar{a}_{ij})$ which satisfy the assertion of the theorem. Now the coefficients $\bar{a}_{ij}$ of $\bar{\xi}(\bar{x},\bar{y})$ are linear homogeneous forms in the coefficients a_{ij} of $\xi(x,y)$. Let $\{F_m\}$ be the set of forms in the a_{ij} consisting of the forms $\bar{F}(\bar{a}_{ij})$, expressed in terms of the a_{ij}, and of the left-hand members of the equations (20) and (22). Let moreover, $\{G_n\}$ be the set of forms in the a_{ij} consisting of the forms $\bar{G}(\bar{a}_{ij})$ (expressed in terms of the a_{ij}) and of the form $a_{0,\nu}$. We assert that the forms $\{F_m\}$ and $\{G_n\}$ satisfy the assertion of our theorem. In the first place, by the definition of the forms F_m, the equations $F_m=0$ must be satisfied if $\xi(x,y)$ can be uniformized by an expansion $y=y_1$ of the type (18). Assume that the coefficients a_{ij} satisfy the equations $F_m=0$ and the inequalities $G_n\neq 0$. The validity of the equations (20) implies that the substitution (19) introduces in $\xi(x,y)$ a factor $\bar{x}^{\lambda}$, where $\lambda \geqq \nu\alpha'_1$. Hence in (20′) the factor $\bar{\xi}(\bar{x},\bar{y})$ contains no negative powers of $\bar{x}$ or of $\bar{y}$. The coefficients $\bar{a}_{ij}$ of $\bar{\xi}$ satisfy, by hypothesis, the equations $\bar{F}(\bar{a}_{ij})=0$ and the inequalities $\bar{G}(\bar{a}_{ij})\neq 0$, relative to the branch (21). Hence the equation $\bar{\xi}(\bar{x},\bar{y})=0$ can be uniformized by an expansion $\bar{y}=\bar{y}_1$ of type (21), and consequently also the equation $\xi(x,y)=0$ can be uniformized by an expansion $y=y_1$ of type (18). The power series $\xi(x,y)$ must be divisible by the irreducible element $\prod_{j=1}^{\nu}(y-y_1^{(j)})$. Let $\xi=\eta(x,y)$, $\prod_{j=1}^{\nu}(y-y_1^{(j)})$. But, by assumption $a_{0,\nu}\neq 0$, i. e. $\xi(x,y)$ contains a term in y^{ν}. Consequently $\eta(x,y)$ must contain a constant term $\neq 0$, since the coefficient of any term y^j, $j<\nu$, in the product $\prod_{j=1}^{\nu}(y-y_1^{(j)})$, is divisible by x. Hence η is a unit, and ξ is irreducible.

Conversely, if $\xi=0$ admits a uniformization $y=y_1$ of type (18) and if ξ is irreducible, we will have $\xi=\eta(x,y)\prod_{j=1}^{\nu}(y-y_1^{(j)})$, where η is a unit. This implies in the first place $a_{0\nu}\neq 0$, and it also implies, as was pointed out before, that the equations $F_m=0$ hold true. The inequalities $G_n(a_{ij})\neq 0$ arising from the inequalities $\bar{G}(\bar{a}_{ij})\neq 0$ must also be satisfied, since the hypothesis that ξ is irreducible in $\mathfrak{k}\{x,y\}$ implies that $\bar{\xi}(\bar{x},\bar{y})$ is irreducible in $\mathfrak{k}\{\bar{x},\bar{y}\}$.

It remains to prove that the elements $\xi(x,y)=\Sigma a_{ij}x^iy^j$ whose coefficients satisfy the linear relations $F_m=0$ form a simple ideal $\mathfrak{P}$. Let us regard the coefficients a_{ij} as elements of the field defined by the equations $F_m(a_{ij})=0$. The inequalities $G_n\neq 0$ are then satisfied, and $\xi=0$ admits a uniformization $y=y_1$ of type (18). As a consequence also $x\xi=0$ and $y\xi=0$ admit the uniformization $y=y_1$, and hence the coefficients of the two power series $x\xi$

and $y\xi$ must satisfy the relations $F_m = 0$. This implies that the relations $F_m(a_{i-1,j}) = 0$, $F_m(a_{i,j-1}) = 0$ are consequences of the relations $F_m(a_{ij}) = 0$, i. e. if ξ is any element in $\mathfrak{P}$, then $x\xi \varepsilon \mathfrak{P}$ and $y\xi \varepsilon \mathfrak{P}$. Hence $\mathfrak{P}$ is an ideal. To prove that $\mathfrak{P}$ is a simple ideal, we observe, that if $\mathfrak{P}$ was a composite ideal, then, by Theorem 10. 1, a suitable quasi-general element t of $\mathfrak{P}$ would be reducible in the algebraic closure of the field of the coefficients of t. Now t, a quasi-general element of $\mathfrak{P}$, has the property that its coefficients satisfy no linear relations other than those which hold for the coefficients of the general element of $\mathfrak{P}$, i. e. only the relations $F_m = 0$. Hence the coefficients of t certainly satisfy the inequalities $G_n \neq 0$, and therefore t could not be reducible, in contradiction with our assumption that $\mathfrak{P}$ is composite, q. e. d.

In order to apply the above theorem, we begin with some preliminary remarks. Let $t = \Sigma t_{ij} x^i y^j$ be the general element of a valuation ideal $\mathfrak{q}$, and let us assume that t is absolutely irreducible (this is certainly the case if $\mathfrak{q}$ is a simple v-ideal, see Theorem 10. 2, Corollary). We will have then $t = \epsilon . \prod_{i=1}^{\nu} (y - y_i)$, where ϵ is a unit and where $y_1, y_2, \cdots, y_\nu$ are the ν determinations of a Puiseux series $y_1 = \sum_{i=1}^{\infty} t_i x^{i/\nu}$, the t_i's being algebraic functions of the t_{ij}. Some of the coefficients t_i may be constants, i. e. elements in $\mathfrak{k}$. Let $t_i = c_i$, $i = 1, 2, \cdots, h$, $c_i \varepsilon \mathfrak{k}$, while t_{h+1} is the first coefficient which is transcendental with respect to $\mathfrak{k}$. The ideal $\mathfrak{q}$ is a valuation ideal for some algebraic valuation B, defined by an expansion $y = \eta = \sum_{i=1}^{\infty} d_i x^{i/\mu}$, $d_i \varepsilon \mathfrak{k}$, and the value of t, i. e. the evaluation of $\mathfrak{q}$, is the exponent of the term of lowest degree in $t(x, \eta) = \epsilon . \prod_{i=1}^{\nu} (y - y_i)$. It is clear that the term of lowest degree in $\eta - y_i$ is the same as the term of lowest degree in $\eta - \Sigma c_i x^{i/\nu} - t_{h+1} x^{(h+1)/\nu}$. It follows that the value of t is not altered if we regard the coefficients $t_{h+1}, t_{h+2}, \cdots$ in the expansion y_1 as entirely independent indeterminates. Now if $\bar{t}(x, y)$ denotes the *direct* product $\epsilon . \prod_{i=1}^{\nu} (y - y_i)$, in which $t_{h+1}, t_{h+2}, \cdots$ are regarded as indeterminates and ϵ is a unit with indeterminate coefficients (i. e. ϵ is the general element of the unit ideal), then $\bar{t}$ is a variable element of $\mathfrak{q}$, since $v(\bar{t}) = v(\mathfrak{q})$, and on the other hand $\bar{t}$ is at least as general a power series as t (i. e. t is a specialization of $\bar{t}$). Since t is the general element of $\mathfrak{q}$, it follows that t can be identified with $\bar{t}$, and hence the coefficients $t_{h+1}, t_{h+2}, \cdots$ in the original Puiseux series y_1 are indeed algebraically independent with respect to $\mathfrak{k}$, and can be regarded as indeterminates. Changing slightly our notation and putting into evidence the coefficients c_i which are different from zero, we re-write the series y_1 as follows:

$$(23)\qquad y_1 = \sum_{i=1}^{k} c_i x^{\alpha_i/\nu} + \sum_{j=0}^{\infty} t_j x^{(\alpha_{k+1}+j)/\nu}, \qquad \begin{array}{l} 0 < \alpha_1 < \alpha_2 < \cdots, \\ c_1 c_2 \cdots c_k \neq 0, \end{array}$$

where $t_0, t_1, t_2,$ are indeterminates. Let δ be the highest common divisor of $\alpha_1, \alpha_2, \cdots, \alpha_{k+1}, \nu$. *We assert that if* $\delta > 1$, *then* $t = \epsilon \prod_{i=1}^{\nu} (y - y_i)$ *cannot be the general element of an ideal.* We shall show, namely, that if $\delta > 1$, then the coefficients t_{ij} of $t = \Sigma t_{ij} x^i y^j$ satisfy non-linear relations (which are not consequences of linear relations). Consider another sample of the series (23),

$$y'_1 = \sum_{i=1}^{k} c_i x^{\alpha_i/\nu} + \sum_{j=0}^{\infty} t'_j x^{(\alpha_{k+1}+j)/\nu},$$

where the t'_j are new indeterminates, and let $t' = \sum t'_{ij} x^i y^j = \epsilon' \prod_{i=1}^{\nu} (y - y'_i)$, where ϵ' is a unit with indeterminate coefficients. Also t' is a general element of $\mathfrak{q}$. Assume that the coefficients t_{ij} (and hence also the coefficients t'_{ij}) satisfy only linear homogeneous relations. Then it is clear that $t + t'$ is also a general element of $\mathfrak{q}$, whence $t + t' = \bar{\epsilon} \prod_{i=1}^{\nu} (y - \bar{y}_i)$, where $\bar{\epsilon}$ is a unit and

$$\bar{y}_1 = \sum_{i=1}^{k} c_i x^{\alpha_i/\nu} + \sum_{j=0}^{\infty} \tau_j x^{(\alpha_{k+1}+j)\nu}.$$

The substitution $y = \bar{y}_1$ must annihilate $t + t'$, i. e. $\epsilon \prod_{i=1}^{\nu} (y - y_i) + \epsilon' \prod_{i=1}^{\nu} (y - y'_i)$. We proceed to express the fact that the coefficients of the first two terms of $\epsilon \prod_{i=1}^{\nu} (\bar{y}_1 - y_i) + \epsilon' \prod_{i=1}^{\nu} (\bar{y}_1 - y'_i)$ vanish. If ω denotes a primitive ν-th root of unity, then

$$y_i = \sum_{j=1}^{k} c_j (\omega^i x^{1/\nu})^{\alpha_j} + t_0 (\omega^i x^{1/\nu})^{\alpha_{k+1}} + \sum_{j=1}^{\infty} t_j (\omega^i x^{1/\nu})^{\alpha_{k+1}+j},$$

whence

$$(24)\qquad \begin{aligned} \bar{y}_1 - y_i = \sum_{j=1}^{k} c_j (1 - \omega^{i\alpha_j}) x^{\alpha_j/\nu} &+ (\tau_0 - t_0 \omega^{i\alpha_{k+1}}) x^{\alpha_{k+1}/\nu} \\ &+ (\tau_1 - t_1 \omega^{i(\alpha_{k+1}+1)}) x^{(\alpha_{k+1}+1)/\nu} + \cdots. \end{aligned}$$

Let δ' be the h. c. d. of $\alpha_1, \alpha_2, \cdots, \alpha_k, \nu$, whence $\delta \equiv 0(\delta')$. We denote by $\Pi'(\bar{y}_1 - y_i)$ the product of the δ' factors $\bar{y}_1 - y_i$ for which $i\delta' \equiv 0(\nu)$, and by $\Pi''(\bar{y}_1 - y_i)$ the product of the remaining factors $\bar{y}_1 - y_i$, so that

$$\Pi'(\bar{y}_1 - y_i)\, \Pi''(\bar{y}_1 - y_i) = \prod_{i=1}^{\nu} (\bar{y}_1 - y_i).$$

If $i\delta' \not\equiv 0(\nu)$, the differences $1 - \omega^{i\alpha_j}$, $j = 1, 2, \cdots, k$, cannot all vanish.

Let σ be the smallest value of j, such that $1-\omega^{i\alpha_\sigma}\neq 0$. Then, by (24), $c_\sigma(1-\omega^{i\alpha_\sigma})x^{\alpha_\sigma/\nu}$, $\sigma\leqq k$, is the term of smallest degree in $\bar{y}_1-y_i$, while the exponent of the next term will be greater than $\alpha_\sigma/\nu+1/\nu$, since, for $j=1,2,\cdots,k$, $\alpha_{j+1}-\alpha_j\equiv 0(\delta)$, whence $\alpha_{j+1}-\alpha_j\geqq\delta>1$. It follows that

$$(25)\qquad \Pi''(\bar{y}_1-y_i)=dx^\lambda+d_1x^{\lambda_1}+\cdots,\ 0\neq d\,\varepsilon\,\mathfrak{k},\ \lambda_1>\lambda+1/\nu.$$

Similarly, replacing the t_j by the t'_j, we will have

$$(26)\qquad \Pi''(\bar{y}_1-y'_i)=dx^\lambda+d'_1x^{\lambda_1}+\cdots.$$

We now consider the product $\Pi'(\bar{y}_1-y_i)$. Here i assumes the values $\nu/\delta', 2\nu/\delta',\cdots,\delta'\nu/\delta'$, whence

$$(27)\qquad \bar{y}_1-y_i=\bar{y}-y_{j\nu/\delta'}=(\tau_0-t_0\omega_1{}^{j\alpha_{k+1}})x^{\alpha_{k+1}/\nu} + (\tau_1-t_1\omega_1{}^{j(\alpha_{k+1}+1)})x^{(\alpha_{k+1}+1)/\nu}+\cdots,\ \omega_1=\omega^{\nu/\delta'},\quad (j=1,2,\cdots,\delta'),$$

where ω_1 is a primitive root of unity of exponent δ'. Taking into account that $(\delta',\alpha_{k+1})=\delta>1$ and letting $\delta'=\delta h$, we find from (27),

$$\Pi'(\bar{y}_1-y_i)=(\tau_0{}^h-t_0{}^h)^\delta x^{\delta'\alpha_{k+1}/\nu}+\delta h(\tau_0{}^h-t_0{}^h)^{\delta-1}\tau_0{}^{h-1}\tau_1x^{(\delta'\alpha_{k+1}+1)/\nu}+\cdots.$$

It follows, by (25), that

$$\prod_{i=1}^{\nu}(\bar{y}_1-y_i)=d(\tau_0{}^h-t_0{}^h)^\delta x^{\lambda+\delta'\alpha_{k+1}/\nu}+d\delta h(\tau_0{}^h-t_0{}^h)^{\delta-1}\tau_0{}^{h-1}\tau_1x^{(\delta'\alpha_{k+1}+1)/\nu+\lambda}+\cdots.$$

The first two terms of $\prod_{i=1}^{\nu}(\bar{y}_1-y'_i)$ are obtained from the above by replacing t_0 by t'_0. Hence we must have

$$\epsilon_{00}(\tau_0{}^h-t_0{}^h)^\delta+\epsilon'_{00}(\tau_0{}^h-t'_0{}^h)^\delta=0,\quad \epsilon_{00}(\tau_0{}^h-t_0{}^h)^{\delta-1}+\epsilon'_{00}(\tau_0{}^h-t'_0{}^h)^{\delta-1}=0,$$

where ϵ_{00} and ϵ'_{00} are the constant terms in ϵ and ϵ' respectively. These equations imply the relations $\tau_0{}^h=t_0{}^h=t'_0{}^h$, in contradiction with the algebraic independence of the t_i and the t'_i. This proves our assertion.

This result shows that the general element $t=\epsilon\prod_{i=1}^{\nu}(y-y_i)$ of the valuation ideal $\mathfrak{q}$ is exactly of the type considered in the proof of Theorem

11. 1, and hence $\mathfrak{q}$ belongs to the class of ideals $\mathfrak{P}$ defined in that theorem. This holds true for any valuation ideal $\mathfrak{q}$ whose general element is absolutely irreducible, in particular for any simple valuation ideal. But it has been shown that the ideals $\mathfrak{P}$ are simple ideals. Hence, *a valuation ideal whose general element is irreducible is simple.* The converse has already been proved before (Theorem 10. 2, Corollary). Reassuming and recalling Theorem 10. 2, we have the following theorem:

THEOREM 11. 2. *A valuation ideal $\mathfrak{P}$ is simple if and only if its general element $t = \Sigma t_{ij}x^iy^j$ is absolutely irreducible. The general element t of $\mathfrak{P}$ is the direct product $t = \epsilon \cdot \prod_{i=1}^{\nu} (y - y_i)$, where ϵ is a unit (with indeterminate coefficients) and $y_1 = \sum_{i=1}^{k} c_i x^{\alpha_i/\nu} + \sum_{j=0}^{\infty} t_j x^{(\alpha_{k+1}+j)/\nu}$, $(\alpha_1, \alpha_2, \cdots, \alpha_{k+1}, \nu) = 1$, the t_j being indeterminates. Given any valuation ideal $\mathfrak{A}$, the factorization of the general element of $\mathfrak{A}$ into irreducible factors yields a factorization of $\mathfrak{A}$ into simple v-ideals, according to the scheme indicated in Theorem* 10. 2.

We are now in position to prove the basic Theorem 5. 2 of Part I:

COROLLARY 11. 2 (Theorem 5. 2). *The transform of a simple v-ideal by a quadratic transformation is a simple v-ideal.*

Proof. It is immaterial whether the theorem is proved for polynomial ideals or for power series ideals, in view of the relationship between these ideals described in section 9. Let $\mathfrak{P}$ be a simple 0-dimensional ideal v-ideal in $\mathfrak{O}^* = \mathfrak{k}\{x, y\}$, belonging to a valuation B^*, and let $x'^{\nu}\mathfrak{P}'$ be the extended ideal of $\mathfrak{P}$ in $\mathfrak{O}'^* = \mathfrak{k}\{x', y'\}$, where $x = x'$, $y = y'x'$ (we assume as usual that $v(y) > v(x)$) and where ν is the integer such that $\mathfrak{P} \equiv 0(\mathfrak{p}^{*\nu})$, $\mathfrak{P} \not\equiv 0(\mathfrak{p}^{*\nu+1})$, $\mathfrak{p}^* = (x, y)$. Here $\mathfrak{P}'$ is necessarily either zero-dimensional or the unit ideal. Let $t(x, y) = \Sigma t_{ij}x^iy^j$ be the general element of $\mathfrak{P}$. By the definition of the integer ν, we must have $t_{ij} = 0$ for all i, j such that $i + j < \nu$, and $t_{ij} \neq 0$ for some i, j such that $i + j = \nu$. Hence $t(x, y) = t(x', x'y') = x'^{\nu}\tau(x', y')$, where $\tau(x', y') = \Sigma \tau_{ij}x'^iy'^j$ is a formal power series in x', y'. Here $\tau_{ij} = t_{i+\nu-j,j}$, if $i + \nu \geqq j$, and $\tau_{ij} = 0$, if $i + \nu < j$. In our special case the general element t of $\mathfrak{P}$ is of the form indicated in Theorem 11. 2. Hence

$$\tau = \epsilon(x, y) \prod_{i=1}^{\nu} (y' - y'_i),$$

where

$$y'_1 = \sum_{i=1}^{k} c_i x'^{(\alpha_i - \nu)/\nu} + \sum_{j=0}^{\infty} t_j x'^{(\alpha_{k+1}+j-\nu)/\nu}.$$

If then $\epsilon'(x', y')$ denote a general unit of $\mathfrak{O}'^*$ (i. e. a unit with indeterminate coefficients), then the direct product $\epsilon'\tau$ is the general element of a simple ideal $\bar{\mathfrak{P}}'$. We have $v(\bar{\mathfrak{P}}') = v(\tau)$ and

$$v(\tau) + v(x'^{\nu}) = v(t) = v(\mathfrak{P}),$$

whence $v(\bar{\mathfrak{P}}') = v(\mathfrak{P}')$ and $\bar{\mathfrak{P}}' \equiv 0(\mathfrak{P}')$.

On the other hand, any element $t^0 = \Sigma t_{ij}{}^0 x^i y^j$ of $\mathfrak{P}$ is obtained from the general element t by a specialization $t_{ij} \to t_{ij}{}^0$. Hence, if

$$t^0(x, y) = x'^{\nu}\tau^0(x', y') = x'^{\nu}\Sigma\tau_{ij}{}^{(0)} x'^i y'^j,$$

then τ^0 is obtained from $\epsilon'\tau$ by the specialization $\epsilon' \to 1$, $\tau_{ij} \to \tau_{ij}{}^0$, and therefore τ^0 is an element of the ideal $\bar{\mathfrak{P}}'$. Since $x'^{\nu}\mathfrak{P}'$ is the extended ideal of $\mathfrak{P}$, a finite number of elements such as τ^0 form a base of $\mathfrak{P}'$. Hence $\mathfrak{P}' \equiv 0(\bar{\mathfrak{P}}')$, and consequently $\mathfrak{P}' = \bar{\mathfrak{P}}'$, q. e. d.

Remark 1. In view of the unicity of the factorization of a v-ideal $\mathfrak{A}$ into simple v-ideals (Theorem 7. 1), the second part of Theorem 11. 2 implies that the factorization of the general element of $\mathfrak{A}$ into irreducible factors yields *the* factorization of $\mathfrak{A}$ into simple v-ideals.

Remark 2. Concerning the characterization of simple v-ideals given in Theorem 11. 2, it is not difficult to show that, conversely, *an ideal $\mathfrak{P}$ whose general element t is of the type described in Theorem* 11. 2, *is a v-ideal* (necessarily simple). For the proof we assume $\alpha_1 > \nu$ and we consider the transform $\mathfrak{P}'$ of $\mathfrak{P}$ by the quadratic transformation $x' = x$, $y' = y/x : \mathfrak{O}'^*\mathfrak{P} = x'^{\nu}\mathfrak{P}'$. Using the notation of the proof above, we find as above the congruence $\mathfrak{P}' \equiv 0(\bar{\mathfrak{P}}')$. The preceding proof of the congruence $\bar{\mathfrak{P}}' \equiv 0(\mathfrak{P}')$ was based upon the fact that $\mathfrak{P}'$ was a v-ideal. But we may proceed without making use of this property of $\mathfrak{P}'$. Let $t'(x', y')$ be the general element of $\mathfrak{P}'$, $t' = \Sigma t'_{ij} x'^i y'^j$, and let $\Sigma c_{ij} t'_{ij} = 0$ be a true linear relation between the t'_{ij}. If $t^{(0)} = \Sigma t_{ij}{}^{(0)} x^i y^j$ is any element in $\mathfrak{P}$, and if $t^{(0)} = x'^{\nu}\tau_0$, $\tau_0 = \Sigma\tau_{ij}{}^{(0)} x'^i y'^j$, then τ_0 is an element of $\mathfrak{P}'$. Hence we must have

$$\Sigma c_{ij}\tau_{ij}{}^{(0)} = \sum_{i+\nu \geqq j} c_{ij} t^{(0)}{}_{i+\nu-j,j} = 0.$$

This relation holds for any element $t^{(0)}$ in $\mathfrak{P}$, consequently we must have $\sum_{i+\nu \geqq j} c_{ij} t_{i+\nu-j,j} = 0$, whence $\sum c_{ij}\tau_{ij} = 0$. This last relation shows that τ *is a variable element of* $\mathfrak{P}'$. But then, in view of Theorem 10. 1, also $\epsilon'\tau$ is a variable element of $\mathfrak{P}'$, since ϵ' is the general element of the unit ideal. Consequently $\bar{\mathfrak{P}}' \equiv 0(\mathfrak{P}')$, whence again $\mathfrak{P}' = \bar{\mathfrak{P}}'$.

Consider the valuation B defined by the branch

$$y = \zeta = \sum_{i=1}^{k} c_i x^{\alpha_i/\nu} + \sum_{j=0}^{\infty} d_j x^{(\alpha_{k+1}+j)/\nu},$$

where $d_0, d_1, \cdots$ are arbitrary constants (in $\mathfrak{k}$) and $d_0 \neq 0$. Let the value of t in this valuation be λ_0/ν, λ_0—an integer. The integer $\lambda_0 = \lambda(\mathfrak{P})$ is independent of the particular constants d_j and is uniquely determined by the ideal $\mathfrak{P}$ and by the auxiliary condition $\alpha_1 \geqq \nu$ which prevents a special position of the axis $x = 0$. If the variables x', y' are used, then the valuation B is defined by the branch

$$y' = \zeta' = \sum_{i=1}^{k} c_i x'^{(\alpha_i-\nu)/\nu} + \sum_{j=0}^{\infty} d_j x^{(\alpha_{k+1}+j-\nu)/\nu}.$$

We have $v(\mathfrak{P}') = v(\mathfrak{P}) - v(x^\nu)$, whence $v(\mathfrak{P}') = (\lambda_0 - \nu^2)/\nu$. If $\alpha_1 - \nu \geqq \nu$, then $\lambda(\mathfrak{P}') = \lambda_0 - \nu^2 < \lambda(\mathfrak{P})$. If $\alpha_1 - \nu < \nu$, then the rôles of the variables x', y' must be interchanged, and putting $\alpha_1 - \nu = \nu'$ we find $v(\mathfrak{P}') = [(\lambda_0 - \nu^2)/\nu] \cdot (\nu/\nu')$, whence again $\lambda(\mathfrak{P}') = \lambda_0 - \nu^2 < \lambda(\mathfrak{P})$. The inequality $\lambda(\mathfrak{P}') < \lambda(\mathfrak{P})$ leads to a complete induction with respect to $\lambda(\mathfrak{P})$. If $\lambda(\mathfrak{P}) = 1$, then $\nu = 1$, since evidently $\lambda(\mathfrak{P}) \geqq \nu$, and hence $\mathfrak{P} = \mathfrak{p}^* = (x, y)$, so that if $\lambda(\mathfrak{P}) = 1$, $\mathfrak{P}$ is a v-ideal. Since $\lambda(\mathfrak{P}') < \lambda(\mathfrak{P})$ we may assume, according to our induction, that $\mathfrak{P}'$ is a v-ideal. Now $x'^\nu \mathfrak{P}'$ is the extended ideal of $\mathfrak{P}$, and from the expression of the general element t of $\mathfrak{P}$ it is seen that the subform of degree ν of t contains only the term y^ν (since $\alpha_1 > \nu$). By Theorem 4. 3, our assertion that $\mathfrak{P}$ is a v-ideal, will follow, provided it is shown that $\mathfrak{P}$ *is the contracted ideal of* $x'^\nu \mathfrak{P}'$. The proof of this is immediate. The general element $\bar{t}$ of the contracted ideal of $x'^\nu \mathfrak{P}'$ is of the form $x'^\nu \bar{\tau}(x', y')$, and we can assert that not only is $\bar{\tau}$ a variable element of $\mathfrak{P}'$ (this is obvious) but also that its coefficients $\bar{\tau}_{ij}$ satisfy those inequalities (see Theorem 11. 1) which insure the irreducibility of $\bar{\tau}$, *because this is true for the general element* $t = x'^\nu \tau(x', y')$ *of the ideal* $\mathfrak{P}$. It follows that $\bar{t}$ is necessarily of the form $\bar{t} = \epsilon(x, y) \prod_{i=1}^{\nu} (y - y_i)$, where $y_1 = \sum c_i x^{\alpha_i/\nu} + \sum_{j=0}^{\infty} t_j x^{(\alpha_{k+1}+j)/\nu}$, whence $\bar{t}$ is a variable element of $\mathfrak{P}$. Since $\mathfrak{P}$ is contained in the contracted ideal of $x'^\nu \mathfrak{P}'$, it follows that $\mathfrak{P}$ coincides with the contracted ideal, q. e. d.

12. The class of complete ideals. It has been pointed out in section 7 that a product of valuation ideals is not always a valuation ideal. The class of ideals (in $\mathfrak{k}[x, y]$ or in $\mathfrak{k}\{x, y\}$) which can be factored into valuation ideals is therefore larger than the class of valuation ideals. We shall call ideals in this class *complete ideals.* The term "complete" is suggested by the notion of complete linear systems in algebraic geometry, these being the linear systems which are defined uniquely by *base conditions,* i. e. by the condition of passing with assigned multiplicities through an assigned set of proper or infinitely near points. It will be seen that complete ideals are those and only those ideals whose elements are subject to given base conditions, and to no other conditions. In other words, the polynomials which belong to a complete ideal and whose degree is not greater than a given integer n, form, for any n, a complete linear system.

By our definition of complete ideals, the class of complete ideals is closed under multiplication. Moreover, by Theorem 7. 1, a complete ideal has a unique factorization into simple complete ideals, these last ones being necessarily valuation ideals. Our next aim is to prove that this class is also closed under the other ideal operations ([,]), (:) (intersection and quotient); not however under addition (+).* The theorem which we wish to prove is the following:

THEOREM 12. 1. *If $\mathfrak{A}$ and $\mathfrak{B}$ are complete ideals and $\mathfrak{C}$ is an arbitrary ideal, then $[\mathfrak{A}, \mathfrak{B}]$ and $\mathfrak{A} : \mathfrak{C}$ are complete ideals.*

This theorem is an immediate consequence of the following:

LEMMA. *Any complete ideal is the intersection of valuation ideals, and, conversely, the intersection of valuation ideals is a complete ideal.*

Proof of the lemma. Let $\mathfrak{A} = [\mathfrak{A}_1, \mathfrak{A}_2, \cdots, \mathfrak{A}_k]$, where each $\mathfrak{A}_i$ is a valuation ideal for some valuation B_i. We may assume that the $\mathfrak{A}_i$ belong to one and the same prime ideal $\mathfrak{p} = (x, y)$. Denote by t the general element of $\mathfrak{A}$ and let $t = t_1 t_2 \cdots t_m$ be the factorization of t into irreducible factors in the ring $\mathfrak{K}\{x, y\}$, where $\mathfrak{K}$ is the algebraic closure of the field of the coefficients of t. Let $\alpha_{ij} = v(t_i)$ be the value of t_i in the valuation B_j $(i = 1, 2, \cdots, m;\ j = 1, 2, \cdots, k)$, and let $\mathfrak{A}_{ij}$ be the v-ideal belonging to B_j such that $v(\mathfrak{A}_{ij}) = \alpha_{ij}$. Since $v(t_i) = v(\mathfrak{A}_{ij})$ in B_j, t_i is a variable element of $\mathfrak{A}_{ij}$, whence t_i is also a variable element of the intersection

* *Example.* The ideals $(y^2, \mathfrak{p}^3)$, $(x^2, \mathfrak{p}^3)$ are v-ideals, but their sum (join) is the *simple* ideal $(x^2, y^2, \mathfrak{p}^3)$, which is obviously not a v-ideal and consequently not complete.

13

$$\mathfrak{B}_i = [\mathfrak{A}_{i1}, \mathfrak{A}_{i2}, \cdots, \mathfrak{A}_{ik}], \qquad (i = 1, 2, \cdots, m).$$

Let τ_i be the general element of $\mathfrak{B}_i$. We have $v(\tau_i) \leqq v(t_i)$ in B_j, since all the linear relations between the coefficients of τ_i are also satisfied by the coefficients of t_i. On the other hand, we have in any of the valuations B_j, $v(\tau_i) = v(\mathfrak{B}_i) \geqq v(\mathfrak{A}_{ij}) = v(t_i)$. Hence $v(\tau_i) = v(t_i)$. Let $\tau = \tau_1\tau_2 \cdots \tau_m$ be the direct product of the elements τ_i. The elements t and τ have the same value in each of the valuations B_j, whence $v(\tau) = v(\mathfrak{A}) \geqq v(\mathfrak{A}_j)$ in B_j. This shows that τ is a variable element of $\mathfrak{A}_j$, whence τ also is a variable element of the intersection $\mathfrak{A}$ of the $\mathfrak{A}_j$. But τ is not less general than t (since τ_i is the general element of the ideal $\mathfrak{B}_i$, of which t_i is a variable element, and since τ is a *direct* product of the τ_i); consequently, since t is the general element of $\mathfrak{A}$, also τ is the general element of $\mathfrak{A}$. We may then identify t with τ, and we deduce *that t is the direct product of its irreducible factors t_i and that t_i is the general element of $\mathfrak{B}_i$.*

We now apply the considerations developed in the proof of Theorem 11.2. For the irreducible element t_i we have the following uniformization: $t_i = \epsilon_i(x, y) \prod_{h=1}^{\nu_i} (y - y_h^{(i)})$, where

$$y_1^{(i)} = \sum_{j=1}^{k} c_{ij} x^{\alpha_{ij}/\nu_i} + \sum_{j=0}^{\infty} \tau_{ij} x^{(\alpha_{i,k+1}+j)/\nu_i},$$

τ_{i0}—transcendental with respect to field $\mathfrak{k}$.

Since the inequalities $v(t) \geqq v(\mathfrak{A}_j)$ in B_j, $j = 1, 2, \cdots, k$, are the only conditions which should be imposed on t, it follows that all the τ_{ij} are indeterminates, as the value of t_i in any algebraic valuation is not altered if we regard the τ_{ij} as indeterminates. Also $\epsilon_i(x, y)$ is to be regarded as a unit element with indeterminate coefficients, and the product of ϵ_i by $\prod_{i=1}^{\nu_i} (y - y_h^{(i)})$ is to be intended as direct. But then we must have necessarily $(\alpha_{i1}, \alpha_{i2}, \cdots, \alpha_{i,k+1}, \nu_i) = 1$ and consequently t_i is the general element of a simple v-ideal (by the preceding section, Remark 2). Hence $\mathfrak{B}_i = \mathfrak{P}^{(i)}$, where $\mathfrak{P}^{(i)}$ is a simple v-ideal. By Theorem 10.1, $t_1 t_2 \cdots t_m$ is a quasi-general element of the ideal $\mathfrak{P}^{(1)}\mathfrak{P}^{(2)} \cdots \mathfrak{P}^{(m)}$. But the coefficients of the direct product Πt_i, i.e. of t, satisfy only linear relations, since t is the general element of $\mathfrak{A}$. Hence t is also the general element of $\mathfrak{P}^{(1)}\mathfrak{P}^{(2)} \cdots \mathfrak{P}^{(m)}$, and consequently $\mathfrak{A} = \mathfrak{P}^{(1)}\mathfrak{P}^{(2)} \cdots \mathfrak{P}^{(m)}$. This proves the first part of the lemma.

To prove the second part of the lemma, let $\mathfrak{A} = \mathfrak{A}_1\mathfrak{A}_2 \cdots \mathfrak{A}_m$ be a complete ideal, where $\mathfrak{A}_1, \mathfrak{A}_2, \cdots, \mathfrak{A}_m$ are simple v-ideals, which we may assume as primary ideals belonging to one and the same prime ideal $\mathfrak{p} = (x, y)$. Let

$\mathfrak{A}_j$ be of kind h_j, and let $h = \max(h_1, h_2, \cdots, h_m)$. The assertion of the lemma is trivial if $m = 1$. We assume that the assertion is true for all integers h' such that $h' < h$ (if $h = 1$, then necessarily $m = 1$). Let $\mathfrak{A}_j$ belong to a valuation B_j. Assuming that $v(y) \geqq v(x)$ in each of the valuation B_j, we apply the quadratic transformation T, $x' = x$, $y' = y/x$ to the polynomial ring $\mathfrak{O} = \mathfrak{k}[x, y]$, getting the ring $\mathfrak{O} = \mathfrak{k}[x', y']$ of polynomials in x', y'. Let $\mathfrak{O}'\mathfrak{A}_j = x^{\rho_j}\mathfrak{A}'_j$, whence $\mathfrak{O}'\mathfrak{A} = x'^{\rho}\mathfrak{A}'_1\mathfrak{A}'_2 \cdots \mathfrak{A}'_m$, $\rho = \sum_{j=1}^{m} \rho_j$, where $\mathfrak{A}_j \equiv 0(\mathfrak{p}^{\rho_j})$ and $\mathfrak{A}_j \not\equiv 0(\mathfrak{p}^{\rho_j+1})$. By Theorem 4. 1, $\mathfrak{A}'_j$ is a v-ideal in $\mathfrak{O}'$, and by Theorems 5. 2 and 5. 3 $\mathfrak{A}'_j$ is a simple v-ideal of kind $h_j - 1$. The product $\prod_{i-1}^{\rho} \mathfrak{A}'_j$ can be first written as the intersection of the partial products consisting of factors $\mathfrak{A}'_j$ which belong to one and the same prime ideal. Then, by our induction, each partial product can be written as the intersection of v-ideals. Let then

$$\mathfrak{A}' = \mathfrak{A}'_1\mathfrak{A}'_2 \cdots \mathfrak{A}'_m = [\mathfrak{B}'_1, \mathfrak{B}'_2, \cdots, \mathfrak{B}'_k],$$

whence

$$\mathfrak{O}'\mathfrak{A} = [x^{\rho}\mathfrak{B}'_1, x^{\rho}\mathfrak{B}'_2, \cdots, x^{\rho}\mathfrak{B}'_k], \tag{28}$$

where $\mathfrak{B}'_1, \mathfrak{B}'_2, \cdots, \mathfrak{B}'_k$ are v-ideals in $\mathfrak{O}'$.

Let σ_i be the smallest integer such that $x'^{\sigma_i}\mathfrak{B}'_i$ is an extended ideal of an ideal in $\mathfrak{O}$, and let $\mathfrak{B}_i$ be the contracted ideal of $x'^{\sigma_i}\mathfrak{B}'_i$. By Theorem 4. 3, $\mathfrak{B}_i$ is a v-ideal and $\mathfrak{B}_i = T^{-1}(\mathfrak{B}'_i)$. We have $\mathfrak{O}'\mathfrak{B}_i = x'^{\sigma_i}\mathfrak{B}'_i$ and $\mathfrak{O}'\mathfrak{A} = x^{\rho}\mathfrak{A}'$. We assert that $\sigma_i \leqq \rho$. In fact, assuming that $\sigma_i > \rho$, we have

$$\mathfrak{O}'\mathfrak{p}^{\sigma_i-\rho}\mathfrak{A} = x'^{\sigma_i}\mathfrak{A}' \subseteqq x'^{\sigma_i}\mathfrak{B}'_i,$$

whence, passing to the contracted ideal of $x'^{\sigma_i}\mathfrak{B}'_i$, $\mathfrak{p}^{\sigma_i-\rho}\mathfrak{A} \equiv 0(\mathfrak{B}_i)$. Now $\mathfrak{A} \equiv 0(\mathfrak{p}^{\rho})$ and $\mathfrak{A} \not\equiv 0(\mathfrak{p}^{\rho+1})$, hence $\mathfrak{p}^{\sigma_i-\rho}\mathfrak{A} \equiv 0(\mathfrak{p}^{\sigma_i})$ and $\not\equiv 0(\mathfrak{p}^{\sigma_i+1})$. Since also $\mathfrak{B}_i$ is in $\mathfrak{p}^{\sigma_i}$ and not in $\mathfrak{p}^{\sigma_i+1}$, the congruence $\mathfrak{p}^{\sigma_i-\rho}\mathfrak{A} \equiv 0(\mathfrak{B}_i)$ implies that the subforms of degree σ_i of the polynomials in $\mathfrak{B}_i$ form a linear system $\Omega(\mathfrak{B}_i)$ of dimension $\geqq \sigma_i - \rho > 0$. This is impossible since, by our definition of $\mathfrak{B}_i (= T^{-1}(\mathfrak{B}'_i))$, the system $\Omega(\mathfrak{B}_i)$ is of dimension zero.

We have therefore $\sigma_i \leqq \rho$, and consequently $x^{\rho}\mathfrak{B}'_i$ is an extended ideal, its contracted ideal being $\mathfrak{p}^{\rho-\sigma_i}\mathfrak{B}_i$ (Theorem 4. 3). Denoting by $\mathfrak{A}^*$ the contracted ideal of $\mathfrak{O}'\mathfrak{A}$, we obtain from (28):

$$\mathfrak{A} \subseteqq \mathfrak{A}^* = [\mathfrak{p}^{\rho-\sigma_1}\mathfrak{B}_1, \mathfrak{p}^{\rho-\sigma_2}\mathfrak{B}_2, \cdots, \mathfrak{p}^{\rho-\sigma_k}\mathfrak{B}_k].$$

Now $\mathfrak{B}_i$ is a valuation ideal for some valuation, and in that valuation the ideal $\mathfrak{p}^{\rho-\sigma_i}\mathfrak{B}_i$ is equivalent to some valuation ideal $\mathfrak{C}_i$. By Corollary 3.3 we have $\mathfrak{p}^{\rho-\sigma_i}\mathfrak{B}_i = [\mathfrak{C}_i, \mathfrak{p}^\rho]$, whence

$$\mathfrak{A} \subseteqq \mathfrak{A}^* = [\mathfrak{C}_1, \mathfrak{C}_2, \cdots, \mathfrak{C}_k, \mathfrak{p}^\rho].$$

Hence $\mathfrak{A}^*$ is the intersection of valuation ideals.[13] By the first part of the lemma, just proved, $\mathfrak{A}^*$ is a complete ideal,

$$\mathfrak{A}^* = \mathfrak{A}^*_1\mathfrak{A}^*_2 \cdots \mathfrak{A}^*_n,$$

where the $\mathfrak{A}^*_j$ are simple v-ideals. $\mathfrak{A}$ and $\mathfrak{A}^*$ have the same extended ideal $x'^\rho\mathfrak{A}'$. If then $T(\mathfrak{A}^*_j) = \mathfrak{A}'^*_j$ we must have

$$x'^\rho\mathfrak{A}'_1\mathfrak{A}'_2 \cdots \mathfrak{A}'_m = \mathfrak{O}'\mathfrak{A} = \mathfrak{O}'\mathfrak{A}^* = x'^\rho\mathfrak{A}'^*_1\mathfrak{A}'^*_2 \cdots \mathfrak{A}'^*_n.$$

By the unique factorization theorem (7.1), it follows, for a proper ordering of the factors, $m = n$, $\mathfrak{A}'_j = \mathfrak{A}'^*_j$, whence also $\mathfrak{A}_j = \mathfrak{A}^*_j$, $\mathfrak{A} = \mathfrak{A}^*$, and consequently $\mathfrak{A}$ is the intersection of v-ideals, q. e. d.

Theorem 12.1 now follows immediately. That $[\mathfrak{A}, \mathfrak{B}]$ is a complete ideal, if $\mathfrak{A}$ and $\mathfrak{B}$ are complete ideals, is now trivial, since, by our lemma, complete ideals can also be defined as intersections of valuation ideals. Consider now the ideal $\mathfrak{A}:\mathfrak{C}$, where $\mathfrak{A}$ is a complete ideal and $\mathfrak{C}$ is an arbitrary ideal. We have $\mathfrak{A} = [\mathfrak{A}_1, \mathfrak{A}_2, \cdots, \mathfrak{A}_k]$, where the $\mathfrak{A}_j$ are valuation ideals, whence $\mathfrak{A}:\mathfrak{C} = [\mathfrak{A}_1:\mathfrak{C}, \mathfrak{A}_2:\mathfrak{C}, \cdots, \mathfrak{A}_k:\mathfrak{C}]$. But by Theorem 2.1, the quotient $\mathfrak{A}_j:\mathfrak{C}$ is again a valuation ideal. Hence $\mathfrak{A}:\mathfrak{C}$ is the intersection of valuation ideals, and therefore is a complete ideal, q. e. d.

COROLLARY 12.2. *If $\mathfrak{A}$ is a complete ideal* (in $\mathfrak{k}[x, y]$ or in $\mathfrak{k}\{x, y\}$) *and if $x'^\rho\mathfrak{A}'$ is the extended ideal $\mathfrak{O}'\mathfrak{A}$ in the ring $\mathfrak{k}[x', y']$ or $\mathfrak{k}\{x', y'\}$, where $x' = x$, $y' = y/x$, then $\mathfrak{A}'$ is a complete ideal and $\mathfrak{A}$ is the contracted ideal of $x'^\rho\mathfrak{A}'$.*

That $\mathfrak{A}'$ is a complete ideal is trivial. The second part of the corollary follows from the relation $\mathfrak{A} = \mathfrak{A}^*$ established in the course of the proof of the second part of the lemma.

---

[13] *Any power of $\mathfrak{p}$ is a valuation ideal.* Consider the 0-dimensional valuation defined by the divisor (x') and by the point $y' = 0$ of this divisor, i. e. if $F(x', y')$ is any rational function of x', y', then put $v(F) = (m, n)$, where x'^m is the highest power of x' which divides F, $F = x'^m F_1(x'y')$, $F_1(0, y') \neq 0$, and where y'^n is the highest power of y' which divides $F_1(0, y')$. Then it is easily seen that the sequence of v-ideals

When a linear system of curves $f = 0$ is subjected to base conditions, it is required that the curves of the system have assigned intersection multiplicities with an assigned set of algebraic branches γ_j. For each branch γ_j the corresponding condition is equivalent to the condition that the value of the polynomial f in the valuation B_j defined by the branch γ_j be not inferior to a given integer; in other words: it is required that $f \equiv 0(\mathfrak{A}_j)$, where $\mathfrak{A}_j$ is a given valuation ideal belonging to B_j. The full set of base conditions is then described by the congruence: $f \equiv 0([\mathfrak{A}_1, \mathfrak{A}_2, \cdots])$, i. e. by the condition that f belong to a given complete ideal. However, the representation of a complete ideal as an intersection of valuation ideals is not unique. For instance, let $\mathfrak{A} = (xy, \mathfrak{p}^3)$, where $\mathfrak{p} = (x, y)$. $\mathfrak{A}$ is not a valuation ideal, since it has only one subform xy and this is not a power of a linear form. Let $\mathfrak{A}_1 = (x^2, xy, \mathfrak{p}^3)$, $\mathfrak{A}_2 = (xy, y^2, \mathfrak{p}^3)$, $\mathfrak{A}'_1 = (x, x^2, xy, \mathfrak{p}^3)$, $\mathfrak{A}'_2 = (y, xy, y^2, \mathfrak{p}^3)$. These 4 ideals are valuation ideals and we have $\mathfrak{A} = [\mathfrak{A}_1, \mathfrak{A}_2] = [\mathfrak{A}'_1, \mathfrak{A}'_2]$. This ambiguity is the algebraic equivalent of the distinction which the geometric theory makes between the *assigned* or *virtual* multiplicities of the curves of a linear system and the *effective* multiplicities of these curves. It is the representation of a complete ideal as a *product* of valuation ideals that is unique and puts into evidence the effective multiplicities of the general curve of a linear system.*

We point out explicitly the following result arrived at in the course of the first part of the above proof: *the general element t of a complete ideal $\mathfrak{A}$ is the direct product of its irreducible factors, and the factorization of t yields the factorization of $\mathfrak{A}$ into simple v-ideals.* This is the generalization to complete ideals of the similar property of valuation ideals given in Theorem 11. 2.

We conclude this section with the definition of an operation which assigns to each ideal $\mathfrak{A}$ in $\mathfrak{O}$ a uniquely determined complete ideal $\mathfrak{A}'$, *the complete ideal determined by* $\mathfrak{A}$. The analogue of this operation in the theory of linear systems is given by the passage from an arbitrary linear system to the corresponding complete linear system. *We define $\mathfrak{A}'$ as the intersection of all the complete ideals containing* $\mathfrak{A}$. Since $\mathfrak{A}$ has a finite length, it is clear that $\mathfrak{A}'$ is also the intersection of a finite number of complete ideals, *hence $\mathfrak{A}'$ itself is a complete ideal*, i. e. *$\mathfrak{A}'$ is the smallest complete ideal containing* $\mathfrak{A}$. The operation comes under the heading of the (′) *operations* studied in other connections by van der Waerden ([9], § 103), Prüfer [8] and others, since it enjoys the following formal properties:

---

in $\mathfrak{k}[x, y]$ consists of all ideals of the form $(x^\lambda y^{\rho-\lambda}, x^{\lambda-1} y^{\rho-\lambda+1}, \ldots, y^\rho, \mathfrak{p}^{\rho+1})$, $\rho \geqq \lambda \geqq 0$, $\rho = 1, 2, 3, \ldots$. For $\lambda = \rho$, we find the ideal $\mathfrak{p}^\rho$.

* See Note at end of Section 12.

1. $(\mathfrak{A}')' = \mathfrak{A}'$. 2. If $\mathfrak{A}_1 \supseteqq \mathfrak{A}_2$, then $\mathfrak{A}'_1 \supseteqq \mathfrak{A}'_2$.
3. $(\mathfrak{A}'_1, \mathfrak{A}'_2)' = (\mathfrak{A}_1, \mathfrak{A}_2)'$. 4. $(\mathfrak{A}'_1\mathfrak{A}'_2)' = (\mathfrak{A}_1\mathfrak{A}_2)'$.
5. $(a)' = (a)$; $((a) \cdot \mathfrak{A})' = (a) \cdot \mathfrak{A}'$.

Proof. 1. Trivial, because $\mathfrak{A}'$ is a complete ideal itself;

2. Self-evident;

3. By 2, $(\mathfrak{A}_1, \mathfrak{A}_2)' \supseteqq (\mathfrak{A}'_1, \mathfrak{A}'_2)$, whence $(\mathfrak{A}_1, \mathfrak{A}_2)' \supseteqq (\mathfrak{A}'_1, \mathfrak{A}'_2)'$. On the other hand $\mathfrak{A}'_1 \supseteqq \mathfrak{A}_1$, $\mathfrak{A}'_2 \supseteqq \mathfrak{A}_2$, whence $(\mathfrak{A}'_1, \mathfrak{A}'_2)' \supseteqq (\mathfrak{A}_1, \mathfrak{A}_2)'$, by 2.

4. By 2, we have $(\mathfrak{A}'_1\mathfrak{A}'_2)' \supseteqq (\mathfrak{A}_1\mathfrak{A}_2)'$. Let $\mathfrak{B}$ be any valuation ideal belonging to some valuation B and containing $\mathfrak{A}_1\mathfrak{A}_2$, and let $\mathfrak{B}_i$, $i = 1, 2$, be the valuation ideal for B which is equivalent to $\mathfrak{A}_i$. We have $v(\mathfrak{B}_1\mathfrak{B}_2) = v(\mathfrak{A}_1\mathfrak{A}_2) \geqq v(\mathfrak{B})$, whence $\mathfrak{B}_1\mathfrak{B}_2 \equiv 0(\mathfrak{B})$. Since $\mathfrak{A}_i \equiv 0(\mathfrak{B}_i)$ and $\mathfrak{B}_i$ is a valuation ideal, hence complete, we have $\mathfrak{A}'_i \equiv 0(\mathfrak{B}_i)$, consequently $\mathfrak{A}'_1\mathfrak{A}'_2 \equiv 0(\mathfrak{B}_1\mathfrak{B}_2)$. Since every complete ideal is the intersection of valuation ideals, it follows that $(\mathfrak{A}_1\mathfrak{A}_2)'$ is the intersection of all the valuation ideals $\mathfrak{B}$ containing $\mathfrak{A}_1\mathfrak{A}_2$. Hence $\mathfrak{A}'_1\mathfrak{A}'_2 \subseteqq (\mathfrak{A}_1\mathfrak{A}_2)'$. As a consequence also $(\mathfrak{A}'_1\mathfrak{A}'_2)' \subseteqq ((\mathfrak{A}_1\mathfrak{A}_2)')$, whence 4 follows.

We observe, however, that $(\mathfrak{A}'_1\mathfrak{A}'_2)' = \mathfrak{A}'_1\mathfrak{A}'_2$, since the product of the complete ideals $\mathfrak{A}'_1, \mathfrak{A}'_2$ is itself complete. Hence 4 can be written as follows:

4′. $\mathfrak{A}'_1\mathfrak{A}'_2 = (\mathfrak{A}_1\mathfrak{A}_2)'$.

5. A principal ideal (a) is itself a complete ideal. If $a = a_1^{\rho_1}a_2^{\rho_2} \cdots a_k^{\rho_k}$, where $a_i(x, y)$ is an irreducible polynomial, then $(a) = (a_1^{\rho_1}) \cap (a_2^{\rho_2}) \cap \cdots \cap (a_k^{\rho_k})$, and evidently $(a_i^{\rho_i})$ is a v-ideal, belonging to the 1-dimensional valuation defined by the divisor (a_i). Hence $(a)' = (a)$. In a similar straightforward manner the relation $((a) \cdot \mathfrak{A})' = (a)\mathfrak{A}'$ can be proved.

THEOREM 12. 3 (*invariance of the operation* (′) *under quadratic transformations*). *If* $\mathfrak{p} = (x, y)$ *and if* $\mathfrak{A} \equiv 0(\mathfrak{p}^\rho)$, $\mathfrak{A} \not\equiv 0(\mathfrak{p}^{\rho+1})$, *then also* $\mathfrak{A}' \equiv 0(\mathfrak{p}^\rho)$, $\mathfrak{A}' \not\equiv 0(\mathfrak{p}^{\rho+1})$. *Moreover, if* $\mathfrak{B}$ *is the transform of* $\mathfrak{A}$ *under the quadratic transformation* T, *then* $\mathfrak{B}'$ *is the transform of* $\mathfrak{A}'$ *under* T, i.e. $\mathfrak{O}'\mathfrak{A} = x'^\rho\mathfrak{B}$ implies $\mathfrak{O}'\mathfrak{A}' = x'^\rho\mathfrak{B}'$.

Proof. Since any power of $\mathfrak{p}$ is a complete ideal (even a valuation ideal) and since $\mathfrak{A}'$ is the smallest complete ideal containing $\mathfrak{A}$, the congruence $\mathfrak{A} \equiv 0(\mathfrak{p}^\rho)$ implies the congruence $\mathfrak{A}' \equiv 0(\mathfrak{p}^\rho)$. Moreover, if $\mathfrak{A} \not\equiv 0(\mathfrak{p}^{\rho+1})$, also $\mathfrak{A}' \not\equiv 0(\mathfrak{p}^{\rho+1})$ since $\mathfrak{A} \equiv 0(\mathfrak{A}')$. Let then $\mathfrak{O}'\mathfrak{A} = x'^{\rho}\mathfrak{B}$, $\mathfrak{O}'\mathfrak{A}' = x'^{\rho}\mathfrak{B}^*$, where $\mathfrak{O}' = \mathfrak{k}[x', y']$ (or $\mathfrak{O}' = \mathfrak{k}\{x', y'\}$), $x' = x$, $y' = y/x$, and where $\mathfrak{B}$ and $\mathfrak{B}^*$ are not divisible by x'. Since $\mathfrak{A} \equiv 0(\mathfrak{A}')$, also $\mathfrak{B} \equiv 0(\mathfrak{B}^*)$, and by Corollary 12. 2, $\mathfrak{B}^*$ is a complete ideal. To prove that $\mathfrak{B}^* = \mathfrak{B}'$, we have to show that if $\mathfrak{B}_1$ is any complete ideal containing $\mathfrak{B}$, then $\mathfrak{B}^* \equiv 0(\mathfrak{B}_1)$. Let σ be the smallest integer such that $x'^{\sigma}\mathfrak{B}_1$ is an extended ideal of an ideal in $\mathfrak{O}$. The same reasoning employed in the proof of the second part of the Lemma, for the derivation of the inequality $\sigma_i \leqq \rho$, can be used also now in order to show that $\sigma \leqq \rho$. It is only necessary to observe that by Corollary 12. 2 the contracted ideal of $x'^{\sigma}\mathfrak{B}_1$ is a complete ideal, say $\mathfrak{C}$. The system $\Omega(\mathfrak{C})$ of the subforms (of degree σ) of $\mathfrak{C}$ can be of dimension greater than zero, only if in the factorization of $\mathfrak{C}$ into simple v-ideals there occurs the factor $\mathfrak{p}$. We would have then $\mathfrak{C} = \mathfrak{p}\mathfrak{D}$, whence $\mathfrak{O}'\mathfrak{D} = x'^{\sigma-1}\mathfrak{B}_1$, in contradiction with the definition of the integer σ. Hence $\Omega(\mathfrak{C})$ is of dimension zero, and it was this value of the dimension of $\Omega(\mathfrak{B}_i)$ that played a rôle in the proof of the Lemma.

Since $\sigma \leqq \rho$ and $\mathfrak{B}_1 \supseteqq \mathfrak{B}$, we have $x'^{\rho}\mathfrak{B}_1 \supseteqq x'^{\rho}\mathfrak{B}$. Now let $\mathfrak{C}$ be the contracted ideal of $x'^{\sigma}\mathfrak{B}_1$. $\mathfrak{C}$ is a complete ideal, and its extension ideal is $x'^{\sigma}\mathfrak{B}_1$. Hence the extended ideal of $\mathfrak{p}^{\rho-\sigma}\mathfrak{C}$ is $x'^{\rho}\mathfrak{B}_1$, and consequently (Corollary 12. 2) $\mathfrak{p}^{\rho-\sigma}\mathfrak{C}$ is the contracted ideal of $x'^{\rho}\mathfrak{B}_1$, since $\mathfrak{p}^{\rho-\sigma}$ is also a complete ideal. Since the contracted ideal of $x'^{\rho}\mathfrak{B}$ contains $\mathfrak{A}$, the congruence $x'^{\rho}\mathfrak{B}_1 \supseteqq x'^{\rho}\mathfrak{B}$ implies the congruence $\mathfrak{p}^{\rho-\sigma}\mathfrak{C} \supseteqq \mathfrak{A}$. Hence $\mathfrak{p}^{\rho-\sigma}\mathfrak{C} \supseteqq \mathfrak{A}'$, and passing to the extended ideals in $\mathfrak{O}'$, we find $x'^{\rho}\mathfrak{B}_1 \supseteqq x'^{\rho}\mathfrak{B}^*$, i. e. $\mathfrak{B}_1 \supseteqq \mathfrak{B}^*$, q. e. d.

NOTE.—For those not familiar with the geometric terminology we give here the definition of the effective multiplicities on the basis of the present treatment. We associate with each simple v-ideal $\mathfrak{P}_{k+1}$ of kind $k+1$, belonging to the prime ideal $\mathfrak{P}_1 = \mathfrak{p} = (x, y)$, a point 0_{k+1} in the k-th neighborhood of the point $0_1(0,0)$ of the (x,y)-plane. Let $\mathfrak{P}_1, \mathfrak{P}_2, \cdots, \mathfrak{P}_k$ be the simple v-ideals of kind $1, 2, \cdots, k$ determined by $\mathfrak{P}_{k+1}$ and preceding it (Theorem 6. 1), and let $0_1, 0_2, \cdots, 0_k$ be the associated points. We proceed to define the *set of base points* of the ideal $\mathfrak{P}_{k+1}$ or the symbol

$$B(\mathfrak{P}_{k+1}) = (0_1^{r_1} 0_2^{r_2} \cdots 0_k^{r_k} 0_{k+1}), \qquad r_i > 0,$$

and we shall say that r_i is the *effective multiplicity* of the ideal $\mathfrak{P}_{k+1}$ at $0_i (r_{k+1} = 1)$. We set $B(\mathfrak{P}_1) = (0_1)$. For any k we define $B(\mathfrak{P}_{k+1})$ by induction with respect to k. We know, by Theorem 6. 2, that if B is any valuation for which $\mathfrak{P}_{k+1}$ is a v-ideal, the v-ideals for B which precede $\mathfrak{P}_{k+1}$ are independent of B. Let $\mathfrak{q}_\lambda$ be the v-ideal which is followed immediately by $\mathfrak{P}_{k+1}$ and let $\mathfrak{q}_\lambda = \mathfrak{P}_1^{a_1}\mathfrak{P}_2^{a_2}\cdots\mathfrak{P}_k^{a_k}$. Assuming that the symbol $B(\mathfrak{P}_i)$ has already been defined for all $i < k+1$, we put

$$B(\mathfrak{P}_{k+1}) = [B(\mathfrak{P}_1)]^{a_1}[B(\mathfrak{P}_2)]^{a_2} \cdots [B(\mathfrak{P}_k)]^{a_k} O_{k+1},$$

where, if

$$B(\mathfrak{P}_{i+1}) = (0_1^{s_1} 0_2^{s_2} \cdots 0_i^{s_i} 0_{i+1})$$

and

$$B(\mathfrak{P}_{j+1}) = (0_1^{t_1} 0_2^{t_2} \cdots 0_j^{t_j} 0_{j+1}),$$

then

$$B(\mathfrak{P}_i) B(\mathfrak{P}_j) = (0_1^{s_1+t_1} 0_2^{s_2+t_2} \cdots).$$

To check this definition against the customary geometric definition, we point out two implications of our definition:

(1) *$\mathfrak{P}_{k+1}$ is exactly divisible by $\mathfrak{p}^{r_1}$*, i.e. $\mathfrak{P}_{k+1} \equiv 0(\mathfrak{p}^r)$, $\mathfrak{P}_{k+1} \not\equiv 0(\mathfrak{p}^{r+1})$. The assertion is true for $k = 0$. Assuming that is true for $\mathfrak{P}_{i+1}$, $i = 0, 1, \cdots, k-1$ and that $\mathfrak{P}_{i+1}$ is exactly divisible by $\mathfrak{p}^{r_{i1}}$, then

$$B(\mathfrak{P}_{i+1}) = (0_1^{r_{i1}} \cdots), \qquad r_1 = \alpha_1 r_{01} + \cdots + \alpha_k r_{k-1,1}$$

and from this we conclude, in view of $\mathfrak{q}_\lambda = \mathfrak{P}_1^{\alpha_1} \mathfrak{P}_2^{\alpha_2} \cdots \mathfrak{P}_k^{\alpha_k}$, that $\mathfrak{q}_\lambda$ is exactly divisible by $\mathfrak{p}^{r_1}$. Now $\mathfrak{p}\mathfrak{q}_\lambda \equiv 0(\mathfrak{P}_{k+1})$, since $\mathfrak{P}_{k+1}$ is a maximal subideal of $\mathfrak{q}_\lambda$. Hence $\mathfrak{P}_{k+1}$ is divisible at most by $\mathfrak{p}^{r_1+1}$. If $\mathfrak{P}_{k+1}$ was divisible by $\mathfrak{p}^{r_1+1}$, then the system of subforms $\Omega(\mathfrak{p}\mathfrak{q}_r)$ would have been a subsystem of $\Omega(\mathfrak{P}_{k+1})$. This is impossible, since $\Omega(\mathfrak{p}\mathfrak{q}_\lambda)$ is of dimension $\geqq 1$, while $\Omega(\mathfrak{P}_{k+1})$ is of dimension 0. Hence $\mathfrak{P}_{k+1}$ is divisible exactly by $\mathfrak{p}^{r_1}$. We have thus proved that in the general polynomial $f(x, y)$ in $\mathfrak{P}_{k+1}$ the terms of lowest degree are of degree r_1, i.e. *the curve $f = 0$ has at 0_r an r_1-fold point (while no curve $f = 0$, $f \varepsilon \mathfrak{P}_{k+1}$, has at 0_1 a multiplicity less than r_1).*

(2) *If $\mathfrak{P}'_i$ is the transform of $\mathfrak{P}_{i+1}$ by a quadratic transformation T having at 0_1 a fundamental point, then $B(\mathfrak{P}'_k) = (0'^{r_2}_1 \cdots 0'^{r_k}_{k-1} 0'_k)$, where $0'_i$ is the point associated with $\mathfrak{P}'_i$.* To prove this, we first observe that from the fact, just proved above, that $\mathfrak{q}_\lambda$ and $\mathfrak{P}_{k+1}$ are both divisible exactly by the same power, $\mathfrak{p}^{r_1}$, of $\mathfrak{p}$, it follows that $\Omega(\mathfrak{P}_{k+1})$ is a subsystem of $\Omega(\mathfrak{q}_\lambda)$. Since $\mathfrak{P}_{k+1}$ is a maximal subideal of $\mathfrak{q}_\lambda$, the dimension of $\Omega(\mathfrak{q}_\lambda)$ cannot exceed the dimension of $\Omega(\mathfrak{P}_{k+1})$ augmented by 1. But $\Omega(\mathfrak{P}_{k+1})$ is of dimension 0, while $\Omega(\mathfrak{q}_\lambda)$ is of dimension α_1. Hence $\alpha_1 \leqq 1$. Let $T(\mathfrak{q}_\lambda) = x'^{r_1} \mathfrak{q}'_\mu$, where $\mathfrak{q}'_\mu = \mathfrak{P}'^{\alpha_2}_1 \mathfrak{P}'^{\alpha_3}_2 \cdots \mathfrak{P}'^{\alpha_k}_{k-1}$. We assert that *$\mathfrak{q}'_\mu$ is the immediate predecessor of $\mathfrak{P}'_k$.* If $\alpha_1 = 0$, the assertion follows immediately from Theorem 4.4, since we have in this case $T^{-1}(\mathfrak{q}'_\mu) = \mathfrak{q}_\lambda$. Let now $\alpha_1 = 1$. Assume that $\mathfrak{q}'_\mu$ is not the immediate predecessor of $\mathfrak{P}'_k$ and let $\mathfrak{q}'_s$ be a v-ideal between $\mathfrak{q}'_\mu$ and $\mathfrak{P}'_k$. Let $\mathfrak{q}_m = T^{-1}(\mathfrak{q}'_\mu) = \mathfrak{P}_2^{\alpha_2} \cdots \mathfrak{P}_k^{\alpha_k}$, $\mathfrak{q}_r = T^{-1}(\mathfrak{q}'_s)$, whence $\mathfrak{q}_m \supset \mathfrak{q}_r \supset \mathfrak{P}_{k+1}$ (by Theorem 4.4). We have $\mathfrak{q}_r \equiv 0(\mathfrak{q}_m) \equiv 0(\mathfrak{p}^{r_1-1})$, and also $\mathfrak{q}_r \not\equiv 0(\mathfrak{p}^{r_1+1})$, since $\mathfrak{P}_{k+1} \equiv 0(\mathfrak{q}_r)$. We also have $\mathfrak{p}\mathfrak{q}_m = \mathfrak{q}_\lambda \equiv 0(\mathfrak{q}_r)$, since $\mathfrak{q}_\lambda$ is the immediate predecessor of $\mathfrak{P}_{k+1}$. If $\mathfrak{q}_r$ was divisible by $\mathfrak{p}^{r_1}$, then $\Omega(\mathfrak{q}_\lambda)$ would be a subsystem of $\Omega(\mathfrak{q}_r)$, and this is impossible, since $\Omega(\mathfrak{q}_\lambda)$ is of dimension 1, while $\Omega(\mathfrak{q}_r)$ must be of dimension 0. Hence $\mathfrak{q}_r$ is divisible exactly by $\mathfrak{p}^{r_1-1}$. Now we have $v(\mathfrak{q}_r) > v(\mathfrak{q}_m)$, $v(\mathfrak{p}\mathfrak{q}_r) > v(\mathfrak{p}\mathfrak{q}_m) = v(\mathfrak{q}_\lambda)$, and consequently $\mathfrak{p}\mathfrak{q}_r \equiv 0(\mathfrak{P}_{k+1})$. This is a contradiction, since both $\mathfrak{p}\mathfrak{q}_r$ and $\mathfrak{P}_{k+1}$ are divisible exactly by $\mathfrak{p}^{r_1}$ and $\Omega(\mathfrak{p}\mathfrak{q}_r)$ is of dimension 1. It is thus proved that $\mathfrak{q}'_\mu$ is the immediate predecessor of $\mathfrak{P}'_k$. As a consequence we have

$$B(\mathfrak{P}'_k) = [B(\mathfrak{P}'_1)]^{\alpha_2} [B(\mathfrak{P}'_2)]^{\alpha_3} \cdots [B(\mathfrak{P}'_{k-1})]^{\alpha_k} 0'_k.$$

From this relation our statement follows immediately by induction with respect to k. Our result implies that *if the general curve $f = 0$, $f \varepsilon \mathfrak{P}_{k+1}$, passes through $0_1, 0_2, \cdots, 0_{k+1}$ with effective multiplicities $r_1, r_2, \cdots, r_k, 1$, then its transform by* *the quadratic transformation T passes through the points $O'_1, O'_2, \cdots, O'_k$ with effective multiplicities $r_2, r_3, \cdots, r_k, 1$.* The identity between our definition of effective multiplicities and the customary geometric definition is thus fully proved. It is hardly necessary to add that the definition of the symbol $B(\mathfrak{A})$ for any complete ideal amounts to postulating the relation $B(\mathfrak{B}\mathfrak{C}) = B(\mathfrak{B})B(\mathfrak{C})$.

13. Simple v-ideals and divisors of the second kind. Let $\mathfrak{P}$ be a divisor of the field $\Sigma = \mathfrak{k}(x, y)$, i. e. an homomorphism of Σ upon a field Σ' of dimension 1 (and the symbol ∞), and let us assume that $\mathfrak{P}$ is of second kind for the ring $\mathfrak{O} = \mathfrak{k}[x, y]$, i. e. that the prime ideal $\mathfrak{p}$ determined in $\mathfrak{O}$ by the divisor $\mathfrak{P}$ is zero-dimensional, say $\mathfrak{p} = (x, y)$ (we exclude the case in which x or y are mapped upon ∞). The points of the Riemann surface of the field Σ' define a set of valuations $\{B_\alpha\}$ of the field Σ, all of rank 2. Let $\{\mathfrak{q}_i^{(\alpha)}\}$ be the Jordan sequence of v-ideals in $\mathfrak{O}$ belonging to the valuation B_α. The ideal $\mathfrak{q}_1^{(\alpha)} = \mathfrak{p}$ is independent of α. There may be other values of i such that $\mathfrak{q}_i^{(\alpha)}$ is independent of α, and, in particular, there may occur in the sequence $\{\mathfrak{q}_i^{(\alpha)}\}$ simple ideals independent of α. Their number is necessarily finite, because the simple v-ideals in the sequence $\{\mathfrak{q}_i^{(\alpha)}\}$ determine the sequence completely (Theorem 6. 2) and since the valuations B_α are all distinct. Let $\mathcal{P}_\rho$ be the last simple v-ideal, of kind ρ, which occurs in all the sequences $\{\mathfrak{q}_i^{(\alpha)}\}$. The simple ideal $\mathcal{P}^{(\alpha)}_{\rho+1}$, of kind $\rho + 1$, will then vary with α. We have thus associated with every divisor of Σ, of the second kind, with respect to $\mathfrak{O}$, a simple v-ideal $\mathcal{P}_\rho$ in $\mathfrak{O}$. If we apply ρ successive quadratic transformations, getting a polynomial ring $\mathfrak{k}[X, Y]$ of the new indeterminates X, Y, the ideal $\mathcal{P}^{(\alpha)}_{\rho+1}$ is transformed into a prime 0-dimensional ideal $\mathfrak{p}^{(\alpha)} = (X, Y - c^{(\alpha)})$, and the constant $c^{(\alpha)}$ must vary as α varies. As a consequence, the divisor $\mathfrak{P}$ is of the first kind with respect to the ring $\mathfrak{k}[X, Y]$, and the corresponding prime ideal in this ring is necessarily the 1-dimensional ideal (X). This shows that there exists a divisor $\mathfrak{P}$ for any preassigned simple v-ideal $\mathcal{P}_\rho$ and that $\mathfrak{P}$ is uniquely determined by $\mathcal{P}_\rho$. *We have then a one to one correspondence between the divisors of the second kind (with respect to the ring $\mathfrak{k}[x, y]$) and the 0-dimensional simple v-ideals in $\mathfrak{k}[x, y]$.* The field Σ' upon which Σ is mapped by $\mathfrak{P}$ coincides with the field $\mathfrak{k}(Y)$ and is therefore purely transcendental.

It is important to point out that as the valuation B_α runs through the set $\{B_\alpha\}$ determined by the points of the field Σ', the set $\{\mathcal{P}^{(\alpha)}_{\rho+1}\}$ will include *all* the simple v-ideals $\mathcal{P}_{\rho+1}$ of kind $\rho + 1$, such that $\mathcal{P}_{\rho+1}$ and $\mathcal{P}_\rho$ belong together to one and the same valuation. This follows from the fact, that all such ideals $\mathcal{P}_{\rho+1}$ are transformed by ρ successive quadratic transformations into the above ideals $\mathfrak{p}^{(\alpha)}$.

The preceding considerations refer to the field $\mathfrak{k}(x, y)$ of rational functions in x, y, but can be immediately extended to the field Σ^* of meromorphic functions in x, y. If we put $x = x$, $y = x'y'$, every holomorphic function $\xi(x, y)$ assumes the form $x'^{\rho}(f_1(y') + x'f_2(y') + \cdots)$. If η is another holomorphic function of x, y, and $\eta = x'^{\sigma}(\phi_1(y') + x'\phi_2(y') + \cdots)$, we map the function ξ/η upon 0, ∞ or $f_1(y')/\phi_1(y')$, according as $\rho > \sigma$, $\rho < \sigma$ or $\rho = \sigma$. This mapping defines an homomorphism of the field Σ^* upon the purely transcendental field $\mathfrak{k}(y')$. We regard this homomorphism as a divisor of the second kind of Σ^*, since the prime ideal in $\mathfrak{k}\{x, y\}$ determined by this divisor is the 0-dimensional ideal $\mathfrak{p}^* = (x, y)$. We associate this divisor with the simple ideal $\mathfrak{p}^*$. In the same manner we may associate with any simple ideal $\mathfrak{P}_{h+1}$ of kind $h + 1$ in $\mathfrak{k}\{x, y\}$ a divisor of Σ^*, in which Σ^* is mapped upon a purely transcendental field of one variable. We do this by first applying h successive quadratic transformations, getting a ring $\mathfrak{O}^*_h = \mathfrak{k}\{x_h, y_h\}$, which contains $\mathfrak{k}\{x, y\}$ and in which the ideal $\mathfrak{P}_{h+1}$ corresponds to a simple v-ideal of kind 1, i. e. to the ideal (x_h, y_h). This ideal defines a divisor of the field Σ^*_h of meromorphic functions of x_h, y_h, mapping Σ^*_h upon the field $\mathfrak{k}(y'_h)$ of rational functions of y'_h $(= y_h/x_h)$. In this homomorphism the subfield Σ^* of Σ^*_h is mapped upon the entire field $\mathfrak{k}(y'_h)$, since x_h, y_h are rational functions of x and y. We associate the divisor of Σ^*, obtained in this manner, with the simple v-ideal $\mathfrak{P}_{h+1}$. In exactly the same manner as for the field of rational functions of x and y, it is shown that the correspondence between all the divisors of Σ^*, defined by homomorphic mappings of Σ upon fields of dimension one with respect to $\mathfrak{k}$ and of second kind with respect to $\mathfrak{k}\{x, y\}$, and the simple v-ideals in $\mathfrak{k}\{x, y\}$ is $(1, 1)$, and that consequently any such divisor is purely transcendental (i. e. the field upon which Σ^* is mapped by a divisor of the second kind is necessarily purely transcendental).

We point out explicitly, that if a divisor of Σ^* *of the first kind with respect to* $\mathfrak{k}\{x, y\}$ is defined as an homomorphic mapping of Σ^* upon a field Ω, such that the prime ideal determined in $\mathfrak{k}\{x, y\}$ by the divisor is one-dimensional, then Ω is necessarily the field of all meromorphic functions of one indeterminate. Thus for fields of meromorphic functions Σ^* the classification of the divisors into two kinds is not merely a relative classification with respect to the ring $\mathfrak{k}\{x, y\}$, but rather a classification in terms of the properties of the field Σ^* itself. That this should be so is only natural, in view of the privileged rôle which the ring of holomorphic functions plays in the field Σ^*.

We conclude with one final remark, which finds an application in the proof of the well-known algebro-geometric theorem, that a pencil of curves on an algebraic surface is necessarily linear, if the pencil has a base point at a simple point of the surface. This remark will be elaborated in a joint note by Dr. O. Schilling and the present author. At this place we wish only to observe that the proof of this theorem is based upon the following assertion:

Given a meromorphic function $f(x,y)/\phi(x,y)$, *and assuming that the elements* f *and* ϕ *of* $\mathfrak{k}\{x,y\}$ *are relatively prime and that neither is a unit, then there exists a divisor* $\mathfrak{P}$ *of second kind of the field* Σ^* *of meromorphic functions, such that* f/ϕ *is mapped upon a transcendental element of the image field.* This assertion can be readily proved as follows. Consider the ideal $\mathfrak{A} = (f, \phi)$ in $\mathfrak{k}\{x,y\}$. This ideal is zero-dimensional, since f and ϕ are relatively prime and are both contained in the ideal (x,y). Let $\mathfrak{A}'$ be the complete ideal determined by $\mathfrak{A}$. Let us first consider the case in which $\mathfrak{A}' = \mathfrak{p}^* = (x,y)$. In this case we assert that the required divisor $\mathfrak{P}$ is the one associated with $\mathfrak{p}^*$. In fact, assume that f/ϕ is mapped by $\mathfrak{P}$ upon a constant c. We may assume $c = 0$, replacing $f - c\phi$ by f. Under this hypothesis f/ϕ will have positive value in all the zero-dimensional valuations B_a defined by the points of the Riemann surface of $\mathfrak{P}$, and hence, since $\phi \equiv 0(\mathfrak{p}^*)$, f must belong, for any α, to the valuation ideal $\mathfrak{q}_2^{(a)}$ which follows $\mathfrak{p}^*$ in the Jordan sequence of v-ideals belonging to B_a. As α varies, $\mathfrak{q}_2^{(a)}$ can be any maximal subideal of $\mathfrak{p}^*$ (see section 8) and is a simple ideal (of kind 2). Now for *some* α we will have $\phi \equiv 0(\mathfrak{q}_2^{a})$. In fact, it is sufficient to consider the valuation defined by an irreducible branch of the analytical locus $\phi = 0$. For this valuation it is true that the element ϕ is contained in *all* the 0-dimensional v-ideals $\mathfrak{q}_i$ belonging to the valuation, whence also in $\mathfrak{q}_2$. It follows then that for some α, both f and ϕ are contained in $\mathfrak{q}_2^{(a)}$. But this is impossible, since $\mathfrak{q}_2^{(a)}$ is a complete ideal which does not contain the complete ideal $\mathfrak{A}' = \mathfrak{p}^*$ determined by (f, ϕ).

In the general case, let $\mathfrak{A}' = \mathcal{P}_1^{a_1}\mathcal{P}_2^{a_2}\cdots\mathcal{P}_k^{a_k}$, where $\mathcal{P}_i$ is a simple v-ideal of kind h_i. Let $h = \max\{h_i\}$, and let us apply the quadratic transformation $x' = x$, $y' = y/x$. If $\mathfrak{A} \equiv 0(\mathfrak{p}^{*\rho})$, $\mathfrak{A} \not\equiv 0(\mathfrak{p}^{*\rho+1})$, then by Theorem 12. 3, also $\mathfrak{A}' \equiv 0(\mathfrak{p}^{*\rho})$, $\mathfrak{A}' \not\equiv 0(\mathfrak{p}^{*\rho+1})$, and putting $f = x'^{\rho} f_1$, $\phi = x'^{\rho}\phi_1$, $\mathfrak{A}_1 = (f_1, \phi_1)$, the complete ideal $\mathfrak{A}'_1$ determined by $\mathfrak{A}_1$ is the transform $T(\mathfrak{A}')$, i. e. the ideal

$$\mathfrak{A}'_1 = \mathcal{P}'^{a_1}_1\mathcal{P}'^{a_2}_2\cdots\mathcal{P}'^{a_k}_k.$$

Here each $\mathcal{P}'_i$ is of kind $h_i - 1$, so that $\max\{h_i - 1\} = h - 1$. Since the theorem has already been proved for $h = 1$ ($\mathfrak{A}' = \mathfrak{p}^*$), we may assume that the theorem is true for $h - 1$. We may even assume that for the function f_1/ϕ_1 the divisor whose existence is stated in our assertion is the divisor associated with a factor $\mathcal{P}'_s$, such that $\mathcal{P}'_s$ is of maximum kind $h - 1$. Then it follows immediately that the divisor of the field Σ^* associated with the factor $\mathcal{P}_s$ satisfies the assertion.

THE JOHNS HOPKINS UNIVERSITY.

REFERENCES.

1. Enriques, F. and Chisini, O., *Lezioni sulla teoria geometrica delle equazioni e delle funzioni algebriche*, vol. 2.
2. Krull, W., "Idealtheorie," *Ergebnisse der Mathematik und ihrer Grenzgebiete*, IV, 3.
3. Lasker, E., "Zur Theorie der Moduln und Ideale," *Mathematische Annalen*, vol. 60 (1905).
4. Macaulay, F. S., "The theorem of residuation," *Proceedings of the London Mathematical Society* (1), vol. 31 (1900).
5. Macaulay, F. S., "Algebraic theory of modular systems," *Cambridge Tracts in Mathematics*, vol. 19 (1916).
6. Macaulay, F. S., "Modern algebra and polynomial ideals," *Proceedings of the Cambridge Philosophical Society* (1), vol. 30 (1934).
7. Ostrowski, A., "Über einige Lögungen der Funktionalgleichung $\phi(x) \cdot \phi(y) = \phi(x \cdot y)$," *Acta Mathematica*, vol. 41 (1917).
8. Prüfer, H., "Untersuchungen über die Teilbarkeitseigenschaften in Körpern," *Journal für reine und angewandte Mathematik*, vol. 168 (1932).
9. van der Waerden, B. L., *Moderne Algebra*, vol. 2.
10. Zariski, O., "Algebraic surfaces," *Ergebnisse der Mathematik und ihrer Grenzgebiete*, III, 5.

ON THE LINEARITY OF PENCILS OF CURVES ON ALGEBRAIC SURFACES.*

By O. F. G. SCHILLING [1] and O. ZARISKI.

The object of this note is to give an arithmetical proof of the following often used theorem: "*If a pencil of curves on an algebraic surface has a base point at a simple point of the surface then the pencil is either a linear system or its curves are cut out by hypersurfaces* $(\phi + \lambda\psi)^p$." The essential feature of our approach consists in eliminating the difficulties which arise from the possible singularities of the curves at the simple base point. The interpretation of the pencil of curves as a rational transform of the given surface allows us to apply a theorem proved by one of us.[2]

Let K be a field of algebraic functions of two variables over an algebraically closed field k. An algebraic surface f in the affine n-dimensional space S_n over k is said to be a model of K in S_n if the quotient field of $k[x_1, \cdots, x_n]/\mathfrak{p}(f)$ which is determined by the prime ideal $\mathfrak{p}(f)$ defining the surface f, is isomorphic with K. Thus f is described by the order $k[\xi_1, \cdots, \xi_n] = \mathfrak{O}$ in K where $\xi_i = x_i \bmod \mathfrak{p}(f)$. The 0-dimensional prime ideals $\mathfrak{p}$ of $k[x_1, \cdots, x_n]$ which divide $\mathfrak{p}(f)$ correspond to the points P of the surface f. A point P with the coördinates $\{a_1, \cdots, a_n\}$ is called a simple point of f if

(i) the ideal $(\xi_1 - a_1, \cdots, \xi_n - a_n)$ is a 0-dimensional prime ideal $\mathfrak{p}$ in the integral closure of $\mathfrak{O}$ and

(ii) it is possible to choose two algebraically independent elements, say ξ_1, ξ_2 among $\xi_1, \xi_2, \cdots, \xi_n$ such that the ideal $(\xi_1 - a_1, \xi_2 - a_2)$ is divisible by $\mathfrak{p}$ but not divisible by any primary ideal belonging to $\mathfrak{p}$.

It can be shown that all elements of $\mathfrak{O}$ can be expanded in formal power series of $u = \xi_1 - a_1$, $v = \xi_2 - a_2$ with coefficients in k.[3] Hence the elements of $\mathfrak{O}$ are contained in the ring of holomorphic functions $\{\sum_{i,j \geqq 0} a_{ij}u^i v^j\}$ which itself is contained in the field of all formal meromorphic functions of u, v:

* Received February 7, 1938.

[1] Johnston Scholar of the Johns Hopkins University for 1937-1938.

[2] O. Zariski, "Polynomial ideals defined by infinitely near base points," § 13, *American Journal of Mathematics*, vol. 60 (1938).

[3] O. Zariski, "Some results in the arithmetic theory of algebraic functions of several variables," *Proceedings of the National Academy of Sciences*, vol. 23 (1937).

$$k\{u, v\} = \{(\Sigma a_{ij}u^iv^j)(\Sigma b_{ij}u^iv^j)^{-1}\}.$$

An irreducible algebraic system Σ_r of curves on the surface f is given by an irreducible algebraic correspondence between f and a r-dimensional algebraic variety V_r such that to a generic point on V_r there corresponds a curve $C \subset \Sigma_r$ on f. A pencil Σ_1 of curves on f is an irreducible algebraic system Σ_1 such that there passes through a generic point P of f exactly one curve C in Σ_1.[4] This definition of a pencil Σ_1 is equivalent to the following: the function field K_1 belonging to the variety V_1 defining Σ_1 is a rational transform of the surface f, i. e. K_1 is isomorphic with a 1-dimensional subfield $\bar{K}_1$ of K.

If V_r is a linear r-dimensional space and if the curves of Σ_r are cut out by hypersurfaces $\phi = \sum_{i=0}^{r} \lambda_i\phi_i = 0$, ϕ_i being forms in the imbedding space of f, then Σ_r is called a linear system. In a linear system one usually omits the fixed curves which are cut out by all hypersurfaces ϕ.

After these preliminary remarks we proceed to the proof of the

THEOREM. *If a pencil of curves Σ_1 on an algebraic surface f has a base point at a simple point of f then Σ_1 is either a linear system or its curves are cut out by hypersurfaces $(\phi + \lambda\psi)^p = 0$.*

Proof. Let P be the simple base point of the pencil Σ_1. Since P is assumed to be a simple point of the surface f, there exist functions u, v in $\mathfrak{O}$ defining a field of meromorphic functions $k\{u, v\}$ which contains a subfield $K^* \cong K$. Moreover, $u = v = 0$ at P. Consequently, the field $\bar{K}_1$ which belongs to the pencil Σ_1 has also an isomorphic map K^*_1 in $k\{u, v\}$. Thus each element $a^* \varepsilon K^*$ is represented by a ratio $\dfrac{\alpha(u, v)}{\beta(u, v)}$ of holomorphic functions $\alpha(u, v)$, $\beta(u, v)$. Moreover, there exists a function $\dfrac{\alpha(u, v)}{\beta(u, v)} = A^* \varepsilon K^*_1$ such that

$$\alpha(0, 0) = \beta(0, 0) = 0$$

when u and v assume the constant values and $0, 0$, respectively, at the given point P. The existence of such a function A^* is a consequence of the assumption that P is a simple base point of Σ_1, i. e. that there corresponds to P the whole curve V_1 under the correspondence between f and V_1. In fact, let us assume that the surface f is given in a 3-dimensional affine space $k[x_1, x_2, x_3]$ and that V_1 is given in an n-dimensional projective space $k[y_0, y_1, \cdots, y_n]$. Since K_1 is a rational transform of K we have relations of the following type

[4] For these definitions see for example O. Zariski, "Algebraic surfaces," Chapters II, V, *Ergebnisse der Mathematik und ihrer Grenzgebiete* (Berlin, 1935).

5

$$P(x_1, x_2, x_3) y_i - Q_i(x_1, x_2, x_3) y_0 = 0 \qquad (i = 1, 2, \cdots, n)$$

where $P(x_1, x_2, x_3)$ and $Q_i(x_1, x_2, x_3)$ are polynomials in x_1, x_2, x_3. Using the imbedding of K as K^* in $k\{u, v\}$ we obtain

$$P_i(u, v) y_i - Q_i(u, v) y_0 = 0 \qquad (i = 1, 2, \cdots, n)$$

where $P_i(u, v)$ and $Q_i(u, v)$ are relatively prime holomorphic functions in u, v. These equations can be considered as relations which are contained in the ideal $\mathfrak{C}$ of relations defining the correspondence. The assumption that P be a base point of Σ_1 implies then that

$$P_i(0, 0) y_i - Q_i(0, 0) y_0 \equiv 0 \text{ in } K^*_1.$$

We may suppose that y_0 is different from 0, then

$$P_i(0, 0) \frac{y_i}{y_0} - Q_i(0, 0) \equiv 0.$$

Consequently, since at least one function y_i/y_0 of the field $K^*_1 \cong K_1$ does not lie in k,

$$P_i(0, 0) = Q_i(0, 0) = 0,$$

or y_i/y_0 is a function having the desired properties. Now we are in a position to apply a result of the general theory of valuation ideals stating that for each function $A^* = \frac{\alpha(u, v)}{\beta(u, v)}$, where $\alpha(0, 0) = \beta(0, 0) = 0$, and α, β are relatively prime, there exists a prime divisor $\mathfrak{P}$ of $k\{u, v\}$ which maps $k\{u, v\}$ upon a purely transcendental field $k(t)$ in which the map $A^* * \mathfrak{P}$ of A^* is a transcendental quantity with respect to k.[5] Consequently, the field $K^*_1 \cong K_1$ is mapped upon a transcendental subfield of $k(t)$. Hence K^*_1 is itself a purely transcendental subfield, for the divisor $\mathfrak{P}$ acts as an isomorphic mapping on K^*_1, since K^*_1 and its map have the same degree of transcendentality. We have

$$K^*_1 \cong \bar{K}_1 = k(\lambda) \subset K$$

where λ is the ratio $\frac{p(x_1, x_2, x_3)}{q(x_1, x_2, x_3)}$ of relatively prime polynomials $p(x_1, x_2, x_3)$ and $q(x_1, x_2, x_3)$ in $k[x_1, x_2, x_3]$.

We observe that we do not change the nature of the algebraic pencil Σ_1 if we use instead of the original variety V_1 the birationally equivalent curve $k[\lambda]$ for the definition of Σ_1.

Now it remains to be shown that Σ_1 is a linear system cut out by surfaces

[5] See note 2, *loc. cit.*, p. 203

$\lambda q(x_1, x_2, x_3) - p(x_1, x_2, x_3) = 0$ or a system of curves cut out by the surfaces $(\lambda q(x_1, x_2, x_3) - p(x_1, x_2, x_3))^\rho = 0,\ \rho > 1$. Consider for this purpose the ideal

$$\mathfrak{A} = (\lambda(x_1, x_2, x_3) - p(x_1, x_2, x_3), f) = (\lambda p - q, f)$$

in the ring $k[x_1, x_2, x_3, \lambda]$ where $(f) = \mathfrak{p}(f)$ denotes the prime ideal defining the surface f. According to a well-known theorem of Macauly the ideal $\mathfrak{A}$ is unmixed of dimension 2, thus

$$\mathfrak{A} = [\mathfrak{q}_1, \mathfrak{q}_2, \cdots, \mathfrak{q}_s]$$

where the ideals $\mathfrak{q}_i$ are 2-dimensional primary ideals with the associated prime ideals $\mathfrak{P}_i$. The contracted ideal $\bar{\mathfrak{A}} = \mathfrak{A} \cap k[x_1, x_2, x_3]$ of $\mathfrak{A}$ is equal to (f). Hence

$$\bar{\mathfrak{A}} = (f) = [\bar{\mathfrak{q}}_1, \bar{\mathfrak{q}}_2, \cdots, \bar{\mathfrak{q}}_s]$$

where $\bar{\mathfrak{q}}_i = \mathfrak{q}_i \cap k[x_1, x_2, x_3]$. This representation implies that one component $\bar{\mathfrak{q}}_i$ must be equal to (f); let $\bar{\mathfrak{q}}_1$ be such a component.

We consider next an arbitrary element $F(x_1, x_2, x_3, \lambda)$ lying in $\mathfrak{q}_1$, then

$$q^\sigma F(x_1, x_2, x_3, \lambda) = A(x_1, x_2, x_3, \lambda)(\lambda q - p) + B(x_1, x_2, x_3).$$

Since $F(x_1, x_2, x_3, \lambda)$ and $\lambda p - q$ both lie in $\mathfrak{q}_1$ we get

$$B(x_1, x_2, x_3) \subset \mathfrak{q}_1,$$

consequently

$$B(x_1, x_2, x_3) \equiv 0 \pmod{\bar{\mathfrak{q}}_1}$$
$$\equiv 0 \pmod{f}.$$

There therefore exists a common exponent $\tau > 0$ such that

$$q^\tau F(x_1, x_2, x_3, \lambda) \equiv 0 \pmod{\mathfrak{A}}$$

for any element $F \subset \mathfrak{q}_1$, because $\mathfrak{q}_1$ has a finite base. Hence

$$q^\tau \mathfrak{q}_1 \equiv 0 \pmod{\mathfrak{q}_2},$$

consequently

$$q^\tau \equiv 0 \pmod{\mathfrak{q}_2},$$

for $\mathfrak{A}$ is an unmixed ideal and consequently all components $\mathfrak{q}_i$ have the same dimension. Therefore

$$q \equiv 0 \pmod{\mathfrak{P}_2}$$

and also

$$q \equiv 0 \pmod{\bar{\mathfrak{P}}_2}.$$

Since $\lambda = p/q$ we have $q \not\equiv 0(f)$, thus

$$\bar{\mathfrak{q}}_2 \not\equiv 0(f) \quad \text{and} \quad \bar{\mathfrak{P}}_2 \not\equiv 0(f).$$

Therefore $\bar{\mathfrak{q}}_2 \supset (f)$, consequently the ideals $\bar{\mathfrak{q}}_2, \cdots, \bar{\mathfrak{q}}_3$ are 0- or 1-dimensional ideals. They must be 1-dimensional, for $\bar{\mathfrak{q}}_2 k[x_1, x_2, x_3, \lambda] \subset \mathfrak{q}_2$ and hence

$$\dim \mathfrak{q}_2 = 2 \leq \dim \bar{\mathfrak{q}}_2 k[x_1, x_2, x_3, \lambda]$$
$$= \dim \bar{\mathfrak{q}}_2 + 1,$$

or

$$\dim \bar{\mathfrak{q}}_2 = 1.$$

This relation between the dimensions of $\mathfrak{q}_2$ and $\bar{\mathfrak{q}}_2$ shows that the components $\bar{\mathfrak{q}}_2, \cdots, \bar{\mathfrak{q}}_s$ do not depend on λ, i. e. they are extended ideals of $\bar{\mathfrak{q}}_2, \cdots, \bar{\mathfrak{q}}_s$. In geometric terms, the curves $\mathfrak{q}_2, \cdots, \mathfrak{q}_s$ correspond to the entire line $k[\lambda]$ under the algebraic correspondence. Such fixed components shall be left out in the definition of a linear system, and consequently $\mathfrak{q}_1 = \mathfrak{q}$ is the ideal defining Σ_1. According to the properties of primary ideals we have

$$(\lambda q - p)^\rho \subset \mathfrak{q},$$

i. e. Σ_1 is cut out by the hypersurfaces $(\lambda q - p)^\rho = 0$ where fixed components are omitted.

We remark that if q lies in one of the components $\mathfrak{q}_i \neq \mathfrak{q}$ then also $p \subset \mathfrak{q}_i$ for $\lambda q - p \subset \mathfrak{A}$; hence $p = q = 0$ occur among the equations defining the correspondence $\mathfrak{q}$.

THE JOHNS HOPKINS UNIVERSITY.

SOME RESULTS IN THE ARITHMETIC THEORY OF ALGEBRAIC VARIETIES.*

By Oscar Zariski.

Introduction. In Part I we treat systematically some basic questions of the theory of singularities of an algebraic variety. The main results bear upon various characterizations of simple points of a variety. We begin with an ideal theoretic definition of a simple point (section 2). This definition is given of terms of relative unramified prime zero-dimensional ideals. We next characterize a simple point by the existence of a local uniformization satisfying certain conditions (sections 3 and 4). At the same time we obtain a characterization of a simple point by an intrinsic (non-relative) property of the corresponding prime zero-dimensional ideal (Theorem 3.2). Finally we connect the singular points of a variety with the properties of the different (section (5-8)) and we exhibit an ideal whose variety is the manifold of singular points of the given variety (sections 8-9).

In Part II we derive some properties of the conductor of a finite integral domain $\mathfrak{o}$ with respect to its integral closure $\mathfrak{o}^*$. Those properties concern the usual question of the decomposition in $\mathfrak{o}^*$ of a prime ideal in $\mathfrak{o}$. The result is complete in the case that $\mathfrak{o}$ is generated by the independent variables and a primitive element (sections 11-12). Also here it was possible to obtain an intrinsic result valid in $\mathfrak{o}^*$ (section 13).

The contents of Parts I and II are, in the main, generalizations of well-known theorems in the arithmetic theory of algebraic functions of one variable.

The contributions of Parts III and IV are new also in the case of functions of one variable. Here we introduce the concept of a *normal* variety, both in the affine and in the projective space, and we are led to a geometric interpretation of the operation of integral closure. The importance of normal varieties is due to their following two properties: 1) *the singular manifold of a normal V_r is of dimension $\leqq r - 2$* (in particular a normal curve (V_1) is free from singularities); 2) *the system of hyperplane sections of a normal V_r is complete.* There is a definite class of normal varieties associated with and birationally equivalent to a given variety V_r. This class is obtained by a process of integral closure carried out in a suitable fashion for varieties in projective spaces. It turns out that the varieties of this class are those on

* Received October 7, 1938.

which the hyperplanes cut out the complete systems $|hC|$, sufficiently high multiples of the system $|C|$ of hyperplane sections of the given V_r. These results seem to point to a fruitful arithmetic approach to questions of the birational theory of varieties.

The special birational transformations effected by the operation of integral closure, and the properties of normal surfaces, play an essential rôle in our arithmetic proof for the reduction of singularities of an algebraic surface. This proof will be published in the July issue of the *Annals of Mathematics.*

Although the underlying field of coefficients is supposed throughout to be of characteristic zero, the proofs remain valid for any characteristic, provided only separable extensions are being considered. The separability is no restriction from the birational point of view, or even from the projective point of view.

Many of the results of Parts I, II, and III have been announced by the author without proof in a Note of the *Proceedings of the National Academy of Sciences* [10].

I. Simple and multiple points of an algebraic variety.

1. We consider an algebraic irreducible r-dimensional variety V_r, in an affine space $S_n(x_1, \cdots, x_n)$, over an algebraically closed ground field K of characteristic zero. Let $\xi_1, \cdots, \xi_n$ be the coördinates of the general point of V_r and let $\Sigma = \mathsf{K}(\xi_1, \cdots, \xi_n)$ be the field of rational functions on V_r, of degree of transcendency r over K. We denote by $\mathfrak{o}$ the ring $\mathsf{K}[\xi_1, \cdots, \xi_n]$, whose elements are polynomials in the ξ's. The defining ideal of V_r in the polynomial ring $\mathsf{K}[x] (= \mathsf{K}[x_1, \cdots, x_n])$ is the prime r-dimensional ideal $\mathfrak{p}'_r$, consisting of all polynomials $f(x_1, \cdots, x_n)$ such that $f(\xi_1, \cdots, \xi_n) = 0$; $\mathfrak{o}$ is the ring of residual classes $\mathsf{K}[x]/\mathfrak{p}'_r$ and Σ is the quotient field of $\mathfrak{o}$.

The polynomials in $\mathsf{K}[x]$ which vanish at a given point $P(a_1, \cdots, a_n)$ of S_n form a prime 0-dimensional ideal $\mathfrak{p}'_0 = (x_1 - a_1, \cdots, x_n - a_n)$. The point P is on V_r if and only if $\mathfrak{p}'_r \subset \mathfrak{p}'_0$. The homomorphism between $\mathsf{K}[x]$ and $\mathsf{K}[\xi]$ sets up a one-to-one correspondence between the prime 0-dimensional ideals $\mathfrak{p}'_0$ in $\mathsf{K}[x]$ which contain $\mathfrak{p}'_r$ and the prime 0-dimensional ideals $\mathfrak{p}_0$ in $\mathfrak{o}$ $(= \mathsf{K}[\xi])$; here $\mathfrak{p}_0 = \mathfrak{p}'_0/\mathfrak{p}'_r$. Thus there is a one-to-one correspondence betwen the points of V_r (in the affine S_n) and the prime 0-dimensional ideals in $\mathfrak{o}$: if $P(a_1, \cdots, a_n)$ is a point on V_r, then the ideal $\mathfrak{p}_0 = (\xi_1 - a_1, \cdots, \xi_n - a_n)$ in $\mathfrak{o}$ is prime and 0-dimensional, and conversely. If P is not on V_r, then $(\xi_1 - a_1, \cdots, \xi_n - a_n)$ is the unit ideal.

If $\mathfrak{p}_0$ is any prime 0-dimensional ideal in $\mathfrak{o}$, then the ring $\mathfrak{o}/\mathfrak{p}_0$ contains a field $\mathsf{K}^* \cong \mathsf{K}$ (if $c_1, c_2 \subset \mathsf{K}$ and $c_1 \neq c_2$, then $c_1 - c_2 \not\equiv 0(\mathfrak{p}_0)$, since $\mathfrak{p}_0$ is not the unit ideal), and moreover every element of $\mathfrak{o}/\mathfrak{p}_0$ is algebraic over K^*. Since K^* is assumed algebraically closed, it follows that $\mathfrak{o}/\mathfrak{p}_0$ coincides with the field K^*.[1] Thus every element ω in $\mathfrak{o}$ satisfies a congruence of the form $\omega \equiv c(\mathfrak{p}_0)$. We shall say that ω *has the value* c *at* P. In particular, $\xi_i \equiv a_i(\mathfrak{p}_0)$ and to $\mathfrak{p}_0$ there corresponds the point $P(a_1, \cdots, a_n)$ on V_r.

2. DEFINITION. *A point $P(a_1, \cdots, a_n)$ on V_r is said to be a simple point of V_r, if there exist, in $\mathfrak{o}$, r elements $\eta_1, \cdots, \eta_r$ such that the ideal $\mathfrak{A} = \mathfrak{o} \cdot (\eta_1, \cdots, \eta_r)$ is divisible by (is contained in) $\mathfrak{p}_0$ $(= \mathfrak{o} \cdot (\xi_1 - a_1, \cdots, \xi_n - a_n))$ but is not divisible by any proper primary ideal belonging to $\mathfrak{p}_0$.*

The condition in the above definition is equivalent to the following: $\mathfrak{p}_0$ *must be an isolated component of the ideal* $\mathfrak{A}$. In fact, $\mathfrak{A} \equiv 0(\mathfrak{p}_0)$ and if $\mathfrak{p}_0$ were not an isolated component of $\mathfrak{A}$, then $\mathfrak{A}$ must be divisible by some prime ideal $\mathfrak{p}$ of dimension > 0 and contained in $\mathfrak{p}_0$ (since, by hypothesis, no primary ideal belonging to $\mathfrak{p}_0$ can be a component of $\mathfrak{A}$). This, however, is impossible, since it can be easily shown that any such prime ideal $\mathfrak{p}$ is also a multiple of some primary ideal belonging to $\mathfrak{p}_0$.[2]

We proceed to derive properties of a simple point which we shall have occasion to use in the sequel and which will also bear upon the geometric content of our definition.

Let $P(a_1, \cdots, a_n)$ be a simple point of V_r and let, according to our definition of a simple point, $\eta_1, \cdots, \eta_r$ be r elements in $\mathfrak{o}$ such that the ideal $\mathfrak{p}_0 = (\xi_1 - a_1, \cdots, \xi_n - a_n)$ is an isolated component of the ideal $\mathfrak{A} = (\eta_1, \cdots, \eta_r)$. We may assume, without loss of generality, that P is at the origin of coördinates, whence $\mathfrak{p}_0 = (\xi_1, \cdots, \xi_n)$.

The elements η_i are polynomials in the ξ's. In each of these polynomials the constant term is zero, since $\eta_i \equiv 0(\mathfrak{p}_0)$. Let

$$(1) \qquad \eta_i = c_{i1}\xi_1 + \cdots + c_{in}\xi_n + \text{terms of degree} \geqq 2.$$

We assert that *the r linear forms $\sum_{j=1}^{n} c_{ij}\xi_j$ are linearly independent modulo $\mathfrak{p}_0^2$.* To see this, we observe that if this were not the case, then also the r elements η_i would be linearly dependent modulo $\mathfrak{p}_0^2$, since $\eta_i - \sum_{j=1}^{n} c_{ij}\xi_j \equiv 0(\mathfrak{p}_0^2)$. Let, say,

---

[1] That $\mathfrak{o}/\mathfrak{p}_0$ is a field follows also from the fact that $\mathfrak{p}_0$ has no divisors other than $\mathfrak{p}_0$ and the unit ideal.

[2] In the homomorphism $\mathfrak{o} \simeq \bar{\mathfrak{o}} = \mathfrak{o}/\mathfrak{p}$, the ideal $\mathfrak{p}_0$, as a divisor of $\mathfrak{p}$, is mapped upon a 0-dimensional ideal $\bar{\mathfrak{p}}_0$ in $\bar{\mathfrak{o}}$, and the primary ideals belonging to $\bar{\mathfrak{p}}_0$ correspond to primary ideals in $\mathfrak{o}$ belonging to $\mathfrak{p}_0$ and containing $\mathfrak{p}$.

$$d_1\eta_1 + \cdots + d_r\eta_r \equiv 0(\mathfrak{p}_0^2),$$

where $d_1, \cdots, d_r$ are in K and are not all zero. Let $d_r \neq 0$. It follows from the above congruence that

$$(2) \qquad \eta_r \equiv 0(\eta_1, \cdots, \eta_{r-1}, \mathfrak{p}_0^2).$$

We consider the ideal $\mathfrak{B} = (\eta_1, \cdots, \eta_{r-1})$. We observe that $\mathfrak{B}$ is not the unit ideal, since $\mathfrak{B} \subseteq \mathfrak{A} \subseteq \mathfrak{p}_0$. Hence, by a well known theorem [3] the minimal (non-imbedded) prime ideals of $\mathfrak{B}$ are all of dimension not less than 1. Since $\mathfrak{B} \subseteq \mathfrak{p}_0$, $\mathfrak{p}_0$ must divide at least one of the minimal prime ideals of $\mathfrak{B}$, say $\mathfrak{p}'$. It follows, as has been pointed out in footnote 2, that there exist primary ideals belonging to $\mathfrak{p}_0$ (and distinct from $\mathfrak{p}_0$) which divide $\mathfrak{p}'$. There will also then exist a maximal primary ideal $\mathfrak{q}_0$ with this same property, i. e. a primary ideal $\mathfrak{q}_0$ belonging to $\mathfrak{p}_0$ and such that there are no ideals between $\mathfrak{p}_0$ and $\mathfrak{q}_0$. Now it is well known that each maximal primary ideal of $\mathfrak{p}_0$ is a divisor of $\mathfrak{p}_0^2$.[4] Hence $\mathfrak{p}_0^2 \equiv 0(\mathfrak{q}_0)$, and since also $\mathfrak{B} \equiv 0(\mathfrak{p}') \equiv 0(\mathfrak{q}_0)$, it follows, by (2), $\eta_r \equiv 0(\mathfrak{q}_0)$. Hence $\mathfrak{A} \equiv 0(\mathfrak{q}_0)$, in contradiction with the hypothesis that $\mathfrak{p}_0$ is an isolated component of $\mathfrak{A}$.

In view of the linear independence of the forms $\sum_{j=1}^{n} c_{ij}\xi_j \bmod \mathfrak{p}_0^2$, it follows, *a fortiori*, that the matrix (c_{ij}) is of rank r. Hence by means of a non-singular linear homogeneous transformation of the coördinates ξ_i of the general point of V_r, it can be arranged that the elements η_i have the following form:

$$(3) \qquad \eta_i = \xi_i + f_i(\xi_1, \cdots, \xi_n), \qquad (i = 1, 2, \cdots, r),$$

where f_i is a polynomial whose terms are all of degree $\geqq 2$. Now let ω be any element in $\mathfrak{p}_0$. Since $\mathfrak{p}_0$ is an isolated component of $\mathfrak{A}$, it follows that there exist an element α in $\mathfrak{o}$, such that $\alpha \not\equiv 0(\mathfrak{p}_0)$ and $\alpha\omega \equiv 0(\mathfrak{A})$. Since $\alpha \equiv c(\mathfrak{p}_0)$, $c \neq 0$, $c \subset K$, this implies that $\omega \equiv 0(\mathfrak{A}, \mathfrak{p}_0^2)$ i. e.

$$\omega = A_1\eta_1 + \cdots + A_r\eta_r + \beta, \qquad A_i \subset \mathfrak{o}, \ \beta \equiv 0(\mathfrak{p}_0^2).$$

Replacing the η_i by their expressions in (3) and observing that β, as an ele-

---

[3] See [2], p. 43 and the references on p. 45 to Macaulay and van der Waerden.

[4] See, for instance, [1], theorem 2, p. 529. Although the assertion is there proved only for rings in which the "weak" Doppelkettensatz holds true, the proof of this theorem, as well as of Theorem 3 and Corollary on p. 529, *loc. cit.*, carry over to *zero-dimensional* ideals in arbitrary finite integral domains. On the other hand, our assertion is practically trivial if we observe that it obviously holds in the polynomial ring $K[x_1, \cdots, x_n]$ for 0-dimensional ideals and if we consider the homomorphism between $K[x]$ and $K[\xi]$.

ment of $\mathfrak{p}_0^2$, can be expressed as a polynomial in the ξ's in which all the terms are of degree $\geqq 2$, we see that our arbitrary element ω in $\mathfrak{p}_0$ can be put in the form:

$$\omega = c_1\xi_1 + \cdots + c_r\xi_r + g(\xi_1, \cdots, \xi_n), \tag{4}$$

where g contains only terms of degree $\geqq 2$. Or, in other words: there exist constants $c_1, \cdots, c_r$ such that

$$\omega \equiv c_1\xi_1 + \cdots + c_r\xi_r(\mathfrak{p}_0^2). \tag{5}$$

THEOREM 1. *Let* $\omega_1, \omega_2, \cdots, \omega_r$ *be elements of* $\mathfrak{p}_0$; $\omega_i \equiv \sum_{j=1}^{r} c_{ij}\xi_j(\mathfrak{p}_0^2)$. *A necessary and sufficient condition that* $\mathfrak{p}_0$ *be an isolated component of the ideal* $(\omega_1, \cdots, \omega_r)$ *is that the determinant* $|c_{ij}|$ *be different from zero.*

Proof. Assume $|c_{ij}| \neq 0$. Given any element ω in $\mathfrak{p}_0$, it is then possible to find constants $d_1, \cdots, d_r$, such that $\omega \equiv d_1\omega_1 + \cdots + d_r\omega_r(\mathfrak{p}_0^2)$. Hence $\mathfrak{p}_0 = (\omega_1, \cdots, \omega_r, \mathfrak{p}_0^2)$, and this implies that $\mathfrak{p}_0$ is an isolated component of the ideal $(\omega_1, \cdots, \omega_r)$ (in view of the fact that $\mathfrak{p}_0^2$ is a multiple of every maximal primary ideal belonging to $\mathfrak{p}_0$; see footnote 4).

Assume $|c_{ij}| = 0$. There exist then constants $d_1, \cdots, d_r$, not all zero, such that $d_1\omega_1 + \cdots + d_r\omega_r \equiv 0(\mathfrak{p}_0^2)$. If, for instance, $d_r \neq 0$, then it follows that $\omega_r \equiv 0(\omega_1, \cdots, \omega_{r-1}, \mathfrak{p}_0^2)$. This congruence is analogous to the congruence (2), encountered above, and leads therefore to a similar conclusion, e. g. that the ideal $(\omega_1, \cdots, \omega_r)$ is a multiple of some maximal primary ideal $\mathfrak{q}_0$ belonging to $\mathfrak{p}_0$, q. e. d.

In particular, if we take for $\omega_1, \cdots, \omega_r$ linear forms $\bar{\xi}_1, \cdots, \bar{\xi}_r$ in $\xi_1, \cdots, \xi_n$, then, for non-special values of the coefficients of these forms, the determinant $|c_{ij}|$ will be different from zero. Moreover, by a well known "normalization theorem,"[5] for non-special values of these coefficients the elements $\bar{\xi}_1, \cdots, \bar{\xi}_r$ have the property that they are algebraically independent[6] and that every element in $\mathfrak{o}$ is integrally dependent on $\bar{\xi}_1, \cdots, \bar{\xi}_r$. Hence our definition of a simple point P can now be completed by the following remark: the r elements $\eta_1, \cdots, \eta_r$ enjoying the property that $\mathfrak{p}_0[= (\xi_1, \cdots, \xi_n)]$

[5] See [2], p. 41.

[6] The algebraic independence is already implied by the non-vanishing of the determinant $|c_{ij}|$, as a consequence of the fact that $\mathfrak{p}_0$ is then an isolated component of the ideal $(\bar{\xi}_1, \cdots, \bar{\xi}_r)$. Namely, if the $\bar{\xi}$'s are algebraically dependent, it is permissible to assume that $\bar{\xi}_r$ is integrally dependent on $\bar{\xi}_1, \cdots, \bar{\xi}_{r-1}$. In view of the algebraic closure of the ground field, this implies that if $\mathfrak{p}$ is any minimal prime ideal of the ideal $(\bar{\xi}_1, \cdots, \bar{\xi}_{r-1})$ which is a multiple of $\mathfrak{p}_0$, then $\bar{\xi}_r$, and hence also the ideal $(\bar{\xi}_1, \cdots, \bar{\xi}_r)$, is divisible by $\mathfrak{p}$.

(the point P being at the origin) is an isolated component of the ideal $(\eta_1, \cdots, \eta_r)$, can be chosen in such a manner that, in addition, every element in $\mathfrak{o}$ be integrally dependent on $\eta_1, \cdots, \eta_r$; in particular, r suitable linear forms in $\xi_1, \cdots, \xi_n$ will meet both these requirements. We make the necessary transformation of the coördinates ξ_i and we assume from now on that these linear forms are $\xi_1, \cdots, \xi_r$ respectively.

3. Let again $\eta_1, \cdots, \eta_r$ be r elements in $\mathfrak{o}$ with the property that $\mathfrak{p}_0$ is an isolated component of the ideal $(\eta_1, \cdots, \eta_r)$. As has been pointed out above (footnote 6), the η's are algebraically independent. We introduce the ring $\mathsf{K}\{\eta\} = \mathsf{K}\{\eta_1, \cdots, \eta_r\}$ of all formal power series in $\eta_1, \cdots, \eta_r$ with coefficients in K. Any element in $\mathsf{K}\{\eta\}$ can be written in the form

$$\psi_0 + \psi_1 + \cdots + \psi_m + \cdots, \quad \text{or} \quad \phi_m + R_{m+1},$$

where ψ_i is a form of degree i in $\eta_1, \cdots, \eta_r$ and where $\phi_m = \psi_0 + \psi_1 + \cdots + \psi_m$.

THEOREM 2. *There exists an isomorphic mapping of the ring $\mathfrak{o}$ upon a subring of $\mathsf{K}\{\eta\}$,[7] with the following property: if ω is any element in $\mathfrak{o}$ and if $\psi_0 + \psi_1 + \cdots$ is the corresponding element in $\mathsf{K}\{\eta\}$, then, for all m,*

$$(6) \qquad \omega \equiv \phi_m(\mathfrak{p}_0^{m+1}), \qquad \phi_m = \psi_0 + \psi_1 + \cdots + \psi_m.$$

Proof. We shall first prove the theorem in the case when $\eta_1, \cdots, \eta_r$ coincide with $\xi_1, \cdots, \xi_r$ respectively. Its validity in the general case will then be an immediate consequence.

We first show that given any element ω in $\mathfrak{o}$ there exists a polynomial $\phi_m(\xi_1, \cdots, \xi_r)$ of degree $\leqq m$, such that the congruence (6) holds true. The assertion is trivial for $m = 0$, and has been proved for $m = 1$ (see congruence (5)). We assume that the assertion is true for $m = i$ and we prove it for $m = i + 1$. Let then $\omega \equiv \phi_i(\mathfrak{p}_0^{i+1})$, where $\phi_i = \phi_i(\xi_1, \cdots, \xi_r)$ is a polynomial of degree $\leqq i$. We can write:

$$\omega = \phi_i + f_{i+1}(\xi_1, \cdots, \xi_n) + \cdots + f_m(\xi_1, \cdots, \xi_n),$$

where f_j is a form of degree j. Let

$$\xi_{r+i} \equiv \sum_{j=1}^{r} c_{ij}\xi_j(\mathfrak{p}_0^2), \qquad (i = 1, 2, \cdots, n-r).$$

[7] In a more precise language: there exists an isomorphic mapping of $\mathfrak{o}$ upon a subring of $\mathsf{K}\{u_1, \cdots, u_r\}$ (the u's being parameters) in which $\eta_i \to u_i$, etc. What we did is to identify the η's with the u's.

Then it is clear that

$$f_{i+1}(\xi_1,\cdots,\xi_n)\equiv f_{i+1}(\xi_1,\cdots,\xi_r,\sum_{j=1}^{r}c_{1j}\xi_j,\cdots,\sum_{j=1}^{r}c_{n-r,j}\xi_j)\,(\mathfrak{p}_0^{i+2}),$$

and that moreover $f_j(\xi_1,\cdots,\xi_n)\equiv 0(\mathfrak{p}_0^{i+2})$, if $j>i+1$. Hence, if we put

$$(7)\qquad \begin{cases}\psi_{i+1}=f_{i+1}(\xi_1,\cdots,\xi_r,\sum_{j=1}^{r}c_{1j}\xi_j,\cdots,\sum_{j=1}^{r}c_{n-r,j}\xi_j)\\ \phi_{i+1}=\phi_i+\psi_{i+1},\end{cases}$$

then

$$\omega\equiv\phi_{i+1}(\mathfrak{p}_0^{i+2}),$$

which proves our assertion.

We next show that, given the element ω, the polynomial ϕ_m in (6) is uniquely determined. We shall prove this by induction with respect to m, since for $m=0$ the assertion is trivial. Let us assume that there exist two polynomials, $\phi_m(\xi_1,\cdots,\xi_r)$ and $\phi'_m(\xi_1,\cdots,\xi_r)$, both of degree $\leqq m$, such that $\omega\equiv\phi_m(\mathfrak{p}_0^{m+1})$, $\omega\equiv\phi'_m(\mathfrak{p}_0^{m+1})$. Let $\phi_m=\phi_{m-1}+\psi_m$, $\phi'_m=\phi'_{m-1}+\psi'_m$, where ϕ_{m-1} and ϕ'_{m-1} are of degree $\leqq m-1$ and ψ_m and ψ'_m are forms of degree m. Since ψ_m and ψ'_m are in $\mathfrak{p}_0^m$, it follows that $\omega\equiv\phi_{m-1}(\mathfrak{p}_0^m)$ and $\omega\equiv\phi'_{m-1}(\mathfrak{p}_0^m)$. By our induction we conclude that $\phi_{m-1}=\phi'_{m-1}$. Hence $\phi_m-\phi'_m=\psi_m-\psi'_m=g_m$, and $g_m\equiv 0(\mathfrak{p}_0^{m+1})$, where g_m is a form in $\xi_1,\cdots,\xi_r$, not identically zero, of degree m.

Let us denote the ideal $(\xi_1,\cdots,\xi_r)$ by $\mathfrak{A}$. By hypothesis, $\mathfrak{p}_0$ is an isolated component of $\mathfrak{A}$. Moreover, since every element in $\mathfrak{o}$ is integrally dependent on $\xi_1,\cdots,\xi_r$, $\mathfrak{A}$ is unmixed, all its components being necessarily zero-dimensional. Let $\mathfrak{A}=[\mathfrak{p}_0,\mathfrak{q}_1,\cdots,\mathfrak{q}_s]$. Since all the components are zero-dimensional, no two of them have common divisors. Hence their intersection coincides with their product, and we can write $\mathfrak{A}=\mathfrak{p}_0\mathfrak{q}_1\cdots\mathfrak{q}_s$. Now $g_m=0(\mathfrak{A}^m)$, i. e. $g_m\subset\mathfrak{p}_0^m\mathfrak{q}_1^m\cdots\mathfrak{q}_s^m=[\mathfrak{p}_0^m,\mathfrak{q}_1^m,\cdots,\mathfrak{q}_s^m]$. On the other hand $g_m\equiv 0(\mathfrak{p}_0^{m+1})$, whence $g_m\subset[\mathfrak{p}_0^{m+1},\mathfrak{q}_1^m,\cdots,\mathfrak{q}_s^m]=\mathfrak{p}_0^{m+1}\mathfrak{q}_1^m\cdots\mathfrak{q}_s^m$. Hence

$$g_m\equiv 0(\mathfrak{p}_0\cdot\mathfrak{A}^m).$$

This last congruence shows that g_m can be expressed as a form $f(\xi_1,\cdots,\xi_r)$ in $\xi_1,\cdots,\xi_r$, of degree m, *with coefficients in* $\mathfrak{p}_0$. Consider the norm $N(f)$ of f over $K(\xi_1,\cdots,\xi_r)$. Clearly $N(f)=g_m^\nu$, where $\nu=[\Sigma:K(\xi_1,\cdots,\xi_r)]$, the relative degree of Σ over $K(\xi_1,\cdots,\xi_r)$. On the other hand $N(f)$ is a form of degree νm in $\xi_1,\cdots,\xi_r$, and its coefficients are polynomials in $\xi_1,\cdots,\xi_r$ (since the coefficients of f are integrally dependent on $\xi_1,\cdots,\xi_r$) which vanish for $\xi_1=\cdots=\xi_r=0$ (since the coefficients of f are in $\mathfrak{p}_0$).[8]

[8] All this can been seen as follows. Let $u_1,u_2,\cdots,u_M$ be the various power products $\xi_1^{i_1}\cdots\xi_r^{i_r}$, $i_1+i_2+\cdots+i_r=m$. Then f is of the form $f=u_1\alpha_1+u_2\alpha_2+\cdots$

We have therefore the following result: the form $g_m{}^\nu$ in $\xi_1, \cdots, \xi_r$, of degree $m\nu$, is equal to a polynomial in $\xi_1, \cdots, \xi_r$, having no terms of degree $< m\nu + 1$. This is in contradiction with the algebraic independence of $\xi_1, \cdots, \xi_r$. The uniqueness of the polynomial ϕ_m in (6) is thus established.

The formula (7) shows that if $\omega \equiv \phi_m(\mathfrak{p}_0{}^{m+1})$ and $\omega \equiv \phi_{m+1}(\mathfrak{p}_0{}^{m+2})$, then $\phi_{m+1} - \phi_m$ is a form ψ_{m+1} in $\xi_1, \cdots, \xi_r$, of degree $m + 1$. *We let correspond to* ω *the power series* $\psi = \psi_0 + \psi_1 + \cdots$:

$$(8) \qquad \omega \to \psi = \sum_{i=0}^{\infty} \psi_i(\xi_1, \cdots, \xi_r).$$

It is now a simple matter to show that (8) defines an isomorphic mapping of $\mathfrak{o}$ upon a subring H of $\mathsf{K}\{\xi_1, \cdots, \xi_r\}$.

(a) First, let ω and ω' be distinct elements of $\mathfrak{o}$, and let

$$\omega \to \psi = \sum_{i=0}^{\infty} \psi_i, \qquad \omega' \to \psi' = \sum_{i=0}^{\infty} \psi'_i.$$

Assume that $\psi_i = \psi'_i$, for all i. Since

$$\omega - \sum_{i=0}^{m} \psi_i \equiv 0(\mathfrak{p}_0{}^{m+1}) \text{ and } \omega' - \sum_{i=0}^{m} \psi'_i \equiv 0(\mathfrak{p}_0{}^{m+1}),$$

it would follow that $\omega - \omega' \equiv 0(\mathfrak{p}_0{}^{m+1})$, for all m. This is impossible if $\omega \neq \omega'$, since zero is the only element which is common to all the powers of $\mathfrak{p}_0$. Hence, *to two distinct elements of* $\mathfrak{o}$ *there correspond two distinct power series.*

(b) We have $\omega + \omega' - \sum_{i=0}^{m} (\psi_i + \psi'_i) \equiv 0(\mathfrak{p}_0{}^{m+1})$, hence

$$\omega + \omega' \to \sum_{i=0}^{\infty} (\psi_i + \psi'_i) = \psi + \psi'.$$

$+ u_M\alpha_M$, $\alpha_j \equiv \mathfrak{o}(\mathfrak{p}_0)$, $j = 1, 2, \cdots, M$. Let $f^{(1)}, f^{(2)}, \cdots, f^{(\nu-1)}$ be the conjugates of f over $\mathsf{K}(\xi_1, \cdots, \xi_r)$; $f^{(i)} = u_1\alpha_1{}^{(i)} + u_2\alpha_2{}^{(i)} + \cdots + u_M\alpha_M{}^{(i)}$, where $\alpha_j{}^{(1)}, \cdots, \alpha_j{}^{(\nu-1)}$ are the conjugates of α_j. The conjugates of any α_j are integrally dependent on $\xi_1, \cdots, \xi_r$. Hence $N(f)$ $(= ff^{(1)} \cdots f^{(\nu-1)})$ is a form of degree νm in $\xi_1, \cdots, \xi_r$ with coefficients which are polynomials in $\xi_1, \cdots, \xi_r$. We consider the least Galois extension of $\mathsf{K}(\xi_1, \cdots, \xi_r)$ which contains the field Σ. We also consider the smallest ring $\mathfrak{o}^*$ containing $\mathfrak{o}$ and its conjugate rings, i. e. the ring $\mathfrak{o}^* = \mathsf{K}[\xi_1, \cdots, \xi_n;\ \xi_1{}^{(1)}, \cdots, \xi_n{}^{(1)};\ \cdots;\ \xi_1{}^{(\nu-1)}, \cdots, \xi_n{}^{(\nu-1)}]$. Let $\mathfrak{p}_0{}^*$ be any prime ideal divisor of $\mathfrak{p}_0$ in $\mathfrak{o}^*$. The coefficients of $N(f)$, considered as a form of degree νm in $\xi_1, \cdots, \xi_r$, clearly belong to the ideal $\mathfrak{o}^*(\alpha_1, \cdots, \alpha_M)$. Since $\alpha_j \equiv 0(\mathfrak{p}_0) \equiv 0(\mathfrak{p}_0{}^*)$, all these coefficients belong to $\mathfrak{p}_0{}^*$. On the other hand, these coefficients are in $\mathsf{K}[\xi_1, \cdots, \xi_r]$, and it is clear that the intersection $\mathsf{K}[\xi_1, \cdots, \xi_r] \cap \mathfrak{p}^*{}_0$ is the prime 0-dimensional ideal $(\xi_1, \cdots, \xi_r)$. Hence the coefficients of $N(f)$ are polynomials in $\xi_1, \cdots, \xi_r$, without constant term.

(c) Let $\psi\psi' = g = g_0 + g_1 + \cdots$. Clearly, $\sum_{i=0}^{m} g_i$ differs from $\sum_{i=0}^{m} \psi_i \cdot \sum_{i=0}^{m} \psi'_i$ by terms of degree $> m$, whence

$$\sum_{i=0}^{m} \psi_i \cdot \sum_{i=0}^{m} \psi'_i - \sum_{i=0}^{m} g_i \equiv 0(\mathfrak{p}_0{}^{m+1}). \tag{9}$$

On the other hand, we have

$$\omega\omega' - \sum_{i=0}^{m} \psi_i \cdot \sum_{i=0}^{m} \psi'_i = (\omega - \sum_{i=0}^{m} \psi_i)\omega' + (\omega' - \sum_{i=0}^{m} \psi'_i) \sum_{i=0}^{m} \psi_i,$$

whence

$$\omega\omega' - \sum_{i=0}^{m} \psi_i \cdot \sum_{i=0}^{m} \psi'_i \equiv 0(\mathfrak{p}_0{}^{m+1}). \tag{9'}$$

Adding (9) and (9′), we obtain:

$$\omega\omega' - \sum_{i=0}^{m} g_i \equiv 0(\mathfrak{p}_0{}^{m+1}),$$

and this shows that to $\omega\omega'$ there corresponds the power series $\psi\psi'$. This completes the proof of Theorem 2 for the case when $\eta_1, \cdots, \eta_r$ coincide with $\xi_1, \cdots, \xi_r$.

The general case can now be settled in a few words. Let us consider the power series which correspond to $\eta_1, \cdots, \eta_r$ in our mapping of $\mathfrak{o}$ upon the subring H of $\mathsf{K}\{\xi_1, \cdots, \xi_r\}$. The constant term in each of these power series is zero, since $\eta_i \equiv 0(\mathfrak{p}_0)$. Let

$$\eta_i \to \eta^*_i = c_{i1}\xi_1 + \cdots + c_{ir}\xi_r + \text{terms of higher degree},\ \eta^*_i \subset H, \qquad (i = 1, 2, \cdots, r). \tag{10}$$

Since $\mathfrak{p}_0$ is an isolated component of the ideal $(\eta_1, \cdots, \eta_r)$, it follows by Theorem 1, that the determinant $| c_{ij} |$ is different from zero. It is therefore possible to "solve" the relations (10) for $\xi_1, \cdots, \xi_r$ and to express each ξ_i as a power series in $\eta^*_1, \cdots, \eta^*_r$. In other terms: the ring $\mathsf{K}\{\xi_1, \cdots, \xi_r\}$ can be mapped isomorphically upon the ring $\mathsf{K}\{\eta_1, \cdots, \eta_r\}$. In this mapping the subring H of $\mathsf{K}\{\xi_1, \cdots, \xi_r\}$ is mapped upon a subring H' of $\mathsf{K}\{\eta_1, \cdots, \eta_r\}$ and $\mathfrak{o} \cong H'$. Let ω be any element in $\mathfrak{o}$ and let

$$\omega \to \sum_{i=0}^{\infty} g_i = g, \qquad g \subset H',$$

where g_i is a form in $\eta_1, \cdots, \eta_r$ of degree i. The two power series in

$\mathsf{K}\{\xi_1,\cdots,\xi_r\}$ which correspond to ω and to $\sum_{i=0}^{m} g_i$ respectively, differ by terms of degree $> m$. Hence $\omega - \sum_{i=0}^{m} g_i \equiv 0(\mathfrak{p}_0^{m+1})$, q. e. d.

In the sequel we shall feel free to identify any element ω with its corresponding power series $\sum_{i=0}^{\infty} g_i$ in $\eta_1,\cdots,\eta_r$, and we shall write $\omega = \sum_{i=0}^{\infty} g_i$. We shall say that $\sum_{i=0}^{\infty} g_i$ is *the expansion of* ω *in a power series of* $\eta_1,\cdots,\eta_r$. We shall also refer to the isomorphic mapping of $\mathfrak{o}$ upon the subring H of $\mathsf{K}\{\eta_1,\cdots,\eta_r\}$ as a *uniformization at* P. The elements $\eta_1,\cdots,\eta_r$ shall be called *uniformizing parameters.*

4. Geometrically speaking, Theorem 2, in the special case $\eta_i = \xi_i$, says that the variety V_r possesses at the point $P(0,\cdots,0)$ a linear analytical branch, whose tangent S_r at P has only the point P in common with the S_{n-r} given by the equation $\xi_1 = \cdots = \xi_r = 0$. We have not yet shown that there are no other branches at P, and that consequently P is indeed a simple point in the ordinary geometric sense.[9]

[9] Actually the existence of a linear analytical branch at P is implied already by a part of Theorem 2. Namely, if we only knew that there exists an isomorphic mapping of $\mathfrak{o}$ upon a subring of $\mathsf{K}\{\xi_1,\cdots,\xi_r\}$ such that the congruence (6) holds true *for* $m = 0$, we could already associate with this mapping a linear branch at $P(0,\cdots,0)$. It will follow from Theorem 4 that it is the validity of the congruence (6) *for all* m that implies that there are no other branches through P. Thus *the existence of an isomorphic mapping of* $\mathfrak{o}$ *upon a subring of* $\mathsf{K}\{\xi_1,\cdots,\xi_r\}$ *together with the validity of the congruence* (6) *for all* m *characterizes a simple point of* V_2. In this connection we wish to call attention to the following question: *assuming* $\mathfrak{o}$ *to be integrally closed* (i. e. assuming that V_2 is *normal* in the affine space (section 14)), *is it true that the neighborhood of every point of* V_2 *is an analytically irreducible variety?* For algebraic functions of one variable the answer is well-known to be affirmative. It seems to us that for functions of several variables the question presents serious difficulties. An equivalent ideal theoretic formulation of the question is the following: Let ω be a primitive element of $\mathfrak{o}$ over $\mathsf{K}(\xi_1,\cdots,\xi_r)$, $\mathfrak{o}$ being assumed to be the integral closure of $\mathsf{K}[\xi_1,\cdots,\xi_r]$ in Σ. Let $F(\xi_1,\cdots,\xi_r,\omega) = 0$ be the defining (irreducible) equation for ω. Over the field of meromorphic functions of $\xi_1,\cdots,\xi_r$ the polynomial F factors: $F_1F_2\cdots F_h$, where F_h is a polynomial in ω with coefficients in $\mathsf{K}\{\xi_1,\cdots,\xi_r\}$ and leading coefficient 1. It is not difficult to associate with each factor F_i a definite prime 0-dimensional ideal in $\mathfrak{o}$ which is a divisor of $\mathfrak{o}\cdot(\xi_1,\cdots,\xi_r)$ (see next section). The question is: do there correspond to distinct factors distinct ideals? The methods used in this connection for algebraic functions of one variable and for fields of algebraic numbers (see [4]) break down in the case of algebraic functions of more than one variable. If the answer to this question was affirmative, then the following could be proved: *if* $\mathfrak{o}$ *is integrally closed and if there exists an isomorphic mapping of* $\mathfrak{o}$ *upon a subring of* $\mathsf{K}\{\xi_1,\cdots,\xi_r\}$ *such that* (6) *holds true for* $m = 0$, *then* (6) *holds true for*

We will show, however, that there exists a non-singular projection V'_r of V_r into an $S_{r+1}(y_1, \cdots, y_{r+1})$, such that if $f(y_1, \cdots, y_{r+1}) = 0$ is the equation of V'_r, then the point P is projected into a point P' of V'_r at which not all the derivatives f_{y_i} vanish and which is therefore a simple point of V'_r, in the ordinary sense of this term.

In addition to this we shall also wish to show that the existence of a uniformization at P with the property described in Theorem 2 (i. e. congruence (6), for all m) implies that P is a simple point of V_r (in the sense of our definition). Accordingly, we consider a point P of V_r, which we shall assume to be the origin (whence the corresponding prime 0-dimensional ideal in $\mathfrak{o}$ is $\mathfrak{p}_0 = (\xi_1, \cdots, \xi_n)$) and we make the following assumptions: *there exist in* $\mathfrak{o}$ *r uniformizing parameters* $\eta_1, \cdots, \eta_r$ for the point P, i. e. there exists an isomorphic mapping of $\mathfrak{o}$ upon a subring of $K\{\eta_1, \cdots, \eta_r\}$ in which every element of $\mathfrak{p}_0$ is mapped upon a power series in $\eta_1, \cdots, \eta_r$ which has no constant term (congruence (6), for $m = 0$!). We also assume that *this uniformization has the property that if to an element* ω *in* $\mathfrak{o}$ *there corresponds the power series* $\psi_0 + \psi_1 + \cdots$, *then* $\omega \equiv \psi_0 + \cdots + \psi_m (\mathfrak{p}_0^{m+1})$, *for all* m (congruence (6), for all m). When these assumptions are satisfied, we shall say briefly that *there exists a uniformization of the whole neighborhood of P.*

THEOREM 3. *If* $\eta_1, \cdots, \eta_r$ *are uniformizing parameters for a uniformization of the whole neighborhood of the point* $P(0, \cdots, 0)$, *then* $\mathfrak{p}_0$ *is an isolated component of the ideal* $\mathfrak{o} \cdot (\eta_1, \cdots, \eta_r)$ *and hence P is a simple point of* V_r.

Proof. By hypothesis, any element ω in $\mathfrak{o}$ satisfies a congruence of the form $\omega \equiv c_0 + c_1\eta_1 + \cdots + c_r\eta_r (\mathfrak{p}_0^2)$, and $c_0 = 0$ if $\omega \equiv 0 (\mathfrak{p}_0)$. It follows immediately that $\mathfrak{p}_0 = (\eta_1, \cdots, \eta_r, \mathfrak{p}_0^2)$, and this implies (see proof of Theorem 1) that $\mathfrak{p}_0$ is an isolated component of the ideal $\mathfrak{o} \cdot (\eta_1, \cdots, \eta_r)$, q. e. d.

We observe that we have used in the proof only the congruence $\omega \equiv \psi_0 + \psi_1 (\mathfrak{p}_0^2)$. We can therefore state the preceding theorem in the following stronger form:

THEOREM 3. 1. *If there exist r elements* $\eta_1, \cdots, \eta_r$ *in* $\mathfrak{o}$ *such that every element* ω *in* $\mathfrak{o}$ *satisfies a congruence of the form* $\omega \equiv c_0 + c_1\eta_1 + \cdots + c_r\eta_r (\mathfrak{p}_0^2)$ *(congruence (6), for* $m = 0, 1$ *only!), then* $\mathfrak{p}_0$ *is an isolated component of the ideal* $\mathfrak{o} \cdot (\eta_1, \cdots, \eta_r)$ *and P is a simple point of* V_r.

---

all m, and hence P is a simple point. Geometrically speaking, this statement signifies just this: the analytical irreducibility of the neighborhood of P together with the existence of a linear branch through P characterize P as a simple point.

The assumption in Theorem 3.1 implies that the elements $\eta_1, \cdots, \eta_r$ form a modular basis of the ring $\mathfrak{p}_0/\mathfrak{p}_0^2$ considered as a K-module. Moreover, since $\mathfrak{p}_0$ is, according to this theorem, an isolated component of $\mathfrak{o} \cdot (\eta_1, \cdots, \eta_r)$, we know, from section 2, that $\eta_1, \cdots, \eta_r$ are linearly independent modulo $\mathfrak{p}_0^2$. *Hence* $\mathfrak{p}_0/\mathfrak{p}_0^2$ *is a* K-*module of rank* r. Conversely, if $\mathfrak{p}_0/\mathfrak{p}_0^2$ is of rank $\leqq r$, then there exist r elements in $\mathfrak{o}$, say $\eta_1, \cdots, \eta_r$, such that every other element ω in $\mathfrak{p}_0$ satisfies a congruence of the form $\omega \equiv c_1\eta_1 + \cdots + c_r\eta_r(\mathfrak{p}_0^2)$. Hence $\mathfrak{p}_0 = (\eta_1, \cdots, \eta_r, \mathfrak{p}_0^2)$, and from this it follows, as in section 2, that $\mathfrak{p}_0$ is an isolated component of the ideal $\mathfrak{o} \cdot (\eta_1, \cdots, \eta_r)$ and that $\eta_1, \cdots, \eta_r$ are linearly independent mod $\mathfrak{p}_0^2$. Hence P is a simple point and necessarily $\mathfrak{p}_0/\mathfrak{p}_0^2$ is of rank exactly r. We therefore have the following

THEOREM 3.2. *If P is a point on V_r and $\mathfrak{p}_0$ is the corresponding prime 0-dimensional ideal in $\mathfrak{o}$, a necessary and sufficient condition that P be a simple point is that the* K-*module* $\mathfrak{p}_0/\mathfrak{p}_0^2$ *be of rank* r.

5. Let $P(0, \cdots, 0)$ be a simple point of V_r. We assume, as in the preceding section, that a linear homogeneous transformation of the coördinates ξ_i has already been performed, so as to make $\xi_1, \cdots, \xi_r$ uniformizing parameters for a uniformization of the whole neighborhood of the point P, and, moreover, that every element in $\mathfrak{o}$ is integrally dependent on $\mathsf{K}[\xi_1, \cdots, \xi_r]$.

The field $\Sigma = \mathsf{K}(\xi_1, \cdots, \xi_n)$ is an algebraic extension of $\mathsf{K}(\xi_1, \cdots, \xi_r)$. Let m be the relative degree of Σ over $\mathsf{K}(\xi_1, \cdots, \xi_r)$. If ω is an element in $\mathfrak{o}$, then

$$N(z - \omega) = F(z; \xi_1, \cdots, \xi_r) = z^m + A_1z^{m-1} + \cdots + A_m,$$

where $A_1, \cdots, A_m$ are polynomials in $\xi_1, \cdots, \xi_r$. Reducing the equation $F(\omega; \xi_1, \cdots, \xi_r) = 0$ modulo $\mathfrak{p}_0$, we find $F(c, 0, \cdots, 0) = 0$, where $\omega \equiv c(\mathfrak{p}_0)$. Hence c is a root of $F(z, 0, \cdots, 0)$.

Since every element in $\mathfrak{o}$ is integrally dependent on $\mathsf{K}[\xi_1, \cdots, \xi_r]$, the ideal $\mathfrak{o} \cdot (\xi_1, \cdots, \xi_r)$ is unmixed and zero-dimensional, and $\mathfrak{p}_0$ is one of its components. Let $[\mathfrak{p}_0, \mathfrak{q}_1, \cdots, \mathfrak{q}_s]$ be the decomposition of the ideal $\mathfrak{o} \cdot (\xi_1, \cdots, \xi_r)$ into primary components and let $\mathfrak{p}_0, \mathfrak{p}_1, \cdots, \mathfrak{p}_s$ be the associated prime zero-dimensional ideals.

THEOREM 4. *Under the assumption that $\xi_1, \cdots, \xi_r$ are uniformizing parameters for a uniformization of the whole neighborhood of the simple point $P(0, \cdots, 0)$ on V_r and that every element in $\mathfrak{o}$ is integrally dependent on $\xi_1, \cdots, \xi_r$, there exist elements ω in $\mathfrak{o}$ such that $F'_\omega(\omega; \xi_1, \cdots, \xi_r) \not\equiv 0(\mathfrak{p}_0)$, where $F(z; \xi_1, \cdots, \xi_r)$ is the norm of $z - \omega$ over* $\mathsf{K}(\xi_1, \cdots, \xi_r)$. *Such elements ω are characterized by the following condition: if ω has the* *value c at P* (i. e. *if* $\omega \equiv c(\mathfrak{p}_0)$) *and if* $\mathfrak{p}_1, \cdots, \mathfrak{p}_s$ *are the prime* 0-*dimensional ideals, other than* $\mathfrak{p}_0$, *which divide the ideal* $\mathfrak{o}\cdot(\xi_1, \cdots, \xi_r)$, *then* $\omega \not\equiv c(\mathfrak{p}_i)$, $i = 1, 2, \cdots, s$.

We give here two proofs of this theorem. The first is based upon the *factorization of* $F(z; \xi_1, \cdots, \xi_r)$ in the ring $K\{\xi_1, \cdots, \xi_r\}[z]$ and upon higher congruences to which this factorization leads.[10]

The second proof makes use of the least Galois extension field which contains our field Σ and its conjugates over $K(\xi_1, \cdots, \xi_r)$.

First proof. Elements ω in $\mathfrak{o}$ satisfying the conditions $\omega \equiv c(\mathfrak{p}_0)$, $\omega \not\equiv c(\mathfrak{p}_i)$, $i = 1, 2, \cdots, s$, certainly exist. Namely, since no two of the ideals $\mathfrak{p}_0, \mathfrak{p}_1, \cdots, \mathfrak{p}_s$ have common divisors, it is well known that given any $s + 1$ elements $\alpha_0, \alpha_1, \cdots, \alpha_s$ in $\mathfrak{o}$, there exist elements ω in $\mathfrak{o}$ satisfying the congruences $\omega \equiv \alpha_i(\mathfrak{p}_i)$, $i = 0, 1, \cdots, s$. To satisfy the above conditions, we have only to take for $\alpha_0, \cdots, \alpha_s$ any $s + 1$ constants in K, say $c_0, c_1, \cdots, c_s$, such that $c_0 \neq c_i$, $i = 1, 2, \cdots, s$.

It is not difficult to prove that *there also exist primitive elements* ω *over* $K(\xi_1, \cdots, \xi_r)$ satisfying the above conditions (as a matter of fact, we shall prove later on that *these conditions imply that* ω *is a primitive element*). To show this we consider an element ζ in $\mathfrak{o}$ satisfying the congruences $\zeta \equiv d_i(\mathfrak{p}_i)$, where $d_0, d_1, \cdots, d_s$ are *distinct* constants. Let ζ' be some primitive element in $\mathfrak{o}$ and let $\zeta' \equiv d'_i(\mathfrak{p}_i)$. If t is a parameter, then the discriminant of $N(z - \zeta' - t\zeta)$ is a polynomial $D(t; \xi_1, \cdots, \xi_r)$ which does not vanish identically in t, since $D(0, \xi_1, \cdots, \xi_r)$ is the discriminant of $N(z - \zeta')$ and does not vanish. Let $A_{ij}(t) = (d'_i - d'_j) + t(d_i - d_j)$, $i, j = 0, 1, 2, \cdots, s$, $i \neq j$. Since $d_i \neq d_j$, if $i \neq j$, none of these polynomials vanishes identically. Let t_0 be a value of t (in K) such that $D(t_0, \xi_1, \cdots, \xi_r) \neq 0$, $A_{ij}(t_0) \neq 0$. Then the element $\omega = \zeta' + t_0\zeta$ is obviously a primitive element. Moreover, we have $\omega \equiv d'_i + t_0 d_i(\mathfrak{p}_i)$ and $d'_i + t_0 d_i \neq d'_j + t_0 d'_j$, if $i \neq j$, since $A_{ij}(t_0) \neq 0$.

Let then ω be a primitive element in $\mathfrak{o}$ over $K(\xi_1, \cdots, \xi_r)$, $\omega \equiv c_i(\mathfrak{p}_i)$, $c_i \neq c_0$, $i = 1, 2, \cdots, s$. Let

$$N(z - \omega) = F(z; \xi_1, \cdots, \xi_r) = F(z; \xi).$$

[10] This proof follows in part a well-known pattern of the theory of algebraic functions of one variable and of algebraic number fields. See [4]. The generalization is made possible by the existence of the uniformization. In the case of algebraic functions of one variable and of algebraic number fields the proof is based upon the fact that the ring of polynomials in one variable and the ring of rational integers are principal ideal rings.

Here F is a polynomial in z and $\xi_1,\cdots,\xi_r$, of degree m $(=[\Sigma:\mathsf{K}(\xi_1,\cdots,\xi_r)])$ in z and with leading coefficient 1.

Let

(11) $$F=F_1(z;\xi)\cdot F_2(z;\xi)\cdots F_h(z;\xi)$$

be the decomposition of F into prime factors in the ring $\mathsf{K}\{\xi_1,\cdots,\xi_r\}[z]$ of polynomials in z with coefficients in the ring of formal power series in $\xi_1,\cdots,\xi_r$. Since the discriminant of F is different from zero, the prime factors F_i are distinct. Moreover, each factor F_i, of degree m_i in z, may be assumed to have leading coefficient 1. We have $m=m_1+\cdots+m_h$.

Since $F_j(z;\xi)$ is a prime element in $\mathsf{K}\{\xi\}[z]$, it follows by Weierstrass' preparation theorem [11] that the polynomial $F_j(z;0,0,\cdots,0)$ in $\mathsf{K}[z]$ cannot have two distinct roots. Hence

(12) $$F_j(z,0,\cdots,0)=(z-d_j)^{m_j},\qquad d_j\subset\mathsf{K}.$$

The h constants $d_1,d_2,\cdots,d_h$ (not necessarily distinct) are the roots of $F(z,0,\cdots,0)$:

(12′) $$F(z;0,\cdots,0)=\prod_{j=1}^{h}(z-d_j)^{m_j}.$$

Every one of the constants c_i, $i=0,1,\cdots,s$, is equal to some d_j. In fact, reducing the equation $F(\omega;\xi_1,\cdots,\xi_r)$ modulo $\mathfrak{p}_i$, we find $F(c_i;0,\cdots,0)=0$, i. e. c_i is a root of $F(z;0,\cdots,0)$. *We shall have occasion to prove later* (footnote 12) *that, conversely, each of the constants d_j is equal to some c_i.* At any rate, if the constants $c_0,c_1,\cdots,c_s$ were distinct, then necessarily $h\geqq s+1$.

Let $\omega=\phi=\phi_0+\phi_1+\cdots$ be the expansion of ω into a power series in $\xi_1,\cdots,\xi_r$. Here $\phi_0=c_0$. Since the mapping $\omega\rightarrow\phi$ of $\mathfrak{o}$ upon the subring H of $\mathsf{K}\{\xi_1,\cdots,\xi_r\}$ is an isomorphism, the equation $F(\omega;\xi)=0$ implies that ϕ is a root of $F(z;\xi)$. Hence $z-\phi$ is a factor of F. Let, say

$$F_1(z;\xi)=z-\phi,$$

whence $m_1=1$ and

(13) $$F(z,0,\cdots,0)=(z-c_0)\prod_{j=2}^{h}(z-d_j)^{m_j},\quad m_2+\cdots+m_h=m-1.$$

We wish to show that the constants $d_2,\cdots,d_h$ are all different from c_0, i. e. that c_0 is a simple root of $F(z,0,\cdots,0)$. This will imply that

[11] See [5], p. 261-262.

$$F'_\omega(\omega; \xi_1, \cdots, \xi_r) \not\equiv 0(\mathfrak{p}_0),$$

as was asserted in our theorem. For the proof, let us denote by $F_{j\rho}(z; \xi)$ the polynomial in z and in $\xi_1, \cdots, \xi_r$ obtained from $F_j(z; \xi)$ by omitting all terms of degree $> \rho$ in $\xi_1, \cdots, \xi_r$. It is clear that $F(z; \xi) - \prod_{j=1}^{h} F_{j\rho}(z; \xi)$ is a polynomial which does not contain terms of degree $< \rho + 1$ in $\xi_1, \cdots, \xi_r$. It follows that

$$(14) \qquad \prod_{j=1}^{h} F_{j\rho}(\omega; \xi) \equiv 0(\xi_1, \cdots, \xi_r)^{\rho+1},$$

or also, since $\mathfrak{o} \cdot (\xi_1, \cdots, \xi_r) = \mathfrak{p}_0 \mathfrak{q}_1 \cdots \mathfrak{q}_s$,

$$(14') \qquad \prod_{j=1}^{h} F_{j\rho}(\omega; \xi) \equiv 0(\mathfrak{p}_0^{\rho+1} \mathfrak{q}_1^{\rho+1} \cdots \mathfrak{q}_s^{\rho+1}).$$

Now suppose that one of the constants $d_2, \cdots, d_h$ coincides with $c_0 (= d_1)$, and let, for instance $d_h = c_0$. We show that this hypothesis leads to a contradiction. It is clear that $F_{j\rho}(z; 0, \cdots, 0) = (z - d_j)^{m_j}$. Hence $F_{j\rho}(\omega; \xi) \equiv (\omega - d_j)^{m_j}(\mathfrak{p}_i)$, $i = 0, 1, \cdots, s$. In particular, for $j = h$, we have

$$F_{h\rho}(\omega; \xi) \equiv (c_i - d_h)^{m_h}(\mathfrak{p}_i), \qquad (i = 0, 1, \cdots, s).$$

Since $d_h = c_0$ and $c_0 \neq c_i$, $i = 1, 2, \cdots, s$, we conclude that

$$(15) \qquad F_{h\rho}(\omega; \xi) \not\equiv 0(\mathfrak{p}_i), \qquad (i = 1, 2, \cdots, s).$$

From this we derive, in view of (14′), that

$$\prod_{j=1}^{h-1} F_{j\rho}(\omega; \xi) \equiv 0(\mathfrak{q}_1^{\rho+1} \cdots \mathfrak{q}_s^{\rho+1}).$$

On the other hand we have

$$(15') \qquad F_{1\rho}(\omega; \xi) = \omega - (\phi_0 + \phi_1 + \cdots + \phi_\rho) \equiv 0(\mathfrak{p}_0^{\rho+1}).$$

Consequently, if we put

$$(16) \qquad G_\rho(\omega; \xi) = \prod_{j=1}^{h-1} F_{j\rho}(\omega; \xi),$$

then

$$G_\rho(\omega; \xi) \equiv 0(\mathfrak{p}_0^{\rho+1} \mathfrak{q}_1^{\rho+1} \cdots \mathfrak{q}_s^{\rho+1}),$$

or [12]

---

[12] If d_h was distinct from any one of the constants c_i, $i = 0, 1, \cdots, s$, then (15) would hold true for $i = 0, 1, \cdots, s$ and hence (16′) would follow directly, independently

$$(16') \qquad G_\rho(\omega;\xi) \equiv 0(\xi_1,\cdots,\xi_r)^{\rho+1}, \qquad (\rho = 0, 1, \cdots).$$

In view of (16′) it must be possible to express $G_\rho(\omega;\xi)$ as a form of degree $\rho+1$ in $\xi_1,\cdots,\xi_r$, with coefficients which are elements in $\mathfrak{o}$. Now any element in $\mathfrak{o}$ is an integral function of $\xi_1,\cdots,\xi_r$ and hence can be expressed in the form $g(\omega;\xi)/D(\xi)$, where g is a polynomial and D is the discriminant of $F(z;\xi)$. Hence we may write $G_\rho(\omega;\xi)$ in the form:

$$(17) \qquad G_\rho(\omega;\xi) = g_\rho(\omega;\xi)/D(\xi),$$

where g_ρ is a polynomial in ω, of degree $\leqq m-1$, with coefficients which are polynomials in $\xi_1,\cdots,\xi_r$ having no terms of degree $< \rho+1$. Now $G_\rho(\omega;\xi)$, according to (16), is of degree $m - m_h \leqq m-1$ in ω. Hence (17) must be an identity in $\omega, \xi_1,\cdots,\xi_r$. This is impossible, if ρ is sufficiently high. Namely, if the terms of lowest degree in D are of degree q, then the coefficient of the highest power of ω in $D\cdot G_\rho(\omega;\xi)$ begins with terms of degree q (since the leading coefficient of G_ρ is 1), while $g_\rho(\omega;\xi)$ has no terms of degree $<\rho+1$ in $\xi_1,\cdots,\xi_r$. Hence, if $\rho+1>q$, the relation (17) is impossible.

We have thus shown that there exist elements ω in $\mathfrak{o}$ with the property that, if $F(z;\xi) = N(z-\omega)$, then $F'_\omega(\omega;\xi) \not\equiv 0(\mathfrak{p}_0)$. Namely, we have also shown that if $\omega \equiv c_i(\mathfrak{p}_i)$, $i = 0, 1,\cdots, s$, then ω certainly enjoys this property, provided it is a primitive element over $K(\xi_1,\cdots,\xi_r)$ and provided $c_0 \neq c_i$, $i = 1, 2, \cdots, s$.

We now wish to prove that the first provision is a consequence of the second. Let ζ be any element in $\mathfrak{o}$, $\zeta \equiv b_i(\mathfrak{p}_i)$, $i = 0, 1, \cdots, s$, and let us assume that $b_0 \neq b_i$, $i = 1, 2, \cdots, s$. We have to show that ζ is a primitive element.

We fix in $\mathfrak{o}$ a primitive element ω such that $\omega \equiv c_i(\mathfrak{p}_i)$ and $c_0, c_1, \cdots, c_s$ are *distinct constants.* Let $N(z-\omega) = F(z;\xi)$ and let us again consider the factorization (11) of F in $K\{\xi_1,\cdots,\xi_r\}[z]$. Let $K^*\{\xi_1,\cdots,\xi_r\}$ denote the field of *meromorphic functions* of $\xi_1,\cdots,\xi_r$ and let Σ_i be the algebraic extension of this field defined by the irreducible equation $F_i(\omega;\xi) = 0$. Our field $\Sigma = K(\xi_1,\cdots,\xi_r,\omega)$ can be regarded as a subfield of Σ_i.[13] Let

---

of (15′), i.e. independently of the assumption that P is a simple point. The remainder of the proof is based on the impossibility of the congruences (16′). *Hence this also proves that any d_j must coincide with some c_i.*

[13] The following are well-known facts (see [7], p. 47). The ring of residual classes of $K^*\{\xi_1,\cdots,\xi_r\}[z]$ modulo F contains the field Σ and is the direct sum $\Sigma_1+\cdots+\Sigma_h$ of the fields Σ_i; each field Σ_i contains a subfield $\Sigma_i^{(0)} \cong \Sigma$, and the decomposition $a = a_1 + a_2 + \cdots + a_h$, $a \subset \Sigma$, $a_i \subset \Sigma_i^{(0)}$ sets up an isomorphism $a_i \to a_j$ between $\Sigma_i^{(0)}$

$N_i(z-\zeta)$ denote the norm of $z-\zeta$ over $K^*\{\xi_1,\cdots,\xi_r\}$, when ζ is regarded as an element of Σ_i. Since, in Σ, ζ is an integral function of $\xi_1,\cdots,\xi_r$, it follows that $N_i(z-\zeta)$ is a polynomial in z with coefficients in $\mathsf{K}\{\xi_1,\cdots,\xi_r\}$ and leading coefficient 1. Moreover, if $N(z-\zeta)$ denotes the usual norm of $z-\zeta$, when ζ is regarded as an element of Σ, then

$$N(z-\zeta)=\prod_{i=1}^{h} N_i(z-\zeta).$$

In particular, since $F_1(z;\zeta)=z-\phi$, $\phi\subset K\{\xi_1,\cdots,\xi_r\}$, we have $N_1(z-\zeta)=z-\psi$, where $\psi=\psi_0+\psi_1+\cdots$ (the expansion of ζ) and where $\psi_0=b_0$.

Now let us assume that ζ is not a primitive element. That $N(z-\zeta)$ is the power of a polynomial of degree $<m$, whence one of the factors $N_i(z-\zeta)$, $i\neq 1$, must be divisible by $z-\psi$. Let it be $N_2(z-\zeta)$. Then $N_2(z-\zeta)$ is necessarily a power of $z-\psi$:

$$N_2(z-\zeta)=(z-\psi)^{m_2}. \tag{18}$$

We consider in Σ_i the ring $\mathfrak{o}_i=\mathsf{K}\{\xi_1,\cdots,\xi_r\}\cdot\mathfrak{o}$ and the ideal $\mathfrak{p}_0^{(i)}=\mathfrak{o}_i\cdot\mathfrak{p}_0$. It is clear that $\mathfrak{p}_0^{(i)}$ contains the ideal of the non-units in $\mathsf{K}\{\xi_1,\cdots,\xi_r\}$ and that $\mathfrak{o}_i/\mathfrak{p}_0^{(i)}\cong\mathsf{K}$. Thus $\mathfrak{p}_0^{(i)}$ is prime and every element α in $\mathfrak{o}_i$ satisfies a congruence of the form $\alpha\equiv c(\mathfrak{p}_0^{(i)})$, $c\subset\mathsf{K}$. By (18), ζ, as an element of the field Σ_2, satisfies the relation $\zeta-\psi(\xi_1,\cdots,\xi_r)=0$, whence $\zeta\equiv\psi_0(\mathfrak{p}_0^{(2)})$, i. e. $\zeta\equiv b_0(\mathfrak{p}_0^{(2)})$, since $\psi_0=b_c$. We consider the contracted ideal $\mathfrak{p}_0^{(2)}\cap\mathfrak{o}$. This ideal contains the elements $\xi_1,\cdots,\xi_r$. Moreover, it also contains $\omega-d_2$, since the equation $F_2(\omega;\xi)=0$ yields the congruence $(\omega-d_2)^{m_2}\equiv 0(\mathfrak{p}_0^{(2)})$ (by (12′)). Now $\mathfrak{p}_0,\mathfrak{p}_1,\cdots,\mathfrak{p}_s$ are the only prime ideals in $\mathfrak{o}$ which divide the ideal $\mathfrak{o}\cdot(\xi_1,\cdots,\xi_r)$. Since $\mathfrak{p}_0^{(2)}\cap\mathfrak{o}$ is a prime 0-dimensional ideal, it must therefore coincide with one of the ideals $\mathfrak{p}_0,\cdots,\mathfrak{p}_s$. It cannot coincide with $\mathfrak{p}_0$, since $d_2\neq c_0$; therefore $\omega-d_2\not\equiv 0(\mathfrak{p}_0)$. Hence $\mathfrak{p}_0^{(2)}\cap\mathfrak{o}$ is one of the ideals $\mathfrak{p}_i$, $i\neq 0$. Let, say, $\mathfrak{p}_0^{(2)}\cap\mathfrak{o}=\mathfrak{p}_1$. We have then $\zeta\equiv b_0(\mathfrak{p}_1)$, and this is in contradiction with our hypothesis $b_0\neq b_i$, $i\neq 0$. Hence ζ is a primitive element, as was asserted.

To complete the proof of the theorem, we still have to show that, conversely, $F'_\omega(\omega;\xi_1,\cdots,\xi_r)\not\equiv 0(\mathfrak{p}_0)$ implies $c_0\neq c_i$, $i=1,2,\cdots,s$, where, again $F(z;\xi_1,\cdots,\xi_r)=N(z-\omega)$ and $\omega\equiv c_i(\mathfrak{p}_i)$. The hypothesis $F'_\omega(\omega;\xi)\not\equiv 0(\mathfrak{p}_0)$ signifies that c_0 is a simple root of $F(z;0,\cdots,0)$. Hence

and $\Sigma_j^{(0)}$. In particular we have $\omega=\omega_1+\cdots+\omega_h$, where $F_i(\omega_i;\xi_1,\cdots,\xi_r)=0$. There can be no confusion if, *for a fixed i*, we identify ω with ω_i and Σ with $\Sigma_i^{(0)}$.

2

the discriminant $D(\xi_1, \cdots, \xi_r)$ of $F(z;\xi)$ does not vanish, and consequently ω is a primitive element. Let ζ be an element in $\mathfrak{o}$ which assumes *distinct* values $b_0, b_1, \cdots, b_s$ at $\mathfrak{p}_0, \mathfrak{p}_1, \cdots, \mathfrak{p}_s$, and let, in our preceding notations, $N(z-\zeta) = G(z;\xi)$, $N_i(z-\zeta) = G_i(z;\xi)$. Here $G_1(z;\xi) = z - b_0 - \psi_1 - \psi_2 - \cdots$. Since $z - b_1$ is a factor of $G(z, 0, \cdots, 0)$, one of the polynomials $G_i(z; 0, \cdots, 0)$, $i \neq 1$, must be a power of $z - b_1$. Let, say, $G_2(z; 0, \cdots, 0) = (z - b_1)^{m_2}$. Then it follows that $\zeta \equiv b_1(\mathfrak{p}_0^{(2)})$ and we conclude as before that $\mathfrak{p}_0^{(2)} \circ \mathfrak{o} = \mathfrak{p}_1$. Now $\omega \equiv d_2(\mathfrak{p}_0^{(2)})$, hence $\omega \equiv d_2(\mathfrak{p}_1)$, i. e. $d_2 = c_1$. Since c_0 is a simple root of $F(z; 0, \cdots, 0)$, the remaining roots $d_2, \cdots, d_h$ are all distinct from d_1 $(= c_0)$. Consequently $c_1 \neq c_0$. Similarly $c_2 \neq c_0, \cdots, c_s \neq c_0$, and this completes the proof of our theorem.

6. *Second proof.* Let $\xi_i = \xi_i^{(1)}; \xi_i^{(2)}, \cdots, \xi_i^{(m)}$ be the conjugates of ξ_i over $K(\xi_1, \cdots, \xi_r)$ $(i = r+1, \cdots, n)$ and let Σ^* be the Galois extension field obtained by adjoining to $K(\xi_1, \cdots, \xi_r)$ the elements $\xi_{r+1}, \cdots, \xi_n$ and their conjugates. Let $\mathfrak{o}_j = K[\xi_1, \cdots, \xi_r, \xi_{r+1}^{(j)}, \cdots, \xi_n^{(j)}]$ and let $\mathfrak{o}^*$ be the smallest ring in Σ^* which contains the m conjugate rings $\mathfrak{o}_1, \mathfrak{o}_2, \cdots, \mathfrak{o}_m$.[14]

By the isomorphism $\xi_i^{(1)} \rightarrow \xi_i^{(j)}$ between $\mathfrak{o}$ $(= \mathfrak{o}_1)$ and $\mathfrak{o}_j$, the ideal $\mathfrak{p}_0$ in $\mathfrak{o}$ is carried into a prime 0-dimensional ideal $\mathfrak{p}_0^{(j)}$ in $\mathfrak{o}_j$. Again the m ideals $\mathfrak{p}_0^{(j)}$, in general, need not be distinct. However, under the hypothesis of the theorem we prove that not only are they distinct, but that also *any two of the extended ideals* $\mathfrak{o}^*\mathfrak{p}_0^{(j)}$ *in* $\mathfrak{o}^*$ *have no common divisors.*

Let us consider, for instance, $\mathfrak{o}^*\mathfrak{p}_0^{(1)}$ and $\mathfrak{o}^*\mathfrak{p}_0^{(2)}$. These ideals are unmixed and 0-dimensional, since every element in $\mathfrak{o}^*$ is integrally dependent on $\mathfrak{o}_j$. Hence if they have a common divisor, they also have a common prime 0-dimensional divisor, say $\mathfrak{p}^*_0$. Let $\omega = \omega_1$ be a primitive element in $\mathfrak{o}$ over $K(\xi_1, \cdots, \xi_r)$ and let ω_2 be its conjugate in $\mathfrak{o}_2$. By Theorem 2 we have $\omega_1 \equiv \psi_0 + \psi_1 + \cdots + \psi_{i-1}(\mathfrak{p}_0^{(1)i})$, where $\omega_1 = \psi_0 + \psi_1 + \cdots$ is the expansion of ω_1 into a power series of $\xi_1, \cdots, \xi_r$. Applying a substitution of the Galois group of $\Sigma^*/K(\xi_1, \cdots, \xi_r)$ which carries $\mathfrak{o}_1$ into $\mathfrak{o}_2$, we get: $\omega_2 \equiv \psi_0 + \psi_1 + \cdots + \psi_{i-1}(\mathfrak{p}_0^{(2)i})$. Since $\mathfrak{p}^*_0$ is a common divisor of $\mathfrak{p}_0^{(1)}$ and $\mathfrak{p}_0^{(2)}$ it follows $\omega_1 \equiv \psi_0 + \cdots + \psi_{i-1}(\mathfrak{p}^{*i}_0)$ and $\omega_2 \equiv \psi_0 + \cdots + \psi_{i-1}(\mathfrak{p}^{*i}_0)$, whence

$$\omega_1 - \omega_2 \equiv 0(\mathfrak{p}^{*i}_0), \qquad (i = 1, 2, \cdots).$$

The validity of this congruence *for all* i implies $\omega_1 = \omega_2$, in contradiction with the hypothesis that ω_1 is a primitive element. This proves that $\mathfrak{o}^*\mathfrak{p}_0^{(1)}$ and $\mathfrak{o}^*\mathfrak{p}_0^{(2)}$ have no common divisors.

Let

$$\mathfrak{o}^*\mathfrak{p}_0^{(j)} = \mathfrak{q}_1^{(j)} \cdots \mathfrak{q}_\sigma^{(j)} = [\mathfrak{q}_1^{(j)}, \cdots, \mathfrak{q}_\sigma^{(j)}] \tag{19}$$

[14] These m rings $\mathfrak{o}_j$ are not necessarily distinct.

be the decomposition of $\mathfrak{o}^*\mathfrak{p}_0^{(j)}$ into maximal primary (0-dimensional) components. Let $\mathfrak{q}_a^{(j)}$ belong to the prime ideal $\mathfrak{p}_a^{(j)}$. The prime ideals $\mathfrak{p}_1^{(j)}, \cdots, \mathfrak{p}_\sigma^{(j)}$ are the conjugates of the ideals $\mathfrak{p}_1^{(1)}, \cdots, \mathfrak{p}_\sigma^{(1)}$ respectively, under a substitution of the Galois group of Σ^* which carries $\mathfrak{o}_1$ into $\mathfrak{o}_j$. Since any two of the ideals $\mathfrak{o}^*\mathfrak{p}_0^{(j)}$ have no common divisors, *the* σm *ideals* $\mathfrak{p}_a^{(j)}$ *are all distinct.*

Let now $\omega = \omega_1$ be an element of $\mathfrak{o}$ $(= \mathfrak{o}_1)$ such that $\omega \equiv c_0(\mathfrak{p}_0)$ and $\omega \not\equiv c_0(\mathfrak{p}_i)$, $i = 1, 2, \cdots, s$. Since $\omega \equiv c_0(\mathfrak{p}_0^{(1)})$, it follows from (19) that

$$(20) \qquad \omega - c_0 \equiv 0(\mathfrak{p}_a^{(1)}), \qquad (\alpha = 1, 2, \cdots, \sigma).$$

We have $\mathfrak{o} \cdot (\xi_1, \cdots, \xi_r) \subseteqq \mathfrak{p}_0^{(1)}$, whence $\mathfrak{o}^*(\xi_1, \cdots, \xi_r) \subseteqq \mathfrak{p}_a^{(1)}$, $\alpha = 1, 2, \cdots, \sigma$. Since the ideal $\mathfrak{o}^*(\xi_1, \cdots, \xi_r)$ is invariant under all the substitutions of the Galois group of Σ^*, it follows $\mathfrak{o}^*(\xi_1, \cdots, \xi_r) \subseteqq \mathfrak{p}_a^{(j)}$, i. e. any of the $m\sigma$ ideals $\mathfrak{p}_a^{(j)}$ is a divisor of $\mathfrak{o}^*(\xi_1, \cdots, \xi_r)$. Since $\mathfrak{p}_0, \mathfrak{p}_1, \cdots, \mathfrak{p}_s$ are the prime ideals of $\mathfrak{o} \cdot (\xi_1, \cdots, \xi_r)$, any $\mathfrak{p}_a^{(j)}$ must be a divisor of one of these ideals. Now if $j \neq 1$, then $\mathfrak{p}_a^{(j)}$ is not a divisor of $\mathfrak{p}_0$ $(= \mathfrak{p}_0^{(1)})$, since the prime ideals $\mathfrak{p}_a^{(1)}$ of $\mathfrak{o}^*\mathfrak{p}_0^{(1)}$ are all distinct from any $\mathfrak{p}_a^{(j)}$, $j \neq 1$. Hence, if $j \neq 1$, we must have, for some $\beta \neq 0$, $\mathfrak{p}_\beta \subseteqq \mathfrak{p}_a^{(j)}$. Since, by assumption, $\omega_1 - c_0 \not\equiv 0(\mathfrak{p}_\beta)$ if $\beta \neq 0$, it follows

$$(21) \qquad \omega_1 - c_0 \not\equiv 0(\mathfrak{p}_a^{(j)}), \qquad (j = 2, \cdots, m;\ \alpha = 1, 2, \cdots).$$

We now consider one of the ideals $\mathfrak{p}_a^{(1)}$, say $\mathfrak{p}_1^{(1)}$. Let $\omega_1, \omega_2, \cdots, \omega_m$ be the conjugates of $\omega = \omega_1$; here $\omega_j \subset \mathfrak{o}_j$ and we do not yet assert that $\omega_1, \cdots, \omega_m$ are distinct, since we do not know whether ω is a primitive element. A substitution of the Galois group of Σ^* which carries $\mathfrak{o}_1$ into $\mathfrak{o}_2$, will carry ω_1 into ω_2 and also some ring $\mathfrak{o}_j$, $j \neq 1$, into $\mathfrak{o}_1$. Hence it will carry some ideal $\mathfrak{p}_a^{(j)}$, $j \neq 1$, into $\mathfrak{p}_1^{(1)}$. Applying this substitution to (21) we obtain:

$$\omega_2 - c_0 \not\equiv 0(\mathfrak{p}_1^{(1)}).$$

In a similar manner we get $\omega_j - c_0 \not\equiv 0(\mathfrak{p}_1^{(1)})$, $j = 3, \cdots, m$. Reassuming and recalling (20), we have:

$$(22) \quad \omega_1 - c_0 \equiv 0(\mathfrak{p}_1^{(1)}),\ \omega_2 - c_0 \not\equiv 0(\mathfrak{p}_1^{(1)}), \cdots, \omega_m - c_0 \not\equiv 0(\mathfrak{p}_1^{(1)}).$$

From (22) it follows that $\omega_1 \neq \omega_j$, $j \neq 1$, whence ω *is a primitive element.* Moreover, let $N(z - \omega) = F(z; \xi_1, \cdots, \xi_r) = (z - \omega_1) \cdots (z - \omega_m)$, and let $\omega_j = c_j(\mathfrak{p}_1^{(1)})$, $j = 2, \cdots, m$. If we reduce the coefficients of $F(z; \xi_1, \cdots, \xi_r)$ mod $\mathfrak{p}_1^{(1)}$, we must replace $\xi_1, \cdots, \xi_r$ by zeros. Hence $F(z; 0, \cdots, 0) = (z - c_0)(z - c_2) \cdots (z - c_m)$. Since $c_j \neq c_0$, it follows that c_0 is a simple root of $F(z; 0, \cdots, 0)$. Hence $F'_{\omega}(\omega; \xi_1, \cdots, \xi_r) \not\equiv 0(\mathfrak{p}_0)$, as was asserted.

Conversely, let us assume that c_0 is a simple root of $F(z; 0, \cdots, 0)$. This implies that $\omega_j \not\equiv c_0(\mathfrak{p}_\alpha^{(1)})$, $j \neq 1$, $\alpha = 1, 2, \cdots, \sigma$, since $\omega_1 \equiv c_0(\mathfrak{p}_\alpha^{(1)})$. Applying the substitutions of the Galois group of Σ^* which carry ω_j into ω_1, we get (21), and from this, retracing our reasoning which led to (21), we conclude that $\omega \not\equiv c_0(\mathfrak{p}_i)$, $i = 1, 2, \cdots, s$. Thus all the assertions of our theorem are proved.

7. It is clear that Theorem 4 remains true if we replace $\xi_1, \cdots, \xi_r$ by any other set of uniformizing parameters $\eta_1, \cdots, \eta_r$ in $\mathfrak{o}$, such that every element in $\mathfrak{o}$ is integrally dependent on $\eta_1, \cdots, \eta_r$. Let now $\eta_1, \cdots, \eta_r$ be any set of r algebraically independent elements in $\mathfrak{o}$, such that every element in $\mathfrak{o}$ is integrally dependent on $\eta_1, \cdots, \eta_r$. Let $F'_{\omega}(\omega; \eta_1, \cdots, \eta_r)$ be the different of an element ω in $\mathfrak{o}$, where $F(z; \eta_1, \cdots, \eta_r)$ is the norm $N(z - \omega)$ over $\mathsf{K}(\eta_1, \cdots, \eta_r)$. We consider the ideal $Z_{\eta_1, \ldots, \eta_r} = Z_{(\eta)}$ generated by the differents F'_{ω} as ω varies arbitrarily in $\mathfrak{o}$ (the h. c. d. of the ideals $\mathfrak{o} \cdot F'_{\omega}$, ω-arbitrary in $\mathfrak{o}$). Let $\mathfrak{p}_0$ be a prime zero-dimensional ideal in $\mathfrak{o}$, P—the corresponding point of V_r, and let $\eta_i \equiv c_i(\mathfrak{p}_0)$. If $\mathfrak{p}_0$ is an isolated component of the ideal $\mathfrak{o} \cdot (\eta_1 - c_1, \cdots, \eta_r - c_r)$, then P is a simple point of V_r and $\eta_1 - c_1, \cdots, \eta_r - c_r$ are uniformizing parameters for the whole neighborhood of P. It follows then, by Theorem 4, that for some ω in $\mathfrak{o}$ it is true that $F'_{\omega} \not\equiv 0(\mathfrak{p}_0)$. Hence $Z_{(\eta)} \not\equiv 0(\mathfrak{p}_0)$.

Conversely, let us assume that $Z_{(\eta)} \not\equiv 0(\mathfrak{p}_0)$. There exists then an element ω in $\mathfrak{o}$ for which $F'_{\omega} \not\equiv 0(\mathfrak{p}_0)$, i. e. such that $\omega \equiv d_0(\mathfrak{p}_0)$ and d_0 is a simple root of $F(z; c_1, \cdots, c_r)$. Let $\mathfrak{A} = \mathfrak{o} \cdot (\eta_1 - c_1, \cdots, \eta_r - c_r) = [\mathfrak{q}_0, \mathfrak{q}_1, \cdots, \mathfrak{q}_s]$, where $\mathfrak{q}_0, \mathfrak{q}_1, \cdots, \mathfrak{q}_s$ are all 0-dimensional and $\mathfrak{q}_0$ belongs to $\mathfrak{p}_0$. If we reduce the equation $F(\omega; \eta_1, \cdots, \eta_r) = 0$ modulo $\mathfrak{A}$, we get $F(\omega; c_1, \cdots, c_r) \equiv 0(\mathfrak{A})$ or

$$(23) \qquad (\omega - d_0) \prod_{i=1}^{m-1} (\omega - d_i) \equiv 0(\mathfrak{A}), \qquad d_i \neq d_0, \text{ if } i \neq 0,$$

where m is the relative degree of Σ over $\mathsf{K}(\eta_1, \cdots, \eta_r)$. *Since* $\omega - d_i \not\equiv 0(\mathfrak{p}_0)$, $i \neq 0$, *it follows by* (23), *that* $\omega - d_0 \equiv 0(\mathfrak{q}_0)$.

Let now ζ *be any element in* $\mathfrak{p}_0$. We form the norm $N(z - \bar{\zeta})$, where $\bar{\zeta} = t\zeta + \omega$, t—a parameter. This norm is a polynomial $G(z; t, \eta_1, \cdots, \eta_r)$ which for $t = 0$ becomes $F(z; \eta_1, \cdots, \eta_r)$. Moreover $G'_{\bar{\zeta}}(\bar{\zeta}; t, \eta_1, \cdots, \eta_r) = H(t)$, where $H(t)$ is a polynomial in t with coefficients in $\mathfrak{o}$. Now $H(0) = F'_{\omega} \not\equiv 0(\mathfrak{p}_0)$, hence the coefficients of $H(t)$ are not all $\equiv 0(\mathfrak{p}_0)$. As a consequence, if t_0 is a non-special value of t, and if $\bar{\zeta}_0 = t_0\zeta + \omega$, then

$G'_{0\bar{\zeta}_0} \not\equiv 0(\mathfrak{p}_0)$, where $G_0(z;\eta_1,\cdots,\eta_r) = N(z-\bar{\zeta}_0)$. Since $\bar{\zeta}_0 \equiv d_0(\mathfrak{p}_0)$, it follows, as above for ω, that $\bar{\zeta}_0 \equiv d_0(\mathfrak{q}_0)$, i. e. $t_0\zeta + \omega \equiv d_0(\mathfrak{q}_0)$. Since also $\omega \equiv d_0(\mathfrak{q}_0)$ and since we may assume $t_0 \neq 0$, we conclude that $\zeta \equiv 0(\mathfrak{q}_0)$. *Thus every element of* $\mathfrak{p}_0$ *is in* $\mathfrak{q}_0$, i. e. $\mathfrak{p}_0 = \mathfrak{q}_0$, whence $\mathfrak{p}_0$ *is an isolated component of the ideal* $\mathfrak{o}\cdot(\eta_1 - c_1,\cdots,\eta_r - c_r)$. We have therefore proved that $\mathfrak{p}_0$ *is an isolated component of* $\mathfrak{o}\cdot(\eta_1 - c_1,\cdots,\eta_r - c_r)$ *if and only if* $Z_{(\eta)} \not\equiv 0(\mathfrak{p}_0)$.

If we then denote by Z the h. c. d. of all ideal $Z_{(\eta)}$, as $\eta_1,\cdots,\eta_r$ vary arbitrarily in $\mathfrak{o}$ (*subject to the only condition that every element in* $\mathfrak{o}$ *be integrally dependent on* $\eta_1,\eta_2,\cdots,\eta_r$), we conclude with the following

THEOREM 5. *If P is a point on V_r and if $\mathfrak{p}_0$ is the corresponding prime 0-dimensional ideal in $\mathfrak{o}$, a necessary and sufficient condition that P be a simple point is that $Z \not\equiv 0(\mathfrak{p}_0)$.*

As a corollary we have that *the manifold of singular points of V_r is algebraic, of dimension $\leqq r-1$.* It is the manifold given by the ideal Z.

8. Actually the singular manifold can be defined by an ideal which is possibly a multiple of Z and in whose construction intervene only the coördinates $\xi_1,\cdots,\xi_n$ of the general point of V_r. Namely, let P be a simple point of V_r, $\mathfrak{p}_0$—the corresponding ideal in $\mathfrak{o}$, and let $\xi_i \equiv c_i(\mathfrak{p}_0)$. We know from section 2 that if $\bar{\xi}_i = u_{i1}\xi_1 + \cdots + u_{in}\xi_n$, $i = 1,2,\cdots,r$, and $\xi_i \equiv d_i(\mathfrak{p}_0)$, then $\mathfrak{p}_0$ is an isolated component of the ideal $\mathfrak{o}\cdot(\bar{\xi}_1 - d_1,\cdots,\xi_r - d_r)$, provided the coefficients u_{ij} are non-special. Moreover, every element in $\mathfrak{o}$ is then integrally dependent on $\bar{\xi}_1,\cdots,\bar{\xi}_r$. Hence P is a simple point if and only if $Z_{\bar{\xi}_1,\ldots,\bar{\xi}_r} \not\equiv 0(\mathfrak{p}_0)$, *for general values of the coefficients* u_{ij}. Now, assume that $Z_{\bar{\xi}_1,\ldots,\bar{\xi}_r} \not\equiv 0(\mathfrak{p}_0)$, i. e. there exists an element ω in $\mathfrak{o}$ such that $F'_\omega \not\equiv 0(\mathfrak{p}_0)$, where $F(z;\bar{\xi}_1,\cdots,\bar{\xi}_r) = N(z-\omega)$, the norm of $z-\omega$ relative to the field $\mathsf{K}(\bar{\xi}_1,\cdots,\bar{\xi}_r)$. We assert that *there also exists such an element ω which is linear and homogeneous in $\xi_1,\cdots,\xi_n$.* In fact, let

$$\mathfrak{o}\cdot(\bar{\xi}_1 - d_1,\cdots,\bar{\xi}_r - d_r) = [\mathfrak{p}_0,\mathfrak{q}_1,\cdots,\mathfrak{q}_s]$$

be the decomposition of $\mathfrak{o}\cdot(\bar{\xi}_1 - d_1,\cdots,\bar{\xi}_r - d_r)$ into primary (0-dimensional) components, and let $\xi_i \equiv c_{ij}(\mathfrak{p}_j)$, $i = 1,2,\cdots,n$, $j = 0,1,\cdots,s$, where $\mathfrak{q}_j$ belongs to the prime ideal $\mathfrak{p}_j$. Since $\mathfrak{p}_j$ and $\mathfrak{p}_0$ determine two distinct points of V_r, if $j \neq 0$, namely $P_j(c_{1j},\cdots,c_{nj})$ and $P(c_{10},\cdots,c_{n0})$, it follows that constants $\bar{v}_1,\cdots,\bar{v}_n$ can be found in such a manner that

$$\bar{v}_1c_{1j} + \cdots + \bar{v}_nc_{nj} \neq \bar{v}_1c_{10} + \cdots + \bar{v}_nc_{n0}, \qquad (j = 1,2,\cdots,s).$$

Let $\bar{\xi} = \bar{v}_1\xi_1 + \cdots + \bar{v}_n\xi_n$. The value of the element $\bar{\xi}$ at P_0 is then distinct from its value at any of the points $P_1, \cdots, P_s$, and hence by Theorem 4 it follows that $F'_{\bar{\xi}}(\bar{\xi}; \bar{\xi}_1, \cdots, \bar{\xi}_r) \not\equiv 0(\mathfrak{p}_0)$, where $F(z; \bar{\xi}_1, \cdots, \bar{\xi}_r) = N(z - \bar{\xi})$. This proves our assertion.

From these considerations we conclude immediately with the following

THEOREM 6. *Let* $\bar{\xi}_i = u_{i1}\xi_1 + \cdots + u_{in}\xi_n$, $i = 1, 2, \cdots, r+1$, *be* $r+1$ *linear forms with indeterminate coefficients* u_{ij}, *and let* $F(\bar{\xi}_1, \cdots, \bar{\xi}_{r+1}) = 0$ *be the irreducible algebraic relation between the* $\bar{\xi}_i$. *Let* $\mathfrak{Z}'$ *be the ideal whose basis consists of the coefficients* $f_\sigma(\xi_1, \cdots, \xi_n)$ *of the various power products of the* u_{ij} *in the polynomial* $F'_{\bar{\xi}_{r+1}}$ *(the* $\bar{\xi}$*'s having been replaced in* $F'_{\bar{\xi}_{r+1}}$ *by the corresponding linear forms in the* ξ*'s). The submanifold of* V_r *defined by this ideal* $\mathfrak{Z}'$ *is the manifold of singular points of* V_r.

9. Let $\eta_1, \cdots, \eta_r$ be elements in $\mathfrak{o}$ such that every element in $\mathfrak{o}$ is integrally dependent on $K[\eta_1, \cdots, \eta_r]$ and let $a_1, \cdots, a_r$ be arbitrary constants. The ideal $\mathfrak{A} = \mathfrak{o} \cdot (\eta_1 - a_1, \cdots, \eta_r - a_r)$ is unmixed and zero dimensional,

$$\mathfrak{A} = [\mathfrak{q}_0, \mathfrak{q}_1, \cdots, \mathfrak{q}_s] = \mathfrak{q}_0\mathfrak{q}_1 \cdots \mathfrak{q}_s,$$

where $\mathfrak{q}_i$ is a primary ideal belonging to the zero-dimensional prime ideal $\mathfrak{p}_i$. Let ω be an element in $\mathfrak{o}$ and let $G(z; \eta_1, \cdots, \eta_r) = N(z - \omega)$ be the norm of $z - \omega$ with respect to the field $K(\eta_1, \cdots, \eta_r)$. We have proved in sections 5, 6 that if $\mathfrak{q}_i \neq \mathfrak{p}_i$ and if $\omega \equiv c_i(\mathfrak{p}_i)$, then c_i is necessarily a multiple root of $G(z; a_1, \cdots, a_r)$. The following theorem gives a lower bound for the multiplicity of the root c_i:

THEOREM 7. *If* $\mathfrak{q}_i$ *belongs to the exponent* ρ_i, *then the multiplicity of the root* c_i *is* $\geqq \rho_i$.

Proof. Let $\omega_1, \cdots, \omega_\mu$ be an integral base of $\mathfrak{o}$ over the polynomial ring $K[\eta_1, \cdots, \eta_r]$. We adjoin to the ground field K the indeterminates $u_1, u_2, \cdots, u_\mu$ and we consider the element $\omega_u = u_1\omega_1 + \cdots + u_\mu\omega_\mu$. Let

$$N(z - \omega_u) = F(z; \eta_1, \cdots, \eta_r; u_1, \cdots, u_\mu) = F(z; \eta; u)$$

be the norm of $z - \omega_u$ with respect to the field $K(\eta; u)$. It is clear that if $\omega_i \equiv c_{ij}(\mathfrak{p}_j)$, $j = 0, 1, \cdots, s$, then $u_1c_{1j} + \cdots + u_\mu c_{\mu j}$ is a root of $F(z; a_1, \cdots, a_r; u_1, \cdots, u_\mu)(= F(z; a; u))$. It has been proved by van der Waerden [8] that, conversely, every root of $F(z; a; u)$ is a linear form $u_1d_1 + \cdots + u_\mu d_\mu$, $d_j \subset K$, and that $(a_1, \cdots, a_r, d_1, \cdots, d_\mu)$ is a point of the variety whose general point has coördinates $\eta_1, \cdots, \eta_r, \omega_1, \cdots, \omega_\mu$. In other words: $d_1, \cdots, d_\mu$ must be the values of $\omega_1, \cdots, \omega_\mu$ at one of the ideals $\mathfrak{p}_0, \mathfrak{p}_1, \cdots, \mathfrak{p}_s$. As a consequence, $F(z; a; u)$ is a product of factors $z - (u_1 c_{1j} + \cdots + u_\mu c_{\mu j})$:

$$F(z; a; u) = \prod_{j=0}^{s} (z - c_{1j}u_1 - \cdots - c_{\mu j}u_\mu)^{\sigma_j}.$$

Since $\omega_1, \cdots, \omega_\mu$ is a base of $\mathfrak{o}$, it follows that

$$\mathfrak{p}_j = (\eta_1 - a_1, \cdots, \eta_r - a_r, \ \omega_1 - c_{1j}, \cdots, \omega_\mu - c_{\mu j}).$$

Hence for $j \neq j'$, *the factors* $z - c_{1j}u_1 - \cdots - c_{\mu j}u_\mu$ and $z - c_{1j'}u_1 - \cdots - c_{\mu j'}u_\mu$ *are distinct.*

We have $F(\omega_u; \eta; u) = 0$. Reducing the η modulo $\mathfrak{A}$ we get the congruence: $F(\omega_u; a; u) \equiv 0(\mathfrak{A})$, or

$$(24) \qquad \prod_{j=0}^{s} (\omega_u - c_{1j}u_1 - \cdots - c_{\mu j}u_\mu)^{\sigma_j} \equiv 0(\mathfrak{A}).$$

This congruence should be intended in the following sense: if ω_u is replaced by $u_1\omega_1 + \cdots + u_\mu\omega_\mu$, then on the left-hand side we get a polynomial in $u_1, \cdots, u_\mu$, and the coefficients—element of $\mathfrak{o}$—of the various power products of the u's belong to $\mathfrak{A}$. Now, for a given j, we have

$$\omega_u - c_{1j}u_1 - \cdots - c_{\mu j}u_\mu = u_1(\omega_1 - c_{1j}) + \cdots + u_\mu(\omega_\mu - c_{\mu j}),$$

and the coefficients of this linear form in the u's do not all belong to $\mathfrak{p}_{j'}$, if $j' \neq j$. Hence, by (24),

$$[u_1(\omega_1 - c_{1j}) + \cdots + u_\mu(\omega_\mu - c_{\mu j})]^{\sigma_j} \equiv 0(\mathfrak{q}_j).$$

This implies that all the power products of degree σ_j in $\omega_1 - c_{1j}, \cdots, \omega_\mu - c_{\mu j}$ are in $\mathfrak{q}_j$. Since $\omega_1 - c_{1j}, \cdots, \omega_\mu - c_{\mu j}$ and the elements $\eta_1 - a_1, \cdots, \eta_r - a_r$ form a base of $\mathfrak{p}_j$, and since the elements $\eta_1 - a_1, \cdots, \eta_r - a_r$ are also in $\mathfrak{q}_j$, we conclude that $\mathfrak{p}_j^{\sigma_j} \equiv 0(\mathfrak{q}_j)$. *Hence* $\sigma_j \geqq \rho_j$.

Now let ω be any element in $\mathfrak{o}$,

$$\omega = P_1\omega_1 + \cdots + P_\mu\omega_\mu, \qquad P_i \subset \mathsf{K}[\eta_1, \cdots, \eta_r],$$

and let $\omega \equiv c_j(\mathfrak{p}_j)$, $P_i \equiv P_{i0}(\mathfrak{A})$, $P_{i0} \subset \mathsf{K}$. Then

$$N(z - \omega) = F(z; \eta_1, \cdots, \eta_r; \ P_1, \cdots, P_\mu) = G(z; \eta_1, \cdots, \eta_r),$$

whence

$$G(z; a_1, \cdots, a_r) = F(z; a_1, \cdots, a_r; \ P_{10}, \cdots, P_{\mu 0})$$
$$= \prod_{j=0}^{s} (z - c_{1j}P_{10} - \cdots - c_{\mu j}P_{\mu 0})^{\sigma_j}.$$

Since $c_{1j}P_{10} + \cdots + c_{\mu j}P_{\mu 0} = c_j$, our theorem follows.

Do there exist elements ω for which c_j is a root of multiplicity exactly ρ_j? The answer: not necessarily (except when $\rho_j = 1$, i. e. $\mathfrak{p}_j = \mathfrak{q}_j$, according to Theorem 4). Here is an example. Let $\eta^3{}_3 = \eta_1{}^2\eta_2$ be the defining equation of an algebraic surface and let $\eta_4 = \eta_3{}^2/\eta_1$. Consider the ring $\mathfrak{o} = \mathsf{K}[\eta_1, \eta_2, \eta_3, \eta_4]$. Every element of $\mathfrak{o}$ is integrally dependent on η_1, η_2, since $\eta_4{}^3 = \eta_1\eta_2{}^2$ (incidentally, it is not difficult to see that $\mathfrak{o}$ is integrally closed). The ideal $\mathfrak{A} = (\eta_1, \eta_2)$ is primary, $\mathfrak{A} = \mathfrak{q}$, and its prime ideal is $\mathfrak{p}_0 = (\eta_1, \eta_2, \eta_3, \eta_4)$. The exponent of $\mathfrak{A}$ is 2, since

$$\eta_3{}^2 = \eta_1\eta_4, \quad \eta_4{}^2 = \eta_2\eta_3, \quad \eta_3\eta_4 = \eta_1\eta_2.$$

On the other hand, the field $\mathsf{K}(\eta_1, \eta_2, \eta_3, \eta_4)$ is in the present case of relative degree 3 over $\mathsf{K}(\eta_1, \eta_2)$. Since $\mathfrak{A}$ itself is primary, it follows that if ω is any element of $\mathfrak{o}$ and if $F(z; \eta_1, \eta_2) = N(z - \omega)$, then $F(z, 0, 0)$ must have a triple root.

II. Properties of the conductor.

10. Let $\mathfrak{o}$, $\mathfrak{o}^*$ be two finite integral domains in Σ. We assume that $\mathfrak{o}$ is a subring of $\mathfrak{o}^*$ and that Σ is the quotient field of $\mathfrak{o}$. The conductor of $\mathfrak{o}$ with respect to $\mathfrak{o}^*$, in symbols $\mathfrak{c}(\mathfrak{o}, \mathfrak{o}^*) = \mathfrak{c}$, is, by definition, the largest ideal in $\mathfrak{o}$ which is also an ideal in $\mathfrak{o}^*$. This implies that $\mathfrak{c}$ is the totality of all elements ξ in $\mathfrak{o}$ such that $\xi\mathfrak{o}^* \subset \mathfrak{o}$. Every element in $\mathfrak{o}^*$ can therefore be written in the form of a quotient η/ξ, $\eta \subset \mathfrak{o}$, and ξ—any element in $\mathfrak{c}$.

It is well known that $\mathfrak{c} \neq (0)$ *if and only if every element in* $\mathfrak{o}^*$ *is integrally dependent on* $\mathfrak{o}$. The proof is immediate. Namely, assume $\mathfrak{c} \neq (0)$ and let ξ be an element in $\mathfrak{c}$, different from zero. The elements in $\mathfrak{o}$ are integrally dependent on the ring of polynomials $\mathsf{P} = \mathsf{K}[\eta_1, \cdots, \eta_r]$, where $\eta_1, \cdots, \eta_r$ are suitable elements in $\mathfrak{o}$. Hence $\mathfrak{o}$ is a finite P-module (since the totality of all elements of Σ which are integrally dependent on P is also a finite P-module; see [7], p. 94). Let $\omega_1, \cdots, \omega_m$ be a P-basis for $\mathfrak{o}$. Then $\mathfrak{o}^*$ is contained in the finite P-module $(\omega_1/\xi, \cdots, \omega_m/\xi)$ and hence $\mathfrak{o}^*$ itself is a finite P-module. Consequently, every element in $\mathfrak{o}^*$ is integrally dependent on P.

Conversely, assume that every element in $\mathfrak{o}^*$ is integrally dependent on $\mathfrak{o}$, whence also on P (in view of the transitivity of integral dependence). Then $\mathfrak{o}^*$ is a finite P-module. Let $\omega^*{}_1, \cdots, \omega^*{}_m$ be a P-basis for $\mathfrak{o}^*$. We can write each $\omega^*{}_i$ in the form ω_i/ξ, where ω_i, $\xi \subset \mathfrak{o}$, ξ being a common denominator (since $\mathfrak{o}^*$ is contained in the quotient ring of $\mathfrak{o}$). Hence $\xi\mathfrak{o}^* \subset \mathfrak{o}$, $\xi \subset \mathfrak{c}(\mathfrak{o}, \mathfrak{o}^*)$, $\xi \neq 0$, q. e. d.

Let $\mathfrak{p}_0$ be a prime 0-dimensional ideal in $\mathfrak{o}$. We assume that $\mathfrak{c}(\mathfrak{o}, \mathfrak{o}^*) \neq (0)$ and we consider the extended ideal $\mathfrak{o}^*\mathfrak{p}_0$. Since the elements in $\mathfrak{o}^*$ are integrally dependent on $\mathfrak{o}$, the ideal $\mathfrak{o}^*\mathfrak{p}_0$ is unmixed and zero-dimensional.

THEOREM 8. *If* $\mathfrak{c}(\mathfrak{o}, \mathfrak{o}^*) \not\equiv 0(\mathfrak{p}_0)$, *then* $\mathfrak{o}^*\mathfrak{p}_0$ *is a prime ideal in* $\mathfrak{o}^*$.

Proof. Let ξ be an element in $\mathfrak{c}$, not in $\mathfrak{p}_0$, and let $\xi \equiv d(\mathfrak{p}_0)$, $d \neq 0$, $d \subset \mathsf{K}$. For any element ω^* in $\mathfrak{o}^*$ we have a relation of the form $\xi\omega^* = \omega \subset \mathfrak{o}$. Reducing this relation modulo $\mathfrak{o}^*\mathfrak{p}_0$ we find $d\omega^* \equiv c(\mathfrak{o}^*\mathfrak{p}_0)$, where $\omega \equiv c(\mathfrak{p}_0)$. Since $d \neq 0$, we conclude that the ring of residual classes $\mathfrak{o}^*/\mathfrak{o}^*\mathfrak{p}_0$ is a field simply isomorphic to K. Hence $\mathfrak{o}^*\mathfrak{p}_0$ is prime (and zero-dimensional).

11. In this and in the following sections we shall derive several properties of the conductor in the special case when $\mathfrak{o}^*$ is the integral closure of $\mathsf{P} = \mathsf{K}[\eta_1, \cdots, \eta_r]$ and $\mathfrak{o} = \mathsf{K}[\eta_1, \cdots, \eta_r, \omega]$. Here $\eta_1, \cdots, \eta_r$ are algebraically independent elements of Σ and ω $(\subset \mathfrak{o}^*)$ is a primitive element of Σ with respect to the field $\mathsf{K}(\eta_1, \cdots, \eta_r)$. Let n be the relative degree $[\Sigma : \mathsf{K}(\eta_1, \cdots, \eta_r)]$, and let $\mathfrak{c}(\omega)$ be the conductor $\mathfrak{c}(\mathfrak{o}; \mathfrak{o}^*)$.

Let $\bar{\mathfrak{p}}$ be an arbitrary zero-dimensional ideal in P. We may assume that $\bar{\mathfrak{p}} = \mathsf{P} \cdot (\eta_1, \cdots, \eta_r)$.

THEOREM 9. *A necessary and sufficient condition that* $1, \omega, \cdots, \omega^{n-1}$ *form a* K*-basis of the* K*-module* $\mathfrak{o}^*/\mathfrak{o}^*\bar{\mathfrak{p}}$, *is that the contracted ideal* $\mathfrak{c}(\omega) \cap \mathsf{P}$ *should not be divisible by* $\bar{\mathfrak{p}}$.

Proof. That the condition is sufficient is trivial. In fact, assuming $\mathfrak{c}(\omega) \cap \mathsf{P} \not\equiv 0(\bar{\mathfrak{p}})$, let g be an element in $\mathfrak{c}(\omega) \cap \mathsf{P}$ but not in $\bar{\mathfrak{p}}$. Let ζ be an arbitrary element in $\mathfrak{o}^*$ and let

$$g\zeta = g_0 + g_1\omega + \cdots + g_{n-1}\omega^{n-1}, \qquad g_i \subset \mathsf{P}.$$

If we reduce $g, g_0, \cdots, g_{n-1}$ modulo $\bar{\mathfrak{p}}$, we get, in view of $g \not\equiv 0(\bar{\mathfrak{p}})$:

$$\zeta \equiv c_0 + c_1\omega + \cdots + c_{n-1}\omega^{n-1}(\mathfrak{o}^*\bar{\mathfrak{p}}), \tag{25}$$

as was asserted.

Somewhat more difficult is the proof that the condition is necessary. We shall make use of the following well-known relation between $\mathfrak{c}(\omega)$, the complementary module e of $\mathfrak{o}^*$ and the different G'_ω of ω:

$$\mathfrak{c}(\omega) = \mathsf{e}G'_\omega.^{15} \tag{26}$$

[15] As was pointed out by Schmeidler [6], this relation holds true for fields of algebraic functions of several variables, the proof being the same as in the case of algebraic functions of one variable.

We assume then that for every element ζ in $\mathfrak{o}^*$ a congruence such as (25) holds true. We can write (25) as follows:

$$(25') \quad \zeta = c_0 + c_1\omega + \cdots + c_{n-1}\omega^{n-1} + \eta_1 A_1 + \cdots + \eta_r A_r, \quad A_i \subset \mathfrak{o}^*.$$

If we apply the congruence (25), taking as ζ any of the elements $A_1, \cdots, A_r$, we derive from (25′) a congruence of the form:

$$\zeta \equiv f_1(\omega; \eta_1, \cdots, \eta_r)\,(\mathfrak{o}^*\bar{\mathfrak{p}}^2),$$

where $f_1 = f_{10} + f_{11}\omega + \cdots + f_{1,n-1}\omega^{n-1}$, and f_{1i} is a polynomial of first degree in $\eta_1, \cdots, \eta_r$. Applying repeatedly this procedure we get more generally the following congruence:

$$(27) \qquad \zeta \equiv f_{\rho-1}(\omega; \eta_1, \cdots, \eta_r)\,(\mathfrak{o}^*\bar{\mathfrak{p}}^\rho),$$

where ρ is an arbitrary integer and $f_{\rho-1}$ is a polynomial, of degree $\leqq n-1$ in ω and of degree $\leqq \rho - 1$ in $\eta_1, \cdots, \eta_r$.

We now consider the complementary module e' of $\mathfrak{o}$, i. e. the set of all elements η in Σ such that $T(\eta\xi)$ (trace of $\eta\xi$ with respect to the field $\mathsf{K}(\eta_1, \cdots, \eta_r)$) is in P if ξ is in $\mathfrak{o}$. It is well known that the elements

$$1/G'_\omega,\ \omega/G'_\omega, \cdots, \omega^{n-1}/G'_\omega,$$

form a P-basis for e'. Here G'_ω is the different of ω; $G(\eta_1, \cdots, \eta_r; \omega) = 0$ is the irreducible equation for ω.

In view of the existence of a finite P-basis in $\mathfrak{o}^*$ and in e', it follows that the traces of *all* the products $\eta\zeta$, $\eta \subset \mathsf{e}'$, $\zeta \subset \mathfrak{o}^*$, can be written as rational functions in $\eta_1, \cdots, \eta_r$ with the same denominator. Let $h(\eta) = h(\eta_1, \cdots, \eta_r)$ be this common denominator. Let us fix an element ζ in $\mathfrak{o}^*$ and an element η in e' and let $T(\zeta\eta) = g(\eta)/h(\eta)$. We apply the congruence (27). The trace of the product $f_{\rho-1}\cdot\eta$ is a polynomial $\psi(\eta)$ in $\eta_1, \cdots, \eta_r$, since $\eta \subset \mathsf{e}'$. The trace of the product $(\zeta - f_{\rho-1})\cdot\eta$ will be, by (27), of the form $g_\rho(\eta)/h(\eta)$, where $g_\rho \equiv 0(\bar{\mathfrak{p}}^\rho)$. We have then the relation: $g/h = \psi + g_\rho/h$, or $g = h\psi + g_\rho$, or finally the congruence

$$(28) \qquad g(\eta) \equiv 0(h(\eta), \bar{\mathfrak{p}}^\rho).$$

We separate in the polynomial $h(\eta)$ the factors which belong to $\bar{\mathfrak{p}}$ (i. e. those which vanish for $\eta_1 = \cdots = \eta_r = 0$) from the remaining factors. We write then $h(\eta) = \sigma(\eta)h_1(\eta)$, where $h_1(0, \cdots, 0) \neq 0$ and $\sigma(\eta)$ is the product of the factors which are $\equiv 0(\bar{\mathfrak{p}})$. If such factors are not present in h, then we put $\sigma(\eta) = 1$. We proceed to show that from the fact that the congruence (28) holds true for any positive integer ρ, it necessarily follows that *$g(\eta)$ is divisible by $\sigma(\eta)$.*

It is clear that in (28) it is permissible to divide through g and h by any common factor of g and h, provided—if that common factor begins with terms of lowest degree ρ_0—that we replace ρ by $\rho - \rho_0$. Hence, to prove our assertion it is sufficient to show that *g and h cannot be relatively prime unless $\sigma(\eta) = 1$.*

Performing, if necessary, a preliminary linear homogeneous transformation on the elements $\eta_1, \cdots, \eta_r$, we may assume that $h(\eta)$, considered as polynomial in η_r, has leading coefficient 1. Since $g = h\psi + g_\rho$, we deduce then that the resultant R of g and h (considered as polynomials in η_r) coincides with the resultant of g_ρ and h. Let

$$h = \eta_r^\nu + A_1(\eta_1, \cdots, \eta_{r-1})\eta_r^{\nu-1} + \cdots + A_\nu(\eta_1, \cdots, \eta_{r-1}),$$
$$g_\rho = B_0(\eta_1, \cdots, \eta_{r-1})\eta_r^\mu + \cdots + B_\mu(\eta_1, \cdots, \eta_{r-1}).$$

Since $g_\rho \equiv 0(\mathfrak{p}^\rho)$, it follows that $B_\mu, B_{\mu-1}, \cdots, B_{\mu-\rho+1}$ begin with terms of lowest degree not less than $\rho, \rho - 1, \cdots, 1$ respectively. Now let us assume that $\sigma(\eta) \neq 1$. Then A_ν begins with terms of lowest degree $\geqq 1$. We apply a theorem on the resultant proved in [9], p. 250. If we attach to A_ν the weight 1, to $B_\mu, B_{\mu-1}, \cdots, B_{\mu-\rho+1}$ the weights $\rho, \rho - 1, \cdots, 1$ respectively, and to the remaining coefficients the weight zero, then—as a special case of the theorem referred to—every term in the resultant R is of weight $\geqq \rho$. It follows that R, as a polynomial in $\eta_1, \cdots, \eta_{r-1}$, begins with terms of lowest degree $\geqq \rho$. Since ρ is an arbitrary integer, it follows that R is identically zero. Hence h and g have a common factor, q. e. d.[16]

---

[16] Our assertion can also be proved without making use of the properties of the resultant. Assume that g and h are relatively prime. Then the intersection of the ideals $(h, \bar{\mathfrak{p}}^\rho)$, $\rho = 1, 2, \cdots$ is at most $(r-2)$-dimensional, since both g and h belong to it. If then $\bar{\eta}_1, \ldots, \bar{\eta}_{r-2}$ are linear forms in $\eta_1, \cdots, \eta_r$ with non special coefficients, then the intersection of the ideals $(h, \bar{\eta}_1, \cdots, \bar{\eta}_{r-2}, \bar{\mathfrak{p}}^\rho)$ is at most zero-dimensional. We may assume that $\bar{\eta}_1, \cdots, \bar{\eta}_{r-2}$ coincide with $\eta_1, \cdots, \eta_{r-2}$ respectively, and we denote by $\mathfrak{B}$ the intersection of the ideals $(h, \eta_1, \cdots, \eta_{r-2}, \bar{\mathfrak{p}}^\rho)$, $\rho = 1, 2, \cdots$. $\mathfrak{B}$ is at most 0-dimensional. If $\sigma \neq 1$, then $h \equiv 0(\bar{\mathfrak{p}})$ and hence $\mathfrak{B} \equiv 0(\bar{\mathfrak{p}})$. Consequently, $\mathfrak{B}$ is 0-dimensional and one of its prime ideals is $\bar{\mathfrak{p}}$. If q is the exponent of the corresponding primary component of $\mathfrak{B}$, then (see [7], p. 49) $\bar{\mathfrak{p}}^q \equiv 0(\mathfrak{B}, \bar{\mathfrak{p}}^{q+1}) \equiv 0(h, \eta_1, \cdots, \eta_{r-2}, \bar{\mathfrak{p}}^\rho, \bar{\mathfrak{p}}^{q+1})$. Since ρ is arbitrary, this implies $\bar{\mathfrak{p}}^q \equiv 0(h, \eta_1, \cdots, \eta_{r-2}, \bar{\mathfrak{p}}^{q+1})$. It is immediately seen that this congruence leads to the following absurd conclusion: if $h = h_j + h_{j+1} + \cdots$, where h_j is a form of degree j, and if $h'_j = h'_j(\eta_{r-1}, \eta_r)$ denotes the sum of terms in h_j which depend only on η_{r-1} and η_r, then every form in η_{r-1}, η_r, of degree q, is divisible by h'_j (or is identically zero, if $h'_j = 0$).

One can also observe that the congruence (28) implies that the hypersurface $g = 0$ has at the origin a contact of infinite order with every algebraic curve which passes

Since h is the least common denominator of all the traces $T(\zeta\eta)$, $\zeta \subset \mathfrak{o}^*$, $\eta \subset \mathfrak{e}'$, *we conclude by the above result that necessarily* $h \not\equiv 0(\bar{\mathfrak{p}})$. Now if $T(\zeta\eta) = g/h$, then $T(\zeta \cdot \eta h) = g$. From this it follows $h\eta \subset \mathfrak{e}$, η arbitrary in $\mathfrak{e}'$. In particular $h \cdot (1/G'_\omega) \subset \mathfrak{e}$, i. e. $h \subset \mathfrak{e}G'_\omega$. Hence, by (26), $h \equiv 0(\mathfrak{c}(\omega))$, and since $h \not\equiv 0(\bar{\mathfrak{p}})$, our theorem is proved.

12. We are now in position to prove the converse of Theorem 8, always under the assumption: $\mathfrak{o} = K[\eta_1, \cdots, \eta_r, \omega]$, $\mathfrak{o}^*$-integral closure of $\mathfrak{o}$.

THEOREM 10. *If* $\mathfrak{p}_0$ *is a prime zero-dimensional ideal in* $\mathfrak{o}$ *and if* $\mathfrak{o}^*\mathfrak{p}_0$ *is prime in* $\mathfrak{o}^*$, *then* $\mathfrak{c}(\mathfrak{o}, \mathfrak{o}^*) \not\equiv 0(\mathfrak{p}_0)$.

Proof. We may assume that $\eta_i \equiv 0(\mathfrak{p}_0)$, $i = 1, 2, \cdots, r$, and also that $\omega \equiv 0(\mathfrak{p}_0)$. Then $\mathfrak{p}_0 = (\eta_1, \eta_2, \cdots, \eta_r, \omega)$. Let V_r be the variety given in S_{r+1} by the defining equation $G(\eta_1, \cdots, \eta_r, \omega) = 0$ and let A be the point $(0, 0, \cdots, 0)$ of V_r which corresponds to $\mathfrak{p}_0$. Let M be the subvariety of V_r defined by the conductor $\mathfrak{c}(\mathfrak{o}, \mathfrak{o}^*)$. A general line on A will intersect V_r, outside of A, in points which are not on M. Subjecting the elements $\eta_1, \cdots, \eta_r, \omega$ to a preliminary linear homogeneous transformation, we may assume that the line $\eta_1 = \cdots = \eta_r = 0$ is general in the sense just specified. Let $[\mathfrak{q}_0, \mathfrak{q}_1, \cdots, \mathfrak{q}_s]$ be decompositions of the ideal $\mathfrak{o}(\eta_1, \cdots, \eta_r)$ into 0-dimensional primary components, $\mathfrak{p}_0, \mathfrak{p}_1, \cdots, \mathfrak{p}_s$—the corresponding prime ideals. Our assumption concerning the line $\eta_1 = \cdots = \eta_r = 0$ implies that $\mathfrak{c}(\mathfrak{o}, \mathfrak{o}^*) \not\equiv 0(\mathfrak{p}_i)$, $i = 1, 2, \cdots, s$. Let $\omega \equiv c_i(\mathfrak{p}_i)$. The constants $c_0 = 0$, $c_1, \cdots, c_s$ are distinct, since $\mathfrak{p}_i = (\eta_1, \cdots, \eta_r, \omega - c_i)$. By Theorem 8 it follows that the ideals

$$\mathfrak{p}^*_i = \mathfrak{o}^*\mathfrak{p}_i = \mathfrak{o}^*(\eta_1, \cdots, \eta_r, \omega - c_i), \qquad (i = 1, 2, \cdots, s)$$

are prime. By our assumption, the ideal

$$\mathfrak{p}^*_0 = \mathfrak{o}^*\mathfrak{p}_0 = \mathfrak{o}^*(\eta_1, \cdots, \eta_r, \omega)$$

is also prime. From this it follows:

$$(29) \quad [\mathfrak{p}^*_0, \mathfrak{p}^*_1, \cdots, \mathfrak{p}^*_s] = \mathfrak{p}^*_0\mathfrak{p}^*_1 \cdots \mathfrak{p}^*_s \equiv \mathfrak{o}^*(\eta_1, \cdots, \eta_r, \prod_{i=0}^{s} (\omega - c_i)).$$

Let now ζ be an arbitrary element in $\mathfrak{o}^*$ and let $\zeta \equiv d_i(\mathfrak{p}^*_i)$, $i = 0, 1, \cdots s$. We can find a polynomial $f(\omega)$, with constant coefficients (of degree $\leqq s$ in ω), such that $f(c_i) = d_i$. For such a polynomial $f(\omega)$ we will have $\zeta - f(\omega) \equiv 0([\mathfrak{p}^*_0, \cdots, \mathfrak{p}^*_s])$, or, in view of (29),

$$\zeta = f(\omega) + \alpha^*_1\eta_1 + \cdots + \alpha^*_r\eta_r + \beta^* \cdot \prod_{i=0}^{s} (\omega - c_i), \quad \alpha^*_i, \beta^* \subset \mathfrak{o}^*.$$

through the origin and lies on the hypersurface $h = 0$. Hence every such irreducible algebraic curve must also lie on the hypersurface $g = 0$, which shows that the two hypersurfaces $g = 0$ and $h = 0$ must have a common component.

Applying to the elements α^*_i and β^* the considerations just applied to ζ, we find that ζ satisfies a congruence of the form

$$\zeta \equiv f_1(\omega; \eta_1, \cdots, \eta_r)(\mathfrak{p}^*{}_0{}^2\mathfrak{p}^*{}_1{}^2 \cdots \mathfrak{p}^*{}_s{}^2),$$

where f_1 is a polynomial (of degree $\leqq 1$ in $\eta_1, \cdots, \eta_r$). More generally, we find

$$\zeta \equiv f_\rho(\omega; \eta_1, \cdots, \eta_r)(\mathfrak{p}^*{}_0{}^{\rho+1} \cdots \mathfrak{p}^*{}_s{}^{\rho+1}),$$

where f_ρ is a polynomial in $\omega, \eta_1, \cdots, \eta_r$, whose degree in ω may be assumed to be $\leqq n-1$. Now if $\rho+1$ is sufficiently high, then $\mathfrak{p}^*{}_i{}^{\rho+1} \equiv 0(\mathfrak{o}^*\mathfrak{q}_i)$.[17] Hence, for ρ sufficiently high, we will have

$$\zeta \equiv f_\rho(\omega; \eta_1, \cdots, \eta_r)(\mathfrak{o}^*(\eta_1, \cdots, \eta_r)),$$

or, letting $f_\rho(\omega; 0, \cdots, 0) = c_0 + c_1\omega + \cdots + c_{n-1}\omega^{n-1}$, $c_i \subset \mathsf{K}$,

$$\zeta \equiv c_0 + c_1\omega + \cdots + c_{n-1}\omega^{n-1}(\mathfrak{o}^*\bar{\mathfrak{p}}).$$

Such a congruence holds true for any element ζ in $\mathfrak{o}^*$. Hence, by Theorem 9, we must have $\mathfrak{c}(\mathfrak{o}, \mathfrak{o}^*) \frown \mathsf{P} \not\equiv 0(\bar{\mathfrak{p}})$. This implies $\mathfrak{c}(\mathfrak{o}, \mathfrak{o}^*) \not\equiv 0(\mathfrak{p}_i)$, $i = 0, 1, \cdots, s$, in particular $\mathfrak{c}(\mathfrak{o}, \mathfrak{o}^*) \not\equiv 0(\mathfrak{p}_0)$, which proves our assertion.

13. In the preceding section we have characterized in $\mathfrak{o}$ the prime zero-dimensional divisors $\mathfrak{p}_0$ of the conductor $\mathfrak{c}(\mathfrak{o}, \mathfrak{o}^*)$, by means of the decomposition of the extended ideal $\mathfrak{o}^*\mathfrak{p}_0$. Namely, $\mathfrak{p}_0$ is a divisor of $\mathfrak{c}(\mathfrak{o}, \mathfrak{o}^*)$ if and only if the zero-dimensional ideal $\mathfrak{o}^*\mathfrak{p}_0$ is not prime in $\mathfrak{o}^*$. This is a relative characterization. It is of interest to give an intrinsic characterization in $\mathfrak{o}^*$ of those zero-dimensional ideals $\mathfrak{p}^*{}_0$ which are extended ideals $\mathfrak{o}^*\mathfrak{p}_0$ of ideals $\mathfrak{p}_0$ in some subrings of $\mathfrak{o}^*$, such as $\mathfrak{o}$. The question is then the following. We consider subrings $\mathfrak{o}$ of $\mathfrak{o}^*$ which satisfy the following conditions: (1) $\mathfrak{o}^*$ is the integral closure of $\mathfrak{o}$; (2) $\mathfrak{o}$ is generated by $r+1$ elements, $\mathfrak{o} = \mathsf{K}[\eta_1, \cdots, \eta_{r+1}]$; (3) $\mathfrak{o}$ and $\mathfrak{o}^*$ have the same quotient field. Given a prime zero-dimensional ideal $\mathfrak{p}^*{}_0$ in $\mathfrak{o}^*$, we ask: under what condition does there exist a subring $\mathfrak{o}$ such that $\mathfrak{p}^*{}_0 = \mathfrak{o}^*\mathfrak{p}_0$, where $\mathfrak{p}_0 = \mathfrak{p}^*{}_0 \frown \mathfrak{o}$? We proceed to prove that such a subring $\mathfrak{o}$ exists if and only if *the rank of the* K*-module* $\mathfrak{p}^*{}_0/\mathfrak{p}^*{}_0{}^2$ *is not greater than* $r+1$.[18]

Proof. Consider, quite generally, any zero-dimensional ideal $\mathfrak{p}^*{}_0$ in $\mathfrak{o}^*$, and let ρ be the rank of the ring $\mathfrak{p}^*{}_0/\mathfrak{p}^*{}_0{}^2$, considered as a K-module. Let

---

[17] Since $\mathfrak{o}^*\mathfrak{p}_i = \mathfrak{p}_i{}^*$, the ideal $\mathfrak{o}^*\mathfrak{q}_i$ is primary and belongs to the ideal $\mathfrak{p}_i{}^*$. Namely, in the first place $\mathfrak{o}^*\mathfrak{q}_i$ is zero-dimensional. Let $\mathfrak{p}'^*$ be one of its prime ideals, and let $\mathfrak{p}'^* \frown \mathfrak{o} = \mathfrak{p}'$. Also $\mathfrak{p}'$ is prime and zero-dimensional. Since $\mathfrak{p}'^* \supseteqq \mathfrak{q}_i$, we have $\mathfrak{p}' \supseteqq \mathfrak{q}_i$, whence $\mathfrak{p}' = \mathfrak{p}_i$. Consequently $\mathfrak{p}'^* \supseteqq \mathfrak{o}^*\mathfrak{p}' = \mathfrak{o}^*\mathfrak{p}_i = \mathfrak{p}^*{}_i$, i. e., $\mathfrak{p}'^* = \mathfrak{p}^*{}_i$.

[18] Hence, in particular, if the point which corresponds to $\mathfrak{p}_0{}^*$ on the variety determined by the ring $\mathfrak{o}^*$, is simple, because then the rank is r (see Theorem 3.2).

$\omega_1, \cdots, \omega_\rho$ be a K-basis of $\mathfrak{p}^*_0/\mathfrak{p}^*_0{}^2$. We have evidently: $(\omega_1, \cdots, \omega_\rho, \mathfrak{p}^*_0{}^2) = \mathfrak{p}^*_0$. From this we deduce (see section 2) that $\mathfrak{p}^*_0$ is an isolated component of the ideal $\mathfrak{o}^*(\omega_1, \cdots, \omega_\rho)$. Since $\mathfrak{p}^*_0$ is zero-dimensional, we must have $\rho \geqq r$ (see footnote 3).

Now let us assume that the rank ρ is $\leqq r+1$. Let $\mathfrak{o}^* = \mathsf{K}[\xi_1, \cdots, \xi_n]$ and let us assume, as usual, that $\mathfrak{p}^*_0 = (\xi_1, \cdots, \xi_n)$. We can then take as a K-basis for $\mathfrak{p}^*_0/\mathfrak{p}^*_0{}^2$ ρ linear forms in the ξ's. We may assume that $\xi_1, \cdots, \xi_\rho$ form such a K-basis. Let $\xi_i \equiv \sum_{j=1}^{\rho} c_{ij}\xi_j(\mathfrak{p}^*_0{}^2)$, $i = 1, 2, \cdots, n$. The rank of the matrix (c_{ij}) is ρ. Given ρ forms $\eta_j = \sum_{i=1}^{n} u_{ji}\xi_i$, they will form a K-basis for $\mathfrak{p}^*_0/\mathfrak{p}^*_0{}^2$ if and only if the determinant of the ρ by ρ matrix $(u_{ji}) \cdot (c_{i\nu})$ is $\neq 0$. We consider the two possible cases: $\rho = r$, $\rho = r+1$.

Let $\rho = r$. We choose the coefficients u_{ji} so that the above determinant be $\neq 0$ and that, in addition, every element of $\mathfrak{o}^*$ be integrally dependent on $\eta_1, \cdots, \eta_r$. Then $\mathfrak{p}^*_0$ is an isolated component of the *zero-dimensional* ideal $\mathfrak{o}^*(\eta_1, \cdots, \eta_r)$. Let $\mathfrak{p}^*_1, \cdots, \mathfrak{p}^*_s$ be the other prime ideals of $\mathfrak{o}^*(\eta_1, \cdots, \eta_r)$. We choose in $\mathfrak{o}^*$ a primitive element ω (with respect to the field $\mathsf{K}(\eta_1, \cdots, \eta_r)$) such that $\omega \equiv 0(\mathfrak{p}^*_0)$, $\omega \not\equiv 0(\mathfrak{p}^*_i)$, $i = 1, 2, \cdots, s$. Then $\mathfrak{p}^*_0 = \mathfrak{o}^*(\eta_1, \cdots, \eta_r, \omega)$ and the ring $\mathfrak{o} = \mathsf{K}[\eta_1, \cdots, \eta_r, \omega]$ satisfied all our conditions.

Let $\rho = r+1$. We first choose the coefficients of the first r rows of the matrix $(u_{ji}) \cdot (c_{i\nu})$ in such a manner that their matrix be of rank r and that every element in $\mathfrak{o}^*$ be integrally dependent on $\eta_1, \cdots, \eta_r$. With this choice of the first r rows we are certain that when $u_{r+1,1}, \cdots, u_{r+1,n}$ have non-special values, then the determinant of the $(r+1)$-row matrix $(u_{ji})(c_{i\nu})$ is $\neq 0$. Let $[\mathfrak{q}^*_0, \mathfrak{q}^*_1, \cdots, \mathfrak{q}^*_s]$ be the decomposition of the *zero-dimensional* ideal $\mathfrak{o}^*(\eta_1, \cdots, \eta_r)$ into primary components. Let $\mathfrak{p}^*_0, \mathfrak{p}^*_1, \cdots, \mathfrak{p}^*_s$ be the corresponding prime ideals. Again, for non-special values of the coefficients $u_{r+1,i}$, the element $\eta_{r+1} = \sum_{i=1}^{n} u_{r+1,i}\xi_i$ will satisfy the conditions: $\eta_{r+1} \not\equiv 0(\mathfrak{p}_i)$, $i = 1, 2, \cdots, s$ (since the n elements ξ_j are not all $\equiv 0(\mathfrak{p}_i)$, if $i \neq 0$). Finally, for non-special coefficients $u_{r+1,i}$ the element η_{r+1} will be a primitive element of $\mathfrak{o}^*$ with respect to the field $\mathsf{K}(\eta_1, \cdots, \eta_r)$. If we choose the coefficients $u_{r+1,i}$ so as to satisfy all these conditions, it will follow that $\mathfrak{o}^*(\eta_1, \cdots, \eta_r, \eta_{r+1}) = \mathfrak{p}^*_0$ and that the ring $\mathfrak{o} = \mathsf{K}[\eta_1, \cdots, \eta_{r+1}]$ satisfies all our requirements, q. e. d.[19]

---

[19] The example at the end of section 9 illustrates the possibility $\rho > r+1$. In that example we have $r = 2$, $\mathfrak{p}_0{}^* = (\eta_1, \eta_2, \eta_3, \eta_4)$, and it is easily seen that the four elements η_i are linearly independent modulo $\mathfrak{p}_0{}^{*2}$. Hence $\rho = 4$.

III. Normal varieties in the affine space.

14. In part I we have given several characterizations of simple and multiple points of a V_r in S_n, mostly of ideal theoretic nature. We have seen that these characterizations are not so much properties of the set of coördinates $\xi_1, \cdots, \xi_n$ of the general point of V_r, as of the whole ring $\mathfrak{o} = K[\xi_1, \cdots, \xi_n]$, namely of the zero-dimensional ideals in $\mathfrak{o}$. If $\eta_1, \cdots, \eta_m$ is another set of elements in $\mathfrak{o}$ such that $\mathfrak{o} = K[\eta_1, \cdots, \eta_m]$, then the η's are the coördinates of the general point of a variety W_r in an S_m, birationally equivalent to V_r. The points of V_r and W_r are in $(1,1)$ correspondence without exceptions (everything being confined to the points in the affine space, i. e. at finite distance), since their points are in $(1,1)$ correspondence with the prime zero-dimensional ideals in $\mathfrak{o}$. To simple points of V_r correspond simple points of W_r, and conversely. Topologically speaking, V_r and W_r (points at infinity excluded) are homeomorphic loci ("open" or "relative" circuits). We shall say that V_r and W_r are *integrally equivalent*, alluding to the fact that they are both defined in terms of one and the same integral domain $\mathfrak{o}$ in the field Σ. Thus $\mathfrak{o}$ determines a *class* of birationally equivalent varieties any two of which are integrally equivalent.

We shall say that an algebraic variety V_r in an affine S_n is *normal in the affine space*, if the coördinates $\xi_1, \cdots, \xi_n$ of the general point of V_r give rise to an integral domain $\mathfrak{o} = K[\xi_1, \cdots, \xi_n]$ which is *integrally closed in its quotient field.* In the sequel we will speak of *the* normal variety determined by an integrally closed finite integral domain $\mathfrak{o}$ in the field Σ, meaning by this any one of the integrally equivalent varieties determined by $\mathfrak{o}$.

15. We prove now the following

LEMMA. *Let V_r be an irreducible r-dimensional algebraic variety in S_n, $\mathfrak{o} = K[\xi_1, \cdots, \xi_n]$ the corresponding integral domain. If V_ρ is a ρ-dimensional irreducible algebraic subvariety of V_r and $\mathfrak{p}_\rho$ is the corresponding ρ-dimensional prime ideal in $\mathfrak{o}$, then a sufficient condition that V_ρ be a non-singular manifold of V_r* (i. e. *that not all points of V_ρ be multiple points of V_r*) *is that there should exist $r-\rho$ elements $\eta_1, \cdots, \eta_{r-\rho}$ in $\mathfrak{o}$ such that $\mathfrak{p}_\rho$ is an isolated component of the ideal $\mathfrak{o}\cdot(\eta_1, \cdots, \eta_{r-\rho})$.*

This lemma generalizes the property which we have used in section 2 for the definition of a simple point $(\rho = 0)$.

For the proof we consider the ring of residual classes $\bar{\mathfrak{o}} = \mathfrak{o}/\mathfrak{p}_\rho$. If $\bar{\xi}_1, \cdots, \bar{\xi}_n$ denote the elements of $\bar{\mathfrak{o}}$ which correspond to $\xi_1, \cdots, \xi_n$ respectively in the homomorphism $\mathfrak{o} \cong \bar{\mathfrak{o}}$, then it is clear that $\bar{\xi}_1, \cdots, \bar{\xi}_n$ are the coördinates of the general point of V_ρ. We take a point P of V_ρ which does not lie on either one of the following two subvarieties of V_ρ (both of dimension $\leqq \rho - 1$): (1) the variety of singular points of V_ρ (the points of V_ρ which are multiple *for* V_ρ); (2) the variety in which V_ρ intersects the possible other components of the variety V' defined by the ideal $\mathfrak{o} \cdot (\eta_1, \cdots, \eta_{r-\rho})$ (since $\mathfrak{p}_\rho$ is an isolated component of $\mathfrak{o} \cdot (\eta_1, \cdots, \eta_{r-\rho})$, V_ρ itself is one of the irreducible components of V'). Let $\bar{\mathfrak{p}}_0$ and $\mathfrak{p}_0$ be the prime 0-dimensional ideals in $\bar{\mathfrak{o}}$ and $\mathfrak{o}$ respectively which correspond to the point P, regarded either as a point of V_ρ or as a point of V_r. Clearly $\mathfrak{p}_\rho \equiv 0(\mathfrak{p}_0)$ and $\mathfrak{p}_0 \rightarrow \bar{\mathfrak{p}}_0$ in the homomorphism between $\mathfrak{o}$ and $\bar{\mathfrak{o}}$. Since P is a simple point for V_ρ, it follows that the rank of the K-module $\bar{\mathfrak{p}}_0/\bar{\mathfrak{p}}_0^2$ is ρ (Theorem 3.2). Let $\bar{\eta}_{r-\rho+1}, \cdots, \bar{\eta}_r$ be a K-module basis of $\bar{\mathfrak{p}}_0/\bar{\mathfrak{p}}_0^2$, and let $\eta_{r-\rho+1}, \cdots, \eta_r$ be any set of ρ elements in $\mathfrak{o}$ belonging to the residual classes $\bar{\eta}_{r-\rho+1}, \cdots, \bar{\eta}_r$ respectively. Any element $\bar{\omega}$ in $\bar{\mathfrak{o}}$ satisfies a congruence of the form $\bar{\omega} \equiv d_0 + \sum_{i=1}^{\rho} d_i \bar{\eta}_{r-\rho+i} (\bar{\mathfrak{p}}_0^2)$, $d_i \subset K$. Since the largest ideal in $\mathfrak{o}$ which is mapped upon $\bar{\mathfrak{p}}_0^2$ in the homomorphism $\mathfrak{o} \simeq \bar{\mathfrak{o}}$ is the ideal $(\mathfrak{p}_0^2, \mathfrak{p}_\rho)$ it follows that any element ω of $\mathfrak{o}$ satisfies a congruence of the form

$$\omega \equiv d_0 + \sum_{i=1}^{p} d_i \eta_{r-\rho+i} (\mathfrak{p}_0^2, \mathfrak{p}_\rho). \tag{30}$$

Now, in view of our hypothesis concerning the ideal $\mathfrak{o} \cdot (\eta_1, \cdots, \eta_{r-\rho})$ and the point P, it follows that $\mathfrak{p}_0$ does not divide any of the primary components of $\mathfrak{o} \cdot (\eta_1, \cdots, \eta_{r-\rho})$ distinct from $\mathfrak{p}_\rho$. Hence there exists an element α in $\mathfrak{o}$ such that $\alpha \not\equiv 0(\mathfrak{p}_0)$ and $\alpha \mathfrak{p}_\rho \equiv 0(\eta_1, \cdots, \eta_{r-\rho})$. Hence for any element ζ in $\mathfrak{p}_\rho$ we have: $\alpha\zeta = A_1\eta_1 + \cdots + A_{r-\rho}\eta_{r-\rho}$, where $A_i \subset \mathfrak{o}$. If then $\alpha \equiv c_0(\mathfrak{p}_0)$, $c_0 \neq 0$ and $A_j \equiv c_j(\mathfrak{p}_0)$, $c_0, \cdots, c_{r-\rho} \subset K$, it follows immediately, since the elements $\zeta, \eta_1, \cdots, \eta_{r-\rho}$ are in $\mathfrak{p}_0$, that a congruence of the form

$$c_0\zeta \equiv c_1\eta_1 + \cdots + c_{r-\rho}\eta_{r-\rho} (\mathfrak{p}_0^2),$$

holds true for any element ζ in $\mathfrak{p}_\rho$. From this, in view of (30), we conclude the r elements $\eta_1, \cdots, \eta_r$ form a K-module basis for $\mathfrak{p}_0/\mathfrak{p}_0^2$, which is then of rank r. Hence P is a simple point of V_r. Thus it is proved that V_ρ contains points which are simple for V_r, q. e. d.

The above lemma implies as an immediate consequence the following important property of normal varieties:

THEOREM 11. *The singular manifold of a normal variety V_r in an affine space is of dimension $\leqq r - 2$.*

Proof. It is sufficient to prove that every $(r-1)$-dimensional subvariety V_{r-1} of V_r is non-singular. Let $\mathfrak{p}_{r-1}$ be the prime ideal of V_{r-1} in $\mathfrak{o}$. Since

$\mathfrak{o}$ is integrally closed, the primary ideals belonging to the *minimal* ideal $\mathfrak{p}_{r-1}$ are the symbolic powers $\mathfrak{p}^{(i)}_{r-1}$ of $\mathfrak{p}_{r-1}$, $i = 1, 2, \cdots$.[20] Let η be an element in $\mathfrak{p}_{r-1}$ but not in $\mathfrak{p}^{(2)}_{r-1}$. Then, by the principal ideal theorem in the integrally closed ring $\mathfrak{o}$, it follows that $\mathfrak{p}_{r-1}$ is an isolated component of $\mathfrak{o} \cdot \eta$. Our theorem now follows from the above lemma in the case $\rho = r - 1$.

16. If V_r is not normal, whence the corresponding ring $\mathfrak{o} = \mathsf{K}[\xi_1, \cdots, \xi_n]$ is not integrally closed, the passage to the integral closure $\mathfrak{o}^*$ of $\mathfrak{o}$ defines a birational transformation of V_r into the normal variety V^*_r defined by $\mathfrak{o}^*$ (or into any other variety V'_r integrally equivalent to V^*_r). We shall say that V^*_r is *the derived normal variety* of V_r. For lack of a better word we shall use the term " integral closure " to denote the birational transformation which carries V_r into V^*_r. By Theorem 11, we may say that *the effect of the integral closure is the elimination of all singular manifolds of dimension* $r - 1$. Thus for algebraic curves, *the integral closure transformation resolves all the singularities of the curve* (singularities of finite distance).[21] For an algebraic surface the integral closure resolves the multiple curves (at finite distance). *A normal surface in the affine space has only a finite number of singularities* (at finite distance).

One more remark concerning the birational correspondence between V_r and V^*_r. If P is a point of V_r, $\mathfrak{p}_0$—the corresponding 0-dimensional prime ideal in $\mathfrak{o}$, then $\mathfrak{o}^*\mathfrak{p}_0$ is also 0-dimensional and unmixed. If $\mathfrak{o}^*\mathfrak{p}_0 = [\mathfrak{q}^*_1, \cdots, \mathfrak{q}^*_s]$, where each $\mathfrak{q}^*_i$ belongs to a prime 0-dimensional ideal $\mathfrak{p}^*_i$ in $\mathfrak{o}^*$, and if P^*_i is the point on V^*_r determined by $\mathfrak{p}^*_i$, then to the point P of V_r there correspond on V_r the points $P^*_1, \cdots, P^*_s$. On the other hand, every prime 0-dimensional ideal $\mathfrak{p}^*_0$ in $\mathfrak{o}^*$ determines uniquely its contracted ideal $\mathfrak{p}_0 = \mathfrak{o} \cap \mathfrak{p}^*_0$ in $\mathfrak{o}$. Hence to a point on V^*_r there corresponds a unique point on V_r. By Theorem 8, if $\mathfrak{p}_0$ does not divide the conductor of $\mathfrak{o}$ relative to $\mathfrak{o}^*$, then $s = 1$ and $\mathfrak{o}^*\mathfrak{p}_0$ is prime in $\mathfrak{o}^*$. Now assume that P is a simple point of V_r, and let $\xi_1, \cdots, \xi_r$ be uniformizing parameters for the whole neighborhood of P, where we assume that $\xi_i \equiv 0(\mathfrak{p}_0)$ and that every element of $\mathfrak{o}$ (whence also of $\mathfrak{o}^*$) is integrally dependent on $\xi_1, \cdots, \xi_r$. By Theorem 4, there exists an element ω in $\mathfrak{o}$ such that $F'_\omega(\omega; \xi_1, \cdots, \xi_r) \not\equiv 0(\mathfrak{p}_0)$, where $F(z; \xi_1, \cdots, \xi_r)$ is the norm of $z - \omega$ over $\mathsf{K}(\xi_1, \cdots, \xi_r)$. Since $\mathfrak{p}_0 = \mathfrak{p}^*_i \cap \mathfrak{o}$, it follows that $F'_\omega(\omega; \xi_1, \cdots, \xi_r) \not\equiv 0(\mathfrak{p}^*_i)$, and hence, by Theorem 5, we conclude that P^*_i *is a simple point of* V^*_r and that $\mathfrak{p}^*_i$ is an isolated component of the ideal $\mathfrak{o}^*(\xi_1, \cdots, \xi_r)$. Since $\mathfrak{o}^*\mathfrak{p}_0$ divides this last ideal, it follows

[20] See van der Waerden [7], p. 105.

[21] See Muhly and Zariski [3].

3

a fortiori that each $\mathfrak{p}^*_i$ is an isolated component of $\mathfrak{o}^*\mathfrak{p}_0$, whence $\mathfrak{q}^*_i = \mathfrak{p}^*_i$ and $\mathfrak{o}^*\mathfrak{p}_0 = [\mathfrak{p}^*_1, \cdots, \mathfrak{p}^*_s]$. Now at each point P^*_i we have a definite isomorphic mapping τ_i of $\mathfrak{o}^*$ upon a subring H_i of the ring of formal power series of $\xi_1, \cdots, \xi_r$. *These s mappings τ_i must coincide on* $\mathfrak{o}$. In fact, since $\xi_1, \cdots, \xi_r$ are uniformizing parameters for the whole neighborhood of P on V_r, we have a mapping τ of $\mathfrak{o}$ upon a subring H of $K\{\xi_1, \cdots, \xi_r\}$. Let ω be any element of $\mathfrak{o}$ and let

$$\tau\omega = \phi_0 + \phi_1(\xi_1, \cdots, \xi_r) + \phi_2(\xi_1, \cdots, \xi_r) + \cdots \subset H.$$

Then $\omega \equiv \phi_0 + \phi_1 + \cdots + \phi_m(\mathfrak{p}_0^{m+1})$, m arbitrary, and consequently also $\omega \equiv \phi_0 + \phi_1 + \cdots + \phi_m(\mathfrak{p}^*{}_0^{m+1})$. If $\tau_i\omega \rightarrow \phi_0^{(i)} + \phi_1^{(i)} + \phi_2^{(i)} + \cdots \subset H_i$, then also $\omega \equiv \phi_0^{(i)} + \phi_1^{(i)} + \cdots + \phi_m^{(i)}(\mathfrak{p}^*{}_0^{m+1})$. In view of the uniqueness of the polynomial $\phi_0^{(i)} + \cdots + \phi_m^{(i)}$, it follows $\phi_m^{(i)} = \phi_m$, and this proves our assertion $\tau = \tau_i$ on $\mathfrak{o}$, $i = 1, 2, \cdots, s$. Since $\mathfrak{o}$ and $\mathfrak{o}^*$ have the same quotient field Σ, it follows that these mappings also coincide on $\mathfrak{o}^*$, *whence necessarily* $s = 1$.

Reassuming, we have the following

THEOREM 12. *If V^*_r is the derived normal variety of V_r (in the affine space), then to each point of V^*_r there corresponds a unique point of V_r, while to every point P of V_r there correspond at most a finite number of points of V^*_r. This number can be greater than one only if P is a singular point of V_r and lies on the subvariety defined on V_r by the conductor of* $\mathfrak{o}$ *relative to* $\mathfrak{o}^*$.[22]

This theorem shows that the birational transformation between V_r and V^*_r is free from fundamental points on either variety.

IV. Normal varieties in the projective space.

17. A normal variety in the affine space may have singularities at infinity. Concerning these, Theorem 11 gives us no information whatever. It may very well happen that a normal V_r has a singular V_{r-1} at infinity. Hence, from a projective—and consequently also from an algebro-geometric point of view—Theorem 11 is not significant. Also the relationship between a V_r and its derived normal variety V^*_r in an affine space has no invariantive character from an algebro-geometric standpoint. Thus, for instance, the birational

[22] The preceding proof shows that every point of the variety defined by the conductor is necessarily a singular point of V_r. The converse is of course not always true. On the other hand, it should be noted that to P there may correspond a unique point of V_r^* even if P lies on the variety of the conductor. Namely, we may have $\mathfrak{o}^*\mathfrak{p}_0 = \mathfrak{q}^*$, $\mathfrak{q}^*$ primary (not prime).

correspondence between V_r and V^*_r may have fundamental loci at infinity. We shall therefore now deal with the extension of the notion of a normal variety to projective spaces.

Let $y_0, y_1, \cdots, y_n$ be homogeneous coördinates in an n-dimensional projective space P_n. The quotients $x_i = y_i/y_0$, $i = 1, 2, \cdots, n$, are point coördinates in the affine S_n consisting of those points of P_n which are not on the hyperplane $y_0 = 0$. Let V_r be an irreducible r-dimensional variety in P_n, *not in the hyperplane* $y_0 = 0$, and let $\xi_1, \cdots, \xi_n$ be the coördinates of the general point of V_r in S_n. The field $\Sigma = K(\xi_1, \cdots, \xi_n)$ is of degree of transcendency r over K. The variety V_r is defined in P_n by the H-ideal (*homogeneous ideal*) $\mathfrak{P}$ in $K[y_0, \cdots, y_n]$, generated by all forms $f(y_0, \cdots, y_n)$ such that $f(1, \xi_1, \cdots, \xi_n) = 0$.

Now suppose that we choose as hyperplane at infinity another hyperplane, say $c_0y_0 + c_1y_1 + \cdots + c_ny_n = 0$, which does not contain the variety V_r, and let S'_n be the affine space consisting of the points of P_n which are not on this hyperplane. If, say, $c_i \neq 0$, then the n quotients $y_j/(c_0y_0 + \cdots + c_ny_n)$, $j = 0, 1, \cdots, i-1, i+1, \cdots, n$, can be taken as coördinates in S'_n. The coördinates of the general point of V_r, considered in S'_n, are then

$$\xi'_0 = 1/(c_0 + c_1\xi_1 + \cdots + c_n\xi_n), \quad \xi'_j = \xi_j/(c_0 + c_1\xi_1 + \cdots + c_n\xi_n), \quad j \neq i.$$

The ring of polynomials in the ξ'_j contains also the quotient

$$\xi'_i = \xi_i/(c_0 + c_1\xi_1 + \cdots + c_n\xi_n),$$

since $c_0\xi'_0 + \cdots + c_n\xi'_n = 1$ and $c_i \neq 0$. Thus, inasmuch as we regard V_r as a variety in the affine space S'_n, it determines in Σ the integral domain

$$(31) \quad \mathfrak{o}' = K[1/(c_0 + c_1\xi_1 + \cdots + c_n\xi_n), \\ \xi_1/(c_0 + c_1\xi_1 + \cdots + c_n\xi_n), \cdots, \xi_n/(c_0 + c_1\xi_1 + \cdots + c_n\xi_n)].$$

We go back to the homogeneous coördinates. If we regard $y_0, \cdots, y_n$ as non-homogeneous coördinates in an affine S_{n+1}, then the homogeneous ideal $\mathfrak{P}$ defines an $(r+1)$-dimensional irreducible variety W_{r+1} in S_{n+1}. The coördinates $\xi^*_0, \xi^*_1, \cdots, \xi^*_n$ of the general point of W_{r+1} are the residual classes in $K[y]/\mathfrak{P}$ containing $y_0, y_1, \cdots, y_n$ respectively. If we imagine P_n as being the hyperplane at infinity of S_{n+1}, then W_{r+1} is the hypercone which projects V_r from the origin $y_0 = \cdots = y_n = 0$.

Let $\Sigma^* = K(\xi^*_0, \xi^*_1, \cdots, \xi^*_n)$ be the field of rational functions on W_{r+1}. Its subfield $K(\xi^*_1/\xi^*_0, \cdots, \xi^*_n/\xi^*_0)$ is simply isomorphic with the field $\Sigma = K(\xi_1, \cdots, \xi_n)$, and in this isomorphism to ξ^*_i/ξ^*_0 there corresponds the element ξ_i. This follows immediately from the relationship between the homogeneous ideal $\mathfrak{P}$ in $K[y]$ and the defining ideal of V_r in $K[x]$. If then we identify ξ^*_i/ξ^*_0 with ξ_i, we may regard Σ as a subfield of Σ^*. The degree of transcendency of Σ^* is one greater than that of Σ. Since $\xi^*_i = \xi_i \xi^*_0$ (in view of our identification), we have $\Sigma^* = \Sigma(\xi^*_0)$. Hence Σ^* is a *simple transcendental extension* of Σ.[23] Thus we have an invariantive relationship between Σ and Σ^*, independent on the particular projective model V_r of the field Σ. We note the existence of the group of relative automorphisms τ of Σ^* with respect to Σ; these are all of the form:

$$(32) \qquad \tau: \quad \xi_i \to \xi_i; \; \xi^*_0 \to \frac{\alpha\xi^*_0 + \beta}{\gamma\xi^*_0 + \delta}, \quad \begin{array}{l} \alpha, \beta, \gamma, \delta \subset \Sigma, \\ \alpha\delta - \beta\gamma \neq 0. \end{array}$$

Such a relative automorphism is described by a birational transformation of the hypercone W_{r+1} into itself which leaves invariant each generator.

Consider the special automorphisms τ_t: $\xi_i \to \xi_i$; $\xi^*_0 \to t\xi^*_0$, $t \subset \mathsf{K}$. We shall say that an element ω^* in Σ^* is *homogeneous of degree* ν, if $\tau_t(\omega^*) = t^\nu \omega^*$, t arbitrary in K. *Any homogeneous element* ω^* *of degree* ν *belonging to the ring* $\mathfrak{o}^* = \mathsf{K}[\xi^*_0, \cdots, \xi^*_n]$ *is a form of degree* ν *in* $\xi^*_0, \cdots, \xi^*_n$. In fact, let $\omega^* = f_\rho(\xi^*) + f_{\rho+1}(\xi^*) + \cdots + f_\sigma(\xi^*)$, where $f_i(\xi^*)$ is a form of degree i in $\xi^*_0, \cdots, \xi^*_n$ and $f_\rho \neq 0$. We must have:

$$\begin{aligned} \tau_t(\omega^*) &= t^\rho f_\rho(\xi^*) + t^{\rho+1} f_{\rho+1}(\xi^*) + \cdots + t^\sigma f_\sigma(\xi^*) \\ &= t^\nu f_\rho(\xi^*) + t^\nu f_{\rho+1}(\xi^*) + \cdots + t^\nu f_\sigma(\xi^*), \quad t\text{—arbitrary.} \end{aligned}$$

This implies $\rho = \nu$, $f_{\rho+1} = \cdots = f_\sigma = 0$.

The elements of Σ are homogeneous of degree 0. Conversely, *any element of* Σ^* *which is homogeneous of degree zero, is an element of* Σ. Namely, it is clear that the elements of $\Sigma^* = \Sigma(\xi^*_0)$ which are left invariant by all the relative automorphisms $\xi^*_0 \to t\xi^*_0$ are necessarily elements of Σ.

In terms of the field Σ^* we are now in position to define quite generally the notion of the *homogeneous coördinates of the general point of an algebraic variety* V_r *in the projective space* P_n:

Any set of $n+1$ elements in Σ^*, say $\zeta^*_0, \zeta^*_1, \cdots, \zeta^*_n$, are homogeneous coördinates of the general point of V_r in P_n if:

(a) $\zeta^*_i / \zeta^*_0 = \xi_i$;

(b) The field $\mathsf{K}(\zeta^*_0, \zeta^*_1, \cdots, \zeta^*_n)$ is a transcendental extension of Σ.

The condition (b) implies that each element ζ^*_i is transcendental with respect to Σ. A particular set of homogeneous coördinates is given by the elements $\xi^*_0, \xi^*_1, \cdots, \xi^*_n$. The most general set of homogeneous coördinates $\zeta^*_0, \zeta^*_1, \cdots, \zeta^*_n$ (relative to our fixed coördinate system $y_0, \cdots, y_n$ in P_n) is obtained by multiplying the coördinates ξ^*_i by any element σ^* in Σ^*, pro-

[23] That $\xi_0{}^*$ is transcendental with respect to Σ $(= \mathsf{K}(\xi_1{}^*/\xi_0{}^*, \cdots, \xi_n{}^*/\xi_0{}^*)$, follows also from the fact that all the algebraic relations between $\xi_0{}^*, \xi_1{}^*, \cdots, \xi_n{}^*$ are homogeneous (or consequences of homogeneous relations).

vided, however, that $\sigma^*\xi^*_i$ be transcendental over Σ. Thus, we may also say that any set $\zeta^*_0, \cdots, \zeta^*_n$ of homogeneous coördinates of the general point of V_r is of the form: $\rho^*, \rho^*\xi_1, \cdots, \rho^*\xi_n$, ρ^* an element of Σ^* which is transcendental over Σ. An immediate consequence of this is the following: *the correspondence* $\xi^*_i \rightarrow \zeta^*_i$ *sets up an isomorphism between the two rings* $K[\xi^*]$ and $K[\zeta^*]$.

18. Definition. *Let* $\xi^*_0, \xi^*_1, \cdots, \xi^*_n$ *(or* $\sigma^*\xi^*_0, \cdots, \sigma^*\xi^*_n, \sigma^* \subset \Sigma^*$*) be homogeneous coördinates of the general point of a* V_r *in* P_n. *The variety* V_r *shall be said to be normal (in* P_n*), if the ring* $\mathfrak{o}^* = K[\xi^*_0, \cdots, \xi^*_n]$ *(or the ring* $K[\sigma^*\xi^*_0, \cdots, \sigma^*\xi^*_n]$*) is integrally closed in its quotient field.*

It is clear that the above defining property of a normal variety V_r is independent of the choice of the factor of proportionality σ^*, in view of the remark at the end of the preceding section.

When we speak in the sequel of a normal variety, it will be understood that the variety is normal in the projective space.

Let us take any given hyperplane $c_0y_0 + \cdots + c_ny_n = 0$ in P_n as hyperplane at infinity, and let S'_n be the corresponding affine space. We assert that *if* V_r *is normal, then it is also normal in the affine space* S'_n (provided, of course, that V_r does not lie entirely in the preassigned hyperplane at infinity). We have to show that the ring (31), or what is the same, that the ring

$$\mathfrak{o}' = K[\xi^*_0/\sum_{i=0}^{n} c_i\xi^*_i, \xi^*_1/\sum_{i=0}^{n} c_i\xi^*_i, \cdots, \xi^*_n/\sum_{i=0}^{n} c_i\xi^*_i]$$

is integrally closed in Σ. Let $\sum_{i=0}^{n} c_i\xi^*_i = \eta^*$, and let ω be an element in Σ which depends integrally on $\mathfrak{o}'$. Then it is clear that there exists an integer ρ such that $\omega \cdot (\eta^*)^\rho$ depends integrally on $\mathfrak{o}^* = K[\xi^*_0, \cdots, \xi^*_n]$. Hence $\omega \cdot (\eta^*)^\rho \subset \mathfrak{o}^*$,

$$\omega \cdot (\eta^*)^\rho = f(\xi^*_0, \cdots, \xi^*_n),$$

where f is a polynomial. Now $\omega \cdot (\eta^*)^\rho$ is an homogeneous element of degree ρ. Hence f is necessarily a form of degree ρ. It follows that $\omega = f/(\eta^*)^\rho$ is a polynomial in $\xi^*_0/\eta^*, \cdots, \xi^*_n/\eta^*$. Hence $\omega \subset \mathfrak{o}'$, and this proves our assertion.

Theorem 11′. *The manifold of singular points of a normal variety* V_r *is of dimension* $\leqq r - 2$.

This follows from Theorem 11 and from the fact that our V_r has just been proved to be normal in the affine space S'_n, *for any choice of the hyperplane at infinity.*

In this connection we wish to point out that *if, conversely, a* V_r *has the property that it is normal in the affine space* S'_n, *for every choice of the hyper-*

plane at infinity, then it does not yet necessarily follow that V_r is normal in the projective space P_n. We prove namely the following

THEOREM 13. *In order that a V_r in P_n be normal in the affine sense, for every choice of the hyperplane at infinity in P_n, it is necessary and sufficient that the conductor of the ring $\mathfrak{o}^* = K[\xi^*_0, \cdots, \xi^*_n]$ with respect to the integral closure $\bar{\mathfrak{o}}^*$ of $\mathfrak{o}^*$ (in the quotient field of $\mathfrak{o}^*$) divide a power of the 0-dimensional prime ideal $\mathfrak{p}^*_0 = (\xi^*_0, \cdots, \xi^*_n)$ (the vertex of the hypercone W_{r+1}).*

If $\mathfrak{o}^*$ is not itself integrally closed, the conductor must be then a primary ideal belonging to $\mathfrak{p}^*_0$.

Proof. The condition is sufficient. For let the conductor $\mathfrak{c}(\mathfrak{o}^*, \bar{\mathfrak{o}}^*)$ be a primary ideal $\mathfrak{q}^*$ belonging to $\mathfrak{p}^*_0$ and let σ be the exponent of $\mathfrak{q}^*$. If η^* is any linear form in $\xi^*_0, \cdots, \xi^*_n$ and if, as before, ω is any element in Σ which is integrally dependent on $\mathfrak{o}' = K[\xi^*_0/\eta^*, \cdots, \xi^*_n/\eta^*]$, then, for some integer ρ, the product $\omega \cdot (\eta^*)^\rho$ is in $\bar{\mathfrak{o}}^*$. Hence $\omega \cdot (\eta^*)^{\rho+\sigma} \subset \mathfrak{o}^*$, since $(\eta^*)^\sigma$ is in $\mathfrak{c}(\mathfrak{o}^*, \bar{\mathfrak{o}}^*)$. We conclude, as before, that $\omega \cdot (\eta^*)^{\rho+\sigma}$ is a form of degree $\rho + \sigma$ in $\xi^*_0, \cdots, \xi^*_n$ and that consequently $\omega \subset \mathfrak{o}'$, i. e. $\mathfrak{o}'$ is integrally closed in Σ.

The condition is necessary. Suppose that $\mathfrak{o}'$ is integrally closed in Σ. Let us consider any *homogeneous element* ω^* of $\bar{\mathfrak{o}}^*$. Let ω^* be homogeneous of degree ν, and let

$$(33) \quad \omega^{*m} + a_1(\xi^*_0, \cdots, \xi^*_n)\omega^{*m-1} + \cdots + a_m(\xi^*_0, \cdots, \xi^*_n) = 0$$

be an equation of smallest degree for ω^* over $\mathfrak{o}^*$, with leading coefficient 1. The equation remains true if we apply any automorphism: $\xi^*_i \rightarrow t\xi^*_i$, $\omega^* \rightarrow t^\nu\omega^*$. The resulting equation must be an identity in t. Equating to zero the coefficient of $t^{\nu m}$, we get an equation similar to (33) but homogeneous of degree νm. Hence we may assume that $a_1, \cdots, a_m$ are homogeneous of degrees $\nu, 2\nu, \cdots, \nu m$ respectively, i. e. a_i is a form of degree $i\nu$ in $\xi^*_0, \cdots, \xi^*_n$.

From (33) it now follows that the element $\omega^*/(\eta^*)^\nu$, which is homogeneous of degree zero and is therefore an element of Σ, is integrally dependent on $\mathfrak{o}' = K[\xi^*_0/\eta^*, \cdots, \xi^*_n/\eta^*]$. Hence $\omega^*/(\eta^*)^\nu \subset \mathfrak{o}'$, i. e.

$$\omega^*/(\eta^*)^\nu = f(\xi^*_0/\eta^*, \cdots, \xi^*_n/\eta^*), \ f \text{ a polynomial.}$$

Clearing the denominator, we find

$$\omega^* \cdot (\eta^*)^\rho = g(\xi^*_0, \cdots, \xi^*_n), \qquad \rho \geqq 0,$$

where g is a form of degree $\rho + \nu$. Thus, given any linear form η^* in $\xi^*_0, \cdots, \xi^*_n$, there exists an integer ρ (perhaps depending on η^*) such that $\omega^* \cdot (\eta^*)^\rho \subset \mathfrak{o}^*$. In particular, let $\omega^* \cdot (\xi^*_i)^{\rho_i} \subset \mathfrak{o}^*$ and let $\rho' = 1 + \sum_{i=0}^{n} (\rho_i - 1)$.

Then it is clear that $\omega^* \cdot (\eta^*)^{\rho'} \subset \mathfrak{o}^*$ for any linear form η^* in $\xi^*_0, \cdots, \xi^*_n$. Hence $\omega^* \cdot \mathfrak{p}^*_0{}^{\rho'} \subset \mathfrak{o}^*$.

Now it is not difficult to see that every element in $\bar{\mathfrak{o}}^*$ is a sum of homogeneous elements of $\bar{\mathfrak{o}}^*$.[24] Since $\bar{\mathfrak{o}}^*$ possesses a finite $\mathfrak{o}^*$-basis, it follows that there also exists an $\mathfrak{o}^*$-basis in $\bar{\mathfrak{o}}^*$ consisting of homogeneous elements, say $\omega^*_1, \cdots, \omega^*_q$. We have just proved that for each ω^*_i we can find an integer σ_i such that $\omega^*_i \cdot \mathfrak{p}^*_0{}^{\sigma_i} \subset \mathfrak{o}^*$. If we put $\sigma = \max(\sigma_1, \cdots, \sigma_q)$, then we will have $\bar{\mathfrak{o}}^* \cdot \mathfrak{p}^*_0{}^{\sigma} \subset \mathfrak{o}^*$. Hence $\mathfrak{c}(\mathfrak{o}^*, \bar{\mathfrak{o}}^*) \supset \mathfrak{p}^*_0{}^{\sigma}$, and this proves our theorem.

We shall have occasion to point out later (see footnote 26) that there actually exist varieties which satisfy the condition of the last theorem and which yet are not normal.

19. There is an important connection between the notion of a normal variety V_r and the concept of a complete linear system of V_{r-1}'s on V_r. We begin with some simple remarks concerning *the order of* V_r.

Let $\xi^*_0, \xi^*_1, \cdots, \xi^*_n$ be homogeneous coördinates of the general point of V_r and let $\xi_i = \xi^*_i/\xi^*_0$ be the non-homogeneous coördinates. The order ν of V_r is ordinarily defined as the number of distinct intersections of V_r with a *general* $(n - r)$-dimensional subspace of P_n. Let

$$\eta_i = u_{i0} + u_{i1}\xi_1 + \cdots + u_{in}\xi_n, \qquad (i = 1, 2, \cdots, r),$$

where the u_{ij}'s are indeterminates. Clearly, the above definition of the order ν of V_r is equivalent with the following: ν is the relative degree of $\Sigma(u_{ij})$ with respect to the field $\mathsf{K}(\eta_1, \cdots, \eta_r; u_{ij})$ (see [7], p. 82). Now, with respect to this last field, there always exists in $\Sigma(u_{ij})$ a primitive element of the form $\eta_{r+1} = u_{r+1,0} + u_{r+1,1}\xi_1 + \cdots + u_{r+1,n}\xi_n$, where the coefficients $u_{r+1,i}$ are "non-special" constants in K. It follows that if all the u_{ij}'s are indeterminates, the irreducible equation $G(\eta_1, \cdots, \eta_{r+1}; u_{ij})$ between $\eta_1, \cdots, \eta_r, \eta_{r+1}$ is of degree ν in η_{r+1}. For reasons of symmetry it also must be of degree ν in each of the variables $\eta_1, \cdots, \eta_r$. Finally, since it is permissible to operate on $\eta_1, \cdots, \eta_r, \eta_{r+1}$ by a non-singular linear transformation, it follows that G *must be of degree* ν *in all the arguments* $\eta_1, \cdots, \eta_{r+1}$.

This is so as long as the u_{ij}'s are indeterminates. Now we specialize $u_{ij} \to u_{ij}^0 \subset \mathsf{K}$, $\eta_i \to \eta_i^0$. If the polynomial $G(\eta_1^0, \cdots, \eta_{r+1}^0; u_{ij}^0)$ in $\mathsf{K}[\eta_1^0, \cdots, \eta_{r+1}^0]$ does not vanish identically, then we get an algebraic rela-

[24] Since $\mathfrak{o}^* = \mathsf{K}[\xi_0^*, \xi_1^*, \cdots, \xi_n^*] = \mathsf{K}[\xi_0^*, \xi_0^*\xi_1, \cdots, \xi_0^*\xi_n] \subset \Sigma[\xi_0^*]$ and since $\Sigma[\xi_0^*]$ is integrally closed in $\Sigma^* = \Sigma(\xi_0^*)$, it follows that any element in $\bar{\mathfrak{o}}^*$ is a polynomial in ξ_0^* with coefficients in Σ. Let now ω^* be any element in $\bar{\mathfrak{o}}^*$, $\omega^* = \alpha_0 + \alpha_1\xi_0^* + \cdots, \alpha_s\xi_0^{*s}$, $\alpha_i \subset \Sigma$, and let $t_0 t_1, \cdots, t_s$ be $s + 1$ distinct constants. The automorphisms τ_{t_i}: $\xi_j^* \to t_i\xi_j^*$ leave $\mathfrak{o}^*$ invariant. Hence the $s + 1$ elements $\omega_i^* = \tau_{t_i}\omega^* = \alpha_0 + t_i\alpha_1\xi_0^* + \cdots t^s\alpha_s\xi_0^{*s}$ also belong to $\bar{\mathfrak{o}}^*$. From this it follows that the homogeneous components $\alpha_0, \alpha_1\xi_0^*, \cdots, \alpha_s\xi_0^{*s}$ of ω^* are in $\bar{\mathfrak{o}}^*$.

tion between $\eta_1{}^0, \cdots, \eta_{r+1}{}^0$ which is of degree $\leqq \nu$. In the contrary case we still get an equation of degree $\leqq \nu$ between $\eta_1{}^0, \cdots, \eta_{r+1}{}^0$, provided that we specialize the u_{ij} one at a time and divide, when possible, by factors $u_{ij} - u_{ij}{}^0$. Consequently, *any* $r + 1$ *linear polynomials* $\eta_1{}^0, \cdots, \eta_{r+1}{}^0$ *in the* ξ_i *satisfy an algebraic relation of degree* $\leqq \nu$. In particular, if r *of the elements* $\eta_j{}^0$ *are algebraically independent, then the irreducible algebraic relation between* $\eta_1{}^0, \cdots, \eta_{r+1}{}^0$ *is of degree* $\leqq \nu$. It is equal to ν, when the coefficients $u_{ij}{}^0$ are not special.

Let V'_r be another algebraic variety, birationally equivalent to V_r and lying in a projective n'-space $P_{n'}$, $n' > n$. Let $\xi^{*\prime}_0, \cdots, \xi^{*\prime}_{n'}$ be the homogeneous coördinates of the general point of V'_r. Here the $\xi^{*\prime}_j$ as well as the ξ^*_i are elements of the field Σ^*, a simple transcendental extension of Σ. The following is self-evident: V_r *is a projection of* V'_r *if* $\xi^*_0, \cdots, \xi^*_n$ *are proportional to linear forms in* $\xi^{*\prime}_0, \cdots, \xi^{*\prime}_{n'}$ *with coefficients in* K. Assuming, as it is permissible, that the ξ^*_i are linearly independent, we may so choose the coördinate system in $P_{n'}$ that $\xi^*_0, \cdots, \xi^*_n$ be proportional to $\xi^{*\prime}_0, \cdots, \xi^{*\prime}_n$.

We now define: The system of hyperplane sections of a V_r in P_n is said to be *complete*, if V_r is not the projection of a V'_r *of the same order* as V_r and *belonging* to a space $P_{n'}$ of dimension n' greater than n.[25]

When we say that V'_r *belongs* to $P_{n'}$ we mean that V'_r does not lie in any subspace of $P_{n'}$. Similarly we suppose that V_r belongs to P_n.

We now prove the following

THEOREM 14. *The system of hyperplane sections of a normal* V_r *is complete.*[26]

---

[25] The usual procedure is to give directly the general definition of a complete linear system of V_{r-1}'s on V_r. The property on which our definition is based becomes then a consequence of the definition as applied to the special case when the system of V_{r-1}'s is the system of hyperplane sections. We reverse the procedure. Dealing with an arbitrary linear system $|V_{r-1}|$ of V_{r-1}'s on a V_r, we would first transform our V_r into a V'_r on which the system $|V_{r-1}|$ is cut out by the hyperplanes, and then our definition is applicable, *provided* V_r *and* V'_r *are birationally equivalent.* The case in which the correspondence between V_r and V'_r is $(\alpha, 1)$, $\alpha > 1$, arises when the system $|V_{r-1}|$ on V_r is composite with an involution of degree α. This case would then require a separate treatment.

[26] This explains our term "normal." In the algebro-geometric literature a variety is called normal if its system of hyperplane sections is complete. However, it should be pointed out that while a variety, normal in our (arithmetic) sense, is also normal in the above geometric sense, the converse is not true. For instance, a curve may be normal in the geometric sense and still have singularities (example: a plane quartic of genus 2). Such a curve cannot be normal in the arithmetic sense, in view of Theorem 11′, section 18.

On the other hand, a curve may be free from singularities in the projective space and not be normal in the geometric sense—for instance—a rational space quartic. Such

Proof. Assume that V_r, of order ν and belonging to P_n, is the projection of birationally equivalent V'_r of the same order ν, lying in a $P_{n'}$, $n' > n$. Let $\xi^*_0, \cdots, \xi^*_n$ and $\xi^{*\prime}_0, \cdots, \xi^{*\prime}_{n'}$ be the homogeneous coördinates of the general point of V_r and of V'_r respectively. Then $\xi^*_0, \cdots, \xi^*_n$ are proportional to $\xi^{*\prime}_0, \cdots, \xi^{*\prime}_n$. Let $\xi^*_i/\xi^{*\prime}_i = \rho^* \subset \Sigma^*$, $i = 0, 1, \cdots, n$. The $n' + 1$ elements $\xi^*_0, \xi^*_1, \cdots, \xi^*_n, \rho^*\xi^{*\prime}_{n+1}, \cdots, \rho^*\xi^{*\prime}_{n'}$ are proportional to the coördinates $\xi^{*\prime}_j$, and moreover generate the field Σ^*, since $\Sigma^* = \mathsf{K}(\xi^*_0, \cdots, \xi^*_n)$. Hence these $n' + 1$ elements can also be taken as homogeneous coördinates of the general point of V'_r. We may therefore assume that $\xi^*_i = \xi^{*\prime}_i$, $i = 0, 1, \cdots, n$. We drop the primes in the remaining $n' - n$ coördinates, so that now the coördinates of the general point of V'_r are $\xi^*_0, \cdots, \xi^*_n, \xi^*_{n+1}, \cdots, \xi^*_{n'}$.

To prove our theorem we have to show that V'_r actually belongs to P_n, i. e. *that* $\xi^*_{n+1}, \cdots, \xi^*_{n'}$ *are linearly dependent on* $\xi^*_0, \cdots, \xi^*_n$. Let $\xi_i = \xi^*_i/\xi^*_0$. Subject to a preliminary linear transformation on $\xi_1, \cdots, \xi_n$ we may assume that $\xi_1, \cdots, \xi_r$ are algebraically independent and that the relative degree $[\Sigma\colon \mathsf{K}(\xi_1, \cdots, \xi_r)]$ is equal to ν. Let us now consider one of the elements $\xi_{n+1}, \cdots, \xi_{n'}$, say the element ξ_{n+1}. We can find constants $c_1, \cdots, c_n, c_{n+1}$; $c_{n+1} \neq 0$, such that $\bar{\xi}_{n+1} = c_1\xi_1 + \cdots + c_{n+1}\xi_{n+1}$ is a primitive element of Σ with respect to the field $\mathsf{K}(\xi_1, \cdots, \xi_r)$. Let then

$$A_0(\xi_1, \cdots, \xi_r)\bar{\xi}^{\nu}_{n+1} + A_1(\xi_1, \cdots, \xi_r)\bar{\xi}^{\nu-1}_{n+1} + \cdots + A_\nu(\xi_1, \cdots, \xi_r) = 0$$

be the irreducible equation for $\bar{\xi}_{n+1}$ over $\mathsf{K}(\xi_1, \cdots, \xi_r)$. Since V'_r is also of order ν, the above equation must be of degree ν in all the arguments $\xi_1, \cdots, \xi_r, \bar{\xi}_{n+1}$. Hence A_0 is a constant, and therefore $\bar{\xi}_{n+1}$ is integrally dependent on $\mathsf{K}[\xi_1, \cdots, \xi_r]$. Since V_r is normal, the ring $\mathsf{K}[\xi_1, \cdots, \xi_n]$ is integrally closed, and consequently $\bar{\xi}_{n+1} \subset \mathsf{K}[\xi_1, \cdots, \xi_n]$. Since

$$\bar{\xi}_{n+1} = c_1\xi_1 + \cdots c_n\xi_n + c_{n+1}\xi_{n+1}, \qquad c_{n+1} \neq 0,$$

it follows that ξ_{n+1} is in $\mathsf{K}[\xi_1, \cdots, \xi_n]$. Passing to the homogeneous coördinates we conclude that there exists an integer h_0 such that

$$\xi^*_{n+1}(\xi^*_0)^{h_0} \subset \mathsf{K}[\xi^*_0, \cdots, \xi^*_n].$$

In a similar fashion we show the existence of an integer h_i such that

$$\xi^*_{n+1} \cdot \xi^{*h_i}_i \subset \mathsf{K}[\xi^*_0, \cdots, \xi^*_n], \qquad (i = 0, 1, \cdots, n).$$

If then $h = \max(h_0, h_1, \cdots, h_n)$, then

$$\xi^*_{n+1} \cdot \xi^{*h}_i \subset \mathsf{K}[\xi^*_0, \cdots, \xi^*_n], \qquad (i = 0, 1, \cdots, n).$$

---

a curve is normal in the affine space, for every choice of the hyperplane at infinity (see footnote 22), but since it is not normal in the geometric sense, it is *a fortiori* not normal in the arithmetic sense.

It follows that if we put $\lambda = (h-1)(n+1)+1$, and if η^* is an arbitrary form in $\xi^*_0, \cdots, \xi^*_n$, of degree λ, then $\xi^*_{n+1}\eta^* \subset K[\xi^*_0, \cdots, \xi^*_n]$. Since $\xi^*_{n+1}\eta^*$ is homogeneous of degree $\lambda+1$ (with respect to the automorphism $\xi^*_i \to t\xi_i$), it follows that $\xi^*_{n+1}\eta^*$ is a form in $\xi^*_0, \cdots, \xi^*_n$, of degree $\lambda+1$. Let $\omega_1, \cdots, \omega_N$ be the various power products of $\xi^*_0, \cdots, \xi^*_n$, of degree λ. If we apply the above result to $\eta^* = \omega_i$, we find:

$$\xi^*_{n+1}\omega_i = h_{i1}\omega_1 + \cdots + h_{iN}\omega_N, \qquad (i = 1, 2, \cdots, N),$$

where the h_{ij} are linear forms in $\xi^*_0, \cdots, \xi^*_n$. Hence $| h_{ij} - \delta_{ij}\xi^*_{n+1} | = 0$, where $\delta_{ij} = 0$ or 1 according as $i \neq j$ or $i = j$. It follows that ξ^*_{n+1} is integrally dependent on $K[\xi^*_0, \cdots, \xi^*_n]$, whence $\xi^*_{n+1} \subset K[\xi^*_0, \cdots, \xi^*_n]$, since V_r is normal. Now ξ^*_{n+1} is homogeneous of degree 1, *whence* ξ^*_{n+1} *is necessarily a linear form in* $\xi^*_0, \cdots, \xi^*_n$. In a similar fashion it follows that $\xi^*_{n+2}, \cdots, \xi^*_{n'}$ are linear forms in $\xi^*_0, \cdots, \xi^*_n$, and this completes the proof of our theorem.

20. We now proceed to establish the *existence* of normal varieties in any given class of birationally equivalent varieties. Specifically we shall show that for any given V_r in P_n it is possible to define a class of *derived normal varieties*—an extension of the analogous notion in affine spaces (see section 16).

Let V_r be an irreducible variety in P_n, and let $\Sigma = K(\xi_1, \cdots, \xi_n)$, $\Sigma^* = K(\xi^*_0, \xi^*_1, \cdots, \xi^*_n)$, where the ξ^*_i are the homogeneous coördinates of the general point of V_r, and $\xi_i = \xi^*_i/\xi^*_0$. Let $\mathfrak{o}^* = K[\xi^*_0, \cdots, \xi^*_n]$ and let $\bar{\mathfrak{o}}^*$ be the integral closure of $\mathfrak{o}^*$ in Σ^*. $\bar{\mathfrak{o}}^*$ is a finite integral domain, say

$$\bar{\mathfrak{o}}^* = K[\zeta^*_1, \cdots, \zeta^*_h].$$

Every element in $\bar{\mathfrak{o}}^*$ is a sum of homogeneous elements also belonging to $\bar{\mathfrak{o}}^*$, and an homogeneous element of $\bar{\mathfrak{o}}^*$ which is not a constant is necessarily of positive degree. Hence we may assume that each ζ^*_j is homogeneous and of degree $\delta_j > 0$. The integers $\delta_1, \cdots, \delta_h$ are not necessarily distinct; let $d_1, d_2, \cdots, d_q$ be the distinct integers among them and let d be their l. c. m. We consider in $\bar{\mathfrak{o}}^*$ the homogeneous elements whose degree is a given multiple of d, say σd. Every such element is a sum of power products of the ζ^*_j, each power product being an homogeneous element of $\bar{\mathfrak{o}}^*$, of degree σd. If in a given power product of the ζ^*_j there are g_1 factors of degree of homogeneity d_1, g_2 factors—of degree of homogeneity d_2, etc., then we must have

$$g_1d_1 + g_2d_2 + \cdots + g_qd_q = \sigma d. \tag{34}$$

Thus the determination of the homogeneous elements of $\bar{\mathfrak{o}}^*$ of degree σd depends on finding all the non-negative solutions $g_1, \cdots, g_q$ of the above diophantine equation. Now suppose that $\sigma \geqq q$. Then, for at least one of the integers $g_1, \cdots, g_q$ we must have $g_i \geqq d/d_i$. If, say, $g_1 \geqq d/d_1$, then $(g_1 - d/d_1, g_2, \cdots, g_q)$ is a solution in non-negative integers of the equation

(34) with σ replaced by $\sigma - 1$, and the system $(g_1, g_2, \cdots, g_q)$ is the sum of the two systems

$$(g_1 - d/d_1, g_2, \cdots, g_q), \qquad (d/d_1, 0, \cdots, 0),$$

of which the second is a non-negative solution of the equation (34) for $\sigma = 1$. By repeated application of this reduction we conclude that if $\sigma \geqq q$, then every solution of (34) in non-negative integers g_j can be expressed as the sum of a non-negative solution of the equation

$$g_1 d_1 + \cdots + g_q d_q = (q-1)d$$

and of $\sigma - q + 1$ non-negative solutions of the equation $g_1 d_1 + \cdots g_q d_q = d$. If we now put $\delta = m \cdot (q-1)d$, m—an integer $\geqq 1$, we can assert that *every non-negative solution of the diophantine equation*

$$g_1 d_1 + \cdots + g_q d_q = \rho\delta, \ \rho\text{—an arbitrary positive integer,}$$

is the sum of ρ *non-negative solutions of the diophantine equation*

$$g_1 d_1 + \cdots + g_q d_q = \delta.$$

This property is shared by any integer δ which is a multiple of $(q-1)d$. Actually it is very likely that d itself enjoys this property, but we have no proof for this conjecture.

Now consider any integer δ with the above property. Let $\omega^*_0, \omega^*_1, \cdots, \omega^*_m$, denote all the possible power products of $\zeta^*_1, \cdots, \zeta^*_h$, whose degree of homogeneity in $\bar{\mathfrak{o}}^*$ is equal to δ. By our choice of δ it follows that any power product of $\zeta^*_1, \cdots, \zeta^*_h$ whose degree of homogeneity in $\bar{\mathfrak{o}}^*$ is a multiple $\rho\delta$ of δ is necessarily a power product of $\omega^*_0, \cdots, \omega^*_m$, of degree ρ. Hence, *every element in* $\bar{\mathfrak{o}}^*$, *homogeneous of degree* $\rho\delta$, ρ*—an arbitrary positive integer, can be expressed as a form of degree* ρ *in* $\omega^*_0, \cdots, \omega^*_m$.

We shall call *character of homogeneity* of our V_r in P_n any integer δ which enjoys this last mentioned property. Any multiple of $(q-1)d$ is certainly a character of homogeneity of V_r.

Let δ be a character of homogeneity of V_r and let, as before, $\omega^*_0, \omega^*_1, \cdots, \omega^*_m$ be all the possible power products of $\zeta^*_1, \cdots, \zeta^*_h$, which are homogeneous elements of $\bar{\mathfrak{o}}^*$, of degree δ. The elements ω^*_i can be regarded as the homogeneous coördinates of the general point of an algebraic irreducible variety V'_r in a projective space $P_m(y'_0, y'_1, \cdots, y'_m)$. *The variety* V'_r *is birationally equivalent to* V_r. Namely, the quotients ω^*_i/ω^*_0, $i = 1, 2, \cdots, m$, are homogeneous elements of Σ^*, of degree zero, and hence are elements of Σ. Consequently $K(\omega^*_1/\omega^*_0, \cdots, \omega^*_m/\omega^*_0) \subseteqq \Sigma$. On the other hand, we have

$$\xi_i = \xi^*_i/\xi^*_0 = \xi^*_i \xi^{*\,\delta-1}_0/\xi^{*\,\delta}_0.$$

The elements $\xi^*_i \xi^{*\,\delta-1}_0$ and $\xi^{*\,\delta}_0$ are homogeneous of degree δ, and consequently can be expressed as linear forms in $\omega^*_0, \omega^*_1, \cdots, \omega^*_m$. It follows that

$$(35) \qquad \xi_i = \frac{c_0{}^{(i)} + c_1{}^{(i)}\omega^*{}_1/\omega^*{}_0 + \cdots + c_m{}^{(i)}\omega^*{}_m/\omega^*{}_0}{c_0{}^{(0)} + c_1{}^{(0)}\omega^*{}_1/\omega^*{}_0 + \cdots + c_m{}^{(0)}\omega^*{}_m/\omega^*{}_0},$$

and consequently $\Sigma \subseteqq \mathsf{K}(\omega^*{}_1/\omega^*{}_0, \cdots, \omega^*{}_m/\omega^*{}_0)$. Hence the two fields $\mathsf{K}(\xi_1, \cdots, \xi_n)$ and $\mathsf{K}(\omega^*{}_1/\omega^*{}_0, \cdots, \omega^*{}_m/\omega^*{}_0)$ coincide, and this proves that V_r and V'_r are birationally equivalent. Note that the equations of the birational transformation between V_r in $P_n(y_0, \cdots, y_n)$ and V'_r in $P_m(y'_0, \cdots, y'_m)$ are, by (35), of the form

$$(36) \qquad \rho y_i = c_0{}^{(i)}y'_0 + c_1{}^{(i)}y'_1 + \cdots + c_m{}^{(i)}y'_m, \qquad (i = 0, 1, \cdots, n),$$
ρ—a factor of proportionality.

The linearity of these equations signifies that *V_r is a projection of V'_r*

We assert that the variety V'_r is normal (in its ambient projective space P_m). We have to show that the ring $\mathsf{K}[\omega^*{}_0, \cdots, \omega^*{}_m]$ is integrally closed in its quotient field. We first point out that since $\omega^*{}_0, \cdots, \omega^*{}_m$ are homogeneous elements of degree δ, every element ω^* in Σ^*, which depends integrally on $\omega^*{}_0, \cdots, \omega^*{}_m$, is a sum of homogeneous elements whose degrees are multiples of δ and which are also integrally dependent on $\omega^*{}_0, \cdots, \omega^*{}_m$. In view of the transitivity of integral dependence, the homogeneous components of ω^* are in $\bar{\mathfrak{o}}^*$. Hence they are forms in $\omega^*{}_0, \cdots, \omega^*{}_m$, and consequently $\omega^* \subset \mathsf{K}[\omega^*{}_0, \cdots, \omega^*{}_m]$, q. e. d.

21. In the construction of the normal variety V'_r there occur arbitrary elements, for instance the character of homogeneity δ. We thus get a whole class of normal varieties associated with V_r. Any variety of this class shall be called a *derived normal variety of V_r*. We wish to investigate the relationship between any two derived normal varieties of V_r.

A first arbitrary element in our construction is the choice of the elements $\zeta^*{}_1, \cdots, \zeta^*{}_h$, such that $\bar{\mathfrak{o}}^* = \mathsf{K}[\zeta^*{}_1, \cdots, \zeta^*{}_h]$. This choice affects the elements $\omega^*{}_0, \cdots, \omega^*{}_m$, which are the various power products of the $\zeta^*{}_j$, of degree δ. However, since the elements $\omega^*{}_i$ always form a linear base for the homogeneous elements of degree δ in $\bar{\mathfrak{o}}^*$, it is clear that *two derived normal varieties of V_r belonging to one and the same character of homogeneity of V_r are projectively equivalent.*

Let now V'_r and V''_r be two derived normal varieties of V_r belonging to two distinct characters of homogeneity δ' and δ'' respectively. Let P_μ and P_ν be their ambient projective spaces respectively. Finally, let $\omega'_0, \cdots, \omega'_\mu$ (or $\omega''_0, \cdots, \omega''_\nu$) be the homogeneous coördinates of the general point of V'_r (or V''_r). We observe that if δ is a character of homogeneity of V_r, then any multiple of δ is a character of homogeneity. We may therefore consider the derived normal variety M_r of V_r belonging to the character of homogeneity $\delta'\delta''$. Let $\omega^*{}_0, \cdots, \omega^*{}_m$ be the homogeneous coördinates of M_r, its ambient space being a P_m. The ω^* are forms of degree δ'' in $\omega'_0, \cdots, \omega'_\mu$, and every form of degree δ'' in $\omega'_0, \cdots, \omega'_\mu$ is necessarily a linear form in $\omega^*_0, \cdots, \omega^*_m$. It follows that M_r *is obtained from* V'_r *by referring projectively the hypersurfaces of order* δ'' *in* P_μ *to the hyperplanes in* P_m. (We assume that $\omega'_0, \cdots, \omega'_\mu$ are linearly independent, so that V'_r does not lie in a subspace of P_μ; we make similar hypotheses for V''_r in P_ν and M_r in P_m). In a similar manner it follows that M_r is obtained from V''_r by referring the hypersurfaces of P_ν of degree δ' to the hyperplanes of P_m. We have therefore the following

THEOREM 15. *The birational correspondence between* V'_r *and* V''_r *has the property that the linear system of sections of* V'_r *with the hypersurfaces of order* δ'' *of its ambient space is transformed into the linear system cut out on* V''_r *by the hypersurfaces of order* δ' *of its ambient space* P_ν.

This result implies, in particular, that the correspondence between V'_r and V''_r is $(1,1)$ *without exceptions.* It is free from fundamental elements on either variety. We connect up this result with the notion *of the quotient ring at a point of a* V_r.

Let $\xi^*_0, \cdots, \xi^*_n$ be the homogeneous coördinates of the general point of a V_r in P_n, and let $A(a_0, a_1, \cdots, a_n)$ be a given point of V_r. We define as the *quotient ring of the point* A, in symbols: $Q(A)$, the set of all elements $f(\xi^*)/g(\xi^*)$ in Σ (f, g—forms of like degree) such that $g(a) \neq 0$. In other words, $Q(A)$ consists of all elements of Σ which have a definite finite value at the point A. The quotient ring $Q(A)$ is independent of the choice of coördinates in P_n. In particular, if, say, $a_0 \neq 0$, and if we pass to non-homogeneous coördinates $\xi_i = \xi^*_i/\xi^*_0$, $x_i^0 = a_i/a_0$, then $Q(A)$ is the quotient ring of the prime ideal $\mathfrak{p}_0 = (\xi_1 - x_1^0, \cdots, \xi_n - x_n^0)$ in the ring $\mathfrak{o} = K[\xi_1, \cdots, \xi_n]$.

Let now W_r be another variety in a P_m, birationally equivalent to V_r, and let $\eta^*_0, \cdots, \eta^*_m$ be the homogeneous coördinates of the general point of W_r. Let

$$(37) \qquad \begin{cases} \rho\eta^*_i = f_i(\xi^*_0, \cdots, \xi^*_n), & (i = 0, 1, \cdots, m), \\ \sigma\xi^*_j = \phi_j(\eta^*_0, \cdots, \eta^*_m), & (j = 0, 1, \cdots, n), \end{cases}$$

be the equations of the birational transformation between V_r and W_r (the f_i-forms of like degree; similarly for the ϕ_j). Assume that the point A on V_r is not fundamental for the transformation. Then the quantities $f_i(a_0, a_1, \cdots, a_n)$ are not all zero, and there corresponds to A a unique point $B(b_0, b_1, \cdots, b_m)$ on W_r, where $\rho b_i = f_i(a_0, a_1, \cdots, a_n)$. If $g(\eta^*_0, \cdots, \eta^*_m)$ is a form such that $g(b_0, \cdots, b_m) \neq 0$, then it is clear that $g(f_0, \cdots, f_m)$ will be a form in $\xi^*_0, \cdots, \xi^*_n$ which is $\neq 0$ at A. It follows that $Q(B) \subseteq Q(A)$. If we also assume that B is not a fundamental point on W_r, then we may conclude likewise that $Q(A) \subseteq Q(B)$, whence $Q(A) = Q(B)$.

Finally, we point out that if $\mathfrak{o} = K[\xi_1, \cdots, \xi_n]$ is integrally closed in its quotient field Σ, and if $\mathfrak{p}$ is any prime ideal in $\mathfrak{o}$, then the quotient ring $\mathfrak{o}_{\mathfrak{p}}$ is also intgrally closed in Σ. Reassuming, we may state the following

THEOREM 16. *The quotient ring of any point P of a normal V_r is integrally closed in the field of rational functions on V_r. The birational correspondence between the points of two derived normal varieties V'_r and V''_r of one and the same V_r, is $(1,1)$ without exceptions. The quotient rings of any two corresponding points P', P'' of V'_r and V''_r respectively, coincide.*

We conclude with a final important remark which clears up geometrically the relationship between a V_r and its derived normal varieties V'_r.

Let V'_r belong to the character of homogeneity δ and let $\omega^*_0, \cdots, \omega^*_m$ be the homogeneous coördinates of the general point of V'_r. Any form in $\xi^*_0, \cdots, \xi^*_n$ of degree δ, is necessarily a linear form in $\omega^*_0, \cdots, \omega^*_m$. Hence in the birational correspondence between V_r and V'_r, to the sections of V_r with the hypersurfaces of order δ, there correspond hyperplane sections of V'_r. In other words: if we denote by V_{r-1} the hyperplane sections of V_r, then the system of hyperplane sections of V'_r is the complete system $| \delta V_{r-1} |$. Thus we have the following

THEOREM 17. *The derived normal varieties V'_r of a given V_r are those on which the hyperplanes cut out the complete system $| \delta V_{r-1} |$, where δ is a character of homogeneity of V_r.*

THE JOHNS HOPKINS UNIVERSITY.

REFERENCES

1. Grell, H., "Zur Theorie der Ordnungen in algebraischen Zahl- und Funktionenkörpen," *Mathematische Annalen*, vol. 97 (1927).
2. Krull, W., "Idealtheorie," *Ergebnisse der Mathematik und ihrer Grenzgebiete*, IV, 3, Berlin, Springer (1935).
3. Muhly, H. T. and Zariski, O., "The resolution of singularities of an algebraic curve," *American Journal of Mathematics*, vol. 61, no. 1 (January, 1939).
4. Ore, O., "Uber den Zusammenhang zwischen den definierenden Gleichungen und der Idealtheorie in algebraischen Körpen, I," *Mathematische Annalen*, vol. 96 (1926).
5. Rückert, W., "Zum Eliminationsproblem der Potenzreihenideale," *Mathematische Annalen*, vol. 107 (1933).
6. Schmeidler, W., "Grundlagen einer Theorie der algebraischen Funktionen mehrerer Veränderlichen," *Mathematische Zeitschrift*, vol. 28 (1928).
7. van der Waerden, *Moderne Algebra*, II.
8. van der Waerden, "Zur algebraischen Geometrie III. Ueber irreduzibile algebraische Mannigfaltigkeiten," *Mathematische Annalen*, vol. 108 (1933).
9. Zariski, O.," Generalized weight properties of the resultant of $n+1$ polynomials in n indeterminates," *Transactions of the American Mathematical Society*," vol. 41, no. 2 (March, 1937).
10. Zariski, O., "Some results in the arithmetic theory of algebraic functions of several variables," *Proceedings of the National Academy of Sciences*, vol. 23, no. 7 (July, 1937).

ALGEBRAIC VARIETIES OVER GROUND FIELDS OF CHARACTERISTIC ZERO.*

By Oscar Zariski.

Introduction. In an earlier paper (see footnote [18]) we have derived a number of characteristic properties of *simple points* of an algebraic r-dimensional variety V_r. There the *ground field* K (field of coefficients, or field of constants) was assumed throughout to be *algebraically closed.* In the present paper we generalize our results to any V_r defined by a field Σ of algebraic functions over an arbitrary ground field K of characteristic zero. *We do not assume that* K *is maximally algebraic in* Σ.

Our generalization has an immediate application to *simple subvarieties* of V_r, of any dimension. This application is given in the last part (V) of the paper. An irreducible s-dimensional subvariety V_s of V_r can be treated as a point $\bar{P}$, provided we pass to a new ground field $\bar{\mathsf{K}}$—a suitable *transcendental* extension of K in Σ—and regard our V_r as an $(r-s)$-dimensional variety $\bar{V}_{r-s}$ over $\bar{\mathsf{K}}$. From our definitions it will follow that V_s is simple for V_r if and only if $\bar{P}$ is simple for $\bar{V}_{r-s}$. The properties of the simple point $\bar{P}$ yield corresponding properties of the simple V_s. It is this application that should justify (in the eyes of a geometer) our consideration of ground fields which are not algebraically closed.

Let $\xi_1, \cdots, \xi_n$ be the coördinates of the general point of V_r and let $\mathfrak{o}$ denote the ring $\mathsf{K}[\xi_1, \cdots, \xi_n]$. An irreducible V_s on V_r is given by a prime s-dimensional ideal $\mathfrak{p}$ in $\mathfrak{o}$. Let $\mathfrak{J}$ be the *quotient ring* of V_s ($\mathfrak{J} = \mathfrak{o}_\mathfrak{p}$, $a/b \in \mathfrak{J}$ if $a, b \in \mathfrak{o}$, $b \not\equiv 0(\mathfrak{p})$) and let $\mathfrak{P} = \mathfrak{J} \cdot \mathfrak{p}$ be the prime ideal of non units of $\mathfrak{J}$. We define a *simple* V_s by the condition that *there exist* $r-s$ elements $\eta_1, \cdots, \eta_{r-s}$ in $\mathfrak{J}$ such that $\mathfrak{J}(\eta_1, \cdots, \eta_{r-s}) = \mathfrak{P}$. The elements η_i are referred to as *uniformizing parameters along* V_s, or *of* V_s. Our main result concerns the characterization of a simple V_s and of its uniformizing parameters with the aid of the different F'_ω of primitive elements ω in $\mathfrak{o}$. In this characterization we start with an arbitrary set of r elements $\zeta_1, \cdots, \zeta_r$ in $\mathfrak{o}$ such that $\mathfrak{o}$ is integrally dependent on $\mathsf{K}[\zeta_1, \cdots, \zeta_r]$. Let F'_ω be the different of an element ω in $\mathfrak{o}$ if $\zeta_1, \cdots, \zeta_r$ are taken as the independent variables. Just as a matter of arrangement of the indices it is permissible to assume that $\zeta_1, \cdots, \zeta_s$ are algebraically independent mod $\mathfrak{p}$. Let $f_i(\zeta_1, \cdots, \zeta_s; \zeta_{s+i}) \equiv 0 \pmod{\mathfrak{p}}$ be the irreducible congruence mod $\mathfrak{p}$ which ζ_{s+i} satisfies over $\mathsf{K}(\zeta_1, \cdots, \zeta_s)$ $(i = 1, 2, \cdots, r-s)$. We show that *if there exists an element* ω *in* $\mathfrak{o}$ *such*

* Received September 28, 1939.

that $F'_\omega \not\equiv 0(\mathfrak{p})$, *then* V_s *is simple and the* $r - s$ *elements* $f_i(\zeta_1, \cdots, \zeta_s; \zeta_{s+i})$ *are uniformizing parameters of* V_s; *and conversely.*

An almost immediate consequence of this result is that *the quotient ring* $\mathfrak{J}$ *of a simple* V_s *is integrally closed in* Σ.

The burden of the proofs rests naturally on the case of simple points. We consider the *residue class field* $K_\mathfrak{p}$ of a point P, i. e. the field $K_\mathfrak{p} = \mathfrak{o}/\mathfrak{p}$. This field is a finite algebraic extension of K. Let K^* be the least normal extension of K which contains $K_\mathfrak{p}$. Upon extending the ground field K to K^*, a new variety V^*_r is obtained, and on V^*_r the point P splits into a finite number of points $P^*_1, \cdots, P^*_h$. The most difficult step of the theory is the proof that P *is simple for* V_r *if and only if the points* P^*_i *are simple for* V^*_r *and if the quotient ring* $\mathfrak{J}$ *of* P *contains the relative algebraic closure of* K *in* Σ. With the aid of this result, the various theorems concerning the simple point P can be readily deduced from the corresponding theorems concerning the points P^*_i.

This reduction succeeds because at each point P^*_i we have a very special state of affairs, namely *the residue class field at each point* P^*_i *coincides with the new ground field* K^*. This is therefore a special case of our problem: it is characterized by the condition $K_\mathfrak{p} = K$. This special case is treated first (Part III). Here we pass directly from K to *the algebraically closed field determined by* K. It is shown that this ground field extension does not cause any splitting of the point P. We then use the results already established in the case of an algebraically closed ground field.

The method just outlined necessitates a preliminary study of *the splitting of prime ideals in* $\mathfrak{o}$ *under algebraic extensions of the ground field* (Part I). We could not take over directly the results established in this connection by van der Waerden and Krull, because these authors have only dealt with the special case in which K *is maximally algebraic in* Σ.

The systematic study of simple points and of simple subvarieties undertaken in this paper is a necessary preliminary to the problem of local uniformization on algebraic varieties which we shall treat in a forthcoming paper.

I. Normal ground field extensions.

1. Let Σ be a field and let K be a subfield of Σ, *of characteristic zero.* The field K shall be referred to as the *ground field.* We consider a normal algebraic extension field K^* of K and we wish to show how this extension of the ground field defines a corresponding extension field of Σ, which we shall denote by Σ^*, or by $K^*\Sigma$.

Let Ω be the algebraically closed field determined by K and let K' be the relative algebraic closure of K in Σ, i. e. the field consisting of all those elements of Σ which are algebraic over K. The fields K′ and K* can be imbedded in Ω. This imbedding is defined to within relative automorphisms of K′ and K* over K, but since K* is a normal extension of K, the intersection of K′ and K* is a subfield of K′ which is independent of the imbedding. Let this subfield be denoted by Δ.

The elements of Σ^* shall be the formal finite sums $\xi^* = a^*_1\xi_1 + \cdots a^*_h\xi_h$,[1] $a^*_i \subset$ K*, $\xi_i \subset \Sigma$, h-arbitrary. Addition, subtraction and multiplication are defined formally in an obvious fashion. We need a rule for identifying two formal sums, and for this it is sufficient to give a rule for identifying a formal sum ξ^* with the zero element of Σ^*. Let $\xi^* = a^*_1\xi_1 + \cdots + a^*_h\xi_h$ and let $b^*_1, \cdots, b^*_n$ ($b^*_i \subset$ K*) be an independent Δ-basis of the algebraic extension field $\Delta(a^*_1, \cdots, a^*_h)$ of Δ. If we substitute formally into the sum $a^*_1\xi_1 + \cdots + a^*_h\xi_h$ the expressions of $a^*_1, \cdots, a^*_h$ in terms of $b^*_1, \cdots, b^*_n$ (linear forms with coefficients in Δ), we get an expression of the form $b^*_1\eta_1 + \cdots + b^*_n\eta_n$, $\eta_i \subset \Sigma$. To indicate this substitution we write: $\xi^* \to b^*_1\eta_1 + \cdots + b^*_n\eta_n$. *We identify the element ξ^* with the zero element of Σ^*, if and only if $\eta_1 = 0, \cdots, \eta_n = 0$.* It is self-evident that this identification rule is independent of the choice of the base $b^*_1, \cdots, b^*_n$. More generally, let $c^*_1, \cdots, c^*_m$ be a set of elements of K* which are such that: (1) they are linearly independent over Δ; (2) the a^*_i can be expressed as linear forms of the c^*_j with coefficients in Δ. The elements c^*_j need not belong to the field $\Delta(a^*_1, \cdots, a^*_h)$. By condition (2) we get, through formal substitution: $\xi^* \to c^*_1\zeta_1 + \cdots + c^*_m\zeta_m$. *We assert that $\xi^* = 0$ if and only if $\zeta_1 = \cdots = \zeta_m = 0$.* For the proof, let $d^*_1, \cdots, d^*_\nu$ be an independent Δ-basis of $\Delta(b^*_1, \cdots, b^*_n, c^*_1, \cdots, c^*_m)$ and let $\xi^* \to d^*_1\omega_1 + \cdots + d^*_\nu\omega_\nu$. It is clear that $b^*_1\eta_1 + \cdots + b^*_n\eta_n \to d^*_1\omega_1 + \cdots + d^*_\nu\omega_\nu$ and also $c^*_1\zeta_1 + \cdots + c^*_m\zeta_m \to d^*_1\omega_1 + \cdots + d^*_\nu\omega_\nu$. If $b^*_i = \sum_{j=1}^{\nu} k_{ij}d^*_j$, $k_{ij} \subset \Delta$, then the matrix (k_{ij}) is or rank n, since $b^*_1, \cdots, b^*_n$ are linearly independent over Δ, and moreover $\omega_i = \sum_{j=1}^{n} k_{ji}\eta_j$. Similarly, if $c^*_i = \sum_{j=1}^{\nu} l_{ij}d^*_j$, then the matrix (l_{ij}) is of rank m, and we have $\omega_i = \sum_{j=1}^{m} l_{ji}\zeta_j$. Hence, if $\xi^* = 0$, i. e. if $\eta_1 = \cdots = \eta_n = 0$, then $\omega_1 = \cdots = \omega_\nu = 0$, and since (l_{ij}) is of rank m, it follows that $\zeta_1 = \cdots = \zeta_m = 0$. Conversely, if $\zeta_1 = \cdots = \zeta_m = 0$, then $\omega_1 = \cdots = \omega_\nu = 0$, i. e. $\sum_{j=1}^{n} k_{ji}\eta_j = 0$, $i = 1, 2, \cdots, \nu$, and since the matrix (k_{ij}) is of rank n, it follows that $\eta_1 = \cdots = \eta_n = 0$, i. e. $\xi^* = 0$.

---

[1] We use small Greek letters for elements of Σ and small Latin letters for elements of K. The same letters with an asterisk denote elements of Σ^* and K* respectively.

As an immediate consequence we have the following: *if* $a^*_1, \cdots, a^*_h$ *are themselves linearly independent over* Δ, *then* $a^*_1\xi_1 + \cdots + a^*_h\xi_h = 0$ *if and only if* $\xi_1 = \cdots = \xi_h = 0$.

It is clear that formal addition and multiplication of the elements of Σ^*, considered as formal sums, is consistent with out identification rule. Hence Σ^* *is a ring.*

LEMMA 1. *Let* θ *be an element of* K^* *and let* $f(\theta) = \theta^g + a_1\theta^{g-1} + \cdots a_g = 0$, $a_i \subset \Delta$, *be the irreducible equation for* θ *over* Δ. *The polynomial* $f(x)$ *remains irreducible in the polynomial ring* $\Sigma[x]$; *in other words: the relative degrees* $[\Delta(\theta):\Delta]$, $[\Sigma(\theta):\Sigma]$ *are the same.*

Proof. Let $\phi(x) = x^m + \omega_1 x^{m-1} + \cdots + \omega_m$, $\omega_i \subset \Sigma$, be an irreducible factor of $f(x)$ in $\Sigma[x]$. Let $\theta^{(1)} = \theta, \theta^{(2)}, \cdots, \theta^{(m)}$ be the roots of $\phi(x)$. Since $\theta \subset K^*$ and since K^* is a normal extension of Δ, all the roots of $f(x)$ are in K^*. Consequently, $\omega_1, \cdots, \omega_m \subset K^*$. Since the ω's are in Σ and are algebraic over K, they must also belong to the field K'. Consequently $\omega_1, \cdots, \omega_m \subset \Delta$, whence $\phi(x) = f(x)$, q. e. d.

By means of this Lemma we now show that Σ^* is an *integral domain.* (i. e. Σ^* has no zero divisors). Let $\xi^*\eta^* = 0$, $\xi^* = a^*_1\xi_1 + \cdots + a^*_m\xi_m$, $\eta^* = b^*_1\eta_1 + \cdots + b^*_n\eta_n$, and let g be the relative degree of the field $\Delta(a^*_1, \cdots, a^*_m, b^*_1, \cdots, b^*_n)$ with respect to Δ. Let θ be a primitive element of this field, satisfying an irreducible equation $F(\theta) = 0$ of degree g, with coefficients in Δ. By our identification rule we have:

$$(1) \qquad \xi^* = \alpha_0 + \alpha_1\theta + \cdots + \alpha_{g-1}\theta^{g-1} = \phi(\theta),$$
$$(1') \qquad \eta^* = \beta_0 + \beta_1\theta + \cdots + \beta_{g-1}\theta^{g-1} = \psi(\theta), \qquad \alpha_i, \beta_j \subset \Sigma,$$

and by the same rule, the relation $\phi(\theta)\cdot\psi(\theta) = 0$ implies that the polynomial $\phi(x)\cdot\psi(x)$ is divisible (in $\Sigma[x]$) by $F(x)$. By Lemma 1, $F(x)$ is irreducible in $\Sigma[x]$. Hence, either $\phi(x)$ or $\psi(x)$ is identically zero, i. e. either $\xi^* = 0$ or $\eta^* = 0$, which shows that Σ^* has no zero divisors.

It now follows immediately that Σ^* *is a field.* In fact, every element ξ^* of Σ^* is of the form (1), for some $\theta \subset K^*$, and, by Lemma 1, Σ^* contains the entire field $\Sigma(\theta)$.

Remark 1. We call attention to the important role which the field Δ plays in the definition of the field Σ^*. It is this field, rather than the ground field K, which really matters in our construction. By definition, Δ is the largest subfield of Σ which can be imbedded in K^*. We would get the same field Σ^* if we took Δ as ground field instead of K.

Of particular importance is the special case $K = K'$ (i. e. K is "*maximally algebraic*" in Σ, or K is *algebraically closed in* Σ). In this case we have $K = \Delta$ for every normal extension of K.

Remark 2. The fields Σ and K^* are subfields of Σ^* [2] and have at least the field K in common. It is not difficult to see that Σ^* is *the smallest field having this property*, i. e. any field Γ with this property contains Σ^*. Our hypothesis is to the effect that the field Γ contains two subfields Σ_1 and K^*_1 simply isomorphic to Σ and K^* respectively, and, moreover, that the field K_1 which corresponds to K in the isomorphism between K^*_1 and K^* is a subfield of Σ_1. It is then clear that the intersection of Σ_1 and K^*_1 must be the field Δ_1 which corresponds to Δ in the isomorphism $\Sigma \cong \Sigma_1$. Using the reasoning of the proof of Lemma 1, it is immediately seen that the join (Σ_1, K^*_1) of the two subfields Σ_1 and K^*_1 of Γ is abstractly isomorphic to the field Σ^*, and that this isomorphism induces the given isomorphisms between Σ_1 and Σ, and between K^*_1 and K^*.

2. Let $\mathfrak{o}$ be an arbitrary subring of Σ, subject to the only condition: $K \subset \mathfrak{o}$. Let $\mathfrak{o}^* = K^*\mathfrak{o}$ be the extended ring in Σ^*, i. e. the ring whose elements are of the form $a^*_1\xi_1 + \cdots + a^*_n\xi_n$, $a^*_i \subset K^*$, $\xi_i \subset \mathfrak{o}$. Let Δ' be the intersection of $\mathfrak{o}$ with Δ. Since Δ is an algebraic extension of K and since $K \subset \mathfrak{o}$, it follows that Δ' is a field.

THEOREM 1. *If* $\Delta' = \Delta$, *then* $\mathfrak{o}^*\mathfrak{A} \cap \mathfrak{o} = \mathfrak{A}$ *for any* $\mathfrak{o}$-*ideal* $\mathfrak{A}$. *In the general case the relation* $\mathfrak{o}^*\mathfrak{A} \cap \mathfrak{o} = \mathfrak{A}$ *still holds true if* $\mathfrak{A}$ *is prime.*

Proof. Let $\xi = a^*_1\xi_1 + \cdots + a^*_n\xi_n$, $a^*_i \subset K^*$, $\xi_i \subset \mathfrak{A}$, be an element of $\mathfrak{o}^*\mathfrak{A} \cap \mathfrak{o}$, and let θ be a primitive element of $\Delta'(a^*_1, \cdots, a^*_n)/\Delta'$. Since $\Delta' \subset \mathfrak{o}$, we can write ξ in the form:

$$\xi = \eta_0 + \eta_1\theta + \cdots + \eta_{g-1}\theta^{g-1}, \tag{2}$$

where $\eta_i \subset \mathfrak{A}$ and where g is the relative degree of $\Delta'(a^*_1, \cdots, a^*_n)$ with respect to Δ'. Under the hypothesis that $\Delta' = \Delta$, the elements $1, \theta, \cdots, \theta^{g-1}$ are linearly independent over Δ, and hence the equation (2) implies that $\xi = \eta_0, \eta_1 = \cdots = \eta_{g-1} = 0$. Hence $\xi \equiv 0(\mathfrak{A})$. This shows that $\mathfrak{o}^*\mathfrak{A} \cap \mathfrak{o} \subseteqq \mathfrak{A}$, and since $\mathfrak{A} \subseteqq \mathfrak{o}^*\mathfrak{A}$, it follows that $\mathfrak{o}^*\mathfrak{A} \cap \mathfrak{o} = \mathfrak{A}$.

In the general case and for a *prime ideal* $\mathfrak{A}$,[3] we proceed as follows. Multiplying (2) by $1, \theta, \cdots, \theta^{g-1}$ respectively, we get relations of the form:

---

[2] More precisely: Σ^* contains two subfields abstractly isomorphic to Σ and K^* respectively, consisting of the elements $a^*_1\xi_1 + \cdots + a^*_m\xi_m$ in which $a^*_1, \cdots, a^*_m \subset K$ or $\xi_1, \cdots, \xi_m \subset K$ respectively.

[3] The following example illustrates the possibility: $\mathfrak{o}^*\mathfrak{A} \cap \mathfrak{o} \neq \mathfrak{A}$, if $\Delta \neq \Delta'$. Let K be the field of rational numbers, $K^* = K(\sqrt{2})$ and let $\Sigma = K^*(x)$. If we regard K as the ground field then the extension $K \to K^*$ does not affect Σ, i. e. we have $\Sigma = \Sigma^*$. Let $\mathfrak{o} = K[x, x \cdot \sqrt{2}]$, $\mathfrak{A} = \mathfrak{o} \cdot x$. Then $\mathfrak{o}^* = K^*[x]$, $\mathfrak{o}^*\mathfrak{A} = \mathfrak{o}^* \cdot x$ and $\mathfrak{o}^*\mathfrak{A} \cap \mathfrak{o} = \mathfrak{o} \cdot (x, x \cdot \sqrt{2}) \neq \mathfrak{A}$. Here the fields Δ and Δ' coincide with K^* and K respectively.

$$\begin{aligned} \xi &= \eta_0 + \eta_1\theta + \cdots + \eta_{g-1}\theta^{g-1}, \\ \xi\theta &= \eta_0^{(1)} + \eta_1^{(1)}\theta + \cdots + \eta_{g-1}^{(1)}\theta^{g-1}, \\ &\cdots\cdots\cdots \\ \xi\theta^{g-1} &= \eta_0^{(g-1)} + \eta_1^{(g-1)}\theta + \cdots + \eta_{g-1}^{(g-1)}\theta^{g-1}, \end{aligned}$$

where all the $\eta_j^{(i)}$ are in $\mathfrak{A}$. Hence $| \eta_i^{(j)} - \delta_i^{(j)}\xi | = 0$, where $\delta_i^{(j)} = 0$ if $i \neq j$, and $\delta_i^{(i)} = 1$. This equation is of the following form:

$$\xi^g + \beta_1\xi^{g-1} + \cdots + \beta_g = 0,$$

where $\beta_i \equiv 0(\mathfrak{A})$, $i = 1, 2, \cdots, g$. Hence $\xi^g \equiv 0(\mathfrak{A})$, and since $\mathfrak{A}$ is prime, we conclude, as in the first part of the proof, that $\xi \equiv 0(\mathfrak{A})$, q. e. d.

An ideal $\mathfrak{A}^*$ in $\mathfrak{o}^*$ is said *to lie over an ideal* $\mathfrak{A}$ in $\mathfrak{o}$ if the relation $\mathfrak{A}^* \cap \mathfrak{o} = \mathfrak{A}$ is satisfied. It has been proved by Krull [4] that over every prime ideal $\mathfrak{p}$ in $\mathfrak{o}$ there lies at least one prime ideal $\mathfrak{p}^*$ in $\mathfrak{o}^*$, provided that $\mathfrak{o}^*$ *be integrally dependent on* $\mathfrak{o}$ (i. e. that each element of $\mathfrak{o}^*$ be integrally dependent on elements of $\mathfrak{o}$). This provision is satisfied in our case, since $\mathfrak{o}^* = \mathsf{K}^*\mathfrak{o}$ and since K^*, as an algebraic extension field of K, is certainly integrally dependent on K ($\mathsf{K} \subset \mathfrak{o}$).

We consider a prime $\mathfrak{o}^*$-ideal $\mathfrak{p}^*$ which lies over $\mathfrak{p}$ and we denote by $\mathsf{K}_\mathfrak{p}$ and $\mathsf{K}^*_{\mathfrak{p}^*}$ the residue class fields of $\mathfrak{p}$ and $\mathfrak{p}^*$ respectively, i. e. the quotient fields of the residue class rings $\mathfrak{o}/\mathfrak{p}$ and $\mathfrak{o}^*/\mathfrak{p}^*$ respectively. Since $\mathfrak{p}^* \cap \mathfrak{o} = \mathfrak{p}$, $\mathsf{K}_\mathfrak{p}$ may be regarded as a subfield of $\mathsf{K}^*_{\mathfrak{p}^*}$. Moreover, K and K^* may be regarded as subfields of $\mathsf{K}_\mathfrak{p}$ and $\mathsf{K}^*_{\mathfrak{p}^*}$ respectively.

LEMMA 2. *$\mathsf{K}^*_{\mathfrak{p}^*}$ is the extension field of $\mathsf{K}_\mathfrak{p}$ obtained by the extension $\mathsf{K} \rightarrow \mathsf{K}^*$ of the ground field K; in symbols: $\mathsf{K}^*_{\mathfrak{p}^*} = \mathsf{K}^* \cdot \mathsf{K}_\mathfrak{p}$.*

We observe that K^* and $\mathsf{K}_\mathfrak{p}$ are subfields of $\mathsf{K}^*_{\mathfrak{p}^*}$ having at least the field K in common. Hence, by Remark 2 of the preceding section, we have: $\mathsf{K}^*_{\mathfrak{p}^*} \supseteq \mathsf{K}^* \cdot \mathsf{K}_\mathfrak{p}$. On the other hand, any element of $\mathfrak{o}^*/\mathfrak{p}^*$ is of the form $a^*_1\bar{\xi}_1 + \cdots + a^*_m\bar{\xi}_m$, $a^*_i \subset \mathsf{K}^*$, $\bar{\xi}_i \subset \mathfrak{o}/\mathfrak{p}$. This shows that the ring $\mathfrak{o}^*/\mathfrak{p}^*$, and hence also its quotient field $\mathsf{K}^*_{\mathfrak{p}^*}$, is contained in the field $(\mathsf{K}^*, \mathsf{K}_\mathfrak{p})$. Hence $\mathsf{K}^*_{\mathfrak{p}^*} = \mathsf{K}^*\mathsf{K}_\mathfrak{p}$, as was asserted.

3. Unramified character of the maximal $\mathfrak{o}$-ideals. We make the following assumption:

The field Δ is a finite extension of K. This assumption is always satisfied if, for instance, K' (the algebraic closure of K in Σ) is itself a finite extension of K.

Under this assumption we prove the following fundamental theorem:

[4] W. Krull, "Zum Dimensionbegriff der Idealtheorie" (Beiträge zur Arithmetik kommutativer Integritätsbereiche, III), *Mathematische Zeitschrift*, vol. 42 (1937), p. 749.

THEOREM 2. *If $\mathfrak{p}$ is a maximal $\mathfrak{o}$-ideal* [5] *then $\mathfrak{o}^*\mathfrak{p}$ is the intersection of the prime $\mathfrak{o}^*$-ideals which lie over $\mathfrak{p}$.*

For polynomial rings $\mathfrak{o} = \mathsf{K}[x_1, \cdots, x_n]$ this theorem is due to van der Waerden.[6] It appears as a special case of a generalized discriminant theorem proved by Krull for any pair of integral domains $\mathfrak{o}$, $\mathfrak{o}^*$ ($\mathfrak{o}^*$-integrally dependent on $\mathfrak{o}$) under the hypothesis that $\mathfrak{o}$ is integrally closed in its quotient field.[7] If we assume, as it is permissible to do, that Σ is the quotient field of $\mathfrak{o}$, then Krull's hypothesis in our special case implies that $\mathsf{K}' \subset \mathfrak{o}$, whence $\Delta = \Delta'$. The special case when $\mathfrak{o}^*$ is obtained from $\mathfrak{o}$ by a separable extension of the ground field has been treated separately by Krull in his report "Idealtheorie" (p. 40). However, also this treatment is based on the tacit assumption that the fields Δ and Δ' coincide. Namely, under this assumption it is permissible to take Δ as ground field, since $\Delta = \Delta' \subset \mathfrak{o}$, i. e. we may put $\Delta = \mathsf{K}$, and then our assumption $\Delta = \Delta'$ becomes: $\mathsf{K}' \cap \mathsf{K}^* = \mathsf{K}$. It follows then, by Lemma 1, that the Galois groups of Σ^*/Σ and of K^*/K coincide (i. e. every relative automorphism of K^* over K can be extended to a relative automorphism Σ^* over Σ; note that Σ^* is at any rate a normal extension of Σ). One defines then in a natural fashion the concepts of *conjugate ideals* and of *invariant ideals* in $\mathfrak{o}^*$. The proof by van der Waerden and its generalization by Krull are then applicable, leading to the following theorem:

THEOREM 2′ (van der Waerden-Krull). *If $\Delta = \Delta'$, and if K^* is a separable extension of Δ, then each invariant $\mathfrak{o}^*$-ideal $\mathfrak{A}^*$ is the extended ideal of its contracted ideal in $\mathfrak{o}$: $\mathfrak{A}^* = \mathfrak{o}^* \cdot (\mathfrak{A}^* \cap \mathfrak{o})$, and for each prime $\mathfrak{o}$-ideal $\mathfrak{p}$* [8] *it is true that $\mathfrak{o}^*\mathfrak{p}$ is the intersection of the prime $\mathfrak{o}^*$-ideals which lie over $\mathfrak{p}$.*

We shall make use of Theorem 2′ in order to prove our more general theorem for maximal ideals.

Let $\bar{\Delta}$ be the least normal extension of K which contains the field Δ, i. e. $\bar{\Delta}$ is the join of Δ and of its conjugate fields over K. By our assumption, $\bar{\Delta}$ is a finite extension of K. We introduce the intermediate field $\bar{\Sigma} = \bar{\Delta}\Sigma$, a finite algebraic extension of Σ, and the intermediate ring $\bar{\mathfrak{o}} = \bar{\Delta}\mathfrak{o}$, so that $\Sigma \subseteqq \bar{\Sigma} \subseteqq \Sigma^*$, $\mathfrak{o} \subseteqq \bar{\mathfrak{o}} \subseteqq \mathfrak{o}^*$. The ground field extension $\mathsf{K} \rightarrow \mathsf{K}^*$ is thus decomposed into two successive normal extensions: $\mathsf{K} \rightarrow \bar{\Delta}$, $\bar{\Delta} \rightarrow \mathsf{K}^*$. We have clearly the relations: $\Sigma^* = \mathsf{K}^*\bar{\Sigma}$, $\mathfrak{o}^* = \mathsf{K}^*\bar{\mathfrak{o}}$.

---

[5] An ideal is maximal (or divisorless) if it is not properly contained in any other ideal, different from the unit ideal.

[6] B. L. van der Waerden, "Eine Verallgemeinerung des Bezoutschen Theorems," § 5, *Mathematische Annalen*, vol. 99 (1928).

[7] W. Krull, "Der allgemeine Discriminantensatz. Unverzweigte Ringerweiterungen" (Beiträge zur Arithmetik kommutativer Integritätsbereiche, VI), *Mathematische Zeitschrift*, vol. 45 (1939).

[8] Not necessarily maximal as in Theorem 2.

13

We assert that *the relative algebraic closure* $\bar{\Delta}'$ *of* $\bar{\Delta}$ *in* $\bar{\Sigma}$ *is the field* $\bar{\Delta}\mathsf{K}'$. To show this, we first observe that $\Delta = \bar{\Delta} \cap \mathsf{K}'$, and hence, by Lemma 1, $[\bar{\Delta} : \Delta] = [\bar{\Sigma} : \Sigma] = g$. Let $\bar{\alpha}$ be an element of $\bar{\Sigma}$ which is algebraic over $\bar{\Delta}$, hence also algebraic over K'. The relative degree $[\mathsf{K}'(\bar{\alpha}) : \mathsf{K}']$ cannot be greater than the relative degree $[\Sigma(\bar{\alpha}) : \Sigma]$, and since $[\Sigma(\bar{\alpha}) : \Sigma] \leqq [\bar{\Sigma} : \Sigma] = g$, it follows that $[\mathsf{K}'(\bar{\alpha}) : \mathsf{K}'] \leqq g$. This last inequality holds true for any element α in $\bar{\Delta}'$, and consequently $[\bar{\Delta}' : \mathsf{K}'] \leqq g$. On the other hand, $\bar{\Delta}'$ contains the field $\bar{\Delta}\mathsf{K}'$, and we have $[\bar{\Delta}\mathsf{K}' : \mathsf{K}'] = [\bar{\Delta} : \Delta] = g$, in view of the relation $\Delta = \bar{\Delta} \cap \mathsf{K}'$ and of Lemma 1. Hence necessarily $\bar{\Delta}' = \bar{\Delta}\mathsf{K}'$, as was asserted.

We now prove the relation:

$$\bar{\Delta} = \mathsf{K}^* \cap \bar{\Delta}\mathsf{K}'. \tag{3}$$

Let a^* be an element of $\mathsf{K}^* \cap \bar{\Delta}\mathsf{K}'$. Since $\Delta = \mathsf{K}^* \cap \mathsf{K}'$, we have $[\mathsf{K}'(a^*) : \mathsf{K}'] = [\Delta(a^*) : \Delta]$. Now we have just proved that $[\bar{\Delta}\mathsf{K}' : \mathsf{K}'] = g$. Since $a^* \subset \bar{\Delta}\mathsf{K}'$, we conclude that $[\Delta(a^*) : \Delta] \leqq g$, for any element a^* in $\mathsf{K}^* \cap \bar{\Delta}\mathsf{K}'$. Hence this last field is of relative degree $\leqq g$ over Δ. Since on the other hand this field contains $\bar{\Delta}$, and since $[\bar{\Delta} : \Delta] = g$, the relation (3) is established.

The relation (3) says that $\bar{\Delta}$ is the intersection of K^* with the algebraic closure of $\bar{\Delta}$ in $\bar{\Sigma}$. The ground field extension $\bar{\Delta} \rightarrow \mathsf{K}^*$ therefore satisfies the condition of Theorem 2′. We therefore know that every prime ideal $\bar{\mathfrak{p}}$ in $\bar{\mathfrak{o}}$ is the intersection of the prime $\mathfrak{o}^*$-ideals which lie over $\mathfrak{p}$. Let us assume that Theorem 2 has already been proved for the ground field extension $\mathsf{K} \rightarrow \bar{\Delta}$ and, moreover, let us assume that there is only a finite number of prime $\bar{\mathfrak{o}}$-ideals which lie over a given maximal prime $\mathfrak{o}$-ideal $\mathfrak{p}$. We will have then:

$$\bar{\mathfrak{o}}\mathfrak{p} = [\bar{\mathfrak{p}}_1, \bar{\mathfrak{p}}_2, \cdots, \bar{\mathfrak{p}}_m].$$

The ideals $\bar{\mathfrak{p}}_i$ are also maximal in $\mathfrak{o}$,[9] hence are two by two free from common divisors. Therefore their intersection coincides with their product:

$$\bar{\mathfrak{o}}\mathfrak{p} = \bar{\mathfrak{p}}_1\bar{\mathfrak{p}}_2 \cdots \bar{\mathfrak{p}}_m. \tag{4}$$

By Theorem 2′ we have

$$\mathfrak{o}^*\bar{\mathfrak{p}}_i = [\mathfrak{p}^*_{i1}, \mathfrak{p}^*_{i2}, \cdots],$$

where $\mathfrak{p}^*_{ij} \cap \bar{\mathfrak{o}} = \bar{\mathfrak{p}}_i$. Since $(\bar{\mathfrak{p}}_i, \bar{\mathfrak{p}}_j) = \bar{\mathfrak{o}}$, if $i \neq j$, we have also $(\mathfrak{o}^*\bar{\mathfrak{p}}_i, \mathfrak{o}^*\bar{\mathfrak{p}}_j) = \mathfrak{o}^*$. Hence the product of the ideals $\mathfrak{o}^*\bar{\mathfrak{p}}_i$ coincides with their intersection, and therefore, by (4),

[9] Since $\mathfrak{p}$ is maximal, the ring $\mathfrak{o}/\mathfrak{p}$ is a field. The integral domain $\bar{\mathfrak{o}}/\bar{\mathfrak{p}}_i$ is integrally dependent on its subfield $\mathfrak{o}/\mathfrak{p}$ and hence is also a field. Consequently $\bar{\mathfrak{p}}_i$ is maximal.

$$\mathfrak{o}^*\mathfrak{p} = \mathfrak{o}^*\tilde{\mathfrak{p}}_1 \cdot \mathfrak{o}^*\tilde{\mathfrak{p}}_2 \cdots \mathfrak{o}^*\tilde{\mathfrak{p}}_m = [\mathfrak{o}^*\tilde{\mathfrak{p}}_1, \cdots, \mathfrak{o}^*\tilde{\mathfrak{p}}_m]$$
$$= [\mathfrak{p}^*_{11}, \mathfrak{p}^*_{12}, \cdots, \mathfrak{p}^*_{m1}, \cdots],$$

which proves Theorem 2.

Thus, to complete the proof of Theorem 2, we have only to prove it for arbitrary *finite* ground field extension $K \to \tilde{\Delta}$, and we have also to show that the number of prime $\tilde{\mathfrak{o}}$-ideals which lie over $\mathfrak{p}$ is finite. This we shall do in the following section.

4. Merely as a matter of notations, we may identify Δ' with K, since $\Delta' \subset \mathfrak{o}$. Let $\mathfrak{p}$ be a maximal $\mathfrak{o}$-ideal and let $K_\mathfrak{p}$ $(= \mathfrak{o}/\mathfrak{p})$ be the residue class field of $\mathfrak{p}$. Let $K_\mathfrak{p} \cap \tilde{\Delta} = \Delta_\mathfrak{p}$,[10] whence $K \subseteqq \Delta_\mathfrak{p} \subseteqq \tilde{\Delta}$, and let $\Delta_\mathfrak{p} = K(\vartheta)$, where ϑ is a primitive element of $\Delta_\mathfrak{p}$ over K. Let $f(\vartheta) = 0$ be the irreducible equation, say of degree m, which ϑ satisfies over K. Since $\vartheta \subset K_\mathfrak{p}$ and $K_\mathfrak{p} = \mathfrak{o}/\mathfrak{p}$ (by hypothesis: $\mathfrak{p}$ is maximal!), there must exist in $\mathfrak{o}$ an element ω such that

(5) $$f(\omega) \equiv 0(\mathfrak{p}).$$

If $\tilde{\mathfrak{p}}$ is a prime $\tilde{\mathfrak{o}}$-ideal which lies over $\tilde{\mathfrak{p}}$, then

(5′) $$f(\omega) \equiv 0(\tilde{\mathfrak{p}}).$$

Since $\tilde{\Delta}$ is normal over K and since one root, $\vartheta = \vartheta_1$, of the polynomial $f(x)$ is in $\tilde{\Delta}$, all its roots are in $\tilde{\Delta}$, whence also in $\tilde{\mathfrak{o}}$. Hence, by (5′), we must have $\omega \equiv \vartheta_i(\tilde{\mathfrak{p}})$, where ϑ_i is one of the roots $\vartheta_1, \cdots, \vartheta_m$ of $f(x)$. Let, say $\omega \equiv \vartheta_1(\mathfrak{p})$. *We assert that*

(6) $$\tilde{\mathfrak{p}} = (\tilde{\mathfrak{o}}\mathfrak{p}, \omega - \vartheta_1).$$

Let θ be a primitive element of $\tilde{\Delta}$ over $K(\vartheta_1)$, and let $[\tilde{\Delta}: K(\vartheta_1)] = n$. Every element $\tilde{\alpha}$ of $\tilde{\mathfrak{o}}$ can be written in the form:

$$\tilde{\alpha} = \tilde{\alpha}_0 + \tilde{\alpha}_1\theta + \cdots + \tilde{\alpha}_{n-1}\theta^{n-1},$$

where

$$\tilde{\alpha}_i = \alpha_{io} + \alpha_{i1}\vartheta_1 + \cdots + \alpha_{i,m-1}\vartheta_1{}^{m-1}, \qquad \alpha_{ij} \subset \mathfrak{o}.$$

Since $\omega \equiv \vartheta_1(\tilde{\mathfrak{p}})$, we have: $\tilde{\alpha}_i \equiv \alpha_{io} + \alpha_{i1}\omega + \cdots + \alpha_{i,m-1}\omega^{m-1}(\tilde{\mathfrak{p}})$. The right-hand side of this congruence is an element of $\mathfrak{o}$. Consequently, in the homomorphism $\tilde{\mathfrak{o}} \cong \tilde{\mathfrak{o}}/\tilde{\mathfrak{p}}$ the elements $\tilde{\alpha}_0, \tilde{\alpha}_1, \cdots, \tilde{\alpha}_{n-1}$ are mapped upon elements of $K_\mathfrak{p}$ $(= \mathfrak{o}/\mathfrak{p})$. Since $K(\vartheta_1) = \Delta_\mathfrak{p} = K_\mathfrak{p} \cap \tilde{\Delta}$, the elements $1, \theta, \cdots, \theta^{n-1}$ are linearly independent not only over $K(\vartheta_1)$, but also over $K_\mathfrak{p}$ (Lemma 1). Consequently $\tilde{\alpha}$ cannot belong to $\tilde{\mathfrak{p}}$, unless all the elements $\tilde{\alpha}_0, \tilde{\alpha}_1, \cdots, \tilde{\alpha}_{n-1}$ belong to $\tilde{\mathfrak{p}}$. We have: $\tilde{\alpha}_i \equiv \alpha_{io} + \alpha_{i1}\omega + \cdots + \alpha_{i,m-1}\omega^{m-1}(\mathfrak{o}\ (\omega - \vartheta_1))$,

---

[10] By $K_\mathfrak{p} \cap \tilde{\Delta}$ is meant the intersection of $\tilde{\Delta}$ (normal finite extension of K) with the relative algebraic closure $K'_\mathfrak{p}$ of K in $K_\mathfrak{p}$, in the same sense as Δ was defined by the relation: $\Delta = K^* \cap K'$. See Section 1.

and if $\bar{\alpha}_i \equiv 0(\bar{\mathfrak{p}})$, then $\alpha_{i0} + \alpha_{i1}\omega + \cdots + \alpha_{i,m-1}\omega^{m-1} \equiv 0(\mathfrak{p})$, since $\mathfrak{p} = \bar{\mathfrak{p}} \cap \mathfrak{o}$. This shows that if $\bar{\alpha}_i \equiv 0(\bar{\mathfrak{p}})$, then $\bar{\alpha}_i \equiv 0(\bar{\mathfrak{o}}\mathfrak{p}, \omega - \vartheta_1)$, and consequently also $\bar{\alpha} \equiv 0(\bar{\mathfrak{o}}\mathfrak{p}, \omega - \vartheta_1)$, which proves the relation (6).

From (6) it follows already that the number of prime $\bar{\mathfrak{o}}$-ideals which lie over $\mathfrak{p}$ is finite, since it cannot be greater than m ($= [\Delta_{\mathfrak{p}} : \mathsf{K}]$). Let these prime ideals be $\bar{\mathfrak{p}}_1, \cdots, \bar{\mathfrak{p}}_h$, and let

$$(6') \qquad \bar{\mathfrak{p}}_i = (\bar{\mathfrak{o}}\mathfrak{p}, \omega - \vartheta_i), \qquad (i = 1, 2, \cdots, h,\ h \leqq m).$$

Since the $\bar{\mathfrak{p}}_i$ are also maximal, we have

$$(7) \qquad [\bar{\mathfrak{p}}_1, \cdots, \bar{\mathfrak{p}}_h] = \bar{\mathfrak{p}}_1 \cdots \bar{\mathfrak{p}}_h = (\bar{\mathfrak{o}}\mathfrak{p}, \prod_{i=1}^{h} (\omega - \vartheta_i)).$$

Let $\xi = \prod_{i=h+1}^{m} (\omega - \vartheta_i)$ and let us consider the ideal $(\bar{\mathfrak{o}}\mathfrak{p}, \xi)$. *We assert that it is the unit ideal.* Namely, in the contrary case let $\bar{\mathfrak{p}}$ be a prime ideal divisor of $(\bar{\mathfrak{o}}\mathfrak{p}, \xi)$. Since $\mathfrak{p} \subset \bar{\mathfrak{p}}$ and $\mathfrak{p}$ is maximal, $\bar{\mathfrak{p}}$ must lie over $\mathfrak{p}$. Hence $\bar{\mathfrak{p}}$ must be one of the ideals $\bar{\mathfrak{p}}_1, \cdots, \bar{\mathfrak{p}}_h$, say $\bar{\mathfrak{p}} = \bar{\mathfrak{p}}_1$. Now since $\xi \equiv 0(\bar{\mathfrak{p}})$, one of the factors $\omega - \vartheta_i$, $i = h+1, \cdots, m$, must belong to $\bar{\mathfrak{p}}_1$, say $\omega - \vartheta_{h+1} \equiv 0(\bar{\mathfrak{p}}_1)$. Hence $\vartheta_1 - \vartheta_{h+1} \equiv 0(\bar{\mathfrak{p}}_1)$, and this is impossible, since $\vartheta_1 \neq \vartheta_{h+1}$ and since $\vartheta_1 - \vartheta_{h+1}$ is an element of the subfield $\bar{\Delta}$ of $\bar{\mathfrak{o}}$.

It is therefore proved that $(\bar{\mathfrak{o}}\mathfrak{p}, \xi) = \bar{\mathfrak{o}}$. Consequently $\prod_{i=1}^{h} (\omega - \vartheta_i) \equiv 0(\bar{\mathfrak{o}}\mathfrak{p})$, since $\xi \cdot \prod_{i=1}^{h} (\omega - \vartheta_i) = f(\omega) \equiv 0(\mathfrak{p})$. Comparing with (7) we find:

$$[\bar{\mathfrak{p}}_1, \cdots, \bar{\mathfrak{p}}_h] = \bar{\mathfrak{o}}\mathfrak{p},$$

as was asserted.

5. It can be shown by examples that Theorem 2 is not generally true for non-maximal ideals.[11] For arbitrary prime $\mathfrak{o}$-ideals some weaker result

[11] Let K be the field of rational numbers, and let $\Sigma = \mathsf{K}(\sqrt{2})(x, y)$, where x, y are independent variables. We put $\mathsf{K}^* = \mathsf{K}(\sqrt{2})$, $\mathfrak{o} = \mathsf{K}[x, y, z]$, where $z = \sqrt{2} \cdot xy$. We have $\Sigma^* = \mathsf{K}^*\Sigma = \Sigma$ and $\mathfrak{o}^* = \mathsf{K}(\sqrt{2})[x, y]$. Let $\mathfrak{p} = \mathfrak{o} \cdot (y^2 - 2, z - 2x)$. Observing that every element of $\mathfrak{o}$ can be put in the form $f(x, y) + z \cdot g(x, y)$, where $f(x, y)$, $g(x, y) \subset \mathsf{K}[x, y]$, it is a straightforward matter to verify that $\mathfrak{p}$ is prime. It is not maximal, since it is contained in the prime ideal $\mathfrak{o}(x, z, y^2 - 2)$. We have $\mathfrak{o}^*\mathfrak{p} = \mathfrak{o}^*(y^2 - 2, \sqrt{2} \cdot x(y - \sqrt{2})) = [\mathfrak{p}^*, \mathfrak{p}^*_0]$, where

$$\mathfrak{p}^* = \mathfrak{o}^*(y - \sqrt{2}), \qquad \mathfrak{p}^*_0 = \mathfrak{o}^*(x, y + \sqrt{2}).$$

The ideal $\mathfrak{p}^*$ lies over $\mathfrak{p}$. In fact, any element $f(x, y) + zg(x, y)$, reduced modulo $\mathfrak{p}$, gives a residue of the form $A(x) + yB(x)$ (since $y^2 \equiv 2(\mathfrak{p})$ and $z \equiv 2x(\mathfrak{p})$). Here $A(x)$ and $B(x)$ are in $\mathsf{K}[x]$. Should this residue belong to $\mathfrak{p}^*$, it is necessary that $A(x) + \sqrt{2} \cdot B(x)$ be identically zero. Hence $A(x) + yB(x)$ is also identically zero, and this shows that $\mathfrak{p}^* \cap \mathfrak{o} = \mathfrak{p}$. However, the ideal $\mathfrak{p}^*_0$ lies over the prime ideal $\mathfrak{o} \cdot (x, z, y^2 - 2)$ which is a proper divisor of $\mathfrak{p}$. It is remarkable that in this example $\mathfrak{o}^*\mathfrak{p}$ possesses even an isolated component $\mathfrak{p}^*_0$ different from $\mathfrak{p}^*$, since $\mathfrak{p}^* \not\equiv 0(\mathfrak{p}^*_0)$.

can be established by the usual artifice of quotient rings. Let $\mathfrak{p}$ be an arbitrary prime $\mathfrak{o}$-ideal and let $\mathfrak{J} = \mathfrak{o}_{\mathfrak{p}}$ be the quotient ring of $\mathfrak{p}$.[12] Let $\mathfrak{J}^* = K^*\mathfrak{J}$. It is well known that the prime ideals in $\mathfrak{J}$ correspond in one to one fashion to $\mathfrak{p}$ and the prime $\mathfrak{o}$-ideals which are contained in $\mathfrak{p}$. If $\mathfrak{p}_1$ and $\mathfrak{P}_1$ are corresponding prime ideals in $\mathfrak{o}$ and $\mathfrak{J}$ respectively, then $\mathfrak{P}_1 = \mathfrak{J} \cdot \mathfrak{p}_1, \mathfrak{p}_1 = \mathfrak{P}_1 \cap \mathfrak{o}$. The prime ideal $\mathfrak{P} = \mathfrak{J} \cdot \mathfrak{p}$ is maximal in $\mathfrak{J}$. By Theorem 2 we have therefore:

$$(8) \qquad \mathfrak{J}^*\mathfrak{P} = [\mathfrak{P}^*_1, \mathfrak{P}^*_2, \cdots],$$

where $\mathfrak{P}^*_1, \mathfrak{P}^*_2, \cdots$ are the prime ideals in $\mathfrak{J}^*$ which lie over $\mathfrak{P}$.

Similarly, it can be shown in a simple manner that the prime ideals in $\mathfrak{J}^*$ correspond, one to one, to the prime ideals in $\mathfrak{o}^*$ which lie over $\mathfrak{p}$ or over prime multiples of $\mathfrak{p}$. The correspondence is again the one of contracted and extended ideals.[13] Let $\mathfrak{P}^*_i \cap \mathfrak{o}^* = \mathfrak{p}^*_i$ and put

$$\mathfrak{o}^*\mathfrak{p} = \mathfrak{m}^*$$
$$[\mathfrak{p}^*_1, \mathfrak{p}^*_2, \cdots] = \mathfrak{m}^*_1.$$

We have evidently: $\mathfrak{J}^*\mathfrak{m}^* = \mathfrak{J}^*\mathfrak{m}^*_1 = [\mathfrak{P}^*_1, \mathfrak{P}^*_2, \cdots]$. *Let us assume that Hilbert's basis theorem holds in* $\mathfrak{o}^*$. From the relation $\mathfrak{J}^*\mathfrak{m}^* = \mathfrak{J}^*\mathfrak{m}^*_1$ follows that for every $\alpha^* \subset \mathfrak{m}^*_1$ there exists an element α in $\mathfrak{o}$ but not in $\mathfrak{p}$, such that $\alpha^*\alpha \subset \mathfrak{m}^*$. By Hilbert's basis theorem there exists then an element β in $\mathfrak{o}$, not in $\mathfrak{p}$, such that $\beta\mathfrak{m}^*_1 \equiv 0(\mathfrak{m}^*)$. *This shows that* $\mathfrak{p}^*_1, \mathfrak{p}^*_2, \cdots$ *are isolated components of* $\mathfrak{m}^*$. Since $\mathfrak{o}^*\mathfrak{p} \cap \mathfrak{o} = \mathfrak{p}$, by Theorem 1, it follows that *the decomposition of* $\mathfrak{o}^*\mathfrak{p}$ *into primary components is of the form*

$$\mathfrak{o}^*\mathfrak{p} = [\mathfrak{p}^*_1, \mathfrak{p}^*_2, \cdots; \mathfrak{q}'^*_1, \mathfrak{q}'^*_2, \cdots],$$

where the prime ideals $\mathfrak{p}'^*_1, \mathfrak{p}'^*_2, \cdots$ *to which* $\mathfrak{q}'^*_1, \mathfrak{q}'^*_2, \cdots$ *belong, lie over proper prime divisors of* $\mathfrak{p}$.

6. The following theorem, which we shall have occasion to use in the sequel, gives a sufficient condition that $\mathfrak{o}^*\mathfrak{p}$ be prime, where $\mathfrak{p}$ is now an arbitrary prime $\mathfrak{o}$-ideal, maximal or not.

---

[12] $\mathfrak{o}_{\mathfrak{p}}$ consists of all quotients α/β, $\alpha, \beta \subset \mathfrak{o}$, $\beta \not\equiv 0(\mathfrak{p})$.

[13] The elements of $\mathfrak{J}^*$ are all of the form α^*/α, $\alpha^* \subset \mathfrak{o}^*$, $\alpha \subset \mathfrak{o}$, $\alpha \not\equiv 0(\mathfrak{p})$. Let $\mathfrak{p}^*$ be a prime $\mathfrak{o}^*$-ideal which lies over a prime multiple of $\mathfrak{p}$, and let $\mathfrak{P}^* = \mathfrak{J}^*\mathfrak{p}^*$. Let α^*/α, β^*/β be two elements in $\mathfrak{J}^*$ whose product is in $\mathfrak{P}^*$. Then $\alpha^*\beta^*/\alpha\beta = \gamma^*/\gamma$, where $\gamma^* \equiv 0(\mathfrak{p}^*)$, and therefore $\gamma\alpha^*\beta^* \equiv 0(\mathfrak{p}^*)$. Since $\mathfrak{p}^* \cap \mathfrak{o} \equiv 0(\mathfrak{p})$, it follows that $\gamma \not\equiv 0(\mathfrak{p}^*)$, and hence either α^* or β^* is in $\mathfrak{p}^*$, i. e. either α^*/α or β^*/β is in $\mathfrak{P}^*$. This shows that $\mathfrak{P}^*$ is prime. Let $\alpha^* \subset \mathfrak{P}^* \cap \mathfrak{o}^*$, $\alpha^* = \beta^*/\beta$, $\beta^* \equiv 0(\mathfrak{p}^*)$. Then $\alpha^*\beta \equiv 0(\mathfrak{p}^*)$, and it follows by the same argument that α^* is in $\mathfrak{p}^*$. This shows that $\mathfrak{P}^* \cap \mathfrak{o}^* = \mathfrak{p}^*$.

If $\mathfrak{p}^* \cap \mathfrak{o} \not\equiv 0(\mathfrak{p})$, then let α be an element of $\mathfrak{o}$ which is in $\mathfrak{p}^*$ but not in $\mathfrak{p}$. Since α is a unit in $\mathfrak{J}^*$, it follows that $\mathfrak{J}^*\mathfrak{p}^* = \mathfrak{J}^*$.

If $\mathfrak{P}^*$ is an arbitrary prime ideal in $\mathfrak{J}^*$, and if $\mathfrak{p}^* = \mathfrak{P}^* \cap \mathfrak{o}^*$, then any element α^*/α in $\mathfrak{P}^*$ is such that α^* is in $\mathfrak{p}^*$, since α is a unit in $\mathfrak{J}^*$. This shows that $\mathfrak{P}^* = \mathfrak{J}^*\mathfrak{p}^*$.

THEOREM 3. *A sufficient condition that* $\mathfrak{o}^*\mathfrak{p}$ *be prime is that* Δ' *be the intersection of the fields* $K_\mathfrak{p}$ *and* K^*.

Proof. Let $\mathfrak{p}^*$ be a prime $\mathfrak{o}^*$-ideal over $\mathfrak{p}$. Every element α^* of $\mathfrak{o}^*$ can be written in the form: $\alpha^* = \alpha_0 + \alpha_1\theta + \cdots + \alpha_{g-1}\theta^{g-1}$, where $\alpha_i \subset \mathfrak{o}$ and θ is an element of K^*, of degree g over Δ'. If $\alpha^* = 0(\mathfrak{p}^*)$, then passing to the residue field $K^*_{\mathfrak{p}^*}$ we find the relation: $\bar{\alpha}_0 + \bar{\alpha}_1\theta + \cdots + \bar{\alpha}_{g-1}\theta^{g-1} = 0$, where $\bar{\alpha}_i \subset K_\mathfrak{p}$. Now if $K^* \cap K_\mathfrak{p} = \Delta'$, then, by Lemma 1, the elements $1, \theta, \cdots, \theta^{g-1}$ are linearly independent over $K_\mathfrak{p}$. Hence $\bar{\alpha}_i = 0$, i. e. $\alpha_i \equiv 0(\mathfrak{p})$. This shows that α^* is contained in $\mathfrak{o}^*\mathfrak{p}$, consequently $\mathfrak{o}^*\mathfrak{p} = \mathfrak{p}^*$, q. e. d.

Let K_0 be the largest subfield of $K_\mathfrak{p}$ which is algebraic over K, i. e. K_0 is the relative algebraic closure of K in $K_\mathfrak{p}$. Let K^* be the least normal extension of K which contains K_0 and let $\mathfrak{o}^* = K^*\mathfrak{o}$. If $\mathfrak{p}^*$ is any prime $\mathfrak{o}^*$-ideal over $\mathfrak{p}$, then, by a result proved in Section 2, we have: $K^*_{\mathfrak{p}^*} = K^* \cdot K_\mathfrak{p}$. In view of our choice of K^*, it follows that this field is algebraically closed in $K^*_{\mathfrak{p}^*}$. Hence if K^*_1 is any normal extension of the new ground field K^*, the condition of Theorem 3 is satisfied and $\mathfrak{p}^*$ will remain prime when we pass from $\mathfrak{o}^*$ to the ring $K^*_1\mathfrak{o}^*$. This is true, in particular, if we pass from K^* to the algebraically closed field determined by K. In other words: *the extension* $K \to K^*$ *causes the maximal splitting of* $\mathfrak{p}$ *into prime ideals.*

II. Algebraic varieties over arbitrary ground fields.

7. Let Σ be a field of algebraic functions of r independent variables, over an arbitrary ground field K of characteristic zero.[14] *We do not assume that* K *is algebraically closed in* Σ. Let $\xi_1, \cdots, \xi_n$ be a set of generators of Σ, i. e. $\Sigma = K(\xi_1, \cdots, \xi_n)$, and let $\mathfrak{o} = K[\xi_1, \cdots, \xi_n]$ be the ring consisting of those elements of Σ which can be expressed as polynomials in $\xi_1, \cdots, \xi_n$. With the elements ξ_i we associate *an irreducible algebraic r-dimensional variety* V_r whose general point has coördinates $\xi_1, \cdots, \xi_n$. A *point* P of V_r shall be associated with a prime zero-dimensional ideal $\mathfrak{p}_0$ in $\mathfrak{o}$. The geometric terms: "variety," "coördinates," "point," are so far purely formal and conventional expressions. To confer upon these terms a geometric reality it is necessary to imbed our V_r in an affine n-dimensional space $S_n{}^\Lambda$ over some field Λ.[15] The field Λ may be either K itself or an algebraic extension of K. Now the residue class field $K_{\mathfrak{p}_0}$ $(= \mathfrak{o}/\mathfrak{p}_0)$ of the prime zero-dimensional ideal $\mathfrak{p}_0$ may very well be a proper extension of K (necessarily algebraic). Hence, in general, there

[14] We assume, of course, that Σ is a *finite* extension of K (of degree of transcendency r).

[15] By the symbol $S_n{}^\Lambda$ we mean an affine n-space in which every point has coördinates $a_1, \cdots, a_n$, $a_i \subset \Lambda$.

will not exist elements $c_1, \cdots, c_n$ in K such that $\xi_i \equiv c_i(\mathfrak{p}_0)$. In such a case our point P is not represented by a geometric point of $S_n{}^K$. On the other hand, if we take for Λ some normal extension of K, for instance the algebraically closed field determined by K, then the results of the preceding sections show that P may be represented in $S_n{}^\Lambda$ by *a set of points*.[16]

More generally, we associate with a prime s-dimensional ideal $\mathfrak{p}_s$ in $\mathfrak{o}$ an *irreducible algebraic s-dimensional subvariety* V_s of V_r. The coördinates of the general point of this V_s are the elements $\bar{\xi}_1, \cdots, \bar{\xi}_n$ upon which the elements $\xi_1, \cdots, \xi_n$ are mapped in the homomorphism $\mathfrak{o} \simeq \mathfrak{o}/\mathfrak{p}_s$. The residue class field $\mathsf{K}_{\mathfrak{p}_s}$ (i. e. the quotient field of $\mathfrak{o}/\mathfrak{p}_s$) is the field of rational functions on V_s: $\mathsf{K}_{\mathfrak{p}_s} = \mathsf{K}(\bar{\xi}_1, \cdots, \bar{\xi}_n)$. Given two irreducible subvarieties V_s and V'_σ of V_r, defined by the prime ideals $\mathfrak{p}_s$ and $\mathfrak{p}'_\sigma$ respectively, we say that V_s belongs to V'_σ if $\mathfrak{p}'_\sigma \equiv 0(\mathfrak{p}_s)$.

8. Let V_s be an irreducible s-dimensional subvariety of V_r and let $\mathfrak{p} = \mathfrak{p}_s$ be the corresponding prime s-dimensionl $\mathfrak{o}$-ideal. We consider the quotient ring $\mathfrak{J} = \mathfrak{o}_\mathfrak{p}$. The ideal $\mathfrak{J} \cdot \mathfrak{p} = \mathfrak{P}$ is prime and maximal in $\mathfrak{J}$ and we have $\mathfrak{P} \cap \mathfrak{o} = \mathfrak{p}$.[17] The quotient ring $\mathfrak{J}_\mathfrak{P}$ is evidently $\mathfrak{J}$ itself, and the residue class field of $\mathfrak{P}$ $(= \mathfrak{J}/\mathfrak{P})$ coincides with $\mathsf{K}_\mathfrak{p}$.

DEFINITION. *V_s is said to be a simple subvariety of V_r if there exist $r - s$ elements $\eta_1, \cdots, \eta_{r-s}$ in $\mathfrak{J}$ such that:*

$$\mathfrak{J} \cdot (\eta_1, \cdots, \eta_{r-s}) = \mathfrak{P}. \tag{9}$$

Elements such as $\eta_1, \cdots, \eta_{r-s}$ shall be referred to in the sequel as *uniformizing parameters along V_s, or at V_s.*

We shall see later that if V_s is simple, then the uniformizing parameters η_i can already be found in the ring $\mathfrak{o}$. Now if $\eta_1, \cdots, \eta_{r-s}$ are in $\mathfrak{o}$, *then* (9) *is equivalent to the condition that $\mathfrak{p}$ itself occur among the maximal primary components of the ideal* $\mathfrak{o} \cdot (\eta_1, \cdots, \eta_{r-s})$,[17] i. e.

$$\mathfrak{o} \cdot (\eta_1, \cdots, \eta_{r-s}) = [\mathfrak{p}, \cdots], \tag{9'}$$

where the right-hand side is a decomposition of $\mathfrak{o} \cdot (\eta_1, \cdots, \eta_{r-s})$ into maximal primary components.

---

[16] This set is finite since the relative algebraic closure of K in $\mathsf{K}_{\mathfrak{p}_0}$ is a finite extension of K. See the footnote [14] and the considerations at the end of Section 6.

[17] Concerning the relationship between the prime ideals in $\mathfrak{o}$ and in $\mathfrak{J}$ see Section 5. To that we add that, more generally, there is a $(1, 1)$ correspondence between the $\mathfrak{J}$-ideal $\mathfrak{A}$ and those $\mathfrak{o}$-ideals $\mathfrak{a}$ which have the property that each maximal primary component of $\mathfrak{a}$ is a multiple of $\mathfrak{p}$. If $\mathfrak{A}$ and $\mathfrak{a}$ are corresponding ideals, then $\mathfrak{A} = \mathfrak{J} \cdot \mathfrak{a}$, $\mathfrak{a} = \mathfrak{A} \cap \mathfrak{o}$. If $\mathfrak{a}$ is an *arbitrary* ideal in $\mathfrak{o}$, then the ideal $\mathfrak{o} \cap \mathfrak{J} \cdot \mathfrak{a}$ differs from $\mathfrak{a}$ only by primary components which are not multiplies of $\mathfrak{p}$ (such primary components are missing in the decomposition of $\mathfrak{o} \cap \mathfrak{J} \cdot \mathfrak{a}$).

Our purpose is to derive from the above definition a number of characteristic properties of simple subvarieties. The results will be on the main generalization of theorems proved by us elsewhere [18] for simple points in the case of an algebraically closed ground field K. Practically all the rest of this paper deals with simple points. Once the results for simple points are established, the extension of these results to simple subvarieties of any dimension is rapidly achieved by the usual artifice of a transcendental extension of the ground field.

Dealing with a point P of V_r, given by a prime zero-dimensional $\mathfrak{o}$-ideal $\mathfrak{p}$, we shall proceed in the following manner. The residue class field $\mathsf{K}_\mathfrak{p}$ is a finite algebraic extension of K. Let K^* be the least normal extension of K which contains $\mathsf{K}_\mathfrak{p}$. We take K^* as new ground field and we pass to the field $\Sigma^* = \mathsf{K}^*\Sigma$ and to the ring $\mathfrak{o}^* = \mathsf{K}^*\mathfrak{o} = \mathsf{K}^*[\xi_1, \cdots, \xi_n]$. Regarded as elements of Σ^* the elements $\xi_1, \cdots, \xi_n$ are the coördinates of the general point of an irreducible variety V^*_r. Let $\mathfrak{p}^*_1, \cdots, \mathfrak{p}^*_h$ be the prime $\mathfrak{o}^*$-ideals which lie over $\mathfrak{p}$ and let $P^*_1, \cdots, P^*_h$ be the corresponding points of V^*_r. We may say that these points P^*_i correspond to the point P, and that P splits into the h points P^*_i of V^*_r. By Lemma 2 (section 2), the residue class field $\mathsf{K}^*_{\mathfrak{p}^*_i}$ coincides with $\mathsf{K}^*\mathsf{K}_\mathfrak{p}$, and since $\mathsf{K}_\mathfrak{p} \subseteqq \mathsf{K}^*$, it follows that $\mathsf{K}^*_{\mathfrak{p}^*_i} = \mathsf{K}^*$. Thus on the new variety V^*_r we now are dealing with points P^*_i $(i = 1, 2, \cdots, h)$ which have the property that for each of them *the residue class field* $\mathsf{K}^*_{\mathfrak{p}^*_i}$ *coincides with the ground field* K^*. If we now pass from K^* to *the algebraically closed field* determined by K, then each prime ideal $\mathfrak{p}^*_i$ remains prime (section 6), and it stands to reason that the results valid in the case of algebraically closed ground fields [18] can therefore be carried over to the points P^*_i of V^*_r. For this reason we study first the special case in which $\mathsf{K}_\mathfrak{p} = \mathsf{K}$. When this special case has been settled, the only thing left to do in the general case will be to study the finite ground field extension $\mathsf{K} \rightarrow \mathsf{K}^*$, where K^* is the least normal extension of K which contains $\mathsf{K}_\mathfrak{p}$.

III. Simple points. Case $\mathsf{K}_\mathfrak{p} = \mathsf{K}$.

9. Let $\mathfrak{p}$ be a prime zero-dimensional ideal in $\mathfrak{o}$ $(= \mathsf{K}[\xi_1, \cdots, \xi_n])$ and let the corresponding point P of V_r be a simple point, with $\eta_1, \cdots, \eta_r$ as uniformizing parameters, and such that the residue class field $\mathsf{K}_\mathfrak{p}$ $(= \mathfrak{o}/\mathfrak{p}$, since $\mathfrak{p}$ is divisorless) coincides with the ground field K. Let K^* be the algebraically closed field determined by K, and let $\Sigma^* = \mathsf{K}^*\Sigma$, $\mathfrak{o}^* = \mathsf{K}^*\mathfrak{o} = \mathsf{K}^*[\xi_1, \cdots, \xi_n]$, $\mathfrak{J}^* = \mathsf{K}^*\mathfrak{J}$, where $\mathfrak{J} = \mathfrak{o}_\mathfrak{p}$. As was pointed out in the preceding section, the

[18] "Some results in the arithmetic theory of algebraic varieties," *American Journal of Mathematics*, vol. 61 (April, 1939), no. 2, pp. 249-294.

ideal $\mathfrak{o}^*\mathfrak{p} = \mathfrak{p}^*$ is prime and lies over $\mathfrak{p}$. It is maximal in $\mathfrak{o}^*$, hence is zero-dimensional, and defines a point P^* of the variety V^*_r.

LEMMA 3. $\mathfrak{J}^* = \mathfrak{o}^*_{\mathfrak{p}^*}$.

Proof. Since the relation $\mathfrak{J}^* \subseteqq \mathfrak{o}^*_{\mathfrak{p}^*}$ is trivial, in view of the relation $\mathfrak{p}^* \cap \mathfrak{o} = \mathfrak{p}$, we have only to show that $\mathfrak{o}^*_{\mathfrak{p}^*}$ is contained in $\mathfrak{J}^*$. Let α^*/β^* be an element of $\mathfrak{o}^*_{\mathfrak{p}^*}$, $\alpha^*, \beta^* \subset \mathfrak{o}^*$, $\beta^* \not\equiv 0(\mathfrak{p}^*)$. Since $\mathfrak{p}^*$ is maximal, the ideal $(\mathfrak{o}^*\mathfrak{p}, \beta^*)$ is the unit ideal, i. e. we have a relation of the form

$$A^*\beta^* = 1 + \xi^*. \tag{10}$$

Here ξ^* is an element of $\mathfrak{o}^*\mathfrak{p}$ and hence can be put in the form:

$$\xi^* = \xi_0 + \xi_1\theta_1 + \cdots + \xi_{g-1}\theta_1^{g-1},$$

where $\xi_i \equiv 0(\mathfrak{p})$ and θ_1 is an element of K^* satisfying an irreducible equation of degree g over K. If $\theta_2, \cdots, \theta_g$ are the conjugates of θ_1 over K and if we multiply (10) by $\prod_{i=2}^{g} (1 + \xi^*_i)$, where $\xi^*_i = \xi_0 + \xi_1\theta_i + \cdots + \xi_{g-1}\theta_i^{g-1}$, we get a relation of the form:

$$B^*\beta^* = 1 + \eta,$$

where $\eta \equiv 0(\mathfrak{p})$ and $B^* \subset \mathfrak{o}^*$. Hence $\alpha^*/\beta^* = \alpha^*B^*/(1 + \eta)$, and since $1 + \eta$ is an element of $\mathfrak{o}$, not in $\mathfrak{p}$, it follows that α^*/β^* belongs to $\mathfrak{J}^*$, q. e. d.

Let $\mathfrak{P}^* = \mathfrak{J}^*\mathfrak{p} = \mathfrak{J}^*\mathfrak{P} = \mathfrak{J}^*\mathfrak{p}^*$, where $\mathfrak{P} = \mathfrak{p} \cdot \mathfrak{o}_\mathfrak{p}$. By (9), we have

$$\mathfrak{J}^*(\eta_1, \cdots, \eta_r) = \mathfrak{P}^*, \tag{11}$$

and since by the preceding Lemma the quotient ring $\mathfrak{o}^*_{\mathfrak{p}^*}$ coincides with $\mathfrak{J}^*$, it follows that *P^* is a simple point of V^*_r and that $\eta_1, \cdots, \eta_r$ are uniformizing parameters at P^**. Since K^* is algebraically closed, we are in position to apply the results of our paper.[18]

Since $\mathsf{K}_\mathfrak{p} = \mathsf{K}$, every element ω of $\mathfrak{J}$ satisfies a congruence of the form: $\omega \equiv c(\mathfrak{P})$, $c \subset \mathsf{K}$. In particular, let $\xi_i \equiv c_i(\mathfrak{P})$, $c_1, \cdots, c_n \subset \mathsf{K}$. The point P is therefore represented by an actual point $(c_1, \cdots, c_n)$ of the affine S_n^K. We shall assume from now on that P is the origin of coördinates in S_n^K, whence $\xi_i \equiv 0(\mathfrak{P})$, $i = 1, 2, \cdots, n$.

By (9), every element ω of $\mathfrak{P}$ can be put in the form: $\omega = A_1\eta_1 + \cdots + A_r\eta_r$, $A_i \subset \mathfrak{J}$. Let $A_i \equiv c_i(\mathfrak{P})$. Then

$$\omega \equiv c_1\eta_1 + \cdots + c_r\eta_r(\mathfrak{P}^2). \tag{12}$$

A congruence such as (12) holds true for any element ω in $\mathfrak{P}$. Since $\mathfrak{P}^* = \mathfrak{J}^*\mathfrak{P}$, it follows from (12) that $\omega \equiv c_1\eta_1 + \cdots + c_r\eta_r(\mathfrak{P}^{*2})$. We have proved in [18] that in this last congruence the coefficients $c_1, \cdots, c_r$ are uniquely determined. From this we conclude immediately with the following:

THEOREM 4. *The coefficients $c_1, \cdots, c_r$ in (12) are uniquely determined and belong to* K. *The elements $\eta_1, \cdots, \eta_r$ are linearly independent mod $\mathfrak{P}^{*2}$ over* K*. *Moreover, we have the following relation:*

$$\mathfrak{P}^{*2} \cap \mathfrak{J} = \mathfrak{P}^2. \tag{13}$$

By a similar argument and observing that $\mathfrak{P}^2 = \mathfrak{J}(\eta_1^2, \eta_1\eta_2, \cdots, \eta_r^2)$, we find that any element ω in $\mathfrak{P}$ satisfies a congruence of the form:

$$\omega \equiv c_1\eta_1 + \cdots + c_r\eta_r + c_{11}\eta_1^2 + c_{12}\eta_1\eta_2 + \cdots + c_{rr}\eta_r^2 (\mathfrak{P}^3),$$

where the coefficients $c_1, \cdots, c_r$ are the same as in (12). Proceeding in the same fashion, we find, more generally, that *for any element ω in $\mathfrak{J}$ there exists a formal integral power series:*

$$\psi_0 + \psi_1 + \cdots + \psi_m + \cdots,$$

where ψ_i is a form of degree i in $\eta_1, \cdots, \eta_r$, with coefficients in K, *such that*

$$\omega \equiv \psi_0 + \psi_1 + \cdots + \psi_m (\mathfrak{P}^{m+1}), \quad m\text{-arbitrary}. \tag{14}$$

Here $\psi_0 = 0$, if and only if $\omega \equiv 0(\mathfrak{P})$. From (14) it follows that $\omega \equiv \psi_0 + \psi_1 + \cdots + \psi_m(\mathfrak{P}^{*m+1})$, and we know that *in this congruence the polynomial $\psi_0 + \psi_1 + \cdots + \psi_m$ is uniquely determined by m.*[18] Hence, if $\omega \equiv 0(\mathfrak{P}^{*m+1})$, then this polynomial must be identically zero, and consequently

$$\mathfrak{P}^{*m} \cap \mathfrak{J} = \mathfrak{P}^m, \quad m\text{-arbitrary}, \tag{15}$$

a generalization of (13).

The result to the effect that the uniformizing parameters at P are also uniformizing parameters at P^*, can be inverted. We show namely that if P is *a simple point and if r elements $\omega_1, \cdots, \omega_r$ in $\mathfrak{o}$ are uniformizing parameters at P^*, then they are also uniformizing parameters at P,* i. e. $\mathfrak{J}^*(\omega_1, \cdots, \omega_r) = \mathfrak{P}^*$ *implies* $\mathfrak{J}(\omega_1, \cdots, \omega_r) = \mathfrak{P}$. Let

$$\omega_i \equiv c_{i1}\eta_1 + \cdots + c_{ir}\eta_r (\mathfrak{P}^{*2}), \tag{16}$$

$c_{ij} \subset$ K. It has been proved ([18], Theorem 1) that the non-vanishing of the determinant $| c_{ij} |$ is a necessary and sufficient condition in order that $\omega_1, \cdots, \omega_r$ be uniformizing parameters at P^*. Hence $| c_{ij} | \neq 0$, and since the c_{ij} are in K we conclude from (16) and (13) that $\eta_1, \cdots, \eta_r$ satisfy congruences of the form:

$$\eta_i \equiv d_{i1}\omega_1 + \cdots + d_{ir}\omega_r (\mathfrak{P}^2), \qquad (i = 1, 2, \cdots, r) \quad d_{ij} \subset \mathsf{K}.$$

Hence, by (12), every element ω in $\mathfrak{P}$ satisfies a congruence of the form: $\omega \equiv e_1\omega_1 + \cdots + e_r\omega_r(\mathfrak{P}^2)$, $e_i \subset$ K. Denote the ideal $\mathfrak{J} \cdot (\omega_1, \cdots, \omega_r)$ by $\mathfrak{A}$. The above relation implies the following relation:

$$(\mathfrak{A}, \mathfrak{P}^2) = \mathfrak{P}. \tag{17}$$

Since $\mathfrak{J}^*\mathfrak{A} = \mathfrak{P}^*$, it follows that $\mathfrak{A}$ has no prime ideal divisors in $\mathfrak{J}$ other than $\mathfrak{P}$.[19] Consequently $\mathfrak{A}$ is a primary ideal, with $\mathfrak{P}$ as associated prime ideal, and this, in view of (17), implies that $\mathfrak{A}$ coincides with $\mathfrak{P}$,[20] as was asserted.

We now are in position to prove the following

THEOREM 5. *There exist uniformizing parameters* $\omega_1, \cdots, \omega_r$ *at* P *which are elements of* $\mathfrak{o}$ *and are such that* $\mathfrak{o}$ *is integrally dependent on the ring* $K[\omega_1, \cdots, \omega_r]$. *Such parameters are furnished, for instance, by linear forms in* $\xi_1, \cdots, \xi_n$ *with non special coefficients in* K.

Proof. Since the original uniformizing parameters $\eta_1, \cdots, \eta_r$ are polynomials in $\xi_1, \cdots, \xi_n$, and since $\mathfrak{P} = \mathfrak{J} \cdot (\xi_1, \cdots, \xi_n)$, we have relations of the form: $\eta_i \equiv \sum_{j=1}^{n} c_{ij}\xi_j (\mathfrak{P}^2)$, $i = 1, 2, \cdots, r$. The r forms $\sum_{j=1}^{n} c_{ij}\xi_j$ are linearly independent mod $\mathfrak{P}^2$ (Theorem 4). Hence if $\xi_i \equiv \sum_{j=1}^{r} e_{ij}\eta_j (\mathfrak{P}^2)$, then the n by r matrix (e_{ij}) must be of rank r. If then we put $\omega_i = \sum_{j=1}^{n} u_{ij}\xi_j$, $i = 1, 2, \cdots, r$, then for non special constants u_{ij} in K the r-row square matrix $(u_{ij})(e_{j\nu})$ will be non singular. The elements $\omega_1, \cdots, \omega_r$ will then be uniformizing parameters at P^*, hence also at P.

In addition, by a well known normalization theorem of E. Noether, for non special u_{ij} the ring $\mathfrak{o}$ will be integrally dependent on $K[\omega_1, \cdots, \omega_r]$. This completes the proof of the theorem.

10. We have seen in the preceding section that if the point P is simple for V_r, then P^* is simple for V^*_r. It can be shown by examples that the converse is not generally true.[21] We prove, however, the following

THEOREM 6. *Under the hypothesis* $K_{\mathfrak{p}} = K$, *a necessary and sufficient condition that* P *be a simple point of* V_r *is that* P^* *be a simple point of* V^*_r *and that* K *be maximally algebraic in* Σ (K *algebraically closed in* Σ).

Proof. The condition is sufficient. For if K is maximally algebraic in Σ,

---

[19] Let $\mathfrak{P}_1$ be a prime ideal divisor of $\mathfrak{A}$. There exists in $\mathfrak{J}^*$ at least one prime ideal, say $\mathfrak{P}^*_1$, which lies over $\mathfrak{P}_1$ (Krull). Since $\mathfrak{P}^*_1$ is a divisor of $\mathfrak{J}^*\mathfrak{A}$, and since $\mathfrak{J}^*\mathfrak{A}$ $(= \mathfrak{P}^*)$ is maximal, necessarily $\mathfrak{P}^*_1 = \mathfrak{P}^*$, whence $\mathfrak{P}_1 = \mathfrak{P}$.

[20] Let ρ be the exponent of $\mathfrak{A}$, i. e. let $\mathfrak{P}^\rho \equiv 0(\mathfrak{A})$, $\mathfrak{P}^{\rho-1} \not\equiv 0(\mathfrak{A})$. Assuming $\rho > 1$, we multiply (17) by $\mathfrak{P}^{\rho-2}$, getting $\mathfrak{P}^{\rho-1} = (\mathfrak{A} \cdot \mathfrak{P}^{\rho-2}, \mathfrak{P}^\rho) \equiv 0(\mathfrak{A})$, a contradiction.

[21] We refer to the example given in the footnote [3]. Let $\mathfrak{p} = \mathfrak{o} \cdot (x, x\sqrt{2})$, $\mathfrak{p}^* = \mathfrak{o}^*(x)$. The point P^* is simple, and x is a uniformizing parameter at P^*. The quotient field $K_{\mathfrak{p}}$ is obviously the field K. However, P is not a simple point, since the ring $\mathfrak{p}/\mathfrak{p}^2$ is a K-module of rank 2, while, according to Theorem 4, the ring $\mathfrak{p}/\mathfrak{p}^2$ for a simple point must be of rank r. In the present case we have $r = 1$.

then the relation $\mathfrak{J}^*\mathfrak{A} \cap \mathfrak{J} = \mathfrak{A}$ holds true for any $\mathfrak{J}$-ideal $\mathfrak{A}$ (Theorem 1). If, in addition, P^* is simple for V^*_r, then from the proof of Theorem 5 it follows that we can find uniformizing parameters $\omega_1, \cdots, \omega_r$ at P^* which are elements of $\mathfrak{J}$. We will have then the relation: $\mathfrak{J}^*(\omega_1, \cdots, \omega_r) = \mathfrak{P}^*$. Since $\mathfrak{P}^* \cap \mathfrak{J} = \mathfrak{P}$ and since, by Theorem 1, $\mathfrak{J}^* \cdot (\omega_1, \cdots, \omega_r) \cap \mathfrak{J} = \mathfrak{J} \cdot (\omega_1, \cdots, \omega_r)$, it follows that $\mathfrak{J} \cdot (\omega_1, \cdots, \omega_r) = \mathfrak{P}$, whence P is a simple point of V_r.

The condition is necessary. Let $\eta_1, \cdots, \eta_r$ be uniformizing parameters at P and let θ be an element of Σ which is algebraically dependent on K. Let $\theta = \alpha/\beta$, $\alpha, \beta \subset \mathfrak{J}$, and let

$$\beta = \psi_\rho(\eta_1, \cdots, \eta_r) + \psi_{\rho+1}(\eta_1, \cdots, \eta_r) + \cdots, \psi_\rho \neq 0,$$

be the power series expansion for β (see preceding section, especially congruence (14)). The coefficients of these power series are in K. Since $\alpha = \theta\beta$ and $\theta \subset \mathsf{K}^*$, the element α will have the following power series expansion: $\alpha = \theta\psi_\rho + \theta\psi_{\rho+1} + \cdots$. Since the coefficient of the form $\theta\psi_\rho$ must also be elements of K, it follows that θ is an element of K. Hence K is algebraically closed in Σ, q. e. d.[22]

Let $\eta_1, \cdots, \eta_r$ be r algebraically independent elements in $\mathfrak{o}$ such that $\mathfrak{o}$ is integrally dependent on the ring $\mathsf{K}[\eta_1, \cdots, \eta_r]$. Let ω be an element of $\mathfrak{o}$ and let $G(\eta_1, \cdots, \eta_r; z)$ be the norm of $z - \omega$ with respect to the field $\mathsf{K}(\eta_1, \cdots, \eta_r)$. Let moreover $\eta_i \equiv c_i(\mathfrak{p})$, $c_i \subset \mathsf{K}$. In the case of an algebraically closed field K we have proved the following ([18], Theorem 4): *a necessary and sufficient condition that P be a simple point and that $\eta_1 - c_1, \cdots, \eta_r - c_r$ be uniformizing parameters at P, is that there should exist an element ω in $\mathfrak{o}$ such that $G'_\omega(\eta_1, \cdots, \eta_r; \omega) \not\equiv 0(\mathfrak{p})$.* Using Theorem 6 we are now in position to extend this result to the case under consideration ($\mathsf{K}_\mathfrak{p} = \mathsf{K}$).

Assume that P is a simple point and that the elements $\eta_1 - c_1, \cdots, \eta_r - c_r$ are uniformizing parameters at P. The elements $\eta_1 - c_1, \cdots, \eta_r - c_r$ are then also uniformizing parameters at P^*, and $\mathfrak{o}^*$ is integrally dependent on the ring $\mathsf{K}^*[\eta_1, \cdots, \eta_r]$. By the quoted theorem, proved for the algebraically closed field K^*, there exists in $\mathfrak{o}^*$ an element ω such that $F'_\omega(\eta_1, \cdots, \eta_r; \omega) \not\equiv 0(\mathfrak{p}^*)$, where $F(\eta_1, \cdots, \eta_r; z)$ is the norm of $z - \omega$ with respect to the field $\mathsf{K}^*(\eta_1, \cdots, \eta_r)$. More specifically we have shown ([18], p. 269) that we may put $\omega = v_1\xi_1 + \cdots + v_n\xi_n$, $v_i \subset \mathsf{K}^*$, provided the coefficients v_i do not satisfy certain linear relations with coefficients in K^*. Hence, we may choose the v_i in K, *and we may therefore assume that ω is an element of $\mathfrak{o}$.* The relation $F'_\omega \not\equiv 0(\mathfrak{p}^*)$ implies at any rate that $F(\eta_1, \cdots, \eta_r; z)$ is irreducible (over

[22] An immediate corollary of Theorem 6 is the following: *if K is not maximally algebraic in Σ then the residue class field $\mathsf{K}_\mathfrak{p}$ for any simple point P of V_r is necessarily a proper algebraic extension of K.* This is a special case of a more general theorem (Theorem 9) proved in Section 14.

K^*) and that ω is a primitive element of $\Sigma^*/\mathsf{K}^*(\eta_1,\cdots,\eta_r)$. By Theorem 6, K is algebraically closed in Σ. Hence, by Lemma 1, the relative degrees $[\Sigma^*:\mathsf{K}^*(\eta_1,\cdots,\eta_r)]$, $[\Sigma:\mathsf{K}(\eta_1,\cdots,\eta_r)]$ are the same. Consequently $F(\eta_1,\cdots,\eta_r;z)$ is also the norm of $z-\omega$ with respect to the field $\mathsf{K}(\eta_1,\cdots,\eta_r)$, and since $F'_\omega\not\equiv 0(\mathfrak{p}^*)$ implies $F'_\omega\not\equiv 0(\mathfrak{p})$, it is thus proved that our condition is necessary.

Conversely, assume $G'_\omega(\eta_1,\cdots,\eta_r;\omega)\not\equiv 0(\mathfrak{p})$, and let $F(\eta_1,\cdots,\eta_r;z)$ be the norm of $z-\omega$ with respect to the field $\mathsf{K}^*(\eta_1,\cdots,\eta_r)$. Clearly F either coincides with G or is a proper factor of G, according as the relative degree $[\Sigma^*:\mathsf{K}^*(\eta_1,\cdots,\eta_r)]$ coincides with or is less than the relative degree $[\Sigma:\mathsf{K}(\eta_1,\cdots,\eta_r)]$. In either case the relation $G'_\omega\not\equiv 0(\mathfrak{p})$ implies the relation $F'_\omega\not\equiv 0(\mathfrak{p}^*)$, and hence P^* is a simple point, with $\eta_1-c_1,\cdots,\eta_r-c_r$ as uniformizing parameters. To prove that P is a simple point of V_r, we have only to show, according to Theorem 6, that K is maximally algebraic in Σ. Let K' be the relative algebraic closure of K in Σ and let $[\mathsf{K}':\mathsf{K}]=g$. Let θ_1 be a primitive element of K' over K, so that $\mathsf{K}'=\mathsf{K}(\theta_1)$. Since the relative degrees $[\Sigma:\mathsf{K}'(\eta_1,\cdots,\eta_r)]$ and $[\Sigma^*:\mathsf{K}^*(\eta_1,\cdots,\eta_r)]$ are the same (Lemma 1), $F(\eta_1,\cdots,\eta_r;z)$ is the also norm of $z-\omega$ with respect to the field $\mathsf{K}'(\eta_1,\cdots,\eta_r)$. Hence F is a polynomial in $\eta_1,\cdots,\eta_r,z$ and θ_1, with coefficients in K: $F=F(\eta_1,\cdots,\eta_r;z;\theta_1)$. If $\theta_2,\cdots,\theta_g$ are the conjugates of θ_1 over K, then

$$G(\eta_1,\cdots,\eta_r;z)=\prod_{i=1}^{g}F(\eta_1,\cdots,\eta_r;z;\theta_i). \tag{18}$$

Let $\omega\equiv c(\mathfrak{p})$, $c\subset\mathsf{K}$ (since $\mathsf{K}_\mathfrak{p}=\mathsf{K}$). If we reduce the equation $F(\eta_1,\cdots,\eta_r;\omega;\theta_1)=0$ modulo $\mathfrak{p}^*$, we get $F(c_1,\cdots,c_r;c;\theta_1)=0$. Hence also $F(c_1,\cdots,c_r;c;\theta_i)=0$, $i=1,2,\cdots,g$, i. e.

$$F(\eta_1,\cdots,\eta_r;\omega;\theta_i)\equiv 0(\mathfrak{p}^*),\qquad (i=1,2,\cdots,g). \tag{18'}$$

Now if g were greater than 1, then it would follow from (18) and (18′) that $G'_\omega\equiv 0(\mathfrak{p}^*)$, i. e. $G'_\omega\equiv 0(\mathfrak{p})$, since $G'_\omega\subset\mathfrak{o}$, a contradiction. Hence $g=1$, i. e. $\mathsf{K}'=\mathsf{K}$, q. e. d.

Remark. Let $\mathfrak{o}\cdot(\eta_1,\cdots,\eta_r)=[\mathfrak{p},\mathfrak{q}_1,\mathfrak{q}_2,\cdots]$ be the decomposition of the ideal $\mathfrak{o}\cdot(\eta_1,\cdots,\eta_r)$ into maximal primary components (see (9′)), where we assume that $\eta_1,\cdots,\eta_r$ are uniformizing parameters at the simple point P and that $\mathfrak{o}$ is integrally dependent on $\mathsf{K}[\eta_1,\cdots,\eta_r]$. This last condition implies that the above primary components are all zero-dimensional. Let $\omega\equiv c(\mathfrak{p})$. For algebraically closed ground fields we have proved ([18], Theorem 4), that the elements ω such that $G'_\omega\not\equiv 0(\mathfrak{p})$ are characterized by the condition: $\omega\not\equiv c(\mathfrak{p}_i)$, $i=1,2,\cdots$, where $\mathfrak{p}_1,\mathfrak{p}_2,\cdots$ are the prime ideals to which $\mathfrak{q}_1,\mathfrak{q}_2,\cdots$ belong respectively. It is clear that this result holds true also in the present case where $\mathsf{K}_{\mathfrak{p}} = \mathsf{K}$. It is sufficient to take into account the relations: $\mathfrak{o}^*(\eta_1, \cdots, \eta_r) = \mathfrak{o}^*\mathfrak{p} \cdot \mathfrak{o}^*\mathfrak{q}_1 \cdot \mathfrak{o}^*\mathfrak{q}_2 \cdots = [\mathfrak{o}^*\mathfrak{p}, \mathfrak{o}^*\mathfrak{q}_1, \mathfrak{o}^*\mathfrak{q}_2, \cdots]$.

IV. Simple points. General case.

11. Let P be a point of V_r, $\mathfrak{p}$ the corresponding prime zero-dimensional ideal in $\mathfrak{o}$. Following the plan outlined in Section 8, we extend our ground field K as follows: we pass from K to the least normal extension K^* which contains the residue class field $\mathsf{K}_{\mathfrak{p}}$. We put: $\Sigma^* = \mathsf{K}^*\Sigma$, $\mathfrak{o}^* = \mathsf{K}^*\mathfrak{o}$, $\mathfrak{I}^* = \mathsf{K}^*\mathfrak{I}$, where $\mathfrak{I} = \mathfrak{o}_{\mathfrak{p}}$. Since K^* is a finite extension of K, there is only a finite number of prime $\mathfrak{o}^*$-ideals which lie over $\mathfrak{p}$, say $\mathfrak{p}^*_1, \cdots, \mathfrak{p}^*_h$, and we have, by Theorem 2:

$$(19) \qquad \mathfrak{o}^*\mathfrak{p} = [\mathfrak{p}^*_1, \cdots, \mathfrak{p}^*_h] = \mathfrak{p}^*_1 \cdots \mathfrak{p}^*_h.$$

We denote by P^*_i the point of V^*_r defined by the prime ideal $\mathfrak{p}^*_i$. For the residue class field $\mathsf{K}^*_{\mathfrak{p}^*_i}$ we have the relation (Lemma 2, Section 2): $\mathsf{K}^*_{\mathfrak{p}^*_i} = \mathsf{K}^* \cdot \mathsf{K}_{\mathfrak{p}} = \mathsf{K}^*$, since $\mathsf{K}_{\mathfrak{p}} \subseteqq \mathsf{K}^*$. Hence *the residue class field at each point P^*_i coincides with the new ground field K^*.* If then P^*_i is a simple point, we have, as far as V^*_r is concerned, the situation studied in the preceding Part III.

We prove the relation $\mathfrak{I}^* = \mathfrak{o}^*_{\mathfrak{p}^*_1} \cap \mathfrak{o}^*_{\mathfrak{p}^*_2} \cap \cdots \cap \mathfrak{o}^*_{\mathfrak{p}^*_h}$.

Proof. It is clear that $\mathfrak{I}^*$ is contained in each of the quotient rings $\mathfrak{o}^*_{\mathfrak{p}^*_i}$. Hence to prove the above relation we have only to show that if η^* is an element which belongs to each quotient ring $\mathfrak{o}^*_{\mathfrak{p}^*_i}$, then $\eta^* \subset \mathfrak{I}^*$. We can write $\eta^* = \alpha^*_i/\beta^*_i$, $i = 1, 2, \cdots, h$, where α^*_i, $\beta^*_i \subset \mathfrak{o}^*$, $\beta^*_i \not\equiv 0(\mathfrak{p}^*_i)$. Since the $\mathfrak{p}^*_i$ are maximal ideals, we can find, for each $i = 1, 2, \cdots, h$, an element ω^*_i in $\mathfrak{o}^*$ satisfying the congruences: $\omega^*_i \equiv 1(\mathfrak{p}^*_i)$, $\omega^*_i \equiv 0(\mathfrak{p}^*_j)$, $j \neq i$. If we put $\gamma^* = \alpha^*_1\omega^*_1 + \cdots + \alpha^*_h\omega^*_h$, $\delta^* = \beta^*_1\omega^*_1 + \cdots + \beta^*_h\omega^*_h$, then $\eta^* = \gamma^*/\delta^*$ and $\delta^* \equiv \beta^*_i \not\equiv 0(\mathfrak{p}^*_i)$. We have thus found for η^* a quotient representation γ^*/δ^* in which the denominator δ^* is not in any of the ideals $\mathfrak{p}^*_i$, $i = 1, 2, \cdots, h$. But then the ideal $(\mathfrak{o}^*\mathfrak{p}, \mathfrak{o}^*\delta^*)$ is the unit ideal, and the rest of the proof is the same as that of Lemma 3 (Section 9).

It now follows immediately that $\mathfrak{I}^*$ has h and only h distinct prime zero-dimensional ideals $\bar{\mathfrak{P}}_1, \bar{\mathfrak{P}}_2, \cdots, \bar{\mathfrak{P}}_h$ which correspond to the ideals $\mathfrak{p}^*_1, \cdots, \mathfrak{p}^*_h$ respectively. Namely, let $\mathfrak{I}^*_i = \mathfrak{o}^*_{\mathfrak{p}^*_i}$ and let $\mathfrak{P}^*_i$ be *the* prime zero-dimensional ideal of $\mathfrak{I}^*_i$:

$$(20) \qquad \mathfrak{P}^*_i = \mathfrak{I}^*_i \cdot \mathfrak{p}^*_i, \qquad \mathfrak{p}^*_i = \mathfrak{o}^* \cap \mathfrak{P}^*_i.$$

Then

$$(21) \qquad \bar{\mathfrak{P}}_i = \mathfrak{P}^*_i \cap \mathfrak{I}^* = \mathfrak{I}^* \cdot \mathfrak{p}^*_i, \qquad (i = 1, 2, \cdots, h).$$

[23]

[23] Let $\bar{\mathfrak{P}}$ be a prime zero-dimensional ideal in $\mathfrak{I}^*$, and let $\bar{\mathfrak{P}} \cap \mathfrak{o}^* = \mathfrak{p}^*$. The prime ideal $\mathfrak{p}^*$ is zero-dimensional and must coincide with one of the ideals $\mathfrak{p}^*_1, \cdots, \mathfrak{p}^*_h$,

The quotient ring $\mathfrak{J}^*_{\bar{\mathfrak{P}}_i}$ contains the quotient ring $\mathfrak{J}^*_i$, since $\mathfrak{o}^* \cap \bar{\mathfrak{P}}_i = \mathfrak{p}^*_i$. On the other hand we have $\mathfrak{J}^*_{\bar{\mathfrak{P}}_i} \subseteq \mathfrak{J}^*_i$, since $\mathfrak{P}^*_i \cap \mathfrak{J}^* = \bar{\mathfrak{P}}_i$. Hence

(22) $$\mathfrak{J}^*_{\bar{\mathfrak{P}}_i} = \mathfrak{J}^*_i = \mathfrak{o}^*_{\mathfrak{p}^*_i}.$$

12. The relations (20-22) are true for any point P, simple or not. Now we assume that P is a simple point and that $\eta_1, \cdots, \eta_r$ are uniformizing parameters at P. We have then $\mathfrak{J} \cdot (\eta_1, \cdots, \eta_r) = \mathfrak{P} = \mathfrak{J} \cdot \mathfrak{p}$. Hence, by (19) and (21), $\mathfrak{J}^* \cdot (\eta_1, \cdots, \eta_r) = \bar{\mathfrak{P}}_1 \cdots \bar{\mathfrak{P}}_h$, and consequently, in view of (22), $\mathfrak{J}^*_i \cdot (\eta_1, \cdots, \eta_r) = \mathfrak{P}^*_i$. This shows that *each of the points P^*_i is a simple point of V^*_r and that $\eta_1, \cdots, \eta_r$ are uniformizing parameters at P^*_i.* The following theorem is in a sense the converse:

THEOREM 7. *If P is a simple point of V_r and if the elements $\omega_1, \cdots, \omega_r$ of $\mathfrak{J}$ are uniformizing parameters at one of the points P^*_i, then they are also uniformizing parameters at P.*

Proof. For the proof, *we first establish the following relation:*

(23) $$\mathfrak{P}^{*m}_i \cap \mathfrak{J} = \mathfrak{P}^m, \qquad m \text{ — an arbitrary integer } \geqq 0.$$

Since $\mathfrak{P}^m \equiv 0(\mathfrak{P}^{*m}_i)$, we have only to show that any element α of $\mathfrak{P}^{*m}_i \cap \mathfrak{J}$ is contained in $\mathfrak{P}^m$. Let $\eta_1, \cdots, \eta_r$ be uniformizing parameters at P. The element α certainly belongs to $\mathfrak{P}$, and since $\mathfrak{P} = \mathfrak{J} \cdot (\eta_1, \cdots, \eta_r)$, we have: $\alpha = A_1\eta_1 + \cdots + A_r\eta_r$, $A_i \subset \mathfrak{J}$. If $A_1, \cdots, A_r$ also belong to $\mathfrak{P}$, then we can put α in the form: $\alpha = \sum_{i,j} A_{ij}\eta_i\eta_j$. Continuing in this manner we will ultimately get for α an expression of the form: $\alpha = \phi_s(\eta_1, \cdots, \eta_r)$, where ϕ_s is a form of degree s in $\eta_1, \cdots, \eta_r$ whose coefficients $A_{(j)} (= A_{j_1 \ldots j_r})$ are elements of $\mathfrak{J}$, and the following are the only two possibilities: (a) *either $s \geqq m$, or* (b) *$s < m$ and not all the elements $A_{(j)}$ are in $\mathfrak{P}$.* In the case (a) we have $\alpha \equiv 0(\mathfrak{P}^m)$, as was asserted. We show that the case (b) leads to a contradiction. Let us denote by $\phi_s^{(0)} (= \phi_s^{(0)}(\eta_1, \cdots, \eta_r))$ the reduced form obtained from $\phi_s(\eta_1, \cdots, \eta_r)$ by reducing the elements $A_{(j)}$ modulo $\mathfrak{P}^*_i$ to elements of K^*.[24] By hypothesis the form $\phi_s^{(0)}$ is not identically zero in $\eta_1, \cdots, \eta_r$. Since $\alpha \equiv 0(\mathfrak{P}^{*m}_i)$ and since $s < m$, we have obviously the congruence:

$$\phi_s^{(0)}(\eta_1, \cdots, \eta_r) \equiv 0(\mathfrak{P}^{*s+1}_i).$$

---

because any element which is not in any one of the ideals $\mathfrak{p}^*_1, \cdots, \mathfrak{p}^*_h$ is a unit in $\mathfrak{J}^*$. Let, say, $\bar{\mathfrak{P}} \cap \mathfrak{o}^* = \mathfrak{p}^*_1$. If α^*/β is an element of $\bar{\mathfrak{P}}$, where $\alpha^* \subset \mathfrak{o}^*$, $\beta \subset \mathfrak{o}$, $\beta \not\equiv 0(\mathfrak{p})$, then $\alpha^* \equiv 0(\mathfrak{p}^*_1)$, since β is a unit in $\mathfrak{J}^*$. Hence $\bar{\mathfrak{P}} = \mathfrak{J}^* \cdot \mathfrak{p}^*_1$. $\mathfrak{J}^*$ contains at least h distinct prime zero-dimensional ideals, namely the ideals $\bar{\mathfrak{P}}_i = \mathfrak{P}^*_i \cap \mathfrak{J}^*$, $i = 1, 2, \cdots, h$. They are distinct, because $\bar{\mathfrak{P}}_i \cap \mathfrak{o}^* = \mathfrak{P}^*_i \cap \mathfrak{o}^* = \mathfrak{p}^*_i$. Hence the h ideals $\mathfrak{J}^*\mathfrak{p}^*_i$, $i = 1, 2, \cdots, h$, are the only prime zero-dimensional ideals in $\mathfrak{J}^*$.

[24] We recall that K^* coincides with the residue class field of $\mathfrak{P}^*_i$.

This congruence is in contradiction with the uniqueness of the polynomial $\psi_0 + \psi_1 + \cdots + \psi_m$ satisfying the congruence (14) (Section 9) (with $\mathfrak{P}$ replaced by $\mathfrak{P}^*_i$). The uniqueness of this polynomial, for a given element ω, shows namely that no form of degree s in $\eta_1, \cdots, \eta_r$, with coefficients in K^*, can belong to $\mathfrak{P}^{*s+1}_i$. The relation (23) is thus proved.

Now let $\omega_1, \cdots, \omega_r$ be elements of $\mathfrak{o}$ which are uniformizing parameters at one of the points P^*_i, say at P^*_1. We have then the relation: $\mathfrak{J}^*_1 \cdot (\omega_1, \cdots, \omega_r) = \mathfrak{P}^*_1$. Let α be any element of $\mathfrak{P}$ and let, by (12) (Section 9):

(24) $$\alpha \equiv c^*_1\omega_1 + \cdots + c^*_r\omega_r(\mathfrak{P}^{*2}_1), \qquad c^*_i \subset \mathsf{K}^*.$$

Let, in particular,

(25) $$\eta_i \equiv c^*_{i1}\omega_1 + \cdots + c^*_{ir}\omega_r(\mathfrak{P}^{*2}_1), \qquad c^*_{ij} \subset \mathsf{K}^*, \quad (i = 1, 2, \cdots, r).$$

Since the η_i are uniformizing parameters at P, we have:

$$\omega_i = A_{i1}\eta_1 + \cdots + A_{ir}\eta_r, \qquad A_{ij} \subset \mathfrak{J}.$$

Let $A_{ij} \equiv d^*_{ij}(\mathfrak{P}^*_1)$. Since the A_{ij} are in $\mathfrak{J}$, *the elements* d^*_{ij} *are not only in* K^* *but also in* $\mathsf{K}_{\mathfrak{p}}$. We have:

(26) $$\omega_i \equiv d^*_{i1}\eta_1 + \cdots + d^*_{ir}\eta_r(\mathfrak{P}^{*2}_1), \qquad (i = 1, 2, \cdots, r).$$

The matrix (c^*_{ij}) in (25) is non-singular, since $\eta_1, \cdots, \eta_r$ are also uniformizing parameters at P^*_1. Comparing with (26) we see that the matrix (c^*_{ij}) is the inverse of the matrix (d^*_{ij}), and *consequently, since* $d^*_{ij} \subset \mathsf{K}_{\mathfrak{p}}$, *we conclude that the* c^*_{ij} *also belong to* $\mathsf{K}_{\mathfrak{p}}$. Now, let $\alpha \equiv d^*_1\eta_1 + \cdots d^*_r\eta_r(\mathfrak{P}^{*2}_1)$. The same argument by which the d^*_{ij} have been proved to belong to $\mathsf{K}_{\mathfrak{p}}$ shows that $d^*_1, \cdots, d^*_r$ belong to $\mathsf{K}_{\mathfrak{p}}$. Since $c^*_i = \sum_{j=1}^{r} d^*_j c^*_{ji}$, it follows that *also the coefficients* c^*_i in (24) *belong to* $\mathsf{K}_{\mathfrak{p}}$. Consequently there exist in $\mathfrak{J}$ elements $A_1, \cdots, A_r$ such that $A_i \equiv c^*_i(\mathfrak{P}^*_1)$, and for such elements the relation (24) implies the following:

$$\alpha \equiv A_1\omega_1 + \cdots + A_r\omega_r(\mathfrak{P}^{*2}_1).$$

Since $\alpha - A_1\omega_1 - \cdots - A_r\omega_r \subset \mathfrak{J}$, this last congruence, in view of (23), implies that:

$$\alpha \equiv A_1\omega_1 + \cdots + A_r\omega_r(\mathfrak{P}^2), \qquad A_i \subset \mathfrak{J}.$$

Such a congruence holds for any element α in $\mathfrak{P}$. Consequently

$$(\mathfrak{J} \cdot (\omega_1, \cdots, \omega_r), \mathfrak{P}^2) = \mathfrak{P},$$

and therefore, by an argument used before (footnotes [19, 20]),

$$\mathfrak{J} \cdot (\omega_1, \cdots, \omega_r) = \mathfrak{P},$$

i. e. $\omega_1, \cdots, \omega_r$ are uniformizing parameters at P, q. e. d.

13. Let $\xi^*_i = \sum_{j=1}^{n} u_{ij}\xi_j$, $i = 1, 2, \cdots, r$, $u_{ij} \subset \mathsf{K}^*$, and let $\xi^*_i \equiv v^*_i(\mathfrak{P}^*_1)$, $v^*_i \subset \mathsf{K}^*$. By Theorem 5, the elements $\xi^*_i - v^*_i$ are uniformizing parameters at P^*_1, provided the u_{ij} are "non special," and moreover, $\mathfrak{o}^*$ is integrally dependent on $\mathsf{K}^*[\xi^*_1, \cdots, \xi^*_r]$. Since the values of the u_{ij} to be avoided are those which satisfy certain algebraic relations, we may choose the u_{ij} in K, and *therefore we may assume that* $\xi^*_1, \cdots, \xi^*_r$ are elements of $\mathfrak{o}$. The constants $v^*_1, \cdots, v^*_r$ are algebraic over K. Let $f_i(v^*_i) = 0$ be the irreducible equation over K which is satisfied by v^*_i, and let us consider the elements $\omega_i = f_i(\xi^*_i)$, $i = 1, 2, \cdots, r$. These are elements in $\mathfrak{o}$ and $\mathfrak{o}$ is integrally dependent on the ring $\mathsf{K}[\omega_1, \cdots, \omega_r]$, since ξ^*_i is integrally dependent on $\mathsf{K}[\omega_i]$. Moreover, $\omega_i \equiv 0(\mathfrak{P})$, since $f_i(\xi^*_i) \equiv f_i(v^*_i)(\mathfrak{P})$ and $f_i(v^*_i) = 0$. Let $v^*_{i1} = v^*_i, v^*_{i2}, \cdots, v^*_{i,g_i}$ be the conjugates of v^*_i over K. We have: $\omega_i = f_i(\xi^*_i) = \prod_{j=1}^{g_i}(\xi^*_i - v^*_{ij})$. Since $\xi^*_i - v^*_{ij} \not\equiv 0(\mathfrak{P}^*_1)$, if $j \neq 1$, the product $\prod_{j=2}^{g_i}(\xi^*_i - v^*_{ij})$ is a unit in the quotient ring $\mathfrak{J}^*_1$. Hence

$$\mathfrak{J}^*_1 \cdot (\omega_1, \cdots, \omega_r) = \mathfrak{J}^*_1 \cdot (\xi^*_1 - v^*_1, \cdots, \xi^*_r - v^*_r) = \mathfrak{P}^*_1,$$

since $\xi^*_1 - v^*_1, \cdots, \xi^*_r - v^*_r$ are uniformizing parameters at P^*_1. It follows that also $\omega_1, \cdots, \omega_r$ are uniformizing parameters at P^*_1, and consequently they are uniformizing parameters also at P (Theorem 7). We have thus proved the following

THEOREM 8. *If P is a simple point, uniformizing parameters $\omega_1, \cdots, \omega_r$ at P can be found in such a fashion as to satisfy the conditions:* (a) $\omega_i \subset \mathfrak{o}$; (b) $\mathfrak{o}$ *is integrally dependent on* $\mathsf{K}[\omega_1, \cdots, \omega_r]$.

This is an extension of Theorem 5, except for that part of Theorem 5 which asserts that the uniformizing parameters may be chosen as linear forms in the ξ_i. This part of the theorem is not valid, of course, in the general case.

14. In this and in the following sections we wish to prove the following important theorem:

THEOREM 9. *The quotient ring $\mathfrak{J}(= \mathfrak{o}_{\mathfrak{p}})$ of a simple point P contains the relative algebraic closure of* K *in* Σ.

We shall need several lemmas. Let K' denote, as usual, the .relative algebraic closure of K in Σ.

LEMMA 4. K' *is contained in the residue class field* $\mathsf{K}_{\mathfrak{p}}$.

Proof. By the assertion $\mathsf{K}' \subseteq \mathsf{K}_{\mathfrak{p}}$ we mean the following. We know that the residue class field $\mathsf{K}^*_{\mathfrak{p}^*_i}$ at each point P^*_i $(i = 1, 2, \cdots, h)$ coincides with the ground field K^*. Since P^*_i is a simple point, it follows (Theorem 6) that K^* is algebraically closed in Σ^*, whence $\mathsf{K}' \subseteq \mathsf{K}^*$. In the homomorphic mapping of $\mathfrak{J}^*_i(= \mathfrak{o}^*_{\mathfrak{p}^*_i})$ upon $\mathsf{K}^*(= \mathfrak{o}^*/\mathfrak{p}^*_i)$, the elements of $\mathfrak{J}$ are mapped

14

upon a set of elements which form a subfield $\mathsf{K}_{\mathfrak{p}}^{(i)}$ of K^*, simply isomorphic to $\mathsf{K}_{\mathfrak{p}}$. The assertion of the lemma is to the effect that $\mathsf{K}' \subseteqq \mathsf{K}_{\mathfrak{p}}^{(i)}$, for $i = 1, 2, \cdots, h$.

The proof is similar to the second part of the proof of Theorem 6. Let $\theta \subset \mathsf{K}'$, $\theta = \alpha/\beta$, $\alpha, \beta \subset \mathfrak{J}$. Let $\beta = \psi_\rho(\eta_1, \cdots, \eta_r) + \cdots$, $\psi_\rho \neq 0$, be the expansion of β *at the point* P^*_i, in terms of the uniformizing parameters $\eta_1, \cdots, \eta_r$ at P. We have $\beta \equiv 0(\mathfrak{P}^*{}_i{}^\rho)$, whence, by (23), $\beta \equiv 0(\mathfrak{P}^\rho)$. We therefore can write β in the form: $\beta = \sum_{(i)} A_{i_1} \cdots {}_{i_r} \eta_1{}^{i_1} \cdots \eta_r{}^{i_r}$, $i_1 + \cdots + i_r = \rho$, $A_{(i)} \subset \mathfrak{J}$. The coefficients $c_{i_1} \cdots {}_{i_r}$ of the form $\psi_\rho(\eta_1, \cdots, \eta_r)$ are obviously the K^*-residues of the elements $A_{(i)}$ mod $\mathfrak{P}^*{}_i$. Since the $A_{(i)}$ are elements of $\mathfrak{J}$, we conclude that the $c_{(i)}$ belong to $\mathsf{K}_{\mathfrak{p}}^{(i)}$. For the element α we will have the expansion: $\alpha = \theta\psi_\rho + \cdots$, and by the same argument we deduce that the coefficients $\theta c_{(i)}$ belonging to $\mathsf{K}_{\mathfrak{p}}^{(i)}$. Since not all the coefficients $c_{(i)}$ are zero, it follows that θ is an element of $\mathsf{K}_{\mathfrak{p}}^{(i)}$, as was asserted.

LEMMA 5. *If $\beta \subset \mathfrak{J}$ and if $\eta_1, \cdots, \eta_r$ are uniformizing parameters at the simple point P, then the power series expansion*

$$\beta = \psi_0 + \psi_1(\eta_1, \cdots, \eta_r) + \cdots \tag{27}$$

of β at P^*_i has all its coefficients in $\mathsf{K}_{\mathfrak{p}}^{(i)}$.

Proof.[25] That ψ_0 is an element of $\mathsf{K}_{\mathfrak{p}}^{(i)}$ is trivial, since $\beta \equiv \psi_0(\mathfrak{P}^*{}_i)$ and $\beta \subset \mathfrak{J}$. We therefore use induction. We assume namely, *for every element β in $\mathfrak{J}$*, that the coefficients of $\psi_0, \psi_1, \cdots, \psi_{m-1}$ are in $\mathsf{K}_{\mathfrak{p}}^{(i)}$, and we prove that also the coefficients of ψ_m are in $\mathsf{K}_{\mathfrak{p}}^{(i)}$.

Let $\phi_\sigma = \sum_{(j)} c_{j_1 \ldots j_r} \eta_1{}^{j_1} \cdots \eta_r{}^{j_r}$, $j_1 + \cdots + j_r = \sigma$, be an arbitrary form of degree σ in $\eta_1, \cdots, \eta_r$, *whose coefficients $c_{(j)}$ are in* $\mathsf{K}_{\mathfrak{p}}^{(i)}$. Let $A_{j_1 \ldots j_r}$ be an element of $\mathfrak{J}$ such that $A_{j_1 \ldots j_r} \equiv c_{j_1 \ldots j_r}(\mathfrak{P}^*{}_i)$. If we put $\alpha = \sum_{(j)} A_{j_1 \ldots j_r} \eta_1{}^{j_1} \cdots \eta_r{}^{j_r}$, then the expansion of α at P^*_i is of the form:

$$\alpha = \phi_\sigma + \text{terms of higher degree.}$$

Let $\alpha = \phi_\sigma + \phi_{\sigma+1} + \cdots$. The form $\phi_{\sigma+\nu}$ depends in an obvious manner on the terms of degree ν of the expansion of the various elements $A_{(j)}$. By our induction, the coefficients of these terms are in $\mathsf{K}_{\mathfrak{p}}^{(i)}$, if $\nu \leqq m - 1$. *Hence the coefficients of $\phi_{\sigma+1}, \cdots, \phi_{\sigma+m-1}$ belong to* $\mathsf{K}_{\mathfrak{p}}^{(i)}$. By the same argument we can find successively elements $\alpha_1, \cdots, \alpha_{m-2}$ in $\mathfrak{J}$ such that:

$$\alpha_j = \phi^{(j)}_{\sigma+j} + \phi^{(j)}_{\sigma+j+1} + \cdots,$$

where the coefficients of the forms $\phi^{(j)}_{\sigma+j}, \cdots, \phi^{(j)}_{\sigma+j+m-1}$ are in $\mathsf{K}_{\mathfrak{p}}^{(i)}$ and

$$\phi_{\sigma+j} + \phi^{(1)}_{\sigma+j} + \cdots + \phi^{(j)}_{\sigma+j} = 0, \qquad (j = 1, 2, \cdots, m-2).$$

[25] We point out that in the course of the proof of the preceding Lemma we have incidentally established the truth of Lemma 5 *for the coefficients of the terms of lowest degree of the expansion of β.*

If we put $\gamma = \alpha + \alpha_1 + \cdots + \alpha_{m-2}$, then the expansion of γ is of the form: $\gamma = \phi_\sigma + g_{\sigma+m-1}$ + terms of higher degree, where $g_{\sigma+m-1}$ is a form of degree $\sigma + m - 1$ in $\eta_1, \cdots, \eta_r$ with coefficients in $\mathsf{K}_{\mathfrak{p}}^{(i)}$. We now take successively for ϕ_σ the forms $\psi_1, \psi_2, \cdots, \psi_{m-1}$ of (27). We get then elements $\gamma_1, \gamma_2, \cdots, \gamma_{m-1}$ such that $\gamma_1 = \psi_1 + g_m$ + terms of degree $> m$, $\gamma_j = \psi_j$ + terms of degree $> m$, $j = 2, \cdots m - 1$. *Here the coefficients of g_m are elements of $\mathsf{K}_{\mathfrak{p}}^{(i)}$*. Let $\psi_0 = \theta_1 \subset \mathsf{K}_{\mathfrak{p}}^{(i)}$ and let

$$\omega = \beta - (\gamma_1 + \gamma_2 + \cdots + \gamma_{m-1}).$$

The element ω has the following expansion at P^*_i:

$$\omega = \theta_1 + (\psi_m - g_m) + \text{terms of degree} > m. \tag{28}$$

Let $f(\theta_1) = 0$ be the irreducible equation, of degree g, which θ_1 satisfies over K, and let $\theta_2, \cdots, \theta_g$ be the conjugates of K. We have $f(\omega) = (\omega - \theta_1)(\omega - \theta_2) \cdots (\omega - \theta_g)$, and from (28) it follows immediately that:

$$f(\omega) \equiv (\psi_m - g_m) f'(\theta_1) \pmod{\mathfrak{P}^{*\,m+1}_i}.$$

Now $f(\omega)$ is an element of $\mathfrak{J}$ and $(\psi_m - g_m)f'(\theta_1)$ is the set of terms of lowest degree in the expansion of $f(\omega)$ at P^*_i. Hence [25] the coefficients of the form $(\psi_m - g_m)f'(\theta_1)$ are in $\mathsf{K}_{\mathfrak{p}}^{(i)}$. Since $\theta_1 \subset \mathsf{K}_{\mathfrak{p}}^{(i)}$, also $f'(\theta_1)$ belongs to $\mathsf{K}_{\mathfrak{p}}^{(i)}$, and since the coefficients of g_m are in $\mathsf{K}_{\mathfrak{p}}^{(i)}$, it follows that the coefficients of ψ_m belong to $\mathsf{K}_{\mathfrak{p}}^{(i)}$, as was asserted.

15. Our next lemma concerns arbitrary integral domains in which every ideal possesses a finite basis (Hilbert's basis theorem). Let $\mathfrak{S}$ be such an integral domain and let $\mathfrak{p}$ be a maximal prime ideal in $\mathfrak{S}$. We consider an arbitrary ideal $\mathfrak{A}$ in $\mathfrak{S}$ and its decomposition into maximal primary components. Let $\mathfrak{q}_1, \mathfrak{q}_2, \cdots, \mathfrak{q}_m$ be the primary components of $\mathfrak{A}$ whose prime ideals $\mathfrak{p}_1, \mathfrak{p}_2, \cdots, \mathfrak{p}_m$ are multiples of $\mathfrak{p}$. Let $\mathfrak{q}'_1, \mathfrak{q}'_2, \cdots$ be the remaining primary components of $\mathfrak{A}$; $\mathfrak{p}'_1, \mathfrak{p}'_2, \cdots,$ — their prime ideals. Thus we have:

$$\mathfrak{A} = [\mathfrak{q}_1, \mathfrak{q}_2, \cdots, \mathfrak{q}_m; \mathfrak{q}'_1, \mathfrak{q}'_2, \cdots]$$
$$\mathfrak{p}_i \equiv 0(\mathfrak{p}), \qquad (i = 1, 2, \cdots, m);$$
$$\mathfrak{p}'_j \not\equiv 0(\mathfrak{p}), \qquad (j = 1, 2, \cdots).$$

Let $\mathfrak{q}$ denote a primary ideal belonging to $\mathfrak{p}$ and let $\underset{\mathfrak{q}}{\Delta}(\mathfrak{A}, \mathfrak{q})$ be *the intersection of all the ideals* $(\mathfrak{A}, \mathfrak{q})$ *as* $\mathfrak{q}$ *runs through the totality of all primary ideals belonging to* $\mathfrak{p}$.

LEMMA 6. $\underset{\mathfrak{q}}{\Delta}(\mathfrak{A}, \mathfrak{q}) = [\mathfrak{q}_1, \mathfrak{q}_2, \cdots, \mathfrak{q}_m]$.

Proof. If we assume that the lemma is true for *primary ideals* $\mathfrak{A}$, then the lemma follows in general. Namely, we have: $\underset{\mathfrak{q}}{\Delta}(\mathfrak{A}, \mathfrak{q}) \subseteqq \underset{\mathfrak{q}}{\Delta}(\mathfrak{q}_i, \mathfrak{q})$, and hence, by your assumption:

$$(29) \qquad \underset{\mathfrak{q}}{\Delta}(\mathfrak{A}, \mathfrak{q}) \subseteqq [\mathfrak{q}_1, \mathfrak{q}_2, \cdots, \mathfrak{q}_m].$$

We put $\mathfrak{A}_1 = [\mathfrak{q}_1, \cdots, \mathfrak{q}_m]$, $\mathfrak{A}'_1 = [\mathfrak{q}'_1, \mathfrak{q}'_2, \cdots]$. Since $\mathfrak{p}$ is maximal and $\mathfrak{A}'_1 \not\equiv 0(\mathfrak{p})$, we have $(\mathfrak{A}'_1, \mathfrak{q}) = \mathfrak{S}$, for any primary ideal $\mathfrak{q}$ belonging to $\mathfrak{p}$. Consequently, we can write: $\mathfrak{A}_1 = (\mathfrak{A}_1\mathfrak{A}'_1, \mathfrak{A}_1\mathfrak{q}) \equiv 0(\mathfrak{A}, \mathfrak{q})$, whence:

$$(29') \qquad [\mathfrak{q}_1, \cdots, \mathfrak{q}_m] = \mathfrak{A}_1 \subseteqq \underset{\mathfrak{q}}{\Delta}(\mathfrak{A}, \mathfrak{q}).$$

From (29) and (29') our lemma follows.

Let now $\mathfrak{A}$ be a primary ideal, and let $\mathfrak{P}$ be the associated prime ideal. We may assume $\mathfrak{P} \equiv 0(\mathfrak{p})$, because in the contrary case the lemma is trivial. We denote by δ our ideal $\underset{\mathfrak{q}}{\Delta}(\mathfrak{A}, \mathfrak{q})$ and we first establish the following relation:

$$(30) \qquad (\delta\mathfrak{p}, \mathfrak{A}) = \delta.$$

Since $\mathfrak{A} \equiv 0(\delta)$, the relation

$$(31) \qquad (\delta\mathfrak{p}, \mathfrak{A}) \equiv 0(\delta)$$

is trivial. Let

$$(32) \qquad (\delta\mathfrak{p}, \mathfrak{A}) = [\mathfrak{q}, \mathfrak{q}'_1, \mathfrak{q}'_2, \cdots]$$

be the decomposition of the ideal $(\delta\mathfrak{p}, \mathfrak{A})$ into maximal primary components, where we assume that the prime ideal $\mathfrak{p}'_i$ associated with $\mathfrak{q}'_i$ is $\neq \mathfrak{p}$, $i = 1, 2, \cdots$, and that $\mathfrak{q}$ is either the unit ideal or belongs to $\mathfrak{p}$. Since $\mathfrak{A} \equiv 0(\mathfrak{q})$, we have, by definition of δ, that $\delta \equiv 0(\mathfrak{q})$. On the other hand $\delta\mathfrak{p} \equiv 0(\mathfrak{q}'_i)$, and since $\mathfrak{p} \not\equiv 0(\mathfrak{p}'_i)$ it follows that $\delta \equiv 0(\mathfrak{q}'_i)$. Hence $\delta \subseteqq (\delta\mathfrak{p}, \mathfrak{A})$, and (30) follows in view of (31).

By means of the relation (30) the proof of the lemma is readily completed. Let $d_1, \cdots, d_\rho$ be a basis for the ideal δ. By (30) we have the following set of relations: $d_i = \sum_{j=1}^{\rho} p_{ij}d_j + a_i$, $i = 1, 2, \cdots, \rho$, where the p_{ij} are in $\mathfrak{p}$ and $a_1, \cdots, a_\rho$ are elements of $\mathfrak{A}$. Hence

$$(33) \qquad \sum_{j=1}^{\rho}(\delta_{ij} - p_{ij})d_j \equiv 0(\mathfrak{A}), \qquad (i = 1, 2, \cdots, \rho),$$

where $\delta_{ij} = 0$ or 1, according as $i \neq j$ or $i = j$. The determinant $\lambda = |\delta_{ij} - p_{ij}|$ is of the form $1 + p$, $p \equiv 0(\mathfrak{p})$. Hence $\lambda \not\equiv 0(\mathfrak{p})$, whence a fortiori $\lambda \not\equiv 0(\mathfrak{P})$. Since $\mathfrak{A}$ is primary, with $\mathfrak{P}$ as associated prime ideal, we conclude from (33) that $d_1, \cdots, d_\rho$ belong to $\mathfrak{A}$. Hence $\delta \equiv 0(\mathfrak{A})$, and since $\mathfrak{A} \equiv 0(\delta)$, it follows that $\underset{\mathfrak{q}}{\Delta}(\mathfrak{A}, \mathfrak{q}) = \delta = \mathfrak{A}$, q. e. d.

16. Now at last we are in position to prove Theorem 9. Let θ be an element of K'. Since Σ is the quotient field of $\mathfrak{J}$, we can write $\theta = \alpha/\beta$, $\alpha, \beta \subset \mathfrak{J}$. We may assume $\beta \equiv 0(\mathfrak{P})$, because otherwise there is nothing to prove. Consider one of the points $P^*_1, \cdots, P^*_h$, say P^*_1. By Lemma 4 there exists an element γ_1 in $\mathfrak{J}$ such that $\gamma_1 \equiv \theta(\mathfrak{P}^*_1)$. Let $\gamma_1 = \theta + \psi_1^{(1)} + \psi_2^{(1)} + \cdots$

be the expansion of γ_1 at P^*_1, in terms of the uniformizing parameters $\eta_1, \cdots, \eta_r$ at P. Here $\psi_i^{(1)}$ is a form of degree i, and its coefficients are in $\mathsf{K}_{\mathfrak{p}}^{(1)}$ (Lemma 5). In particular, the coefficients of the linear form $\psi_1^{(1)}$ are in $\mathsf{K}_{\mathfrak{p}}^{(1)}$, and therefore we can find an element γ'_1 in $\mathfrak{J}$ whose expansion is of the form: $\gamma'_1 = -\psi_1^{(1)} + \cdots$. If we put $\gamma_2 = \gamma_1 + \gamma'_1$, then $\gamma_2 = \theta + \psi_2^{(2)}$ + terms of degree > 2, where $\psi_2^{(2)}$ is a quadratic form in $\eta_1, \cdots, \eta_r$ with coefficients in $\mathsf{K}_{\mathfrak{p}}^{(1)}$. Let γ'_2 be an element of $\mathfrak{J}$ whose expansion is of the form: $\gamma'_2 = -\psi_2^{(2)}$ + terms of degree > 2, and let $\gamma_3 = \gamma_2 + \gamma'_2$. Then γ_3 is an element of $\mathfrak{J}$ whose expansion at P^*_1 is of the form: $\gamma_3 = \theta + \psi_3^{(3)}$ + terms of degree > 3, and again all the coefficients in this expansion belong to $\mathsf{K}_{\mathfrak{p}}^{(1)}$ (Lemma 5). Continuing in this manner, we can find for each positive integer i an element γ_i in $\mathfrak{J}$ whose expansion is of the form: $\gamma_i = \theta + \psi_i^{(i)}$ + terms of degree $> i$. Here $\psi_i^{(i)}$ is a form of degree i in $\eta_1, \cdots, \eta_r$ with coefficients in $\mathsf{K}_{\mathfrak{p}}^{(1)}$. Hence $\gamma_i \equiv \theta(\mathfrak{P}^*{}_1{}^i)$, and since $\alpha = \theta\beta$, it follows that $\alpha - \gamma_i\beta \equiv 0(\mathfrak{P}^*{}_1{}^i)$. Now $\alpha - \gamma_i\beta$ is an element of $\mathfrak{J}$ and we have proved earlier that $\mathfrak{P}^*{}_1{}^i \cap \mathfrak{J} = \mathfrak{P}^i$ (congruence (23), Section 12). Hence

$$\alpha - \gamma_i\beta \equiv 0(\mathfrak{P}^i), \qquad (i = 1, 2, \cdots). \tag{34}$$

Let $\mathfrak{Q}$ be an arbitrary primary ideal belonging to $\mathfrak{P}$ and let ρ be the exponent of $\mathfrak{Q}$. From (34), for $i = \rho$, we deduce $\alpha \equiv 0(\beta, \mathfrak{Q})$, whence

$$\alpha \subseteqq \underset{\mathfrak{Q}}{\Delta}(\beta, \mathfrak{Q}). \tag{35}$$

Since $\mathfrak{P}$ is maximal, we may apply Lemma 6, where we put $\mathfrak{A} = \mathfrak{J} \cdot \beta$, $\mathfrak{p} = \mathfrak{P}$. In applying this lemma we must take into account that every ideal in $\mathfrak{J}$ is a multiple of $\mathfrak{P}$, whence the primary components $\mathfrak{q}'_1, \mathfrak{q}'_2 \cdots$ of $\mathfrak{A}$ are never present in the case under consideration. In other words: in the present case our lemma asserts that the intersection of the ideals $(\beta, \mathfrak{Q})$ is the ideal $\mathfrak{J} \cdot \beta$ itself. Hence, in view of (35), we conclude that $\alpha \subset \mathfrak{J} \cdot \beta$, whence α/β, i. e. θ, is an element of $\mathfrak{J}$. This completes the proof of Theorem 9.

The following theorem is a generalization of Theorem 6:

THEOREM 10. *In order that P be a simple point of V_r it is necessary and sufficient that:* (1) *$P^*_1, \cdots, P^*_h$ be simple points of V^*_r and* (2) *that the quotient ring $\mathfrak{J}(=\mathfrak{o}_{\mathfrak{p}})$ of P contain the relative algebraic closure K' of K in Σ.*

Proof. We have already proved that the conditions are necessary (see Section 12 and Theorem 9). We prove that they are sufficient. The uniformizing parameters at the simple point P^*_1 may be chosen in $\mathfrak{J}$ (Section 13). Let $\eta_1, \cdots, \eta_r$ be such uniformizing parameters. We have

$$\mathfrak{J}^*_1(\eta_1 \cdots, \eta_r) = \mathfrak{P}^*_1, \tag{36}$$

where $\mathfrak{J}^*_1 = \mathfrak{o}^*_{\mathfrak{p}^*_1}$. We now use condition (2). Since $\mathsf{K}' \subset \mathfrak{J}$, we can apply Theorem 2′ to the ring $\mathfrak{J}$ (put $\mathfrak{o} = \mathfrak{J}$, $\Delta = \Delta' = \mathsf{K}'$). Let $\mathfrak{J}^* = \mathsf{K}^*\mathfrak{J}$ and let $\bar{\mathfrak{P}}_2, \cdots, \bar{\mathfrak{P}}_\sigma, \sigma \leqq h$ (see (21)), be the conjugates of the prime ideal $\bar{\mathfrak{P}}_1$ under the relative automorphisms of Σ^* over Σ. The intersection $[\bar{\mathfrak{P}}_1, \cdots, \bar{\mathfrak{P}}_\sigma]$ is an invariant ideal and its contracted ideal in $\mathfrak{J}$ is $\mathfrak{P}$. Hence, by Theorem 2′, $[\bar{\mathfrak{P}}_1, \cdots, \bar{\mathfrak{P}}_\sigma] = \mathfrak{J}^*\mathfrak{P}$, and consequently $\sigma = h$, i. e. *the prime ideals* $\bar{\mathfrak{P}}_1, \cdots, \bar{\mathfrak{P}}_h$ *in* $\mathfrak{J}^*$ *which lie over* $\mathfrak{P}$ *form a complete set of conjugate ideals.* From (22) and (36) it follows that $\bar{\mathfrak{P}}_1$ is a maximal isolated component of the ideal $\mathfrak{J}^*(\eta_1, \cdots, \eta_r)$. Since this last ideal is invariant, it follows that all the ideals $\bar{\mathfrak{P}}_i$, $i = 1, 2, \cdots, h$, are maximal isolated components of $\mathfrak{J}^*(\eta_1, \cdots, \eta_r)$.[26] Taking into account the fact that $\bar{\mathfrak{P}}_1, \cdots, \bar{\mathfrak{P}}_h$ are the only maximal prime ideals in $\mathfrak{J}^*$, it follows that $\mathfrak{J}^*(\eta_1, \cdots, \eta_r) = \bar{\mathfrak{P}}_1 \cdots \bar{\mathfrak{P}}_h = \mathfrak{J}^*\mathfrak{P}$. Hence, by Theorem 1,

$$\mathfrak{J} \cdot (\eta_1, \cdots, \eta_r) = \mathfrak{P},$$

and this shows that P is a simple point, q. e. d.

17. In Section 10 we have extended to the case $\mathsf{K}_\mathfrak{p} = \mathsf{K}$ the theorem on the different G'_ω proved in our paper.[18] We now propose to prove this theorem in the most general case now under consideration.

Let P be a point of V_r, $\mathfrak{p}$—the corresponding prime $\mathfrak{o}$-ideal. Let $\eta_1, \cdots, \eta_r$ be algebraically independent elements of $\mathfrak{p}$ such that $\mathfrak{o}$ is integrally dependent on the ring $\mathsf{K}[\eta_1, \cdots, \eta_r]$. Given an element ω in $\mathfrak{o}$ we denote by $G(\eta_1, \cdots, \eta_r, z)$ the norm of $z - \omega$ with respect to the field $\mathsf{K}(\eta_1, \cdots, \eta_r)$.

THEOREM 11. *A necessary and sufficient condition that P be a simple point and that $\eta_1, \cdots, \eta_r$ be uniformizing parameters at P, is that there exist an element* ω such that $G'_\omega(\eta_1, \cdots, \eta_r; \omega) \not\equiv 0(\mathfrak{p})$.

We first prove that the condition is necessary. If K' is the relative algebraic closure of K in Σ, then since P is a simple point, $\mathsf{K}' \subseteqq \mathsf{K}_\mathfrak{p}$ (Theorem 9, or Lemma 4). Let

$$[\mathsf{K}_\mathfrak{p}: \mathsf{K}'] = m, \quad [\mathsf{K}': \mathsf{K}] = \mu, \quad [\mathsf{K}^*: \mathsf{K}'] = g,$$

whence $[\mathsf{K}_\mathfrak{p}: \mathsf{K}] = m\mu$. Since $\mathsf{K}' \subset \mathfrak{J}$, the ideal $\mathfrak{J}^*\mathfrak{P}$ decomposes into at most m prime ideals $\bar{\mathfrak{P}}_i$, i. e. we have $h \leqq m$ (see Section 4, where we should put $\mathsf{K} = \mathsf{K}' = \Delta$, $\bar{\Delta} = \mathsf{K}^*$, whence $\Delta_\mathfrak{p} \doteq \mathsf{K}_\mathfrak{p}$, since $\mathsf{K}_\mathfrak{p} \subseteqq \mathsf{K}^*$). On the other

[26] Hence $\eta_1, \cdots, \eta_r$ are also uniformizing parameters at $P^*_2, \cdots, P^*_h$. Without the condition $\mathsf{K}' \subset \mathfrak{J}$ this is not always true. For instance, in the example given in footnote [11] let P be the point given by the ideal $\mathfrak{o} \cdot (x, y^2 - 2, z)$. The corresponding points P^*_1, P^*_2 on V^*_r are given by the prime ideals $\mathfrak{o}^*(x, y - \sqrt{2})$, $\mathfrak{o}^*(x, y + \sqrt{2})$ respectively. The elements $\eta_1 = y^2 - 2$, $\eta_2 = z + 2x$ $(= \sqrt{2} \cdot x(y + \sqrt{2}))$ are uniformizing parameters at P^*_1 but not at P^*_2.

hand, since K' is algebraically closed in Σ, the prime ideals of $\mathfrak{J}^*\mathfrak{P}$ form a set of conjugates under the relative automorphisms of Σ^* over Σ. These automorphisms are extensions of the relative automorphisms of K^* over K'. If we now take into account the relation (6), or (6′), of Section 4, we deduce that $h = m$ and that

$$(37) \qquad \mathfrak{J}^*\mathfrak{P} = [\bar{\mathfrak{P}}_1 \cdots, \bar{\mathfrak{P}}_m] = \bar{\mathfrak{P}}_1 \cdots \bar{\mathfrak{P}}_m,$$

$$(37') \qquad \mathfrak{o}^*\mathfrak{p} = [\mathfrak{p}^*_1, \cdots, \mathfrak{p}^*_m] = \mathfrak{p}^*_1 \cdots \mathfrak{p}^*_m.$$

Let $F(\eta, \cdots, \eta_r; z)$ denote the norm of $z - \omega$ with respect to the field $\mathsf{K}^*(\eta_1, \cdots, \eta_r)$, where ω is an element of $\mathfrak{o}^*$. Our theorem is true for each of the simple points $P^*_1, \cdots, P^*_m$, in view of the fact that at each of these points the residue class field coincides with the ground field K^*. Thus, dealing with the point P^*_1, we may assert that there exists an element ω in $\mathfrak{o}^*$ such that

$$(38) \qquad F'_\omega(\eta_1, \cdots, \eta_r; \omega) \not\equiv 0(\mathfrak{p}^*_1).$$

With the aid of the Remark at the end of Section 10 we proceed to make a judicious choice of the element ω. Let $\theta = \theta_1^{(1)}$ be a primitive element of $\mathsf{K}_\mathfrak{p}^{(1)}$ with respect to K [27] and let $\theta_i^{(j)}$, $i = 1, 2, \cdots, m$, $j = 1, 2, \cdots, \mu$, be the conjugates of θ over K. We choose the notations in such a fashion that $\theta_1^{(j)}, \theta_2^{(j)}, \cdots, \theta_m^{(j)}$ is a complete set of conjugates with respect to the intermediate field K'. Let $f(\theta) = 0$ be the irreducible equation, of degree $m\mu$, which θ satisfies over K. Since $\theta \subset \mathsf{K}_\mathfrak{p}^{(1)}$, there exists an element ζ in $\mathfrak{o}$ such that

$$(39) \qquad \zeta \equiv \theta_1^{(1)}(\mathfrak{p}^*_1),$$

whence

$$(39') \qquad f(\zeta) \equiv 0(\mathfrak{p}).$$

Let

$$\mathfrak{o} \cdot (\eta_1, \cdots, \eta_r) = [\mathfrak{p}, \mathfrak{q}_1, \cdots, \mathfrak{q}_\sigma]$$

be the decomposition of the ideal $\mathfrak{o} \cdot (\eta_1, \cdots, \eta_r)$ into primary maximal components (see (9′), Section 8), and let $\mathfrak{p}_1, \cdots, \mathfrak{p}_\sigma$ be prime ideals associated with the primary ideals $\mathfrak{q}_1, \cdots, \mathfrak{q}_\sigma$ respectively. The ideals $\mathfrak{p}_1, \cdots, \mathfrak{p}_\sigma$ are zero-dimensional, since $\mathfrak{o}$ is integrally dependent on the ring $\mathsf{K}[\eta_1, \cdots, \eta_r]$. *We show that there exists an element ω in $\mathfrak{o}$ such that*

$$(40) \qquad f(\omega) \equiv 0(\mathfrak{p}),$$

$$(40') \qquad f(\omega) \not\equiv 0(\mathfrak{p}_i), \qquad (i = 1, 2, \cdots, \sigma).$$

[27] $\mathsf{K}_\mathfrak{p}^{(1)}$ is a subfield of K^*, simply isomorphic to $\mathsf{K}_\mathfrak{p}$, and is contained in the residue class field of the point P^*_1. This field has been first introduced in the proof of Lemma 4 (Section 14). The fields $\mathsf{K}_\mathfrak{p}^{(1)}, \mathsf{K}_\mathfrak{p}^{(2)}, \cdots, \mathsf{K}_\mathfrak{p}^{(m)}$ are conjugate fields over K', not necessarily distinct.

We put, namely, $\omega = \zeta + c\pi$, where $c \subset \mathsf{K}$ and π is an element of $\mathfrak{p}$ but not in $\mathfrak{p}_i$, $i = 1, 2, \cdots, \sigma$. Since $\omega \equiv \zeta(\mathfrak{p})$, the congruence $f(\omega) \equiv 0(\mathfrak{p})$ follows from (39′). On the other hand we have:

$$f(\omega) = f(\zeta + c\pi) = f(\zeta) + c \cdot \pi f'(\zeta) + \cdots + c^{m\mu}\pi^{m\mu}$$

(we assume that the leading coefficient of $f(\theta)$ is 1). This is a polynomial in c with coefficients not all $\equiv 0(\mathfrak{p}_i)$, $i = 1, 2, \cdots, \sigma$, since $\pi^{m\mu} \not\equiv 0(\mathfrak{p}_i)$. Hence we can find the constant c in K so as to satisfy the relation $f(\omega) \not\equiv 0(\mathfrak{p}_i)$, for all $i = 1, 2, \cdots, \sigma$. *We assert that the element ω of $\mathfrak{o}$, constructed in this fashion, necessarily satisfies* (38). Namely, let $\mathfrak{p}^*_{i1}, \mathfrak{p}^*_{i2}, \cdots$ be the prime ideals in $\mathfrak{o}^*$ which lie over $\mathfrak{p}_i$. The ideals $\mathfrak{p}^*_1, \cdots, \mathfrak{p}^*_m, \mathfrak{p}^*_{ij}$ are then the prime ideals associated with the primary components of the ideal $\mathfrak{o}^*(\eta_1, \cdots, \eta_r)$. Since $\mathfrak{p}^*_2, \cdots, \mathfrak{p}^*_m$ are the conjugates of $\mathfrak{p}^*_1$, it follows, by (39), that

(39″) $$\omega \equiv \theta_i^{(1)}(\mathfrak{p}^*_i),$$

whence

(41) $$\omega \not\equiv \theta_1^{(1)}(\mathfrak{p}^*_i), \qquad (i = 2, \cdots, m).$$

Moreover, by (40′), we have $f(\omega) \not\equiv 0(\mathfrak{p}^*_{ij})$, $j = 1, 2, \cdots$, whence, in particular,

(41′) $$\omega \not\equiv \theta_1^{(1)}(\mathfrak{p}^*_{ij}), \qquad (i = 1, 2, \cdots, \sigma;\ j = 1, 2, \cdots).$$

The relations (39″), (41) and (41′) imply (38), in view of the remark at the end of Section 10.

Since K' is algebraically closed in Σ, it follows, by an argument repeatedly used before and based on Lemma 1, that the norm of $z - \omega$ with respect to the field $\mathsf{K}^*(\eta_1, \cdots, \eta_r)$ is the same as the norm of $z - \omega$ with respect to the field $\mathsf{K}'(\eta_1, \cdots, \eta_r)$. Therefore the coefficients of $F(\eta_1, \cdots, \eta_r, z)$ are in K'. If $F_1, \cdots, F_{\mu-1}$ denote the conjugate polynomials of F with respect to K and if $G(\eta_1, \cdots, \eta_r; z)$ is the norm of $z - \omega$ wtih respect to $\mathsf{K}(\eta_1, \cdots, \eta_r)$, then obviously

(42) $$G(\eta_1, \cdots, \eta_r; z) = FF_1 \cdots F_{\mu-1}.$$

If we put $F(0, \cdots, 0; z) = \phi(z)$, then $\phi(\omega) \equiv 0(\mathfrak{p}^*_1)$, since $F(\eta_1, \cdots, \eta_r; \omega) = 0$ and $\eta_i \equiv 0(\mathfrak{p}^*_1)$. Consequently $\phi(z)$ is divisible by $z - \theta_1^{(1)}$ (39″). *But* (38) *implies that $\phi(z)$ is not divisible by $(z - \theta_1^{(1)})^2$*. Since $F(\eta_1, \cdots, \eta_r, z)$ is invariant under the relative automorphism of Σ^* over Σ (its coefficient being in K'), it follows likewise that $\phi(z)$ is divisible by $z - \theta_i^{(1)}$, but is not divisible by $(z - \theta_i^{(1)})^2$, $i = 1, 2, \cdots, m$.

We can also show that $\phi(z)$ is not divisible by $z - \theta_i^{(j)}$, for all $j \neq 1$ and $i = 1, 2, \cdots, m$. In fact, if say $\phi(z)$ was divisible by $z - \theta_1^{(2)}$, then $\theta_1^{(2)}$ would have to be the value of ω at some prime ideal $\mathfrak{p}^*$ belonging to the ideal $\mathfrak{o}^* \cdot (\eta_1, \cdots, \eta_r)$,[28] i. e. $\omega \equiv \theta_1^{(2)}(\mathfrak{p}^*)$. This ideal $\mathfrak{p}^*$ could not be any of the ideals $\mathfrak{p}^*_1, \cdots, \mathfrak{p}^*_m$, since $\omega \equiv \theta_i^{(1)}(\mathfrak{p}^*_i)$, $i = 1, 2, \cdots, m$, and $\theta_1^{(2)} \neq \theta_i^{(1)}$. Hence $\mathfrak{p}^*$ would have to be one of the ideals $\mathfrak{p}^*_{i1}, \mathfrak{p}^*_{i2}, \cdots$, $i = 1, 2, \cdots, \sigma$, considered above. This, however, would be in contradiction with (40′).

The polynomial $\phi(z)$ therefore factors as follows:

$$\phi(z) = \prod_{i=1}^{m} (z - \theta_i^{(1)}) \cdot \psi(z),$$

where $\psi(z)$ and $f(z)$ have no common roots. Let $\phi_j(z) (= F_j(0, \cdots, 0, z))$, $j = 1, 2, \cdots, \mu - 1$, be the conjugates of $\phi(z)$ with respect to K. We will have likewise:

$$\phi_j(z) = \prod_{i=1}^{m} (z - \theta_i^{(j)}) \cdot \psi_j(z), \qquad (j = 1, 2, \cdots, \mu - 1),$$

where again $\psi_j(z)$ and $f(z)$ have no common roots. By (42) we have:

$$G(0, \cdots, 0; z) = \phi(z)\phi_1(z) \cdots \phi_{\mu-1}(z) = f(z) \cdot A(z),$$

where $A(z) (= \psi\psi_1 \cdots \psi_{\mu-1})$ and $f(z)$ are relatively prime. Since $\eta_i \equiv 0(\mathfrak{p})$, we have: $G'_\omega \equiv f'(\omega) \cdot A(\omega) + f(\omega) \cdot A'(\omega)(\mathfrak{p})$, i. e. $G'_\omega \equiv f'(\omega) \cdot A(\omega)(\mathfrak{p})$, since $f(\omega) \equiv 0(\mathfrak{p})$ (40). Now we observe that $f'(\omega) \not\equiv 0(\mathfrak{p})$, since $f(\omega) \equiv 0(\mathfrak{p})$ and f is an irreducible polynomial, and we also note that $A(\omega) \not\equiv 0(\mathfrak{p})$, since $A(z)$ is not divisible by $f(z)$. Hence $G'_\omega \not\equiv 0(\mathfrak{p})$, as was asserted.

18. Continuation of the proof. The condition is sufficient. From (42) we deduce in the first place that $G'_\omega \not\equiv 0(\mathfrak{p})$ implies the relations $F'_\omega \not\equiv 0(\mathfrak{p}^*_i)$, $i = 1, 2, \cdots, h$. Hence from the hypothesis $G'_\omega \not\equiv 0(\mathfrak{p})$ follows at any rate that the points $P^*_1, \cdots, P^*_h$ are simple (by the special case $K_\mathfrak{p} = K$; see Section 10). It remains to prove that $K' \subset \mathfrak{J}$ (Theorem 10). We shall prove the following stronger result: *$\mathfrak{J}$ is integrally closed in Σ.* Since $K \subset \mathfrak{J}$ and K' is an algebraic extension of K, the property of $\mathfrak{J}$ being integrally closed will obviously imply that K' is a subfield of $\mathfrak{J}$. Now to show that $\mathfrak{J}$ is integrally closed in Σ, we consider the complementary module e [29] of the ring $K[\eta_1, \cdots, \eta_r, \omega]$. If ν denotes the relative degree $[\Sigma : K(\eta_1, \cdots, \eta_r)]$, then it is well known that the elements

$$1/G'_\omega, \omega/G'_\omega, \cdots, \omega^{\nu-1}/G'_\omega$$

form a module basis for e with respect to the ring $K[\eta_1, \cdots, \eta_r]$. Since $G'_\omega \not\equiv 0(\mathfrak{p})$, it follows therefore that *e is contained in $\mathfrak{J}$.* Since e contains

[28] This assertion has been proved in the case of an algebraically closed ground field ([18], p. 263, footnote [12]) and therefore is obviously true for any ground field.

[29] The module e consists of those elements ζ of Σ for which the trace $T(\zeta \cdot \alpha)$ is in $K[\eta_1, \cdots, \eta_r]$, *for every element* α in $K[\eta, \cdots, \eta_r, \omega]$.

the integral closure of the ring $\mathsf{K}[\eta_1, \cdots, \eta_r]$ and since $\mathfrak{o}$ is integrally dependent on this ring, we conclude that $\mathfrak{J}$ *contains the integral closure* $\bar{\mathfrak{o}}$ *of* $\mathfrak{o}$. Let $\bar{\mathfrak{p}}$ be the contracted ideal of $\mathfrak{P}$ in $\bar{\mathfrak{o}}$: $\bar{\mathfrak{p}} = \mathfrak{P} \cap \bar{\mathfrak{o}}$. We have then $\mathfrak{J} = \mathfrak{J}_{\mathfrak{P}} \supseteqq \bar{\mathfrak{o}}_{\bar{\mathfrak{p}}}$. On the other hand $\bar{\mathfrak{o}}_{\bar{\mathfrak{p}}} \supseteqq \mathfrak{o}_{\mathfrak{p}}$ (since $\bar{\mathfrak{p}} \cap \mathfrak{o} = \mathfrak{p}$), whence $\bar{\mathfrak{o}}_{\bar{\mathfrak{p}}} \supseteqq \mathfrak{J}$. Consequently $\mathfrak{J} = \bar{\mathfrak{o}}_{\bar{\mathfrak{p}}}$, and therefore $\mathfrak{J}$ is integrally closed (since for the integrally closed ring $\bar{\mathfrak{o}}$ it is true that the quotient ring of any prime $\bar{\mathfrak{o}}$-ideal is integrally closed). This completes the proof of Theorem 11.

As a corollary we have the following

THEOREM 12. *The quotient ring* $\mathfrak{J}$ *of a simple point is integrally closed in its quotient field.*

Remark. If P is a simple point, with $\eta_1, \cdots, \eta_r$ as uniformizing parameters, and if $\mathfrak{o}$ is integrally dependent on $\mathsf{K}[\eta_1, \cdots, \eta_r]$, then *the elements* ω *of* $\mathfrak{o}$ *such that* $G'_\omega \not\equiv 0(\mathfrak{p})$ *are characterized by the relations* (40), (40′), *where the irreducible polynomial* $f(\omega)$ *must be of degree* $m\mu = [\mathsf{K}_{\mathfrak{p}} : \mathsf{K}]$. This follows from the first part of the proof of Theorem 11 (Section 17) and from the remark at the end of Section 10.

In Theorem 11 we have assumed that the elements $\eta_1, \cdots, \eta_r$ are in $\mathfrak{p}$. We now drop this assumption, and we consider the uniquely determined *irreducible* polynomials $f_i(z)$ in $\mathsf{K}[z]$ $(i = 1, 2, \cdots, r)$ such that $f_i(\eta_i) \equiv 0(\mathfrak{p})$. We can easily prove the following stronger theorem:

THEOREM 11′. *A necessary and sufficient condition that* P *be a simple point and that* $f_1(\eta_1), \cdots, f_r(\eta_r)$ *be uniformizing parameters at* P *is that there exist an element* ω *in* $\mathfrak{o}$ *such that* $G'_\omega(\eta_1, \cdots, \eta_r; \omega) \not\equiv 0(\mathfrak{p})$.

That the condition is necessary follows almost immediately from Theorem 11. Namely, let us put $\zeta_i = f_i(\eta_i)$ and let $H(\zeta_1, \cdots, \zeta_r; z)$ denote the norm of $z - \omega$ with respect to the field $\mathsf{K}(\zeta_1, \cdots, \zeta_r)$, while $G(\eta_1, \cdots, \eta_r; z)$ denotes, as before, the norm of $z - \omega$ with respect to the field $\mathsf{K}(\eta_1, \cdots, \eta_r)$. Since this last field contains the field $\mathsf{K}(\zeta_1, \cdots, \zeta_r)$, it is clear that we have an identity (in z) of the form:

$$H(\zeta_1, \cdots, \zeta_r; z) = G(\eta_1, \cdots, \eta_r; z) \cdot A(\eta_1, \cdots, \eta_r; z),$$

where A is a polynomial with coefficients in K. Since $G(\eta_1, \cdots, \eta_r; \omega) = 0$ we have

$$H'_\omega(\zeta_1, \cdots, \zeta_r; \omega) = G'_\omega(\eta_1, \cdots, \eta_r; \omega) \cdot A(\eta_1, \cdots, \eta_r; \omega).$$

Now, by hypothesis, $\zeta_1, \cdots, \zeta_r$ are uniformizing parameters at P, and moreover, $\mathfrak{o}$ depends integrally on the ring $\mathsf{K}[\zeta_1, \cdots, \zeta_r]$, since each element η_i is integrally dependent on $\mathsf{K}[\zeta_i]$. Hence, by Theorem 11, we must have $H'_\omega \not\equiv 0(\mathfrak{p})$, for a suitable element ω in $\mathfrak{o}$. The above identity shows then that we must also have $G'_\omega \not\equiv 0(\mathfrak{p})$, as was asserted.

The proof that the condition is sufficient is direct and follows the lines of the proof of sufficiency given in the special case of Theorem 11. The relation $G'_\omega \not\equiv 0(\mathfrak{p})$ implies the relations $F'_\omega(\eta_1, \cdots, \eta_r; \omega) \not\equiv 0(\mathfrak{p}^*_i)$, for $i = 1, 2, \cdots, h$. Hence $P^*_1, \cdots, P^*_h$ are simple points and $\eta_1 - \theta_1^{(i)}, \cdots, \eta_r - \theta_r^{(i)}$ are uniformizing parameters at P^*_i, where the $\theta_i^{(j)}$ are elements of K^* and $\eta_j \equiv \theta_j^{(i)}(\mathfrak{p}^*_i)$ (Section 10). Since $f_j(\eta_j) \equiv 0(\mathfrak{p}) \equiv 0(\mathfrak{p}^*_i)$, $\theta_j^{(i)}$ must be a root of $f_j(z)$. Since $f_j(z)$ has no repeated roots, it follows that $f_j(\eta_j)$ differs from $\eta_j - \theta_j^{(i)}$ by a factor which is a unit in the quotient ring $\mathfrak{J}^*_i$ of the point P^*_i. Hence $f_1(\eta_1), \cdots, f_r(\eta_r)$, i. e. $\zeta_1, \cdots, \zeta_r$ are also uniformizing parameters at P^*_i $(i = 1, 2, \cdots, h)$. It remains to prove that $\mathsf{K}' \subset \mathfrak{J}$, since from this it will follow that P is a simple point (Theorem 10) and that $\zeta_1, \cdots, \zeta_r$ are uniformizing parameters at P (Theorem 7). The rest of the proof is the same as before, namely it is shown that $\mathfrak{J}$ is integrally closed in Σ.

COROLLARY. *If V_r is a linear space, i. e. if $\mathfrak{o}$ is a polynomial ring, then every point of V_r is simple.*

In fact, if $\mathfrak{o} = \mathsf{K}[\xi_1, \cdots, \xi_r]$, where $\xi_1, \cdots, \xi_r$ are algebraically independent elements, then the norm G of $z - \omega$ with respect to the field $\mathsf{K}(\xi_1, \cdots, \xi_r)$ is $z - \omega$ itself, whence $G'_\omega = 1$.

V. Simple subvarieties.

19. Let V_s be an irreducible *simple* subvariety of V_r, of dimension s. Let $\mathfrak{p}$ be the corresponding s-dimensional prime ideal in $\mathfrak{o}$ and let $\mathfrak{J} = \mathfrak{o}_\mathfrak{p}$. By definition, there exist $r - s$ elements in $\mathfrak{J}$, say $\eta_1, \cdots, \eta_{r-s}$, such that

$$\mathfrak{J} \cdot (\eta_1, \cdots, \eta_{r-s}) = \mathfrak{P} = \mathfrak{J} \cdot \mathfrak{p}. \tag{43}$$

Let $\zeta_1, \cdots, \zeta_s$ be elements of $\mathfrak{J}$ which are algebraically independent mod $\mathfrak{p}$ (with respect to the ground field K), but otherwise arbitrary. We take as new ground field the field $\Omega = \mathsf{K}(\zeta_1, \cdots, \zeta_s)$ and we put $\bar{\mathfrak{o}} = \Omega \cdot \mathfrak{o}$, $\bar{\mathfrak{p}} = \bar{\mathfrak{o}}\mathfrak{p}$. Since Ω is a pure transcendental extension of K, $\bar{\mathfrak{p}}$ is prime. It is of dimension zero, since the ζ's are algebraically independent mod $\mathfrak{p}$. With Ω as new ground field, the elements $\xi_1, \cdots, \xi_n$ define an $(r - s)$-dimensional variety $\bar{V}_{r-s}$ of which they are the coördinates of the general point. The ideal $\bar{\mathfrak{p}}$ defines a point $\bar{P}$ on $\bar{V}_{r-s}$. The quotient ring $\bar{\mathfrak{J}} = \bar{\mathfrak{o}}_{\bar{\mathfrak{p}}}$ coincides with $\mathfrak{J}$, since the elements of Ω are units in $\mathfrak{J}$. Hence, by (43),

$$\bar{\mathfrak{J}} \cdot (\eta_1, \cdots, \eta_{r-s}) = \bar{\mathfrak{J}} \cdot \bar{\mathfrak{p}}. \tag{43'}$$

i. e. *$\bar{P}$ is a simple point of $\bar{V}_{r-s}$.*

Conversely, assume that $\bar{P}$ is a simple point of $\bar{V}_{r-s}$. There will exist then elements $\eta_1, \cdots, \eta_{r-s}$ in $\bar{\mathfrak{J}}$, i. e. in $\mathfrak{J}$, satisfying (43′). The relation (43) is equivalent to (43′). Hence V_s is a simple subvariety of V_r. Thus the *assertions: V_s is a simple subvariety of V_r; $\bar{P}$ is a simple point of $\bar{V}_{r-s}$—are equivalent.*

THEOREM 13. *The quotient ring $\mathfrak{J}(=\mathfrak{o}_{\mathfrak{p}})$ of a simple subvariety is integrally closed in Σ.*

The theorem is an immediate consequence of Theorem 12 and of the fact that $\mathfrak{J}=\bar{\mathfrak{J}}=\bar{\mathfrak{o}}_{\bar{\mathfrak{p}}}$.

THEOREM 14. *If V_s is a simple subvariety of V_r, the uniformizing parameters $\eta_1,\cdots,\eta_{r-s}$ along V_s and the elements $\zeta_1,\cdots,\zeta_s$ algebraically independent mod $\mathfrak{p}$, can be so chosen that they be elements of $\mathfrak{o}$ and that $\mathfrak{o}$ be integrally dependent on the ring $K[\eta_1,\cdots,\eta_{r-s};\zeta_1,\cdots,\zeta_s]$.*

Proof. Let $\bar{\mathfrak{o}}=\mathfrak{o}/\mathfrak{p}=K[\bar{\xi}_1,\cdots,\bar{\xi}_n]$, whence $\bar{\mathfrak{o}}$ is of degree of transcendency s over K. Subject to a preliminary linear homogeneous transformation on $\xi_1,\cdots,\xi_n$, with coefficients u_{ij} in K, we may assume that the following conditions are satisfied.

(a) $\xi_{r+1},\cdots,\xi_n$ are integrally dependent on $K[\xi_1,\cdots,\xi_r]$;

(b) $\bar{\xi}_{s+1},\cdots,\bar{\xi}_n$ are integrally dependent on $K[\bar{\xi}_1,\cdots,\bar{\xi}_s]$;

(c) $f_i(\xi_1,\cdots,\xi_s;\xi_{s+i})$, $i=1,2,\cdots,r-s$, are uniformizing parameters along V_s, where f_i is an irreducible polynomial in $\xi_1,\cdots,\xi_s,\xi_{s+i}$ with coefficients in K.

The possibility of satisfying conditions (a) and (b) is trivial. As to condition (c), we observe that by (b) the elements $\xi_1,\cdots,\xi_s$ are algebraically independent mod $\mathfrak{p}$. We put $\zeta_i=\xi_i$, $i=1,2,\cdots,s$. It then follows from the proof of Theorem 8 (Section 13) that for "non special" u_{ij}, certain polynomials $f_i(\xi_{s+i})$, $i=1,2,\cdots,r-s$, with coefficients in $\Omega(=K(\zeta_1,\cdots,\zeta_s))$ will be uniformizing parameters at the point $\bar{P}$ of $\bar{V}_{r-s}$, hence also along V_s. The values of the u_{ij} to be avoided are those which satisfy certain algebraic relations with coefficients in Ω. Since K contains infinitely many elements, the u_{ij} may be chosen in K. We may assume that the leading coefficient of f_i is 1. If we put $\eta_i=f_i(\xi_1,\cdots,\xi_s;\xi_{s+i})$, $i=1,2,\cdots,r-s$, then $\zeta_1,\cdots,\zeta_s$ $(=\xi_1,\cdots,\xi_s)$ are algebraically independent mod $\mathfrak{p}$ and $\eta_1,\cdots,\eta_{r-s}$ are uniformizing parameters along V_s. Since $\eta_i\equiv 0(\mathfrak{p})$, we find, passing to the ring $\bar{\mathfrak{o}}$: $f_i(\bar{\xi}_1,\cdots,\bar{\xi}_s;\bar{\xi}_{s+i})=0$. Since f_i is irreducible, it follows by condition (b) that the coefficients of f_i are polynomials in $\xi_1,\cdots,\xi_s$. Hence ξ_{s+i} is integrally dependent on $K[\xi_1,\cdots,\xi_s;\eta_i]$, $i=1,2,\cdots,r-s$, whence

$\xi_1, \cdots, \xi_r$ are integrally dependent on the ring $K[\zeta_1, \cdots, \zeta_s; \eta_1, \cdots, \eta_{r-s}]$. This completes the proof of the theorem, in view of condition (a).

20. Let $\eta_1, \cdots, \eta_r$ be algebraically independent elements of $\mathfrak{o}$ such that $\mathfrak{o}$ is integrally dependent on the ring $K[\eta_1, \cdots, \eta_r]$. The residual class ring $\tilde{\mathfrak{o}} = \mathfrak{o}/\mathfrak{p}$ will depend integrally on the ring $K[\tilde{\eta}_1, \cdots, \tilde{\eta}_r]$, and since $\tilde{\mathfrak{o}}$ is of degree of transcendency s over K, s elements $\tilde{\eta}_i$ have to be algebraically independent. Consequently s of the elements η_i are algebraically independent mod $\mathfrak{p}$. We assume that $\eta_1, \cdots, \eta_s$ are algebraically independent mod $\mathfrak{p}$. Let $f_i(\eta_1, \cdots, \eta_s; \eta_{s+i}) \equiv 0(\mathfrak{p})$ be the irreducible congruence which η_{s+i} satisfies over $K[\eta_1, \cdots, \eta_s]$. Let moreover, $F(\eta_1, \cdots, \eta_r; z)$ be the norm of $z - \omega (\omega \subset \mathfrak{o})$, over the field $K(\eta_1, \cdots, \eta_r)$. As a generalization of Theorem 11′, we prove the following

THEOREM 15. *The existence of an element* ω *in* $\mathfrak{o}$ such that $F'_\omega(\eta_1, \cdots, \eta_r; \omega) \not\equiv 0(\mathfrak{p})$ *is a necessary and sufficient condition in order that* V_s *be a simple subvariety of* V_r *and that* $f_1(\eta_1, \cdots, \eta_s, \eta_{s+1}), \cdots, f_{r-s}(\eta_1, \cdots, \eta_s; \eta_r)$ *be uniformizing parameters along* V_s.

The theorem is an immediate consequence of Theorem 11′. It is sufficient to observe that $F(\eta_1, \cdots, \eta_r; z)$ is also the norm of $z - \omega$ with respect to the field $\Omega(\eta_{s+1}, \cdots, \eta_r)$, where $\Omega = K(\eta_1, \cdots, \eta_s)$. Moreover, as was pointed out in the preceding section, the subvariety V_s of V_r and the point $\tilde{P}$ of $\tilde{V}_{r-s}$ (the elements $\eta_1, \cdots, \eta_s$ now play the rôle of $\zeta_1, \cdots, \zeta_s$) are both simple or not simple at the same time, and that uniformizing parameters along V_s are also uniformizing parameters at $\tilde{P}$, and conversely (see (43) and (43′)).

An immediate corollary of Theorem 15 is the following: *if* V_s *contains a simple point* P *of* V_r, *then* V_s *itself is a simple subvariety*. In fact, if $\mathfrak{p}_0$ is the prime zero-dimensional $\mathfrak{o}$-ideal which corresponds to the point P, then $\mathfrak{p} \equiv 0(\mathfrak{p}_0)$, and $G'_\omega \not\equiv 0(\mathfrak{p}_0)$ (Theorem 11′) implies the relation $G'_\omega \not\equiv 0(\mathfrak{p})$.

We can invert this result. We show namely that *a simple subvariety* V_s *contains at least one simple point of* V_r. Using the notations of Theorem 15, if V_s is simple, then $G'_\omega \not\equiv 0(\mathfrak{p})$, for some ω in $\mathfrak{o}$. We can therefore find a prime zero-dimensional divisor $\mathfrak{p}_0$ of $\mathfrak{p}$ such that $G'_\omega \not\equiv 0(\mathfrak{p}_0)$.[30] If P is the point of V_r defined by $\mathfrak{p}_0$, then P is simple (Theorem 11′) and lies on V_s (since $\mathfrak{p} \equiv 0(\mathfrak{p}_0)$).

THE JOHNS HOPKINS UNIVERSITY.

---

[30] If we pass to the ring $\mathfrak{o}/\mathfrak{p}$, then our assertion is equivalent to the following: if $\alpha \neq 0$, then there exists a zero-dimensional prime ideal which does not contain α. The proof of this assertion is straightforward.

PENCILS ON AN ALGEBRAIC VARIETY AND A NEW PROOF OF A THEOREM OF BERTINI

BY

OSCAR ZARISKI(1)

Introduction. A well known theorem of Bertini-Enriques on *reducible* linear systems of V_{r-1}'s on an algebraic V_r (i.e., linear systems in which each element is a reducible V_{r-1}) states that any such system, if free from fixed components, is composite with a pencil. The usual geometric proof of this theorem is based on another theorem of Bertini, to the effect that the *general* V_{r-1} of a linear system cannot have multiple points outside the singular locus of the variety V_r and the base locus of the system. This geometric proof has been subsequently completed and presented by van der Waerden under an algebraic form [3].

In this paper we give a new proof of the theorem of Bertini on reducible linear systems and we also extend this theorem to *irrational* pencils, i.e., pencils of genus $p>0$. Our proof does not make use of the second theorem of Bertini just quoted. In the case of pencils (linear or irrational), we first observe that a pencil $\{W\}$ on V_r is defined by a field P of algebraic functions of one variable which is a subfield of the field Σ of rational functions on V_r. The whole proof is then essentially based on the simple remark that *the pencil* $\{W\}$ *is composite with another pencil* $\{\overline{W}\}$, *defined by a field* $\overline{P}$, *if and only if* P *is a subfield of* $\overline{P}$. This property is a straightforward consequence of the geometric definition of composite pencils. As a matter of fact we prefer to define composite pencils by this property. At any rate, it is then true that *a pencil* $\{W\}$ *is non-composite if and only if the corresponding field* P *is maximally algebraic in* Σ. In the light of this approach to the question, the theorem of Bertini on reducible pencils is almost a direct consequence of the well known fact that an irreducible algebraic variety V, over a ground field K, is absolutely irreducible if K is maximally algebraic in the field of rational functions on V.

In the case of linear systems of dimension >1 the proof is even simpler, provided use is made of a certain lemma (Lemma 5). This lemma is, however, of interest in itself.

A sizable portion of the paper (Part I) is devoted to the development of the concept of a pencil and of the basic properties of pencils in the abstract case of an arbitrary ground field (of characteristic zero).

Presented to the Society, January 1, 1941; received by the editors July 1, 1940.

(1) Guggenheim Fellow.

I. Pencils of V_{r-1}'s on a V_r

1. **Divisors of the first kind.** Let V be an irreducible algebraic r-dimensional variety in a projective n-space, over an arbitrary ground field K of characteristic zero. We shall assume that V is *normal* in the projective space(²). Let Σ be the field of rational functions on V. Since V is normal, any irreducible $(r-1)$-dimensional subvariety Γ of V defines a prime divisor $\mathfrak{P}$ of Σ, i.e., an homomorphic mapping of Σ upon (Σ_1, ∞), where Σ_1 is the field of rational functions on Γ. There is also an associated $(r-1)$-dimensional valuation B of Σ/K, whose valuation ring $\mathfrak{R}$ is the quotient ring $Q(\Gamma)$ of Γ(³). It is well known that B (being of dimension $r-1$) is a *discrete valuation of rank* 1, i.e., its value group is the group of integers. If η is an element of Σ and if its value $v_B(\eta)$ in the valuation B is α (α a positive, negative, or zero integer), we shall say that η has *order α at the prime divisor* $\mathfrak{P}$, or *along the variety* Γ. We shall also say that η *vanishes to order α at* $\mathfrak{P}$, *or along* Γ, if $\alpha>0$, and that η *is infinite to the order* $-\alpha$ *at* $\mathfrak{P}$, *or along* Γ, if $\alpha<0$.

The prime divisors $\mathfrak{P}$ defined as above, by means of irreducible subvarieties of V *of dimension* $r-1$, shall be referred to as *divisors of the first kind* (*with respect to* V(⁴)). Dealing with the given normal variety V, we shall only deal with prime divisors of the first kind with respect to V. Concerning these we state the following well known theorem (Krull [1, p. 137, Vollständigkeitseigenschaft]):

(²) For the definition of normal varieties see our paper [4, p. 279, 283]. Our assumption implies that V is normal in the affine space, *for any choice of the hyperplane at infinity*. It is this weaker condition that really matters in our present treatment, rather than the condition that V be normal in the projective space. The restriction to normal varieties (either in the projective or in the above affine sense) is a sound principle from the standpoint of birational geometry. We have proved, in fact, that normal varieties exist in every class of birationally equivalent varieties. We have also associated with any given variety V a definite class of projectively related normal varieties, the *derived normal varieties of* V [4, p. 292], and therefore results proved for these can be readily restated as results concerning the original V. Finally, we point out that the class of varieties which are normal in the above affine sense includes the class of varieties which are free from singularities (in the projective space).

(³) We choose as hyperplane at infinity any hyperplane which does not contain Γ. Let $\xi_1, \cdots, \xi_n$ be the nonhomogeneous coordinates of the general point of V. Since the subvariety Γ is not at infinity, it is given by a prime ideal $\mathfrak{p}$ of the ring $\mathfrak{o}=K[\xi_1, \cdots, \xi_n]$. This ideal is $(r-1)$-dimensional and is minimal in $\mathfrak{o}$. Since V is normal, the ring $\mathfrak{o}$ is integrally closed in its quotient field Σ. Therefore the *quotient ring* $\mathfrak{o}_{\mathfrak{p}}$ $(=Q(\Gamma))$ is a valuation ring $\mathfrak{R}$. The residue field of the corresponding valuation B is the quotient field of the residue class ring $\mathfrak{o}/\mathfrak{p}$, and hence coincides with Σ_1. The homomorphic mapping of $\mathfrak{R}$ upon Σ_1 defines the prime divisor $\mathfrak{P}$.

(⁴) The notion of a prime divisor *of the second kind with respect to* V is defined as follows. Any prime divisor $\mathfrak{P}$ of Σ is by definition a homomorphic mapping of Σ upon (Σ_1, ∞), where Σ_1 is a field of algebraic functions of $r-1$ variables. By a proper choice of the hyperplane at infinity we may arrange matters so that none of the coordinates ξ_i is mapped upon the symbol ∞. The elements of the ring $\mathfrak{o}$ (see Footnote 3) which are mapped upon the zero element of Σ form then a prime ideal $\mathfrak{p}$ in $\mathfrak{o}$, and this prime ideal defines an irreducible subvariety Γ of V. If Γ is of dimension $r-1$, then our divisor $\mathfrak{P}$ is of the first kind and is uniquely determined by Γ. If Γ is

If an element η of Σ is transcendental over K, *then the set of prime divisors of the first kind along which η is infinite is finite and non-empty.*

A *divisor of the first kind* will be by definition a power product $\mathfrak{A}=\mathfrak{P}_1^{\alpha_1}\cdots\mathfrak{P}_h^{\alpha_h}$, where $\mathfrak{P}_1, \cdots, \mathfrak{P}_h$ are *prime* divisors of the first kind and where $\alpha_1, \cdots, \alpha_h$ are positive, negative, or zero integers. If all the α are positive integers or zero, then $\mathfrak{A}$ is said to be an *integral divisor.*

$\mathfrak{A}$ is said to be a *principal divisor* if there exists an element η in the field Σ such that η has order α_i at $\mathfrak{P}_i$ $(i=1, 2, \cdots, h)$ and order 0 at any other prime divisor of the first kind. Notation: $\mathfrak{A}=(\eta)$.

2. **Definition of a pencil.** Let P be a subfield of Σ *containing the ground field* K and being of *degree of transcendency* 1 *over* K. By means of such a subfield P we proceed to define a collection $\{W\}$ of $(r-1)$-dimensional subvarieties W of V, and namely *one* W for each *place*, or *prime divisor*, of P/K.

Given an $(r-1)$-dimensional irreducible subvariety Γ of V, let $\mathfrak{P}$ be the corresponding prime divisor of Σ/K and let B be the valuation defined by $\mathfrak{P}$. The valuation B induces a valuation B_1 in the field P. The valuation B_1 is either the trivial valuation, in which every element of P (different from 0) has value zero, or B_1 is a non-trivial valuation. In the first case, the mapping of P in the divisor $\mathfrak{P}$ is an isomorphism. In the second case, P is mapped by $\mathfrak{P}$ homorphically upon a field which is algebraic over K. This mapping defines a *place*, or a prime divisor $\mathfrak{p}$ of P/K; it is the prime divisor which is also directly defined by the non-trivial valuation B_1. We say in this second case that the irreducible $(r-1)$-dimensional subvariety Γ, or the corresponding prime divisor $\mathfrak{P}$, *corresponds to the place* $\mathfrak{p}$ *of* P/K.

It is not difficult to see that there is at most a *finite number of irreducible* V_{r-1}*'s which correspond to a given place* $\mathfrak{p}$ *of* P/K. In fact, let t be an element of P whose order at the place $\mathfrak{p}$ is *positive.* Then it is clear that the prime divisors of the first kind which correspond to $\mathfrak{p}$ must be among those prime factors of the principal divisor (t) whose exponents are positive. The number of such prime factors is finite, since, by a previously stated theorem, $1/t$ is infinite only at a finite number of prime divisors of the first kind.

We shall see later that *to each place* $\mathfrak{p}$ *there corresponds at least one irreducible* V_{r-1} on V.

Let $\Gamma_1, \cdots, \Gamma_m$ be the irreducible V_{r-1}'s on V which correspond to the

---

of dimension *less* than $r-1$, then $\mathfrak{P}$ is of the second kind with respect to V. There exist infinitely many prime divisors $\mathfrak{P}$ of the second kind leading to one and the same irreducible subvariety Γ of dimension $<r-1$.

It is well known that a prime divisor $\mathfrak{P}$ of Σ/K defines a discrete valuation B of Σ, of rank 1. The valuation ring $\mathfrak{R}$ of B is the set of all elements of Σ which are mapped upon elements of Σ_1. Our condition on the choice of the hyperplane at infinity implies that $\mathfrak{o}\subset\mathfrak{R}$. The subvariety Γ is called *the center* of the valuation B on V. A prime divisor $\mathfrak{P}$ of Σ/K is of the first kind with respect to V if and only if the center of the valuation defined by $\mathfrak{P}$ is a subvariety of V of *dimension* $r-1$ (and not less).

given place $\mathfrak{p}$ of P/K. Let t be a uniformizing parameter at $\mathfrak{p}$. We attach to each Γ_i *a multiplicity* h_i: namely, h_i *shall be the order to which t vanishes along* Γ_i. It is clear that h_i is a positive integer and that it is independent of the choice of the uniformizing parameter t(5). We regard the variety

$$W_{\mathfrak{p}} = h_1\Gamma_1 + \cdots + h_m\Gamma_m$$

as *the total subvariety* of V which corresponds to the place $\mathfrak{p}$. We define *the pencil* $\{W\}$ as *the totality of all* $W_{\mathfrak{p}}$ *obtained as* $\mathfrak{p}$ *varies on the Riemann surface of* P/K.

3. **Adjunction of indeterminates.** In order to derive the basic properties of pencils, we proceed to give an explicit construction of the pencil $\{W\}$ based on Kronecker's method of indeterminates. Let $\xi_1, \cdots, \xi_n$ be, as before, the nonhomogeneous coordinates of the general point of V(6). We introduce the $r-1$ forms:

$$\eta_i = u_{i1}\xi_1 + \cdots + u_{in}\xi_n, \qquad i = 1, 2, \cdots, r-1,$$

where the $n(r-1)$ elements u_{ij} are indeterminates. These indeterminates we adjoin to the field Σ, getting a field $\Sigma^* = \Sigma(\{u_{ij}\}) = \Sigma(u_{11}, \cdots, u_{r-1,n})$. The field Σ^* is a pure transcendental extension of Σ, of degree of transcendency $n(r-1)$ over Σ. We also consider the fields:

$$P^* = P(\{u_{ij}\}, \{\eta_i\}),$$
$$K^* = K(\{u_{ij}\}, \{\eta_i\}).$$

The elements $\eta_1, \cdots, \eta_{r-1}$ *are algebraically independent over the field* $P(\{u_{ij}\})$. For, if $\eta_1, \cdots, \eta_{r-1}$ and the u_{ij} satisfied an algebraic relation with coefficients in P, then by specializing the u_{ij}, $u_{ij} \to c_{ij} \in K$, we could get an algebraic relation over P between any $r-1$ of the elements ξ_i(7). This is impossible, since Σ is of degree of transcendency $r-1$ over P.

From the algebraic independence of the η_i it follows that Σ^* and P^* have the same degree of transcendency over K, and that P^* is of degree of transcendency 1 over K^*. Hence:

(5) If t' is another uniformizing parameter at $\mathfrak{p}$, then t'/t has order zero at $\mathfrak{p}$, and consequently t'/t has also order zero along Γ_i.

(6) Generally speaking, the proper procedure would have been to use the homogeneous coordinates of the general point of V (Zariski [4, p. 284]), thus avoiding special considerations for *divisors of infinity* (i.e., prime divisors of the first kind, at which at least one of the elements ξ_i is infinite and which therefore correspond to the irreducible components of the section of V with the hyperplane at infinity), However, the use of homogeneous coordinates would have required introductory definitions and proofs concerning homogeneous prime divisors and similar concepts associated with homogeneous coordinates. The size of such an introduction would be out of proportion to the limited object of this paper. We therefore prefer to deal with nonhomogeneous coordinates, also because in the present case the special considerations for the divisors at infinity are very simple and brief.

(7) See van der Waerden [2, lemma on page 17].

Σ^* is an algebraic extension of P^*. *If* K^* *is taken as ground field, then* P^* *is a field of algebraic functions of one variable.*

Given a prime divisor $\mathfrak{p}$ of the field P/K, we shall want to extend $\mathfrak{p}$ to a prime divisor of the field $\mathrm{P}^*/\mathrm{K}^*$. This extension is based on the following lemma, which we shall use also later on in a different connection:

LEMMA 1. *Let* $\Omega^* = \Omega(x_1, \cdots, x_m)$ *be a pure transcendental extension of a field* Ω *(i.e.,* $x_1, \cdots, x_m$ *are algebraically independent over* Ω*). Given a valuation* B *of* Ω*, there exists one and only one extended valuation* B^* *of* Ω^* *such that the* B^**-residues of* $x_1, \cdots, x_m$ *are* $\neq \infty$ *and are algebraically independent over the residue field of* B.

Proof. Let B^* be a valuation of Ω^* satisfying the conditions of the lemma, and let $\eta^* = f(x_1, \cdots, x_m)/g(x_1, \cdots, x_m)$, $f, g \in \Omega[x_1, \cdots, x_m]$, be an arbitrary element of Ω^*. Some of the coefficients of the rational function f/g may have negative values in B. However, if we divide through f and g by a coefficient of minimum value, we get a rational function whose coefficients have *finite* B-residues, *not all zero*. We may assume then that the rational function f/g already satisfies this condition. Let us assume that the B-residues of the coefficients of the denominator g are all zero. Then the B^*-residue of $g(x_1, \cdots, x_m)$ is zero, since the B^*-residues of $x_1, \cdots, x_m$ are $\neq \infty$. On the other hand, the B-residues of the coefficients of the numerator f are necessarily not all zero. Hence the B^*-residue of f is different from zero, since the B^*-residues of $x_1, \cdots, x_m$ are algebraically independent over the residue field of B. Consequently the B^*-residue of η^* is ∞.

On the other hand, assuming that the B-residues of the coefficients of g are not all zero, we conclude in a similar fashion that the B^*-residue of g is $\neq 0$, while the B^*-residue of f is $\neq \infty$. Hence the B^*-residue of η^* is $\neq \infty$.

Hence *the valuation ring* $\mathfrak{R}^*$ *of* B^* *consists of all quotients* f/g *such that the* B*-residues of the coefficients are all finite and the* B*-residues of the coefficients of* g *are not all zero*. This shows that B^* is uniquely determined. On the other hand, the set of all such quotients is a ring $\mathfrak{R}^*$ satisfying the condition that if $\eta^* \not\subset \mathfrak{R}^*$, then $1/\eta^* \subset \mathfrak{R}^*$. Hence $\mathfrak{R}^*$ is a valuation ring. It is then immediately seen that the corresponding valuation B^* of Ω^* is an extension of B (i.e., $\mathfrak{R}^* \cap \Omega$ is the valuation ring of B) and satisfies the condition of the lemma.

We point out that *the value group of* B^* *is the same as the value group of* B and that *the residue field of* B^* *is a pure transcendental extension of the residue field of* B*, the adjoined transcendentals being the* B^**-residues of* $x_1, \cdots, x_m$. The proofs of these assertions are straightforward. We also wish to point out explicitly that B^* *depends on the particular set of generators* $x_1, \cdots, x_m$ of Ω^*/Ω. Thus, the set of generators $cx_1, x_2, \cdots, x_m$, $c \subset \Omega$, defines an extended valuation of Ω^* which is different from B^*, *whenever* $v_B(c) \neq 0$.

We apply the above lemma to the fields P, P^*, of which the second is a pure transcendental extension of P. As generators of P*/P we take the elements u_{ij} and η_i. Let $\mathfrak{p}$ be a prime divisor of P/K, B_1 the corresponding valuation of P, B_1^* the extended valuation of P*. If Δ_1 is the residue field of B_1, then $\Delta_1^* = \Delta_1(\{u_{ij}^*\}, \{\eta_i^*\})$ is the residue field of B_1^*, where the u_{ij}^* and η_i^* are the B_1^*-residues of the u_{ij} and η_i. The field Δ_1 contains a subfield simply isomorphic to K, which we may identify with K. The fields $K(\{u_{ij}^*\}, \{\eta_i^*\})$, $K(\{u_{ij}\}, \{\eta_i\})$ are simply isomorphic, in view of the algebraic independence of the residues u_{ij}^*, η_i^* with respect to Δ_1. Hence the valuation B_1^* defines a *prime divisor* $\mathfrak{p}^*$ of the field P*/K*, i.e., if K* is taken as ground field and if P* is regarded as a field of functions of one variable. *Hence we have associated with each prime divisor* $\mathfrak{p}$ *of* P/K *a prime divisor* $\mathfrak{p}^*$ of P*/K*:

$$\mathfrak{p} \rightarrow \mathfrak{p}^*. \tag{1}$$

Since B_1^* is an extension of B_1, it is clear that *distinct prime divisors of* P/K *extend to distinct prime divisors of* P*/K*.

We now consider the field Σ^*. It is an algebraic extension of P*. Hence the prime divisor $\mathfrak{p}^*$ factors into a power product of prime divisors of Σ^*/K^*. Let

$$\mathfrak{p}^* = \mathfrak{P}_1^{*h_1} \cdots \mathfrak{P}_m^{*h_m}. \tag{2}$$

The following sections are devoted to the proof of the following assertions:

(A) *Each prime divisor* $\mathfrak{P}_i^*$ $(i=1, 2, \cdots, m)$ *induces in* Σ *a prime divisor* $\mathfrak{P}_i$ *of the first kind.*

(B) *The prime divisors* $\mathfrak{P}_1, \cdots, \mathfrak{P}_m$ *are distinct.*

(C) *If* Γ_i *is the irreducible* V_{r-1} *defined on* V *by the prime divisor* $\mathfrak{P}_i$, *then* $h_1\Gamma_1 + \cdots + h_m\Gamma_m$ *is the member* $W_{\mathfrak{p}}$ *of the pencil* $\{W\}$ *which corresponds to the place* $\mathfrak{p}$ *of* P/K (in the sense of the definition given in §2).

4. **The induced prime divisors $\mathfrak{P}_i$.** We consider one of the prime divisors $\mathfrak{P}_i^*$ in (2), say $\mathfrak{P}_1^*$, and we denote by B^* the valuation of Σ^* defined by $\mathfrak{P}_1^*$. In Σ, a subfield of Σ^*, B^* induces a valuation B. Since Σ^* is of degree of transcendency $(r-1)n$ over Σ, *the residue field of* B^* *can be at most of degree of transcendency* $(r-1)n$ *over the residue field of* B[8]. Since the residue field of B^* is an algebraic extension of K*, it is of degree of transcendency $(r-1)n+(r-1)$

(8) The proof of this assertion is immediate. Let, quite generally, Σ^* be an extension field of a field Σ, of degree of transcendency ρ over Σ, and let B and B^* be respectively a valuation of Σ and an extended valuation of Σ^*. Let us assume that the residue field of B^* is of degree of transcendency $\geqq \rho$ over the residue field of B. We consider ρ elements $\omega_1, \cdots, \omega_\rho$ of Σ^* whose B^*-residues $\omega_1^*, \cdots, \omega_\rho^*$ are algebraically independent over the residue field Δ of B. It is clear that $\omega_1, \cdots, \omega_\rho$ are algebraically independent over Σ. Hence, by the remark at the end of the proof of Lemma 1, it follows that the residue field of the valuation induced in the field $\Sigma(\omega_1, \cdots, \omega_\rho)$ by the valuation B^*, is $\Delta(\omega_1^*, \cdots, \omega_\rho^*)$. Now Σ^* is an algebraic extension of the field $\Sigma(\omega_1, \cdots, \omega_\rho)$. Hence the residue field of B^* is an algebraic extensions of $\Delta(\omega_1^*, \cdots, \omega_\rho^*)$, and therefore its degree of transcendency over Δ is exactly ρ.

over K. Hence the degree of transcendency of the residue field of B over K must be not less than $r-1$. On the other hand, $\mathfrak{P}_1^*$ induces in P the non-trivial prime divisor $\mathfrak{p}$. *Hence B, as a non-trivial valuation, is of dimension $r-1$.* It defines in Σ a prime divisor $\mathfrak{P}_1$. Thus *the prime divisors $\mathfrak{P}_1^*, \cdots, \mathfrak{P}_m^*$ induce in Σ/K prime divisors $\mathfrak{P}_1, \cdots, \mathfrak{P}_m$.*

We next show that the *prime divisors $\mathfrak{P}_1, \cdots, \mathfrak{P}_m$ are of the first kind with respect to V.* Let us consider for instance the divisor $\mathfrak{P}_1$. We examine separately two cases, according as the center of $\mathfrak{P}_1$ (i.e., of the corresponding valuation [see Footnote 4]) on V is or is not at finite distance.

First case. The center of $\mathfrak{P}_1$ is at finite distance (i.e., it does not lie in the hyperplane at infinity). In this case the B-residues $\xi_1^*, \cdots, \xi_n^*$ of $\xi_1, \cdots, \xi_n$ are all different from ∞. Let u_{ij}^*, η_i^* be the B^*-residues of the elements u_{ij}, η_i. These B^*-residues are algebraically independent over K. On the other hand we have: $\eta_i^* = u_{i1}^*\xi_1^* + \cdots + u_{in}^*\xi_n^*$. Hence $r-1$ of the elements ξ_i^* must be algebraically independent over K. This shows that the center of $\mathfrak{P}_1$ on V is of dimension $r-1$, whence $\mathfrak{P}_1$ is of the first kind.

Second case. The center of $\mathfrak{P}_1$ is at infinity. In this case some of the elements ξ_i have negative values in B. Without loss of generality we may assume that $0 > v_B(\xi_1) = \min\ (v_B(\xi_1), \cdots, v_B(\xi_n))$. We consider the following projective transformation of coordinates:

$$\xi_1' = 1/\xi_1, \qquad \xi_i' = \xi_i/\xi_1, \qquad i = 2, \cdots, n. \tag{3}$$

With respect to the new nonhomogeneous coordinates ξ_i' the center of $\mathfrak{P}_1$ is at finite distance. To show that $\mathfrak{P}_1$ is of the first kind, we have to show that among the B-residues $\xi_1'^*, \xi_2'^*, \cdots, \xi_n'^*$ of the ξ_i' there are $r-1$ which are algebraically independent over K. To show this we observe that we have:

$$\eta_i = \frac{u_{i1} + u_{i2}\xi_2' + \cdots + u_{in}\xi_n'}{\xi_1'}. \tag{4}$$

Hence passing to the B^*-residues and noting that the B-residue of ξ_1' is zero we find:

$$u_{i1}^* + u_{i2}^*\xi_2'^* + \cdots + u_{in}^*\xi_n'^* = 0. \tag{5}$$

These relations show that the $r-1$ elements u_{i1}^* belong to the field $K(\{u_{ij}^*\}, j>1; \xi_2'^*, \cdots, \xi_n'^*)$. Since the $(r-1)n$ elements u_{ij}^* are algebraically independent over K, $r-1$ of the residues $\xi_2'^*, \cdots, \xi_n'^*$ must be algebraically independent over K, q.e.d.

Thus assertion (A) is fully established.

5. **Proof of** (B). Let us consider one of the divisors $\mathfrak{P}_1, \cdots, \mathfrak{P}_m$, say $\mathfrak{P}_1$, and let, as before, B be the corresponding valuation of Σ, while B^* is the valuation of Σ^* defined by $\mathfrak{P}_1^*$. We know that B^* is an extended valuation of B. We also know that Σ^* is a pure transcendental extension of Σ. By Lemma 1,

B^* is uniquely determined by B if a set of generators of Σ^* over Σ is known whose B^*-residues are algebraically independent over the residue field of B. Since $\mathfrak{P}_1^*, \cdots, \mathfrak{P}_m^*$ are distinct divisors, the assertion (B) will be proved if we show that *it is possible to exhibit such a set of generators which depend only on B*. Let us first suppose that the center of B is at finite distance. The residue field of B is the field $K(\xi_1^*, \cdots, \xi_n^*)$, where the ξ_i^* are the B-residues of the ξ_i. Since the field $K(\{u_{ij}^*\}, \{\eta_i^*\})$ is of degree of transcendency $(r-1)n+(r-1)$ over K and since this field is contained in the field $K(\{u_{ij}^*\}, \xi_1^*, \cdots, \xi_n^*)$, it follows that the $(r-1)n$ residues u_{ij}^* are algebraically independent over the field $K(\xi_1^*, \cdots, \xi_n^*)$ (since this last field is of degree of transcendency $r-1$ over K). Hence, *if the center of B is at finite distance, the* $(r-1)n$ *elements* u_{ij} *form a set of generators of the desired nature.*

Let us now assume that the center of B is at infinity, and let, say, $0>v_B(\xi_1)=\min\ (v_B(\xi_1), \cdots, v_B(\xi_n))$. In this case the B^*-residues u_{ij}^* of the u_{ij} are not algebraically independent over the residue field $K(\xi_2'^*, \cdots, \xi_n'^*)$ of B, in view of (5). However, the elements $u_{ij}, j>1$, and $\eta_1, \cdots, \eta_{r-1}$ also form a set of generators of Σ^* over Σ. We assert that the *B^*-residues of these elements are algebraically independent over* $K(\xi_2'^*, \cdots, \xi_n'^*)$. To show this, we observe that in view of the relations (5), the field

$$\Delta^* = K(\xi_2'^*, \cdots, \xi_n'^*; \{u_{ij}^*\}, j > 1; \eta_1^*, \cdots, \eta_{r-1}^*)$$

contains the subfield $K(\{u_{ij}^*\}, \eta_1^*, \cdots, \eta_{r-1}^*)$. Since this last field is of degree of transcendency $(r-1)n+(r-1)$ over K, and since $K(\xi_2'^*, \cdots, \xi_n'^*)$ is of degree of transcendency $r-1$ over K, it follows that the $(n-1)r$ residues of u_{ij}^* $(j>1)$, $\eta_1^*, \cdots, \eta_{r-1}^*$, are algebraically independent over $K(\xi_2'^*, \cdots, \xi_n'^*)$, q.e.d.

The proof of assertion (B) is now complete.

6. **Proof of** (C). Let $\Gamma_1, \cdots, \Gamma_m$ be the $(r-1)$-dimensional irreducible subvarieties of V which correspond to the prime divisors $\mathfrak{P}_1, \cdots, \mathfrak{P}_m$. Let $W_{\mathfrak{p}}$ be the member of the pencil $\{W\}$ which corresponds to the place $\mathfrak{p}$ of P/K, in the sense of the definition given in §2. Since each prime divisor $\mathfrak{P}_i^*$, and hence also each $\mathfrak{P}_i$, induces in P/K the prime divisor $\mathfrak{p}$, *it follows that* $\Gamma_1, \cdots, \Gamma_m$ *are components of* $W_{\mathfrak{p}}$.

Let t be an element of P which has order 1 at $\mathfrak{p}$. Then t, considered as an element of P*, also has order 1 at $\mathfrak{p}^*$. In view of (2), t has order h_i at $\mathfrak{P}_i^*$. Since the value group of the valuation of Σ defined by $\mathfrak{P}_i$ is the same as the value group of the valuation of Σ^* defined by $\mathfrak{P}_i^*$ (see remark at the end of the proof of Lemma 1), it follows that *t has also order h_i at $\mathfrak{P}_i$*. We conclude that *Γ_i is an h_i-fold component of $W_{\mathfrak{p}}$* (in the sense of the definition of §2).

To complete the proof of (C), we have to show that if Γ is an irreducible component of $W_{\mathfrak{p}}$, then Γ coincides with one of the varieties $\Gamma_1, \cdots, \Gamma_m$. Let $\mathfrak{P}$ be the prime divisor of Σ defined by Γ and let B be the corresponding valuation of Σ. Let B^* be *the* extended valuation of B in Σ^* defined by the following condition: (1) if Γ is at finite distance, then the B^*-residues of the $(r-1)n$ elements u_{ij} are algebraically independent over the residue field of B; (2) if $\mathfrak{P}$ is a divisor at infinity, and if, say, $0>v_B(\xi_1)=\min(v_B(\xi_1), \cdots, v_B(\xi_n))$, then the B^*-residues of the $(r-1)n$ elements $u_{ij}(j>1), \eta_1, \cdots, \eta_{r-1}$, are algebraically independent over the residue field of B.

In view of Lemma 1, the valuation B^* is well defined in either case. We assert that the *B^*-residues of the elements u_{ij}, η_i are algebraically independent over* K.

Proof. *First case. $\mathfrak{P}$ is a divisor at finite distance.* Let u_{ij}^*, η_i^*, ξ_i^* denote B^*-residues of the corresponding elements u_{ij}, η_i, ξ_i. We have $\eta_i^*=u_{i1}^*\xi_1^*+\cdots+u_{in}^*\xi_n^*$. By our construction of B^*, the elements u_{ij}^* are algebraically independent over the field $K(\xi_1^*, \cdots, \xi_n^*)$. The rest of the proof is based on the van der Waerden lemma quoted in Footnote 7 and is identical with the proof that $\eta_1, \cdots, \eta_{r-1}$ are algebraically independent over $P(\{u_{ij}\})$.

Second case. $\mathfrak{P}$ is a divisor at infinity, and $0>v_B(\xi_1)=\min(v_B(\xi_1), \cdots, v_B(\xi_n))$. By our construction of B^*, the residues u_{ij}^*, $j>1$, are algebraically independent over the field $K(\xi_2'^*, \cdots, \xi_n'^*)$, where the ξ_i' are given by the relations (3)([9]). By a specialization argument similar to the one used in the preceding case and based on van der Waerden's lemma, it follows that the $r-1$ forms $u_{i2}^*\xi_2'^*+\cdots+u_{in}^*\xi_n'^*$ are algebraically independent over the field $K(\{u_{ij}^*\}, j>1)$. But these forms are equal to the elements $-u_{i1}^*$, in view of the relations (5). Hence the $(r-1)n$ elements u^* are algebraically independent over K. On the other hand, the residues $\eta_1^*, \cdots, \eta_{r-1}^*$ are, by hypothesis, algebraically independent over the field $K(\xi_2'^*, \cdots, \xi_n'^*; \{u^*\}, j>1)$. Hence *a fortiori* these residues are algebraically independent over the subfield $K(\{u^*\})$, and this proves our assertion.

The fact that the B^*-residues u_{ij}^*, η_i^* are algebraically independent over K, implies that B^* is a valuation of Σ^* *over the ground field* K^*, i.e., *B^* defines a prime divisor $\mathfrak{P}^*$ of Σ^*/K^**. This prime divisor $\mathfrak{P}^*$ lies over a definite prime divisor $^*\mathfrak{p}^*$ of P^*/K^*. To show that $\mathfrak{P}^*$ coincides with one of the prime divisors $\mathfrak{P}_1^*, \cdots, \mathfrak{P}_m^*$, it is only necessary to show that $^*\mathfrak{p}^*=\mathfrak{p}^*$. Now both $\mathfrak{p}^*$ and $^*\mathfrak{p}^*$ induce in the field P/K one and the same prime divisor, namely $\mathfrak{p}$, since, by hypothesis, Γ is a component of $W_{\mathfrak{p}}$. In view of Lemma 1, the relation $^*\mathfrak{p}^*=\mathfrak{p}^*$ will follow if we prove that the residues of the $(n+1)(r-1)$ elements u_{ij}, η_i in the divisor $^*\mathfrak{p}^*$ are algebraically independent over the residue field of the divisor $\mathfrak{p}$. But this is obvious, since this last residue field is algebraic over K and since we have just proved that the B^*-residues of the elements u_{ij}, η_i are algebraically independent over K. This completes the proof of assertion (C).

7. **Order of $W_{\mathfrak{p}}$.** Let $\eta_i=u_{i1}\xi_1+\cdots+u_{in}\xi_n$, $i=1, 2, \cdots, r$, where the

(9) Note that since V is normal, $K(\xi_1'^*, \cdots, \xi_n'^*)$ is the residue field of B, and also that $\xi_1'^*=0$.

nr elements u_{ij} are indeterminates. We define the *order of* V as the relative degree $[\Sigma(\{u_{ij}\}):K(\{u_{ij}\}, \eta_1, \cdots, \eta_r)]$. If we apply this definition to each irreducible component Γ_i of $W_\mathfrak{p}$, we can speak of the *order of* $W_\mathfrak{p}$: it shall be, by definition, the sum of the orders of $\Gamma_1, \cdots, \Gamma_m$, each counted to its multiplicity $h_1, \cdots, h_m$, respectively.

Given a field Ω of algebraic functions of one variable, over a ground field K, and given a prime divisor $\mathfrak{p}$ in Ω/K, by *the degree of* $\mathfrak{p}$ we mean the relative degree $[\Delta:K]$, where Δ is the residue field of $\mathfrak{p}$.

From this definition it follows immediately that the prime divisor $\mathfrak{p}$ of P/K and the extended prime divisor $\mathfrak{p}^*$ of P^*/K^* have the same degree, since the residue field of $\mathfrak{p}^*$ is a pure transcendental extension of the residue field of $\mathfrak{p}$ (see Lemma 1 and the application to the fields P, P*, §3). Let d be the common degree of $\mathfrak{p}$ and $\mathfrak{p}^*$. Let $d_1, \cdots, d_m$ be the degrees of the prime divisors $\mathfrak{P}_1^*, \cdots, \mathfrak{P}_m^*$ respectively. It is well known that the relation (2) implies

$$\nu d = h_1 d_1 + \cdots + h_m d_m, \tag{6}$$

where *ν is the relative degree* $[\Sigma^*:P^*]$. The relation (6) has a geometric interpretation, namely, that *the right-hand side* $h_1d_1 + \cdots + h_{1n}d_{1n}$ *is the order of* $W_\mathfrak{p}$. To prove this, we have to show that *d_i is the order of Γ_i*.

Proof. *First case. Γ_1 is at finite distance.* The nonhomogeneous coordinates of the general point of Γ_1 are $\xi_1^*, \cdots, \xi_n^*$, where the ξ_i^* are the residues of the ξ_i in the valuation B defined by the prime divisor $\mathfrak{P}_1$. The field $\Delta = K(\xi_1^*, \cdots, \xi_n^*)$ is the residue field of B. The residue field of the extended valuation B^* in Σ^* is (§5) $K(\xi_1^*, \cdots, \xi_n^*, \{u_{ij}^*\})$. The degree of $\mathfrak{P}_1^*$ is the relative degree

$$d_1 = [\Delta(\{u_{ij}^*\}:K(\{u_{ij}^*\}, \{\eta_i^*\})].$$

Since the u_{ij}^* are algebraically independent over Δ, the relation above shows that d_1 is also the order of Γ_1.

Second case. Γ_1 is at infinity, and $0 > v_B(\xi_1) = \min(v_B(\xi_1), \cdots, v_B(\xi_n))$. In this case the general point of Γ_1 is $(\xi_1'^*, \xi_2'^*, \cdots, \xi_n'^*)$, where the ξ_i' are defined by the transformation (3). But since $\xi_1'^* = 0$, Γ_1 actually lies in an S_{n-1} (in the hyperplane at infinity of S_n), and the general point of Γ_1 is $(\xi_2'^*, \cdots, \xi_n'^*)$. The field $\Delta = K(\xi_2'^*, \cdots, \xi_n'^*)$ is the residue field of B. By §5, the residue field of B^* is $\Delta(\{u_{ij}^*\}, j>1; \{\eta_i^*\})$, and we have

$$d_1 = [\Delta(\{u_{ij}^*\}, j > 1; \{\eta_i^*\}):K(\{u_{ij}^*\}, \{\eta_i^*\})].$$

Since $\eta_1^*, \cdots, \eta_{r-1}^*$ are algebraically independent over the field $\Delta(\{u_{ij}^*\}, j>1)$, we can also write:

$$d_1 = [\Delta(\{u_{ij}^*\}, j > 1):K(\{u_{ij}^*\}, j > 1; \{u_{i1}^*\})].$$

The assertion that d_1 is the order of Γ_1 now follows from the relations (5).

In view of (6), we may now state the following result: *the quotient*

$$\frac{\textit{order of } W_{\mathfrak{p}}}{\textit{degree of } \mathfrak{p}}$$

is the same for all members of the pencil $\{W\}$. In the particular case *when* K *is algebraically closed*, the degree of $\mathfrak{p}$ is always 1, and hence *all the members of the pencil* $\{W\}$ *have the same order.*

8. **The base points of the pencil** $\{W\}$. A point A of V is said to be a *base point* of the pencil $\{W\}$ *if every* $W_{\mathfrak{p}}$ *is on* A. Let A be a point of V and let $Q(A)=\mathfrak{J}$ be the quotient ring of A. We consider the intersection $R=\mathfrak{J}\cap \mathrm{P}$. We prove the following theorem:

A is a base point of $\{W\}$ if and only if the ring R is an algebraic extension of K. *If A is not a base point of $\{W\}$, then there exists one and only one $W_{\mathfrak{p}}$ which passes through A.*

Proof. Let us first assume that R contains elements which are transcendental over K, whence *R is of degree of transcendency* 1 *over* K. Since V is normal, $\mathfrak{J}$ is integrally closed in Σ. Consequently

(a) *R is integrally closed in* P.

Since every element of P is algebraic over R, it follows by (a) that

(b) P *is the quotient field of R.*

Let ω be an arbitrary element of R. Since $\mathfrak{J}$ is the quotient ring of a zero-dimensional ideal, there exists a polynomial $f(\omega)$, with coefficients in K, such that $f(\omega)$ is a non-unit in $\mathfrak{J}$. Hence $1/f(\omega)\not\subset\mathfrak{J}$, whence $1/f(\omega)$ is an element of P which is not in R. Therefore

(c) *R is a proper subring of* P.

The non-units of $\mathfrak{J}$ form an ideal. Hence also

(d) *The non-units of R form an ideal.*

Since P is a field of algebraic functions of one variable, the properties (a), (b), (c), (d) of R imply that *R is a valuation ring*. Let $\mathfrak{p}$ be the corresponding prime divisor of P/K.

If *W is a member of the pencil and if $W\neq W_{\mathfrak{p}}$, then W does not pass through A*. For, let Γ be any irreducible component of W. The quotient ring $Q(\Gamma)$ is the valuation ring of the valuation defined by Γ (see Footnote 3). Since $W\neq W_{\mathfrak{p}}$, the intersection $Q(\Gamma)\cap \mathrm{P}$ is different from R (this intersection is the valuation ring associated with a prime divisor of P/K different from $\mathfrak{p}$). Hence $Q(\Gamma)\cap \mathrm{P}\not\supset R$, whence $Q(\Gamma)\not\supset\mathfrak{J}$. *Therefore A is not on* Γ. Since this holds true for any component Γ of W, A is not on W, as was asserted.

The variety $W_{\mathfrak{p}}$ passes through A. Let t be a non-unit of R. Then t is also a non-unit of $\mathfrak{J}$. Therefore at least one prime factor in the numerator of the principal divisor (t) must arise from a minimal prime of $\mathfrak{J}$. This minimal prime corresponds to an irreducible $(r-1)$-dimensional variety Γ passing through A. The order of t along Γ is positive, whence the valuation of Σ/K defined by Γ induces in P/K a non-trivial valuation. The corresponding prime divisor of P/K must coincide with $\mathfrak{p}$, since $Q(\Gamma)\supset Q(A)\supset R$. Hence Γ is a component of $W_{\mathfrak{p}}$, and thus $W_{\mathfrak{p}}$ is on A.

Let us suppose now that R is an algebraic extension of K. Let $\mathfrak{p}$ be an *arbitrary* prime divisor of P/K. Let t be an element of P which has positive order at $\mathfrak{p}$ and nonpositive order at any other prime divisor of P/K. It is well known that such elements t exist. The element t is transcendental over K. Hence $1/t\not\subset R$, and consequently $1/t\not\subset \mathfrak{Z}$. This implies that one of the prime factors in the numerator of the principal divisor (t) must arise from an irreducible subvariety Γ (of dimension $r-1$) which passes through A. Since t has positive order along Γ, it follows as before that Γ is a component of some member of the pencil, and this member must be $W_{\mathfrak{p}}$, since $\mathfrak{p}$ is the only prime divisor of P/K at which t has positive order. *Hence $W_{\mathfrak{p}}$ is on A.* Since $\mathfrak{p}$ was an arbitrary prime divisor of P/K, we conclude that A is a base point of the pencil $\{W\}$. This completes the proof of our theorem.

It is not difficult to see that *the base points of the pencil $\{W\}$ form an algebraic subvariety of V, of dimension $\leqq r-2$.* To see this, we fix an element t in P which is transcendental over K. Let $M_1, \cdots, M_s$ and $N_1, \cdots, N_\sigma$ be the irreducible $(r-1)$-dimensional subvarieties of V along which t has respectively positive or negative order, and let H be the intersection of the two varieties $M_1+\cdots+M_s$ and $N_1+\cdots+N_\sigma$. Clearly, H is of dimension $\leqq r-2$. We assert that *H is the base locus of the pencil $\{W\}$*. Namely, if A is a base point of $\{W\}$, then, by the preceding proof, neither t nor $1/t$ can belong to the quotient ring $Q(A)$, since both t and $1/t$ are transcendentals over K. Consequently the point A must lie on at least one of the varieties M_i and also on at least one of the varieties N_j. Therefore A is on H. Conversely, assume that A is on H, and let, say, A be on M_1 and N_1. It is clear that M_1 belongs to some member $W_{\mathfrak{p}}$ of the pencil, such that t has positive order at $\mathfrak{p}$. Similarly N_1 belongs to some member $W_{\mathfrak{p}'}$ of the pencil such that t has negative order at $\mathfrak{p}'$. Hence the point A lies on two *distinct* members of the pencil, $W_{\mathfrak{p}}$ and $W_{\mathfrak{p}'}$, and consequently A is a base point.

Remark. *Pencil with fixed components.* Let Γ_0 be a fixed $(r-1)$-dimensional subvariety of V and let $\{W\}$ be a pencil. It is convenient to regard also the collection of varieties $W'=W+\Gamma_0$ as a pencil. All members of this new pencil $\{W'\}$ have a fixed component, namely Γ_0. The base locus of $\{W'\}$ is naturally no longer of dimension $\leqq r-2$, since Γ_0 itself is a part of the base locus.

The pencils considered heretofore are pencils without fixed components.

II. The theorem of Bertini for pencils

9. **Composite pencils.** Let $\{W\}$, $\{W^*\}$ be two distinct pencils on V, both free from fixed components, and let P, P* be the corresponding subfields of Σ, of degree of transcendency 1 over K.

Definition. *The pencil* $\{W\}$ *is composite with the pencil* $\{W^*\}$ *if* P *is a subfield of* P*.

The geometric significance of this definition is straightforward. If P is a subfield of P*, then P* is an algebraic extension of P, since both have degree of transcendency 1 over K. Given a prime divisor $\mathfrak{p}$ of P/K, it factors in P*/K:

$$\mathfrak{p} = \mathfrak{p}_1^{*\rho_1} \cdots \mathfrak{p}_s^{*\rho_s}, \tag{7}$$

where, if $\mathfrak{p}$ is of degree d and $\mathfrak{p}_i^*$ is of degree d_i^*, then $\rho_1 d_1^* + \cdots + \rho_s d_s^* = \nu d$, $\nu = [P^*:P]$. Consider the varieties $W_{\mathfrak{p}}$, $W_{\mathfrak{p}_i^*}^*$. If Γ^* is an irreducible component of $W_{\mathfrak{p}^*}$, then Γ^* is also a component of $W_{\mathfrak{p}}$, since $\mathfrak{p}_i^*$ lies over $\mathfrak{p}$. Moreover, if Γ^* occurs in $W_{\mathfrak{p}_i^*}^*$ to the multiplicity h_i, it must occur in $W_{\mathfrak{p}}$ to the multiplicity $h_i\rho_i$, since if an element of P has order 1 at $\mathfrak{p}$, it has order ρ_i at $\mathfrak{p}_i^*$. Finally, if Γ is an irreducible component of $W_{\mathfrak{p}}$ and if $\mathfrak{P}$ is the corresponding prime divisor of Σ, then $\mathfrak{P}$ induces in P the prime divisor $\mathfrak{p}$. Therefore $\mathfrak{P}$ must induce in P* one of the prime divisors $\mathfrak{p}_i^*$, and consequently Γ is a component of one of the varieties $W_{\mathfrak{p}^*}^*$. *We conclude that*

$$W_{\mathfrak{p}} = \rho_1 W_{\mathfrak{p}_1^*}^* + \cdots + \rho_s W_{\mathfrak{p}_s^*}^*,$$

i.e., *if the pencil* $\{W\}$ *is composite with the pencil* $\{W^*\}$, *then every* W *consists of a certain number of members of the pencil* $\{W^*\}$.

In general this number $\rho_1 + \cdots + \rho_s$ depends on $W_{\mathfrak{p}}$. But if K is algebraically closed, then $d = d_i^* = 1$, *whence* $\rho_1 + \cdots + \rho_s = \nu$, the *relative degree* $[P^*:P]$(10). For particular prime divisors $\mathfrak{p}$ it may happen that $\rho_1 + \cdots + \rho_s = 1$, i.e., that $W_{\mathfrak{p}}$ itself is a member of the pencil $\{W^*\}$. This can happen even for infinitely many prime divisors $\mathfrak{p}$. But it is easily seen that *there always exist infinitely many prime divisors* $\mathfrak{p}$ *for which* $\rho_1 + \cdots + \rho_s > 1$, so that *in a composite pencil there always exist infinitely many* $W_{\mathfrak{p}}$ *which are reducible over* K(11).

---

(10) There is one case in which our definition of composite pencils must be slightly modified. It is the case in which Σ is a field of functions of one variable. Since Σ itself is of degree of transcendency 1, it can be taken as field P and it therefore defines a pencil of zero-dimensional varieties on the curve V. This pencil is nothing else than the Riemann surface R of Σ, and every other pencil on V would be, according to our definition, composite with R. To avoid this undesirable conclusion, one must define composite pencils on an algebraic curve (i.e., composite *involutions of sets of points*) by the condition that the field P defining the pencil is not a maximal subfield of Σ.

(11) **Proof.** Let x be a transcendental element of P and let $\mathfrak{S}$ and $\mathfrak{S}^*$ denote the rings of integral functions of x in P and in P* respectively. Let ω be a primitive element of $\mathfrak{S}^*$, with respect to P, which is not algebraic over K, and let D be the discriminant of the base $1, \omega, \cdots, \omega^{\nu-1}$. Let $\mathfrak{p}^*$ be a prime $\mathfrak{S}^*$-ideal at which the residue of ω is an element of K and such that $D \not\equiv 0$ ($\mathfrak{p}^*$). There exist infinitely many such prime ideals $\mathfrak{p}^*$ since K contains infinitely many elements and since the numbers of prime divisors of P* which are at infinity with respect to $\mathfrak{S}^*$ is finite. Let finally $\mathfrak{p}^* \cap \mathfrak{S} = \mathfrak{p}$. The assumption $D \not\equiv 0$ ($\mathfrak{p}$) implies that $1, \omega, \cdots, \omega^{\nu-1}$

The condition that $\{W\}$ is composite with $\{W^*\}$ implies, at any rate, that *the field* P *is not maximally algebraic in* Σ, since P^* is a proper algebraic extension of P. Conversely, if P is not maximally algebraic in Σ, and if P^* denotes the relative algebraic closure of P in Σ, then P^* is of degree of transcendency 1 over K, and the pencil $\{W\}$ is composite with pencil $\{W^*\}$ defined by P^*. Hence, a *pencil* $\{W\}$ is *composite* (with some other pencil) *if and only if the field* P *which defines it is not maximally algebraic in* Σ.

10. **Theorem of Bertini (for pencils).** *If a pencil* $\{W\}$, *free from fixed components, is not composite, then all but a finite number of members of the pencil are irreducible varieties (over* K).

For the proof we first establish two very simple lemmas, probably well known.

LEMMA 2. *If* P *is a maximally algebraic subfield of a field* Σ *and if* $\Sigma^* = \Sigma(x_1, \cdots, x_m)$ *is a pure transcendental extension of* Σ, *then* $P(x_1, \cdots, x_m)$ *is maximally algebraic in* Σ^*.

Proof. It is clearly sufficient to prove the lemma for $m=1$. Let then $\Sigma^* = \Sigma(x)$, x a transcendental over Σ, and let $t=f(x)/g(x)$ be an element of $\Sigma(x)$ which is algebraic over $P(x)$. We assume that $f(x)$ and $g(x)$ are relatively prime (in $\Sigma[x]$). We have to prove that t is an element of $P(x)$. Consider first the case $t \in \Sigma$ (whence f and g are elements of Σ). Since t is algebraic over $P(x)$, it must be already algebraic over P, since otherwise x would be algebraic over $P(t)$, and this is impossible, since $P(t) \subset \Sigma$. Hence $t \in P$, since P is maximally algebraic in Σ.

Let us now assume that t is not in Σ. Then t, being a transcendental over Σ, is *a fortiori* transcendental over P. But t is by hypothesis algebraic over $P(x)$. *Consequently* x *is algebraic over* $P(t)$. Let $f(x) = a_0x^n + \cdots + a_n$, $g(x) = b_0x^n + \cdots + b_n$; a_0, b_0 not both zero, and $a_i, b_j \in \Sigma$. Then

$$x^n(b_0t - a_0) + \cdots + (b_nt - a_n) = 0,$$

is the *irreducible* equation for x over $\Sigma(t)$ (since $f(x)$ and $g(x)$ are relatively prime in $\Sigma[x]$). The n roots of this equation are among the conjugates of x over $P(t)$, since $P(t) \subset \Sigma(t)$. *Consequently, the quotients*

$$\zeta_i = \frac{b_it - a_i}{b_0t - a_0}, \qquad i = 1, 2, \cdots, n,$$

are algebraic over $P(t)$.

Without loss of generality we may assume that $b_0 \neq 0$ (we may replace t by $1/t$). If we write down the algebraic equation for ζ_i over $P(t)$, then after

constitute an $\mathfrak{S}_\mathfrak{p}$-module basis for $\mathfrak{S}^*_{\mathfrak{p}^*}$. Hence if $\omega \equiv c\ (\mathfrak{p})$ and if Δ, Δ^* denote the residue fields at $\mathfrak{p}$ and $\mathfrak{p}^*$ respectively ($\Delta = \mathfrak{S}/\mathfrak{p}$, $\Delta^* = \mathfrak{S}^*/\mathfrak{p}^*$), then $\Delta^* = \Delta(c)$, i.e., $\Delta^* = \Delta$, since $c \in K \subset \Delta$. This of course implies that $\mathfrak{S}^*\mathfrak{p} \neq \mathfrak{p}^*$, because $\mathfrak{S}^*\mathfrak{p} = \mathfrak{p}^*$ would imply the relation $[\Delta^* : \Delta] = \nu$.

clearing the denominators we get an algebraic equation for t over Σ. This must be an identity, since t is transcendental over Σ. If we equate to zero the leading coefficient of this equation in t, we find that *b_i/b_0 is algebraic over* P. But since P is maximally algebraic in Σ, it follows that *b_i/b_0 is an element of* P. Hence we may write $g(x)=b_0\cdot g_1(x)$, where $g_1(x)\in \mathrm{P}[x]$.

If also $a_0\neq 0$, then replacing t by $1/t$ we conclude in the same fashion that $f(x)=a_0f_1(x), f_1(x)\in \mathrm{P}[x]$. We have now

$$t = \frac{a_0}{b_0}\cdot\frac{f_1(x)}{g_1(x)}\cdot$$

Since t is algebraic over $\mathrm{P}(x)$ and since $f_1(x)/g_1(x)\in\mathrm{P}(x)$, it follows that a_0/b_0 is algebraic over $\mathrm{P}(x)$. But since $a_0/b_0\in\Sigma$, it follows by the preceding case that $a_0/b_0\in\mathrm{P}$, whence $t\in\mathrm{P}(x)$. If $a_0=0$, then we replace t by $c+t$ where $c\in\mathrm{P}$ and $c\neq 0$. Then $c+t$ is a quotient of two polynomials, $(cg+f)/g$, *of the same degree*. By the preceding case $a_0\neq 0$, it follows that $c+t\in\mathrm{P}(x)$, whence also $t\in\mathrm{P}(x)$. This completes the proof of the lemma.

LEMMA 3. *Let $f(x_1, \cdots, x_n)$ be an irreducible polynomial in $\mathrm{P}[x_1, \cdots, x_n]$, where* P *is a field, and let Σ denote the field $\mathrm{P}(\xi_1, \cdots, \xi_n)$ defined by the equation $f(\xi_1, \cdots, \xi_n)=0$. If* P *is maximally algebraic in Σ, then the polynomial $f(x_1, \cdots, x_n)$ is absolutely irreducible.*

Proof. It is sufficient to prove that f remains irreducible for any finite extension Ω of P. The lemma is trivial when the field Ω is a pure transcendental extension of P: $\Omega=\mathrm{P}(u_1, \cdots, u_m)$. Moreover, in this case, if Σ^* denotes the field $\Omega(\xi_1, \cdots, \xi_n)$, then Σ^* is a pure transcendental extension of Σ, namely: $\Sigma^*=\Sigma(u_1, \cdots, u_m)$. Since P is maximally algebraic in Σ, it follows, by the preceding lemma, that Ω is maximally algebraic in Σ^*. We conclude that *it is sufficient to prove the lemma for finite algebraic extensions of* P. But in this case the lemma follows from a more general result concerning the behavior of prime ideals under algebraic extensions of the ground field (Zariski [5, Theorem 3, p. 198]).

11. **Proof of the theorem of Bertini.** The notation being that of §3, let ζ be any element of Σ^* and let

$$F(\{u_{ij}\}, \{\eta_i\}, \zeta) = 0 \tag{8}$$

be the irreducible equation for ζ over the field P^* $(=\mathrm{P}(\{u_{ij}\}, \{\eta_i\}))$. We may assume that F is an irreducible polynomial in $\mathrm{P}[\{u_{ij}\}, \{\eta_i\}, \zeta]$. Since Σ^* is a pure transcendental extension of Σ and since, by hypothesis, P is maximally algebraic in Σ, it follows, by Lemma 2, that P is also maximally algebraic in Σ^*. Hence, *a fortiori*, P is maximally algebraic in the field $\mathrm{P}(\{u_{ij}\}, \{\eta_i\}, \zeta)$, a subfield of Σ^*. By Lemma 3, we conclude that *the polynomial F is absolutely irreducible.*

Let $a, b, \cdots$ be the coefficients of the general polynomial in the arguments $\{u_{ij}\}, \{\eta_i\}, \zeta$, of the same degree as F ($a, b, \cdots$ are indeterminates). Let $a_0, b_0, \cdots$ be the corresponding coefficients of F ($a_0, b_0, \cdots$ are elements of P). There exists a finite number of finite sets of polynomials in $a, b, \cdots$, with rational coefficients, say

$$\{G_{i1}(a, b, \cdots), G_{i2}(a, b, \cdots), \cdots\}, \qquad i = 1, 2, \cdots, \rho,$$

with the following property, that if $F^*(\{u_{ij}\}, \{\eta_i\}, \zeta)$ is a polynomial, of the same degree as F, with coefficients $\alpha, \beta, \cdots$ in some field Ω, a necessary and sufficient condition that F^* be reducible in some extension of Ω is that $G_{i1}(\alpha, \beta, \cdots) = G_{i2}(\alpha, \beta, \cdots) = \cdots = 0$, for at least one value of i ($i = 1, 2, \cdots, \rho$). Since F is absolutely irreducible, it follows that for each i at least one of the elements

$$G_{i1}(a_0, b_0, \cdots), G_{i2}(a_0, b_0, \cdots), \cdots$$

is different from zero. These are elements of P. If then $\mathfrak{p}$ is a prime divisor of P/K, and if $a_0^*, b_0^*, \cdots$ denote the residues of $a_0, b_0, \cdots$ at $\mathfrak{p}$, then in the first place for all but a finite number of divisors $\mathfrak{p}$ the residues $a_0^*, b_0^*, \cdots$ will be all finite. In the second place, if an additional *finite* set of prime divisors $\mathfrak{p}$ is avoided, the expressions $G_{i1}(a_0^*, b_0^*, \cdots)$, $G_{i2}(a_0^*, b_0^*, \cdots)$ will not be all zero, for any $i = 1, 2, \cdots, \rho$. Hence the resulting relation between the B^*-residues u_i^*, η_i^*, ζ^*:

$$F^*(\{u_{ij}^*\}, \{\eta_i^*\}, \zeta^*) = 0 \tag{9}$$

remains absolutely irreducible, and *in particular is irreducible over the residue field* Δ of $\mathfrak{p}$. Here B^* is the valuation of Σ^* defined by one of the divisors $\mathfrak{P}_i^*$ in (2), say by $\mathfrak{P}_1^*$.

Let now ζ be a primitive element of Σ^* over P*. Then F is of degree ν in ζ, where $\nu = [\Sigma^*: \text{P}^*]$. Again, with the exception of a finite number of divisors $\mathfrak{p}$, the equation (9) will also be of degree ν in ζ^*. Since this equation is irreducible, it follows that the relative degree of the residue field Δ^* of $\mathfrak{P}_1^*$ with respect to the field $\Delta(\{u_i^*\}, \{\eta_i^*\})$ is at least ν. But this last field is the residue field of the prime divisor $\mathfrak{p}^*$ of P*/K*, and the relative degree in question cannot therefore exceed the relative degree $[\Sigma^*: \text{P}^*]$. Hence

$$[\Delta^*: \Delta(\{u_{ij}^*\}, \{\eta_i^*\})] = \nu. \tag{10}$$

Now the relative degree $[\Delta:\text{K}]$ is equal to d, where d is the degree of $\mathfrak{p}$. Since the u_{ij}^*, η_i^* are algebraically independent over Δ, it follows that $\Delta(\{u^*\}, \{\eta_i^*\})$ is also of relative degree d over the field $\text{K}(\{u_{ij}^*\}, \{\eta_i^*\})$. Hence, by (10),

$$[\Delta^*: \text{K}(\{u_{ij}^*\}, \{\eta_i^*\})] = \nu d.$$

But the left-hand side of this relation is the relative degree d_1 of $\mathfrak{P}_1^*$. Hence

$d_1 = \nu d$. Comparing with the relation (6) of §7, we conclude that $m=1$, $h_1=1$, whence $W_{\mathfrak{p}} = \Gamma_1$, and $W_{\mathfrak{p}}$ is an irreducible subvariety of V. Since this holds true for all but a finite number of prime divisors of P/K, the proof of the theorem of Bertini is complete.

12. **Absolutely irreducible members of a non-composite pencil.** In this section we wish to prove the following theorem:

If K *is not maximally algebraic in* Σ, *then every irreducible subvariety of* V *is absolutely reducible. If* K *is maximally algebraic in* Σ, *and if* $\{W\}$ *is a non-composite pencil free from fixed components, then a* $W_{\mathfrak{p}}$ *is absolutely reducible if the degree of* $\mathfrak{p}$ *is* >1, *while, with a finite number of exceptions, all* $W_{\mathfrak{p}}$ *corresponding to prime divisors of degree* 1 *are absolutely irreducible.*

For the proof of this theorem, we first recall the definition of absolute irreducibility. An irreducible variety V in an $S_n(x_1, \cdots, x_n)$, over a ground field K, is given by a prime ideal $\mathfrak{p}$ in the polynomial ring $\mathfrak{J} = K[x_1, \cdots, x_n]$. The variety V is *absolutely irreducible* if $\mathfrak{p}$ remains prime under *any* extension K* of the ground field K; i.e., if $\mathfrak{J}^*\mathfrak{p}$ is prime, $\mathfrak{J}^* = K^*[x_1, \cdots, x_n]$. In the contrary case, V is *absolutely reducible.* We now derive a necessary and sufficient condition for absolute irreducibility:

Lemma 4. *A necessary and sufficient condition in order that* V *be absolutely irreducible is that* K *be maximally algebraic in the field* Σ *of rational functions on* V(12).

Proof. *The condition is sufficient.* If K* is an algebraic extension of K, then the sufficiency of the condition follows from the quoted theorem in [5, p. 198]. If K* is a pure transcendental extension of K, $K^* = K(u_1, \cdots, u_m)$, $\mathfrak{J}^*\mathfrak{p}$ is prime unconditionally. Moreover, if $\mathfrak{J}^*\mathfrak{p} = \mathfrak{p}^*$, then the quotient field Σ^* of $\mathfrak{J}^*/\mathfrak{p}^*$ coincides with the field $\Sigma(u_1, \cdots, u_m)$, a pure transcendental extension of Σ. By Lemma 2, if K is maximally algebraic in Σ, K* will be maximally algebraic in Σ^*. Hence $\mathfrak{p}^*$ remains prime under any algebraic extension of K*.

The condition is necessary. Let θ be an element of Σ which is algebraic over K, and let $F(\theta)=0$ be the irreducible equation for θ over K. Let $\theta_1, \cdots, \theta_{g-1}$ be conjugates of θ over K, where g is the degree of F. We take as field K* the field $K(\theta, \theta_1, \cdots, \theta_{g-1})$. Since $\theta \in \Sigma$, there must exist a quotient ϕ/ψ ($\phi, \psi \in K[x_1, \cdots, x_n]$), $\psi \not\equiv 0$ ($\mathfrak{p}$), such that $\psi^g F(\phi/\psi) \equiv 0$ ($\mathfrak{p}$). This equation factors in $\mathfrak{J}^*$: $(\phi - \theta\psi)(\phi - \theta_1\psi) \cdots (\phi - \theta_{g-1}\psi) \equiv 0$ ($\mathfrak{J}^*\mathfrak{p}$). Now, by hypothesis, $\mathfrak{J}^*\mathfrak{p}$ is prime. Hence one of the factors $\phi - \theta_i\psi$ ($\theta_0 = \theta$) must be in $\mathfrak{J}^*\mathfrak{p}$. Since $\mathfrak{J}^*\mathfrak{p}$ is invariant under all automorphisms of K* over K, it follows that all the g factors $\phi - \theta_i\psi$ must be in $\mathfrak{J}^*\mathfrak{p}$. This implies that g is 1, since $\theta_i \neq \theta_j$ if $i \neq j$ and since $\psi \not\equiv 0$ ($\mathfrak{p}$). Hence $\theta \in K$, and this shows that K is maximally algebraic in Σ.

(12) In Lemma 3 we have proved the sufficiency of this condition in the special case of principal ideals (f).

We now pass to the proof of the theorem stated at the beginning of this section.

Let us first assume that K is not maximally algebraic in Σ (whence V itself is absolutely reducible). *Since V is normal*, the relative algebraic closure $\overline{K}$ of K in Σ must be contained in the *integrally closed* ring $\mathfrak{o}=K[\xi_1, \cdots, \xi_n]$. Hence $\overline{K}$ is also contained in the quotient ring $\mathfrak{o}/\mathfrak{p}$ of any prime $\mathfrak{o}$-ideal $\mathfrak{p}$. Hence K is not maximally algebraic in the field of rational functions on the subvariety defined by $\mathfrak{p}$. Thus we conclude that in the present case *V does not carry at all absolutely irreducible subvarieties.* Note that the hypothesis that V is normal was essential in the proof.

We now consider the more interesting case in which the ground field K is maximally algebraic in Σ. Let $\mathfrak{p}$ be a prime divisor of P/K, of degree $d>1$. We show that *not only $W_\mathfrak{p}$ but also any irreducible component Γ of $W_\mathfrak{p}$ is absolutely reducible.* Namely, the field Δ of rational functions on Γ is the residue field of the prime divisor $\mathfrak{P}$ defined by Γ (since V is normal). Since $\mathfrak{P}$ induces in the field P the given prime divisor $\mathfrak{p}$, Δ contains the residue field of $\mathfrak{p}$. But since $d>1$, this last residue field is a proper algebraic extension of K. Hence K is not maximally algebraic in Δ, and Γ is absolutely reducible.

There remains to consider the prime divisors $\mathfrak{p}$ of degree 1. For these it is not difficult to see that the construction used in the proof of the theorem of Bertini actually leads to varieties $W_\mathfrak{p}$ which are not only irreducible over K but are also absolutely irreducible. In fact, in the first place we have the absolutely irreducible equation (9). The coefficients of this equation are elements of the residue field of $\mathfrak{p}$, i.e., in the present case, elements of K. The absolute irreducibility of (9) implies, by Lemma 4, that K *is maximally algebraic in the field* $K(\{u_{ij}^*\}, \{\eta_i^*\}, \zeta^*)$. In the second place, we may assume (avoiding a finite set of prime divisors $\mathfrak{p}$) that the equation (9) is of degree ν in ζ, where $\nu=[\Sigma^*:P^*]$. Then the field $K(\{u_{ij}^*\}, \{\eta_i^*\}, \zeta^*)$ coincides with the residue field Δ^* of $\mathfrak{P}_1^*$ (since this residue field is at most of relative degree ν with respect to the residue field of $\mathfrak{p}^*$, and since this last residue field is $K(\{u_{ij}^*\}, \{\eta_i^*\})$). We conclude that K is maximally algebraic in Δ^*. Since the field Δ of rational functions on the *irreducible* $W_\mathfrak{p}$ is a subfield of Δ^*, it follows that K is maximally algebraic in Δ and that consequently $W_\mathfrak{p}$ is absolutely irreducible (Lemma 4).

III. Linear systems of arbitrary dimension

13. **Definitions.** *A pencil $\{W\}$ is linear if the corresponding field* P *is a simple transcendental extension of* K (if $\{W\}$ has fixed components, P is the field which defines the pencil obtained from $\{W\}$ by deleting these fixed components).

Let us assume that $\{W\}$ is linear, and let $P=K(t)$, t a transcendental over K, $t\in\Sigma$. From our definition of a pencil, given in §2, it follows that our linear pencil $\{W\}$ can be obtained in the following fashion. Let $\mathfrak{p}$ be a prime divisor of P/K and let us assume that $\mathfrak{p}$ is not the divisor at infinity of (t). Let $f(t^*)=0$ be the irreducible equation over K, of degree d, satisfied by the residue t^* of t at $\mathfrak{p}$. Here d is the degree of $\mathfrak{p}$. The principal divisor $(f(t))$ in Σ can be written in the form:

$$(f(t)) = \mathfrak{A}_{\mathfrak{p}}/\mathfrak{A}_{\infty}^{d}, \tag{11}$$

where $\mathfrak{A}_{\mathfrak{p}}$, $\mathfrak{A}_{\infty}$ are integral divisors (of the first kind) and $\mathfrak{A}_{\infty}$ is the denominator of the principal divisor (t) and is independent of $\mathfrak{p}$. *The member* $W_{\mathfrak{p}}$ *of the pencil* $\{W\}$, outside of possible fixed components, *is the subvariety of* V *which corresponds to the integral divisor* $\mathfrak{A}_{\mathfrak{p}}$. If $\mathfrak{p}$ is the prime divisor where t becomes infinite, then $\mathfrak{A}_{\mathfrak{p}}=\mathfrak{A}_{\infty}$ and $d=1$.

Suppose now that we consider only those prime divisors of P/K *which are of degree* $d=1$. Then $f(t)=c_0t+c_1$, $c_0, c_1\in$K, and we will have

$$(c_0t + c_1) = \frac{\mathfrak{A}_{\mathfrak{p}}}{\mathfrak{A}_{\infty}}\cdot$$

This suggests a possible definition of a linear pencil which is different from ours. Namely one could define $\{W\}$ as the set of all $W_{\mathfrak{p}}$ for which $\mathfrak{p}$ is of degree 1. This is the customary definition in classical algebraic geometry. Unless K is algebraically closed, these usual pencils are proper subsets of our pencils. They are defined by the linear one-parameter family of functions: c_0t+c_1([13]). We use this customary procedure as a basis for the definition of linear systems. This definition is well known. We consider namely a finite set of functions in Σ, say $t_1, \cdots, t_s$, such that $1, t_1, \cdots, t_s$ are linearly independent over K. For arbitrary constants $c_0, c_1, \cdots, c_s$ in K, not all zero, let

$$c_0 + c_1t_1 + \cdots + c_st_s = \mathfrak{A}_{(c)}/\mathfrak{A}_{\infty},$$

where $\mathfrak{A}_{(c)}$ and $\mathfrak{A}_{\infty}$ are integral divisors (of the first kind) and where $\mathfrak{A}_{\infty}$ is independent of the c's. Let $W_{(c)}$ be the $(r-1)$-dimensional subvariety of V which corresponds to $\mathfrak{A}_{(c)}$. The totality of all $W_{(c)}$, as the constants c_i vary in K, is called *a linear system of dimensions s*, and is denoted by $|W|$. It is clear that $\mathfrak{A}_{\infty}$ itself is a particular $\mathfrak{A}_{(c)}$, namely for $c_0=1$, $c_1=\cdots=c_s=0$. It is also clear that we can always assume that no prime factor of $\mathfrak{A}_{\infty}$ occurs in *all* $\mathfrak{A}_{(c)}$ and that by this condition $\mathfrak{A}_{\infty}$ is uniquely determined. *The system* $|W|$ *is then free from fixed components*. If $\mathfrak{A}_{\infty}$ is chosen in this fashion, and if $\mathfrak{L}$ is an arbitrary *fixed* integral divisor, then $\mathfrak{A}_{(c)}\cdot\mathfrak{L}/\mathfrak{A}_{\infty}\cdot\mathfrak{L}$ is another representation of the principal divisor $(c_0+c_1t+\cdots+c_st)$, but it is not in reduced form. The system $|W|$ defined by this representation would have as fixed component the $(r-1)$-dimensional subvariety which corresponds to $\mathfrak{L}$.

(13) All the results proved in the preceding sections continue to hold for these pencils, except the property that through each point of V there passes at least one $W_{\mathfrak{p}}$.

We note that if $|W|$ is free from fixed components and if

$$\mathfrak{A}_\infty = \mathfrak{P}_{\infty 1}^{h_1} \cdots \mathfrak{P}_{\infty m}^{h_m},$$

then each t_i has at $\mathfrak{P}_{\infty j}$ order $\geqq -h_j$, and at least one function t_i must have at $\mathfrak{P}_{\infty j}$ order exactly equal to $-h_j$.

One more remark. If we pass from $t_1, \cdots, t_s$ to another set of functions $\tau_1, \cdots, \tau_s$, by a nonsingular transformation of the form

$$\tau_i = \frac{d_{i0} + d_{i1}t_1 + \cdots + d_{is}t_s}{d_{00} + d_{01}t_1 + \cdots + d_{0s}t_s},$$

then the functions $1, \tau_1, \cdots, \tau_s$ define the same linear system $|W|$. Namely, we will have

$$(b_0 + b_1\tau_1 + \cdots + b_s\tau_s) = \frac{\mathfrak{A}_{(c')}}{\mathfrak{A}_{(d_{0j})}},$$

where $c'_j = b_0d_{0j} + b_1d_{1j} + \cdots + b_sd_{sj}$.

14. **Composite linear systems.** Let $\Omega = K(t_1, \cdots, t_s)$ *and let us assume that* Ω *is of degree of transcendency* 1 *over* K. Let us see what geometric property of $|W|$ corresponds to this assumption. We shall also assume that $|W|$ is free from fixed components.

Let $\{Z\}$ be the pencil (free from fixed components) defined by the field Ω. Let $t = c_0 + c_1t_1 + \cdots + c_st_s$, where $c_0, \cdots, c_s$ are arbitrary but fixed constants. The field $K(t) = P$ defines a pencil $\{W'\}$, free from fixed components. It is clear that each W' is either a W, or becomes a W after a certain fixed component M_0 is added to each W'. Namely, if t has order $-q_i$ at $\mathfrak{P}_{\infty i}$ $(i = 1, 2, \cdots, m)$, and if $\Gamma_{\infty i}$ is the irreducible V_{r-1} which corresponds to $\mathfrak{P}_{\infty i}$, then $W' + (h_1 - q_1)\Gamma_{\infty 1} + \cdots + (h_m - q_m)\Gamma_{\infty m}$ is a W, for each member W' of the pencil $\{W'\}$.

Let the pencil $\{W' + M_0\}$ be denoted by $\{W\}$. We have that P either coincides with Ω or is a subfield of Ω. *Let us first assume* that $\{W\}$ *has no fixed components.* We assert that *in this case* P *is a proper subfield of* Ω, *provided* $s > 1$.

To show this, let us assume the contrary. Each t_i is then an element of $K(t)$, say $t_i = f_i(t)/g_i(t)$, where f_i and g_i are assumed to be relatively prime. Let $t - d$ be a factor of $g_i(t)$, if g_i is of degree $\geqq 1$. Let $\mathfrak{P}$ be a prime divisor of Σ/K which occurs in the numerator of the principal divisor $(t - d)$. Since $\{W\}$ has no fixed components, t, and hence also $t - d$, becomes infinite along each of the varieties $\Gamma_{\infty 1}, \cdots, \Gamma_{\infty m}$. Consequently $\mathfrak{P} \neq \mathfrak{P}_{\infty i}$, $i = 1, 2, \cdots, m$. On the other hand, since $t - d$ is a factor of $g_i(t)$, t_i is infinite at the divisor $\mathfrak{P}$, and this is impossible since t_i can become infinite only at $\mathfrak{P}_{\infty 1}, \cdots, \mathfrak{P}_{\infty m}$.

Hence $g_i(t)$ *is of degree zero*, i.e., $t_i = f_i(t)$. Let ν_i be the degree of $f_i(t)$. Again, since $\{W\}$ is free from fixed components, the order of t at $\mathfrak{P}_{\infty j}$ is exactly $-h_j$. Hence the order of t_i at $\mathfrak{P}_{\infty j}$ is exactly $-h_j\nu$. But this order is at most $-h_j$. Hence $\nu \leqq 1$. Thus each t_i is of the form: $t_i = d_{i0} + d_{i1}t$, and if $s > 1$ this is impossible, since $1, t_1, \cdots, t_s$ are linearly independent over K. This proves our assertion.

Since $\{W\}$ has no fixed components, every W' is a W, and since Ω is a proper algebraic extension of P, *$\{W\}$ is composite with the pencil $\{Z\}$*. We have therefore proved that each W decomposes into a certain number of varieties Z, provided W belongs to some subpencil of $|W|$ which is free from fixed components. But it is clear that there exists such a pencil for each W. Namely, let $W = W_{(c)}$ and let $W_{(c')}$ be a W which has no components in common with $W_{(c)}$. We put

$$t = \frac{c_0 + c_1 t_1 + \cdots + c_s t_s}{c_0' + c_1' t_1 + \cdots + c_s' t_s}.$$

It is then clear that the pencil determined by the field $K(t)$ is contained in the linear system $|W|$, is free from fixed components, and that $W_{(c)}$ and $W_{(c')}$ are members of that pencil.

In conclusion, we have proved that if $s > 1$ and if $K(t_1, \cdots, t_s)$ is of degree of transcendency 1 over K, then each member of the linear system $|W|$ (free from fixed components) decomposes into a certain number of varieties Z. We say that *$|W|$ is composite with the pencil $\{Z\}$*.

15. **The theorem of Bertini for linear systems.** We shall say that a linear system $|W|$ is *reducible* if *each* member W of the system is reducible *over* K. We shall say that $|W|$ is *absolutely reducible*, if each W is absolutely reducible.

THEOREM OF BERTINI. *A reducible linear system $|W|$, free from fixed components, is necessarily composite with a pencil. If* K *is maximally algebraic in Σ, then the assumption that $|W|$ is reducible can be replaced by the weaker assumption that $|W|$ is absolutely reducible.*

For linear systems of dimension 1 this theorem has already been proved in §§11 and 12. In order to prove this theorem for system of dimension $s > 1$, we first prove a lemma.

LEMMA 5. *Let Σ be a field of algebraic functions of r independent variables, $r > 1$, and let the ground field* K *be maximally algebraic in Σ. If $x_1, \cdots, x_\rho$, $1 < \rho \leqq r$ are algebraically independent elements of Σ (over* K*) and if we put*

$$\bar{x}_i = c_{i1}x_1 + \cdots + c_{i\rho}x_\rho, \qquad i = 1, 2, \cdots, \rho - 1,$$

then for non-special constants c_{ij} in K, *the field $K(\bar{x}_1, \cdots, \bar{x}_{\rho-1})$ is maximally algebraic in Σ.*

Proof. If the lemma is true for $\rho = 2$, then for non-special constants c_1, c_2 the field $\bar{K} = K(c_1x_1 + c_2x_2)$ is a maximally algebraic subfield of Σ, and the degree of transcendency of Σ over $\overline{K}$ is $r-1$. If, say, $c_1 \neq 0$, then $x_2, \cdots, x_\rho$ are algebraically independent over $\overline{K}$, and hence replacing K by $\overline{K}$ we achieve a reduction from ρ to $\rho-1$. Therefore it is sufficient to prove the lemma for $\rho=2$.

Let Σ' be the relative algebraic closure of the field $K(x_1, x_2)$ in Σ. We may conduct the proof under the assumption that $\Sigma'=\Sigma$, for if $K(c_1x_1+c_2x_2)$ is maximally algebraic in Σ', then it is also maximally algebraic in Σ. Hence we may assume that $r=2$ and that therefore Σ is an algebraic extension of $K(x_1, x_2)$. Since we are dealing with fields of algebraic functions, Σ is a finite extension of $K(x_1, x_2)$.

Let $\bar{x}=x_1+cx_2$, $c \in K$, and let Ω_c denote the relative algebraic closure of the field $K(\bar{x})$ in Σ. We shall prove that *for all but a finite number of elements c in* K *the field* Ω_c *coincides with* $K(\bar{x})$. This result will establish the lemma.

Let $\Sigma_c = \Omega_c(x_2)$. We have

$$K(x_1, x_2) \subseteq \Sigma_c \subseteq \Sigma.$$

Since Σ is a *finite* algebraic extension of $K(x_1, x_2)$ and since for each c in K the field Σ_c is between $K(x_1, x_2)$ and Σ, it follows that there is only a *finite* number of *distinct* fields Σ_c(14). Therefore, for all but a finite number of elements c in K, it is true that for a given c in K there exists another element d in K, $d \neq c$, such that $\Sigma_c=\Sigma_d$. We proceed to prove that *for any such element c in* K, *the field* Ω_c *coincides with* $K(\bar{x})$.

We may identify x_1+cx_2 and x_1+dx_2 with x_1 and x_2 respectively, since $c \neq d$. The fields Ω_c and Ω_d are now the relative algebraic closures in Σ of the fields $K(x_1)$ and $K(x_2)$ respectively. They shall be denoted by Ω_1 and Ω_2 respectively. Similarly we put $\Sigma_1=\Omega_1(x_2)$, $\Sigma_2=\Omega_2(x_1)$. We have, by hypothesis, $\Sigma_1=\Sigma_2=\Sigma^*$. The field K was assumed to be maximally algebraic in Σ. Hence K is also maximally algebraic in Ω_2. Since Σ^* $(=\Sigma_2)$ is a pure transcendental extension of Ω_2, it follows, by Lemma 2 of §10, that $K(x_1)$ *is maximally algebraic in* Σ^*. But $\Sigma^*=\Sigma_1 \supset \Omega_1$ and Ω_1 is algebraic over $K(x_1)$. Hence $\Omega_1=K(x_1)$, as was asserted.

16. **Proof of the theorem of Bertini.** By a linear homogeneous transformation on $t_1, \cdots, t_s$ we may arrange matters so that t_1 becomes infinite on $\Gamma_{\infty i}$ $(i=1, 2, \cdots, m)$ to the highest order h_i. We consider the pencil $\{W\}$ contained in the system $|W|$ and defined by the functions 1, t_1+ct_i, where $c \in K$. For all but a finite number of constants c in K, t_1+ct_i becomes infinite on $\Gamma_{\infty i}$ to the order h_i. Avoiding the exceptional constants c, we may therefore assert that the pencil $\{W\}$ is free from fixed components and consequently coincides with the pencil determined by the field $K(t_1+ct_i)$. However, the pencil $\{W\}$ contains only those varieties $W_\mathfrak{p}$ which correspond to prime divisors $\mathfrak{p}$ of

(14) Note that K is of characteristic zero and hence that Σ is a separable extension of $K(x_1, x_2)$. If we did not assume that K is of characteristic zero we could not assert that there is only a finite number of fields between $K(x_1, x_2)$ and Σ.

$K(t_1+ct_i)$ *which are of degree* 1. Let K^* be the relative algebraic closure of K in Σ, and let $\{W^*\}$ be the pencil defined by the field $K^*(t_1+ct_i)$. Again we include in $\{W^*\}$ only those varieties $W^*_{\mathfrak{p}^*}$ which correspond to prime divisors $\mathfrak{p}^*$ of $K^*(t_1+ct_i)$ which are of degree 1. If we identify the fields $K(t_1+ct_i)$, $K^*(t_1+ct_i)$ with the fields P, P* respectively, of §9, we notice that if $\mathfrak{p}$ is of degree 1, then $\mathfrak{p}$ extends to a unique prime divisor of $\mathfrak{p}^*$ of P*, i.e., we will have in (7): $s=1$, $\rho_1=1$. *Hence each member of the pencil* $\{W\}$ *is also a member of the pencil* $\{W^*\}$. By hypothesis, each W is reducible over K. Hence, by the theorem of Bertini for pencils, *the field* $K^*(t_1+ct_i)$ *is not maximally algebraic in* Σ. The same conclusion holds true if $K^*=K$ and if each W is absolutely reducible, in view of the theorem of §12. Since this is true for all but a finite number of constants c in K, it follows, by Lemma 5, that t_1 and t_i are algebraically dependent over K^*, and hence also over K. Since this holds for $i=2, \cdots, s$ and since t_1 is a transcendental over K, it follows that the field $K(t_1, \cdots, t_s)$ is an algebraic extension of $K(t_1)$ and is therefore of degree of transcendency 1, q.e.d.

References

1. W. Krull, *Idealtheorie*, Ergebnisse der Mathematik und ihrer Grenzgebiete, vol. 4, no. 3.
2. B. L. van der Waerden, *Moderne Algebra*, vol. 2.
3. B. L. van der Waerden, *Über lineare Scharen von reduziblen Mannigfaltigkeiten*, Mathematische Annalen, vol. 113 (1936).
4. O. Zariski, *Some results in the arithmetic theory of algebraic varieties*, American Journal of Mathematics, vol. 61 (1939).
5. O. Zariski, *Algebraic varieties over ground fields of characteristic zero*, American Journal of Mathematics, vol. 62 (1940).

THE JOHNS HOPKINS UNIVERSITY,
BALTIMORE, MD.
CALIFORNIA INSTITUTE OF TECHNOLOGY,
PASADENA, CALIF.

NORMAL VARIETIES AND BIRATIONAL CORRESPONDENCES

OSCAR ZARISKI

1. **Introduction.** As one advances into the general theory of algebraic varieties, one reluctantly but inevitably reaches the conclusion that there does not exist a general theory of birational correspondences. This may sound too reckless a statement or too harsh a criticism, especially if one thinks of the fundamental role which birational transformations are supposed to have in algebraic geometry. Nevertheless our conclusion is in exact agreement with the facts and it is made with constructive rather than with critical intentions. It is true that the geometers have a fairly good intuitive idea of what happens or what may happen to an algebraic variety when it undergoes a birational transformation; but the only thing they know with any certainty is what happens in a thousand and one special cases. All these special cases—and they include all Cremona transformations—are essentially reducible to one special but very important case, namely, the case in which the varieties under consideration are nonsingular(that is, free from singular points). One can give many reasons for regarding as inadequate any theory which has been developed exclusively for nonsingular varieties. One rather obvious reason is that we have as yet no proof that every variety of dimension greater than 3 can be transformed birationally into a nonsingular variety.[1] But there are other, less transient, reasons. Were such a proof available, it would still be advisable to develop the theory of algebraic varieties, *as far as possible,* without restricting oneself to nonsingular projective models. This certainly would be the correct program of work from an arithmetic standpoint. I have a distinct impression that my friends the algebraists have not much use anyway for the resolution of the singularities. All they want is a general uniformization theorem, and now that they have it, they are content.

The following consideration will perhaps carry greater weight with the geometers. It turns out, as I have found out at some cost to myself, that we have to know a lot more about birational correspondences than we know at present before we can even attempt to carry

An address delivered before the meeting of the Society in Bethlehem, Pa., on December 31, 1941, by invitation of the Program Committee; received by the editors January 22, 1942.

[1] The resolution of the singularities of three-dimensional varieties will be carried out in a forthcoming paper of mine.

out the resolution of the singularities of higher varieties. A general theory of birational correspondences is a necessary prerequisite for such an attempt. I shall have occasion later on to indicate some difficult questions concerning birational correspondences which arise in connection with the resolution of singularities.

2. **Birational correspondences and valuations.** From a formal point of view, there is nothing mysterious about a birational transformation. If $V=V_r^n$ is an irreducible r-dimensional algebraic variety in an n-dimensional projective space, the coordinates $\xi_1, \xi_2, \cdots, \xi_n$ of its general point are algebraic functions of r independent variables, and the field $\Sigma=K(\xi_1, \xi_2, \cdots, \xi_n)$ generated by these functions is *the field of rational functions on* V. Here K denotes the field of constants (in the classical case K is the field of complex numbers). If $V'=V_r'^m$ is another irreducible algebraic variety, with general point $(\xi_1', \xi_2', \cdots, \xi_m')$ and associated field $\Sigma'=K(\xi_1', \xi_2', \cdots, \xi_m')$, then the two varieties V and V' are *birationally equivalent* if the two fields Σ and Σ' are simply isomorphic: $\Sigma/K\cong\Sigma'/K$. A birational transformation is merely the process of passing from one variety to another, birationally equivalent, variety.

The difficulties begin when we wish to associate with this purely formal process a *geometric transformation*, that is, a correspondence between the points of the two varieties. From the equations of the transformation, in which the ξ''s are given as rational functions of the ξ's and vice versa, it is not difficult to conclude that the transformation sets up a (1, 1) correspondence between the non-special points of V and the non-special points of V'. The points of either variety for which the equations of the transformation fail to define corresponding points on the other variety are referred to as special points in the sense that they lie on certain algebraic subvarieties, of dimension less than r. In the classical case, considerations of continuity allow us to complete the definition of the correspondence also for these special points. In the abstract case we use valuation theory instead, as follows:

A *valuation* of the field Σ is an homomorphic mapping v of the multiplicative group $\Sigma-0$ (that is, the element zero excluded) upon an ordered additive abelian group Γ, which satisfies the well known valuation axioms: (1) $v(\omega_1\cdot\omega_2)=v(\omega_1)+v(\omega_2)$; (2) $v(\omega_1\pm\omega_2)\geqq\min\{v(\omega_1), v(\omega_2)\}$; (3) $v(\omega)\neq 0$, for some ω in Σ; (4) $v(c)=0$, for all constants $c\neq 0$. We put $v(0)=+\infty$.

In the case of algebraic functions of one variable, every valuation arises from a branch of our curve V. The value $v(\omega)$ is then *the order of* ω at the branch and is an integer. Positive and negative $v(\omega)$ signify, respectively, that the center of the branch is a zero or a pole of the function ω, while if $v(\omega)=0$ then the function-theoretic value of ω at the center of the branch is a finite constant, different from zero. In this special case, it is clear that from an algebraic standpoint the function-theoretic values of the elements of the field are the cosets of the *valuation ring* $\mathfrak{B}$ $\{\omega\in\mathfrak{B}\leftrightarrow v(\omega)\geqq 0\}$, with respect to its subset consisting of the elements ω such that $v(\omega)>0$. This subset is a prime divisorless ideal $\mathfrak{p}$, and so the cosets form indeed a field; the field of complex numbers, in the classical case. This consideration is independent of the dimension of the field and can therefore be applied directly to the general case. It is therefore always possible to associate with any valuation v of Σ a mapping f of the elements of Σ upon the elements of another field (and the symbol ∞), *the field of residual classes* of $\mathfrak{B}$ mod $\mathfrak{p}$, and we may speak of $f(\omega)$, $\omega\in\Sigma$, as being the *function-theoretic value* of ω (if $v(\omega)<0$, then $f(\omega)=\infty$). This field is the so-called *residue field* of the valuation. However, if $r>1$ then the residue field may be a transcendental extension of the ground field K. Its degree of transcendence s over K, or briefly, *its dimension*, is at most $r-1$, and is referred to as the dimension of the valuation. A zero-dimensional valuation is called *a place* of the field Σ. The function-theoretic values of the elements of Σ at a given place are *constants*, that is, either elements of K or algebraic quantities over K.

Given a valuation v and a projective model V of Σ, with general point $(\xi_1, \xi_2, \cdots, \xi_n)$, it is permissible to assume that the function-theoretic values of the ξ's are different from ∞, since we may subject the coordinates ξ_i to an arbitrary projective transformation. Then the polynomials in the ξ's which have function-theoretic value zero form a prime ideal in the ring of all polynomials in the ξ's. This prime ideal defines an irreducible algebraic subvariety W of V. This subvariety W we call *the center of the valuation* v on the variety V. The dimension of W cannot exceed the dimension of the valuation. In particular, *the center of a place is always a point of* V.

The following geometric picture of a valuation is suggestive, although not entirely adequate. A zero-dimensional valuation, that is, a place, with center at a point P, corresponds to a way of approaching P along some one-dimensional branch, which may be algebraic, analytic, or transcendental. Similarly an s-dimensional valuation with an s-dimensional center W corresponds to a way of approaching W along an $(s+1)$-dimensional branch through W.

After these preliminaries, we define the birational correspondence between two birationally equivalent varieties V and V' as follows:

DEFINITION 1. *Two subvarieties W and W' of V and V', respectively, correspond to each other if there exists a valuation of the field Σ whose center on V is W and whose center on V' is W'.*

Note that our definition does not treat points in any privileged fashion. Any subvariety of V is treated as an element, rather than as a set of points. This procedure is much more convenient than the usual one in which corresponding loci are defined as loci of corresponding points.

In the study of the birational correspondence between V and V', it is found convenient to introduce a third variety $\overline{V}$ which is birationally related to both V and V', the so-called *join of V and V'* (or *the variety of pairs of corresponding points* of V and V'). $\overline{V}$ is defined as follows. We adjoin to Σ a new transcendental η_0 and we regard the $n+1$ quantities $\eta_0, \eta_1=\eta_0\xi_1, \cdots, \eta_n=\eta_0\xi_n$ as the *homogeneous coordinates* of the general point of V. Similarly, the $m+1$ quantities $\eta_0'=\eta_0, \eta_1'=\eta_0\xi_1', \cdots, \eta_m'=\eta_0\xi_m'$ will be the homogeneous coordinates of the general point of V'. The $(n+1)(m+1)$ quantities $\omega_{ij}=\eta_i\eta_j'$ can be regarded as the homogeneous coordinates of the general point of a variety birationally equivalent to V and to V'. This variety is our $\overline{V}$, the join of V and V'. The birational correspondence between $\overline{V}$ and V has the property that to any subvariety of $\overline{V}$ there corresponds a unique subvariety of V; in particular, to every point of $\overline{V}$ there corresponds a unique point of V. Similarly, for $\overline{V}$ and V'. Thus, both V and V' are single-valued transforms of $\overline{V}$. The properties of the birational correspondence between V and V' can be readily derived from the properties of the birational correspondences between $\overline{V}$ and V and between $\overline{V}$ and V'. We therefore replace one of the two varieties V, V', say V', by the join $\overline{V}$, that is, from now on we shall always assume that *V is a single-valued transform of V'.*

3. **Fundamental loci; geometric preliminaries.** On the basis of Definition 1, it is easy to prove that the points of V to which there correspond more than one point on V' constitute an algebraic subvariety F of V, and that every subvariety W of V to which there correspond more than one subvariety W' on V' must lie on F. This variety F is called the *fundamental locus* of the birational correspondence, and every W which lies on F is a *fundamental* variety. This is not our final definition, but it will do for the moment. Note that the fundamental locus on V' is an empty set, in view of our assumption that V is a single-valued transform of V'.

What corresponds on V' to a fundamental variety W of V? To this question we have a complete answer in the case of a nonsingular

V. We have, namely, in this case the following two fundamental theorems (see van der Waerden, *Algebraische Korrespondenzen und rationale Abbildungen*, Mathematische Annalen, vol. 110 (1934)):

A. *If W is an irreducible s-dimensional fundamental variety of V, then the transform of W is an algebraic subvariety of V' whose irreducible components are all of dimension greater than s.*

B. *The transform of the fundamental locus F is a pure $(r-1)$-dimensional subvariety of V'.*

The following examples show that both theorems fail to hold for singular models.

(1) If P is a point of V at which V is *locally reducible*, that is, if in the neighborhood of P the variety V consists of ν $(\nu>1)$ analytical r-dimensional branches, then in *special cases* it turns out that V is the projection of another variety V' on which these ν branches become separated.[2] Then the point P will be the projection of ν distinct points of V', and this contradicts Theorem A.

(2) Let Q be a ruled quadric surface in S_3 and let V be the three-dimensional cone which projects Q from a point O not in S_3. We take another copy of Q, say Q', which we now imagine as being immersed in an S_5. Let l be a line in S_5 which does not meet the S_3 containing the quadric. We set up a (1, 1) projective correspondence between the points P' of l and the lines p of one ruling of the quadric. Let V' be the irreducible three-dimensional variety generated by the planes (P', p), where P' and p are corresponding elements in the above projectivity. It is easy to set up a birational correspondence between V and V' in which to the planes (O, p) there correspond the planes (P', p). There will be no fundamental points on V', while O will be the only fundamental point on V. To the point O there corresponds on V' the line l, in contradiction with Theorem B.

One has the feeling that the second example does more damage than the first, because the first counterexample could be explained away on the basis that the point P, as origin of ν analytical branches, should not be regarded at all as a "point" of the variety, but rather as a point of the ambient projective space at which ν "points" of the variety accidentally happened to come together. This explanation, if stripped of all metaphysics, can have only one mathematical meaning, namely, it means, by implication, that in the general theory of birational correspondence, we should restrict ourselves to varieties

[2] Whether this is true generally, is not at all obvious and, in fact, has never been proved.

which are locally irreducible at each point. What interests us in this argument is the formal admission that some kind of restriction as to the type of varieties to be studied is necessary in the theory of birational correspondences. Whether or not the restriction to locally irreducible varieties is the right one, is a debatable matter. For one thing, it is not certain that in making such a restriction we are not being too stingy, unless we can prove that the branches of a variety can always be separated by the method of projection. This is probably true and should not be too difficult to prove. Actually we regard this restriction as being too generous. For, besides the requirement that the varieties V of our hypothetical restricted class satisfy Theorem A, we find it essential that these varieties also satisfy the following additional condition:

C. *If to a point P of V there corresponds a unique point P' of V', then the birational correspondence, regarded as an analytical transformation, is regular at P.*

The geometric meaning of this condition can be roughly indicated as follows. If this condition and condition A are satisfied for a given variety V, then the analytical structure of the neighborhood of any point P of V cannot be affected by a birational transformation, unless this transformation blows up P into a curve, or a surface and so on (always provided we replace the transform V' of V by the join $\overline{V}$; compare with §2). In particular, it is not possible to simplify any further the type of singularity which V possesses at P without doing a thing as radical as that of spreading out that singular point into a variety of dimension greater than 0. From this point of view, condition C can be looked upon as a sort of maximality condition.

In the case of algebraic curves it follows readily from this geometric interpretation that the only curves which satisfy conditions A and C are the nonsingular curves. But already in the case of surfaces we get a much wider variety of types. For instance, it can be proved that the surfaces in S_3 which satisfy our conditions are the surfaces which have only isolated singularities.

We now proceed to define arithmetically the varieties of our restricted class. We call these varieties *locally normal*. Included among these varieties are those which I have called *normal* varieties. The algebraic operation which plays a fundamental role in our arithmetic approach to the geometric questions just outlined is that of the *integral closure of a ring in its quotient field*. Theorem C follows in a relatively simple fashion from our arithmetic definitions and from some well known theorems in valuation theory. Theorem A lies much deeper and its proof is more difficult. As to Theorem B, it definitely must be sacrificed when we are dealing with singular varieties.

4. **Locally normal and normal varieties.** We use the homogeneous coordinates $\eta_0, \eta_1, \cdots, \eta_n$ of the general point of V (§2) and we define the *quotient ring* $Q(P)$ of any point P of V as the ring of all quotients $f(\eta)/g(\eta)$, where f and g are forms of like degree in $\eta_0, \eta_1, \cdots, \eta_n$ and where $g \neq 0$ at P. In other words, the quotient ring $Q(P)$ consists of all functions in our field Σ which have a definite and finite value at P. In a similar fashion, we define the quotient ring $Q(W)$ of any irreducible algebraic subvariety W of V by the condition that $g(\eta) \neq 0$ on W (that is, that g should not vanish at every point of W).

One is led to the consideration of quotient rings when one examines the equations of a birational correspondence between V and another variety V'. For it is seen immediately that if the nonhomogeneous coordinates $\xi_1', \cdots, \xi_m'$ of the general point of V' belong to the quotient ring $Q(W)$ of a given W on V, then to W there corresponds a unique subvariety W' on V' and moreover $Q(W')$ will be a subring of $Q(W)$. If $Q(W') = Q(W)$, then also W will be the only subvariety of V which corresponds to W'. In this case we say that the birational correspondence is *regular at* W, or *along* W. In particular, if the birational correspondence is regular at a point P of V, then as an analytical transformation it is regular in the neighborhood of P. Therefore, the quotient ring of a point determines uniquely the analytical structure of the neighborhood of the point. In the sequel we shall say that a birational correspondence is *regular* on V if it is regular at each point of V.

DEFINITION 2. *A variety V is locally normal along a subvariety W, if the quotient ring $Q(W)$ is integrally closed in its quotient field* (*that is, in* Σ).

DEFINITION 3. *V is locally normal, if it is locally normal at each point.*

The last definition refers only to points. The reason for this is the following: If V is locally normal at one point P of a subvariety W of V, then it is also locally normal along W. Hence if V is locally normal, in the sense of Definition 3, it is also locally normal along any W.

It is not difficult to show that V is locally normal if and only if the following condition is satisfied: *If $\mathfrak{C}$ is the conductor of the ring $K[\eta_0, \eta_1, \cdots, \eta_n]$ with respect to the integral closure of this ring in its quotient field, then the subvariety of V determined by the* (*homogeneous*) *ideal $\mathfrak{C}$ is empty.* This implies that either $\mathfrak{C}$ has no zeros at all, or its only zero is the trivial one: $(0, 0, \cdots, 0)$. Therefore, $\mathfrak{C}$ is either the unit ideal or is a primary ideal belonging to the irrelevant prime ideal $(\eta_0, \eta_1, \cdots, \eta_n)$.

If $\mathfrak{C}$ is the unit ideal, we say that V is *normal*, that is, we give the following definition:

DEFINITION 4. *A variety V is normal if the ring $K[\eta_0, \eta_1, \cdots, \eta_n]$ is integrally closed in its quotient field.*

It can be proved that the singular manifold of a locally normal r-dimensional variety is of dimension less than or equal to $r-2$ (in particular, *a locally normal curve is nonsingular*). The converse is not generally true, except in the case $r=1$, *since a nonsingular curve is always locally normal.* However, *the converse is true for hypersurfaces*, that is, for V_r's in an S_{r+1}, Thus any surface in S_3 is locally normal if and only if it has a finite number of singularities.

It is important to point out that *nonsingular varieties are always locally normal.*

Our definitions clearly indicate that normal varieties can differ from locally normal varieties only by some property at large, since locally they cannot be distinguished from each other. This difference at large is put into evidence by the following characterization of normal varieties, due to Muhly: *A variety V is normal if and only if the hypersurfaces of its ambient space, of any given order m, cut out on V a complete linear system.* Now it can be proved that locally normal varieties are characterized by the completeness of the above linear systems *for sufficiently high values of m.* This last result, in conjunction with the fact that every nonsingular V is locally normal, contains as a special case of the well known lemma of Castelnuovo concerning nonsingular curves. This lemma plays an important role in Severi's proof of Riemann-Roch's theorem for surfaces.

We have already pointed out that for locally normal varieties Theorems A and C hold true. Moreover it can be shown that these are the only varieties for which these theorems are true. It may be added that there is really no great loss of generality in confining the theory of birational correspondences to locally normal varieties. For it can be shown that any variety V determines uniquely, *to within regular birational transformations*, a locally normal variety V' which is birationally equivalent to V and which is such that: (a) to each point P' of V' there corresponds a unique point P of V, and we have always: $Q(P') \supseteq Q(P)$; (b) to any point P of V there corresponds a finite number of points on V'; (c) if V is locally normal at P, then the birational correspondence between V and V' is regular at P. It is not difficult to show that these properties of the birational correspondence between

V and V' imply that, to within a regular birational transformation, *V is the projection of V' from a center S_k which does not meet V'.*

It would be of interest to characterize the locally normal varieties for which Theorem B holds. It can be proved that the following condition is sufficient for the validity of Theorem B: If P is any point of V and if W is any $(r-1)$-dimensional subvariety through P, then a sufficiently high multiple of W should be *locally* (that is, at P) complete intersection of V with an hypersurface of the ambient space. In terms of ideal theory, this means that in the quotient ring of P a sufficiently high power of any minimal prime ideal should be *quasi-gleich* (in the sense of van der Waerden) to a principal ideal. This condition gives us a good insight into the "real" reason of the validity of Theorem B for nonsingular varieties; for we know that if P is a simple point, then every minimal prime ideal in $Q(P)$ is itself a principal ideal.

5. **Monoidal transformations.** I should now like to discuss briefly a special class of birational correspondences which seem to be very useful in the theory of singularities, whether we deal with the resolution of singularities or with the analysis of the composition of a singularity from the standpoint of infinitely near points. These special transformations are the hyperspace analogue of plane quadratic transformations, and they are therefore of importance also for the general theory of birational correspondences.

When we are dealing with locally normal varieties V, we find it most convenient to define fundamental varieties of a birational correspondence between V and another variety V', as follows:

DEFINITION 5. *An irreducible subvariety W of V is fundamental if a corresponding subvariety W' of V' exists such that $Q(W) \not\supseteq Q(W')$.*

We know from §4 that if $Q(W) \supseteq Q(W')$, then W' is the only subvariety which corresponds to W. Hence if W is fundamental, then the relation $Q(W) \not\supseteq Q(W')$ is true for any W' which corresponds to W. If the birational correspondence has no fundamental points on V and on V', then $Q(W) = Q(W')$, for *any* two corresponding subvarieties W and W', and the transformation is regular on V (and on V'). We do not regard two birational transformations as being essentially distinct if they differ only by a regular birational transformation.

We consider a birational correspondence between two locally normal varieties V and V', and we again restrict ourselves to the case in which the given correspondence has no fundamental points on V'. Then V' is in regular birational correspondence with the join $\overline{V}$ of V'

and V, and we may replace V' by $\overline{V}$. Consequently, we assume that the equations of the birational correspondence between V and V' are of the form:

$$\rho\eta_{ij} = \eta_i\phi_j, \qquad i = 0, 1, \cdots, n; j = 0, 1, \cdots, m,$$

where $\eta_0, \eta_1, \cdots, \eta_n$ are the homogeneous coordinates of the general point of V and where the ϕ_j are forms of like degree in the η's. Here ρ is a factor of proportionality and the η_{ij} are the homogeneous coordinates of the general point of V'. It can be shown that the fundamental locus on V is given by the base manifold of the *linear system* $\lambda_0\phi_0 + \cdots + \lambda_m\phi_m = 0$, provided we first drop all fixed $(r-1)$-dimensional components of the system. In terms of ideal theory, it means that we first write each principal ideal (ϕ_j) as a power product of minimal primes, say $(\phi_j) = \mathfrak{A}\mathfrak{B}_j$, where $\mathfrak{A}$ is the highest common divisor of $(\phi_0), \cdots, (\phi_m)$. Then the fundamental locus is given by the ideal $(\mathfrak{B}_0, \mathfrak{B}_1, \cdots, \mathfrak{B}_m)$. This ideal is of dimension less than or equal to $r-2$.

The special transformations which we wish to discuss are those for which $(\phi_0, \phi_1, \cdots, \phi_m)$ is itself a prime ideal, of dimension $s \leqq r-2$ or differs from a prime ideal by an irrelevant primary component. For the lack of a better name, we call them *monoidal transformations*.[3] The irreducible subvariety W of V defined by the ideal $(\phi_0, \phi_1, \cdots, \phi_m)$ is called the *center* of the transformation. It is not difficult to see that a change of the base of the ideal $(\phi_0, \cdots, \phi_m)$ does not essentially affect the transformation. A *quadratic transformation* is a special case of a monoidal transformation, the center is in that case a point.

The effect of a monoidal transformation consists in that the center W is spread out into an $(r-1)$-dimensional irreducible subvariety W' of V'. Moreover, *points of W which are simple both for V and W, correspond to simple points of W'.* This is the main reason why a monoidal transformation is a useful tool in the resolution of singularities, since while it may conceivably simplify some singular points which lie on its center, it does not introduce new singularities.[4]

There are two outstanding problems concerning monoidal trans-

[3] With some non-essential modifications, and without their projective trimmings, the space Cremona transformations, known as monoidal transformations, are monoidal transformations in our sense.

[4] There is one exception: to a simple point of V which is *singular* for W there may correspond singular points of V! For this reason it is usually advisable to "smooth out" W, that is, to resolve the singularities of W, before one applies the monoidal transformation.

formations which play a role in the problem of resolution of singularities, but which at the same time are decisively of interest in themselves. We proceed to outline these questions.

PROBLEM 1. *Given any birational correspondence between two nonsingular models V and V', and assuming that there are no fundamental points on V', show that the birational transformation can be decomposed into monoidal transformations.*

In other words, the question is to show that for nonsingular models the monoidal transformations form a set of generators of the birational group. It is very likely that this decomposition exists also when only V' is nonsingular.

PROBLEM 2. *Given any birational correspondence between two arbitrary (not necessarily nonsingular) models V and V', and assuming as before that there are no fundamental points on V', show that the fundamental varieties on V can be eliminated by monoidal transformations.*

By this I mean that it is asked to transform V by a sequence of monoidal transformations into another variety V^* such that the birational correspondence between V^* and V' has no fundamental points on V^*.

As to Problem 1, we have a proof in the case of surfaces. In this case, the result can be regarded as a generalization of the well known theorem of Noether on the decomposition of plane Cremona transformations into quadratic transformations, although Noether's theorem is not a special case of this general result. It may be well to clarify the connection between the two results. In the first place, a quadratic Cremona transformation is not at all a quadratic transformation in our sense. Our quadratic transformation has only one ordinary fundamental point, and its inverse has no fundamental points at all, while a plane quadratic transformation and its inverse both have three fundamental points, which in special cases may be infinitely near points. For this reason a plane Cremona transformation can never be a quadratic transformation in our sense. The transform of a plane π under a quadratic transformation in our sense is not a plane, but a certain rational surface M in S_5, or any other surface in regular birational correspondence with M. Of course, an ordinary quadratic transformation between two planes π and π' can be expressed as a product of quadratic transformations in our sense, or more precisely as the product of 3 successive quadratic transformations and of 3 inverses of quadratic transformations. Since our proof takes care of surfaces over abstract fields K, it yields immediately a corresponding result for the fundamental varieties W of dimension $r-2$ in the general case, for the adjunction of certain $r-2$ transcendentals to the ground field will make a surface out of the variety V and a point out of W. In particular, the decomposition into monoidal transformations is thus established for birational correspondences between nonsingular three-dimensional varieties, provided the correpondence has only fundamental curves, but no isolated fundamental points.

In applications of Problem 2, the main interest lies in the elimination of the *simple* fundamental varieties of V. In this case we have a complete proof, provided the resolution theorem is granted for varieties of *dimension two less* than the dimension of V. Thus, in the case of three-dimensional varieties we have to use only a thing as little as that of the resolution of singularities of an algebraic curve. It is clear that Problem 2 is to be viewed as a step in an inductive proof of the general theorem of the resolution of singularities rather than as a problem for the solution of which we first need that general theorem. The really important problem is Problem 1. Its solution seems to be essential for the resolution of singularities of higher varieties. Thus, it is possible to carry out the resolution of singularities of three-dimensional varieties, because we have the theorem of local uniformization for the varieties of this dimension, plus the solution of Problem 1 for surfaces over abstract fields of constants.

THE JOHNS HOPKINS UNIVERSITY

FOUNDATIONS OF A GENERAL THEORY OF BIRATIONAL CORRESPONDENCES

BY

OSCAR ZARISKI

In our papers dealing with the reduction of singularities of an algebraic surface (see [8, 11]), we were forced to devote a good deal of space to certain properties of birational correspondences for which we could find no general proofs in the literature. These properties were of a general character and therefore could not be regarded as part of the reduction proof proper, although they did play an auxiliary role in the proof. A similar situation arose in our reduction proof for three-dimensional varieties (not yet published), but in this case the amount of preliminary general material on birational correspondences used in the proof was even larger and was out of proportion to the length of the reduction proof proper. It thus became increasingly clear that the procedure of treating general questions of birational correspondences only as and when these questions come up in connection with various steps of the reduction process, could no longer be followed in the case of higher varieties. Instead it seemed necessary—and also worthwhile for its own sake—to develop *systematically* in a separate paper the fundamental concepts and theorems of the theory of birational correspondences, and to do this in as general a fashion as possible. This we propose to do in the present paper. We deal here with algebraic varieties, *with or without singularities*, over an arbitrary ground field (of characteristic zero or p).

It is difficult to say which of our results are entirely novel and which are not. Since many of the results hold only for normal varieties, they would appear to be novel inasmuch as our concept of a normal variety is new. On the other hand, most of our results were known for nonsingular models. It is perhaps correct to say that the novelty of the present investigation consists in showing that most of the known properties of birational correspondences between nonsingular varieties remain true more generally for normal varieties.

Of importance for the theory of algebraic functions over arbitrary ground fields of characteristic p is the fact that our construction of normal varieties which we gave in [7] in the case of algebraically closed ground fields of characteristic zero—and which carries over without essential modifications to arbitrary fields of characteristic zero (see II. 2)—can be extended to arbitrary fields of characteristic p (II. 2). This extension is made possible by a theorem of F. K. Schmidt [5] and by the normalization theorem of Emmy

Presented in part to the Society, December 31, 1941 under the title *Normal varieties and birational correspondences* (see [12]); received by the editors September 1, 1942.

Noether. The proof of this last theorem for arbitrary ground fields (and not only for fields containing "sufficiently many" elements; see Krull [2, pp. 41–42]) is given in II.2.

A feature of the treatment is our use of valuation theory. Our very definition of a birational correspondence (II.1) is valuation-theoretic in character, and our proofs are naturally conditioned by this valuation-theoretic approach. The characterization of an integrally closed ring as intersection of valuation rings, and the ideal theory in such a ring, also play an important role in our treatment.

Thanks are due to Irvin Cohen for valuable assistance lent by him during the preparation of this paper.

The following list of contents will give an idea of the individual topics treated.

Contents

Part I. Valuation-theoretic preliminaries

1. **Homogeneous ideals.** Let $\eta_0, \eta_1, \cdots, \eta_n$ be the homogeneous coordinates of the general point of an irreducible r-dimensional algebraic variety V immersed in an n-dimensional projective space. The coordinates η_i are defined to within a linear homogeneous nonsingular transformation with coefficients in the ground field K. By the very definition of homogeneous coordinates (Zariski [7, p. 284]), if $f(\eta)=0$ is an algebraic relation between the η's over K, and if we write f as a sum of forms $f_i(\eta)$ of different degrees, then each $f_i(\eta)$ individually is zero. This is equivalent to saying that the polynomials $f(y)$ in the polynomial ring $K[y_0, y_1, \cdots, y_n]$ (the y's are indeterminates) such that $f(\eta)=0$, form a *homogeneous ideal, that is, an ideal which has a base consisting of forms.*

Let P denote the ring $K[\eta_0, \eta_1, \cdots, \eta_n]$. We shall also consider homogeneous ideals in P, that is, again ideals in P which have a base consisting of forms. These are the ideals on which the homogeneous ideals of $K[y_0, \cdots, y_n]$ are mapped in the homomorphism $K[y] \sim K[\eta]$. The ring P possesses relative automorphisms τ_λ over K, where for any element $\phi(\eta)$ in P we define: $\tau_\lambda(\phi(\eta)) = \phi(\lambda\eta)$, $\lambda \in K$, $\lambda \neq 0$. It is clear that if $\mathfrak{A}$ is a homogeneous ideal in P, then $\tau_\lambda(\mathfrak{A}) = \mathfrak{A}$. Conversely, we have the following theorem:

THEOREM 1. *If* K *has infinitely many elements and if an ideal* $\mathfrak{A}$ *in* P *is such that* $\tau_\lambda(\mathfrak{A}) \subseteq \mathfrak{A}$, *for all* λ *in* K ($\lambda \neq 0$), *then* $\mathfrak{A}$ *is homogeneous.*

Proof. Let $\omega = f(\eta)$ be an arbitrary element of $\mathfrak{A}$ and let $f(\eta) = f_s(\eta) + f_{s+1}(\eta) + \cdots + f_m(\eta)$, where f_i is a form of degree i. We take arbitrarily in K a set of $m-s+1$ *distinct* elements $\lambda_1, \lambda_2, \cdots, \lambda_{m-s+1}$, *all different from zero.* We have, by hypothesis:

$$\tau_{\lambda_i}(\omega) = \lambda_i^s f_s(\eta) + \lambda_i^{s+1} f_{s+1}(\eta) + \cdots + \lambda_i^m f_m(\eta) \equiv 0(\mathfrak{A}).$$

These $m-s+1$ congruences imply the congruences: $f_j(\eta) = 0$, $j = s, s+1, \cdots, m$. Hence $\mathfrak{A}$ is a homogeneous ideal, as was asserted.

THEOREM 2. *If* $\mathfrak{A} = [\mathfrak{q}_1, \mathfrak{q}_2, \cdots, \mathfrak{q}_h]$ *is a normal decomposition of a homogeneous ideal* $\mathfrak{A}$ *into maximal primary components, then each* $\mathfrak{q}_i$ *is either itself homogeneous or—in the case of an embedded component—can be so selected as to be homogeneous.*

Proof. We first consider the case of an infinite ground field K. Let $\mathfrak{p}_i$ be the prime ideal associated with the primary ideal $\mathfrak{q}_i$. The infinitely many automorphisms τ_λ leave $\mathfrak{A}$ invariant and hence must permute the prime ideals $\mathfrak{p}_1, \mathfrak{p}_2, \cdots, \mathfrak{p}_h$. Consequently each of these prime ideals is left invariant by *infinitely many* automorphisms τ_λ. It follows then from the proof of the preceding theorem, that the prime ideals $\mathfrak{p}_i$ of $\mathfrak{A}$ must be all homogeneous.

If $\mathfrak{p}_i$ is an isolated prime ideal of $\mathfrak{A}$, then $\mathfrak{q}_i$ is uniquely determined. Since $\tau_\lambda(\mathfrak{p}_i) = \mathfrak{p}_i$, it follows that also $\tau_\lambda(\mathfrak{q}_i) = \mathfrak{q}_i$, whence $\mathfrak{q}_i$ is homogeneous.

Let now $\mathfrak{p}_i$ be an embedded prime of $\mathfrak{A}$. We apply to $\mathfrak{q}_i$ all the automorphisms τ_λ ($\lambda \in K$, $\lambda \neq 0$) and we denote by $\mathfrak{q}_i^*$ the intersection of all the ideals $\tau_\lambda(\mathfrak{q}_i)$. From the very definition of a primary ideal, it follows immediately that $\mathfrak{q}_i^*$ is a primary ideal and that $\mathfrak{p}_i$ is its associated prime ideal. Moreover, by Theorem 1, $\mathfrak{q}_i^*$ is homogeneous, and since $\mathfrak{A} \subseteq \mathfrak{q}_i^* \subseteq \mathfrak{q}_i$, we find: $\mathfrak{A} = [\mathfrak{q}_1, \mathfrak{q}_2, \cdots, \mathfrak{q}_i^*, \cdots, \mathfrak{q}_h]$, as was asserted.

Let now K be a finite field. We adjoin to K a transcendental u and we put $K^* = K(u)$, $P^* = K^*[\eta_0, \eta_1, \cdots, \eta_n]$ (we assume that u is also a transcendental with respect to the ring P). By the preceding case, our theorem holds for the ring P^* over the new ground field K^*. We can draw from this the conclusion that the theorem also holds for the ring P, provided we first establish certain relations between the ideals in P and the ideals in P^*. If $\mathfrak{B}$ is an ideal in P, it determines in P^* the *extended ideal* $P^*\cdot\mathfrak{B}$. Vice versa, every ideal $\mathfrak{C}^*$ in P^* gives rise to a *contracted ideal* in P, namely the ideal $\mathfrak{C}^*\cap P$ (but $\mathfrak{C}^*$ is not necessarily the extended ideal of its contracted ideal). Note that every element ω^* of P^* can be written in the form:

$$\omega^* = \frac{1}{g(u)}\cdot(\omega_0 u^m + \omega_1 u^{m-1} + \cdots + \omega_m), \tag{1}$$

where $g(u)\in K[u]$ and $\omega_j\in P$. The relations which we wish to establish concern the operations of extension and contraction just described, and are as follows:

a. *An ideal* $\mathfrak{B}^*$ *in* P^* *is the extension of an ideal in* P *if and only if the congruence* $\omega^*\equiv 0(\mathfrak{B}^*)$ *implies* $\omega_j\equiv 0(\mathfrak{B}^*)$, *for* $j=0, 1, \cdots, m$, *where* ω^* *is written in the form* (1).

b. $P^*\cdot\mathfrak{B}\cap P=\mathfrak{B}$.

c. *If* $\mathfrak{B}_1\cap\mathfrak{B}_2=\mathfrak{B}_3$, *then* $P^*\cdot\mathfrak{B}_1\cap P^*\cdot\mathfrak{B}_2=P^*\cdot\mathfrak{B}_3$. The proofs are trivial.

d. *If* $\mathfrak{p}$ *is a prime ideal in* P, *then* $P^*\cdot\mathfrak{p}$ *is also a prime ideal.* This assertion is essentially equivalent to the well known theorem that if R is an integral domain and if u is a transcendental with respect to R, then $R[u]$ is also an integral domain. In the present case R is the ring $P/\mathfrak{p}$.

Let now $\mathfrak{A}$ be a homogeneous ideal in P and let us consider its extended ideal $P^*\mathfrak{A}$ in P^*. It is clear that also $P^*\mathfrak{A}$ is a homogeneous ideal in P^*. Since K^* is an infinite field, we can write

$$P^*\cdot\mathfrak{A} = [\mathfrak{q}_1^*, \mathfrak{q}_2^*, \cdots, \mathfrak{q}_h^*], \tag{2}$$

where each $\mathfrak{q}_i^*$ is a homogeneous primary ideal. If we put $\mathfrak{q}_i=\mathfrak{q}_i^*\cap P$, we find, by property b:

$$\mathfrak{A} = [\mathfrak{q}_1, \mathfrak{q}_2, \cdots, \mathfrak{q}_h].$$

It is obvious that if $\mathfrak{B}^*$ is a homogeneous ideal in P^*, then $\mathfrak{B}^*\cap P$ is also a homogeneous ideal. Hence the ideals $\mathfrak{q}_i$ are all homogeneous. Since they are obviously primary ideals, our theorem follows in view of the unicity theorems concerning the decomposition of ideals into maximal primary components.

In addition to the relations a, b, c, d, we shall also have occasion to use the following property:

e. *If* $\mathfrak{q}$ *is a primary ideal in* P *and if* $\mathfrak{p}$ *is its associated prime ideal, then* $P^*\cdot\mathfrak{q}$ *is also primary and* $P^*\cdot\mathfrak{p}$ *is its associated prime ideal.*

To prove e, we denote by $\mathfrak{p}^*$ the prime ideal $P^*\cdot\mathfrak{p}$ and we consider any prime ideal $\mathfrak{p}_1^*$ of $P^*\cdot\mathfrak{q}$. We have: $\mathfrak{p}_1^*\supseteq P^*\mathfrak{q}\supseteq\mathfrak{q}$, whence $\mathfrak{p}_1^*\cap P\supseteq\mathfrak{q}$. But since the contraction of a prime ideal is also a prime ideal, it follows that $p_1^*\cap P\supseteq\mathfrak{p}$, whence

$$\mathfrak{p}_1^* \supseteq \mathfrak{p}^*. \tag{3}$$

Now let ω^* be an arbitrary element of $\mathfrak{p}_1^*$ and let us write ω^* in the form (1). Since $\mathfrak{p}_1^*$ is a prime ideal of $P^*\cdot\mathfrak{q}$, there exists in P^* an element ξ^* such that $\omega^*\cdot\xi^*\equiv 0\ (P^*\cdot\mathfrak{q})$, $\xi^*\not\equiv 0\ (P^*\cdot\mathfrak{q})$. Let us also write ξ^* in a form similar to (1):

$$\xi^* = \frac{1}{h(u)}\cdot(\xi_0 u^\mu + \xi_1 u^{\mu-1} + \cdots + \xi_\mu), \qquad h(u) \in K[u],\ \xi_j \in P.$$

Since $\xi^*\not\equiv 0(P^*\cdot\mathfrak{q})$, not all the elements $\xi_0, \xi_1, \cdots, \xi_\mu$ are in $\mathfrak{q}$. We may assume that $\xi_0\not\equiv 0(\mathfrak{q})$, since otherwise we can drop the term $\xi_0 u^\mu$ (it is permissible to replace ξ^* by any element of P^* which is congruent to it modulo $P^*\cdot\mathfrak{q}$). Since $\omega^*\xi^*\equiv 0(P^*\mathfrak{q})$, we must have, by a and b: $\omega_0\xi_0\equiv 0(\mathfrak{q})$, and consequently $\omega_0\equiv 0(\mathfrak{p})$, since $\xi_0\not\equiv 0(\mathfrak{q})$. From this we conclude, in view of (3), that the element

$$\frac{1}{g(u)}\cdot\lfloor\omega_1 u^{m-1} + \omega_2 u^{m-2} + \cdots + \omega_m\rfloor$$

also belongs to $\mathfrak{p}_1^*$. Continuing in the same fashion with this new element of $\mathfrak{p}_1^*$, we conclude that $\omega_0, \omega_1, \cdots, \omega_m$ are all in $\mathfrak{p}$. Hence $\omega^*\in P^*\cdot\mathfrak{p}$. Since ω^* is an arbitrary element of $\mathfrak{p}_1^*$, it follows that $\mathfrak{p}_1^*\subseteq\mathfrak{p}^*$, and this, in view of (3), yields the relation: $\mathfrak{p}_1^*=\mathfrak{p}^*$. What we have shown is that $P^*\mathfrak{q}$ has only one prime ideal, namely $\mathfrak{p}^*$, and that is exactly what is asserted in e.

The relation b shows that there is a (1, 1) correspondence between the ideals $\mathfrak{B}$ in P and their extended ideals $P^*\mathfrak{B}$ in P^*, for $P^*\mathfrak{B}_1=P^*\mathfrak{B}_2$ implies $\mathfrak{B}_1=\mathfrak{B}_2$. By the property c, this correspondence is an *isomorphism* with respect to the operation $\cap$ of intersection. It is a straightforward matter to show that this correspondence is also an isomorphism with respect to the other elementary operations on ideals, namely the operation of forming the sum, the product and the quotient of two ideals:

c_1. $P^*\cdot(\mathfrak{A}, \mathfrak{B})=(P^*\mathfrak{A}, P^*\mathfrak{B})$.

c_2. $P^*\cdot(\mathfrak{A}\mathfrak{B})=P^*\mathfrak{A}\cdot P^*\mathfrak{B}$.

c_3. $P^*\cdot(\mathfrak{A}:\mathfrak{B})=P^*\mathfrak{A}:P^*\mathfrak{B}$.

We shall refer to the relations c, c_1, c_2 and c_3 as the *isomorphism relations*. The question whether any one of these relations holds, arises quite generally whenever we deal with ideals in two rings P, P^* such that P is a subring of P^*. For a general treatment of the relationship between the ideals in two such rings, see Grell [1].

The ideal $(\eta_0, \eta_1, \cdots, \eta_n)$ is referred to as the *irrelevant* prime ideal in P. Any primary ideal whose associated prime ideal is the irrelevant prime is also referred to as irrelevant. Any prime homogeneous ideal, other than the irrelevant prime, is of dimension $s+1$, $s\geq 0$, and defines an irreducible s-dimensional subvariety of V. Two homogeneous ideals which differ only by the irrelevant component define one and the same subvariety of V.

2. **Homogeneous and nonhomogeneous coordinates.** A preference for one or the other type of coordinates is in part a matter of taste. However, it can

be claimed that in the study of properties of algebraic varieties, the use of homogeneous coordinates is indicated whenever one deals with properties *in the large*. For instance, the concept of a normal variety (Zariski [7, p. 285]) is defined essentially in terms of homogeneous coordinates. On the contrary, in questions pertaining to *local* properties it is preferable to use nonhomogeneous coordinates. Thus, if our attention is focused on a given subvariety W of V and if, say, $\eta_0 \neq 0$ on W, that is, if η_0 does not belong to the homogeneous prime ideal by which W is defined (whence W does not lie in the hyperplane $y_0=0$), then we may find it convenient to pass to the nonhomogeneous coordinates $\xi_i=\eta_i/\eta_0$, $i=1, 2, \cdots, n$. With respect to these coordinates "*W is at finite distance*"—an expression that we shall use consistently. More generally, if $l=c_0\eta_0+c_1\eta_1+\cdots+c_n\eta_n\neq 0$ on W, $c_i \in \mathrm{K}$, and if, say $c_\alpha \neq 0$, then the quotients η_i/l, $i\neq\alpha$, may be equally well used as nonhomogeneous coordinates ξ_i of the general point of V.

It should be understood that the field Σ of rational functions on V is the field $\mathrm{K}(\xi_1, \xi_2, \cdots, \xi_n)$ generated by the nonhomogeneous coordinates. The field $\mathrm{K}(\eta_0, \eta_1, \cdots, \eta_n)$ is a simple transcendental extension of Σ. The field Σ consists of all quotients $f(\eta)/g(\eta)$, where f and g are forms of like degree; in other words, Σ consists of all elements of the field $\mathrm{K}(\eta_0, \eta_1, \cdots, \eta_n)$ which are homogeneous of degree zero (Zariski [7, p. 284]).

We have also *two coordinate rings*: the ring $\mathrm{K}[\eta_0, \eta_1, \cdots, \eta_n]$ of the homogeneous coordinates, which we have denoted by P, and—for a given choice of the nonhomogeneous coordinates ξ_i—the ring $\mathrm{K}[\xi_1, \xi_2, \cdots, \xi_n]$ which we shall denote by $\mathfrak{o}$. In order to elicit the relationship between the ideals in $\mathfrak{o}$ and the homogeneous ideals in P, we assume for simplicity that $\xi_i=\eta_i/\eta_0$ and we consider the ring $\mathfrak{o}^*=\mathrm{K}(\eta_0)[\xi_1, \xi_2, \cdots, \xi_n]$. Since η_0 is a transcendental with respect to Σ, the two rings $\mathfrak{o}$ and $\mathfrak{o}^*$ are in the same relationship to each other as the two rings P, P^* of the preceding section. Therefore the correspondence between the ideals in $\mathfrak{o}$ and their extended ideals in $\mathfrak{o}^*$ satisfies all the relations a–e (in which P and P^* are naturally to be replaced by $\mathfrak{o}$ and $\mathfrak{o}^*$). We shall denote by M^* the class of $\mathfrak{o}^*$-ideals which are extensions of $\mathfrak{o}$-ideals.

From the pair of rings, $\mathfrak{o}$, $\mathfrak{o}^*$ we pass to the pair of rings $\mathfrak{o}^*$, P. We have: $\mathfrak{o}^*=\mathrm{K}(\eta_0)[\eta_1, \eta_2, \cdots, \eta_n]$. The polynomial ring $\mathrm{K}[\eta_0]$ is a subring of P, and thus $\mathfrak{o}^*$ is a *quotient ring* of P, since $\mathrm{K}[\eta_0]$ is at any rate closed under multiplication(1). There is therefore a (1, 1) correspondence between the ideals $\mathfrak{A}^*$ in $\mathfrak{o}^*$ and those ideals $\mathfrak{A}$ in P which are relatively prime to all elements of $\mathrm{K}[\eta_0]$, that is, which are such that $\mathfrak{A}:\alpha=\mathfrak{A}$, for all α in $\mathrm{K}[\eta_0]$. The correspondence is again that of extension and contraction: $\mathfrak{A}^*=\mathfrak{o}^*\mathfrak{A}$, $\mathfrak{A}=\mathfrak{A}^*\cap P$. Prime or primary ideals $\mathfrak{A}$ go, respectively, into prime or primary ideals $\mathfrak{A}^*$.

(1) For properties of quotient rings used in the text, see Grell [1] and Krull [2, p. 18]. We recall that a quotient ring is defined as follows: if R is an integral domain and if S is a subset of R which is closed under multiplication, then the quotients α/β, $\alpha, \beta \in R$, $\beta \in S$, form a ring. This is the quotient ring R_S. A special important case is the one in which S is the set-theoretic complement of a prime ideal $\mathfrak{p}$ in R, that is, $S=R-\mathfrak{p}$. In this case one writes $R_{\mathfrak{p}}$ instead of R_S.

The isomorphism relations c, c_1, c_2, c_3 of the preceding section, with P^* replaced by $\mathfrak{o}^*$, continue to hold[2]. These are properties which hold quite generally for quotient rings.

It is immediately seen that an $\mathfrak{o}^*$-ideal $\mathfrak{A}^*$ is in the class M^* if and only if the corresponding ideal $\mathfrak{A}$ in P is homogeneous. It is clear that a homogeneous ideal is always relatively prime to any polynomial $f(\eta_0)$ if that polynomial contains a constant term which is different from zero. Hence, for a homogeneous ideal $\mathfrak{A}$ to be relatively prime to each element of $K[\eta_0]$, it is necessary and sufficient that it be relatively prime to η_0. An equivalent condition is that *no prime ideal of* $\mathfrak{A}$ *be a divisor of* (η_0). *We have therefore a* (1, 1) *correspondence between the homogeneous ideals in P which satisfy this last mentioned condition and the ideals in* $\mathfrak{o}$. This conclusion corresponds to the obvious geometrical fact that by passing to nonhomogeneous coordinates η_i/η_0 we lose track of all the subvarieties of V which "are at infinity," that is, which lie in the hyperplane $y_0=0$.

Concretely, the relationship between two corresponding ideals $\mathfrak{a}$ and $\mathfrak{A}$ in $\mathfrak{o}$ and P, respectively, is as follows: a form $f(\eta_0, \eta_1, \cdots, \eta_n)$ belongs to $\mathfrak{A}$ if and only if $f(1, \xi_1, \cdots, \xi_n)$ belongs to $\mathfrak{a}$; a polynomial $\phi(\xi_1, \xi_2, \cdots, \xi_n)$ belongs to $\mathfrak{a}$, if and only if there exists in $\mathfrak{A}$ a form $f(\eta_0, \eta_1, \cdots, \eta_n)$ such that $f(1, \xi_1, \cdots, \xi_n)=\phi(\xi_1, \xi_2, \cdots, \xi_n)$.

3. **The center of a valuation.** Let W be an irreducible subvariety of V. If $\xi_1, \xi_2, \cdots, \xi_n$ are nonhomogeneous coordinates with respect to which W is at finite distance, then W is given by a prime ideal $\mathfrak{p}$ in the ring $\mathfrak{o}$. By the *quotient ring* $Q(W)$ of W we mean the quotient ring $\mathfrak{o}_\mathfrak{p}$. If $\mathfrak{P}$ is the homogeneous prime ideal which corresponds to W in the ring P, then it is easily seen that $Q(W)$ consists of all quotients $f(\eta)/g(\eta)$, where f and g are forms of like degree in $\eta_0, \eta_1, \cdots, \eta_m$ and where $g(\eta)\not\equiv 0(\mathfrak{P})$. This shows, incidentally, that $Q(W)$ is independent of the choice of the nonhomogeneous coordinates.

The relationship between the ideals in $\mathfrak{o}$ and the ideals in $Q(W)$ is the one described in the preceding section for general quotient rings. The (1, 1) correspondence is now between the ideals in $Q(W)$ and those ideals in $\mathfrak{o}$ which are relatively prime to each element[3] of $\mathfrak{o}-\mathfrak{p}$. An ideal in $\mathfrak{o}$ satisfies this condition *if and only if each prime ideal of this ideal is a multiple of* $\mathfrak{p}$. This shows that by passing from the ring $\mathfrak{o}$ (or from the ring P) to the quotient ring $Q(W)$ we lose all irreducible subvarieties of V which do not contain W. For this reason we may regard the ideal theory of $Q(W)$ as that pertaining to the "neighborhood" of W. An important property of the quotient ring $\mathfrak{o}_\mathfrak{p}$ is the

---

(2) However, it is to be pointed out that the set of ideals $\mathfrak{A}$ which correspond to ideals $\mathfrak{A}^*$ in the above correspondence is not in general closed under the operations of multiplication and addition of ideals. It *is* closed under the operations of intersection and quotient formation.

(3) Quite generally, given a quotient ring R_S (see footnote 1), there is a (1, 1) correspondence between the ideals in R_S and the ideals in R which are relatively prime to each element of S.

following: *the non-units in* $\mathfrak{o}_\mathfrak{p}$ *form an ideal*(4). This is the prime ideal in $\mathfrak{o}_\mathfrak{p}$ which corresponds to (that is, is the extension of) the prime ideal $\mathfrak{p}$ in $\mathfrak{o}$.

Let v be a non-trivial valuation of the field Σ over K ($v(c)=0$ if $c \in$ K and $c \neq 0$; $v(\omega) \neq 0$ for some ω in Σ). We consider linear forms l in $\eta_0, \eta_1, \cdots, \eta_n$, with coefficients in K, such that $v(\eta_i/l) \geqq 0$, $i=0, 1, \cdots, n$. If l_0 is one such form, we consider the homogeneous ideal $\mathfrak{P}$ generated by the forms $f(\eta)$ having the following property: if $f(\eta)$ is of degree m, then $v(f/l_0^m)>0$. The ideal $\mathfrak{P}$ is independent of l_0, since if l_1 is another linear form l, then $v(l_1/l_0)=0$. Moreover, $\mathfrak{P}$ is obviously a prime ideal, different from the irrelevant prime. It is also different from the zero ideal, since v is a non-trivial valuation. Consequently $\mathfrak{P}$ defines an irreducible proper subvariety W of V, of dimension at least 0. This subvariety W we call the *center of the valuation* v (on V).

It is clear that any linear form l for which $v(\eta_i/l) \geqq 0$, $i=0, 1, \cdots, n$, is such that $l \neq 0$ on W. Conversely, if l is a linear form in the η's and if $l \neq 0$ on W, then we must have: $v(l/l_0)=0$, whence $v(\eta_i/l)=v(\eta_i/l_0) \geqq 0$. Thus the linear forms l which played an auxiliary role in our definition of the center W of v, turn out simply to be those forms which do not vanish on that center.

In terms of nonhomogeneous coordinates the center W is obtained as follows. Let us consider the nonhomogeneous coordinates $\xi_i=\eta_i/\eta_0$. Should the center W be at finite distance with respect to these coordinates, we must have $\eta_0 \neq 0$ on W. But then, by the remark just made, $v(\eta_i/\eta_0) \geqq 0$, $i=0, 1, \cdots, n$, and *the entire coordinate ring* $\mathfrak{o}$ *must be contained in the valuation ring* R_v *of* v. Conversely, if $\mathfrak{o} \subseteq R_v$, then reversing the above reasoning we see immediately that the center W of v is at finite distance with respect to the nonhomogeneous coordinates ξ_i. If $f(\eta_0, \eta_1, \cdots, \eta_n)$ is a form of degree ν and if $f=0$ on W, then $v(f/\eta_0^\nu)>0$, whence $v(f(1, \xi_1, \cdots, \xi_n))>0$. Conversely, it is seen immediately that every polynomial in $\xi_1, \xi_2, \cdots, \xi_n$ which has positive value in v arises from a form in the η's which vanishes on W. Hence *the elements of* $\mathfrak{o}$ *which have positive value in the given valuation* v *form a prime ideal* $\mathfrak{p}$ *in* $\mathfrak{o}$, *and the center of* v *is the irreducible subvariety of* V *defined by the ideal* $\mathfrak{p}$. This conclusion holds for any choice of the nonhomogeneous coordinates, provided the corresponding coordinate ring is contained in the valuation ring R_v.

The following characterizations of the center of a valuation are useful in applications:

THEOREM 3. *An irreducible subvariety* W *of* V *is the center of a valuation* v *if and only if either one of the following conditions is satisfied*: (1) $Q(W) \subseteq R_v$ *and the non-units of* $Q(W)$ *are non-units of* R_v; (2) $Q(W) \subseteq R_v$ *and* W *is the maximal subvariety of* V *whose quotient ring is contained in* R_v.

Proof. Suppose that W is the center of v. If $f(\eta)/g(\eta) \in Q(W)$, where f and g

(4) Chain theorem rings with the property that their non-units form an ideal have been called by Krull "*Stellenringen*" (see Krull [3]). We propose the translation: "*local rings.*" The quotient ring of any irreducible subvariety of V is a local ring.

are forms of degree m, then $g(\eta)\neq 0$ on W. Therefore $v(g/l_0^m)=0$, while $v(f/l_0^m)\geqq 0$, and consequently $f(\eta)/g(\eta)\in R_v$. Moreover, if f/g is a non-unit in $Q(W)$, then $f=0$ on W. Hence $v(f/l_0^m)>0$, and consequently $v(f/g)>0$, that is, f/g is also a non-unit in R_v. This proves that condition (1) is necessary. To show the necessity and sufficiency of condition (2), let W_1 be another irreducible subvariety of V with the property: $Q(W_1)\subseteq R_v$. Let l_0 be a linear form in $\eta_0, \eta_1, \cdots, \eta_m$ which does not vanish on W_1. Then $\eta_i/l_0\in Q(W_1)\subseteq R_v$, whence $v(\eta_i/l_0)\geqq 0$, $i=0, 1, \cdots, n$. This shows that $l_0\neq 0$ on W. Let now $f(\eta)$ be any form in the η's which vanishes on W. If f is of degree m, then $v(f/l_0^m)>0$, and consequently $l_0^m/f\notin R_v$. Since, by hypothesis, $Q(W_1)\subseteq R_v$, we have *a fortiori*, $l_0^m/f\notin Q(W_1)$, whence $f=0$ on W_1. Thus we find that "$f=0$ on W" implies "$f=0$ on W_1." Therefore $W_1\subseteq W$, and this proves our assertion.

Now it follows immediately that condition (1) is also sufficient. For, if W_1 is any *proper* subvariety of the center W of v, then(5) $Q(W_1)\subset Q(W)$ and there exist non-units in $Q(W_1)$ which are units in $Q(W)$ and which are therefore also units in R_v. This completes the proof of the theorem.

We shall conclude this section with two lemmas which we shall have occasion to use in the sequel.

LEMMA 1. *If W and W_1 are irreducible subvarieties of V, then $W_1\subseteq W$ if and only if $Q(W_1)\subseteq Q(W)$, and*(6) *$W_1\subset W$ if and only if $Q(W_1)\subset Q(W)$.*

The proof is straightforward. If $W_1\subseteq W$ and if W_1 is at finite distance with respect to the nonhomogeneous coordinates ξ_i, then also W is at finite distance, and the corresponding prime ideals $\mathfrak{p}_1$ and $\mathfrak{p}$ are such that $\mathfrak{p}_1\supseteq\mathfrak{p}$. Hence $\mathfrak{o}_{\mathfrak{p}_1}\subseteq\mathfrak{o}_{\mathfrak{p}}$. If $\mathfrak{p}_1\supset\mathfrak{p}$, and if α is an element of $\mathfrak{p}_1$, not in $\mathfrak{p}$, then $1/\alpha\in\mathfrak{o}_{\mathfrak{p}}$, but $1/\alpha\notin\mathfrak{o}_{\mathfrak{p}_1}$, whence $\mathfrak{o}_{\mathfrak{p}_1}\subset\mathfrak{o}_{\mathfrak{p}}$. Conversely, assume that $Q(W_1)\subseteq Q(W)$. If $\mathfrak{P}$ and $\mathfrak{P}_1$ are the homogeneous ideals corresponding respectively to W and W_1, let $g(\eta)$ be an arbitrary form such that $g\neq 0$ on W_1. Let $f(\eta)$ be a form of the same degree as g, such that $f\neq 0$ on W. Then $f/g\in Q(W_1)$, and hence $f/g\in Q(W)$. This implies $g\neq 0$ on W, in view of the assumption that $f\neq 0$ on W. Hence if $g\neq 0$ on W_1 we must also have: $g\neq 0$ on W. This shows that $\mathfrak{P}_1\supseteq\mathfrak{P}$, whence $W_1\subseteq W$, as asserted.

LEMMA 2. *If v and v_1 are two valuations of Σ/K, with centers W and W_1, respectively, and if v is composite with v_1, then $W\subseteq W_1$.*

Proof. A valuation v is composite with another valuation v_1, if v is obtained by combining v_1 with a valuation v' of the residue field Σ_1 of v_1. The manner in which v and v' are to be combined is best described in terms of the homomorphic mapping of Σ upon the residue field of the valuation (together with the symbol ∞), a mapping which is determined by the valuation

(5) See the lemma which follows immediately.

(6) We use the symbol $\subset$ only for *proper* subsets.

and which in its turn determines the valuation uniquely. Let τ_1 be the homomorphic mapping of Σ upon (Σ_1, ∞) determined by v_1, and let τ' be the homomorphic mapping of Σ_1 upon the residue field Σ_1' (and the symbol ∞) of v'. Then v is *the* valuation of Σ determined by the homomorphic mapping $\tau=\tau_1\tau'$ of Σ onto (Σ_1', ∞). For further details, see Krull [2, p. 112].

Now let $f(\eta)$ be a form which vanishes on W_1 and let $g(\eta)$ be a form of the same degree as $f(\eta)$ such that $g\neq 0$ on W *and* on W_1. The quotient $\zeta=f/g$ is a non-unit of $Q(W_1)$. Hence, by Theorem 3, ζ is also a non-unit in R_{v_1}. Consequently $\tau_1(\zeta)=0$, and therefore also $\tau(\zeta)=0$. Hence ζ is a non-unit in R_v, and, in view of our assumption that g is *not* zero on W, this is only possible if $f=0$ on W, q.e.d.

4. **Existence theorems for valuations with a preassigned center.** If R_v is the valuation ring of a given valuation v of Σ/K and if $\mathfrak{P}_v$ denotes the prime ideal of non-units of R_v, then by the dimension of v is meant the degree of transcendency of the residue field $R_v/\mathfrak{P}_v$ (over K). Let W be the center of v on V and let $\mathfrak{J}$ denote the quotient ring $Q(W)$. By Theorem 3 we have: $\mathfrak{P}_v\cap\mathfrak{J}=\mathfrak{m}$, where $\mathfrak{m}$ is the ideal of non-units in $\mathfrak{J}$. Hence $\mathfrak{J}/\mathfrak{m}$ is a subring of $R_v/\mathfrak{P}_v$, and therefore *the dimension of W is not greater than the dimension of v*. If v is of dimension $r-1$, it is called a *divisor*. A divisor is of the *first or of the second kind with respect to V*, according as its center on V is of dimension $r-1$ or less than $r-1$.

THEOREM 4. *Given an s-dimensional irreducible subvariety W of V, there exist valuations of center W, of any dimension ρ, $s\leqq\rho\leqq r-1$.*

Proof. We consider first two special cases: (a) $s=r-1$; (b) $s<r-1$, $\rho=r-1$.

Case (a) $(s=r-1)$. Let $\mathfrak{J}^*$ be the integral closure of $\mathfrak{J}$ in Σ. The $(r-1)$-dimensional prime ideal $\mathfrak{m}$ of $\mathfrak{J}$ may split in $\mathfrak{J}^*$ into several prime ideals $\mathfrak{m}_1^*, \mathfrak{m}_2^*, \cdots, \mathfrak{m}_h^*$, all of the same dimension $r-1$. It is well known that the quotient rings $\mathfrak{J}^*_{\mathfrak{m}_i^*}$ are valuation rings of divisors $v_1, v_2, \cdots, v_h$. The center of each divisor v_i is our preassigned W, and in this fashion all the divisors of center W are obtained.

Case (b) $(s<r-1, \rho=r-1)$. Assuming that W is at finite distance with respect to the nonhomogeneous coordinates ξ_i, let $\mathfrak{o}$ denote, as usual, the ring of these coordinates, and let $\mathfrak{p}$ be the prime ideal of W in $\mathfrak{o}$. Let $\omega_1, \omega_2, \cdots, \omega_m$ be a base of the ideal $\mathfrak{p}$. We select one element among these m elements ω_i and we denote it by ω. We pass to the ring $\mathfrak{o}'=\mathfrak{o}[\omega_1/\omega, \omega_2/\omega, \cdots, \omega_m/\omega]$, and we first prove the following lemma:

LEMMA 3. *For at least one mode of selecting the element ω among the elements $\omega_1, \omega_2, \cdots, \omega_m$, the following relation will be satisfied*: $\mathfrak{o}'\cdot\mathfrak{p}\cap\mathfrak{o}=\mathfrak{p}$.

Proof of the lemma. Since $\omega_i=(\omega_i/\omega)\cdot\omega\in\mathfrak{o}'\cdot\omega$, it follows that the ideal $\mathfrak{o}'\cdot\mathfrak{p}$ coincides with the principal ideal $\mathfrak{o}'\cdot\omega$. Now let us suppose that

$\mathfrak{o}'\cdot\omega\cap\mathfrak{o}\neq\mathfrak{p}$, and let us see what restriction, if any, this assumption imposes on the element ω. Let ζ be an element of $\mathfrak{o}$, which belongs to $\mathfrak{o}'\cdot\omega$ but not to $\mathfrak{p}$. We will have then: $\zeta=\omega\cdot f(\omega_1/\omega, \omega_2/\omega, \cdots, \omega_m/\omega)$, where $f(z)=f(z_1, z_2, \cdots, z_m)\in\mathfrak{o}[z_1, z_2, \cdots, z_m]$. If ν is the degree of f, the above expression for ζ leads to a relation of the form: $\zeta\cdot\omega^{\nu-1}=\phi(\omega_1, \omega_2, \cdots, \omega_m)$, where ϕ is a form of degree ν, with coefficients in $\mathfrak{o}$. This relation implies that $\zeta\cdot\omega^{\nu-1}$ is in $\mathfrak{p}^\nu$. Since $\zeta\not\equiv 0(\mathfrak{p})$, we conclude that *$\mathfrak{p}^\nu:\omega_i^{\nu-1}$ is a proper divisor of* $\mathfrak{p}$.

If our lemma is false, then for each element ω_i, $i=1, 2, \cdots, m$, there must exist an integer ν_i such that $\mathfrak{p}^{\nu_i}:\omega_i^{\nu_i-1}\supset\mathfrak{p}$. If $\sigma=\text{maximum }(\nu_1, \nu_2, \cdots, \nu_m)$, then we will have $\mathfrak{p}^\sigma:\omega_i^{\sigma-1}\supset\mathfrak{p}$, $i=1, 2, \cdots, m$. Hence we have also $\mathfrak{p}^{\sigma+\rho}:\mathfrak{p}^\rho\omega_i^{\sigma-1}\supset\mathfrak{p}$, for any integer ρ. Now if $q=(\sigma-2)\cdot m+1$, then it is clear that $(\mathfrak{p}^\rho\omega_1^{\sigma-1}, \mathfrak{p}^\rho\omega_2^{\sigma-1}, \cdots, \mathfrak{p}^\rho\omega_m^{\sigma-1})=\mathfrak{p}^q$, if $\rho=q-\sigma+1$. Therefore the quotient $\mathfrak{p}^{q+1}:\mathfrak{p}^q$ is the intersection of the ideals $\mathfrak{p}^{q+1}:\mathfrak{p}^{q-\sigma+1}\omega_i^{\sigma-1}$, $i=1, 2, \cdots, m$. But each of these m ideals is, by hypothesis, a proper divisor of $\mathfrak{p}$. Hence also $\mathfrak{p}^{q+1}:\mathfrak{p}^q$ is a proper divisor, and this is impossible since(7) $\mathfrak{p}^{q+1}:\mathfrak{p}^q=\mathfrak{p}$. Our lemma is thus proved.

We therefore may assume that $\mathfrak{o}'\cdot\mathfrak{p}\cap\mathfrak{o}=\mathfrak{p}$. This implies at any rate that the ideal $\mathfrak{o}'\cdot\omega$ is not the unit ideal, whence its minimal prime ideals are all $(r-1)$-dimensional. The relation $\mathfrak{o}'\cdot\mathfrak{p}\cap\mathfrak{o}=\mathfrak{p}$ also implies, and is in fact equivalent to, the assertion that at least one minimal prime ideal of $\mathfrak{o}'\cdot\omega$ must contract to $\mathfrak{p}$. Let $\mathfrak{p}'$ be such a minimal prime.

Now let V' be the projective model whose general point (in nonhomogeneous coordinates) is(8) $(\xi_1, \xi_2, \cdots, \xi_n, \omega_1/\omega, \omega_2/\omega, \cdots, \omega_m/\omega)$, so that $\mathfrak{o}'$ is the ring of the nonhomogeneous coordinates of the general point of V'. Let W' be the $(r-1)$-dimensional subvariety of V' which is defined by the prime ideal $\mathfrak{p}'$. By the preceding case (a) there exists a $(r-1)$-dimensional valuation whose center on V' is W'. Since $\mathfrak{p}'\cap\mathfrak{o}=\mathfrak{p}$, it follows that the center of v on V is W. This establishes our theorem in the case under consideration.

To prove the theorem in the general case, $s<r-1$, $s\leqq\rho<r-1$, we shall keep s and ρ fixed and we shall proceed by induction with respect to r, since, by the case (b), the theorem is true if $r=\rho+1$. We consider an $(r-1)$-dimensional irreducible subvariety W_1 of V which contains W and we denote by v_1 a divisor of center W_1. The residue field Σ_1^* of v_1 is a finite algebraic extension of the field Σ_1 of rational functions on W_1. By our induction there exists a valuation v' of Σ_1, of dimension ρ, whose center on W_1 is W. This valuation v' has at least one extension v^* in Σ_1^*. Compounding v_1 with v^* we get a composite valuation v of Σ, of dimension ρ, whose center is W. This completes the proof of our theorem.

---

(7) Let $\mathfrak{p}^{q+1}:\mathfrak{p}^q=\mathfrak{A}$. We have then: $\mathfrak{A}\mathfrak{p}^q\equiv 0(\mathfrak{p}^{q+1})$. On the other hand $\mathfrak{p}\subseteq\mathfrak{A}$, whence $\mathfrak{A}\cdot\mathfrak{p}^q\supseteq\mathfrak{p}^{q+1}$. Consequently $\mathfrak{A}\cdot\mathfrak{p}^q=\mathfrak{p}^{q+1}=\mathfrak{p}\cdot\mathfrak{p}^q$. From this it follows (see Krull [2, p. 36]) that $\mathfrak{A}$ and $\mathfrak{p}$ have the same radical. Since $\mathfrak{A}\supseteq\mathfrak{p}$, it follows that $\mathfrak{A}=\mathfrak{p}$.

(8) Comparison with section 11 will show that our V' is the transform of V by a monoidal transformation of center W.

The preceding proof does not give an adequate idea of the totality of all valuations having a preassigned center. The valuations obtained in the course of the proof are special in the sense that if their dimension is ρ then the rank of their value group is $r-\rho$. The following theorem gives more information about the arbitrary elements which can be assigned in the construction of valuations with a given center(9):

THEOREM 5. *Given an arbitrary descending chain*(10) $W_0 \supseteq W_1 \supseteq \cdots \supseteq W_{\sigma-1}$ *of irreducible subvarieties of* V *and given any set of integers* $\rho_0, \rho_1, \cdots, \rho_{\sigma-1}$ *such that* $r-1 \geqq \rho_0 > \rho_1 > \cdots > \rho_{\sigma-1}$, $\rho_i \geqq$ *dimension of* W_i, *there exists a sequence of valuations* $v_0, v_1, \cdots, v_{\sigma-1}$ *such that*: (1) v_i *is of dimension* ρ_i, *of rank* $i+1$, *and its center is* W_i; (2) v_i *is compounded with* v_{i-1}.

We first prove the theorem in the following two special cases: (a) $\sigma=1$, $\rho_0=s=$dimension of W_0; (b) $\sigma=1$, $\rho_0>s$.

(a) $\sigma=1$, $\rho_0=s=$dimension W_0. In this case we have to prove the existence of a rank 1 valuation, of dimension s, whose center is a given s-dimensional irreducible subvariety W_0 of V. We shall use nonhomogeneous coordinates $\xi_1, \xi_2, \cdots, \xi_n$ with respect to which W_0 is at finite distance, so that W_0 is given by a prime ideal $\mathfrak{p}$ in the ring $\mathfrak{o}$ of these coordinates. We then adjoin to the ground field s elements of $\mathfrak{o}$ which are algebraically independent on W, that is, algebraically independent mod $\mathfrak{p}$. In this fashion we achieve a reduction to the case $s=0$, so that we may assume that W_0 is a point, say A. It is also permissible to assume(11) that $\xi_{r+1}, \xi_{r+2}, \cdots, \xi_n$ are

(9) Which ordered groups can be preassigned as value groups for valuations of fields of algebraic functions, is a question which has been solved completely by S. MacLane and O. F. G. Schilling in their paper [4].

(10) Note that we do not assume that the chain is strictly descending, that is, that each W_i is a proper subvariety of W_{i-1}.

(11) In the case of infinite ground fields, or of ground fields with "sufficiently many" elements, this assumption can always be realized, in view of the usual proof of Emmy Noether's normalization theorem, by first subjecting the nonhomogeneous coordinates $\xi_1, \xi_2, \cdots, \xi_n$ to a linear transformation with "non-special" coefficients in K. In the case of finite ground fields this is no longer true. However, it will be proved in II.2 that there always exists in the ring $K[\xi_1, \xi_2, \cdots, \xi_n]$ a set of r algebraically independent elements $\zeta_1, \zeta_2, \cdots, \zeta_r$ such that the ring is integrally dependent on the ring $K[\zeta_1, \zeta_2, \cdots, \zeta_r]$. We may then simply include the elements ζ_i among the elements ξ_i, which does not change the ring of nonhomogeneous coordinates, and we may then proceed as in the text.

We could also proceed in the following fashion. We first pass to the field K′ which is generated over K by the coefficients of the linear transformation on the ξ_i mentioned above, and we consider an extension field $\Sigma' = K'\Sigma$ of Σ. The field K′ may be assumed to be an algebraic extension of K. We obtain a new variety V' *over* K′, with the same general point $(\xi_1, \xi_2, \cdots, \xi_n)$ as V. The original subvariety W_0 splits on V' into at most a finite number of varieties, all of the same dimension as W_0. If W_0' is one of them, the proof given in the text leads to a valuation v' of Σ' of dimension s and rank 1, with center W_0'. The valuation v of Σ induced by v' will have center W_0, dimension s and rank **1.**

integrally dependent on $K[\xi_1, \xi_2, \cdots, \xi_r]$. Let A' be the projection(12) of the point A into the linear space of the r independent variables $\xi_1, \xi_2, \cdots, \xi_r$. Let v' be a zero-dimensional and rank 1 valuation of the field $K(\xi_1, \xi_2, \cdots, \xi_r)$ whose center in the above linear space is the point(13) A'. We denote, as usual, by $R_{v'}$ the valuation ring of v', and by $A_1, A_2, \cdots, A_h$ the other points of V, at finite distance and different from A, which project into A'. We can find an element ω in $\mathfrak{o}$ such that $\omega = 0$ at A, $\omega \neq 0$ at A_i, $i = 1, 2, \cdots, h$. Let

$$\omega^m + a_1(\xi_1, \xi_2, \cdots, \xi_r)\omega^{m-1} + \cdots + a_m(\xi_1, \xi_2, \cdots, \xi_r) = 0$$

be the irreducible equation of integral dependence for ω over the ring $K[\xi_1, \xi_2, \cdots, \xi_r]$. Since $\omega = 0$ at A, we must have $a_m = 0$ at A'. Hence a_m is a non-unit in $R_{v'}$. *We assert that $R_{v'}[\omega]$ is a proper ring* (that is, is not a field). We prove this by showing that ω *is a non-unit in this ring*. For suppose that ω is a unit in $R_{v'}[\omega]$. Then we would have: $1 = \omega \cdot g(\omega)$, where $g(\omega)$ is a polynomial with coefficients in $R_{v'}$. Using the above equation of integral dependence and observing that the coefficients a_i are polynomials, hence are elements of $R_{v'}$, we can reduce the degree of $g(\omega)$. We thus find a new relation of the form: $1 = \omega(b_0\omega^{m-1} + b_1\omega^{m-2} + \cdots + b_{m-1})$, where $b_i \in R_{v'}$. Comparing this relation with the above relation of integral dependence, we conclude that $b_0 = -1/a_m$, a contradiction, since a_m is a non-unit in $R_{v'}$.

Since ω is a non-unit in $R_{v'}[\omega]$, there exists at least one valuation v of Σ such that $R_v \supseteq R_{v'}[\omega]$ and such that ω is a non-unit(14) in R_v. Since v' is of rank 1, $R_{v'}$ is a maximal subring of $K(\xi_1, \xi_2, \cdots, \xi_r)$. Hence $R_v \cap K(\xi_1, \xi_2, \cdots, \xi_r) = R_{v'}$, and therefore *the valuation v is an extension of v'* and has the same rank and the same dimension as v', that is, rank 1 and dimension 0. The center of v on V must be a point at finite distance (since $\xi_1, \xi_2, \cdots, \xi_r \in R_{v'} \subseteq R_v$ and hence $\mathfrak{o} \subseteq R_v$, for R_v is integrally closed), and this point must project into the point A'. The center cannot be any of the points $A_1, A_2, \cdots, A_h$, since $\omega \neq 0$ at A_i and this implies that ω is a unit in the quotient ring $Q(A_i)$, while, as we have just seen, ω is a non-unit in R_v (compare with Theorem 3). Hence the center of v is the point A, q.e.d.

(b) Let now $\sigma = 1$, $\rho > s$. We refer to the case (b) of the proof of Theorem 4, where we identify the variety W with our present variety W_0. We had there an $(r-1)$-dimensional prime ideal $\mathfrak{p}'$ in $\mathfrak{o}'$ such that $\mathfrak{p}' \cap \mathfrak{o} = \mathfrak{p}$. From the existence of an $(r-1)$-dimensional prime ideal in $\mathfrak{o}'$ which contracts to $\mathfrak{p}$ follows immediately the existence of prime ideals in $\mathfrak{o}'$ of any dimension ρ,

---

(12) By that we mean that A' is the point which is defined by the contraction of the prime ideal $\mathfrak{p}$ in the polynomial ring $K[\xi_1, \xi_2, \cdots, \xi_r]$.

(13) The existence of v' is proved in the joint paper by MacLane and Schilling [4].

(14) This is implied by the fundamental theorem on principal orders which states that an integrally closed integral domain (*not* a field) is the intersection of the valuation rings which contain it (see Krull [2, p. 111]). It is necessary only to observe that the integral closure of a *proper* ring is also a proper ring. For if a ring R is proper, it contains a non-unit α, and it is seen immediately that $1/\alpha$ cannot be integrally dependent on R.

$r-1 \geqq \rho \geqq s$, which contract to[15] $\mathfrak{p}$. Let $\mathfrak{p}''$ be such a prime ideal and let W'' be the corresponding irreducible ρ-dimensional subvariety of V'. By the preceding case (a), there exists a valuation v, of rank 1 and of dimension ρ, whose center on V' is W''. Since $\mathfrak{p}'' \cap \mathfrak{o} = \mathfrak{p}$, the center of v on V is W, q.e.d.

To prove our theorem in the general case, we first prove this lemma:

LEMMA 4. *If W and W_1 are irreducible subvarieties of V such that $W \subset W_1 \subset V$ and if v_1 is a valuation of center W_1, then there exists a valuation v of center W, which is compounded with v_1.*

Proof. Let Σ_1 be the residue field of the valuation v_1 and let τ_1 denote the homomorphic mapping of Σ onto (Σ_1, ∞) defined by v_1. Since $Q(W) \subset Q(W_1)$ (Lemma 1, I.3) and $Q(W_1) \subseteq R_{v_1}$, it follows that if we put $\mathfrak{J} = Q(W)$, then $\tau_1 \cdot \mathfrak{J} \subseteq \Sigma_1$. The elements of $\mathfrak{J}$ which are mapped into zero under τ_1 are non-units in $Q(W_1)$. Since W is a proper subvariety of W_1, there are non-units in $Q(W)$ which are units in $Q(W_1)$. *Hence $\tau_1 \mathfrak{J}$ is a proper ring.* There exists then at least one valuation v' of Σ_1 such that $R_{v'} \supseteq \tau_1 \cdot \mathfrak{J}$. Let v_2 denote the valuation obtained by compounding v with v', and let W_2 be the center of v_2. By Lemma 2, I.3, we have $W_2 \subseteq W_1$. Since $R_{v'} \supseteq \tau_1 \cdot \mathfrak{J}$, it follows that $R_{v_2} \supseteq \mathfrak{J}$, whence, by Theorem 3 (I.3), we have $W \subseteq W_2$. If $W_2 = W$, then our lemma is proved ($v = v_2$). If, however, W is a proper subvariety of W_2, then we replace v_1 and W_1 by v_2 and W_2 and we repeat the above procedure. Since v_2 is of *smaller dimension* than v_1, this process cannot continue indefinitely, q.e.d.

We now are in position to complete the proof of the theorem in the general case. Since by the special cases (a) and (b) treated above the theorem is true in the case $\sigma = 1$, we assume that the theorem is true for $\sigma = h-1$ and we proceed to prove the theorem for $\sigma = h$. We can therefore assume the existence of the valuation $v_0, v_1, \cdots, v_{\sigma-2}$ and we have only to prove the existence of $v_{\sigma-1}$. The valuation $v_{\sigma-2}$ is of dimension $\rho_{\sigma-2}$ and its center is $W_{\sigma-2}$. We have to prove the existence of a valuation $v_{\sigma-1}$, of dimension $\rho_{\sigma-1}$, which is compounded with $v_{\sigma-2}$, is of rank *one higher* than the rank of $v_{\sigma-2}$ and has center $W_{\sigma-1}$. For simplicity we shall denote $W_{\sigma-2}$, $W_{\sigma-1}$, $\rho_{\sigma-2}$, $\rho_{\sigma-1}$ and $v_{\sigma-2}$ by W_1, W, ρ_1, ρ and v_1, respectively, so that we have now:

$$W_1 \supseteq W, \quad \rho_1 > \rho \geqq \text{dimension } W; \quad \rho_1 \geqq \text{dimension } W_1.$$

We first provide ourselves with a projective model V' on which the center W_1' of v_1 is exactly of dimension[16] ρ_1. We then consider, for auxiliary purposes,

[15] Let $\alpha_1, \alpha_2, \cdots, \alpha_s$ be s elements of $\mathfrak{o}$ which are algebraically independent modulo $\mathfrak{p}$ and let us take as new ground field the field $K_1 = K(\alpha_1, \alpha_2, \cdots, \alpha_s)$. If $\mathfrak{o}_1 = K_1 \cdot \mathfrak{o}$, $\mathfrak{o}_1' = K_1 \cdot \mathfrak{o}'$, $\mathfrak{o}_1 \cdot \mathfrak{p} = \mathfrak{p}_1$, $\mathfrak{o}_1' \cdot \mathfrak{p}' = \mathfrak{p}_1'$, then, *over* K_1, $\mathfrak{p}_1$ is zero-dimensional, $\mathfrak{p}_1'$ is $(r-1-s)$-dimensional, and $\mathfrak{p}_1' \cap \mathfrak{o}_1 = \mathfrak{p}_1$. Let $\mathfrak{p}_1''$ be any $(\rho - s)$-dimensional prime divisor of $\mathfrak{p}_1'$, and let $\mathfrak{p}_1'' \cap \mathfrak{o}' = \mathfrak{p}''$. Then $\mathfrak{p}''$ is of dimension ρ, *over* K, and $\mathfrak{p}'' \cap \mathfrak{o} = \mathfrak{p}$.

[16] As nonhomogeneous coordinates of such a model V' we may take any finite set of generators of Σ which belong to R_{v_1} and such that ρ_1 of these generators have algebraically independent residues in the residue field of v_1.

an arbitrary valuation v which is compounded with v_1 and whose center on V is W (Lemma 4; if $W_1=W$, we put $v=v_1$). Let W' denote the center of v on V', whence $W'\subseteq W_1'$ (Lemma 2, I.3). We next select nonhomogeneous coordinates $\xi_1, \xi_2, \cdots, \xi_n$ for the general point of V and nonhomogeneous coordinates $\zeta_1, \zeta_2, \cdots, \zeta_m$ for the general point of V' in such a fashion that W and W' be at finite distance with respect to these coordinates. Finally we denote by V^* the projective model whose general point is $(\xi_1, \xi_2, \cdots, \xi_n, \zeta_1, \zeta_2, \cdots, \zeta_m)$, and we denote by $\mathfrak{o}$, $\mathfrak{o}'$ and $\mathfrak{o}^*$, respectively, the corresponding rings $K[\xi]$, $K[\zeta]$, $K[\xi, \zeta]$ of the nonhomogeneous coordinates.

Let W^* and W_1^* be the centers on V^* of the valuations v and v_1, respectively. Since W and W' are at finite distance, we have: $\mathfrak{o}\subseteq R_v$, $\mathfrak{o}'\subseteq R_v$, and, by a stronger reason, $\mathfrak{o}\subseteq R_{v_1}$, $\mathfrak{o}'\subseteq R_{v_1}$. Therefore $\mathfrak{o}^*\subseteq R_v$, $\mathfrak{o}^*\subseteq R_{v_1}$, and consequently W^* and W_1^* are at finite distance. Let $\mathfrak{p}, \mathfrak{p}_1$; $\mathfrak{p}', \mathfrak{p}_1'$; $\mathfrak{p}^*, \mathfrak{p}_1^*$ denote the prime ideal of W, W_1; W', W_1'; W^*, W_1^* in $\mathfrak{o}$, $\mathfrak{o}'$ and $\mathfrak{o}^*$ respectively. It is clear that:

$$\mathfrak{p}^* \cap \mathfrak{o} = \mathfrak{p}, \qquad \mathfrak{p}_1^* \cap \mathfrak{o} = \mathfrak{p}_1.$$

Moreover $W^*\subseteq W_1^*$. The above relations show that *any valuation of center* W^* *on* V^* *has* W *as center on* V. We also point out that V^* shares with V' the property that *the center of* v_1 *on that variety is exactly of dimension* ρ_1. This follows immediately from the relation: $\mathfrak{p}_1^*\cap\mathfrak{o}'=\mathfrak{p}_1'$. The variety V' and the auxiliary valuation v have now served their purpose and will not be used any more.

(1) Suppose first that W^* is of dimension at most ρ. By the special case $\sigma=1$, we can find a ρ-dimensional, rank 1 valuation *of the field of rational functions on* W_1^*, having W^* as center. This valuation has at least one extension in the residue field of the valuation v_1. Let v_2 be such an extension. Since the residue field of v_1 is an algebraic extension of the field of rational functions on W_1^*, it follows that also v_2 is of rank 1 and dimension ρ. Compounding v_1 with v_2, we get a valuation v of Σ, of dimension ρ, of rank one higher than v_1. Its center on V^* is W^*, hence its center on V is W. This valuation v is the valuation $v_{\sigma-1}$, whose existence we have claimed in our theorem. (2) Suppose now that W^* is of dimension greater than ρ. Since $\mathfrak{p}^*\cap\mathfrak{o}=\mathfrak{p}$ and since $\mathfrak{p}$ is of dimension at most ρ, it follows that we can find in $\mathfrak{o}^*$ a prime ρ-dimensional ideal which divides $\mathfrak{p}^*$ and which likewise contracts[17] to $\mathfrak{p}$. This prime ideal defines a ρ-dimensional irreducible subvariety W_0^* of V^* which we can use with the same effect instead of W^*, since the two essential conditions: (1) $W_0^*\subseteq W_1^*$, (2) every valuation of center W_0^* has W as center on V, are still satisfied. But now W_0^* has dimension ρ, and we have therefore the case (1) just considered. This completes the proof of Theorem 5.

([17]) See footnote 15.

Part II. General theory of birational correspondences

1. Valuation-theoretic definition of a birational correspondence. Let V and V' be two birationally equivalent r-dimensional irreducible algebraic varieties. The two varieties can be regarded as projective models of one and the same field Σ of algebraic functions([18]), and if they are so regarded there arises a well defined correspondence between the irreducible subvarieties of V (of all possible dimension from 0 to $r-1$ inclusive) and the irreducible subvarieties of V'. It is the *birational correspondence* between V and V', or the *birational transformation* of V into V'. This correspondence, which we shall denote by T, is defined as follows([19]):

Definition 1. *Two irreducible subvarieties W and W' of V and V' respectively (not necessarily of the same dimension) correspond to each other (in symbols: $T(W)=W'$, $T^{-1}(W')=W$), if there exists a valuation v of the field Σ such that the center of v on V is W and the center of v on V' is W'.*

Note that this definition retains its full meaning also when V and V' are *coincident varieties* (as varieties in the projective space). In this case we deal with an automorphism τ of Σ and we have a *birational transformation of W into itself.* It is only necessary to regard the two coincident varieties V and V' as *distinct* projective models of Σ, in the sense that the general point of V is $(\xi_1, \xi_2, \cdots, \xi_n)$ and the general point of V' is $(\tau\xi_1, \tau\xi_2, \cdots, \tau\xi_n)$.

From the results of I.3 and I.4 we deduce immediately the following properties of a birational correspondence:

A. *Given $W\subset V$, there exists at least one $W'\subset V'$ such that $T(W)=W'$* (Theorem 4, I.4).

B. *If $W\subseteq W_1\subset V$ and $W_1'=T(W_1)$, there exists a W' such that $W'=T(W)$ and $W'\subseteq W_1'$. In particular, if to W_1 there corresponds a point P' on V', then P' corresponds to each point of W_1.*

A birational correspondence is, generally speaking, not a (1, 1) correspondence. There may very well exist varieties W on V such that T is not single-valued at W, that is, such that to W there correspond more than one subvariety of V'. Similarly for T^{-1} and V'. These varieties W are exceptional in the sense that they lie on algebraic subvarieties of V (see Theorem 15, II.9). The analysis of these exceptions to the (1, 1) character of a birational correspondence is the main goal of our study. At this stage, however, we wish

([18]) The fields Σ, Σ' of rational functions on V and V', respectively, are isomorphic over **K**. When we say that V and V' are projective models of one and the same field we imply that the fields Σ, Σ' have been identified. The identification is determined to within an automorphism of Σ. When we speak of a birational correspondence we refer to a fixed identification of the two fields.

([19]) From now on *irreducible* subvarieties of V shall be denoted by the letter W, with or without subscripts. Similarly W', W_1', W_i' and so on shall always denote irreducible subvarieties of V'.

to give a very simple but important criterion for the uniqueness of $T(W)$ when W is given:

THEOREM 6. *If $T(W)=W'$ and if $Q(W')\subseteq Q(W)$, then W' is the only subvariety of V' which corresponds to W.*

Proof. By hypothesis, there exists a valuation v_1 whose center on V is W and whose center on V' is W'. Let v be an arbitrary valuation of center W. We have $R_v \supseteq Q(W)$, whence $R_v \supseteq Q(W')$. Every non-unit of $Q(W')$ is a non-unit of R_{v_1} (Theorem 3, I.3), hence it is also a non-unit of $Q(W)$. We conclude that every non-unit of $Q(W')$ is a non-unit in R_v, and our theorem follows from Theorem 3.

2. **The birational correspondence between V and a derived normal model $\overline{V}$.** In our paper [7] we have proved that from any irreducible algebraic variety V it is possible to pass to what we have called a *derived normal model $\overline{V}$ of V*. That proof dealt only with algebraically closed ground fields of characteristic zero. To extend the proof to arbitrary ground fields additional considerations are necessary.

First of all we shall need an extension of the normalization theorem of Emmy Noether to finite ground fields. Given a finite integral domain $K[\xi_1, \xi_2, \cdots, \xi_n]$ of degree of transcendency r over an infinite ground field K, the Noether normalization theorem states that there exist r linear combinations $\xi_i' = \sum_{j=1}^{n} c_{ij}\xi_j$, $i=1, 2, \cdots, r$, with coefficients in K, which are algebraically independent over K and which are such that $\xi_1, \xi_2, \cdots, \xi_n$ are integrally dependent on $\xi_1', \xi_2', \cdots, \xi_r'$. This theorem, as it stands, is not generally true when K is a finite field. In this case we can still assert that elements such as $\xi_1', \xi_2', \cdots, \xi_r'$ can be found in the ring $K[\xi_1, \xi_2, \cdots, \xi_n]$, provided we drop the condition that these elements be linear in the ξ's. The proof of this assertion, as given below, was communicated orally to me by Irvin Cohen.

We shall first consider an homogeneous integral domain $\mathfrak{o}^* = K[\eta_0, \eta_1, \cdots, \eta_n]$, of degree of transcendency $r+1$, that is, one whose generating elements η_i are the homogeneous coordinates of the general point of an r-dimensional variety V.

(a) *If the ideal $(\eta_1, \eta_2, \cdots, \eta_n)$ in $\mathfrak{o}^*$ is irrelevant, then η_0 is integrally dependent on $\eta_1, \eta_2, \cdots, \eta_n$.* For the hypothesis implies that the point $y_0=1$, $y_1=y_2=\cdots=y_n=0$ is not on V, and hence there must exist a form $f(y_0, y_1, \cdots, y_n)$ such that $f(\eta_0, \eta_1, \cdots, \eta_n)=0$ and $f(1, 0, \cdots, 0)\neq 0$. If ρ is the degree of f, the term η_0^ρ must therefore occur in $f(\eta)$, and this proves our assertion.

(b) More generally, *if the ideal $(\eta_{k+1}, \eta_{k+2}, \cdots, \eta_n)$ is irrelevant, then $\eta_0, \eta_1, \cdots, \eta_k$ are integrally dependent on $\eta_{k+1}, \eta_{k+2}, \cdots, \eta_n$.* Proof by induction with respect to k (that is, with respect to the number $k+1$ of elements η_i which do not occur in the set $\eta_{k+1}, \eta_{k+2}, \cdots, \eta_n$). By (a), η_0 is integrally dependent on $\eta_1, \eta_2, \cdots, \eta_n$. Since every element of $\mathfrak{o}^*$ is integrally dependent on $\eta_1, \eta_2, \cdots, \eta_n$, the elements $\eta_{k+1}, \eta_{k+2}, \cdots, \eta_n$ generate in the ring $K[\eta_1, \eta_2, \cdots, \eta_n]$ an ideal of the same dimension as that of the ideal generated by them in $\mathfrak{o}^*$, that is, they generate in $K[\eta_1, \eta_2, \cdots, \eta_n]$ an irrelevant ideal. By our induction it follows that $\eta_1, \eta_2, \cdots, \eta_k$ are integrally dependent on $\eta_{k+1}, \eta_{k+2}, \cdots, \eta_n$, q.e.d.

(c) *If $\omega_0, \omega_1, \cdots, \omega_r$ are forms in $\eta_0, \eta_1, \cdots, \eta_n$, all of the same degree h, and if the ideal $(\omega_0, \omega_1, \cdots, \omega_r)$ is irrelevant, then the η's are integrally dependent on the ω's.* For if $\eta_0^{(h)}, \eta_1^{(h)}, \cdots$ form a linear base for the forms of degree h in the η's and if we include the ω's in this base, then applying (b) to the ring $K[\eta_0^{(h)}, \eta_1^{(h)}, \cdots]$, we conclude that $\eta_0^h, \eta_1^h, \cdots, \eta_n^h$ are integrally dependent on the ω's.

(d) We obviously can select (in many ways) $r+1$ forms $\zeta_0, \zeta_1, \cdots, \zeta_r$ in $\mathfrak{o}^*$ such that the ideal $(\zeta_0, \zeta_1, \cdots, \zeta_r)$ be irrelevant. We can then find exponents σ_i such that the forms $\omega_i = \zeta_i^{\sigma_i}$ be of like degree. Then it follows, by (c), that the η's are integrally dependent on $\omega_0, \omega_1, \cdots, \omega_r$. This completes the proof of the extended "normalization theorem" for homogeneous integral domains.

From an homogeneous integral domain $K[\eta_0, \eta_1, \cdots, \eta_n]$, of degree of transcendency $r+1$, we get an arbitrary integral domain $K[\xi_1, \xi_2, \cdots, \xi_n]$, of degree of transcendency r, by putting $\xi_i = \eta_i/\eta_0$. We apply step (d) above, and we observe that of the $r+1$ forms ζ_i, one, say ζ_0, can be taken arbitrarily. If we put $\zeta_0 = \eta_0$, then ω_0 is a power of η_0, say $\omega_0 = \eta_0^h$. It is then seen immediately that the r elements $\xi_i' = \omega_i/\eta_0^h$, $i = 1, 2, \cdots, r$, are polynomials in the ξ's and that $\xi_1, \xi_2, \cdots, \xi_n$ are integrally dependent on $\xi_1', \xi_2', \cdots, \xi_r'$. This completes the proof.

Let $(\eta_0, \eta_1, \cdots, \eta_n)$ be the general point of V (the coordinates η_i are homogeneous[20]) and let $P = \mathrm{K}[\eta_0, \eta_1, \cdots, \eta_n]$. Let $\overline{P}$ be integral closure of P in its quotient field. We first need to establish *in the most general case* that $\overline{P}$ is a finite P-module. By the normalization theorem of Emmy Noether, let $\zeta_0, \zeta_1, \cdots, \zeta_r$ be $r+1$ algebraically independent elements in P such that every element of P is integrally dependent on $\mathrm{K}[\zeta_0, \zeta_1, \cdots, \zeta_r]$. This last ring is a polynomial ring and we shall denote it by R. Since the field $\Sigma = \mathrm{K}(\eta_0, \eta_1, \cdots, \eta_n)$ is a finite algebraic extension of the quotient field of *the polynomial ring R*, it follows that *the integral closure $\overline{R}$ of R in Σ is a finite R-module*. This result, for arbitrary ground fields, has been proved by F. K. Schmidt [5]. Since $P \subseteq \overline{R}$, it follows that every element of $\overline{P}$ is integrally dependent on $\overline{R}$. But since $\overline{R}$ is a finite R-module and is therefore a chain-theorem ring, it is well known that every element of Σ which is integrally dependent on $\overline{R}$ is also integrally dependent on R, that is, belongs to $\overline{R}$. Hence $\overline{P} \subseteq \overline{R}$, that is, $\overline{P} = \overline{R}$. Thus $\overline{P}$ is a finite R-module, q.e.d.

(20) To avoid repetitions, we stipulate from now on that whenever the subscript in a set of coordinates begins with 0, the coordinates are homogeneous.

Since $\bar{P}$ is a finite P-module, it is a finite integral domain over K. Let $\bar{P}=K[\bar{\zeta}_0, \bar{\zeta}_1, \cdots, \bar{\zeta}_m]$. As in the quoted paper [7] (see p. 290), we may assume also here that the $\bar{\zeta}_i$ are homogeneous elements.

Then by exactly the same procedure as that carried out in our quoted paper we can show that if $\omega_0^*, \omega_1^*, \cdots, \omega_\mu^*$ is a linear K-basis for all the homogeneous elements of $\bar{P}$ of a given degree δ, then for suitable integers δ it will be true that every homogeneous element in $\bar{P}$ whose degree is a multiple $\rho\delta$ of δ, $\rho>0$, is necessarily a form of degree ρ in $\omega_0^*, \omega_1^*, \cdots, \omega_\mu^*$. From that we concluded in the quoted paper that for such an integer δ *the ring* $P^*=K[\omega_0^*, \omega_1^*, \cdots, \omega_\mu^*]$ *is integrally closed in its quotient field*, whence the variety $\bar{V}$ whose general point is $(\omega_0^*, \omega_1^*, \cdots, \omega_\mu^*)$ is normal. This variety $\bar{V}$ we termed a derived normal variety of V. It was pointed out to me by Irvin Cohen that the above conclusion fails to hold true if K is not maximally algebraic in the field Σ of rational functions on V, that is, *if V is not absolutely irreducible* (see [10, Lemma 4, p. 64]). For in this case the elements of Σ which are algebraic over K but are not in K, that is, *the homogeneous elements of degree zero*, are certainly not in the ring P^*. However, it is still true that P^* contains all homogeneous integral quantities (that is, the homogeneous elements of the quotient field of P^* which are integrally dependent of P^*) *of positive degree*. Hence if α^* is any integral quantity in the quotient field of P^*, then the products $\alpha^*\omega_i^*$, $i=0, 1, \cdots, \mu$, belong to P^*, since they are sums of homogeneous integral quantities of positive degree. It follows that *the irrelevant ideal* $P^*(\omega_0^*, \omega_1^*, \cdots, \omega_\mu^*)$ *is the conductor of the ring* P^* *with respect to its integral closure in the quotient field of* P^*. This implies that the variety $\bar{V}$ is *locally normal* in the sense of Definition 3 given later on in this section (see also [7, Theorem 13, p. 286]). For our purpose a locally normal variety is just as effective as a normal variety. We shall continue to call $\bar{V}$ the derived normal model of V, it being understood that if V is not absolutely irreducible then $\bar{V}$ is only locally normal. It may be well to point out at this stage the self-evident fact that if V is not absolutely irreducible, the field Σ does not possess at all normal models *over* K.

For the general theory of birational correspondences it is necessary to establish first some properties of the birational correspondence between a given model V and a derived normal model $\bar{V}$ of V. For it will follow from the properties of this birational correspondence that the properties of a birational correspondence between any two models V, V' can be readily deduced from the properties of the birational correspondence between the derived normal models of V and V'. Therefore, there is no loss of generality if the theory is restricted to normal models. On the other hand, the emphasis on normal models is advantageous, both from a technical and a conceptual standpoint, since in the case of normal varieties the theory of birational correspondences is free from many accidental complications and irrelevant exceptions which one often encounters on non-normal varieties.

Let us therefore consider the case in which one of the two birationally equivalent varieties V, V' is a derived normal model of the other. Let, say, V' be a derived normal model of V and let $(\eta_0, \eta_1, \cdots, \eta_n)$ and $(\eta_0', \eta_1', \cdots, \eta_m')$ be the general points of V and of V', respectively. Let h be the degree of homogeneity of V'. The elements η_i' form then a linear base for the elements of the field $K(\eta_0, \eta_1, \cdots, \eta_n)$ which are integrally dependent on $\eta_0, \eta_1, \cdots, \eta_n$ and which are homogeneous of degree h. Moreover, the ring $P' = K[\eta_0', \eta_1', \cdots, \eta_m']$ is integrally closed (in its quotient field), or at any rate contains all homogeneous integers of positive degree.

It will be convenient to use an auxiliary projective model V^* defined as follows. Let $\eta_0^*, \eta_1^*, \cdots, \eta_s^*$ be a linear base for the forms of degree h in $\eta_0, \eta_1, \cdots, \eta_n$, with coefficients in K. We take as V^* the variety whose general point is $(\eta_0^*, \eta_1^*, \cdots, \eta_s^*)$.

LEMMA 5. *The birational correspondence between V and V^* is* (1, 1) *without exceptions*[21]. *Any two corresponding irreducible subvarieties of V and V' have the same dimension and the same quotient ring.*

Proof. Let W and W^* be two corresponding irreducible subvarieties of V and V^*, respectively. We assume that $\eta_0 \neq 0$ on W. Since the η^*'s constitute a linear base for the forms of degree h in the η's, we may assume that η_0^h is one of the η^*'s, say $\eta_0^h = \eta_0^*$. If v be a valuation of center W and W^*, then $v(\eta_i/\eta_0) \geqq 0$, since $\eta_0 \neq 0$ on W. From this it follows that $v(\eta_j^*/\eta_0^*) \geqq 0$, for $j = 0, 1, \cdots, s$, and hence (see I.3) $\eta_0^* \neq 0$ *on* W^*. Now let ζ be any element of $Q(W)$, say $\zeta = \phi(\eta)/\psi(\eta)$, where ϕ and ψ are forms of like degree and where $\psi(\eta) \neq 0$ on W. We may assume that the common degree of ϕ and ψ is a multiple of h, say ρh, since we can multiply both ϕ and ψ by any power of η_0 without destroying the inequality: $\psi \neq 0$ on W. But if ϕ and ψ are of degree ρh, then they can be expressed as forms of degree ρ in the η^*'s: $\phi(\eta) = \phi^*(\eta^*)$, $\psi(\eta) = \psi^*(\eta^*)$. Since $\psi(\eta) \neq 0$ on W, we have $v(\psi(\eta)/\eta_0^{\rho h}) = 0$. Hence $v(\psi^*(\eta^*)/\eta_0^{*\rho}) = 0$, and this shows that $\psi^*(\eta^*) \neq 0$ on W^*. Since $\zeta = \phi^*(\eta^*)/\psi^*(\eta^*)$, *we conclude that* $\zeta \in Q(W^*)$.

A quite similar argument shows that if $\zeta \in Q(W^*)$, then $\zeta \in Q(W)$. Hence the quotient rings $Q(W)$ and $Q(W^*)$ coincide, and from this our lemma follows in view of Theorem 3 (I.3).

The lemma shows that as far as the study of the birational correspondence between V and V' is concerned, it is permissible to replace V by V^*. Let us see therefore how V' is related to V^*.

Every homogeneous element in $K(\eta_0, \eta_1, \cdots, \eta_n)$, of degree h,—and in particular each element η_i'—can be written as a quotient of two forms in the η's whose degrees are multiples of h. Any such quotient is a quotient of

(21) When we say that a birational correspondence between two varieties V and V' is (1, 1) without exceptions, we mean that it is (1, 1) as a correspondence between the irreducible subvarieties of V and the irreducible subvarieties of V'.

two forms in the η^*'s. Hence $\eta_i' \in K(\eta_0^*, \eta_1^*, \cdots, \eta_s^*)$. Conversely, each element η_i^*, being homogeneous of degree h, is a linear form in the η''s. We therefore conclude that *the two fields* $K(\eta_0^*, \eta_1^*, \cdots, \eta_s^*)$ and $K(\eta_0', \eta_1', \cdots, \eta_m')$ *coincide.*

The elements η_i^*, which as elements of the field $K(\eta_0, \eta_1, \cdots, \eta_n)$ are homogeneous of degree h, as elements of the field $K(\eta_0^*, \eta_1^*, \cdots, \eta_s^*)$ are to be regarded as homogeneous, of degree 1. The same remark applies to the elements η_i'. Moreover, the elements η_i' constitute a linear base for the elements of the field $K(\eta_0^*, \eta_1^*, \cdots, \eta_s^*)$ which are homogeneous of degree 1 and which are integrally dependent on $\eta_0^*, \eta_1^*, \cdots, \eta_s^*$. We conclude from all this that *V' is also a derived normal variety of V^*, of degree of homogeneity* 1. Thus, while not every variety V possesses a derived normal variety of degree of homogeneity 1, we may nevertheless assume—and we do so assume—that we had originally $h=1$; this assumption amounts to replacing V by V^*.

Now that we have $h=1$, it follows that η's are linear combinations of the η''s, whence *V is a projection of the normal model V'.* Moreover, the ring $P'=K[\eta']$ is now either the integral closure of the ring $P=K[\eta]$ in its quotient field or contains at any rate all homogeneous integral quantities of positive degree.

Let now W and W' be corresponding irreducible subvarieties of V and V', respectively. We assume that $\eta_0 \neq 0$ on W and that $\eta_0' = \eta_0$. *I assert that $\eta_0' \neq 0$ on W'.* To see this we have only to show that $v(\eta_i'/\eta_0') \geqq 0$, $i=0, 1, \cdots, m$, for at least one valuation of center W'. We take as v a valuation which has also W as center on V. We write the relation of integral dependence for η_i' over P. It is of the form (see our paper [7, p. 286, equation (33)]):

$$\eta_i'^{\nu} + a_1(\eta)\eta_i'^{\nu-1} + \cdots + a_\nu(\eta) = 0,$$

where $a_j(\eta)$ is a form of degree j in $\eta_0, \eta_1, \cdots, \eta_n$. If we divide this equation by $\eta_0'^{\nu}$, we see that the quotients η_i'/η_0' are integrally dependent on the quotients $\eta_1/\eta_0, \eta_2/\eta_0, \cdots, \eta_n/\eta_0$. Since $\eta_0 \neq 0$ on W, these quotients are in R_v, and this proves our assertion.

Let $\mathfrak{P}$ and $\mathfrak{P}'$ be the prime homogeneous ideals of W and of W' in the ring P and P', respectively. From the fact that $\eta_0 \neq 0$ on W and $\eta_0'(=\eta_0) \neq 0$ on W' and from the very definition of the center of a valuation, it follows immediately that $\mathfrak{P} = \mathfrak{P}' \cap P$. Since the elements of P' are integrally dependent on P, there is only a finite number of prime ideals in P' which contract to $\mathfrak{P}$. These ideals are all homogeneous(22) and of the same dimension as $\mathfrak{P}$. We therefore reach the following conclusion:

To each irreducible subvariety W' of V' there corresponds a unique subvariety W to V, while to each irreducible subvariety W of V there corresponds a finite

(22) The prime ideals $\mathfrak{P}'$ which contract to $\mathfrak{P}$ are the minimal primes of the extended ideal $P' \cdot \mathfrak{P}$ and therefore are homogeneous, by Theorem 2 (I.1).

number of subvarieties W' of V'. Two corresponding varieties W and W' have the same dimension.

We now investigate the relationship between the quotient rings of two corresponding varieties W and W'. We pass to the rings $\mathfrak{o}=K[\xi_1, \xi_2, \cdots, \xi_n]$ and $\mathfrak{o}'=K[\xi_1', \xi_2', \cdots, \xi_m']$ of the nonhomogeneous coordinates $\xi_i=\eta_i/\eta_0$ and $\xi_i'=\eta_i'/\eta_0'$, where $\eta_0=\eta_0'$. Here $\mathfrak{o}'$ is the integral closure of $\mathfrak{o}$. Let $\mathfrak{p}$ be the prime $\mathfrak{o}$-ideal of W. Any W' which corresponds to W will be at finite distance with respect to the coordinates ξ_i' and will be given in $\mathfrak{o}'$ by a prime ideal which contracts to $\mathfrak{p}$. Let $\mathfrak{p}_1', \mathfrak{p}_2', \cdots, \mathfrak{p}_\nu'$ be the prime $\mathfrak{o}'$-ideals which contract to $\mathfrak{p}$, and let $W_1', W_2', \cdots, W_\nu'$ be the corresponding subvarieties of V'. Let

$$\mathfrak{J} = Q(W) = \mathfrak{o}_{\mathfrak{p}}, \qquad \mathfrak{J}_i' = Q(W_i') = \mathfrak{o}_{\mathfrak{p}_i'},$$

and let $\mathfrak{m}$ and $\mathfrak{m}_i'$ denote the ideals of non-units in $\mathfrak{J}$ and $\mathfrak{J}_i'$, respectively. We have $\mathfrak{J}_i'\supseteq\mathfrak{J}$ and $\mathfrak{J}_i'$ is integrally closed. Let $\mathfrak{J}^*$ denote the integral closure of $\mathfrak{J}$. Then $\mathfrak{J}_i'\supseteq\mathfrak{J}^*\supset\mathfrak{o}'$ and we can consider the ideals $\mathfrak{m}_i^*=\mathfrak{m}_i'\cap\mathfrak{J}^*$. *The ideals $\mathfrak{m}_1^*, \mathfrak{m}_2^*, \cdots, \mathfrak{m}_\nu^*$ are distinct,* since $\mathfrak{m}_i^*\cap\mathfrak{o}'=\mathfrak{m}_i'\cap\mathfrak{o}'=\mathfrak{p}_i'$. *The ideals $\mathfrak{m}_i^*$ contract to one and the same ideal in $\mathfrak{J}$, namely to $\mathfrak{m}$,* since $\mathfrak{p}_i'\cap\mathfrak{o}=\mathfrak{p}$ and $\mathfrak{m}=\mathfrak{J}\cdot\mathfrak{p}$. The quotient ring of $\mathfrak{m}_i^*$ in $\mathfrak{J}^*$ is contained in $\mathfrak{J}_i'$ since $\mathfrak{m}_i^*=\mathfrak{m}_i'\cap\mathfrak{J}^*$. On the other hand, the quotient ring $\mathfrak{J}_i'$ is contained in the quotient ring of $\mathfrak{m}_i^*$, since $\mathfrak{m}_i^*\cap\mathfrak{o}'=\mathfrak{p}_i'$. *Hence the quotient ring of $\mathfrak{m}_i^*$ in the ring $\mathfrak{J}^*$ coincides with $\mathfrak{J}_i'$.*

The foregoing properties of $\mathfrak{J}^*$ can also be derived from the general theory of quotient rings (see I.2 (1)). It is only necessary to observe that $\mathfrak{J}^*$ coincides with the quotient ring $\mathfrak{o}_S'$, where $S=\mathfrak{o}-\mathfrak{p}$. From this remark it follows immediately that the ideals $\mathfrak{m}_1^*, \mathfrak{m}_2^*, \cdots, \mathfrak{m}_\nu^*$ are the only prime ideals of $\mathfrak{J}^*$ which contract to $\mathfrak{m}$. The connection between the quotient ring $Q(W)$ and the ν quotient rings $Q(W_i')$ is therefore fully established. Reassuming, we can now state the following theorem:

THEOREM 7. *The birational correspondence between an irreducible algebraic variety V and a derived normal variety V' of V has the following properties:*

(A) *Two corresponding subvarieties W and W' of V and V', respectively, have the same dimension, and we have: $Q(W)\subseteq Q(W')$.*

(B) *Given W', the corresponding W is uniquely determined, while to a given W there corresponds a finite number of varieties W'.*

(C) *If $\mathfrak{J}^*$ denotes the integral closure of the quotient ring $\mathfrak{J}=Q(W)$, and if $\mathfrak{m}_1^*, \mathfrak{m}_2^*, \cdots, \mathfrak{m}_\nu^*$ are the prime ideals in $\mathfrak{J}^*$ which contract to the ideal of non-units in $\mathfrak{J}$, then there are exactly ν varieties $W_1', W_2', \cdots, W_\nu'$ which correspond to W, and for a suitable ordering of the indices we will have $Q(W_i')=\mathfrak{J}^*_{\mathfrak{m}_i^*}$.*

COROLLARY 1. *The birational correspondence between any two derived normal varieties of V is (1, 1) without exceptions, and corresponding subvarieties have the same quotient ring.*

Other corollaries follow from Theorem 7. We first give the following definitions:

DEFINITION 2. *If $Q(W)$ is integrally closed, then V is said to be locally normal at W.*

DEFINITION 3. *If V is locally normal at each of its irreducible subvarieties, then V is said to be a locally normal variety*(23).

A locally normal variety is characterized by the property that it is normal in the affine space for every choice of the nonhomogeneous coordinates. For that it is necessary and sufficient (see our paper [7, Theorem 13, p. 286]) that the conductor of the ring $P=K[\eta_0, \eta_1, \cdots, \eta_n]$ with respect to the integral closure of P be an irrelevant ideal.

COROLLARY 2. *If V is locally normal at W then to W there corresponds a unique subvariety W' of the derived normal model V', and the quotient rings $Q(W)$, $Q(W')$ coincide.*

COROLLARY 3. *If V is locally normal, then the birational correspondence between V and a derived normal variety V' of V is* (1, 1) *without exceptions, and the correspondence preserves quotient rings.*

COROLLARY 4. *The irreducible subvarieties of V to which there corresponds more than one variety on V' all lie on the subvariety C of V which is defined by the conductor $\mathfrak{C}$ of the ring $P=K[\eta_0, \eta_1, \cdots, \eta_n]$ with respect to the integral closure of P. Outside of C, the birational correspondence between V and V' is* (1, 1) *and it preserves quotient rings.*

3. **The fundamental elements of a birational correspondence.** We consider a birational correspondence T between two *locally normal* varieties V and V' (the general case is discussed briefly at the end of this section). Let W be an irreducible subvariety of V.

DEFINITION 4. *We say that W is* (1) *regular,* (2) *irregular, or* (3) *fundamental for T if there exists a W' on V' such that $W'=T(W)$ and, respectively,* (1) $Q(W)=Q(W')$, (2) $Q(W)\supset Q(W')$ or (3) $Q(W)\not\supseteq Q(W')$.

The following theorem is merely a statement of some properties of regular, irregular and fundamental varieties which follow directly from the definition and from Theorem 6 (II.1) and which we shall use very frequently:

THEOREM 8.

(A) *If W is regular or irregular, then to W there corresponds a unique W' on V'.*

(23) For V to be locally normal it is sufficient that V be locally normal at each of its points. For if W is any irreducible subvariety of V and if P is any point of W, then $Q(W)$ is also the quotient ring of a prime ideal in $Q(P)$. If $Q(P)$ is integrally closed, also $Q(W)$ is integrally closed.

(B) *If W is regular for T and if $W'=T(W)$, then W' is of the same dimension as W and is regular for T^{-1}.*

(C) *If W is irregular and $W'=T(W)$, then W' is fundamental for T^{-1}, and the dimension of W' is less than or equal to the dimension of W.*

(D) *If W is fundamental, then $Q(W)\not\supseteq Q(W')$ for any W' which corresponds to W.*

(E) *A necessary and sufficient condition that W not be fundamental is that for a suitable choice of the nonhomogeneous coordinates $\xi_1', \xi_2', \cdots, \xi_m'$ of the general point of V', the ring $\mathfrak{o}'$ of these coordinates be contained in $Q(W)$.*

The proof of (E) is immediate. For assume that $\mathfrak{o}'\subseteq Q(W)$ and let $\mathfrak{p}'$ be the prime $\mathfrak{o}'$-ideal which is the contraction of the ideal of non-units of $Q(W)$. Then if W' is the irreducible subvariety of V' which is defined by $\mathfrak{p}'$, then $Q(W')\subseteq Q(W)$ and by Theorem 3 every valuation of center W on V has center W' on V'.

The sets of regular, irregular and fundamental varieties are mutually exclusive. We now give the following definition:

DEFINITION 5. *A birational correspondence T is regular, if every W is regular for T.*

It is clear that if T is regular, then also T^{-1} is regular (Theorem 8 (B)) A regular birational correspondence is a (1, 1) correspondence, without exceptions, and it preserves quotient rings. We have encountered examples of regular birational correspondences in the preceding section (Lemma 5; Theorem 7, Corollaries 1 and 3).

The next theorem shows that Theorem 8 (A) expresses a *characteristic property* of a non-fundamental W.

THEOREM 9. *If to W there corresponds a unique subvariety W' of V', then W is not fundamental for T (hence is either regular or irregular).*

Proof. Since V is locally normal, the quotient ring $Q(W)$ is integrally closed. By the fundamental theorem on principal orders[14], $Q(W)$ is the intersection of the valuation rings which contain $Q(W)$. Let R_{v_1} be one of these valuation rings. Since $R_{v_1}\supseteq Q(W)$, it follows that the center of v_1 on V is either W or a subvariety W_1 of V which properly contains W (Theorem 3, I.3). In the second case there exists a valuation v which is compounded with v_1 and has center W (Lemma 4, I.4). Since $R_v\subset R_{v_1}$ we can omit R_{v_1} from the set of valuation rings which contain $Q(W)$, without affecting the intersection of these rings. *Hence $Q(W)$ is also the intersection of the valuation rings which belong to valuations of center W.* Since, by hypothesis, all valuations of center W on V have the same center W' on V', it follows that the corresponding valuation rings all contain $Q(W')$. Consequently $Q(W)\supseteq Q(W')$, as was asserted.

COROLLARY. *The dimension of a fundamental variety W cannot exceed* $r-2$.

For if W has dimension $r-1$, then $Q(W)$ is itself a valuation ring, namely the valuation ring of a divisor v. Therefore W is the center of only one valuation, namely of the divisor v.

THEOREM 10. *If W is fundamental, then to W there correspond on V′ infinitely many varieties W′.*

Proof. We shall prove that if to W there corresponds on V' only a finite number of varieties, then W is not fundamental.

Let $W_1', W_2', \cdots, W_h'$ be the irreducible subvarieties of V' which correspond to W. If v is any valuation of center W, then R_v must contain at least one of the h quotient rings $Q(W_i')$. Hence R_v contains the intersection of these quotient rings. Since $Q(W)$ is the intersection of all R_v, it follows that $Q(W)$ *contains the intersection of the h quotient rings* $Q(W_i')$.

We can find a form $\phi(\eta_0', \eta_1', \cdots, \eta_m')$, of a sufficiently high degree ν, such that $\phi\neq 0$ on W_i', $i=1, 2, \cdots, h$. We pass from V' to the variety V_1' whose general point is defined by a linear K-basis of the forms of degree ν in $\eta_0', \eta_1', \cdots, \eta_m'$. By Lemma 5, V' and V_1' are in regular birational correspondence, hence we may replace in our proof V' by V_1'. We may therefore assume that ϕ is one of the elements η_i', say $\phi=\eta_0'$. From the fact that $\eta_0'\neq 0$ on W_i', $i=1, 2, \cdots, h$, it follows that the ring $\mathfrak{o}'$ of the nonhomogeneous coordinates $\xi_i'=\eta_i'/\eta_0'$ is contained in each quotient ring $Q(W_i')$. Since the intersection of the rings $Q(W_i')$ is contained in $Q(W)$, our theorem follows from Theorem 8 (E).

We shall now discuss briefly the general case in which V and V' are not locally normal. Let $\overline{V}$ and $\overline{V}'$ be derived normal varieties of V and V' respectively. Let W be an irreducible subvariety of V and let $\overline{W}_1, \overline{W}_2, \cdots, \overline{W}_h$ be the irreducible varieties on $\overline{V}$ which correspond to W (Theorem 7 (B)). We shall denote by $\overline{T}$ the birational correspondence between $\overline{V}$ and $\overline{V}'$.

DEFINITION 6. *The variety W is regular for T, if each* $\overline{W}_i$, $i=1, 2, \cdots, h$, *is regular for* $\overline{T}$; *W is fundamental for T, if at least one of the varieties* $\overline{W}_i$ *is fundamental for* $\overline{T}$; *W is irregular for T, if it is neither regular nor fundamental, that is, if no* $\overline{W}_i$ *is fundamental for* $\overline{T}$ *and if at least one* $\overline{W}_i$ *is irregular for* $\overline{T}$.

Of the theorems proved in this section for locally normal varieties, Theorems 9 and 10 continue to hold in the general case. The validity of Theorem 10 is obvious. As to Theorem 9, the proof is as follows. If T is single-valued at W, say $T(W)=W'$, then any irreducible subvariety of $\overline{V}'$ which corresponds to $\overline{W}_i$ under $\overline{T}$ ($i=1, 2, \cdots, h$) must be among the irreducible subvarieties of $\overline{V}'$ which correspond to W' in the birational correspondence between V' and $\overline{V}'$. Hence to each $\overline{W}_i$ there can correspond on $\overline{V}'$ only a finite number of varieties. Therefore no $\overline{W}_i$ is fundamental (Theorem 10) for $\overline{T}$, and therefore, by definition, W is not fundamental for T.

In particular, if $T(W)=W'$ and if $Q(W)\supseteq Q(W')$, then W is not fundamental. This follows from Theorem 6, II.1.

On the other hand, other results established for locally normal varieties do not generalize to varieties which are not locally normal. For instance, parts (A) and (B) of Theorem 8 cease to be true in the general case. Also the defining property of a non-fundamental variety used in Definition 4 ceases to be a property of non-fundamental varieties in the general case, that is, if W is not fundamental that does not mean that there must exist a W' such that $W'=T(W)$ and $Q(W)\supseteq Q(W')$. Also the condition stated in Theorem 8 (E) is sufficient, but no longer necessary.

Note that according to Definition 6 the birational correspondence between a variety V and derived normal variety of V is free from fundamental elements on either variety.

We shall agree to use Definition 5 of regular birational correspondences also in the case of varieties which are not locally normal.

4. **A question of terminology.** At this stage it becomes necessary to point out and to discuss the difference between our terminology and the terminology used heretofore in the literature. This difference concerns the meaning of the term "fundamental" and our use of the new term "irregular."

In the case of algebraic surfaces it is the sense of the old terminology that both points *and curves* can be fundamental: a point P is fundamental if it is transformed into a curve Γ', and any such curve Γ', which is then the transform of a point, is "fundamental." As far as the notion of a fundamental point is concerned this is in agreement with our terminology, from Theorem 8 (A) and Theorem 10. However, by Theorem 9, corollary, a curve on an algebraic surface can never be fundamental in our sense. The "fundamental" curves in the sense of the old terminology are irregular curves in our sense.

The reasons for our terminology—or better—the inadequacy of the old terminology([24]) become apparent in the case of higher varieties. Let us consider, for instance, a birational correspondence T between two 3-dimensional varieties V and V'. Again, according to the old terminology we may have "fundamental" loci of all dimensions from 0 to 2. As far as fundamental points and "fundamental" surfaces are concerned, the situation is the same as in the case of algebraic surfaces: there is complete agreement on fundamental points, while according to our terminology there are definitely no "fundamental" surfaces, but only irregular surfaces. It is, however, the use of the term "fundamental curve" that brings out some significant facts.

I can find no clear-cut definition of a fundamental curve in the literature. This much is certain: if a curve Γ is such that $T(\Gamma)$ is a surface, or if Γ corresponds to each point of another curve, then in the old terminology (and

([24]) The best justification for our terminology is its own logical consistency. We call fundamental a variety W if and only if the birational correspondence T is infinitely many-valued at W. Otherwise W is either regular or irregular.

also in our terminology) Γ is fundamental (respectively, of the "first" or of the "second kind"). Suppose, however, that the transform of Γ is a single point. I am not certain whether or not such a curve is fundamental in the sense of the old terminology. If it is, then the terminology is confusing, since we are dealing here with a curve at which the birational transformation is single-valued. If it is not, then the terminology is inconsistent, in view of the use of the term "fundamental" surface, since in both cases we are dealing with a W such that $T(W)$ is *unique* and is of *lower* dimension than W.

It is quite possible that in the old terminology no special name has ever been assigned to a curve Γ such that $T(\Gamma)$ is a single point. If that is the case, then this is probably due to the fact that, as a rule, only nonsingular models have been considered in the literature. If the three-dimensional varieties V and V' are nonsingular, then a curve Γ which is transformed into a point necessarily lies on a surface which is transformed into a curve (see II.10, Theorem 17, corollary). Thus, such a curve Γ always lies on a "fundamental" surface, and there seemed to be no compelling reason for giving these curves a special name. However, in the case of singular models it may very well happen that a curve Γ whose dimension is *lowered* by the birational transformation T and at which T is *single-valued* (these two properties imply that W is irregular; see Theorem 8 (B) and Theorem 9) does not lie on any surface having the same properties (compare with Theorem 17, II.10). Some term for such a curve is necessary, and the term "fundamental" we reject for reasons given above.

5. **The join of two birationally equivalent varieties.** Let $(\eta_0, \eta_1, \cdots, \eta_n)$ and $(\eta_0', \eta_1', \cdots, \eta_m')$ be the general points, respectively, of V and of V', where V and V' are our two birationally equivalent varieties. Since the quotients η_i'/η_0' are rational functions of the quotients η_j/η_0, the η''s are proportional to forms of like degree in the η's:

$$\begin{aligned}(4)\quad &\eta_0' : \eta_1' : \cdots : \eta_m' \\ &= \phi_0(\eta_0, \eta_1, \cdots, \eta_n) : \phi_1(\eta_0, \eta_1, \cdots, \eta_n) : \cdots : \phi_m(\eta_0, \eta_1, \cdots, \eta_n).\end{aligned}$$

DEFINITION 7. *The irreducible algebraic variety V^* whose general point has the $(n+1)(m+1)$ products $\eta_i\phi_j$ as homogeneous coordinates is called the join of V and V'.*

We shall denote the products $\eta_i\phi_j$ by η_{ij} and the quotients η_i/η_0, η_j'/η_0' and η_{ij}/η_{00} by ξ_i, ξ_j' and ξ_{ij}, respectively. We have then:

$$(5)\qquad \xi_{i0} = \xi_i, \qquad \xi_{0j} = \xi_j', \qquad \xi_{ij} = \xi_i\xi_j', \qquad i, j \neq 0,$$

and from these relations it follows that V^* is birationally equivalent to V (and to V'). Moreover, if we take as nonhomogeneous coordinates of the general point of V^* the quotients of the η_{ij}'s by a fixed η_{ij}, say by η_{00}, then the ring of these coordinates is, by (5), the join of the two rings of nonhomogeneous coordinates relative to V and V'. In symbols: if

$$\mathfrak{o} = K[\xi_1, \xi_2, \cdots, \xi_n], \qquad \mathfrak{o}' = K[\xi_1', \xi_2', \cdots, \xi_m'],$$
$$\mathfrak{o}^* = K[\xi_{10}, \xi_{20}, \cdots, \xi_{nm}],$$

then $\mathfrak{o}^* = (\mathfrak{o}, \mathfrak{o}')$.

THEOREM 11. *In the birational correspondence T^* between V and V^* there corresponds to any irreducible subvariety W^* of V^* a unique subvariety W of V. If a given W on V is not fundamental for T, then it is regular for T^*. Similarly for V' and V^*.*

Proof. We may assume that W^* is at finite distance with respect to the nonhomogeneous coordinates ξ_{ij}. If v is any valuation of center W^*, then $R_v \supseteq Q(W^*) \supset \mathfrak{o}^* \supseteq \mathfrak{o}$, whence the center W of v on V is at finite distance with respect to the nonhomogeneous coordinates ξ_i. Similarly for V', W' and the ξ_j'. But then, if $\mathfrak{p}^*$ is the prime $\mathfrak{o}^*$-ideal of W^*, the prime ideals $\mathfrak{p}$ and $\mathfrak{p}'$ of W and W', in the rings $\mathfrak{o}$ and $\mathfrak{o}^*$, respectively, are necessarily the contracted ideals of $\mathfrak{p}^*$, that is, $\mathfrak{p} = \mathfrak{p}^* \cap \mathfrak{o}$, $\mathfrak{p}' = \mathfrak{p}^* \cap \mathfrak{o}'$, and consequently W and W' are uniquely determined by W^*. Notice that the quotient rings $Q(W)$ and $Q(W')$ are subrings of $Q(W^*)$:

$$Q(W) \subseteq Q(W^*), \qquad Q(W') \subseteq Q(W^*). \tag{6}$$

To prove the second part of the theorem, let $T(W) = W'$ and let us assume that W and W' are at finite distance with respect to the nonhomogeneous coordinates ξ_i and ξ_j'. Let v be any valuation of center W and W' on V and V', respectively. The center W^* of v on V^* will be at finite distance with respect to the coordinates ξ_{ij}, since $R_v \supset \mathfrak{o}$, $R_v \supset \mathfrak{o}'$ and therefore $R_v \supset \mathfrak{o}^*$.

Let us first consider the case in which V is locally normal. Since W is not fundamental for T, we have $Q(W) \supseteq Q(W') \supset \mathfrak{o}'$. Therefore $Q(W) \supset \mathfrak{o}^*$, that is, $\mathfrak{o}_{\mathfrak{p}} \supset \mathfrak{o}^*$. Since $\mathfrak{p}^* \cap \mathfrak{o} = \mathfrak{p}$, it follows that $\mathfrak{o}_{\mathfrak{p}}$ contains the ring $\mathfrak{o}^*_{\mathfrak{p}^*}$, that is, $Q(W) \supseteq Q(W^*)$. Hence, by (6), $Q(W) = Q(W^*)$, whence W is regular for T^*.

To prove the theorem in the general case we first observe that if $\bar{V}$ and $\bar{V}'$ denote derived normal varieties of V and of V', respectively, then any derived normal variety $\bar{V}^*$ of the join of V and V' is in regular birational correspondence with any derived normal variety of the join of $\bar{V}$ and $\bar{V}'$. The proof is straightforward and consists in the obvious remark that the integral closure of the ring $(\mathfrak{o}, \mathfrak{o}')$ is the same as integral closure of the ring $(\mathfrak{o}_1, \mathfrak{o}_1')$, where $\mathfrak{o}_1$ and $\mathfrak{o}_1'$ are the integral closures of $\mathfrak{o}$ and of $\mathfrak{o}'$, respectively. Now let $\bar{W}$ be any of the irreducible subvarieties of $\bar{V}$ which correspond to W. To prove that W is regular for T^* we have only to prove (Definition 6, II.3) that $\bar{W}$ is regular for the birational correspondence between $\bar{V}$ and $\bar{V}^*$. Since W, by hypothesis, is not fundamental for T, it follows (Definition 6) that $\bar{W}$ is not fundamental for the birational correspondence between $\bar{V}$ and $\bar{V}'$. Since $\bar{V}$ is locally normal, it follows by the case just considered that $\bar{W}$ is regular for the birational correspondence between $\bar{V}$ and the join of $\bar{V}$ and $\bar{V}'$. Since a derived normal variety of this join is, by the remark made above, in regular birational correspondence with $\bar{V}^*$, it follows that $\bar{W}$ is regular for the birational correspondence between $\bar{V}$ and $\bar{V}^*$, as was asserted.

COROLLARY. *If P and P' are corresponding points of V and V', then there is only a finite number of points P^* on V^* which correspond to both P and P'. If V (or V') is locally normal, then the number of such points P^* can be greater than* 1 *only if P (or P') is a fundamental point of T (or of T^{-1}).*

For if we identify the varieties W, W' and W^* of the preceding proof with the points P, P' and P^*, respectively, we see that $\mathfrak{p}^*$ must be a zero-dimensional prime divisor of the ideal $\mathfrak{o}^* \cdot (\mathfrak{p}, \mathfrak{p}')$. Since this ideal is pure zero-dimensional, the number of possible ideals $\mathfrak{p}^*$ is finite. The second half of the corollary follows directly from the second half of the preceding theorem.

If the ground field K is algebraically closed, the ideal $\mathfrak{o}^* \cdot (\mathfrak{p}, \mathfrak{p}')$ is itself prime, whenever $\mathfrak{p}$ and $\mathfrak{p}'$ are both zero-dimensional. Hence if K is algebraically closed, then not only does every point P^* of V^* determine uniquely a pair of *corresponding* points P, P' of V and V', respectively, but, conversely, every such pair determines uniquely a point P^* on V^*. For this reason the join V^* is often referred to in the literature as the *variety of pairs of corresponding points of V and V'.*

If K is not algebraically closed then P^* need not be uniquely determined by P and P'. The following is an example(25). Let K be the field of real numbers and let $\Sigma = K(x, y)$, where x and y are indeterminates. We take as V and V' two planes given—in nonhomogeneous coordinates—by the general points (x, y_1) and (x_1, y), respectively, where $y_1 = y(x^2+1)$ and $x_1 = x(y^2+1)$. Here we have: $\mathfrak{o} = K[x, y_1]$, $\mathfrak{o}' = K[x_1, y]$, $\mathfrak{o}^* = (\mathfrak{o}, \mathfrak{o}') = K[x, y]$. Let $\mathfrak{p} = (x^2+1, y_1)$, $\mathfrak{p}' = (x_1, y^2+1)$. These ideals are prime and zero-dimensional in their respective rings and they represent corresponding points of the two planes V and V'. However, the ideal $\mathfrak{o}^* \cdot (\mathfrak{p}, \mathfrak{p}')$ is now the intersection of the following prime ideals: $\mathfrak{p}_1^* = (x^2+1, y-x)$, $\mathfrak{p}_2^* = (x^2+1, y+x)$.

The join V^* of two locally normal varieties V and V' need not be locally normal. One may often find it convenient to pass from V^* to a derived normal variety $\bar{V}^*$ of V^*. We may call $\bar{V}^*$ the *normal join* of V and V'.

(25) It should not be too difficult to find necessary and sufficient conditions in order that a given pair of corresponding points P, P' determine uniquely a point P^* of the join V^*. The following is a sufficient condition. *If $\Delta = \mathfrak{o}/\mathfrak{p}$ and $\Delta' = \mathfrak{o}'/\mathfrak{p}'$ are the residue fields of the points P and P' respectively, then a least field Δ^*/K containing Δ/K and Δ'/K should exist such that its relative degree over* K *is the product of the relative degrees of Δ and of Δ' over* K. (It is not difficult to see that this product is the maximum value for the relative degree of Δ^* over K, and if that maximum is reached, then there exists, to within relative isomorphisms, only one least field which contains Δ and Δ'.) However, the above condition is not sufficient. In fact, no condition can be both necessary and sufficient which does not take into account the quotient rings $Q(P)$ and $Q(P')$ themselves, besides the residue fields Δ and Δ'.

The usefulness of the join V^* is due to the possibility of deriving properties of the birational correspondence between V and V' by first passing from V to V^* and then from V^* to V'. In each of these two steps we are dealing with a birational correspondence between two varieties *which has no fundamental elements on one of the varieties* (*on* V^*). Birational correspondences of this sort are easier to handle, and they in fact play an important role in the general theory and in applications.

6. **Further properties of fundamental varieties.**

THEOREM 12. *Given an irreducible subvariety W of V there exists an algebraic subvariety of V' which we shall denote by $T[W]$ and which has the following properties*:

A. *Each irreducible component of $T[W]$ corresponds to W.*

B. *Each irreducible subvariety W' of V' which corresponds to W lies on $T[W]$.*

The variety $T[W]$ shall be referred to in the sequel as the *transform of W*.

Proof. Suppose that the theorem is true for V and the join V^* of V and V'. Then we show that it is also true for V and V'. For let T^* denote, as before, the birational correspondence between V and V^* and let $T^*[W]=W_1^*+W_2^*+\cdots+W_h^*$, where each W_i^* is irreducible. To each W_i^* there corresponds on V' a unique irreducible variety W_i'. Since, by hypothesis, $T^*(W)=W_i^*$, it follows that each of the varieties $W_1', W_2', \cdots, W_h'$ corresponds to W. On the other hand, let W' be any irreducible subvariety of V' which corresponds to W, and let W^* be an irreducible subvariety of V^* which corresponds to both W and W'. By hypothesis, $W^*\subseteq T^*(W)$, say $W^*\subseteq W_1^*$. Passing to the corresponding subvarieties W' and W_1' of V', we conclude that $W'\subseteq W_1'$, that is, $W'\subseteq W_1'+W_2'+\cdots+W_h'$. Hence $T[W]=W_1'+W_2'+\cdots+W_h'$, where, of course, some of the h varieties W_i' may be embedded, so that the number of irreducible components of $T[W]$ may actually be less than h.

We now prove the theorem for V and for the join V^*. We use nonhomogeneous coordinates. As nonhomogeneous coordinates for V^* we can use the coordinates ξ_{ij} of the preceding section, since the $(n+1)(m+1)$ systems of coordinates $\eta_{ij}/\eta_{\alpha\beta}$ (α and β are fixed for each system) cover the entire projective space in which V^* is embedded. To prove the existence of the transform $T^*[W]$, it will be sufficient therefore to exhibit that part L^* of $T^*[W]$ which is at finite distance with respect to the coordinates ξ_{ij}. Let $\mathfrak{o}$, $\mathfrak{o}'$ and $\mathfrak{o}^*$ have the same meaning as in the preceding section. If W is not at finite distance with respect to the coordinates ξ_i, then *no* W^* which corresponds to W on V^* can be at finite distance with respect to the coordinates ξ_{ij} (since $\mathfrak{o}\subseteq\mathfrak{o}^*$). Hence in this case L^* is empty. If W is at finite distance, it is given by a prime ideal $\mathfrak{p}$ in $\mathfrak{o}$. Let $\mathfrak{p}_1^*, \mathfrak{p}_2^*, \cdots, \mathfrak{p}_h^*$ be those *minimal* prime ideals of $\mathfrak{o}^*\cdot\mathfrak{p}$ which contract to $\mathfrak{p}$, and let W_i^* be the irreducible subvariety of V^*

which is defined by $\mathfrak{p}_i^*$. I assert that $L^* = W_1^* + W_2^* + \cdots + W_h^*$. For in the first place, each W_i^* corresponds to W. In the second place, if $W^* = T^*(W)$ and if W^* is at finite distance with respect to the coordinates ξ_{ij}, W^* is given by a prime $\mathfrak{o}^*$-ideal $\mathfrak{p}^*$, such that $\mathfrak{p}^* \cap \mathfrak{o} = \mathfrak{p}$. Since $\mathfrak{p} \subseteq \mathfrak{p}^*$, we have $\mathfrak{o}^*\mathfrak{p} \subseteq \mathfrak{p}^*$, whence $\mathfrak{p}^*$ must divide some minimal ideal $\mathfrak{p}'^*$ of $\mathfrak{o}^*\mathfrak{p}$. Since $\mathfrak{p} \subseteq \mathfrak{p}'^* \subseteq \mathfrak{p}^*$, it follows that $\mathfrak{p} \subseteq \mathfrak{p}'^* \cap \mathfrak{o} \subseteq \mathfrak{p}$, that is $\mathfrak{p}'^* \cap \mathfrak{o} = \mathfrak{p}$. Therefore $\mathfrak{p}'^*$ is one of the ideals $\mathfrak{p}_1^*, \mathfrak{p}_2^*, \cdots, \mathfrak{p}_h^*$, and since $\mathfrak{p}^* \supseteq \mathfrak{p}'^*$, we conclude that $W^* \subseteq W_1^* + W_2^* + \cdots + W_h^*$. This completes the proof.

COROLLARY 1. *If the birational correspondence between V and V' has no fundamental elements on V', then to a fundamental variety W on V there corresponds on V' at least one variety of higher dimension than W; in other words, $T[W]$ is of higher dimension than W.*

For if there are no fundamental elements on V' and if $W' = T(W)$ then dimension $W' \geqq$ dimension W (Theorem 8, (B) and (C), II.3). If every W' which corresponds to W were of the same dimension as W, then $T[W]$ would be pure ρ-dimensional, where $\rho =$ dimension W. But then the irreducible components of $T[W]$ would be the only subvarieties of V' which correspond to W, and since the number of these components is finite, the corollary follows, by Theorem 10, II.3.

As a consequence, we have the following *characterization of fundamental varieties*:

COROLLARY 2. *If V and V' are arbitrary birationally equivalent varieties and if T and T^* denote, respectively, the birational correspondence between V and V' and the birational correspondence between V and V^*, then a given irreducible subvariety W of V is fundamental for T if and only if $T^*[W]$ is of higher dimension than W.*

For W is fundamental for T if and only if it is fundamental for T^* (Theorem 11, II.5) and since T^* has no fundamental elements on V^*.

In addition to the transform $T[W]$ we shall also have occasion to consider what we call the *total transform of W* and that we shall denote by $T\{W\}$. By that we mean *the locus of points of V' which correspond to points of W*. That $T\{W\}$ is an algebraic variety is seen as follows. As in the case of $T[W]$, so also here it is sufficient to show that $T^*\{W\}$ is algebraic, where T^* is, as usual, the birational correspondence between V and the join V^* of V and V'. Now if we consider that part L_1^* of $T^*\{W\}$ which is at finite distance with respect to the nonhomogeneous coordinates ξ_{ij} we see immediately that *L_1^* is the algebraic subvariety of V^* which is defined by the ideal $\mathfrak{o}^* \cdot \mathfrak{p}$*. For a point P^* of V^*, at finite distance with respect to the ξ_{ij}'s, corresponds to a point P on W, if and only if the corresponding 0-dimensional prime $\mathfrak{o}^*$-ideal $\mathfrak{p}^*$ satisfies the relation: $\mathfrak{p}^* \cap \mathfrak{o} \supseteq \mathfrak{p}$, that is, if and only if $\mathfrak{p}^* \supseteq \mathfrak{o}^* \cdot \mathfrak{p}$.

The irreducible components of $T^*\{W\}$, at finite distance, correspond to the minimal prime ideals of $\mathfrak{o}^*\cdot\mathfrak{p}$. We have seen that the irreducible components of $T^*[W]$, at finite distance, correspond to those minimal prime ideals of $\mathfrak{o}^*\mathfrak{p}$ which contract to $\mathfrak{p}$. Hence $T^*[W]$ *lies on* $T^*\{W\}$, and from this follows immediately that *also* $T[W]$ *lies on* $T\{W\}$. This result can also be deduced directly from property B, stated in II.1. In the same fashion one sees immediately that $T\{W\}$ also has the following property: *If* $W_1 \subseteq W$, *and if* $W_1' = T(W_1)$, *then* $W_1' \subseteq T\{W\}$.

We point out explicitly that $T[W]$ *may very well be a proper subvariety* of $T\{W\}$. For instance, if T is a plane Cremona transformation and if W is a curve containing a fundamental point P, then $T\{W\} = T[W] + \Gamma'$, where Γ' is the irregular ("fundamental" in the old terminology) curve which corresponds to the point P. Here $T[W]$ is either a curve (if W is regular) or a fundamental point (if W is irregular)(26). Quite generally, we have the following theorem:

THEOREM 13. *Any irreducible component of the total transform* $T\{W\}$ *which is not a component of the transform* $T[W]$ *must correspond to a proper subvariety* W_1 *of* W. *Moreover, if* V *is locally normal, then* W_1 *must be a fundamental variety.*

Proof. In the proof of Theorem 12 we have seen that if $T^*[W] = W_1^* + W_2^* + \cdots + W_h^*$ and if W_i' is *the* subvariety of V' which corresponds to W_i^*, then $T[W] = W_1' + W_2' + \cdots + W_h'$. In a similar fashion it is seen immediately that if $T^*\{W\} = W_1^* + W_2^* + \cdots + W_h^* + W_{h+1}^* + \cdots$, then $T\{W\} = W_1' + W_2' + \cdots + W_h' + W_{h+1}' + \cdots$. Now let W_0' be an irreducible component of $T\{W\}$ which does not belong to $T[W]$. Then W_0' must correspond to an irreducible component W_0^* of $T^*\{W\}$ which does not belong to $T^*[W]$. Now by definition of $T^*\{W\}$, each point of W_0^* must correspond to some point of W. Since to a point of V^* there corresponds a unique point of V, it follows that the subvariety W_0 of V which corresponds to W_0^* must lie on W. It must be a *proper* subvariety of W, since $W_0^* \not\subseteq T^*[W]$. Now both W_0' and W_0 correspond to W_0^*, whence they correspond to each other. It remains to prove that W_0 is fundamental for T. If W_0 were not funda-

(26) This example shows therefore that the ideal $\mathfrak{o}^*\mathfrak{p}$ *may very well possess minimal prime ideals which contract in* $\mathfrak{o}$ (not to $\mathfrak{p}$ but) *to proper divisors of* $\mathfrak{p}$. In this connection we wish to correct a statement on p. 135 in Krull's Ergebnisse report *Idealtheorie*. The theorem stated on that page consists of three parts, and in the first part it is asserted, among other things, that $\tilde{\mathfrak{p}} = \mathfrak{p}\cdot\tilde{\mathfrak{J}}$. This (and only this) assertion is incorrect. The rings $\mathfrak{J}$ and $\tilde{\mathfrak{J}}$ play there the role of our rings $\mathfrak{o}$ and $\mathfrak{o}^*$. It is quite true that there is only one prime ideal $\tilde{\mathfrak{p}}$ which lies over $\mathfrak{p}$ (that is, such that $\tilde{\mathfrak{p}} \cap \mathfrak{J} = \mathfrak{p}$), and it is also true that $\mathfrak{p}$ is a minimal prime of the ideal $\tilde{\mathfrak{J}}\cdot\mathfrak{p}$. The equality $\tilde{\mathfrak{J}}_{\tilde{\mathfrak{p}}} = \mathfrak{J}_{\mathfrak{p}}$ shows clearly that the case under consideration corresponds to a *regular* W. Nevertheless, $\tilde{\mathfrak{J}}\cdot\mathfrak{p}$ may have minimal prime ideals other than $\tilde{\mathfrak{p}}$, as was pointed out above. Already the quadratic transformation $x' = x$, $y' = y/x$ may serve very well as a source of simple counterexamples. The formal source of that incorrect statement made in Krull's report is the erroneous assertion made earlier on the same page (line 12) to the effect that if the rank equals m then $L_{\mathfrak{p}}$ is a field.

mental, then by a stronger reason W would not be fundamental, since $Q(W)\supset Q(W_0)$ (see Theorem 8 (E)). But then we would have $T[W_0]\subset T[W]$, a contradiction.

If V is not locally normal, one passes to the derived normal varieties of V and of V^* and the rest of the proof is straightforward.

COROLLARY. *If no subvariety of W is fundamental, and if V is locally normal, then the total transform of W coincides with the transform of W.*

We make one more remark about the transforms $T[W]$ and $T\{W\}$. It is clear that in no case can these varieties be empty. However, it may very well happen that with respect to a given system of nonhomogeneous coordinates either $T[W]$ or even $T\{W\}$ is entirely at infinity. Referring to the join V^* and to the rings $\mathfrak{o}$ and $\mathfrak{o}^*$ considered above, we see that $T^*[W]$ is at infinity *if $\mathfrak{o}^*$ does not contain prime ideals which contract to $\mathfrak{p}$* (in the terminology of Krull: *$\mathfrak{p}$ is lost in $\mathfrak{o}^*$*, see [2, p. 134]). If also $T^*\{W\}$ is entirely at infinity, then *the ideal $\mathfrak{o}^*\cdot\mathfrak{p}$ is the unit ideal.*

7. **The main theorem.** If the birational correspondence between V and V' has no fundamental elements on V', and if W is fundamental for T, then, by Theorem 12, Corollary 1, the transform $T[W]$ has at least one component of higher dimension than W. In the general case, that is, when V is an arbitrary variety, that is the best result one may claim, since it is quite possible for $T[W]$ to possess components which have the same dimension as W. However, in the case in which *V is locally normal at W* we have the following important theorem:

MAIN THEOREM. *If W is an irreducible fundamental variety on V of a birational correspondence T between V and V' and if T has no fundamental elements on V', then—under the assumption that V is locally normal at W—each irreducible component of the transform $T[W]$ is of higher dimension than W.*

COROLLARY. *In the more general case in which T has fundamental elements on both V and V' and under the assumption that W is fundamental for T and that V is locally normal at W, the transform $T^*[W]$ of W on the join V^* of V and V' has the property that all its irreducible components are of higher dimension than W.*

We shall first give the main theorem another formulation which is more directly algebraic. Since V is locally normal at W, it is permissible to replace V by a derived normal variety of V (Theorem 7, Corollary 2, II.2). Hence we assume that V is a normal variety. Since there are no fundamental elements on V', the birational correspondence between V' and V^* is regular (Theorem 11, II.5). Hence it is permissible to replace V' by V^*. When these preparations are carried out then the main theorem expresses a feature of the the relationship between the ideals in the two rings $\mathfrak{o}$ and $\mathfrak{o}^*$ considered in II.5, that is, we have to prove the following theorem:

THEOREM 14. *Let* $\mathfrak{o}$ *and* $\mathfrak{o}^*$ *be two finite integral domains with the same quotient field* Σ, *where we assume that the ring* $\mathfrak{o}$ *is integrally closed and that it is a subring*(27) *of* $\mathfrak{o}^*$. *Let* $\mathfrak{p}$ *be a prime ideal in* $\mathfrak{o}$ *and let* $\mathfrak{p}^*$ *be a prime ideal* $\mathfrak{o}^*$ *which lies over* $\mathfrak{p}$, *that is, such that* $\mathfrak{p}^* \cap \mathfrak{o} = \mathfrak{p}$. *If* $\mathfrak{p}$ *and* $\mathfrak{p}^*$ *have the same dimension, then*

(1) *either the quotient rings* $\mathfrak{o}_{\mathfrak{p}}$ *and* $\mathfrak{o}^*_{\mathfrak{p}^*}$ *coincide*(28) (*and in this case* $\mathfrak{p}^*$ *is obviously the only prime* $\mathfrak{o}^*$-*ideal which lies over* $\mathfrak{p}$) *or*

(2) $\mathfrak{p}^*$ *is not minimal with respect to the property of lying over* $\mathfrak{p}$, *that is, there exists in* $\mathfrak{o}^*$ *another prime ideal which also lies over* $\mathfrak{p}$ *and which is a proper multiple of* $\mathfrak{p}^*$.

The proof of this theorem is rather long and will be developed in this and in the following section. That part of the proof which is contained in this section consists of three steps: (a) a reduction to a simpler special case; (b) a reference to the theorem of Krull which we have already mentioned(26); (c) a lemma.

(a) Let $\mathfrak{o}'^*$ be the integral closure of $\mathfrak{o}^*$ in Σ. It is well known that over every prime ideal $\mathfrak{p}^*$ in $\mathfrak{o}^*$ there lies at least one prime ideal $\mathfrak{p}'^*$ in $\mathfrak{o}'^*$, and that the number of such ideals $\mathfrak{p}'^*$ is finite. Moreover $\mathfrak{p}^*$ and $\mathfrak{p}'^*$ have the same dimension and it is clear that if $\mathfrak{p}^* \cap \mathfrak{o} = \mathfrak{p}$, then $\mathfrak{o}_{\mathfrak{p}} \subseteq \mathfrak{o}^*_{\mathfrak{p}^*} \subseteq \mathfrak{o}'^*_{\mathfrak{p}'^*}$. From this it follows that if our theorem is true for the pair of rings $\mathfrak{o}$ and $\mathfrak{o}'^*$, then it is also true for $\mathfrak{o}$ and $\mathfrak{o}^*$. *Hence we may assume that* $\mathfrak{o}^*$ *is integrally closed.*

Since $\mathfrak{o}$ and $\mathfrak{o}^*$ are finite integral domains, $\mathfrak{o}^*$ is a finite ring extension of $\mathfrak{o}$. We prefer to think of $\mathfrak{o}^*$ as a ring obtained from $\mathfrak{o}$ by a *finite* number of simple ring extensions, *each ring extension being followed up by the operation of integral closure.* Let therefore:

$$\mathfrak{o}_1 = \mathfrak{o}[\alpha_1], \quad \mathfrak{o}_1' = \text{integral closure of } \mathfrak{o}_1;$$

$$\mathfrak{o}_2 = \mathfrak{o}_1'[\alpha_2], \quad \mathfrak{o}_2' = \text{integral closure of } \mathfrak{o}_2;$$

$$\cdots\cdots\cdots\cdots\cdots\cdots\cdots\cdots$$

$$\mathfrak{o}_m = \mathfrak{o}_{m-1}'[\alpha_m], \quad \mathfrak{o}_m' = \mathfrak{o}^* = \text{integral closure of } \mathfrak{o}_m.$$

Let us assume that our theorem is true when $m=1$. Then we show, by induction with respect to m, that the theorem is true for any value of m. Let $\mathfrak{p}^* \cap \mathfrak{o}_{m-1}' = \mathfrak{p}_{m-1}'$. Since $\mathfrak{p}_{m-1}' \cap \mathfrak{o} = \mathfrak{p}$, we have: dimension $\mathfrak{p}^* \geqq$ dimension $\mathfrak{p}_{m-1}' \geqq$ dimension $\mathfrak{p}$. If $\mathfrak{p}$ and $\mathfrak{p}^*$ have the same dimension, then it follows that also $\mathfrak{p}_{m-1}'$ has the same dimension as $\mathfrak{p}$. Now let us also assume that no proper prime multiple of $\mathfrak{p}^*$ lies over $\mathfrak{p}$. Then no proper prime multiple of $\mathfrak{p}^*$ can lie over $\mathfrak{p}_{m-1}'$, and therefore, by the case $m=1$, we have that $\mathfrak{o}^*_{\mathfrak{p}^*}$ *coincides with the quotient ring of* $\mathfrak{p}_{m-1}'$ *in* $\mathfrak{o}_{m-1}'$. Consequently, there is a (1, 1) correspondence between the prime multiples of $\mathfrak{p}^*$ in $\mathfrak{o}^*$ and the prime multiples

(27) It will be seen from the proof that these assumptions can be weakened as follows: $\mathfrak{o}$ *has a finite degree of transcendency over the ground field and is a finite discrete principal order* (see Krull [2, p. 104]); $\mathfrak{o}^*$ *is a finite ring extension of* $\mathfrak{o}$.

(28) This case arises when W is not fundamental, therefore regular.

of $\mathfrak{p}'_{m-1}$ in $\mathfrak{o}'_{m-1}$. (See I.2.) Therefore it is equally true that no prime multiple of $\mathfrak{p}'_{m-1}$ lies over $\mathfrak{p}$. By our induction, we conclude that *also* $\mathfrak{o}_\mathfrak{p}$ *coincides with the quotient ring of* $\mathfrak{p}'_{m-1}$ *in* $\mathfrak{o}'_{m-1}$. Hence $\mathfrak{o}_\mathfrak{p}=\mathfrak{o}^*_\mathfrak{p}$, as was asserted.

We therefore have only to prove our theorem in the following special case: $\mathfrak{o}^*$ *is the integral closure of a ring* $\mathfrak{o}'$, *where* $\mathfrak{o}'$ *is a simple ring extension of* $\mathfrak{o}$: $\mathfrak{o}'=\mathfrak{o}[\alpha]$.

(b) In this special case we shall make use of a theorem stated in Krull [2, p. 135], to which we have already referred in the preceding section[26]. We write the principal fractional ideal $\mathfrak{o}\cdot\alpha$ as a quotient of two integral ideals: $\mathfrak{o}\cdot\alpha=\mathfrak{z}/\mathfrak{n}$, where $\mathfrak{z}$ and $\mathfrak{n}$ are symbolic power products of minimal prime ideals in $\mathfrak{o}$, without common factors. Krull distinguishes three cases: (1) $\mathfrak{n}\equiv 0(\mathfrak{p})$, $\mathfrak{z}\not\equiv 0(\mathfrak{p})$; (2) $\mathfrak{n}\not\equiv 0(\mathfrak{p})$; (3) $\mathfrak{n}\equiv 0(\mathfrak{p})$, $\mathfrak{z}\equiv 0(\mathfrak{p})$.

In the first case the element $1/\alpha$ is a non-unit in $\mathfrak{o}_\mathfrak{p}$ and from this it follows immediately that $\mathfrak{p}$ is lost in $\mathfrak{o}'$, that is, no prime ideal in $\mathfrak{o}'$ contracts to $\mathfrak{p}$. But then $\mathfrak{p}$ is also lost in $\mathfrak{o}^*$, since $\mathfrak{o}^*$ is the integral closure of $\mathfrak{o}'$, and in this case there is nothing to prove.

In the second case α is contained in the quotient ring $\mathfrak{o}_\mathfrak{p}$, whence $\mathfrak{o}^*\subset\mathfrak{o}_\mathfrak{p}$, since $\mathfrak{o}_\mathfrak{p}$ is integrally closed. From this we conclude that the two rings $\mathfrak{o}_\mathfrak{p}$, $\mathfrak{o}^*_\mathfrak{p}$. coincide[29]. This is the alternative (1) of the theorem.

The really significant case is the third one. In this case Krull's result is to the effect that $\mathfrak{o}'\cdot\mathfrak{p}$ *is a prime ideal* $\mathfrak{p}'$, *that* $\mathfrak{p}'$ *lies over* $\mathfrak{p}$ *and that the dimension of* $\mathfrak{p}'$ *is one greater than the dimension of* $\mathfrak{p}$. We shall make use of this result.

(c) In addition to the above result which concerns the relationship between the ideal theory in $\mathfrak{o}$ and in $\mathfrak{o}'$, we shall have to make use, in a very essential fashion, of the following property of the conductor $\mathfrak{C}$ of $\mathfrak{o}'$ with respect to $\mathfrak{o}^*$:

LEMMA 6. *Each prime* $\mathfrak{o}^*$*-ideal* $\mathfrak{p}^*$ *of the conductor* $\mathfrak{C}$ *has the property that it contracts in* $\mathfrak{o}$ *to a prime ideal of lower dimension.*

Proof. The lemma implies in particular that $\mathfrak{C}$ has no *zero-dimensional* prime ideals. Let us assume that this particular consequence of the lemma has been established and let us show that then the lemma follows by the usual device of ground field extension.

Let $\mathfrak{p}^*$ be a prime $\mathfrak{o}^*$-ideal and let $\mathfrak{p}^*\cap\mathfrak{o}=\mathfrak{p}$. We assume that $\mathfrak{p}$ and $\mathfrak{p}^*$ have the same dimension, say dimension s. It shall now be shown that $\mathfrak{p}^*$ cannot be a prime ideal of $\mathfrak{C}$.

We select in $\mathfrak{o}$ a set of s elements $\zeta_1, \zeta_2, \cdots, \zeta_s$ which are algebraically independent modulo $\mathfrak{p}$. We adjoin these elements to the ground field K getting a new ground field $K_1=K(\zeta_1, \zeta_2, \cdots, \zeta_s)$ and also the new rings: $\mathfrak{o}_1=K_1\cdot\mathfrak{o}$, $\mathfrak{o}'_1=K_1\cdot\mathfrak{o}'$, $\mathfrak{o}^*_1=K_1\cdot\mathfrak{o}^*$. We point out that $\mathfrak{o}^*_1$ is a quotient ring of $\mathfrak{o}^*$, namely $\mathfrak{o}^*_1=\mathfrak{o}^*_S$, where S is the set of all polynomials in $\zeta_1, \zeta_2, \cdots, \zeta_s$, with coeffi-

[29] The proof is exactly the same as the proof which in II.2 led us to the conclusion that the quotient ring of $\mathfrak{m}^*_i$ in the ring $\mathfrak{J}^*$ coincides with $\mathfrak{J}'_i$.

cients in K. Similarly we have: $\mathfrak{o}_1 = \mathfrak{o}_S$. We therefore can apply the properties of the correspondence between the ideals in a given ring R and a quotient ring R_S, as described in I.1. We find then that $\mathfrak{p}_1 = \mathfrak{o}_1 \cdot \mathfrak{p}$ and $\mathfrak{p}_1^* = \mathfrak{o}_1^* \cdot \mathfrak{p}^*$ are prime ideals and $\mathfrak{C}_1 = K_1\mathfrak{C}$ is the conductor of $\mathfrak{o}_1'$ with respect to $\mathfrak{o}_1^*$. We have $\mathfrak{o}_1' = \mathfrak{o}_1[\alpha]$ and it is clear that $\mathfrak{o}_1^*$ is the integral closure of $\mathfrak{o}_1'$. Since $\mathfrak{p}_1^*$ is *zero-dimensional* over the ground field K_1, it follows, by our assumption, that $\mathfrak{C}_1 : \mathfrak{p}_1^* = \mathfrak{C}_1$. On the other hand we have $\mathfrak{C}_1 : \mathfrak{p}_1^* = K_1 \cdot (\mathfrak{C} : \mathfrak{p}^*)$. Hence *the two ideals* $\mathfrak{C}$ *and* $\mathfrak{C} : \mathfrak{p}^*$ *have the same extended ideal in* $\mathfrak{o}_1^*$. It follows (see II.1([3])) that these two ideals can only differ by primary components whose associate prime ideals contain polynomials in $\zeta_1, \zeta_2, \cdots, \zeta_s$. Since the ζ's are algebraically independent modulo $\mathfrak{p}^*$, we conclude that $\mathfrak{p}^*$ is not among the prime ideals of $\mathfrak{C}$, as was asserted.

The proof of the lemma is thus reduced to the matter of proving that the conductor $\mathfrak{C}$ does not possess zero-dimensional prime ideals.

Let $\mathfrak{p}^*$ be a zero-dimensional prime ideal in $\mathfrak{o}^*$. We have to prove that $\mathfrak{C} : \mathfrak{p}^* = \mathfrak{C}$. Let ζ^* be an arbitrary element of $\mathfrak{C} : \mathfrak{p}^*$. We denote by $f(x)$ the irreducible polynomial in $K[x]$ such that $f(\alpha) \equiv 0(\mathfrak{p}^*)$. We have then: $\zeta^* f(\alpha) \equiv 0(\mathfrak{C})$, whence $\zeta^* f(\alpha) \cdot \eta^* \in \mathfrak{o}'$, *for any element* η^* *in* $\mathfrak{o}^*$. Hence we may write

$$(7) \qquad \zeta^* \eta^* \cdot f(\alpha) = G(\alpha) = \omega_0 \alpha^\nu + \omega_1 \alpha^{\nu-1} + \cdots + \omega_\nu, \qquad \omega_i \in \mathfrak{o}.$$

We divide through $G(x)$ by $f(x)$:

$$G(x) = A(x) f(x) + R(x),$$

where all polynomials are in $\mathfrak{o}[x]$ and where $R(x)$ is of degree at most $m-1$, if m is the degree of $f(x)$. We now rewrite (7) as follows (notice that $f(\alpha) \neq 0$, since α is not in $\mathfrak{o}$ and since $\mathfrak{o}$ is integrally closed):

$$(8) \qquad \zeta^* \eta^* = A(\alpha) + \frac{R(\alpha)}{f(\alpha)} \cdot$$

Let v be an arbitrary valuation of Σ whose valuation ring R_v contains $\mathfrak{o}$. If $v(\alpha) \geqq 0$, then $\mathfrak{o}[\alpha] \in R_v$, and also $\mathfrak{o}^* \subset R_v$. Hence, by (8), we have $R(\alpha)/f(\alpha) \in R_v$. If $v(\alpha) < 0$, then $v(R(\alpha)) > v(f(\alpha))$, since R is of less degree than f and since $v(f(\alpha)) = mv(\alpha)$ (the leading coefficient of $f(\alpha)$ is an element of K). Hence also in this case $R(\alpha)/f(\alpha)$ is contained in R_v. Since this holds true for any valuation v such that $\mathfrak{o} \subset R_v$ and since $\mathfrak{o}$ is integrally closed, we conclude that $R(\alpha)/f(\alpha) \in \mathfrak{o}$. Hence, by (8), $\zeta^* \eta^* \in \mathfrak{o}'$, *for any element* η^* *in* $\mathfrak{o}^*$. Consequently $\zeta^* \in \mathfrak{C}$. Since ζ^* was an arbitrary element of $\mathfrak{C} : \mathfrak{p}^*$, it follows that $\mathfrak{C} : \mathfrak{p}^* = \mathfrak{C}$. This completes the proof of the lemma.

8. **Continuation of the proof of the main theorem.** To prove the main theorem, or better, the equivalent Theorem 14, we shall proceed as follows. We assume that we have the special case described in the preceding section under (a). Let $\mathfrak{p}^*$ be a prime ideal in $\mathfrak{o}^*$ and let $\mathfrak{p}^* \cap \mathfrak{o} = \mathfrak{p}$. We shall also assume

that we are dealing with the significant case $\mathfrak{n}\equiv 0(\mathfrak{p})$, $\mathfrak{z}\equiv 0(\mathfrak{p})$, in which case we have the result of Krull as stated in the preceding section under (b). We shall prove that if $\mathfrak{p}$ and $\mathfrak{p}^*$ have the same dimension, then $\mathfrak{p}^*$ contains properly another prime ideal $\mathfrak{p}_1^*$ with the property: $\mathfrak{p}_1^*\cap\mathfrak{o}=\mathfrak{p}$. This is the second alternative of Theorem 14.

We divide the proof into two parts, according as $\mathfrak{C}\not\equiv 0(\mathfrak{p}^*)$ or $\mathfrak{C}\equiv 0(\mathfrak{p}^*)$.

First case: $\mathfrak{C}\not\equiv 0(\mathfrak{p}^*)$. Let $\mathfrak{p}^*\cap\mathfrak{o}'=\mathfrak{p}_1'$. Since $\mathfrak{C}\not\equiv 0(\mathfrak{p}_1')$, it follows in an elementary fashion from the very definition of the conductor, that $\mathfrak{p}^*$ is the only prime ideal in $\mathfrak{o}^*$ which contracts to $\mathfrak{p}_1'$ and that the quotient rings $\mathfrak{o}^*_{\mathfrak{p}^*}$, $\mathfrak{o}'_{\mathfrak{p}'_1}$ coincide. We have $\mathfrak{p}_1'\cap\mathfrak{o}=\mathfrak{p}$, whence $\mathfrak{p}_1'$ must be a divisor of the prime ideal $\mathfrak{p}'=\mathfrak{o}'\cdot\mathfrak{p}$. It must be *a proper divisor* of $\mathfrak{p}'$, since $\mathfrak{p}_1'$ is of the same dimension as $\mathfrak{p}$, while by Krull's result $\mathfrak{p}'$ is of dimension one greater than $\mathfrak{p}$. Now since $\mathfrak{C}\not\equiv 0(\mathfrak{p}_1')$, we have *a fortiori* $\mathfrak{C}\not\equiv 0(\mathfrak{p}')$. Hence there is a unique prime ideal $\mathfrak{p}'^*$ in $\mathfrak{o}^*$ which contracts to $\mathfrak{p}'$ and we have $\mathfrak{o}^*_{\mathfrak{p}'^*}=\mathfrak{o}'_{\mathfrak{p}'}$. Since we have also $\mathfrak{o}^*_{\mathfrak{p}^*}=\mathfrak{o}'_{\mathfrak{p}'_1}$ and since $\mathfrak{p}_1'$ is a proper divisor of $\mathfrak{p}'$, it follows that also $\mathfrak{p}^*$ *is a proper divisor of* $\mathfrak{p}'^*$. Since $\mathfrak{p}'^*\cap\mathfrak{o}=\mathfrak{p}'\cap\mathfrak{o}=\mathfrak{p}$, our proof is complete.

Second case: $\mathfrak{C}\equiv 0(\mathfrak{p}^*)$. In this case $\mathfrak{p}^*$ is either a prime ideal of the conductor $\mathfrak{C}$ or properly contains a prime ideal of $\mathfrak{C}$. Since, by hypothesis, $\mathfrak{p}^*$ contracts in $\mathfrak{o}$ to an ideal $\mathfrak{p}$ of the same dimension as $\mathfrak{p}^*$, the first possibility is excluded by our lemma. Hence $\mathfrak{p}^*$ properly contains a prime ideal $\mathfrak{p}_1^*$ of $\mathfrak{C}$. Let $\mathfrak{p}_1^*\cap\mathfrak{o}=\mathfrak{p}_1$ and let

$$\text{dimension } \mathfrak{p} = s, \qquad \text{dimension } \mathfrak{p}_1 = s_1.$$

Since $\mathfrak{p}^*\supset\mathfrak{p}_1^*$, we have $\mathfrak{p}\supseteq\mathfrak{p}_1$. If $\mathfrak{p}_1=\mathfrak{p}$ then the theorem is proved, since we have now a proper multiple $\mathfrak{p}_1^*$ of $\mathfrak{p}^*$ which also contracts to $\mathfrak{p}$. We assume therefore that $\mathfrak{p}\supset\mathfrak{p}_1$, whence $s_1>s$. By our lemma, $\mathfrak{p}_1^*$ is of greater dimension than $\mathfrak{p}_1$; by Krull's theorem, the dimension of $\mathfrak{p}_1^*$ is at most one greater than the dimension of $\mathfrak{p}_1$. Hence the dimension of $\mathfrak{p}_1^*$ is s_1+1. Our proof would be complete if we could show that there exists a prime ideal in $\mathfrak{o}^*$, *between* $\mathfrak{p}^*$ and $\mathfrak{p}_1^*$ (and different from $\mathfrak{p}^*$) which contracts to $\mathfrak{p}$. This we proceed to show.

We pass to the residue class rings $\mathfrak{O}=\mathfrak{o}/\mathfrak{p}_1$, $\mathfrak{O}^*=\mathfrak{o}^*/\mathfrak{p}_1^*$. Both rings are finite integral domains and $\mathfrak{O}$ is a subring of $\mathfrak{O}^*$. The first ring is of degree of transcendency s_1, while $\mathfrak{O}^*$ is of degree of transcendency s_1+1. In the homomorphisms $\mathfrak{o}\sim\mathfrak{O}$, $\mathfrak{o}^*\sim\mathfrak{O}^*$, the prime ideals $\mathfrak{p}$ and $\mathfrak{p}^*$ are mapped, respectively, onto prime ideals $\mathfrak{P}$ and $\mathfrak{P}^*$, of the same dimension s, $s<s_1$, and we have: $\mathfrak{P}^*\cap\mathfrak{O}=\mathfrak{P}$. What we have to prove is the existence in $\mathfrak{O}^*$ of a prime ideal which is a proper multiple of $\mathfrak{P}^*$ and which contracts in $\mathfrak{O}$ to the ideal $\mathfrak{P}$. The assertion that such an ideal exists is equivalent to the assertion that $\mathfrak{P}^*$ *is not a minimal prime of the extended ideal* $\mathfrak{O}^*\cdot\mathfrak{P}$. To prove this we pass to the quotient rings $\mathfrak{J}=\mathfrak{O}_{\mathfrak{P}}$, $\mathfrak{J}^*=\mathfrak{O}^*_{\mathfrak{P}^*}$. If $\mathfrak{m}$ and $\mathfrak{m}^*$ denote the ideals of non-units in these two rings, then our problem is to prove that $\mathfrak{m}^*$ is not a minimal prime ideal of the extended ideal $\mathfrak{J}^*\cdot\mathfrak{m}$. The proof of this will complete the proof of the main theorem.

Since $\mathfrak{J}$ is of degree of transcendency s_1, we can find s_1-s elements $\zeta_1, \zeta_2, \cdots, \zeta_{s_1-s}$ in $\mathfrak{J}$ such that the ideal $\mathfrak{A}=\mathfrak{J}\cdot(\zeta_1, \zeta_2, \cdots, \zeta_{s_1-s})$ be exactly s-dimensional. Since the ideal of non-units $\mathfrak{m}$ in $\mathfrak{J}$ is also s-dimensional, it follows that $\mathfrak{A}$ will then be necessarily a primary ideal, with $\mathfrak{m}$ as associated prime. Now consider the ideal $\mathfrak{A}^*=\mathfrak{J}^*\cdot(\zeta_1, \zeta_2, \cdots, \zeta_{s_1-s})$. Since $\mathfrak{J}^*$ is of degree of transcendency s_1+1 and since $\mathfrak{A}^*$ is not the unit ideal (since $\zeta_i \in \mathfrak{A}\equiv 0(\mathfrak{m})$, whence $\mathfrak{A}^*\equiv 0(\mathfrak{m}^*)$), every minimal prime of $\mathfrak{A}^*$ is of dimension at least $s+1$. Let $\mathfrak{P}^*$ be a minimal prime of $\mathfrak{A}^*$. Since $\mathfrak{A}^*\equiv 0(\mathfrak{P}^*)$, we have $\mathfrak{P}^*\cap\mathfrak{J}\supseteq\mathfrak{A}$, whence $\mathfrak{P}^*\cap\mathfrak{J}=\mathfrak{m}$, for $\mathfrak{A}$ is primary and its associated prime is the ideal $\mathfrak{m}$ of non-units. Hence $\mathfrak{P}^*\supseteq\mathfrak{J}^*\mathfrak{m}$, and this shows that $\mathfrak{m}^*$ is not a minimal prime of $\mathfrak{J}^*\cdot\mathfrak{m}$, since $\mathfrak{P}^*$ is a proper multiple of $\mathfrak{m}^*$ (dimension $\mathfrak{m}^*=s$, dimension $\mathfrak{P}^*\geqq s+1$), q.e.d.

In the main theorem we have assumed that the birational correspondence has no fundamental elements on V'. In the general case of an arbitrary pair of birationally equivalent varieties V and V' we may apply the main theorem to V and to the join V^* of V and V'. If we then take into account Theorem **11** of II.5 we deduce the following corollary which expresses the local character of the main theorem:

COROLLARY. *If W is an irreducible subvariety of V at which V is locally normal and if W is fundamental for the birational correspondence T between V and some other variety V', then each irreducible component of $T[W]$ which is not fundamental for T^{-1} is of higher dimension than W.*

9. **The fundamental locus of a birational correspondence.** The forms $\phi_0(\eta), \phi_1(\eta), \cdots, \phi_m(\eta)$ which are proportional to the coordinates $\eta_0', \eta_1', \cdots, \eta_m'$ of the general point of V' (see equations (4), II.5) define a linear system of forms:

$$\phi_\lambda = \lambda_0\phi_0 + \lambda_1\phi_1 + \cdots + \lambda_m\phi_m.$$

We shall allow the parameters λ_i to take arbitrary values (not all zero) in the relative algebraic closure K$'$ of K in Σ, that is, the λ's shall be elements of Σ which are either in K or algebraic over K. We shall also assume that V and V' are locally normal varieties. Under this assumption it is permissible to identify K with K$'$, since any ring of nonhomogeneous coordinates of the general point of a locally normal variety is integrally closed and consequently contains K$'$. We therefore assume that K itself is algebraically closed in Σ.

The principal ideal (ϕ_λ) in the ring K $[\eta_0, \eta_1, \cdots, \eta_m]$ is $(r-1)$-dimensional. Since V is locally normal, the conductor of this ring with respect to its integral closure is a primary irrelevant ideal (or the unit ideal). Hence, to within an irrelevant component which we shall disregard, the ideal (ϕ_λ) is quasi-gleich to a product of symbolic powers of minimal prime (homogeneous) ideals. In particular, let

(9) $$(\phi_i) = \mathfrak{M}\mathfrak{A}_i, \qquad i = 0, 1, \cdots, m,$$

where $\mathfrak{A}_0, \mathfrak{A}_1, \cdots, \mathfrak{A}_m$ have no common factor. Then $\mathfrak{M}$ will be the h.c.d. of all principal ideals (ϕ_λ), that is, we will have: $(\phi_\lambda) = \mathfrak{M}\cdot\mathfrak{A}_{(\lambda)}$. The ideal $\mathfrak{A}_{(\lambda)}$ defines a pure $(r-1)$-dimensional subvariety $\mathfrak{C}_{(\lambda)}$ of V, which may be reducible and in which each irreducible component is counted to a definite multiplicity (equal to the exponent of the corresponding prime factor of $\mathfrak{A}_{(\lambda)}$). As the λ's vary in K, the variety $C_{(\lambda)}$ varies and describes a *linear system* $|C|$ *of* $(r-1)$-*dimensional varieties on* V, free from fixed components since $\mathfrak{A}_0, \mathfrak{A}_1, \cdots, \mathfrak{A}_m$ have no common factor. We have in particular the members $C_0, C_1, \cdots, C_m$ of $|C|$ which correspond to the ideals $\mathfrak{A}_0, \mathfrak{A}_1, \cdots, \mathfrak{A}_m$.

Let F be the algebraic subvariety of V defined by the ideal $\mathfrak{F} = (\mathfrak{A}_0, \mathfrak{A}_1, \cdots, \mathfrak{A}_m)$, or rather, by the radical of this ideal. The variety F is of dimension at most $r-2$ and is common to $C_0, C_1, \cdots, C_m$. We show that *F is the base manifold of the linear system* $|C|$, that is, that *F lies on each* $C_{(\lambda)}$. (This is not obvious, because of the presence of the factor $\mathfrak{M}$.)

Let $\mathfrak{p}$ be a minimal ideal of $\mathfrak{F}$ and let us show that the assumption $\mathfrak{A}_{(\lambda)} \not\equiv 0(\mathfrak{p})$ leads to a contradiction. Let β be an element of $\mathfrak{A}_{(\lambda)}$ not in $\mathfrak{p}$. Since $\phi_i \equiv 0(\mathfrak{M})$, it follows[30], by (9), that $\beta\phi_i \equiv 0(\phi_{(\lambda)})$, $i = 0, 1, \cdots, m$. Let $\beta\phi_i = \beta_i\phi_{(\lambda)}$, $\beta_i \in K[\eta]$. This relation can be written as follows: $\beta\mathfrak{A}_i = \beta_i\mathfrak{A}_{(\lambda)}$. By hypothesis $\mathfrak{A}_i \equiv 0(\mathfrak{p})$, but $\mathfrak{A}_{(\lambda)} \not\equiv 0(\mathfrak{p})$. Hence $\beta_i \equiv 0(\mathfrak{p})$, and therefore the relations $\beta\phi_i = \beta_i\phi_{(\lambda)}$ yield relations of the form

$$\beta\phi_i = \beta_{i0}\phi_0 + \beta_{i1}\phi_1 + \cdots + \beta_{im}\phi_m, \qquad i = 0, 1, \cdots, m,$$

where $\beta_{ij} \equiv 0(\mathfrak{p})$. From this we conclude that the determinant $\Delta = |\beta_{ij} - \delta_{ij}\beta|$ vanishes ($\delta_{ij} = 0$ if $i \neq j$, $\delta_{ii} = 1$), and this is impossible since $\Delta \equiv \pm\beta^m(\mathfrak{p})$, whence $\Delta \not\equiv 0(\mathfrak{p})$.

THEOREM 15. *The base manifold F of the linear system $|C|$ is also the fundamental locus of the birational transformation T, that is, F has the property that any irreducible subvariety W of V which is fundamental for T lies on F, and conversely.*

Proof. Let $\mathfrak{p}$ be the homogeneous prime ideal in the ring $K[\eta]$ which corresponds to W.

Assume $W \not\subseteq F$. Then at least one of the $m+1$ varieties C_i does not contain F. Let, say, $F \not\subseteq C_0$, whence $\mathfrak{A}_0 \not\equiv 0(\mathfrak{p})$. We introduce the nonhomogeneous coordinates $\xi_i' = \eta_i'/\eta_0'$ of the general point of V' and we denote, as usual,

[30] Strictly speaking, the congruences $\beta \equiv 0(\mathfrak{A}_{(\lambda)})$, $\phi_i \equiv 0(\mathfrak{M})$ do not necessarily imply that $\beta\phi_i$ is a multiple of $\phi_{(\lambda)}$, since the equation (9) is only true to within an irrelevant component. However, in view of the regularity and the (1, 1) character of the birational correspondence between a locally normal variety and its derived normal variety, it is permissible to give the proof under the assumption that V and V' are not only locally normal but normal. If V is normal, then the equation (9) is exact. The same remark applies to the other theorems proved in this section.

the ring $K[\xi_1', \xi_2', \cdots, \xi_m']$ by $\mathfrak{o}'$. We have $(\xi_i') = \mathfrak{A}_i/\mathfrak{A}_0$, whence $\xi_i' \in Q(W)$. Thus the entire ring $\mathfrak{o}'$ is contained in the quotient ring $Q(W)$, and this implies that *W is not fundamental.*

Conversely, *assume that W is not fundamental.* There corresponds then to W a unique variety W' on V' and we have $Q(W') \subseteq Q(W)$. We may assume that $\eta_0' \neq 0$ on W', and then we will have $\mathfrak{o}' \subset Q(W')$, whence *a fortiori* $\mathfrak{o}' \subset Q(W)$. The fact that the quotients ϕ_i/ϕ_0 all belong to $Q(W)$ leads immediately to the conclusion that $\mathfrak{A}_0 \not\equiv 0(\mathfrak{p})$. *Hence W does not lie on F*, q.e.d.

While Theorem 15 gives full information about the location of the fundamental elements of a birational correspondence, the following theorem tells us where the *irregular* varieties are located:

THEOREM 16. *If an irreducible subvariety W' of V' is irregular for T^{-1} then W' lies on the total transform $T\{F\}$ of the fundamental locus F of T. Conversely, if W' lies on $T\{F\}$, then it is either irregular or fundamental for T^{-1}.*

The proof is immediate. For if W' is irregular, and if $W = T^{-1}(W')$, then W is fundamental (Theorem 8 (C), II.3), whence $W \subseteq F$. Consequently, $W' = T(W) \subseteq T\{F\}$. Conversely, if $W' \subseteq T\{F\}$, then W' must correspond to some irreducible subvariety W of F. Since W is fundamental, W' cannot be regular, q.e.d.

The linear system $|C|$ which is defined by the linear family of forms $\phi_{(\lambda)}$ has always played an important part in the study of the birational correspondence T with which this system is associated. Theorem 15 is one illustration of the geometric connection between $|C|$ and T. Another property of the linear system $|C|$ which follows in a straightforward fashion from the definition is the following: *the birational correspondence T transforms the linear system $|C|$ into the system of hyperplane sections of V'.* This statement should be intended in the following sense: *a general* member $C_{(\lambda)}$ of $|C|$ is regular and $T(C_{(\lambda)})$ is the section $\Gamma'_{(\lambda)}$ of V' with the hyperplane $\lambda_0 y_0' + \lambda_1 y_1' + \cdots + \lambda_m y_m' = 0$. However, for special values of the λ's, it may very well happen that $C_{(\lambda)}$, or $\Gamma'_{(\lambda)}$, or both, contain irreducible components which are irregular and which therefore correspond to fundamental varieties.

10. **Isolated fundamental varieties.** We assume as before that V is locally normal and we keep the notation V^* for the join of V and V', and T^* for the birational correspondence between V and V^*.

DEFINITION 8. *Let $W_1^*, W_2^*, \cdots, W_h^*$ be the irreducible components of the total transform $T^*\{F\}$, where F is the fundamental locus of T^* (and hence also of T) on V. The irreducible subvarieties $F_i = T^{*-1}(W_i^*)$ of V are called the isolated fundamental varieties of T (and also of T^*).*

It is clear that the isolated fundamental varieties F_i lie on the fundamental locus F. It is also not difficult to see that *the irreducible components of F are among the isolated fundamental varieties.* For let W be an irreducible component of F and let W^* be an irreducible component of the transform $T^*[W]$. We have $W^* \subseteq T^*\{F\}$, since $T^{*-1}(W^*) = W \subseteq F$. Consequently W^* lies on one of the varieties W_i^*, say $W^* \subseteq W_1^*$. We have then $T^{*-1}(W^*) \subseteq T^{*-1}(W_1^*) = F_1$, that is, $W \subseteq F_1$. Since W is a component of F, we conclude that $F_1 = W$.

It is important to point out that in addition to the irreducible components of the fundamental locus F there may exist other "embedded" isolated fundamental varieties, which are proper subvarieties of the irreducible components of F. Thus in the three-dimensional case we may have a fundamental curve Γ on V to which there corresponds a surface on V^*, and on that fundamental curve Γ there may exist some special point P to which there also corresponds a surface on V^*. This point P must be regarded as an isolated fundamental point, although it is embedded in the fundamental curve Γ. The term "isolated" refers not to the position of the point P with respect to the fundamental locus F but to its role in the birational correspondence T.

By the main theorem each irreducible component W_i^* of $T^*\{F\}$ is of higher dimension than the corresponding isolated fundamental variety F_i. Under certain conditions it is possible to assert that *each W_i^* is of dimension $r-1$*. We proceed to find such conditions.

Let η_{ij} denote as usual the homogeneous coordinates of the general point of the join V^*, where $\eta_{ij} = \eta_i \phi_j$ (see II.5 and II.9). Let us consider quite generally an arbitrary homogeneous ideal $(g_0, g_1, \cdots, g_h)$ in the ring $K[\eta_0, \eta_1, \cdots, \eta_n]$, where each g_i is a form, say of degree ν_i. We put

$$(10) \qquad g_{ij} = g_i(\eta_{0j}, \eta_{1j}, \cdots, \eta_{nj}), \qquad i = 0, 1, \cdots, h; j = 0, 1, \cdots, m.$$

The $(m+1)(h+1)$ forms g_{ij} in the η_{ij}'s generate a homogeneous ideal in the ring $K[\eta_{00}, \eta_{10}, \cdots, \eta_{nm}]$.

Lemma 7. *If N is the subvariety of V defined by the ideal $(g_0, g_1, \cdots, g_h)$ and if N^* is the subvariety of V^* defined by the ideal $(g_{00}, \cdots, g_{hm})$, then N^* is the total transform of N, that is, $N^* = T^*\{N\}$.*

Proof. Let W and W^* be two corresponding irreducible subvarieties of V and V^*, respectively. We have to show that $W^* \subseteq N^*$, if and only if $W \subseteq N$.

Assume that $W \subseteq N$. Without loss of generality we may assume that $\eta_0 \neq 0$ on W. Then if v denotes a valuation of centers W and W^*, we will have:

$$(11) \qquad v(\eta_i/\eta_0) \geqq 0, \qquad v(g_i/\eta_0^{\nu_i}) > 0.$$

Also without loss of generality we may assume that $v(\phi_i/\phi_0) \geqq 0$, for $i = 1, 2, \cdots, m$. We will have then $v(\eta_{ij}/\eta_{00}) = v(\eta_i/\eta_0) + v(\phi_j/\phi_0) \geqq 0$. By (10), we can write:

$$(12) \qquad g_{ij} = g_i \phi_j^{\nu_i},$$

and hence $g_{ij}/\eta_{00}^{\nu_i} = g_i/\eta_0^{\nu_i} \cdot (\phi_j/\phi_0)^{\nu_i}$. Consequently $v(g_{ij}/\eta_{00}^{\nu_i}) > 0$, by (11), and this shows that $W^* \subseteq N^*$.

Conversely, assume that $W^* \subseteq N^*$. Then if $\eta_{00} \neq 0$ on W^*, we find that $\eta_0 \neq 0$ on W (since $\eta_i/\eta_0 = \eta_{i0}/\eta_{00}$). On the other hand we have: $g_{i0}/\eta_{00}^{\nu_i} = g_i/\eta_0^{\nu_i}$, whence $v(g_i/\eta_0^{\nu_i}) > 0$, that is, $W \subseteq N$, q.e.d.

We now apply the lemma to the case in which N is given by the ideal $(\phi_0, \phi_1, \cdots, \phi_m)$. Then N^* is given by the ideal $(\phi_{00}, \phi_{10}, \cdots, \phi_{mm})$, where

$$\phi_{ij} = \phi_i(\eta_{0j}, \eta_{1j}, \cdots, \eta_{nj}).$$

The relations (12) now yield: $\phi_{ij} = \phi_i \phi_j^{\nu}$, where ν is the common degree of the form ϕ_i. From these relations we deduce the following: $\phi_{ij}^{\nu+1} = \phi_{ii}\phi_{jj}^{\nu}$, and consequently the two ideals $(\phi_{00}, \phi_{10}, \cdots, \phi_{mm})$ and $(\phi_{00}, \phi_{11}, \cdots, \phi_{mm})$ have the same radical. Therefore N^* is also defined by the ideal $(\phi_{00}, \phi_{11}, \cdots, \phi_{mm})$. Now we have $\phi_{ii}/\phi_{jj} = (\phi_i/\phi_j)^{\nu+1} = \eta_{ki}^{\nu+1}/\eta_{kj}^{\nu+1}$, for any $i, j = 0, 1, \cdots, m$; $k = 0, 1, \cdots, n$. Therefore each irreducible component of N^* at which $\eta_{kj} \neq 0$, for some k and j, is also a component of the principal ideal (ϕ_{jj}) and is therefore $(r-1)$-dimensional. *Consequently N^* is pure $(r-1)$-dimensional.*

In view of the formulas (9) of II.9, the variety of the ideal $(\phi_0, \phi_1, \cdots, \phi_m)$ consists of the $(r-1)$-dimensional variety of the ideal $\mathfrak{M}$ and of the fundamental locus F. We therefore can assert that *$T^*\{F\}$ is pure $(r-1)$-dimensional if $\mathfrak{M}$ is the unit ideal.* The hypothesis $\mathfrak{M} = (1)$ implies that *each member $C_{(\lambda)}$ of the linear system $|C|$ associated with the birational correspondence is complete intersection of V with a hypersurface of the ambient projective space,* namely with the hypersurface $\lambda_0\phi_0(y_0, y_1, \cdots, y_n) + \lambda_1\phi_1(y_0, y_1, \cdots, y_n) + \cdots + \lambda_m\phi_m(y_0, y_1, \cdots, y_n) = 0$.

Conversely, let us assume that each $C_{(\lambda)}$ is *complete intersection*. Then in particular C_0 is complete intersection, whence the ideal $\mathfrak{A}_0$ is a principal ideal, say $\mathfrak{A}_0 = (\psi_0)$. We have: $\phi_i\psi_0/\phi_0 = \mathfrak{M}\mathfrak{A}_i \cdot \mathfrak{A}_0/\mathfrak{M}\mathfrak{A}_0 = \mathfrak{A}_i$, that is, $\phi_i\psi_0/\phi_0$ is an integral ideal. Consequently the quotients $\phi_i\psi_0/\phi_0$ are forms in the η's, say $\phi_i\psi_0/\phi_0 = \psi_i$. The forms $\psi_0, \psi_1, \cdots, \psi_m$ are proportional to the forms $\phi_0, \phi_1, \cdots, \phi_m$, and the linear system $|C|$ is also defined by the linear family of forms $\lambda_0\psi_0 + \lambda_1\psi_1 + \cdots + \lambda_m\psi_m$. If we use the ψ's instead of the ϕ's, we will have $\mathfrak{M} = (1)$, and we reach again the conclusion that $T^*\{F\}$ is pure $(r-1)$-dimensional.

We can go a step further. Let us point out that if we define a projective model V'_ρ by the condition that the homogeneous coordinates of its general point be given by a linear base of the forms of degree ρ in $\eta'_0, \eta'_1, \cdots, \eta'_m$ then V' and V'_ρ are in regular birational correspondence (Lemma 5, II.2). The transition from V' to V'_ρ is equivalent to passing from the linear system $|C|$ to the least linear system which contains as members all sets of ρ C's. Hence, by the preceding result, we conclude that *if a sufficiently high multiple of a C is complete intersection, then $T^*\{F\}$ is pure $(r-1)$-dimensional.*

In order to conclude with a similar result of a local character, let F_1 be any isolated fundamental variety of T, and let us assume that C is *locally, at F_1,* complete intersection. By that we mean that some hypersurface cuts V

along C and along a residual variety *which does not contain* F_1. If we replace the ϕ's by a suitable set of proportional forms, we may arrange matters so that the variety of the ideal $\mathfrak{M}$ does not contain F_1. Since $N=M+F$, it is clear that $T^*\{N\}=T^*[M]+T^*\{F\}$, where we write $T^*[M]$ instead of $T^*\{M\}$, since $T^*\{M\}-T^*[M]$ lies on $T^*\{F\}$ (Theorem 13, II.6). Since $F_1\not\subseteq M$, no component of $T^*[F_1]$ can lie on $T^*[M]$. It follows that the irreducible components of $T^*\{F\}$ which correspond to the isolated fundamental variety F_1 are also components of $T^*\{N\}$, and hence are $(r-1)$-dimensional. The same conclusion is reached if we assume that some sufficiently high multiple of C is complete intersection locally at F_1.

The above results refer to V and to the join of V and V'. In particular, if the birational correspondence T has no fundamental elements on V' then V' may play the role of V^*, since V' and V^* are then in regular birational correspondence. We reassume our results in the following theorem:

THEOREM 17. *If a birational correspondence T between two locally normal r-dimensional varieties V and V' has no fundamental elements on V' and if F denotes the fundamental locus of T on V, then an irreducible component of $T\{F\}$ is of dimension $r-1$, provided the corresponding isolated fundamental variety F_1 has the property that the members of the linear system $|C|$ associated with T, or their sufficiently high multiples, are complete intersections locally, at F_1.*

COROLLARY. *To an isolated simple fundamental variety there always corresponds an $(r-1)$-dimensional variety on V'* (see van der Waerden [6, p. 154]).

For locally, at a simple subvariety of V, every $(r-1)$-dimensional subvariety of V is complete intersection[31].

11. **Monoidal transformations.** Given a homogeneous ideal $\mathfrak{A}$ in the ring $K[\eta_0, \eta_1, \cdots, \eta_n]$ of homogeneous coordinates of the general point of V, it is possible to associate with $\mathfrak{A}$ an infinite set S of birational transforms of V such that: (1) the birational correspondence between V and any variety V' of the set has no fundamental elements on V' and such that (2) any two varieties of the set are in regular birational correspondence. The varieties V' of the set S shall be defined as follows. Let us take a base of $\mathfrak{A}$ consisting of forms of least possible degrees, and let a be the highest degree of the forms in that base. We define V' by its general point $(\phi_0, \phi_1, \cdots, \phi_m)$, where the ϕ's form a linear base for the forms of a given degree ν in $\mathfrak{A}$ and where we impose on ν the condition: $\nu\geqq a+1$. For $\nu=a+1, a+2, \cdots$, we get an infinite set of models V', and this is our set S. We shall denote these models by $V'_{a+1}, V'_{a+2}, \cdots$.

(31) For the case of algebraically closed ground fields of characteristic zero see our paper [8, p. 664]. There the proof is given explicitly for surfaces only, but actually exactly the same proof applies to higher varieties. For ground fields which are not algebraically closed or which are of characteristic p, the statement can be derived from the following result obtained by Irvin Cohen in his dissertation: if the characteristic of a complete p-series ring coincides with the characteristic of its residue field, the ring is a power series ring over a field.

First of all it is clear that each V'_ν *is birationally equivalent to* V. For $\mathfrak{A}$ contains at least one form ψ of degree $\nu-1$ so that the products $\eta_0\psi, \eta_1\psi, \cdots, \eta_n\psi$ can be identified with $n+1$ of the ϕ's. This shows that the quotients ϕ_i/ϕ_0 generate the field Σ.

Since $\mathfrak{A}$ has a basis consisting of forms of degree at most a, it follows that if $(\phi_0, \phi_1, \cdots, \phi_m)$ is a basis for the forms in $\mathfrak{A}$ of degree ν, then the products $\eta_i\phi_j$ constitute a basis for the forms in $\mathfrak{A}$ which are of degree $\nu+1$. This holds true also for $\nu=a$. From this it follows that $V_{\nu+1}$ is the join of V and V_ν, provided $\nu \geqq a+1$. Therefore the birational correspondence between V and V_ν has no fundamental points on V_ν, provided $\nu \geqq a+2$. But then $V_{\nu+1}$, *the join of* V *and* V_ν, *is a regular birational transform of* V_ν, always provided that $\nu \geqq a+2$. As for the case $\nu=a+1$, we can still regard V_{a+1} as the join of V and V_a, although in this case V_a need not be birationally equivalent to V. At any rate, the proof that the birational correspondence between V and V_{a+1} has no fundamental elements on V_{a+1} is exactly the same as that given for the join in II.5.

Thus we may say that a given homogeneous ideal $\mathfrak{A}$ in the ring $K[\eta_0, \eta_1, \cdots, \eta_n]$ determines, to within a regular birational transformation, a birational transform V' of V such that the birational correspondence between V and V' has no fundamental elements on V'.

Let N be the subvariety of V defined by the ideal $\mathfrak{A}$. It is quite clear that if $\nu \geqq a$, then the ideal generated by our base $(\phi_0, \phi_1, \cdots, \phi_m)$ differs from $\mathfrak{A}$ only by an irrelevant component. Hence, by Theorem 15 of II.9, we conclude that *if* N *is of the dimension at most* $r-2$, *then* N *is the fundamental locus* F *of the birational correspondence between* V *and* V'. In particular, if N *is empty*, that is, if $\mathfrak{A}$ *is an irrelevant ideal, then* V' *is a regular transform of* V.

If, however, N is of dimension $r-1$, then the fundamental locus F will consist of the irreducible components of N which are of dimension less than $r-1$ and possibly of some proper subvarieties of the $(r-1)$-dimensional components of N. Thus, even in the case in which N is pure $(r-1)$-dimensional, it may very well happen that F is not empty. According to Theorem 15, this will happen if the residual intersections of the hypersurfaces $\phi_i=0$ $(i=0, 1, \cdots, m)$ with V, outside N, have a base manifold on N.

Let now $\mathfrak{A}_1$ be another homogeneous ideal in the ring $K[\eta_0, \eta_1, \cdots, \eta_n]$. If W is an irreducible subvariety of V given by a prime ideal $\mathfrak{p}$, we shall say that $\mathfrak{A}$ and $\mathfrak{A}_1$ *coincide locally* at W, if the two ideals differ only by primary components whose associated prime ideals are not multiples of $\mathfrak{p}$. In other words, $\mathfrak{A}$ and $\mathfrak{A}_1$ coincide locally at W if they give rise to one and the same ideal in the quotient ring of W (see I.1).

LEMMA 8. *If* $\mathfrak{A}$ *and* $\mathfrak{A}_1$ *coincide locally at an irreducible subvariety* W *of* V *and if* V' *and* V'_1 *are the birational transforms of* V *which are determined (to within a regular birational transformation), respectively, by* $\mathfrak{A}$ *and by* $\mathfrak{A}_1$, *then* *any irreducible subvariety of V' which corresponds to W is regular for the birational correspondence between V' and V_1'.*

Proof. Let $\mathfrak{A}^*$ be the ideal which is obtained from either $\mathfrak{A}$ or $\mathfrak{A}_1$ by the omission of all primary components whose associated prime ideals are not multiples of $\mathfrak{p}$. Let V^* be the birational transform of V determined by the ideal $\mathfrak{A}^*$. It is sufficient to prove the lemma for $\mathfrak{A}$ and $\mathfrak{A}^*$, and for $\mathfrak{A}_1$ and $\mathfrak{A}^*$. We shall prove it, for instance, for $\mathfrak{A}$ and $\mathfrak{A}^*$.

Let W' be an irreducible subvariety of V' which corresponds to W. We have to prove that W' is regular for the birational correspondence between V' and V^*. Let $\phi_0, \phi_1, \cdots, \phi_m$ be a linear base of the forms of degree ν which belong to $\mathfrak{A}$. Since $\mathfrak{A}\subseteq\mathfrak{A}^*$, we may complete this base to a linear base $\phi_0, \phi_1, \cdots, \phi_m, \phi_{m+1}, \cdots, \phi_{m+\mu}$ for the forms of degree ν which belong to the ideal $\mathfrak{A}^*$. We take ν sufficiently high, so that $(\phi_0, \phi_1, \cdots, \phi_m)$ and $(\phi_0, \phi_1, \cdots, \phi_m, \phi_{m+1}, \cdots, \phi_{m+\mu})$ are the general points, respectively, of V' and V^*.

Let v be any valuation whose center on V' is W' and whose center on V is W. We may assume that $v(\phi_i/\phi_0)\geqq 0$, $i=1, 2, \cdots, m$ whence W' is at finite distance with respect to the nonhomogeneous coordinates $\xi_i' = \phi_i/\phi_0$, $i=1, 2, \cdots, m$, of the general point of V'. Let W^* be the center of v on V^*. By our definition of the ideal $\mathfrak{A}^*$, there exists a form $g(\eta_0, \eta_1, \cdots, \eta_n)$ such that $g\cdot\mathfrak{A}^*\equiv 0(\mathfrak{A})$ and such that $g\neq 0$ on W. We have then, in particular: $g\phi_{m+j}=A_0\phi_0+A_1\phi_1+ \cdots +A_m\phi_m$, $j=1, 2, \cdots, \mu$, where $A_0, A_1, \cdots, A_m$ are forms in $\eta_0, \eta_1, \cdots, \eta_n$, of the same degree as g. We write:

$$\phi_{m+j}/\phi_0 = \frac{A_0}{g} + \frac{A_1}{g}\cdot\frac{\phi_1}{\phi_0} + \cdots + \frac{A_m}{g}\cdot\frac{\phi_m}{\phi_0}. \tag{13}$$

Since $g\neq 0$ on W we have $v(A_i/g)\geqq 0$, $i=0, 1, \cdots, m$. Since also $v(\phi_i/\phi_0)\geqq 0$, it follows from the above relation (13) that $v(\phi_{m+j}/\phi_0)\geqq 0$. Hence W^* is at finite distance with respect to the nonhomogeneous coordinates $\xi_1', \xi_2', \cdots, \xi_{m+\mu}'$ of the general point of V^*, where $\xi_i' = \phi_i/\phi_0$. Since the ring $K[\xi_1', \xi_2', \cdots, \xi_m']$ is a subring of the ring $K[\xi_1', \xi_2', \cdots, \xi_{m+\mu}']$, it follows that $Q(W')\subseteq Q(W^*)$.

On the other hand, since $g\neq 0$ on W, the quotients A_i/g belong to the quotient ring $Q(W)$. Since $Q(W)\subseteq Q(W')$ and since also the quotients ϕ_i/ϕ_0, $i=1, 2, \cdots, m$, are in $Q(W')$, we conclude, by (13), that the entire ring $K[\xi_1', \xi_2', \cdots, \xi_{m+\mu}']$ is contained in $Q(W')$. From this it follows immediately that $Q(W^*)\subseteq Q(W')$, whence $Q(W^*)=Q(W')$, q.e.d.

COROLLARY. *If $\mathfrak{A}$ and $\mathfrak{A}_1$ differ only by an irrelevant primary component, then V' and V_1' are in regular birational correspondence.*

From the above general consideration we pass to the special case which interests us, namely to the case in which the given homogeneous ideal $\mathfrak{A}$ is a prime ideal $\mathfrak{p}$, of dimension s, $0 \leqq s \leqq r-2$. Let W be the irreducible subvariety of V defined by $\mathfrak{p}$. The birational transformation T determined by the ideal $\mathfrak{p}$ (that is, by a linear base of forms of sufficiently high degree in $\mathfrak{p}$) is called a *monoidal transformation of center* W. In the special case when W is a point P the transformation is called *quadratic* (of center P). The birational transform V' of V, under a monoidal transformation of given center, is determined to within a regular birational correspondence. The center W of a monoidal transformation is the fundamental locus of the transformation. Moreover, from Theorem 17, II.10, it follows that in the present case $T\{W\}$ is pure $(r-1)$-dimensional. However, it should be pointed out that $T\{W\}$ may very well be reducible and—this is significant—*some components of* $T\{W\}$ *may correspond to proper subvarieties of* W. In other words, *the center* W *of a monoidal transformation is not necessarily the only isolated fundamental variety of the transformation*(32). We shall see presently that this complication arises only if W carries some singular points of V or if W itself has singularities.

Of special importance in applications are monoidal transformations with *simple center*, that is, with center at a simple subvariety W of V. The special case of à quadratic transformation with simple center has been considered in our paper [11]. The results established there carry over to monoidal transformations with simple center, in view of the following considerations. Let the ground field K be extended by the adjunction of s elements of $Q(W)$ which are algebraically independent on W. With respect to this new ground field K_1, the variety W becomes a (simple) point and the monoidal transformation T becomes a quadratic transformation. Therefore certain properties of the monoidal transformation T, over K, can be deduced from corresponding properties of the quadratic transformation over K_1. However, only such properties of T can be deduced in this fashion as concern *W as a whole*. What happens to special points or special subvarieties of W requires new considerations. For instance, we have proved in the quoted paper [11] that if T is a quadratic transformation with simple center P, then the transform $T[P]$ (which, since P is a point, coincides with the total transform $T\{P\}$) is an irreducible, simple and $(r-1)$-dimensional subvariety of V' and, moreover, that every

(32) Here is an example. Let V be the quadric hypersurface $u^2=yz$ in the 4-dimensional space of the variables x, y, z, u. This hypersurface has the double line $y=z=u=0$. Let W be the line $x=y=u=0$. As nonhomogeneous coordinates of the monoidal transform V' of V we can take the elements $x, y, z, u, x/u, y/u$. Let $\mathfrak{o}'$ be the ring of these coordinates and let $\mathfrak{o}=K[x, y, z, u]$. We have $\mathfrak{o}'=\mathfrak{o}[x/u, y/u]=K[x_1, y_1, z]$, where $x_1=x/u$, $y_1=y/u$. Here $\mathfrak{p}=(x, y, u)$ is the prime ideal of W and we have $\mathfrak{o}'\cdot\mathfrak{p}=\mathfrak{o}'\cdot u=\mathfrak{o}'\cdot y_1 z$, that is, that part of $T\{W\}$ which is at finite distance consists of two planes: $y_1=0$ and $z=0$ (note that the *affine* model V' is in regular birational correspondence with the affine space of the variables x_1, y_1, z). The first plane corresponds to W (since $\mathfrak{o}'\cdot y_1 \cap \mathfrak{o}=\mathfrak{p}$). But the plane $z=0$ corresponds to the point $x=y=z=u=0$. This point is imbedded in W, but according to our terminology must be regarded as an isolated fundamental point.

point of $T[P]$ is likewise simple for V'. Now when we pass from the ground field K to the ground field K_1, we lose all those components of $T\{W\}$ which cannot be regarded as varieties over K_1, that is, all those components of $T\{W\}$ which correspond to proper subvarieties of W (since on any proper subvariety of W the s elements which have been adjointed to K are algebraically dependent). Consequently, the correct extrapolation of the above result concerning quadratic transformations to monoidal transformations is the following:

THEOREM 18. *If the center W of a monoidal transformation T is a simple subvariety of V, then the transform*(33) $T[W]$ *of W is an irreducible, simple, $(r-1)$-dimensional subvariety of V', and every irreducible subvariety W' of $T[W]$ is also simple for V', provided $W'=T(W)$.*

The total transform $T\{W\}$ may possess components which are not components of $T[W]$ (even if W is simple(32)), and concerning those components we can assert nothing. Likewise $T[W]$ may contain points which are singular for V. Thus, if V is three-dimensional and if W is a curve, then $T[W]$ is a surface which may carry, in addition to a finite number of singular points of V, also a finite number of singular curves of V, but each such curve must correspond to a point of W.

The following theorem will show, among other things, that these complications can arise only from points or subvarieties of W which are singular for V or for W.

THEOREM 19. *Let W_1 be an irreducible subvariety of W, of dimension s_1. If W_1 is simple both for V and W, then $T[W_1]$ lies on $T[W]$, is irreducible, is of dimension $r-1-s+s_1$ and is simple both for V' and for $T[W]$. Moreover, every irreducible subvariety of $T[W_1]$ which corresponds to W_1 is likewise simple for V', $T[W]$ and also for $T[W_1]$.*

Proof. By the usual device of ground field extension we can achieve a reduction to the case $s_1=0$. Therefore we assume that W_1 is a point P of W, simple both for V and W. It is then possible to select uniformizing parameters $t_1, t_2, \cdots, t_r$ at P in such a fashion that W be locally, at P, complete intersection of the $r-s$ hypersurfaces(34) $t_1=0, t_2=0, \cdots, t_{r-s}=0$. Then

(33) Not the *total* transform $T\{W\}$!

(34) **Proof.** Quite generally, the uniformizing parameters $t_1, t_2, \cdots, t_{r-\rho}$ of a simple ρ-dimensional subvariety L of V have the following property: *if $g(t_1, t_2, \cdots, t_{r-\rho})=0$ is a true homogeneous relation between these parameters, with coefficients in the quotient ring $Q(L)$, then all these coefficients must be zero on L, that is, they are non-units of $Q(L)$.* (See [9, p. 202, (15) and p. 207, (23)].) In view of this property and also because $Q(L)$ is a chain theorem ring in which the non-units form an ideal, the quotient ring of a simple subvariety is a p-series ring (*p-Reihenring*) in the sense of Krull [3]. We shall therefore apply properties of p-series rings due to Krull.

Let $\mathfrak{J}$ denote the quotient ring of P and let $\mathfrak{m}$ be the ideal of non-units in $\mathfrak{J}$. If $\tau_1, \tau_2, \cdots, \tau_r$ are uniformizing parameters of P, then we have $\mathfrak{m}=\mathfrak{J}\cdot(\tau_1, \tau_2, \cdots, \tau_r)$. If α is any element of $\mathfrak{J}$ and if $\alpha\equiv 0(\mathfrak{m}^h)$, $\alpha\not\equiv 0(\mathfrak{m}^{h+1})$, then α can be written as a form $g_h(\tau_1, \tau_2, \cdots, \tau_r)$, of degree h,

$t_1, t_2, \cdots, t_{r-s}$ will be uniformizing parameters for W (that is, the ideal generated by $t_1, t_2, \cdots, t_{r-s}$ in $Q(W)$ will be the prime ideal of non-units), and the ideal generated by the same elements in $Q(P)$ will be the prime ideal of W in $Q(P)$.

The uniformizing parameters $t_1, t_2, \cdots, t_{r-s}$ of W are proportional to certain forms $\psi_1, \psi_2, \cdots, \psi_{r-s}$ in the homogeneous coordinates $\eta_0, \eta_1, \cdots, \eta_n$ of the general point of V. Since these uniformizing parameters belong to $Q(P)$, it follows that the factor of proportionality can be so selected that the ideal generated by the forms $(\psi_1, \psi_2, \cdots, \psi_{r-s})$ coincide locally at P with the prime ideal of W. Hence by Lemma 8 we can replace the transformation T by the birational transformation defined by the ideal(35) $(\psi_1, \psi_2, \cdots, \psi_{r-s})$. Therefore we may assume that T, instead of being our original monoidal transformation of center W, is the birational transformation which carries V into the variety V' whose general point is $(\eta_{01}, \eta_{11}, \cdots, \eta_{n,r-s})$, where(36) $\eta_{ij} = \eta_i\psi_j$.

Without loss of generality we may assume that the point P (and hence also W) is at finite distance with respect to the nonhomogeneous coordinates $\xi_i = \eta_i/\eta_0$, $i = 1, 2, \cdots, n$. Let Γ' denote an irreducible component of $T[P]$. For some value of h, $h = 1, 2, \cdots, r-s$, it will be true that Γ' is at finite distance with respect to the nonhomogeneous coordinates

$$\xi_{ij}^{(h)} = \eta_i\psi_j/\eta_0\psi_h$$

---

with coefficients in $\mathfrak{J}$. If the coefficients of this form are replaced by their residues mod $\mathfrak{m}$, one obtains a form in $\tau_1, \tau_2, \cdots, \tau_r$ with coefficients in the residue field $\mathfrak{J}/\mathfrak{m}$. The property of uniformizing parameters stated above implies that this form is uniquely determined by the element α. This form is called by Krull the *leading form* of α [3, p. 207].

It is a straightforward matter to show that r elements $t_1, t_2, \cdots, t_r$ are uniformizing parameters of P, that is, $\mathfrak{m} = \mathfrak{J}\cdot(t_1, t_2, \cdots, t_r)$, if and only if the leading forms of $t_1, t_2, \cdots, t_r$ are linear and linearly independent.

Let $\mathfrak{p}$ denote the prime ideal of W in $\mathfrak{J}$ and let $\mathfrak{J}^* = \mathfrak{J}/\mathfrak{p}$, $\mathfrak{m}^* = \mathfrak{m}/\mathfrak{p}$. Then $\mathfrak{J}^*$ is the quotient ring of the point P, *regarded as a point of* W, and $\mathfrak{m}^*$ is the ideal of non-units of $\mathfrak{J}^*$. Since, by hypothesis, P is a simple point of W, there exist s elements in $\mathfrak{J}^*$, say $t^*_{r-s+1}, t^*_{r-s+2}, \cdots, t^*_r$ such that $\mathfrak{m}^* = \mathfrak{J}^*\cdot(t^*_{r-s+1}, t^*_{r-s+2}, \cdots, t^*_r)$. Let $t_{r-s+1}, t_{r-s+2}, \cdots, t_r$ be elements of $\mathfrak{J}$ whose $\mathfrak{p}$-residues are $t^*_{r-s+1}, t^*_{r-s+2}, \cdots, t^*_r$, respectively. We will have then: $\mathfrak{m} = \mathfrak{J}\cdot(\mathfrak{p}, t_{r-s+1}, t_{r-s+2}, \cdots, t_r)$. From this relation we draw the following consequences. In the first place it follows that the ideal $\mathfrak{p}$ must contain $r-s$ elements whose leading forms are linear and linearly independent. Let $t_1, t_2, \cdots, t_{r-s}$ be such $r-s$ elements of $\mathfrak{p}$. If $\mathfrak{J}'$ is the perfect closure of $\mathfrak{J}$ (see Krull [3, p. 217]), then it is a straightforward matter to show that the ideal $\mathfrak{J}'\cdot(t_1, t_2, \cdots, t_{r-s})$ is prime. Therefore also the ideal $\mathfrak{J}\cdot(t_1, t_2, \cdots, t_{r-s})$ is prime, since it is the contraction of the ideal $\mathfrak{J}'\cdot(t_1, t_2, \cdots, t_{r-s})$ (Krull [3, Theorem 15]). Since the leading ideal of $\mathfrak{J}\cdot(t_1, t_2, \cdots, t_{r-s})$ is of dimension s, it follows (Krull [3, Theorem 8]) that also $\mathfrak{J}\cdot(t_1, t_2, \cdots, t_{r-s})$ is of dimension s. *Consequently this ideal coincides with* $\mathfrak{p}$. We have therefore: $\mathfrak{m} = \mathfrak{J}\cdot(t_1, t_2, \cdots, t_{r-s}, t_{r-s+1}, \cdots, t_r)$, $\mathfrak{J}\cdot\mathfrak{p} = \mathfrak{J}\cdot(t_1, t_2, \cdots, t_{r-s})$, q.e.d.

(35) Note that if two ideals coincide locally at some W they also coincide locally at any W_1 such that $W \subseteq W_1$.

(36) Since our new transformation behaves locally at P as the given monoidal transformation, we could refer to our new transformation as being *locally monoidal at P*.

of the general point of V'. Let $\mathfrak{o}_h'$ denote the ring of these nonhomogeneous coordinates, and let $\mathfrak{o}$ denote, as usual, the ring $K[\xi_1, \xi_2, \cdots, \xi_n]$. Since $\psi_j/\psi_h = t_j/t_h$, we find:

$$\mathfrak{o}_h' = \mathfrak{o}[t_1/t_h, t_2/t_h, \cdots, t_{r-s}/t_h]. \tag{14}$$

Without loss of generality we may assume that Γ' is at finite distance with respect to the nonhomogeneous coordinates $\xi_{ij}^{(1)}$. For simplicity we shall drop the subscript 1 in the symbol $\mathfrak{o}_1'$, that is, we shall use the symbol $\mathfrak{o}'$ to denote the ring $\mathfrak{o}_1'$.

Let $\mathfrak{p}_0$ denote the prime zero-dimensional $\mathfrak{o}$-ideal of the point P. We shall denote by $\mathfrak{p}$ the prime $\mathfrak{o}$-ideal of W. For clarity of exposition we divide our proof into several steps.

(1) We shall show first that *the ideal* $\mathfrak{o}'\cdot\mathfrak{p}_0$ *is prime*. Let $\mathfrak{J} = \mathfrak{o}_{\mathfrak{p}_0}$ denote the quotient ring of P and let $\mathfrak{J}' = \mathfrak{J}\cdot\mathfrak{o}'$. The ring $\mathfrak{J}'$ is a quotient ring of $\mathfrak{o}'$, namely $\mathfrak{J}' = \mathfrak{o}_S'$ where $S = \mathfrak{o} - \mathfrak{p}_0$. Therefore, in view of the relationship between the ideals in a ring and the ideals in its quotient ring, *the ideal* $\mathfrak{o}'\cdot\mathfrak{p}_0$ *is prime if and only if* $\mathfrak{J}'\cdot\mathfrak{p}_0$ *is prime*([37]). We prefer to deal with the ring $\mathfrak{J}'$ and to show that $\mathfrak{J}'\cdot\mathfrak{p}_0$ *is prime.*

Since $\mathfrak{J}\cdot\mathfrak{p}_0 = \mathfrak{J}\cdot(t_1, t_2, \cdots, t_r)$ and $t_i = t_1\cdot t_i/t_1$, $i = 2, 3, \cdots, r-s$, that is, $t_2, t_3, \cdots, t_{r-s}$ are multiples of t_1 in $\mathfrak{o}'$, it follows that

$$\mathfrak{J}'\cdot\mathfrak{p}_0 = \mathfrak{J}'\cdot(t_1, t_{r-s+1}, \cdots, t_r). \tag{15}$$

Any element α in $\mathfrak{J}'$ can be written in the form:

$$\alpha = \phi_\rho(t_1, t_2, \cdots, t_{r-s})/t_1^\rho,$$

where ϕ_ρ is a form of degree ρ in $t_1, t_2, \cdots, t_{r-s}$, with coefficients in $\mathfrak{J}$. Let β be another element in $\mathfrak{J}'$,

$$\beta = \psi_\sigma(t_1, t_2, \cdots, t_{r-s})/t_1^\sigma,$$

and let us assume that $\alpha\beta \equiv 0(\mathfrak{J}'\cdot\mathfrak{p}_0)$. We will have a relation of the form:

$$\begin{aligned}\phi_\rho(t_1, t_2, \cdots, t_{r-s})&\cdot\psi_\sigma(t_1, t_2, \cdots, t_{r-s})\cdot t_1^\mu \\ &= [t_1 f_\mu(t_1, t_2, \cdots, t_{r-s}) + t_{r-s+1} f_\mu^{(1)}(t_1, t_2, \cdots, t_{r-s}) + \cdots \\ &\qquad + t_r f_\mu^{(s)}(t_1, t_2, \cdots, t_{r-s})]t_1^{\rho+\sigma},\end{aligned}$$

where $f_\mu, f_\mu^{(1)}, \cdots, f_\mu^{(s)}$ are forms of degree μ, with coefficients in $\mathfrak{J}$. The right-hand side of this relation is a form of degree $\rho+\sigma+\mu+1$ in $t_1, t_2, \cdots, t_r$, with

([37]) We have a (1, 1) isomorphic correspondence between the ideals in $\mathfrak{J}'$ and those ideals in $\mathfrak{o}'$ all prime ideals of which contract in $\mathfrak{o}$ to $\mathfrak{p}_0$ or to multiples of $\mathfrak{p}_0$. Now since $\mathfrak{p}_0$ is a maximal ideal, every prime ideal of $\mathfrak{o}'\cdot\mathfrak{p}_0$ contracts to $\mathfrak{p}_0$. Hence $\mathfrak{o}'\cdot\mathfrak{p}_0$ and $\mathfrak{J}'\cdot\mathfrak{p}_0$ are corresponding ideals in the above correspondence.

coefficients in $\mathfrak{J}$. Hence by a well known property of uniformizing parameters[34] either all the coefficients of ϕ_ρ or all the coefficients of ψ_σ must be elements of $\mathfrak{J}\cdot\mathfrak{p}_0$. Suppose that all the coefficients of ϕ_ρ are in $\mathfrak{J}\cdot\mathfrak{p}_0$. Since $\mathfrak{J}\cdot\mathfrak{p}_0=\mathfrak{J}\cdot(t_1, t_2, \cdots, t_r)$, we will have for ϕ_ρ an expression of the form: $\phi_\rho=t_1\phi_\rho^{(1)}(t_1,t_2,\cdots,t_{r-s})+t_2\phi_\rho^{(2)}(t_1,t_2,\cdots,t_{r-s})+\cdots+t_r\phi_\rho^{(r)}(t_1,t_2,\cdots,t_{r-s})$, where the ϕ_ρ^j are again forms of degree ρ, with coefficients in $\mathfrak{J}$. Since $t_2, \cdots, t_{r-s}$ are multiples of t_1 in $\mathfrak{J}'$, we conclude immediately that ϕ_ρ/t_1^ρ is contained in the ideal $\mathfrak{J}'(t_1, t_{r-s+1}, \cdots, t_r)$, that is, in view of (15), $\alpha\equiv 0(\mathfrak{J}'\cdot\mathfrak{p}_0)$. This shows that $\mathfrak{J}'\cdot\mathfrak{p}_0$ is a prime ideal, as was asserted[38].

(2) Let $\mathfrak{o}'\mathfrak{p}_0=\mathfrak{p}'$. *We assert* that *the* $\mathfrak{p}'$*-residues of* $t_2/t_1, \cdots, t_{r-s}/t_1$ *are algebraically independent* (over K). For a relation of algebraic dependence between these residues would imply a relation of the form:

$$\phi_\rho(t_1, t_2, \cdots, t_{r-s})/t_1^\rho = \sum_{i=1}^{r} t_i g_\sigma^{(i)}(t_1, t_2, \cdots, t_{r-s})/t_1^\sigma,$$

where the $g_\sigma^{(i)}$ and ϕ_ρ are forms in $t_1, t_2, \cdots, t_{r-s}$, with coefficients in $\mathfrak{o}$, and *where the coefficients of* ϕ_ρ *are not all in* $\mathfrak{p}_0$. Such a relation, cleared of the denominators, is in contradiction with the property of uniformizing parameters stated above[34].

From the fact that $\mathfrak{o}'\cdot\mathfrak{p}_0$ is prime, follows that Γ' *is the only irreducible component of* $T[P]$ *which is at finite distance with respect to the coordinates* $\xi_{ij}^{(1)}$.

The fact that the $\mathfrak{p}'$-residue of $t_2/t_1, \cdots, t_{r-s}/t_1$ are algebraically independent, in conjunction with the fact that $\mathfrak{p}'\cap\mathfrak{o}$ is the *zero-dimensional* ideal $\mathfrak{p}_0$, *implies that* Γ' *is of dimension* $r-s-1$. Moreover, the algebraic independence of the quotients $t_2/t_1, t_3/t_1, \cdots, t_{r-s}/t_1$ implies in particular that they do not belong to $\mathfrak{p}'$. Hence these quotients are units in the quotient ring $Q(\Gamma')$. But then also $t_i/t_h\in Q(\Gamma')$, for $i, h=1, 2, \cdots, r-s$, whence the rings $\mathfrak{o}_h'$ (see (14)) belong to $Q(\Gamma')$. This shows that Γ' is *at finite distance also with respect to the nonhomogeneous coordinates* $\xi_{ij}^{(h)}$, for $h=1, 2, \cdots, r-s$. Consequently, Γ' is the only irreducible component of $T[P]$, that is, $T[P]$ is irreducible: $T[P]=\Gamma'$.

(3) Let C' denote the irreducible $(r-1)$-dimensional variety $T[W]$. We are interested in the quotient rings of Γ' and of C'. On the basis of the preceding considerations we find immediately that every element of $Q(\Gamma')$ is of the form: $f_\rho(t_1, t_2, \cdots, t_{r-s})/g_\rho(t_1, t_2, \cdots, t_{r-s})$, where f_ρ and g_ρ are forms of like degree ρ, with coefficients in $\mathfrak{o}$, and where the coefficients of g_ρ are not all in $\mathfrak{p}_0$. Similarly, making use of the remark in footnote 38, or also directly from the properties of what we have called "p-adic divisor" in [11], we conclude that the elements of $Q(C')$ are all of the form $f_\rho(t_1, t_2, \cdots, t_{r-s})/g_\rho(t_1, t_2, \cdots, t_{r-s})$,

(38) Exactly the same proof could be applied toward proving the following: if $\mathfrak{p}$ is the prime deal of W in $\mathfrak{o}$ and if $\mathfrak{J}=Q(W)$, $\mathfrak{J}'=\mathfrak{J}\cdot\mathfrak{o}'$, then the ideal $\mathfrak{J}'\cdot\mathfrak{p}(=\mathfrak{J}'\cdot t_1)$ is prime. From this we could conclude that $T[W]$ is irreducible (as asserted in Theorem 18) in exactly the same fashion as we concluded in the text that $T[P]$ is irreducible.

with the same conditions on f_ρ and g_ρ as above, except that now the coefficients of g_ρ *must not all be* in $\mathfrak{p}$, where $\mathfrak{p}$ is the prime ideal of W. Since $P\subseteq W$, it follows that $Q(\Gamma')\subseteq Q(C')$, that is, $T[P]$ *lies on* $T[W]$ (that is, Γ' lies on C').

Moreover, from the preceding considerations (see relation (15)), it follows that the prime ideal of non-units in $Q(\Gamma')$ has the basis $t_1, t_{r-s+1}, \cdots, t_r$, consisting of $s+1$ elements. Since Γ' is of dimension $r-s-1$, it follows that Γ' *is a simple subvariety of* V', and that $t_1, t_{r-s+1}, \cdots, t_r$ are uniformizing parameters of Γ'. In a similar fashion we find that t_1 is a uniformizing parameter of the $(r-1)$-dimensional variety C', and since t_1 is among the uniformizing parameters of Γ', we conclude that Γ' *is a simple subvariety of* C'.

(4) To complete the proof of our theorem we have only to show that every point P' of Γ' is simple for V', C' and Γ'. To show that P' is simple for V' we have to exhibit r uniformizing parameters at P'. Let Δ be the residue class field of the point P, that is, let $\Delta=\mathfrak{o}/\mathfrak{p}$. Similarly let Δ' be the residue class field of P'. Here Δ and Δ' are finite algebraic extensions of K, and $\Delta'\supseteq\Delta$ since $Q(P')\supseteq Q(P)$. Without loss of generality we may assume that P' is at finite distance with respect to the ring $\mathfrak{o}'$ of nonhomogeneous coordinates ($\mathfrak{o}'=\mathfrak{o}_1'$, see (14)), and is therefore given in $\mathfrak{o}'$ by a prime zero-dimensional ideal $\mathfrak{p}_0'$.

Let $c_1, c_2, \cdots, c_n$ be the P-residues of $\xi_1, \xi_2, \cdots, \xi_n$ respectively ($c_i\in\Delta$) and let $d_2, d_3, \cdots, d_{r-s}$ be the P'-residues of $t_2/t_1, t_3/t_1, \cdots, t_{r-s}/t_1 (d_i\in\Delta')$. The element d_i will be the root of an irreducible polynomial $f_i(z)$ with coefficients in Δ. We replace the coefficients of $f_i(z)$ by arbitrary but fixed elements of $\mathfrak{o}$ of which they are residues. Let $F_i(z)$ be the polynomial with coefficients in $\mathfrak{o}$ thus obtained, $i=2, 3, \cdots, r-s$. *If we assume that the polynomials* $f_i(z)$ *are all separable* then we can conclude as in [11, p. 590] that the r elements

$$(16)\qquad t_1, F_2(t_2/t_1), \cdots, F_{r-s}(t_{r-s}/t_1), t_{r-s+1}, \cdots, t_r$$

are uniformizing parameters at P'. Since these elements include the uniformizing parameters $t_1, t_{r-s+1}, \cdots, t_r$ of Γ' and the uniformizing parameter t_1 of C', the proof is complete.

However, if some or all of the polynomials $f_i(z)$ are inseparable, then the elements (16) are no longer uniformizing parameters at P' (compare with footnote 41). We shall therefore give here a new proof which applies both to the separable and non-separable case. We consider the residue class ring $\mathfrak{o}^*=\mathfrak{o}'/\mathfrak{p}'$, where $\mathfrak{p}'$ is the prime $\mathfrak{o}'$-ideal of Γ'. Let $z_2, z_3, \cdots, z_{r-s}$ be the $\mathfrak{p}'$-residues of $t_2/t_1, t_3/t_1, \cdots, t_{r-s}/t_1$. Since $\mathfrak{p}'\cap\mathfrak{o}=\mathfrak{p}_0$ and since we have shown earlier in this section that $z_2, z_3, \cdots, z_{r-s}$ are algebraically independent over K, it follows from (14), for $h=1$, that $\mathfrak{o}^*$ is a *polynomial ring*(39) *over* Δ:

(39) This shows incidentally that the field of rational functions on Γ' is a pure transcendental extension of Δ, whence Γ' *is a rational variety* over Δ.

$$\mathfrak{o}^* = \Delta[z_2, z_3, \cdots, z_{r-s}].$$

The rest of the proof will be based on the following lemma:

LEMMA 9. *In a polynomial ring* $P_n = \Delta[x_1, x_2, \cdots, x_n]$ *over an arbitrary field* Δ *every prime zero-dimensional ideal possesses a base consisting of n elements*(40).

Proof of the lemma. Since the lemma is trivially true for $n=1$, we proceed by induction with respect to n. Let $\mathfrak{p}$ be a prime zero-dimensional ideal in P_n and let $f(z)$ be the irreducible polynomial in $\Delta[z]$ such that $f(x_n) \equiv 0(\mathfrak{p})$. The residue class ring $P_{n-1}^* = P_n/f(x_n)$ is obviously a polynomial ring $\Delta^*[x_1, x_2, \cdots, x_{n-1}]$ over the field $\Delta^* = \Delta(\alpha)$, where α is a root of $f(z)$. The ideal $\mathfrak{p}^* = \mathfrak{p}/f(x_n)$ is prime and zero-dimensional in P_{n-1}^*. By our induction, there exist $n-1$ elements $\omega_1^*, \omega_2^*, \cdots, \omega_{n-1}^*$ in P_{n-1}^* such that $\mathfrak{p}^* = P_{n-1}^*(\omega_1^*, \omega_2^*, \cdots, \omega_{n-1}^*)$. Let $\omega_1, \omega_2, \cdots, \omega_{n-1}$ be elements of P_n whose residues modulo $f(x_n)$ are, respectively, $\omega_1^*, \omega_2^*, \cdots, \omega_{n-1}^*$. Then it is clear that $\mathfrak{p} = (\omega_1, \omega_2, \cdots, \omega_{n-1}, f(x_n))$, q.e.d.(41).

We now apply our lemma. In the homomorphism $\mathfrak{o}' \sim \mathfrak{o}^*$ the prime $\mathfrak{o}'$-ideal $\mathfrak{p}_0'$ of the point P' is mapped upon a prime zero-dimensional $\mathfrak{o}^*$-ideal $\mathfrak{p}_0^*$. By the lemma we have

$$\mathfrak{p}_0^* = (\zeta_2^*, \zeta_3^*, \cdots, \zeta_{r-s}^*).$$

Let $\zeta_2', \zeta_3', \cdots, \zeta_{r-s}'$ be elements of $\mathfrak{o}'$ whose $\mathfrak{p}'$-residues are respectively $\zeta_2^*, \zeta_3^*, \cdots, \zeta_{r-s}^*$. Then we have

$$\mathfrak{p}_0' = (\mathfrak{p}', \zeta_2', \zeta_3', \cdots, \zeta_{r-s}'). \tag{17}$$

Let $\mathfrak{J}_1'$ be the quotient ring of P'. Since $Q(P') \supset Q(P) = \mathfrak{J}$ and since $Q(P') \supset \mathfrak{o}'$, the ring $\mathfrak{J}_1'$ contains all the rings previously considered, that is, the rings $\mathfrak{o}$, $\mathfrak{o}'$, $\mathfrak{J}$ and $\mathfrak{J}'$ ($\mathfrak{J}' = \mathfrak{J}\cdot\mathfrak{o}'$). We have $\mathfrak{p}' = \mathfrak{o}'\mathfrak{p}_0$, whence $\mathfrak{J}_1' \cdot \mathfrak{p}' = \mathfrak{J}_1' \cdot \mathfrak{p}_0 = \mathfrak{J}_1' \cdot \mathfrak{J} \cdot \mathfrak{p}_0 = \mathfrak{J}_1' \cdot (t_1, t_2, \cdots, t_r) = \mathfrak{J}_1' \cdot (t_1, t_{r-s+1}, \cdots, t_r)$ (since $t_i/t_1 \in \mathfrak{o}' \in \mathfrak{J}_1'$, $i = 2, 3, \cdots, r-s$). Substituting into (17) we find:

$$\mathfrak{J}_1' \cdot \mathfrak{p}_0' = \mathfrak{J}_1' \cdot (t_1, \zeta_2, \zeta_3, \cdots, \zeta_{r-s}, t_{r-s+1}, t_{r-s+2}, \cdots, t_r).$$

---

(40) This lemma implies that every point P of an affine (or of a projective) space over Δ is simple. The lemma gives, however, a stronger result, since it shows that uniformizing parameters at P can be so selected that they generate the prime ideal of P not only in $Q(P)$ but also in the polynomial ring. In other words: *every point P of an affine n-space is complete intersection of n hypersurfaces.* This result can be extended without any difficulties to projective spaces by a similar inductive argument.

(41) In the case of ground fields of characteristic zero we have used instead of the above lemma the following property of the polynomial ring P_n: if $f_i(z)$ is the irreducible polynomial in $\Delta[z]$ such that $f_i(x_i) \equiv 0(\mathfrak{p})$, then the ideal $(f_1(x_1), f_2(x_2), \cdots, f_n(x_n))$ is the intersection of prime zero-dimensional ideals, one of which is of course the ideal $\mathfrak{p}$ itself. In this case the n polynomials $f_i(x_i)$ are uniformizing parameters for the point defined by $\mathfrak{p}$. This reasoning applies also in the case in which the polynomials $f_i(x_i)$ are separable.

This exhibits r uniformizing parameters at P', and since these parameters include the uniformizing parameters of Γ' and that of C', the proof is now complete.

COROLLARY. *If all points of W are simple both for W and for V, then $T[W]=T\{W\}$ (T—a monoidal transformation of center W) and $T[W]$ is irreducible, $(r-1)$-dimensional and all its points are simple both for $T[W]$ and V'. Moreover, if W is of dimension s, then $T[W]$ is covered by an s-dimensional algebraic system $\{\Gamma'\}$ of $(r-s-1)$-dimensional varieties Γ' in $(1, 1)$ correspondence with the points of W. Each Γ' is irreducible, rational and free from singularities, and through each point of $T[W]$ there passes a unique Γ'.*

REFERENCES

1. H. Grell, *Beziehungen zwischen den Idealen verschiedener Ringe*, Math. Ann. vol. 97 (1927).
2. W. Krull, *Idealtheorie*, Ergebnisse der Mathematik und ihrer Grenzgebiete, IV 3.
3. ———, *Dimensionstheorie in Stellenringen*, J. Reine Angew. Math., vol. 179 (1938).
4. S. MacLane and O. F. G. Schilling, *Zero-dimensional branches of rank one on algebraic varieties*, Ann. of Math. (2) vol. 40 (1939).
5. F. K. Schmidt, *Über die Erhaltung der Kettensätze der Idealtheorie bei beliebigen endlichen Körpererweiterungen*, Math. Zeit. vol. 41 (1936).
6. B. L. van der Waerden, *Algebraische Korrespondenzen und rationale Abbildungen*, Math. Ann. vol. 110 (1934).
7. O. Zariski, *Some results in the arithmetic theory of algebraic varieties*, Amer. J. Math. vol. 61 (1939).
8. ———, *The reduction of the singularities of an algebraic surface*, Ann. of Math. vol. 40 (1939).
9. ———, *Algebraic varieties over ground fields of characteristic zero*, Amer. J. Math. vol. 62 (1940).
10. ———, *Pencils on an algebraic variety and a new proof of a theorem of Bertini*, Trans. Amer. Math. Soc. vol. 50 (1941).
11. ———, *A simplified proof for the resolution of singularities of an algebraic surface*, Ann. of Math. (2) vol. 43 (1942).
12. ———, *Normal varieties and birational correspondences*, Bull. Amer. Math. Soc. vol. 48 (1942).

THE JOHNS HOPKINS UNIVERSITY,
BALTIMORE, MD.

THE THEOREM OF BERTINI ON THE VARIABLE SINGULAR POINTS OF A LINEAR SYSTEM OF VARIETIES

BY

OSCAR ZARISKI

1. **Pure transcendental extensions of the ground field.** Let V/k be an irreducible r-dimensional algebraic variety over a given ground field k. We assume that V/k is immersed in an n-dimensional projective space and we denote by $x_1, x_2, \cdots, x_n$ the nonhomogeneous coördinates of the general point of V/k. Let $u_1, u_2, \cdots, u_m$ be indeterminates with respect to the field $k(x)$ $[=k(x_1, x_2, \cdots, x_n)]$ of rational functions on V/k. We adjoin these indeterminates to the field $k(x)$ and we denote by K the field $k(u)$ $[=k(u_1, u_2, \cdots, u_m)]$. This subfield K of the field $k(x, u)$ we take as new ground field, and over this new ground field we consider the variety V/K defined by the same general point $(x_1, x_2, \cdots, x_n)$ as V/k. The varieties V/k and V/K are of the same dimension r over their respective ground fields k and K. We shall say that the variety V/K is *the extension of the variety* V/k under the ground field extension $k \rightarrow K$.

By precisely the same argument, every irreducible subvariety W/k of V/k has as extension an irreducible subvariety W/K of V/K, of the same dimension as W/k. If $\bar{x}_1, \bar{x}_2, \cdots, \bar{x}_n$ are the nonhomogeneous coördinates of the general point of W/k, then $\bar{x}_1, \bar{x}_2, \cdots, \bar{x}_n$ are also the nonhomogeneous coördinates of the general point of W/K. Moreover, $u_1, u_2, \cdots, u_m$ are indeterminates with respect to the field $k(\bar{x})$, that is, they are algebraically independent over this field.

Not every irreducible subvariety of V/K is the extension of a subvariety of V/k, but every such subvariety W^*/K defines an irreducible subvariety W/k of V/k, which we shall refer to as *the contraction of* W^*/K and which is obtained as follows. Let $x_1^*, x_2^*, \cdots, x_n^*$ be the nonhomogeneous coördinates of the general point of W^*/K. Since $W^*/K \subseteq V/K$, the ring $K[x_1^*, x_2^*, \cdots, x_n^*]$ is a homomorphic image of the ring $K[x_1, x_2, \cdots, x_n]$. Therefore, also the ring $k[x_1^*, x_2^*, \cdots, x_n^*]$ is a homomorphic image of the ring $k[x_1, x_2, \cdots, x_n]$. Therefore there is a unique irreducible subvariety W/k of V/k, whose general point $(\bar{x}_1, \bar{x}_2, \cdots, \bar{x}_n)$ is defined by the condition that the rings $k[\bar{x}_1, \bar{x}_2, \cdots, \bar{x}_n]$ and $k[x_1^*, x_2^*, \cdots, x_n^*]$ be simply isomorphic and that $\bar{x}_i, x_i^*$ $(i=1, 2, \cdots, n)$ be corresponding elements in the isomorphism. This variety W/k shall be termed the contraction of W^*/K.

LEMMA 1. *Let W/k and W^*/K be irreducible subvarieties of V/k and of V/K respectively, and let $(\bar{x})$, (x^*) $[(\bar{x})=(\bar{x}_1, \bar{x}_2, \cdots, \bar{x}_n), (x^*)=(x_1^*, x_2^*, \cdots, x_n^*)]$ be their general points.*

Presented to the Society, April 29, 1944; received by the editors March 20, 1944.

(a) *W/k is the contraction of W/K.*

(b) *If W/k is the contraction of W^*/K, then $W^*/K \subseteq W/K$.*

(c) *A necessary and sufficient condition that W^*/K be the extension of a subvariety of V/k is that $u_1, u_2, \cdots, u_m$ be algebraically independent over the field $k(x_1^*, x_2^*, \cdots, x_n^*)$.*

Proof. (a) Trivial.

(b) By hypothesis, the rings $k[\bar{x}]$ and $k[x^*]$ are simply isomorphic. Since $u_1, u_2, \cdots, u_m$ are algebraically independent over the field $k(\bar{x})$, it follows that $K[x^*]$ is a homomorphic image of $K[\bar{x}]$, that is, $W^*/K \subseteq W/K$.

(c) The condition is necessary in virtue of the very definition of W/K. Let us suppose that the condition is satisfied. Then every algebraic relation between the elements $x_1^*, x_2^*, \cdots, x_n^*$ over K is a consequence of algebraic relations between $x_1^*, x_2^*, \cdots, x_n^*$ over k. If then W/k is the contraction of W^*/K, then it follows in view of isomorphism $k[x^*] \simeq k[\bar{x}]$ that the two rings $K[x^*]$ and $K[\bar{x}]$ are simply isomorphic. This shows that $W^*/K = W/K$.

We denote the quotient ring of an irreducible subvariety W/k of V/k by $Q_V(W/k)$. Similarly, the quotient ring of W/K, regarded as a subvariety of V/K, shall be denoted by $Q_V(W/K)$. By definition, W/k is a simple subvariety of V/k if the ideal $\mathfrak{m}$ of non-units of $Q_V(W/k)$ has a base consisting of $r-s$ elements, where s is the dimension of W/k(1). The elements of such a base are referred to as *uniformizing parameters of* $W(V/k)$. If $t_1, t_2, \cdots, t_\rho$, $\rho = r-s$, are uniformizing parameters of $W(V/k)$, and if ω is any element of $\mathfrak{m}$ which is exactly divisible by $\mathfrak{m}^\nu$ [that is, $\omega \equiv 0(\mathfrak{m}^\nu)$, $\omega \not\equiv 0(\mathfrak{m}^{\nu+1})$], then $\omega = \phi_\nu(t_1, t_2, \cdots, t_\rho)$, where ϕ_ν is a form of degree ν with coefficients in $Q_V(W/k)$ but not all in $\mathfrak{m}$. If these coefficients are replaced by their residues mod $\mathfrak{m}$, we obtain a form of degree ν in ρ indeterminates, with coefficients in the residue field of W/k, not all zero. This is called the *leading form* of ω. It is easy to show that ρ elements of $Q_V(W/k)$ are uniformizing parameters of $W(V/k)$ if and only if their leading forms are linear and linearly independent(2).

We use the notation of the preceding lemma and we assume that W/k

---

(1) See O. Zariski, *Algebraic varieties over ground fields of characteristic zero*, Amer. J. Math. vol. 62 (1940) p. 199.

(2) *Proof.* If $\omega_1, \omega_2, \cdots, \omega_\rho$ are non-units in $Q_V(W/k)$, then $\omega_i = \sum_{j=1}^{\rho} A_{ij} t_j$, where the A_{ij} are elements of $Q_V(W/k)$. Let A be the matrix $\|A_{ij}\|$ and $\bar{A}$ be the matrix $\|\bar{A}_{ij}\|$, where $\bar{A}_{ij}$ is the $\mathfrak{m}$-residue of A_{ij}. Suppose that the leading forms of $\omega_1, \omega_2, \cdots, \omega_\rho$ are linear and linearly independent. Then $|\bar{A}| \neq 0$, that is, $|A| \not\equiv 0(\mathfrak{m})$, and therefore $t_1, t_2, \cdots, t_\rho$ can be expressed as linear forms in $\omega_1, \omega_2, \cdots, \omega_\rho$, with coefficients which are elements of the matrix A^{-1}. These elements are in $Q_V(W/k)$, and hence $\omega_1, \omega_2, \cdots, \omega_\rho$ form a basis for $\mathfrak{m}$.

Conversely, assume that $\omega_1, \omega_2, \cdots, \omega_\rho$ are uniformizing parameters of $W(V/k)$. We can then write: $t_i = \sum_{j=1}^{\rho} B_{ij} \omega_j$, and hence $t_i = \sum_{j=1}^{\rho} C_{ij} t_j$ where $C = BA$, $C = \|C_{ij}\|$, $B = \|B_{ij}\|$. Since every element of $Q_V(W/k)$ has a unique leading form, it follows from the relations $t_i = \sum_{j=1}^{\rho} C_{ij} t_j$ that modulo $\mathfrak{m}$ the matrix C is the unit matrix. Hence $|B| \cdot |A| \not\equiv 0(\mathfrak{m})$, whence the leading forms of $\omega_1, \omega_2, \cdots, \omega_\rho$ are linear and linearly independent.

is the contraction of W^*/K. If W/k is of dimension s, then W^*/K is of dimension $s-\sigma$, $\sigma \geqq 0$, by part (b) of Lemma 1, that is, the field $K(x^*)$ is of degree of transcendency $s-\sigma$ over $K[=k(u)]$. Since $k(x^*) \simeq k(\bar{x})$ and since $k(\bar{x})$ is of degree of transcendency s over k, it follows that the field $k(u, x^*)$ is of degree of transcendency $m-\sigma$ over $k(x^*)$. If then $U_1, U_2, \cdots, U_m$ are indeterminates over $k(x^*)$, the polynomials in the polynomial ring $k(x^*)[U_1, U_2, \cdots, U_m]$ which vanish after the specialization $U_i \rightarrow u_i$ form a prime ideal $\mathfrak{P}$ of dimension $m-\sigma$. The ideal of non-units in the quotient ring of this polynomial ideal has a base of σ elements[3]. Let $G_1(U), G_2(U), \cdots, G_\sigma(U)$ be such a base, where we may assume, without loss of generality, that the $G_i(U)$ are polynomials in the U's (with coefficients in $k(x^*)$): $G_i(U) = G_i(U, x^*)$. We put

$$\tau_i = G_i(u, x), \qquad i = 1, 2, \cdots, \sigma. \tag{1}$$

LEMMA 2. (a) *If W/K, of dimension s, is simple for V/K, then W/k is simple for V/k, and if $t_1, t_2, \cdots, t_\rho$ $(\rho = r-s)$ are uniformizing parameters for $W(V/k)$, they are also uniformizing parameters for $W(V/K)$.*

(b) *If W/k, of dimension s, is the contraction of W^*/K which is of dimension $s-\sigma$, $\sigma \geqq 0$, and if W/k is simple for V/k, with $t_1, t_2, \cdots, t_\rho$ as uniformizing parameters, then also W^*/K is simple for V/K, and the elements $t_1, t_2, \cdots, t_\rho, \tau_1, \tau_2, \cdots, \tau_\sigma$ are uniformizing parameters of $W^*(V/K)$.*

Proof. (a) We put $\mathfrak{J} = Q_V(W/k)$, $\bar{\mathfrak{J}} = Q_V(W/K)$ and we denote by $\mathfrak{m}$ and $\bar{\mathfrak{m}}$ respectively the ideals of non-units in $\mathfrak{J}$ and in $\bar{\mathfrak{J}}$. The elements of $\bar{\mathfrak{J}}$ are all of the form $\phi(u)/\psi(u)$, where $\phi(u), \psi(u) \in \mathfrak{J}[u_1, u_2, \cdots, u_m]$ and where not all the coefficients of $\psi(u)$ are in $\mathfrak{m}$. The following relations are therefore obvious:

$$\bar{\mathfrak{m}} = \bar{\mathfrak{J}} \cdot \mathfrak{m}, \tag{2}$$

$$\mathfrak{m} = \bar{\mathfrak{m}} \cap \mathfrak{J}. \tag{2'}$$

Let W/K be simple for V/K. It follows then directly from (2) that already the ideal $\mathfrak{m}$ must contain ρ elements which—regarded as elements of $Q_V(W/K)$—have leading forms which are linear and linearly independent. There exists therefore a set of uniformizing parameters of $W(V/K)$ which consists of elements of $\mathfrak{m}$. Let $t_1, t_2, \cdots, t_\rho$ be such a set. If ω is any element of $\mathfrak{m}$, we have by (2'):

$$\omega \cdot \psi(u) = \phi_1(u) \cdot t_1 + \phi_2(u) \cdot t_2 + \cdots + \phi_\rho(u) \cdot t_\rho, \tag{3}$$

where $\psi(u), \phi_i(u) \in \mathfrak{J}[u]$ and where at least one coefficient of $\psi(u)$ does not belong to $\mathfrak{m}$. The presence of such a coefficient and the fact that $u_1, u_2, \cdots, u_m$ are algebraically independent over $\mathfrak{J}$ have, by (3), the consequence that ω

[3] See O. Zariski, *Foundations of a general theory of birational correspondences*, Trans. Amer. Math. Soc. vol. 53 (1943) p. 541.

belongs to ideal $\mathfrak{J}\cdot(t_1, t_2, \cdots, t_\rho)$. Hence $\mathfrak{m}=\mathfrak{J}(t_1, t_2, \cdots, t_\rho)$, and W/k is simple for V/k, with $t_1, t_2, \cdots, t_\rho$ as uniformizing parameters. This and relation (2) complete the proof of part (a) of the lemma.

(b) Let $\mathfrak{J}^*$ denote the quotient ring $Q_V(W^*/K)$ and let $\mathfrak{m}^*$ be the ideal of non-units in $\mathfrak{J}^*$. We assume that W/k, the contraction of W^*/K, is simple for V/k, with $t_1, t_2, \cdots, t_\rho$ as uniformizing parameters. The elements of $\mathfrak{J}^*$ are all of the form $\phi(x, u)/\psi(x, u)$, where ϕ and ψ are polynomials with coefficients in k and where $\psi(x^*, U)\not\equiv 0 \pmod{\mathfrak{P}}$. Since $W^*/K\subseteq W/K$, the variety W/K is defined locally at W^*/K by a prime ideal $\mathfrak{p}^*$ in the quotient ring $\mathfrak{J}^*$. This ideal consists of those quotients $\phi(x, u)/\psi(x, u)$ for which $\phi(x^*, U)=0$ (or in equivalent form: $\phi(\bar{x}, u)=0$). It is therefore clear that the residue class ring $\mathfrak{J}^*/\mathfrak{p}^*$ coincides with the quotient ring of the prime ideal $\mathfrak{P}$. Hence we have by (1):

$$\mathfrak{J}^*\cdot\mathfrak{m}^* = \mathfrak{J}^*\cdot(\mathfrak{p}^*, \tau_1, \tau_2, \cdots, \tau_\sigma). \tag{4}$$

Now if $\phi(x^*, U)=0$, that is, if $\phi(\bar{x}, u)=0$, then $\phi(x, u)\equiv 0(\bar{\mathfrak{m}})$, whence $\phi(x, u)\equiv 0(\overline{\mathfrak{J}}\cdot\mathfrak{m})$. Consequently $\mathfrak{p}^*=\mathfrak{J}^*\cdot(t_1, t_2, \cdots, t_\rho)$, and therefore, by (4),

$$\mathfrak{J}^*\cdot\mathfrak{m}^* = \mathfrak{J}^*\cdot(t_1, t_2, \cdots, t_\rho, \tau_1, \tau_2, \cdots, \tau_\sigma). \tag{5}$$

This completes the proof of part (b) of the lemma.

The above lemma implies that the *singular locus of V/K is the extension of the singular locus of V/k*, in the sense that the irreducible components of the singular locus of V/K are extensions of the irreducible components of the singular locus of V/k.

2. The general member of a linear system of V_{r-1}'s on V/k. Let $f_0(x), f_1(x), \cdots, f_m(x)$ be $m+1$ elements in $k[x]$ which are linearly independent over *the field of constants*, that is, over the algebraic closure of the ground field k in the field $k(x)$ of rational functions on V/k. The $m+1$ polynomials $f_i(x)$ determine uniquely a set of $m+1$ integral divisors $\mathfrak{A}_i$ of the field $k(x)$ with the following properties: (1) each $\mathfrak{A}_i$ is a divisor of the first kind with respect to a derived normal model V' of V/k(4); (2) $f_i/f_j=\mathfrak{A}_i/\mathfrak{A}_j$; (3) $\mathfrak{A}_0, \mathfrak{A}_1, \cdots, \mathfrak{A}_m$ are relatively prime. Each divisor $\mathfrak{A}_i$ determines on V/k (and—if V/k is locally normal(5)—is determined by) a pure $(r-1)$-dimensional subvariety F_i/k, whose irreducible components are counted to well-defined multiplicities. The irreducible components of F_i correspond to the prime factors of $\mathfrak{A}_i$, but if V/k is not locally normal, two distinct prime factors of $\mathfrak{A}_i$ may very well correspond to one and the same irreducible component of F_i/k. However, we agree to keep separate the identity of the irre-

(4) For the definition of a derived normal model, see O. Zariski, *Some results in the arithmetic theory of algebraic varieties*, Amer. J. Math. vol. 61 (1939) p. 292. For a discussion of divisors of the first kind, see O. Zariski, *Pencils on an algebraic variety and a new proof of a theorem of Bertini*, Trans. Amer. Math. Soc. vol. 50 (1941) p. 49.

(5) For a definition of locally normal varieties, see (3), p. 512.

ducible components of F_i/k relative to distinct prime factors of $\mathfrak{A}_i$. This means that we actually regard F_i as a subvariety of the normal derived variety V' rather than of V/k. In this connection we may add the remark that an $(r-1)$-dimensional irreducible subvariety of V/k may arise from more than one prime divisor only if it is singular for V/k.

If $\lambda_0, \lambda_1, \cdots, \lambda_m$ are arbitrary elements of k, then there exists a unique divisor $\mathfrak{A}(\lambda)$, of the first kind with respect to the normal variety V', such that

$$(6) \qquad [\lambda_0 f_0(x) + \lambda_1 f_1(x) + \cdots + \lambda_m f_m(x)]/f_0(x) = \mathfrak{A}(\lambda)/\mathfrak{A}_0.$$

Let $F(\lambda)$ be the $(r-1)$-dimensional subvariety of V/k (or—better—of V') defined by $\mathfrak{A}(\lambda)$. As the λ's vary in k, $F(\lambda)$ varies in a *linear system* $|F(\lambda)|$. The varieties $F_0, F_1, \cdots, F_m$ are particular members of the system.

We proceed to associate with the linear system $|F(\lambda)|$ an irreducible $(r-1)$-dimensional subvariety F^*/K of the variety V/K considered in the preceding section. We define F^* by the following conditions:

(a) The nonhomogeneous coördinates $\eta_1, \eta_2, \cdots, \eta_n$ of the general point of F^* shall satisfy the relation:

$$(7) \qquad f_0(\eta) + u_1 f_1(\eta) + \cdots + u_m f_m(\eta) = 0.$$

(b) The rings $k[x]$ and $k[\eta]$ shall be isomorphic, and in the isomorphism the elements x_i and η_i shall correspond to each other.

We have to show that: (1) there exists an F^*/K, of dimension $r-1$, satisfying conditions (a) and (b); (2) F^*/K is uniquely determined by these two conditions and by the condition that it be of dimension $r-1$; (3) F^*/K is a subvariety of V/K.

We start by introducing another copy $k(\eta)$ of the field $k(x)$, that is, we assume that $k(\eta) \simeq k(x)$, $\eta_i \rightarrow x_i$. We then adjoin to $k(\eta)$ a set of $m-1$ indeterminates $v_1, v_2, \cdots, v_{m-1}$ and we put[6]

$$v_m = -[f_0(\eta) + v_1 f_1(\eta) + \cdots + v_{m-1} f_{m-1}(\eta)]/f_m(\eta).$$

LEMMA 3. *The elements $v_1, v_2, \cdots, v_m$ are algebraically independent over k.*

Proof. The elements $v_1, v_2, \cdots, v_m$ are elements of the polynomial ring $k(\eta)[v_1, v_2, \cdots, v_{m-1}]$. If these elements were algebraically dependent over k, the specialization $v_1 \rightarrow 1$, $v_j \rightarrow 0$, $j = 2, 3, \cdots, m-1$, would lead to a true relation[7] of algebraic dependence of $[f_0(\eta) + f_1(\eta)]/f_m(\eta)$ over k. This contradicts our hypothesis that $f_0(\eta), f_1(\eta), \cdots, f_m(\eta)$ are linearly independent over the algebraic closure of k in $k(\eta)$.

In view of Lemma 3 we can identify $v_1, v_2, \cdots, v_m$ with $u_1, u_2, \cdots, u_m$ respectively. We have then an algebraic variety F^*/K, with general point $(\eta_1, \eta_2, \cdots, \eta_n)$, and conditions (a) and (b) are satisfied.

(6) Note that $f_m(\eta) \neq 0$ since the elements $f_0(\eta), f_1(\eta), \cdots, f_m(\eta)$ were assumed to be linearly independent.

(7) See B. L. van der Waerden, *Moderne Algebra*, vol. 2, Hilfsatz on p. 17.

The field of rational functions on F^*/K is the field $k(\eta, u_1, u_2, \cdots, u_{m-1})$, of degree of transcendency $m-1$ over $k(\eta)$, and the field $k(\eta)$ is of degree of transcendency r over k. Hence the field $k(\eta, u_1, u_2, \cdots, u_{m-1})$ is of degree of transcendency $r+m-1$ over k. The subfield K is of degree of transcendency m over k, and hence $k(\eta, u_1, u_2, \cdots, u_{m-1})$ is of degree of transcendency $r-1$ over K. This shows that F^*/K is of dimension $r-1$.

The above construction of F^*/K shows that if F'^*/K is another irreducible $(r-1)$-dimensional variety, with general point $(\eta_1', \eta_2', \cdots, \eta_n')$ satisfying conditions (a) and (b), then the rings $K[\eta]$ and $K[\eta']$ are necessarily simply isomorphic. Hence F^*/K is uniquely determined.

Finally, every algebraic relation between $x_1, x_2, \cdots, x_n$, with coefficients in K, is a consequence of algebraic relations between $x_1, x_2, \cdots, x_n$ with coefficients in k (since $u_1, u_2, \cdots, u_m$ are algebraically independent over $k(x)$). Hence every such relation remains a true relation after the specialization $x_i \rightarrow \eta_i$, since $k(\eta) \simeq k(x)$. Hence the ring $K[\eta]$ is a homomorphic image of the ring $K[x]$, that is, F^*/K lies on V/K. Thus our three assertions are established.

We shall call the $(r-1)$-dimensional variety F^*/K—*the general member of the linear system* $|F(\lambda)|$.

We note that *the contraction of* F^*/K *is the entire variety* V/k: this follows from the isomorphism $k[\eta] \simeq k[x]$. Also, as a consequence of part (b) of Lemma 2, F^*/K is simple for V/K.

3. **The base locus of the linear system** $|F(\lambda)|$. An irreducible subvariety W/k of V/k is a base variety of the linear system $|F(\lambda)|$ if it lies on every $F(\lambda)$. It is known[8] that W/k is a base variety if it lies on each F_i, $i=0, 1, 2, \cdots, m$.

THEOREM 1. *If* W/k *is a base variety of the linear system* $|F(\lambda)|$, *then* W/K *lies on the general member* F^*/K *of the system*; *and conversely.*

Proof. Without loss of generality we may assume that W/k is at finite distance with respect to the nonhomogeneous coördinates $x_1, x_2, \cdots, x_n$ of the general point of V/k. We denote by $\bar{x}_1, \bar{x}_2, \cdots, \bar{x}_n$ the nonhomogeneous coördinates of the general point of W/k. Let

$$f_i(x) = \mathfrak{M}\mathfrak{A}_i, \qquad i = 0, 1, \cdots, m, \tag{8}$$

be the divisor decomposition of $f_i(x)$ on the derived normal model V' of V/k. Here $\mathfrak{A}_0, \mathfrak{A}_1, \cdots, \mathfrak{A}_m$ are the integral divisors considered in the preceding section, while $\mathfrak{M}$ is a fractional divisor whose denominator consists entirely of divisors at infinity. If $\mathfrak{o}$ denotes the ring of nonhomogeneous coördinates of the normal variety V', then the divisors $\mathfrak{A}_i$ may be identified with certain pure $(r-1)$-dimensional ideals in $\mathfrak{o}$, ideals which we shall continue to denote by $\mathfrak{A}_0, \mathfrak{A}_1, \cdots, \mathfrak{A}_m$.

(8) See (3), p. 528.

Let W/k be a base variety of the linear system $|F(\lambda)|$. That means that W/k lies on the subvariety of V' defined by the ideal $(\mathfrak{A}_0, \mathfrak{A}_1, \cdots, \mathfrak{A}_m)$. To prove that W/K lies on F^*/K we have to show the following: *if $H(u, \eta)=0$, where H is a polynomial with coefficients in k, then $H(u, \bar{x})=0$.* Actually we shall prove the following stronger result: *if $H(u, \eta)=0$, then all the coefficients of $H(u, x)$, regarded as a polynomial in $u_1, u_2, \cdots, u_m$, belong to the ideal $(\mathfrak{A}_0, \mathfrak{A}_1, \cdots, \mathfrak{A}_m)$.*

Let $X_1, X_2, \cdots, X_n$ denote indeterminates and let

$$f(u, X) = f_0(X) + u_1 f_1(X) + \cdots + u_m f_m(X).$$

The elimination of u_m between $H(u, X)$ and $f(u, X)$ leads to an identity of the form:

$$[f_m(X)]^q \cdot H(u, X) = Q(u, X) \cdot f(u, X) + R(u_1, u_2, \cdots, u_{m-1}, X), \tag{9}$$

where all polynomials have coefficients in k. By hypothesis, we have $H(u, \eta)=0$, hence $R(u_1, u_2, \cdots, u_{m-1}, \eta)=0$, since $f(u, \eta)=0$. But since $u_1, u_2, \cdots, u_{m-1}$ are algebraically independent over $k(\eta)$ (by definition of F^*/K) and since $k(\eta) \simeq k(x)$, we conclude that $R(u, x)=0$. The specialization $X \to x$ in the above identity yields therefore the following relation:

$$H(u, x) = f(u, x) \cdot Q(u, x)/[f_m(x)]^q. \tag{10}$$

On the left we have a polynomial in the indeterminates $u_1, u_2, \cdots, u_m$, with coefficients in $k[x]$, hence in $\mathfrak{o}$. On the right we have a product of two polynomials in the same indeterminates, with coefficients respectively in $\mathfrak{o}$ and in the quotient field of $\mathfrak{o}$. Since $\mathfrak{o}$ is integrally closed, the generalized Kronecker lemma[9] is applicable. Since the coefficients of $f(u, x)$ are $f_0(x), f_1(x), \cdots, f_m(x)$, it follows from this lemma of Kronecker and from (8) that the coefficients of $H(u, x)$ must indeed all belong to the ideal $(\mathfrak{A}_0, \mathfrak{A}_1, \cdots, \mathfrak{A}_m)$.

To prove the second part of our theorem, we show that if W/k is not a base variety of the linear system $|F(\lambda)|$ then W/K does not lie on F^*/K. Without loss of generality we may assume then that W/k does not lie on the variety defined by the ideal $\mathfrak{A}_0$. Under this assumption, the quotients $f_i(x)/f_0(x)$, $i=1, 2, \cdots, m$, belong to the quotient ring $Q_V(W/k)$. Hence we may write: $f_i(x)/f_0(x) = \phi_i(x)/\phi_0(x)$, where $\phi_0(\bar{x}) \neq 0$. By (7) we have the following true relation between $\eta_1, \eta_2, \cdots, \eta_n$:

$$\phi_0(\eta) + u_1\phi_1(\eta) + \cdots + u_m\phi_m(\eta) = 0.$$

This relation, however, is destroyed by the specialization $\eta \to \bar{x}$, since $\phi_0(\bar{x}) \neq 0$ and since $u_1, u_2, \cdots, u_m$ are indeterminates over $k(\bar{x})$. Hence W/K does not lie on F^*/K. This completes the proof of the theorem.

(9) W. Krull, *Idealtheorie*, Ergebnisse der Mathematik und ihrer Grenzgebiete, vol. 4, no. 3, p. 125.

We consider an irreducible *simple* subvariety W/k of V/k. If Φ is a pure (not necessarily irreducible) $(r-1)$-dimensional subvariety of V/k containing W/k, Φ is given in the quotient ring $Q_V(W/k)$ by a principal ideal (ω). We say that W/k is simple for Φ if the leading form of ω in $Q_V(W/k)$ is of degree 1. It is well known([10]) that if Φ is irreducible, this definition is equivalent to our usual definition of simple subvarieties of Φ.

Suppose now that the simple variety W/k is a base variety of the linear system $|F(\lambda)|$. We then say that W/k is *a singular base variety of* $|F(\lambda)|$ if it is singular for each $F(\lambda)$.

LEMMA 4. *In order that W/k be a singular base variety of the linear system $|F(\lambda)|$ it is sufficient that it be singular for $F_0, F_1, \cdots, F_m$.*

Proof. Let Δ denote the partial power product of those prime factors of $\mathfrak{M}$ which represent $(r-1)$-dimensional varieties containing W/k, and let us write $\mathfrak{M}=\Delta\mathfrak{N}$. Since Δ is defined in $Q_V(W/k)$ by a principal ideal, we can find a polynomial $g(x)$ with coefficients in k such that $g(x)=\Delta\cdot\mathfrak{N}'$, where each prime divisor which occurs in $\mathfrak{N}'$ represents an $(r-1)$-dimensional subvariety of V/k which does *not* pass through W/k. We have $f_i(x)/g(x)=\mathfrak{N}\mathfrak{A}_i/\mathfrak{N}'$, and hence $f_i(x)/g(x)$ *belongs to the quotient ring* $Q_V(W/k)$. Moreover, $F(\lambda)$ will be defined in this quotient ring by the principal ideal

$$\left(\frac{f_0(x)+\lambda_1 f_1(x)+\cdots+\lambda_m f_m(x)}{g(x)}\right).$$

Now if W/k is singular for F_i, $i=0, 1, \cdots, m$, then the leading form of $f_i(x)/g(x)$ is of degree greater than 1. Therefore also the leading form of $[f_0(x)+\lambda_1 f_1(x)+\cdots+\lambda_m f_m(x)]/g(x)$ will be greater than 1 for all λ, q.e.d.

4. **The theorem of Bertini.** From the preceding considerations we can now derive a well known theorem of Bertini. For reasons explained in the next section, our formulation of this theorem is different from the classical formulation. We first prove the following theorem:

THEOREM 2. *A base variety W/k of $|F(\lambda)|$ which is simple for V/k is a singular base variety if and only if W/K is singular for the general member F^*/K of the linear system $|F(\lambda)|$.*

Proof. The proof of Lemma 4 shows that we can write $f_i(x)/g(x) =\bar{f}_i(x)/\bar{g}(x)$, where $\bar{g}(x)\neq 0$ on W, $\bar{f}_i(x)=\overline{\mathfrak{M}}\mathfrak{A}_i$, and where no prime factor of $\overline{\mathfrak{M}}$ represents a variety passing through W/k. Since the polynomials $\bar{f}_i(x)$ are proportional to the polynomials $f_i(x)$, the linear system $|F(\lambda)|$ and the general member F^*/K are equally well defined by the $\bar{f}_i(x)$ as by the $f_i(x)$. Hence we may assume that no prime factor of $\mathfrak{M}$ represents a variety passing

([10]) If the leading form of ω is linear, then ω is a member of a set of uniformizing parameters $\omega_1(=\omega), \omega_2, \cdots, \omega_\rho$ of $W(V/k)$ (see footnote 2). If $\bar{\omega}_2, \bar{\omega}_3, \cdots, \bar{\omega}_\rho$ are the Φ-residues of $\omega_2, \omega_3, \cdots, \omega_\rho$, then these $\rho-1$ elements are uniformizing parameters of $W(\Phi)$.

through W/k. Under this assumption the elements $f_0(x), f_1(x), \cdots, f_m(x)$ are relatively prime *locally* at W/k. By Theorem 1, and by our assumption that W/k is a base variety of $|F(\lambda)|$, it follows at any rate that W/K lies on F^*/K. We come back to equation (10); it is an identity in the polynomial ring $k(x)[u_1, u_2, \cdots, u_m]$. The coefficients of the polynomial $f(u, x)\cdot Q(u, x)$ are divisible by $[f_m(x)]^q$ in $k[x]$, hence a fortiori in $Q_V(W/k)$. But the coefficients of $f(u, x)$ are $f_0(x), f_1(x), \cdots, f_m(x)$, and these are relatively prime in $Q_V(W/k)$. Hence the coefficients of $Q(u, x)$ must be divisible by $[f_0(x)]^q$ in $Q_V(W/k)$. *Therefore $H(u, x)$ is divisible by $f(u, x)$ in $Q_V(W/k)$.* But $H(u, x)$ is an arbitrary polynomial which vanishes on F^*/K, that is, such that $F(u, \eta)=0$. *We conclude therefore that F^*/K is defined in $Q_V(W/K)$ by the principal ideal $(f(u, x))$.*

If $t_1, t_2, \cdots, t_p$ are uniformizing parameters of $W(V/k)$, they are also uniformizing parameters of $W(V/K)$. By the results just obtained, W/K is simple for F^*/K if and only if the leading form of $f(u, x)$ is of degree 1. Since this degree is equal to the minimum of the degrees of the leading forms of $f_0(x), f_1(x), \cdots, f_m(x)$, our theorem follows from Lemma 4.

THEOREM OF BERTINI. *Let W^*/K be an irreducible subvariety of F^*/K and let W/k be its contraction. If W^*/K is singular for F^*/K, then W/k is either singular for V/k or is a base variety of the linear system $|F(\lambda)|$.*

Proof. We shall show that if W/k is simple for V/k and is not a base variety of $|F(\lambda)|$, then W^*/K is simple for F^*/K. We denote by $x_1^*, x_2^*, \cdots, x_n^*$ the nonhomogeneous coördinates of the general point of W^*/K. Since W/k is not a base variety of $|F(\lambda)|$, we may assume that the elements $f_i(x)$, $i=0, 1, \cdots, m$, are not all zero on W (see proof of Theorem 2), that is, that $f_i(\bar{x})\neq 0$ for some i, where $(\bar{x})$ is the general point of W/k. Since $k[\bar{x}]\simeq k[x^*]$, also the $f_i(x^*)$ are not all zero. The polynomial $f(U, x^*)$ is zero for $U=u$, hence it belongs to the polynomial ideal $\mathfrak{P}$ considered in §1. It is linear in the U's and is not identically zero. Hence $f(U, x^*)$ can be taken as one of the σ polynomials $G_i(U, x^*)$, and $f(u, x)$ can be taken as one of the elements τ_i defined by (1). In other words, *$f(u, x)$ can be taken as one of a set of uniformizing parameters of $W^*(V/K)$.* Since F^*/K is defined in the quotient ring $Q_V(W^*/K)$ by the principal ideal $(f(u, x))$, it follows that W^*/K is simple for F^*/K, q.e.d.

5. **The geometric content of the theorem of Bertini.** In the classical formulation of Bertini's theorem there intervenes the notion of variable singular points of the varieties belonging to a linear system. The theorem is then stated in the following terms: *a variety V_{r-1} which varies in a linear system on a V_r cannot have variable singular points outside the singular locus of V_r and outside the base locus of the linear system.* We proceed to show that if the characteristic of the ground field k is equal to zero, then this classical formulation of Bertini's theorem is an easy consequence of our formulation of this

theorem as given in the preceding section; while if the characteristic of k is different from zero, then the classical formulation of Bertini's theorem cannot be maintained, for in that form the theorem is false.

With reference to the linear system $|F(\lambda)|$ introduced in §2, we must understand by a variable point P of $F(\lambda)$ the composite concept consisting of: (1) a point whose coördinates $x_1^*, x_2^*, \cdots, x_n^*$ are algebraic functions of $u_1, u_2, \cdots, u_m$ (over k) satisfying the relation:

$$(11)\qquad f(u, x^*) = f_0(x^*) + u_1f_1(x^*) + \cdots + u_mf_m(x^*) = 0$$

and which are such that $k[x] \sim k[x^*]$.

(2) An *arbitrary* specialization $u_i \to \lambda_i$, $x_i^* \to x_i(\lambda) = x_i^\lambda$, where $\lambda_i \in k$ and where the $x_i(\lambda)$ are algebraic quantities over k. Since $k[x] \sim k[x^*]$, the point (x^λ) lies on V, and it is clear by (11) and (9) that (x^λ) belongs to $F(\lambda)$. If we wish to insist that P be actually a variable point in the set-theoretic sense, we must add the condition that the field $k(x_1^*, x_2^*, \cdots, x_n^*)$ be of degree of transcendency not less than 1 over k.

The coördinates $x_1^*, x_2^*, \cdots, x_n^*$ define a general point of an irreducible zero-dimensional variety W^*/K which lies on F^*/K; and conversely, any irreducible zero-dimensional subvariety W^*/K of F^*/K defines a variable point of the variable member $F(\lambda)$ of the linear system $|F(\lambda)|$. The contraction W/k of W^*/K is the geometric locus of the variable point (x^λ).

We shall now assume that k is of characteristic zero. We suppose that the variety W/k is simple for V/k and that it is not a base variety of $|F(\lambda)|$. Then by the theorem of Bertini, as formulated in §4, W^*/K is simple for F^*/K. Let $H_1(u, X), H_2(u, X), \cdots, H_N(u, X)$ be a base of the prime ideal in the polynomial ring $K[X_1, X_2, \cdots, X_n]$ which defines the variety F^*/K. Then it is well known(11) that the Jacobian matrix

$$(12)\qquad \|\partial H_i/\partial X_j\|, \qquad i = 1, 2, \cdots, N; j = 1, 2, \cdots, n,$$

must be of rank $n-r+1$ on W^*. We now use the identity (9). Since $R(u_1, u_2, \cdots, u_{m-1}, x) = 0$, it follows from (9) that if $h_1(X), h_2(X), \cdots, h_g(X)$ is a base of the prime ideal of V/k in $k[X]$, then $[f_m(X)]^q H(u, X)$ belongs to the ideal generated by $f(u, X), h_1(X), \cdots, h_g(X)$ in $K[X]$. Here the integer q depends on $H(u, X)$, and $H(u, X)$ is any polynomial which vanishes on F^*/K. For a suitable integer q it will then be true that the products

$$[f_m(X)]^q H_i(u, X), \qquad i = 1, 2, \cdots, N,$$

(11) W. Schmeidler, *Über die Singularitäten algebraischer Gebilde*, Math. Ann. vol. 81 (1920) p. 227. That the definition of simple points by means of nonvanishing Jacobians is equivalent—in the case of characteristic zero—to our intrinsic definition by means of uniformizing parameters will be proved by us in a forthcoming paper in these Transactions. In the same paper we shall extend this equivalence to ground fields of characteristic $p \neq 0$ by using certain mixed Jacobians, which, in addition to partial derivatives with respect to the coördinates of the point, involve also partial derivatives with respect to certain parameters which appear in the coefficients of the equations of the variety.

belong to the ideal generated by $f(u, X)$, $h_1(X)$, $h_2(X)$, $\cdots$, $h_g(X)$ in the polynomial ring $K[X]$.

Since W/k is not a base variety and is simple for V/k, we may suppose without loss of generality that $f_m(x)\neq 0$ on W/k (see proof of Theorem 2). We therefore conclude, since the matrix (12) is of rank $n-r+1$ on W^*/K, that the matrix

$$J(u, X) = \|\partial f(u, X)/\partial X_j, \partial h_i(X)/\partial X_j\|, \quad i = 1, 2, \cdots, g; j = 1, 2, \cdots, n, \tag{13}$$

is of rank not less than $n-r+1$ on W^*/K. But the matrix consisting of the last g columns of the matrix (13) is of rank not greater than $n-r$ on W/k, for V/k is of dimension r. *Hence the matrix* (13) *is exactly of rank* $n-r+1$ *on* W^*/K, that is, the matrix $J(u, x^*)$ is of rank $n-r+1$. From this it follows that in the m-dimensional linear space of $u_1, u_2, \cdots, u_m$ (over k) the points $(\lambda_1, \lambda_2, \cdots, \lambda_m)$, such that $J(\lambda, x(\lambda))$ is of rank less than $n-r+1$, lie on an algebraic variety L of dimension less than m. If then $F(\lambda)$ is any member of the linear system $|F(\lambda)|$ such that the point (λ) does not lie on the above algebraic variety L, the point x^λ is a simple point of $F(\lambda)$. We have therefore shown that if the locus W/k of a variable point x^λ of $|F(\lambda)|$ is a simple subvariety of V/k and is not a base variety of the linear system $|F(\lambda)|$, then x^λ cannot be a variable singular point of $F(\lambda)$. This is precisely the theorem of Bertini in its classical formulation.

If k is of characteristic p it is not difficult to construct any number of counterexamples. For instance, if k is an algebraically perfect field, then every member of the pencil of curves $x^p-\lambda y^p=0$ in the (x, y)-plane consists entirely of p-fold points. In the following counterexample the curves of the pencil are absolutely irreducible: $x^p+y^2-2\lambda y=0$. These curves have a variable double point $x=(\lambda^2)^{1/p}$, $y=\lambda$, for we have $x^p+y^2-2\lambda y=(x-(\lambda^2)^{1/p})^p+(y-\lambda)^2$.

THE JOHNS HOPKINS UNIVERSITY,
BALTIMORE, MD.

THE CONCEPT OF A SIMPLE POINT OF AN ABSTRACT ALGEBRAIC VARIETY

BY

OSCAR ZARISKI

CONTENTS

Presented to the Society, August 23, 1946; received by the editors August 15, 1946.

INTRODUCTION

In [11] and [12](1) we have developed the concept of a simple point of an algebraic variety over an arbitrary ground field of characteristic zero. Our analysis of this concept has led us to the following characteristic property of a simple point which was novel and of intrinsic character (from the standpoint of regular birational transformations), since it referred to the quotient ring of the point and did not involve the ambient space of the variety:

A. *A point P of an r-dimensional irreducible algebraic variety V is a simple point of V if and only if the ideal of nonunits in the quotient ring of P has a basis of r elements.*

This property of a simple point played an essential role in our study of other questions, for instance in the problem of local uniformization, and for this reason we found it convenient to use A as a definition of simple points(2).

On the other hand we have the following classical and time-honored definition of simple points:

B. *If* $((f_1(x_1, x_2, \cdots, x_n), f_2(x_1, x_2, \cdots, x_n), \cdots, f_\nu(x_1, x_2, \cdots, x_n))$ *is a basis of the defining prime ideal of an r-dimensional irreducible algebraic variety V in an* S_n, *then a point P of V is simple for V if and only if the Jacobian matrix* $\partial(f_1, f_2, \cdots, f_\nu)/\partial(x_1, x_2, \cdots, x_n)$ *is of rank* $n-r$ *at P.*

If the ground field κ is of characteristic zero or a perfect field of characteristic $p\neq 0$, the two definitions A and B are equivalent (see §7.2, Theorem 7 and corollary), but in the case of nonperfect fields the two definitions may very well cease to be equivalent. We illustrate this fact by some examples.

Example 1. If a is an element of κ such that $a^{1/p}\notin\kappa$, then the polynomial $f(x, y)=x^p+y^p-a$ is irreducible over κ (p=characteristic of κ, $p\neq 0$). The partial derivatives $\partial f/\partial x$, $\partial f/\partial y$ vanish identically; therefore in the sense of Definition B *all the points of the irreducible curve* $f(x, y)=0$ *are singular.* On the other hand it is easily seen that in the sense of Definition A *all the points of the curve are simple.* To see this one either shows directly that in the quotient ring of any point of the curve the nonunits form a principal ideal, or one observes that we are dealing here with a *normal* curve, since the coördinate ring $\kappa[\xi, \eta]$ (where ξ and η are the coördinates of the general point of the

(1) Numbers in brackets refer to the references cited at the end of the paper.

(2) See [12, p. 199]. This definition is restated in the present paper in terms of vector spaces (Definition 1, §3.1).

curve) is integrally closed(3).

Example 2. The field $\kappa(\xi, \eta)$ of rational functions on the curve of the preceding example is not separably generated over κ. Now, quite generally, let V be an irreducible algebraic variety over κ and let $\mathcal{J}(V)$ denote the field of rational functions on V. It can be shown that *if $\mathcal{J}(V)$ is not separably generated over κ then all the points of V are singular in the sense of Definition* B (see §8.1). On the other hand we prove (Theorem 3 and corollary, §4.2) that *every irreducible variety carries points which are simple in the sense of Definition* A.

Example 3. Our attention has been called by Chevalley to the following example: $f(x, y) = y^2 + x^p - a = 0$, $p \neq 2$, $\alpha = a^{1/p} \notin \kappa$. We have $\partial f/\partial x = 0$, $\partial f/\partial y = 2y$, and hence the curve has one singular point $(\alpha, 0)$ in the sense of Definition B. Again we are dealing here with a normal curve, and hence, according to Definition A, all the points of the curve are simple. This example is significant because not only is the field of functions on the given curve separably generated over κ, but the curve is even absolutely irreducible (that is, the polynomial $y^2 + x^p - a$ remains irreducible upon any extension of the ground field).

In §7.2 we shall prove (Theorem 7) that points which are simple in the sense of Definition B are also simple in the sense of Definition A. There is ample evidence in the present paper, as well as in previous papers of ours, in support of the thesis that it is the more general concept of a simple point, as defined in A, that constitutes the natural generalization of the classical concept of simple point. The considerations expounded below will serve the twofold purpose of reviewing this evidence and of clarifying the underlying ideas and the motivation of our present *systematic treatment of the general concept of a simple point in algebraic geometry.*

(a) The case of algebraic *curves* is particularly illuminating and deserves special consideration. A normal algebraic curve C is a true projective model of the Riemann surface of the field $\mathcal{J}(C)$ (in the sense of Dedekind-Weber), for in the first place the points of such a curve are in 1-1 correspondence with the prime divisors of the field $\mathcal{J}(C)$, and in the second place the quotient ring of any point P of C coincides with the set of functions in $\mathcal{J}(C)$ which have non-negative order at the corresponding divisor $\mathfrak{p}$ (in other words: that quotient ring is a valuation ring). Now as long as we remain within the field $\mathcal{J}(C)$ there is no good reason why certain prime divisors of this field be called singular, and there is even less reason why *all* the prime divisors of $\mathcal{J}(C)$ be regarded as singular whenever $\mathcal{J}(C)$ is not separably generated over κ. From this point of view it would appear that no point of a normal curve deserves to be branded as singular.

(b) On the other hand it must be observed that the curves of Examples 1 and 3 do have at their "singular" points (in the sense of Definition B) a *singular behavior with respect to suitable extensions of the ground field* κ. If we

(3) For the concept of a normal variety see [11], especially Theorem 11′, and [13, p. 506].

pass to the field $\kappa_1=\kappa(\alpha)$, then the curve of Example 1 becomes a p-fold line: $x^p+y^p-a=(x+y-\alpha)^p=0$, and therefore all the points of the given curve become p-fold points upon the ground field extension $\kappa\rightarrow\kappa_1$. A similar situation prevails at the point $(\alpha, 0)$ of the curve of Example 3, and this time the phenomenon is even more apparent because of the absolute irreducibility of the curve: the curve $y^2+x^p-a=0$ remains irreducible over κ_1, but since its equation takes the form $y^2+(x-\alpha)^p=0$ it follows that the point $(\alpha, 0)$ is singular, in the sense of Definition A, if viewed from the level of the new ground field κ_1. These examples show that *points which are simple in the sense of Definition* A *may become singular when the ground field is extended.* We prove in §10.2 (Theorem 13) that *a point which is simple in the sense of Definition* A *is also simple in the sense of Definition* B *if and only if it remains simple under any extension of the ground field.* For this reason we use the term "*absolutely simple*" (proposed by André Weil) to designate points which are simple in the sense of Definition B (see Definition 2, §10.2). From now on we shall use the term "simple" in the sense of Definition A.

(c) We have proved in [13] that simple points behave under birational transformations in the abstract case in much the same way as they do in the classical case. In [14] we have extended to arbitrary ground fields the well known theorem of Bertini on the variable singular points of a linear system (whereas the old proofs of this theorem invariably make use of the usual differential conditions for simple points). Thus we have two significant instances of classical questions which involve the concept of a simple point and in which it turns out that the final results have nothing to do with the Jacobian criterion B. Our work on the problems of local uniformization and the resolution of singularities contains a considerable amount of material which remains valid for ground fields of characteristic $p\neq 0$, provided simple points are intended in the sense of Definition A. The final results in these problems are definitely false if by "simple" we mean "absolutely simple" (see Example 2 above). That the difficulties still to be overcome in the case $p\neq 0$ are merely of a technical nature and are not caused by any flaw in our formulation of the general concept of a simple point is strongly indicated by the fact that these difficulties already arise in the case of perfect ground fields when the two definitions A and B are, as we know, still equivalent.

(d) The Jacobian criterion for absolutely simple points implies that the points of an algebraic variety V which are not absolutely simple are those which satisfy a certain system of algebraic equations. Hence these points form an algebraic manifold, just as in the classical case the singular manifold is always algebraic. The only difference is that in the case of nonperfect fields the points which are not absolutely simple may fill up the entire variety V. In this respect the concept of an absolutely simple point presents no new problem.

We face an entirely different situation when we deal with the general concept of a simple point, since Definition A is strictly of a *local character*; it does not give a *global criterion* which would characterize singular points by means of algebraic equations. Therefore in the case of nonperfect fields two questions can be formulated: (1) *is the singular manifold of an algebraic variety V algebraic*? (2) *is this manifold always a proper subset of V*? The present investigation originated in these two questions, and we answer both in the affirmative. The second question can be settled already by local considerations. That is done in §4 of Part I (see Theorem 3 and corollary in §4.2). In Part I we study the concept of a simple point by local methods only and we go as far as those methods permit.

Much more difficult is the proof that the singular manifold is algebraic. The proof is achieved by deriving an algebraic criterion for simple points in which there occur certain mixed Jacobian matrices (see Theorem 11 and corollary, §9.6). In these matrices there appear derivatives of two types: (1) ordinary derivatives with respect to the variable coördinates $x_1, x_2, \cdots, x_n$; (2) derivatives which arise from abstract differentiations in κ over κ^p. Our general Jacobian criterion reduces of course to the classical criterion B if κ is a perfect field.

Although so far we spoke only of points, actually this paper deals with subvarieties of any dimension of a given variety V. In principle, by a well known reduction to the zero-dimensional case (§2.2), the case of higher subvarieties should not present new features, but in point of fact it does lead to specific results which are of interest. For an indication of the nature of these results, as well as of other results not mentioned in this introduction, we refer the reader to the table of contents.

It is with pleasure that the author takes this opportunity of gratefully acknowledging the stimulating discussions and the lively correspondence which he has had with André Weil and in the course of which the ideas embodied in the present work gradually took shape in his mind.

PART I. THE LOCAL THEORY

1. Notation and terminology. We fix an arbitrary abstract field κ and we refer to it as *ground field* (or *field of coefficients*). The algebraically closed field determined by κ shall be denoted by $\bar{\kappa}$ and shall be referred to as the *field of constants*.

A *point P* is an ordered n-tuple $(\alpha_1, \alpha_2, \cdots, \alpha_n)$ of constants α_i (that is, $\alpha_i \in \bar{\kappa}$). However, we stipulate that two n-tuples (α) and (β) represent one and the same point (over κ) if (and only if) they are conjugate over κ, that is, if there exists a κ-isomorphism of the field $\kappa(\alpha_1, \alpha_2, \cdots, \alpha_n)$ onto the field $\kappa(\beta_1, \beta_2, \cdots, \beta_n)$ in which to α_i there corresponds β_i $(i=1, 2, \cdots, n)$. Hence a point (over κ), if viewed from the level of the field $\bar{\kappa}$, is actually a complete set of conjugate points with respect to κ. The totality of all points is called the *linear n-dimensional space over* κ and is denoted by S_n^κ. The superscript indicating the field will be used only if a field other than κ is used temporarily as ground field. Consequently, in the case of the given ground field κ we shall write S_n instead of S_n^κ.

If $x_1, x_2, \cdots, x_n$ are indeterminates, an ideal $\mathfrak{A}$ in the polynomial ring $\kappa[x_1, x_2, \cdots, x_n]$ defines an algebraic variety $V = V/\kappa$ in S_n over κ, the *zero manifold* of $\mathfrak{A}$. We shall denote this variety by $\mathcal{V}(\mathfrak{A})$. Conversely, every algebraic variety V defines uniquely an ideal, namely the ideal consisting of those polynomials $f(x)$ which vanish at every point of V. We shall denote this ideal by $\boldsymbol{I}(V)$. If $V = \mathcal{V}(\mathfrak{A})$, then $\boldsymbol{I}(V) =$ Radical of $\mathfrak{A}$ (Hilbert's Nullstellensatz) [15] and $\mathcal{V}(\boldsymbol{I}(V)) = V$.

For an *irreducible* variety V the ideal $\boldsymbol{I}(V)$ is prime and shall be denoted by $\boldsymbol{p}(V)$. If $\mathfrak{p} = \boldsymbol{p}(V)$ and if ξ_i denotes the $\mathfrak{p}$-residue of x_i, then the ordered n-tuple $(\xi_1, \xi_2, \cdots, \xi_n)$ is a *general point* of V. We mean by *the coördinate ring of* V the ring $\kappa[\xi_1, \xi_2, \cdots, \xi_n]$, and we denote this ring by $\mathcal{R}[V]$. Similarly we denote by $\mathcal{J}(V)$ the field $\kappa(\xi_1, \xi_2, \cdots, \xi_n)$ of rational functions on V (the quotient field of $\mathcal{R}[V]$).

Let V be irreducible and let W be an irreducible algebraic subvariety of V. Then $\boldsymbol{p}(V) \subset \boldsymbol{p}(W)$ and $\boldsymbol{p}(W)/\boldsymbol{p}(V)$ is a prime ideal in the ring $\mathcal{R}[V]$. We denote this ideal by $\boldsymbol{p}(W/V)$. If $(\eta_1, \eta_2, \cdots, \eta_n)$ is the general point of W then we have $\mathcal{R}[V]/\boldsymbol{p}(W/V) = \kappa[\eta_1, \eta_2, \cdots, \eta_n]$.

We denote by $Q(W/V)$ the *quotient ring of* W *on* V, that is, the quotient ring of the prime ideal $\boldsymbol{p}(W/V)$ with respect to the ring $\mathcal{R}[V]$. The ring $Q(W/V)$ is a local ring in the sense of Krull [7]. Its maximal ideal shall be denoted by $\boldsymbol{m}(W/V)$; this ideal consists of all nonunits of $Q(W/V)$. The residue field $Q(W/V)/\boldsymbol{m}(W/V)$ of this local ring coincides with the field $\mathcal{J}(W)$.

2. The local vector space $\mathcal{M}(W/V)$.

2.1. *The mapping* $\mathfrak{m} \to \mathfrak{m}/\mathfrak{m}^2$. We set $\mathfrak{o} = Q(W/V)$, $\mathfrak{m} = \boldsymbol{m}(W/V)$, $\Delta = \mathfrak{o}/\mathfrak{m} = \mathcal{J}(W)$ and we consider the ring $\mathfrak{m}/\mathfrak{m}^2$. This ring, *as an additive group*, can be regarded as a vector space over Δ if, for $\bar{u}$ in $\mathfrak{m}/\mathfrak{m}^2$ and $\bar{\delta}$ in Δ, we define the product $\bar{\delta}\bar{u}$ as follows:

$$\bar{\delta}\bar{u} = \mathfrak{m}^2\text{-residue of } \delta u,$$

where δ is any element of $\mathfrak{o}$ whose $\mathfrak{m}$-residue is $\bar{\delta}$ and u is any element of $\mathfrak{m}$ whose $\mathfrak{m}^2$-residue is $\bar{u}$. It is immediately seen that the product $\bar{\delta}\bar{u}$ is uniquely determined by $\bar{\delta}$ and $\bar{u}$. We call this vector space *the local vector space of* V *at* W and we denote this space by $\mathcal{M}(W/V)$. We denote by τ the mapping $u \to \bar{u}$ of $\mathfrak{m}$ onto $\mathcal{M}(W/V)$:

$$(1) \qquad \tau\colon\ u \to \bar{u}, \qquad u \in \mathfrak{m}, \qquad \bar{u} = \mathfrak{m}^2\text{-residue of } u.$$

Let $v, v_1, v_2, \cdots, v_g$ be elements of $\mathfrak{m}$ and let $\bar{v} = \tau v$, $\bar{v}_i = \tau v_i$. If $\delta_1, \delta_2, \cdots, \delta_g$ are elements of $\mathfrak{o}$ and if $\bar{\delta}_i$ is the $\mathfrak{m}$-residue of δ_i then the following relations are equivalent: $\bar{v} = \sum_{i=1}^{g} \bar{\delta}_i \bar{v}_i$, $v \equiv \sum_{i=1}^{g} \delta_i v_i \pmod{\mathfrak{m}^2}$. Hence $\bar{v}$ is linearly dependent on $\bar{v}_1, \bar{v}_2, \cdots, \bar{v}_g$ if and only if v belongs to the ideal

$\mathfrak{o}\cdot(v_1, v_2, \cdots, v_g)+\mathfrak{m}^2$. In particular, *if the elements* $v_1, v_2, \cdots, v_g$ *form a basis for the ideal* $\mathfrak{m}$, *then the vectors* $\bar{v}_1, \bar{v}_2, \cdots, \bar{v}_g$ *span the entire vector space* $\mathfrak{M}(W/V)$. Consequently the dimension of this vector space is finite.

Conversely, let us assume that the g vectors $\bar{v}_i$ span the entire vector space $\mathfrak{M}(W/V)$. We show then that *the g elements v_i form a basis for the ideal* $\mathfrak{m}$. For if we denote by $\mathfrak{A}$ the ideal $\mathfrak{o}\cdot(v_1, v_2, \cdots, v_g)$, then it follows, by assumption, that $\mathfrak{m}=\mathfrak{A}+\mathfrak{m}^2$. Hence $\mathfrak{m}^2=\mathfrak{A}\mathfrak{m}+\mathfrak{m}^3\subseteq\mathfrak{A}+\mathfrak{m}^3$ and therefore $\mathfrak{m}=\mathfrak{A}+\mathfrak{m}^3$. This implies $\mathfrak{m}^2=\mathfrak{A}\mathfrak{m}+\mathfrak{m}^4\subseteq\mathfrak{A}+\mathfrak{m}^4$ and therefore $\mathfrak{m}=\mathfrak{A}+\mathfrak{m}^4$. In like fashion we find $\mathfrak{m}=\mathfrak{A}+\mathfrak{m}^i$ for any integer i. Since $\mathfrak{o}$ is a local ring, we have $\bigcap_{i=1}^{\infty}(\mathfrak{A}+\mathfrak{m}^i)=\mathfrak{A}$ ([7, Theorem 2]). Hence $\mathfrak{A}=\mathfrak{m}$, and this proves our assertion.

We say that a basis $(u_1, u_2, \cdots, u_s)$ of $\mathfrak{m}$ is *minimal* if no proper subset of this basis is a basis of $\mathfrak{m}$. From the preceding results it follows that $(u_1, u_2, \cdots, u_s)$ *is a minimal basis of* $\mathfrak{m}$ *if and only if the corresponding vectors* $\bar{u}_1, \bar{u}_2, \cdots, \bar{u}_s$ *form a (independent) basis of the vector space* $\mathfrak{M}(W/V)$. All minimal bases of $\mathfrak{m}$ have therefore the same number of elements, this number being equal to the dimension of $\mathfrak{M}(W/V)$.

Let $(u_1, u_2, \cdots, u_s)$ be a minimal basis of $\mathfrak{m}$, so that s = dimension of $\mathfrak{M}(W/V)$. We have $u_i=f_i(\xi)/g_i(\xi)$, where $f_i(\xi), g_i(\xi)\in\mathfrak{R}[V]$ and $g_i(x)\neq 0$ on W (that is, $g_i(\xi)\notin \mathfrak{p}(W/V)$). Since $g_i(\xi)$ is a unit in $\mathfrak{o}$, also $f_1(\xi)$, $f_2(\xi), \cdots, f_s(\xi)$ form a minimal basis of $\mathfrak{m}$. Therefore we may assume without loss of generality that $u_i\in\mathfrak{R}[V]$, $i=1, 2, \cdots, s$. Let r = dimension of V and ρ = dimension of W. The finite integral domain $R=\mathfrak{R}[V]$ has degree of transcendency r over κ. It is then well known (Krull [5, p. 43]) that every isolated prime ideal of the ideal $R\cdot(u_1, u_2, \cdots, u_s)$ is of dimension not less than $r-s$. Among the isolated prime ideals there is the ideal $\mathfrak{p}=\mathfrak{p}(W/V)$, since $\mathfrak{o}\cdot(u_1, u_2, \cdots, u_s)=\mathfrak{m}=\mathfrak{o}\cdot\mathfrak{p}$, and this prime ideal $\mathfrak{p}$ is of dimension ρ. Hence $\rho\geqq r-s$ or $s\geqq r-\rho$, that is, we have the following result: *the dimension of the local vector space* $\mathfrak{M}(W/V)$ *satisfies the inequality*

$$\dim \mathfrak{M}(W/V) \geqq \dim V - \dim W. \tag{2}$$

In particular, if W is a point P of V, then we have:

$$\dim \mathfrak{M}(P/V) \geqq \dim V. \tag{2'}$$

Two procedures will be used frequently in the sequel for the actual determination of a basis of the vector space $\mathfrak{M}(W/V)$: (1) *ground field extensions*; (2) *insertion of varieties between* W *and* V. The lemmas given below are intended to introduce these procedures.

2.2. *Reduction to dimension zero.* Let $(\eta_1, \eta_2, \cdots, \eta_n)$ be the general point of W. If for a given integer ν, $1\leqq\nu\leqq n$, the κ-homomorphism $\kappa[\xi_1, \xi_2, \cdots, \xi_n]\sim\kappa[\eta_1, \eta_2, \cdots, \eta_n]$ which carries ξ_i into η_i, $i=1, 2, \cdots, n$, induces an isomorphism between $\kappa[\xi_1, \xi_2, \cdots, \xi_\nu]$ and $\kappa[\eta_1, \eta_2, \cdots, \eta_\nu]$, then it is permissible to identify $\xi_1, \xi_2, \cdots, \xi_\nu$ with $\eta_1, \eta_2, \cdots, \eta_\nu$ respectively. We shall express this assumption and the fact that the identification has actually been performed by writing: $\xi_i=\eta_i$, $i=1, 2, \cdots, \nu$.

LEMMA 1. *Let* $\xi_i=\eta_i$, $i=1, 2, \cdots, \nu$, *and let* κ^* *denote the field* $\kappa(\xi_1, \xi_2, \cdots, \xi_\nu)$. *If* V^* and W^* *are the varieties (over* κ^**) in* $S_{n-\nu}^{\kappa^*}$ *having respectively* $(\xi_{\nu+1}, \xi_{\nu+2}, \cdots, \xi_n)$ *and* $(\eta_{\nu+1}, \eta_{\nu+2}, \cdots, \eta_n)$ *as general points, then the quotient ring* $\mathfrak{o}=Q(W/V)$ *coincides with the quotient ring* $\mathfrak{o}^*=Q(W^*/V^*)$ *(and hence also the vector spaces* $\mathfrak{M}(W/V)$, $\mathfrak{M}(W^*/V^*)$ *coincide).*

Proof. The fact that the identification $\xi_i=\eta_i$, $i=1, 2, \cdots, \nu$, is permissible, signifies that 0 is the only element common to $\kappa[\xi_1, \xi_2, \cdots, \xi_\nu]$ and $\mathfrak{p}$ $(=\boldsymbol{p}(W/V))$. *Hence the quotient ring* $\mathfrak{o}$ *contains the entire field* κ^*. It is clear that $\mathfrak{o}\in\mathfrak{o}^*$. On the other hand we have: $R^*=\mathfrak{R}[V^*]=\kappa^*R$, $\mathfrak{p}^*=\boldsymbol{p}(W^*/V^*)=\kappa^*\mathfrak{p}$, and every element of κ^*R which is not in $\kappa^*\mathfrak{p}$ is evidently a unit in $R_\mathfrak{p}$ $(=\mathfrak{o})$. Hence $\mathfrak{o}^*=R^*_{\mathfrak{p}^*}\subseteq\mathfrak{o}$, and therefore $\mathfrak{o}^*=\mathfrak{o}$, as was asserted.

The above lemma includes as a special case the well known *reduction to dimension zero*. Namely, the dimension of W being ρ, we may assume that $\eta_1, \eta_2, \cdots, \eta_\rho$ are algebraically independent over κ. In that case necessarily also $\xi_1, \xi_2, \cdots, \xi_\rho$ are algebraically independent over κ, and the identification $\xi_i=\eta_i$, $i=1, 2, \cdots, \rho$, is permissible. The variety V^* of the lemma is now a variety of dimension $n-\rho$ (over $\kappa^*=\kappa(\xi_1, \xi_2, \cdots, \xi_\rho)$), immersed in a linear space $S_{n-\rho}^{\kappa^*}$, while W^*/κ^* is now a *point* of V^*, because $\eta_{\rho+1}, \eta_{\rho+2}, \cdots, \eta_n$ are algebraic quantities over the new ground field κ^*.

2.3. *The linear transformation* $\mathfrak{M}(W/V)\rightarrow\mathfrak{M}(W/V')$. In the "insertion" procedure we insert an irreducible variety V' between W and V: $W\subset V'\subset V$, and we consider the two vector spaces $\mathfrak{M}=\mathfrak{M}(W/V)$, $\mathfrak{M}'=\mathfrak{M}(W/V')$. These two spaces have the same field of scalars, namely $\mathcal{J}(W)$ $(=\Delta)$. Let $\mathfrak{o}'=Q(W/V')$, $\mathfrak{m}'=\boldsymbol{m}(W/V')$. We have a mapping τ' of $\mathfrak{m}'$ onto $\mathfrak{M}'$ similar to the mapping τ of $\mathfrak{m}$ onto $\mathfrak{M}$ (see (1)):

$$\tau': \quad u'\rightarrow \bar{u}', \qquad u'\in\mathfrak{m}', \qquad \bar{u}' = \mathfrak{m}'^2\text{-residue of } u'.$$

If $(\xi_1', \xi_2', \cdots, \xi_n')$ is the general point of V'/κ, there is a definite κ-homomorphism ψ of $R=\mathfrak{R}[V]$ onto $R'=\mathfrak{R}[V']$ which carries ξ_i into ξ_i' (since $V'\subset V$), and $\mathfrak{p}=\boldsymbol{p}(W/V)$ is the full inverse image of $\mathfrak{p}'=\boldsymbol{p}(W/V')$ under ψ^{-1}. It follows that ψ can be extended (in a unique fashion) to a homomorphism of $\mathfrak{o}=R_\mathfrak{p}$ onto $\mathfrak{o}'=R'_{\mathfrak{p}'}$. We denote this extended homomorphism by the same letter ψ. It is clear that $\psi\mathfrak{m}=\mathfrak{m}'$.

We consider in $\mathfrak{R}[V]$ the ideal $\mathfrak{p}_1=\boldsymbol{p}(V'/V)$. This ideal is the nucleus of the homomorphism ψ of R onto R'. Since $W\subset V'$, $\mathfrak{p}_1$ is contained in $\mathfrak{p}$, and therefore the extended ideal $\mathfrak{P}_1=\mathfrak{o}\cdot\mathfrak{p}_1$ is a prime ideal contained in $\mathfrak{m}$. The ideal $\mathfrak{P}_1$ is the nucleus of the homomorphism ψ of $\mathfrak{o}$ onto $\mathfrak{o}'$. If $\mathfrak{A}$ is any $\mathfrak{o}$-ideal contained in $\mathfrak{m}$, then $\tau\mathfrak{A}$ is a linear subspace of $\mathfrak{M}$. We consider in particular the subspace $\tau\mathfrak{P}_1$ of $\mathfrak{M}$. This subspace is spanned by the vectors belonging to the subset $\tau\mathfrak{p}_1$ of $\mathfrak{M}$.

LEMMA 2. *The transformation* $\phi = \tau^{-1}\psi\tau'$ *is single-valued and is a linear transformation of* $\mathfrak{M} = \mathfrak{M}(W/V)$ *onto* $\mathfrak{M}' = \mathfrak{M}(W/V')$. *The subspace of* $\mathfrak{M}$ *which is annihilated by* ϕ *is the space spanned by the vectors belonging to* $\tau\mathfrak{p}_1$ (*where* $\mathfrak{p}_1$ *is the ideal* $\boldsymbol{p}(V'/V)$ *of* V' *in* $\mathfrak{R}[V]$), *or also the space* $\tau\mathfrak{P}_1$ (*where* $\mathfrak{P}_1 = \mathfrak{o}\cdot\mathfrak{p}_1$).

Proof. Since τ and τ' are respectively transformations of $\mathfrak{m}$ onto $\mathfrak{M}$ and of $\mathfrak{m}'$ onto $\mathfrak{M}'$, and since $\psi\mathfrak{m} = \mathfrak{m}'$, ψ *is a transformation of* $\mathfrak{M}$ *onto* $\mathfrak{M}'$. If $\bar{u}$ is any element of $\mathfrak{M}$, then $\tau^{-1}\bar{u}$ is a residue class $u + \mathfrak{m}^2$ in $\mathfrak{m}$ modulo $\mathfrak{m}^2$; $\psi(u+\mathfrak{m}^2) = \psi u + \psi(\mathfrak{m}^2) = u' + \mathfrak{m}'^2$, since ψ is a homomorphism and since $\psi\mathfrak{m} = \mathfrak{m}'$; and finally $\tau'(u' + \mathfrak{m}'^2) = \tau' u' = \bar{u}'$. *Hence* ψ *is single-valued.* Since τ, ψ and τ' are homomorphisms of additive groups, it follows that ϕ *is a homomorphism of the additive group* $\mathfrak{M}$ *onto the additive group* $\mathfrak{M}'$. If $\bar{\delta} \in \Delta$ $(= \mathfrak{J}(W))$ and $\bar{u} \in \mathfrak{M}$, let u be some element in $\tau^{-1}\bar{u}$ and *let* δ be some element in $\mathfrak{o}$ whose $\mathfrak{m}$-residue is $\bar{\delta}$. We have, by the definition of $\mathfrak{M}$ and τ: $(\tau\phi)\delta u = \phi(\bar{\delta}\bar{u})$. On the other hand, $\bar{\delta}$ is also the $\mathfrak{m}'$-residue of $\psi(\delta)$, and therefore $(\psi\tau')\delta u = \tau'(\psi(\delta)\psi(u)) = \bar{\delta}\cdot(\psi\tau')u$. Since $\tau\phi = \psi\tau'$ we conclude that $\phi(\bar{\delta}\bar{u}) = \bar{\delta}\cdot\phi\bar{u}$ *which shows that* ϕ *is linear.* Finally, we have $(\tau\phi)u = \phi(\bar{u})$ and $(\tau\phi)u = (\psi\tau')u$. Hence $\phi(\bar{u}) = 0$ if and only if $(\psi\tau')u = 0$, that is, if and only if $\psi u \in \mathfrak{m}'^2$. Since the nucleus of the homomorphism ψ of $\mathfrak{o}$ onto $\mathfrak{o}'$ is the ideal $\mathfrak{P}_1$ defined above, we conclude that $\phi(\bar{u}) = 0$ if and only if $u \in \mathfrak{m}^2 + \mathfrak{P}_1$, and this shows that the subspace of $\mathfrak{M}$ annihilated by ϕ is the space $\tau\mathfrak{P}_1$. This completes the proof of the lemma.

COROLLARY. *If* $u_1, u_2, \cdots, u_\nu$ *are elements of* $\mathfrak{m}$ *such that* $\psi u_1, \psi u_2, \cdots, \psi u_\nu$ *form a basis of* $\mathfrak{m}'$, *and if* $u_{\nu+1}, u_{\nu+2}, \cdots, u_{\nu+\mu}$ *form a basis of* $\mathfrak{P}_1$, *then the* $\nu+\mu$ *elements* u_i *form a basis of* $\mathfrak{m}$.

3. Simple points.

3.1. *Definition of simple loci.* The inequality (2′) assigns a lower limit to the dimension of the local vector space at any point P of a given variety V. This lower limit is the dimension r of V. More generally, by (2), the difference $r-\rho$ is a lower limit for the dimension of the local vector space at any ρ-dimentional irreducible subvariety W of V. We speak of *simple points* and of *simple subvarieties* of V when these lower limits are reached, that is, we give the following definition:

DEFINITION 1. *A point* P *of* V *is simple if* $\dim \mathfrak{M}(P/V) = r = \dim V$. *More generally, a* ρ*-dimensional irreducible subvariety* W *of* V *is simple if* $\dim \mathfrak{M}(W/V) = r-\rho$.

According to this definition, a ρ-dimensional irreducible subvariety W of V is simple if the maximal ideal $\boldsymbol{m}(W/V)$ of the quotient ring $Q(W/V)$ has a basis of $r-\rho$ elements (such a basis is then necessarily minimal). If W is simple, the $r-\rho$ elements of any minimal basis of $\boldsymbol{m}(W/V)$ shall be called *uniformizing parameters* of W.

W is *singular* for V if it is not simple.

We observe that our definition of simple loci has an intrinsic character, since the ambient linear space S_n does not intervene at all in the definition. Whether a given irreducible subvariety W of V is or is not simple for V depends entirely on the structure of the quotient ring $Q(W/V)$. This puts in evidence the ***invariantive*** character of the concept of simple loci with respect to *regular transformations*: if W is simple for V and if V is transformed into another variety V^* by a birational transformation which is regular at W (see [13, p. 513]), then to W there corresponds on V^* a (unique) subvariety W^* which is also simple for V^* (and has the same quotient ring as W).

3.2. *Geometric aspects of the definition.* In this section we shall discuss some geometric aspects of our definition of a simple point. For that purpose we shall examine the variety V, together with a given point P on it, in relation to the ambient linear space S_n in which V is immersed.

We have shown elsewhere that every point P of S_n is simple (see [13], Lemma 9, p. 541 and first footnote on that page). It is easy to detect in our proof of this result a specialization of the general set-up dealt with in Lemma 2. Namely, we apply that lemma to the following case: $V=S_n$, $W=P=P(\alpha_1, \alpha_2, \cdots, \alpha_n)$ and $V'=$the variety having $(\alpha_1, x_2, x_3, \cdots, x_n)$ as general point. The ideal $\mathfrak{p}(V'/V)$ is in this case the principal ideal $(f(x_1))$ in $\kappa[x_1, x_2, \cdots, x_n]$, where $f(x_1)$ is an irreducible polynomial in $\kappa[x_1]$ such that $f(\alpha_1)=0$. Hence the space annihilated by the linear transformation ϕ is at most of dimension 1, and therefore dim $\mathfrak{M}(P/S_n) \leqq 1+\dim \mathfrak{M}(P/V')$. If we now replace in Lemma 1 the varieties V and W by V' and P respectively, the lemma is applicable if we set $\nu=1$, since in the present case we have $\xi_1=\eta_1=\alpha_1$. Therefore we conclude that $\mathfrak{M}(P/V')=\mathfrak{M}(P^*/S_{n-1}^{\kappa^*})$, where $\kappa^*=\kappa(\alpha_1)$ and P^* is the point $(\alpha_2, \alpha_3, \cdots, \alpha_n)$ over κ^*. We have therefore: dim $\mathfrak{M}(P/S_n) \leqq 1+\dim \mathfrak{M}(P^*/S_{n-1}^{\kappa^*})$. Applying the same argument to the point P^*, and repeating this procedure n times, we shall get ultimately the inequality: dim $\mathfrak{M}(P/S_n) \leqq n$. Hence necessarily dim $\mathfrak{M}(P/S_n)=n$, and therefore P is a simple point of S_n.

We shall now proceed by analogy with the classical case, where κ is the field of complex numbers, or, more generally, an algebraically closed field. In that case the coordinates $\alpha_1, \alpha_2, \cdots, \alpha_n$ of any point P in S_n are in κ, and the n differences $x_i-\alpha_i$ form a minimal base of $\mathfrak{m}(P/S_n)$, that is, they are uniformizing parameters of P. The field of scalars Δ of the vector space $\mathfrak{M}(P/S_n)$ is the field κ itself, since $\Delta=\kappa(\alpha_1, \alpha_2, \cdots, \alpha_n)$. Hence, this vector space can be identified with the space of all linear forms in $x_1-\alpha_1, x_2-\alpha_2, \cdots, x_n-\alpha_n$ with coefficients in κ. If $u=\sum_{i=1}^n c_i(x_i-\alpha_i)+$terms of degree greater than 1 in the $x_i-\alpha_i$ is any polynomial in $\kappa[x_1, x_2, \cdots, x_n]$ which vanishes at P, then under the mapping τ of $\mathfrak{m}(P/S_n)$ onto $\mathfrak{M}(P/S_n)$ we get $\tau u=\sum_{i=1}^n c_i(x_i-\alpha_i)$. The classroom definition of a simple point of a hypersurface H, given by an equation $u=0$, takes therefore the following form: *a point P of the hypersurface H is simple if and only if the vector τu in $\mathfrak{M}(P/S_n)$ is different from zero.*

At a simple point P the vector τu fixes the position of the tangent hyperplane at P of the hypersurface H.

The above italicized statement can be of course easily proved in the abstract case on the basis of our definition of simple points. However, since we have defined simple points only for irreducible varieties, we must suppose that we are dealing with an irreducible hypersurface. We apply Lemma 2 to the case $V=S_n$, $V'=H$, $W=P$. Since the ideal $\boldsymbol{p}(V'/V)$ is now the principal ideal (u), the subspace of $\mathfrak{M}(P/S_n)$ which is annihilated by ϕ is spanned by the single vector τu. Consequently the dimension of the vector space $\mathfrak{M}(P/H)$ is n or $n-1$, according as $\tau u=0$ or $\tau u\neq 0$. Since H is of dimension $r=n-1$, the statement follows.

In the case of a reducible hypersurface H we shall define simple points just by the above criterion $\tau u\neq 0$. Let $u=\prod_{i=1}^{g} u_i^{\nu_i}$ be the decomposition of u into irreducible factors, and let H_i denote the hypersurface $u_i=0$. The condition $\tau u\neq 0$ signifies that $u\in\mathfrak{m}$, $u\notin\mathfrak{m}^2$, where $\mathfrak{m}=\boldsymbol{m}(P/S_n)$. That is equivalent to the following set of conditions: (1) one, and only one, of the polynomials $u_1, u_2, \cdots, u_g$ is in $\mathfrak{m}$; (2) if, say, $u_1\in\mathfrak{m}$, then $u_1\notin\mathfrak{m}^2$ and $\nu_1=1$. Or in geometric language: P is a simple point of H if and only if the following conditions are satisfied: (1) only one of the irreducible components H_i of the hypersurface H goes through P; (2) if, say, H_1 contains P, then P is a simple point of H_1 and H_1 is a simple component of H.

If P is a simple point of our hypersurface H, we define as *tangent hyperplane of H at P* the one-dimensional subspace of $\mathfrak{M}(P/S_n)$ spanned by the vector τu. A true picture of the set of all hyperplanes through P is the bundle of hyperplanes in an affine n-dimensional space over Δ $(=\mathcal{J}(P)=\kappa(\alpha_1, \alpha_2, \cdots, \alpha_n))$ with center at the origin.

Given ν hypersurfaces H_i: $u_i=0$, $i=1, 2, \cdots, \nu$, passing simply through P, we say that their tangent hyperplanes are linearly independent if the vectors $\tau u_1, \tau u_2, \cdots, \tau u_\nu$ in $\mathfrak{M}(P/S_n)$ are linearly independent. The following result is well known [3, corollary on p. 87]:

LEMMA 3. *If P is a simple point of an irreducible r-dimensional algebraic variety V and if $u_1, u_2, \cdots, u_\nu$ are elements of $\boldsymbol{m}(P/V)$ such that the corresponding vectors $\tau u_1, \tau u_2, \cdots, \tau u_\nu$ of $\mathfrak{M}(P/V)$ are linearly independent, then the ideal $\mathfrak{p}=(u_1, u_2, \cdots, u_\nu)$ in $Q(P/V)$ is prime and $(r-\nu)$-dimensional.*

COROLLARY. *The assumptions being as in Lemma 3, the irreducible $(r-\nu)$-dimensional subvariety W of V passing through P which is defined by the ideal $\mathfrak{p}$ is simple for V.*

For if $\mathfrak{o}=Q(P/V)$, then $Q(W/V)=\mathfrak{o}_\mathfrak{p}$ and $\boldsymbol{m}(W/V)=\mathfrak{o}_\mathfrak{p}\cdot\mathfrak{p}=\mathfrak{o}_\mathfrak{p}\cdot(u_1, u_2, \cdots, u_\nu)$. Hence $\dim \mathfrak{M}(W/V)\leqq\nu=r-\dim W$, and therefore, by (2), $\dim \mathfrak{M}(W/V)=\nu$.

For $V=S_n$ it follows from this lemma that if ν hypersurfaces H_i pass simply through a point P and have at P linearly independent tangent hyperplanes, then locally, at P, the complete intersection W of the hypersurfaces H_i is irreducible and of dimension $n-\nu$. We shall say then that W is locally, at P, *regular intersection* of the hypersurfaces H_i (the "regularity" assumption signifies not only that W is complete intersection of H_i locally at P, but also that the tangent hyperplanes of the hypersurfaces H_i at P are linearly independent).

After these preliminaries, we can express the geometric content of our definition of simple points by means of the following result:

THEOREM 1. *A point P of an r-dimensional algebraic irreducible variety V in S_n is simple for V, if and only if V is locally at P a regular intersection of $n-r$ hypersurfaces.*

Proof. If in Lemma 2 we identify W, V' and V with P, V and S_n respectively and if we take into account the inequality (2′) (§2.1), we conclude (independently of the assumption that P is simple for V) that the ideal $\mathfrak{p}(V)$ in $\kappa[x_1, x_2, \cdots, x_n]$ cannot contain more than $n-r$ polynomials u_i such that the corresponding vectors τu_i in $\mathfrak{M}(P/S_n)$ are linearly independent. The point P is simple for V if and only if the maximum $n-r$ is reached. Hence P is simple for V if and only if there exist $n-r$ hypersurfaces H_i containing V and having at P linearly independent tangent hyperplanes. The complete intersection of the H_i, locally at P, being of dimension r and containing V, it must coincide with V, q.e.d.

3.3. *Local ideal bases at a simple point.* In the preceding discussion the linear space S_n played the privileged role of a universal ambient space. We now replace the S_n by a fixed n-dimensional irreducible variety $\boldsymbol{S}$ (immersed in some linear space of a higher dimension). We consider a fixed *simple* point P of $\boldsymbol{S}$ and we proceed to develop some aspects of the local ideal theory of $\boldsymbol{S}$ at P similar to those developed above for the S_n.

We denote by R the coordinate ring $\mathfrak{R}[\boldsymbol{S}]$ and by $\mathfrak{o}$ the quotient ring $Q(P/\boldsymbol{S})$. Given any ideal $\mathfrak{A}$ in R, we say that a set of elements $\omega_1, \omega_2, \cdots, \omega_s$ in R is *a local P-basis of* $\mathfrak{A}$ if this set is a basis for the ideal $\mathfrak{o}\cdot\mathfrak{A}$; and that it is a *minimal* local P-basis if no proper subset of the set $(\omega_1, \omega_2, \cdots, \omega_s)$ is a local P-basis.

We observe that if $\mathfrak{B}$ is any ideal in $\mathfrak{o}$ and if $\mathfrak{m}$ is the maximal ideal of $\mathfrak{o}$, then the residue class ring $\mathfrak{B}/\mathfrak{m}\mathfrak{B}$, as an additive group, can be regarded as a vector space $\mathfrak{N}$ over the field $\Delta=\mathfrak{o}/\mathfrak{m}=$residue field of P. This is true in any local ring $\mathfrak{o}$, and the definition of that vector space is the same as the one given in §2.1 for the special case $\mathfrak{B}=\mathfrak{m}$.

By exactly the same reasoning which was employed in that special case it can be proved that $(u_1, u_2, \cdots, u_\nu)$ is a basis of $\mathfrak{B}$ if and only if the corresponding vectors $\bar{u}_1, \bar{u}_2, \cdots, \bar{u}_\nu$ span the entire space $\mathfrak{N}$, and that $(u_1, u_2, \cdots, u_\nu)$ is a minimal basis of $\mathfrak{B}$ if and only if the vectors $\bar{u}_1, \bar{u}_2, \cdots, \bar{u}_\nu$

form an independent basis of $\mathcal{N}$. Hence all minimal bases of $\mathfrak{B}$ have the same number of elements.

In particular, we conclude that all minimal local P-bases of a given ideal $\mathfrak{A}$ in R have the same number of elements.

We now extend and further elaborate Theorem 1 as follows:

THEOREM 2. *If V is an irreducible algebraic subvariety of $\mathbf{S}$ passing through P, then P is simple for V if and only if either one of the following two conditions is satisfied:*

(1) *The dimension of V being r, the ideal $\mathbf{p}(V/\mathbf{S})$ contains $n-r$ elements $u_1, u_2, \cdots, u_{n-r}$ such that the corresponding vectors $\tau u_1, \tau u_2, \cdots, \tau u_{n-r}$ in $\mathcal{M}(P/\mathbf{S})$ are linearly independent.*

(2) *The ideal $\mathbf{p}(V/\mathbf{S})$ possesses a local P-basis $u_1, u_2, \cdots, u_s$ such that the vectors $\tau u_1, \tau u_2, \cdots, \tau u_s$ are linearly independent.*

Moreover, if condition (1) *is satisfied, then the $n-r$ elements u_i form necessarily a minimal local P-basis of $\mathbf{p}(V/\mathbf{S})$, and any minimal local P-basis of $\mathbf{p}(V/\mathbf{S})$ necessarily consists of elements $v_1, v_2, \cdots, v_{n-r}$ of R such that the vectors $\tau v_1, \tau v_2, \cdots, \tau v_{n-r}$ are independent. If condition* (2) *holds then* $\dim V = n-s$ *and $(u_1, u_2, \cdots, u_s)$ is necessarily a minimal local P-basis of $\mathbf{p}(V/\mathbf{S})$.*

Proof. We have $P \subseteq V \subseteq \mathbf{S}$ and $\dim \mathcal{M}(P/\mathbf{S}) = n$. Therefore, by Lemma 2, $\dim \mathcal{M}(P/V) = \dim V$ if and only if $\mathbf{p}(V/\mathbf{S})$ contains $n-r$ elements u_i such as specified in condition (1) (no more than $n-r$ such elements can belong to $\mathbf{p}(V/\mathbf{S})$ because $\dim \mathcal{M}(P/V)$ is always greater than or equal to $\dim V$). Hence condition (1) is both necessary and sufficient.

By Lemma 3, the $n-r$ elements of condition (1) generate in $\mathfrak{o}$ a prime ideal of dimension r, and this ideal is contained in the ideal $\mathfrak{o}\cdot\mathfrak{A}$, where $\mathfrak{A} = \mathbf{p}(V/\mathbf{S})$. Since also $\mathfrak{o}\cdot\mathfrak{A}$ is prime and of dimension r, it follows that $(u_1, u_2, \cdots, u_{n-r})$ is a local P-basis of $\mathbf{p}(V/\mathbf{S})$. This shows that condition (2) is necessary. On the other hand, if this condition is satisfied, then again, by Lemma 3, we have necessarily $\dim V = n-s$, and by Lemma 2 we find $\dim \mathcal{M}(P/V) = n-s$. Therefore condition (2) is also sufficient.

No subset of set $(u_1, u_2, \cdots, u_{n-r})$ of condition (1) can be a local P-basis of $\mathbf{p}(V/\mathbf{S})$ for otherwise V would be of dimension greater than r.

To complete the proof of the theorem we proceed as follows. Assuming that condition (1) is satisfied, let $\mathfrak{B} = \mathfrak{o}\cdot\mathbf{p}(V/\mathbf{S})$. We have $\mathfrak{B}\mathfrak{m} \subseteq \mathfrak{m}^2 \cap \mathfrak{B}$. On the other hand, if ω is any element of $\mathfrak{m}^2 \cap \mathfrak{B}$, then $\omega = \delta_1 u_1 + \delta_2 u_2 + \cdots + \delta_{n-r} u_{n-r}$, where $\delta_i \in \mathfrak{o}$. Since $\omega \in \mathfrak{m}^2$ we have $\bar{\delta}_1 \bar{u}_1 + \bar{\delta}_2 \bar{u}_2 + \cdots + \bar{\delta}_{n-r} \bar{u}_{n-r} = 0$, where $\bar{\delta}_i = \mathfrak{m}$-residue of δ_i and $\bar{u}_i = \tau u_i$. Since the vectors τu_i are linearly independent, it follows that $\bar{\delta}_i = 0$, that is, $\delta_i \in \mathfrak{m}$, $i = 1, 2, \cdots, n-r$. Hence $\omega \in \mathfrak{B}\mathfrak{m}$. We have thus proved that $\mathfrak{B}\mathfrak{m} = \mathfrak{m}^2 \cap \mathfrak{B}$. *But this implies that the vector space $\mathfrak{B}/\mathfrak{B}\mathfrak{m}$ can be regarded as a subspace of the vector space* $\mathfrak{m}/\mathfrak{m}^2 = (\mathcal{M}(P/\mathbf{S}))$. Since we know that the elements of a minimal basis of $\mathfrak{B}$ give rise to independent vectors of $\mathfrak{B}/\mathfrak{B}\mathfrak{m}$, we conclude that the elements of such a basis must also give rise to independent vectors of $\mathfrak{M}(P/\mathbf{S})$. This completes the proof of the theorem.

3.4. *Simple zeros of ideals.* The criterion for simple points given by condition (2) of Theorem 2 expresses a property of the ideal $\boldsymbol{p}(V/\mathbf{S})$ which continues to have a meaning if instead of $\boldsymbol{p}(V/\mathbf{S})$ we are dealing with an arbitrary ideal $\mathfrak{A}$ in R $(=\mathfrak{R}[\mathbf{S}])$. This leads to a definition of a *simple zero* of an ideal $\mathfrak{A}$, and in particular of a simple point of a reducible algebraic variety:

DEFINITION 2. *Given an arbitrary ideal $\mathfrak{A}$ in $\mathfrak{R}[\mathbf{S}]$ and given a simple point $P(\alpha)$ of $\mathbf{S}$, we say that (α) is a simple zero of $\mathfrak{A}$ if (α) is a zero of $\mathfrak{A}$ and if $\mathfrak{A}$ possesses a local P-basis $(u_1, u_2, \cdots, u_s)$ such that the vectors $\tau u_1, \tau u_2, \cdots, \tau u_s$ in $\mathfrak{M}(P/\mathbf{S})$ are independent. In particular, given an arbitrary subvariety V of $\mathbf{S}$ (not necessarily irreducible), we say that P is simple for V if (α) is a simple zero of $\boldsymbol{I}(V/\mathbf{S})$.*

Concerning this definition the following observations should be made. By Lemma 3, the elements $u_1, u_2, \cdots, u_s$ generate in $\mathfrak{o}$ a prime ideal $\mathfrak{P}$ of dimension $n-s$. If $\mathfrak{p}=\mathfrak{P}\cap\mathfrak{R}[\mathbf{S}]$ then we have $\mathfrak{o}\cdot\mathfrak{A}=\mathfrak{P}=\mathfrak{o}\cdot\mathfrak{p}$. Moreover, since $(u_1, u_2, \cdots, u_s)$ is also a local basis of $\mathfrak{p}$, it follows from Theorem 2 that *P is a simple point of the $(n-s)$-dimensional irreducible variety $\mathcal{V}(\mathfrak{p})$.* We thus see that the condition that (α) be a simple zero of $\mathfrak{A}$ signifies that *locally, at P, the ideal $\mathfrak{A}$ does not differ essentially from a prime ideal $\mathfrak{p}$ (that is, we have $\mathfrak{o}\cdot\mathfrak{A}=\mathfrak{o}\cdot\mathfrak{p}$) and that the variety $\mathcal{V}(\mathfrak{p})$ of this prime ideal has a simple point at P.* We also point out that, in terms of a normal decomposition of $\mathfrak{A}$ into primary components, the relation $\mathfrak{o}\cdot\mathfrak{A}=\mathfrak{o}\cdot\mathfrak{p}$ expresses the fact that $\mathfrak{p}$ itself is one of the primary (necessarily isolated) components of $\mathfrak{A}$ and that $\mathfrak{p}$ is the only prime ideal of $\mathfrak{A}$ which is contained in the prime ideal $\boldsymbol{p}(P)$ of the point P.

We must point out explicitly that in the second part of our definition we speak of an algebraic variety V in a *strictly set-theoretic sense.* There is no question of multiple components or of embedded components; V is simply a set of points. This is clearly indicated by the fact that our definition is in terms of the ideal $\boldsymbol{I}(V)$, this ideal being its own radical and therefore a finite intersection of prime ideals. The observations just made on the simple zeros of an arbitrary ideal show that, according to our definition, a point P of an algebraic variety V is simple *if and only if V has only one irreducible component through P and this component has a simple point at P.*

For simple zeros of an ideal $\mathfrak{A}$ one could formulate results similar to those stated in Theorem 2. For later applications we shall state explicitly the following theorem:

THEOREM 2′. *If (α) is a zero of an r-dimensional isolated prime ideal $\mathfrak{p}$ of $\mathfrak{A}$ (that is, if the point $P(\alpha)$ belongs to an r-dimensional irreducible component of the variety $\mathcal{V}(\mathfrak{A})$) and if $(\omega_1, \omega_2, \cdots, \omega_\nu)$ is a local P-basis of $\mathfrak{A}$, then (α) is a simple zero of $\mathfrak{A}$ if and only if exactly $n-r$ of the vectors $\tau\omega_1, \tau\omega_2, \cdots, \tau\omega_\nu$ in $\mathfrak{M}(P/\mathbf{S})$ are linearly independent.*

Proof. From the preceding considerations we know that in order that (α) be a simple zero of the ideal $\mathfrak{A}$ it is necessary and sufficient that two conditions be satisfied:

(a) *We must have* $\mathfrak{o}\cdot\mathfrak{A}=\mathfrak{o}\cdot\mathfrak{p}$. This condition is equivalent to the following: $(\omega_1, \omega_2, \cdots, \omega_\nu)$ *is a local P-basis of* $\mathfrak{p}$.

(b) *P must be a simple point of the variety* $\mathcal{V}(\mathfrak{p})$.

If (α) is a simple zero of $\mathfrak{A}$, then by (a) the set $(\omega_1, \omega_2, \cdots, \omega_\nu)$ contains a minimal local P-basis of $\mathfrak{p}$. Let $(\omega_1, \omega_2, \cdots, \omega_\sigma)$ be such a basis. By (b) and by the second part of Theorem 2, we must have $\sigma=n-r$, and the vectors $\tau\omega_1, \tau\omega_2, \cdots, \tau\omega_{n-r}$ must be linearly independent. By Lemma 3 no more than $n-r$ of the ν vectors $\tau\omega_i$ can be independent, since dim $\mathfrak{p}=r$.

Conversely, if $n-r$ of the vectors $\tau\omega_i$, say $\tau\omega_1, \tau\omega_2, \cdots, \tau\omega_{n-r}$, are independent, then by condition (1) of Theorem 2, P is a simple point of $\mathcal{V}(\mathfrak{p})$. By Lemma 3, the $n-r$ elements $\omega_1, \omega_2, \cdots, \omega_{n-r}$ form then a local P-basis of $\mathfrak{p}$. Hence, a fortiori the ν elements $\omega_1, \omega_2, \cdots, \omega_\nu$ form a local P-basis of $\mathfrak{p}$. Therefore conditions (a) and (b) are satisfied, and (α) is a simple zero of $\mathfrak{A}$.

4. **Simple subvarieties.**

4.1. *Generalization of the preceding results.* If W is a ρ-dimensional irreducible variety in S_n, the reduction to the zero-dimensional case (§2.2) has the effect of replacing S_n by an $S^{\kappa^*}_{n-\rho}$ and W by a point P^* $(=W^*)$ in this $S^{\kappa^*}_{n-\rho}$. Every irreducible r-dimensional variety V in S_n *containing* W is then replaced by an $(r-\rho)$-dimensional irreducible variety V^*/κ^* in $S^{\kappa^*}_{n-\rho}$ containing P^*. Since this reduction has no effect on the quotient ring $Q(W/V)$ (Lemma 1), the results of the preceding section are either valid or can be properly interpreted if we deal with W instead of with a point P in S_n. We shall briefly state the corresponding results using W instead of P.

Every variety W in S_n is a simple subvariety of S_n, that is, dim $\mathfrak{M}(W/S_n)=n-\rho$. A set of uniformizing parameters of W in S_n consists of $n-\rho$ elements. If W is a *simple* subvariety of an n-dimensional irreducible variety $\boldsymbol{S}$ and if $u_1, u_2, \cdots, u_\nu$ are elements of $\boldsymbol{m}(W/\boldsymbol{S})$ such that the corresponding vectors $\tau u_1, \tau u_2, \cdots, \tau u_\nu$ in $\mathfrak{M}(W/\boldsymbol{S})$ are independent, then the ideal $(u_1, u_2, \cdots, u_\nu)$ in $Q(W/\boldsymbol{S})$ is prime and of dimension $n-\nu$ (see Lemma 3). Here the dimension is intended with respect to the given ground field κ. After the reduction to the zero-dimensional case the dimension $n-\nu$ of Lemma 3 becomes $(n-\rho)-\nu$, but this is the dimension of the ideal $(u_1, u_2, \cdots, u_\nu)$ in $Q(W/\boldsymbol{S})$ *with respect to a field* κ^* which is of degree of transcendency ρ over κ (see §2.2, Lemma 1). If $\mathfrak{o}=Q(W/\boldsymbol{S})$ and if $\mathfrak{A}$ is any ideal in $\mathfrak{R}[\boldsymbol{S}]$, then any set of elements of $\mathfrak{R}[\boldsymbol{S}]$ which is a basis of $\mathfrak{o}\cdot\mathfrak{A}$ is called a *local W-basis of* $\mathfrak{A}$. All *minimal* local W-bases of a given ideal $\mathfrak{A}$ in $\mathfrak{R}[\boldsymbol{S}]$ have the same number of elements. Theorem 2 remains valid if P is replaced by W and $\mathfrak{M}(P/\boldsymbol{S})$ by $\mathfrak{M}(W/\boldsymbol{S})$. Thus W is simple for V (V irreducible) if and only if either one of the following conditions is satisfied: (1) dim $V=r$ and the ideal $\boldsymbol{p}(V/\boldsymbol{S})$ contains $n-r$ elements u_i such that the vectors $\tau u_1, \tau u_2, \cdots, \tau u_{n-r}$ in $\mathfrak{M}(W/\boldsymbol{S})$

are independent; (2) the ideal $\mathfrak{p}(V/S)$ possesses a local W-basis $(u_1, u_2, \cdots, u_s)$ such that the s vectors τu_i are independent. (Note that by the reduction to the zero-dimensional case the difference $n-r$ remains invariant: $n-r = (n-\rho)-(r-\rho)$; moreover we have $\dim V^*/\kappa^* = (n-\rho)-s = (n-s)-\rho$, whence $\dim V/\kappa = n-s$.) If (η) is the general point of W, then Definition 2 reads as follows: (η) is a simple zero of an ideal $\mathfrak{A}$ in $R = \mathfrak{R}[S]$ if $\mathfrak{A}$ possesses a local W-basis $(u_1, u_2, \cdots, u_s)$ such that $\tau u_1, \tau u_2, \cdots, \tau u_s$ are independent vectors (the reduction to dimension zero leads to elements $u_1^*, u_2^*, \cdots, u_s^*$ in $\kappa^* R$, but since $\kappa^* \subseteq Q(W/S)$ these u_i^* can be replaced by elements in R). In particular, if V is any algebraic variety (not necessarily irreducible) and if $W \subseteq V$ then *W is simple for V if and only if W belongs to only one irreducible component of V and W is simple for that component.*

4.2. *Singular subvarieties and singular points.* The customary geometric way of looking at singular subvarieties of dimension greater than 0 is given by the following statement, which we shall formulate as a theorem:

THEOREM 3. *An irreducible subvariety W of a variety V is singular for V if and only if all the points of W are singular for V.*

This theorem follows as an immediate consequence from an algebraic criterion for simple loci which will be proved later on (see §9.6) and which makes use of derivatives and of certain *mixed Jacobian matrices.* However, it is of interest from a methodological point of view to prove Theorem 3 without making use of differentiation arguments. Such a proof, based directly on our definition of simple loci, must be essentially a local argument and as such is likely to contain elements of interest for the general theory of local or semi-local rings.

Before we proceed with the proof of Theorem 3, we point out the following corollary of this theorem:

COROLLARY. *Every irreducible variety V carries simple points.*

In fact, if we regard V as a subvariety of itself then we find that the vector space $\mathfrak{M}(V/V)$ is of dimension zero (for $Q(V/V)$ is the entire field $\mathcal{J}(V)$ and $\mathfrak{m}(V/V)$ is the zero ideal). Hence by Definition 1 (where we have now: $W = V$, $\rho = r = \dim V$), V itself is simple for V.

Thus Theorem 3 implies that the singular manifold of an irreducible V is a *proper* subset of V. In the course of the proof of Theorem 3 we shall actually establish the following stronger result: *if W is any proper algebraic subvariety of V, then $V-W$ carries at least one simple point of V* (see Lemma 4 below). That is as far as we were able to go by local arguments. For the proof that the singular manifold of V is *algebraic* we have to fall back on the *global* algebraic criterion for simple loci (§9.6).

We proceed to prove Theorem 3 under the assumption that V is irreducible. The generalization to arbitrary varieties V will then be straightforward.

4.3. *Proof of Theorem* 3. We first show that the condition stated in Theorem 3 is necessary. We assume therefore that W contains a simple point of V, and we must prove that then W itself is simple for V. Let P be a point of W which is simple for V and let ρ be the dimension of W. *Suppose that our assertion that W is simple for V has already been proved in the case when W is a curve* ($\rho=1$). We then show that the assertion follows by induction with respect to ρ. We fix on W an irreducible subvariety W_1 passing through P and of dimension $\rho-1$. By our induction we have then that W_1 is a simple subvariety of V. By a reduction to dimension zero we replace W_1 by a *point* W_1^* over a suitable extension κ^* of the ground field κ. By this reduction the varieties W and V are replaced respectively by a *curve* W^* and by a variety V^* of dimension $r-\rho+1$ over κ^*. We have $W_1^*\subset W^*\subseteq V^*$ and $Q(W_1^*/V^*)=Q(W_1/V)$. Hence, since W_1 is simple for V, the point W_1^* is simple for V^*. Since the point W_1^* belongs to the curve W^* and since we have assumed that the case $\rho=1$ has already been settled, it follows that W^* is a simple curve on V^*. But then also W is simple for V, since $Q(W/V)=Q(W^*/V^*)$.

Suppose now that W is a curve. We shall first consider the case in which the point P is simple not only for V but also for W. We apply Lemma 2 of §2.3 taking for V' and W respectively the curve W and the point P. In the present case we have dim $\mathfrak{M}(P/W)=1$ (since P is simple for W) and dim $\mathfrak{M}(P/V)=r=$dimension of V. Hence the dimension of the space annihilated by the linear transformation ϕ of $\mathfrak{M}(P/V)$ onto $\mathfrak{M}(P/W)$ must be $r-1$. This means that the ideal $\mathfrak{p}(W/V)$ must contain $r-1$ elements u_i such that the vectors τu_i in $\mathfrak{M}(P/V)$ are independent. By Lemma 3, the $r-1$ elements $u_1, u_2, \cdots, u_{r-1}$ generate a prime *one-dimensional* ideal in $Q(P/V)$, and the corresponding curve through P must therefore coincide with W. Hence by the corollary to Lemma 3, W is simple for V.

There remains the case in which P is a singular point of the curve W. In this case we apply to V successive quadratic transformations the effect of which is to resolve the singularity P of the *curve* W. Let V' and W' be respectively the transforms of V and W under the product of the successive quadratic transformations. Since the center of each quadratic transformations is a *point*, the quotient ring $Q(W/V)$ is not affected by these transformations, that is, we have $Q(W/V)=Q(W'/V')$. On the other hand W' carries at least one point P' of V' which is simple for *both* V' *and* W', namely any of the points of W' into which the singular points P of W has been resolved. Consequently, by the preceding case, W' is simple for V'. Since $Q(W/V)=Q(W'/V')$, it follows that also W is simple for V.

4.4. *Continuation of the proof*. The sufficiency of the condition stated in Theorem 3 is included in the following lemma:

LEMMA 4. *If W is an irreducible simple subvariety of V, of dimension ρ, $0\leqq\rho\leqq r=\dim V$, and if S is a proper algebraic subvariety of W, then $W-S$ contains at least one simple point of V.*

Proof of the lemma. We shall prove this lemma by induction with respect to r. If $r=1$, V is a curve, and there exists then at most a finite number of points P on V such that $Q(P/V)$ is not integrally closed. For all other points P of V the ring $Q(P/V)$ is a discrete valuation ring, $\boldsymbol{m}(P/V)$ is a principal ideal, and therefore P is a simple point of V. What we have shown is that a curve V has at most a finite number of singular points, and this establishes the lemma for $r=1$ (the permissible values of ρ being 0 and 1).

We now assume that the lemma is true for varieties V of dimension less than r and we proceed to prove the lemma for a given variety V of dimension r.

We first consider the case $\rho<r$. We select in $R=\mathfrak{R}[V]$ a set of $r-\rho$ uniformizing parameters $t_1, t_2, \cdots, t_{r-\rho}$ of W, and we consider the ideal $R\cdot(t_1, t_2, \cdots, t_{r-\rho})$ and a normal decomposition of this ideal in primary components. By the definition of uniformizing parameters and by known relations between ideals in R and in the quotient ring $Q(W/V)$, the prime ideal $\mathfrak{p}=\boldsymbol{p}(W/V)$ must be itself one of the primary (necessarily isolated) components of the ideal $R\cdot(t_1, t_2, \cdots, t_{r-\rho})$. Let $\mathfrak{q}_1, \mathfrak{q}_2, \cdots, \mathfrak{q}_h$ be the other primary components, and let $\mathfrak{p}_1, \mathfrak{p}_2, \cdots, \mathfrak{p}_h$ be their associated prime ideals. Let W_i be the irreducible subvariety of V defined by $\mathfrak{p}_i$. Since $\mathfrak{p}_i \not\subseteq \mathfrak{p}$, we have $W_i \not\supseteq W$, and therefore if we denote by S_1 the intersection of W with the variety $\bigcup_{i=1}^{h} W_i$, then S_1 is a *proper* algebraic subvariety of W. Therefore also $S\cup S_1$ is a proper algebraic subvariety of W. Since $\rho<r$, it follows by our induction that there exists at least one point on W which is simple for W and does not belong to $S\cup S_1$. *We shall now prove that any simple point of W which does not belong to S_1 is necessarily a simple point of V.* This will establish the lemma for varieties V of dimension r, provided $\rho<r$.

Let then P be a point of W such that $P\notin S_1$. Since $P\notin W_i$, $i=1, 2, \cdots, h$, we can find an element ω_i in $\mathfrak{q}_i$ such that $\omega_i \notin \boldsymbol{m}(P/V)$. If $\omega=\prod_{i=1}^{h}\omega_i$ then ω is a unit in $Q(P/V)$ and we have moreover $\omega\cdot\mathfrak{p}\subseteq R\cdot(t_1, t_2, \cdots, t_{r-\rho})$. Since $t_i\in\mathfrak{p}$ this shows that $(t_1, t_2, \cdots, t_{r-\rho})$ is a local P-basis of $\mathfrak{p}$. We now use Lemma 2 (where W, V', and V are now respectively P, W, and V). The subspace of $\mathfrak{M}(P/V)$ which is spanned by the vectors belonging to $\tau\mathfrak{p}$ is in this case spanned by the $r-\rho$ vectors τt_i. Hence $\dim \mathfrak{M}(P/V) \leqq r-\rho + \dim \mathfrak{M}(P/W)$. Since P is simple for W, we have $\dim \mathfrak{M}(P/W)=\rho$, and hence $\dim \mathfrak{M}(P/V)\leqq r$. Hence necessarily $\dim \mathfrak{M}(P/V)=r$, and P is a simple point of V, as was asserted.

There remains to be considered the case $\rho=r$, that is, $W=V$. The consideration of the conductor of the coordinate ring $R=\mathfrak{R}[V]$ with respect to the integral closure of R in its quotient field $\mathcal{J}(V)$ shows that there exists at most a finite number of $(r-1)$-dimensional irreducible subvarieties H of V such that the quotient ring $Q(H/V)$ is not integrally closed. There exist therefore irreducible subvarieties H of V, of dimension $r-1$, such that $Q(H/V)$ is integrally closed *and* $H\not\supseteq S$. Let H_0 be such a subvariety of V. Since $Q(H_0/V)$

is integrally closed, this quotient ring is a discrete valuation ring and therefore H_0 *is simple for* V. Since $H_0 \not\supseteq S$, $H_0 \cap S$ is a proper algebraic subvariety of H_0. By the preceding case $\rho < r$ (namely $\rho = r-1$) we can find at least one point P on H_0 which is simple for V and does not belong to $H_0 \cap S$. We have thus shown that there exist simple points of V which are not in S. This completes the proof of Theorem 3 for an *irreducible* V.

4.5. *The case of a reducible variety.* The case of a reducible V can now be readily taken care of. Let $V_1, V_2, \cdots, V_g$ be the irreducible components of V. We assume first that *W contains a simple point P of V.* Then P belongs to only one of the irreducible components V_i, say to V_1, and P is necessarily a simple point of V_1 (Definition 2, §3.4). From $P \notin V_i$, $i = 2, 3, \cdots, g$, follows $W \not\subseteq V_i$ whence $W \subseteq V_1$. Since W carries the simple point P of V_1, it follows, by the irreducible case of Theorem 3, that W is simple for V_1. Thus W belongs to only one irreducible component of V and is simple for that component. Hence W is simple for V.

Assume now that W is simple for V. For a suitable labeling of the irreducible components V_i of V we may assume that $W \subseteq V_1$, $W \not\subseteq V_i$, $i = 2, 3, \cdots, g$, and that W is simple for V_1. Let S be the intersection of W with $\bigcup_{i=2}^{g} V_i$. Then S is a proper algebraic subvariety of W, and therefore by Lemma 4 there exists a point P on $W - S$ which is simple for V_1. This point P does not belong to any other component V_i, $i \neq 1$. Hence P is simple for V. This completes the proof of Theorem 3.

5. Simple loci and regular rings.

5.1. *The identity of the two concepts.* A special class of local rings, called *regular*, has been first introduced and studied by Krull [7], and the theory of these rings has been further developed by Chevalley [2] and I. S. Cohen [3](4). Let $\mathfrak{o}$ be a local ring, $\mathfrak{m}$ the maximal ideal in $\mathfrak{o}$, and let $(t_1, t_2, \cdots, t_s)$ be a minimal basis of $\mathfrak{m}$. The local ring $\mathfrak{o}$ is called regular if the following condition is satisfied: *in any homogeneous relation $f(t_1, t_2, \cdots, t_s) = 0$ between $t_1, t_2, \cdots, t_s$, with coefficients in $\mathfrak{o}$, the coefficients necessarily all belong to $\mathfrak{m}$.* An equivalent condition is the following: *if $\phi(t_1, t_2, \cdots, t_s) \in \mathfrak{m}^{\nu+1}$, $\phi =$ a form of degree ν with coefficients in $\mathfrak{o}$, then all the coefficients of ϕ must belong to $\mathfrak{m}$.* That the second of these two conditions implies the first is trivial. That the first condition implies the second follows immediately from the fact that every element of $\mathfrak{m}^{\nu+1}$ can be expressed as a form of degree $\nu+1$ in $t_1, t_2, \cdots, t_s$ with coefficients in $\mathfrak{o}$, and hence also as a form of degree ν in $t_1, t_2, \cdots, t_s$, *with coefficients in* $\mathfrak{m}$.

If $\mathfrak{o}$ is an arbitrary local ring, then the ring $\mathfrak{m}/\mathfrak{m}^2$, as an additive group, can be regarded as a vector space over the residue field $\Delta = \mathfrak{o}/\mathfrak{m}$ as field of scalars. We denote this vector space by $\mathfrak{M}_1$. The dimension of $\mathfrak{M}_1$ is equal to the number s of elements $u_1, u_2, \cdots, u_s$ in any minimal basis of $\mathfrak{m}$. More

(4) The term "regular" is due to Chevalley. The original term used by Krull was "p-Reihenringen."

generally, $\mathfrak{m}^\nu/\mathfrak{m}^{\nu+1}$ can be regarded as a vector space over Δ. We shall denote this space by $\mathfrak{M}_\nu$. The $C_{s+\nu-1,\nu}$ power products $u_1^{t_1}u_2^{t_2}\cdots u_s^{t_s}$ of degree ν, reduced modulo $\mathfrak{m}^{\nu+1}$, form a basis of $\mathfrak{M}_\nu$, but not necessarily an independent one. *The ring $\mathfrak{o}$ is regular if and only if the above basis is independent, that is, if and only if* dim $\mathfrak{M}_\nu = C_{s+\nu-1,\nu}$ *for all positive integers ν.*

THEOREM 4. *Given an irreducible algebraic variety V, an irreducible algebraic subvariety W of V is simple for V if and only if the quotient ring $Q(W/V)$ is regular.*

Proof. By the reduction to dimension zero (§2.2) it is sufficient to prove the theorem for points of V. We assume then that $W=P=$ a point of V. Let $\mathfrak{o}=Q(P/V)$, $\mathfrak{m}=\mathfrak{m}(P/V)$, $s=$ dim $\mathfrak{M}(P/V)$, and $r=$ dim V. We know that $s\geqq r$. *We have to show that $s=r$ if and only if $\mathfrak{o}$ is a regular ring.*

(1) *Assume $s>r$.* Let $(t_1, t_2, \cdots, t_s)$ be a minimal basis of $\mathfrak{m}$. It is always possible to choose r elements in that basis in such a fashion that the $\mathfrak{o}$-ideal generated by these elements is of dimension 0. We may assume that $t_1, t_2, \cdots, t_r$ are such elements. Then the $\mathfrak{o}$-ideal $\mathfrak{q}=\mathfrak{o}\cdot(t_1, t_2, \cdots, t_r)$ is primary, and its associated prime ideal is $\mathfrak{m}$. Since $t_s\in\mathfrak{m}$, some power of t_s, say t_s^h, belongs to $\mathfrak{q}$. There exists an integer $\nu\geqq 0$ such that $t_s^h\in\mathfrak{q}\mathfrak{m}^\nu$, $t_s^h\notin\mathfrak{q}\mathfrak{m}^{\nu+1}$ (such an integer exists since $\bigcap_{i=1}^{\infty}\mathfrak{m}^i=(0)$ and $t_s\neq 0$; by $\mathfrak{q}\cdot\mathfrak{m}^0$ we mean $\mathfrak{q}$). Every element of $\mathfrak{m}^\nu$ is expressible as a form of degree ν in $t_1, t_2, \cdots, t_s$ with coefficients in $\mathfrak{o}$, while every element in $\mathfrak{q}$ is expressible as a linear form in $t_1, t_2, \cdots, t_r$ with coefficients in $\mathfrak{o}$. Since $t_s\in\mathfrak{q}\mathfrak{m}^\nu$, we therefore have a relation of the form:

$$F(t) = t_s^h - \phi_1(t_1, t_2, \cdots, t_r)\psi_\nu(t_1, t_2, \cdots, t_s) = 0,$$

where ϕ_1 and ψ_ν are forms with coefficients in $\mathfrak{o}$, of degrees 1 and ν respectively. If $h<\nu+1$ then the relation $t_s^h\in\mathfrak{q}\mathfrak{m}^\nu$ implies $t_s^h\in\mathfrak{m}^{\nu+1}\subseteq\mathfrak{m}^{h+1}$, and therefore $\mathfrak{o}$ is not regular. Assume now that $h\geqq\nu+1$. Since $t_s^h\notin\mathfrak{q}\mathfrak{m}^{\nu+1}$ the coefficients of ϕ_1 as well as the coefficients of ψ_ν are not all in $\mathfrak{m}$. If $h>\nu+1$ then we have $\phi_1(t)\psi_\nu(t)=t_s^h\in\mathfrak{m}^h\subseteq\mathfrak{m}^{\nu+2}$, where, as we have just seen, $\phi_1\psi_\nu$ is a form of degree $\nu+1$ in $t_1, t_2, \cdots, t_s$ whose coefficients are not all in $\mathfrak{m}$. Hence again $\mathfrak{o}$ is not regular. If finally $h=\nu+1$ then $F(t)$ is a form of degree h in $t_1, t_2, \cdots, t_s$, and since the product $\phi_1\psi_\nu$ does not contain a term depending on t_s only (ϕ_1 is linear homogeneous in $t_1, t_2, \cdots, t_r$), it follows that the coefficient of t_s^h in $F(t)$ is 1. The relation $F(t)=0$ shows then that $\mathfrak{o}$ is not regular. We have therefore shown that if $s>r$ then the ring $\mathfrak{o}$ is not regular.

(2) *Assume $s=r$.* To prove that in this case $\mathfrak{o}$ is a regular ring we first consider the case in which the residue field Δ $(=\mathfrak{o}/\mathfrak{m})$ is infinite.

If $f(z_1, z_2, \cdots, z_r)$ is a form in indeterminates z_i with coefficients in $\mathfrak{o}$ we shall denote by $\bar{f}(z_1, z_2, \cdots, z_r)$ the form with coefficients in Δ which is obtained when the coefficients of f are replaced by their $\mathfrak{m}$-residues in Δ. To show that $\mathfrak{o}$ is regular we have to show the following: if $f(t_1, t_2, \cdots, t_r)=0$ then $\bar{f}(z_1, z_2, \cdots, z_r)=0$. We consider an arbitrary linear nonsingular homogeneous transformation

(3) $$z_i' = \sum_{j=1}^{r} \bar{a}_{ij} z_j, \qquad i = 1, 2, \cdots, r; \bar{a}_{ij} \in \Delta.$$

With any such transformation we can associate (in more than one way) a transformation

(4) $$t_i' = \sum_{j=1}^{r} a_{ij} t_j, \qquad i = 1, 2, \cdots, r; a_{ij} \in \mathfrak{o},$$

of the given minimal basis $(t_1, t_2, \cdots, t_r)$ of $\mathfrak{m}$ into another minimal basis $(t_1', t_2', \cdots, t_r')$ of $\mathfrak{m}$. We have only to take for the a_{ij} elements of $\mathfrak{o}$ whose $\mathfrak{m}$-residues are the $\bar{a}_{ij}$ (that the t''s also form a basis of $\mathfrak{m}$ follows from the assumption $|\bar{a}_{ij}| \neq 0$, which implies $|a_{ij}| \notin \mathfrak{m}$, that is, $|a_{ij}|$ is a unit in $\mathfrak{o}$). It is clear that if $f(t_1, t_2, \cdots, t_r)$ is a form in the t's with coefficients in $\mathfrak{o}$, and if $\phi(t_1', t_2', \cdots, t_r')$ is the transform of $f(t)$ under the transformation (4), then $\bar{\phi}(z_1', z_2', \cdots, z_r')$ is the transform of $\bar{f}(z_1, z_2, \cdots, z_r)$ under the transformation (3). Given $f(t_1, t_2, \cdots, t_r)$, homogeneous of degree ν, *and assuming that* $\bar{f}(z_1, z_2, \cdots, z_r) \neq 0$, we can find a linear transformation (3) such that in $\bar{\phi}(z_1', z_2', \cdots, z_r')$ the coefficient of $z_r'^{\nu}$ is not equal to 0 (since Δ is an infinite field). We then can match this transformation by a corresponding transformation (4) in such a fashion that the coefficient of $t_r'^{\nu}$ in $\phi(t_1', t_2', \cdots, t_r')$ is not in $\mathfrak{m}$, hence is a unit in $\mathfrak{o}$. The upshot of the preceding argument is then this: if we have a homogeneous relation $f(t_1, t_2, \cdots, t_r) = 0$ of degree ν between the t's, with coefficients in $\mathfrak{o}$, and if we assume that $\bar{f}(z_1, z_2, \cdots, z_r) \neq 0$ we can assume (by first changing, if necessary, the minimal basis of $\mathfrak{m}$) that the coefficient of t_r^{ν} in f is a unit e in $\mathfrak{o}$. But then the relation $f(t) = 0$ implies that $t_r^{\nu} \in \mathfrak{o} \cdot (t_1, t_2, \cdots, t_{r-1})$, $\mathfrak{m}^{\nu} \subseteq \mathfrak{o} \cdot (t_1, t_2, \cdots, t_{r-1})$, whence the ideal $\mathfrak{o} \cdot (t_1, t_2, \cdots, t_{r-1})$ is zero-dimensional (having $\mathfrak{m}$ as associated prime ideal). This is impossible, since all minimal primes of the ideal $\mathfrak{o} \cdot (t_1, t_2, \cdots, t_{r-1})$ are of dimension not less than $r - (r-1) = 1$. This contradiction has been obtained because we have assumed that $\bar{f}(z_1, z_2, \cdots, z_r) \neq 0$. Hence we must have $\bar{f}(z_1, z_2, \cdots, z_r) = 0$ whenever $f(t_1, t_2, \cdots, t_r) = 0$, and this shows that $\mathfrak{o}$ is a regular ring.

If Δ is a finite field, we adjoin to the field $\mathcal{J}(V)$ an indeterminate u and we take as new ground field the field $\kappa^* = \kappa(u)$. We denote by V^* and P^* respectively the variety over κ^* having the same general point as V and the point over κ^* having the same coördinates as P. We have dim $V^*/\kappa^* =$ dim $V/\kappa = r$, $R^* = \mathfrak{R}[V^*] = \kappa^* R$ (where $R = \mathfrak{R}[V]$) and $\mathfrak{p}(P^*/V^*) = R^* \cdot \mathfrak{p}(P/V)$. From these facts it follows immediately that if $\mathfrak{o}^* = Q(P^*/V^*)$ and $\mathfrak{m}^* = \boldsymbol{m}(P^*/V^*)$, then

(5) $$\mathfrak{m}^* = \mathfrak{o}^* \mathfrak{m},$$

(5′) $$\mathfrak{m} = \mathfrak{m}^* \cap \mathfrak{o}.$$

From (5) it follows that our minimal basis $(t_1, t_2, \cdots, t_r)$ of $\mathfrak{m}$ is also a basis of $\mathfrak{m}^*$—necessarily minimal, since dim $V^* = r$. The residue field $\Delta^* = \mathfrak{o}^*/\mathfrak{m}^*$ contains the *infinite* new ground field κ^*. Hence, by the preceding proof, $\mathfrak{o}^*$ is a regular ring. Therefore if $f(t_1, t_2, \cdots, t_r) = 0$ is a homogeneous relation with coefficients in $\mathfrak{o}$, these coefficients are all in $\mathfrak{m}^*$. But then, in view of (5′), these coefficients must all belong to $\mathfrak{m}$. This completes the proof of the theorem.

5.2. *Unique local factorization at simple loci.* The regular rings which occur in algebraic geometry as quotient rings of simple subvarieties are "unramified" in the sense of I. S. Cohen [3, p. 88], for they contain a field (namely the ground field κ). The theorems proved by Cohen for unramified regular rings are therefore valid for quotient rings of simple subvarieties. Thus if W is an irreducible ρ-dimensional simple subvariety of an irreducible r-dimensional variety V and if $\mathfrak{o} = Q(W/V)$, then the completion $\mathfrak{o}^*$ of $\mathfrak{o}$ (with respect to the powers of the maximal ideal $\mathfrak{m}$ in $\mathfrak{o}$) is isomorphic to the ring of formal power series in $r - \rho$ indeterminates over the residue field $\Delta = \mathfrak{o}/\mathfrak{m}$ [3, Theorem 15, p. 88], and therefore [3, Theorem 18, p. 94], $\mathfrak{o}^*$ *is a unique factorization domain.* The following result is a consequence of a general theorem on local rings due to Chevalley: *if W is an arbitrary (not necessarily simple) irreducible subvariety of V, then any prime ideal $\mathfrak{p}$ in the quotient ring $\mathfrak{o} = Q(W/V)$ is unramified in the completion $\mathfrak{o}^*$ of $\mathfrak{o}$, that is, $\mathfrak{o}^* \cdot \mathfrak{p}$ is a finite intersection of prime ideals in $\mathfrak{o}^*$* (see C. Chevalley, *Intersections of algebraic and algebroid varieties,* Trans. Amer. Math. Soc. vol. 57 (1945) p. 9, last sentence of Lemma 9, and p. 11, Theorem 1). Using this result and the fact that $\mathfrak{o}^*$ is a unique factorization domain, we can now prove the following theorem:

THEOREM 5. *The quotient ring $\mathfrak{o}$ of a simple subvariety is a unique factorization domain.*

Proof. Let $\mathfrak{p}$ be a minimal prime ideal in $\mathfrak{o}$. To prove the theorem we have only to show that $\mathfrak{p}$ is a principal ideal. Let

$$\mathfrak{o}^* \cdot \mathfrak{p} = \bigcap_{i=1}^{h} \mathfrak{p}_i^*, \tag{6}$$

where we may assume of course that no $\mathfrak{p}_i^*$ is superfluous. It has been shown by Chevalley [2, Propositions 5 and 6, p. 699] for the completion $\mathfrak{o}^*$ of an arbitrary local ring $\mathfrak{o}$ that if $\mathfrak{p}$ is a prime ideal in $\mathfrak{o}$ then every prime ideal of $\mathfrak{o}^* \cdot \mathfrak{p}$ contracts to $\mathfrak{p}$. Using this result, we have therefore: $\mathfrak{p}_i^* \cap \mathfrak{o} = \mathfrak{p}$, $i = 1, 2, \cdots, h$.

We now show that *each prime ideal $\mathfrak{p}_i^*$ is minimal in $\mathfrak{o}$.* Consider for instance the ideal $\mathfrak{p}_1^*$. Let ω be a fixed element of $\mathfrak{p}$, $\omega \neq 0$. Since $\omega \in \mathfrak{p}_1^*$, some isolated prime ideal of $\mathfrak{o}^* \cdot \omega$ must be contained in $\mathfrak{p}_1^*$. Let $\mathfrak{p}^*$ be such an isolated prime ideal of $\mathfrak{o}^* \cdot \omega$, $\mathfrak{p}^* \subseteq \mathfrak{p}_1^*$. We have $\mathfrak{p}^* \cap \mathfrak{o} \subseteq \mathfrak{p}_1^* \cap \mathfrak{o} = \mathfrak{p}$ and $\mathfrak{p}^* \cap \mathfrak{o} \neq (0)$ since $\omega \in \mathfrak{p}^*$. Hence $\mathfrak{p}^* \cap \mathfrak{o} = \mathfrak{p}$ since $\mathfrak{p}$ is minimal in $\mathfrak{o}$. Therefore $\mathfrak{o}^*\mathfrak{p} \subseteq \mathfrak{p}^*$ and consequently $\mathfrak{p}^*$ *must contain* one of the prime ideals $\mathfrak{p}_i^*$, $i=1, 2, \cdots, h$. However, since we know already that $\mathfrak{p}^*$ in its turn is contained in $\mathfrak{p}_1^*$ and since $\mathfrak{p}_1^*$ is not superfluous in (6), we *conclude that* $\mathfrak{p}^*=\mathfrak{p}_1^*$. Hence $\mathfrak{p}_1^*$ is an isolated prime ideal of the principal ideal $\mathfrak{o}^*\cdot\omega$, and therefore it is necessarily minimal in $\mathfrak{o}^*$, as was asserted.

So far we have made no use of the assumption that $\mathfrak{o}$ is the quotient ring of a simple subvariety. Using this assumption we know then that $\mathfrak{o}^*$ is a unique factorization domain. Therefore the ideal $\mathfrak{o}^*\cdot\mathfrak{p}$ must be a principal ideal, since we have just proved that $\mathfrak{o}^*\cdot\mathfrak{p}$ is the intersection of *minimal* prime ideals. Hence $\mathfrak{p}$ must contain an element u such that $\mathfrak{o}^*\cdot\mathfrak{p}=\mathfrak{o}^*\cdot u$. It is known that every ideal $\mathfrak{A}$ in $\mathfrak{o}$ satisfies the relation $\mathfrak{o}^*\cdot\mathfrak{A}\cap\mathfrak{o}=\mathfrak{A}$ (Krull [7, Theorem 15]). Hence $\mathfrak{p}=\mathfrak{o}^*\cdot\mathfrak{p}\cap\mathfrak{o}=\mathfrak{o}^*\cdot u\cap\mathfrak{o}=\mathfrak{o}\cdot u$. This completes the proof.

5.3. *The abstract analogue of Theorem 3 and an example of F. K. Schmidt.* We shall conclude this section with a discussion of the first part of the proof of Theorem 3 (§4.3) from the standpoint of the general theory of regular local rings. The necessity of the condition stated in Theorem 3 is a strictly local fact. For let P be a point of W, let $\mathfrak{o}=Q(P/V)$ and let $\mathfrak{p}$ be the prime ideal in $\mathfrak{o}$ defined by $W(\mathfrak{p}=\mathfrak{o}\cdot\boldsymbol{p}(W/V))$. Then in view of Theorem 4 of §5.1, the "only if" part of Theorem 3 can be stated as follows: "*if* $\mathfrak{o}$ *is a regular ring, then also the ring* $\mathfrak{o}_{\mathfrak{p}}$ (that is, $Q(W/V)$) *is regular.*" The question arises whether this statement is true for an *arbitrary* regular local ring $\mathfrak{o}$ and for *any* prime ideal $\mathfrak{p}$ in $\mathfrak{o}$ (the statement is trivial for minimal prime ideals $\mathfrak{p}$ in view of the fact that any regular local ring is an integrally closed domain; see [7, Theorem 6]). At present it is only known that the statement is true if the regular ring $\mathfrak{o}$ is complete (Cohen [3, Theorem 20, p. 97]). In addition to calling attention to this unsolved question, we wish to analyze the proof of the "only if" part of Theorem 3 in order to point out why that proof fails in the general case where we have an arbitrary regular local ring $\mathfrak{o}$. The first part of that proof consists in a reduction to the case dim $W=\rho=1$, and that reduction can be carried out just as it stands also in the general case. In the second part of the proof we have first considered the case in which P is a simple point of W. The corresponding assumption in the general case can be expressed by saying that the *residue class ring* $\mathfrak{o}/\mathfrak{p}$ is a regular ring (hence a discrete valuation ring, since $\mathfrak{p}$ is of dimension 1). Also in this case our proof of the regularity of $\mathfrak{o}_{\mathfrak{p}}$ remains valid in the general case.

The critical phase in our proof is that in which the local ring $\mathfrak{o}/\mathfrak{p}$ (of dimension 1) *is not regular*; in other words: $\mathfrak{o}/\mathfrak{p}$ *is a primary integral domain which is not integrally closed.* In this case we have applied to the neighborhood of the point P on V, that is, to the local ring $\mathfrak{o}$, successive quadratic transformations, the effect of which was to resolve the singularity of the curve W at P, that is, to transform $\mathfrak{o}/\mathfrak{p}$ into a regular ring. It is this step that may fail to work in the general case. The following free presentation of an example by F. K. Schmidt [9] will illustrate this possibility.

Let κ be a field of characteristic $p \neq 0$ and let Σ' be the field $\kappa(x, t)$ in two independent variables x and t over κ. Let $\xi_1, \xi_2, \cdots, \xi_n, \cdots$ be an infinite sequence of elements of κ such that the formal power series

$$\xi_1 + \xi_2 x + \xi_3 x^2 + \cdots$$

is transcendental over $\kappa(x)$(5). Then the analytical branch in the (x, t)-plane defined by

$$t = (\xi_1 + \xi_2 x + \cdots)^p \tag{7}$$

is not algebraic, and therefore it defines a zero-dimensional, discrete, ***rank*** **1** valuation v' of Σ'. Let $\mathfrak{o}'$ be the valuation ring of v'. We now consider the field $\Sigma = \Sigma'(\tau) = \kappa(x, \tau)$, where $\tau = t^{1/p}$, and the ring

$$\mathfrak{o} = \mathfrak{o}'[\tau] = \mathfrak{o}' \cdot 1 + \mathfrak{o}' \cdot \tau + \cdots + \mathfrak{o}' \cdot \tau^{p-1}. \tag{8}$$

Let v be the extension of v' to Σ (the extension is unique, for Σ is purely inseparable over Σ'). It is clear by (7) that v is defined by the following analytical branch in the (x, τ)-plane:

$$\tau = \xi_1 + \xi_2 x + \cdots . \tag{9}$$

Since $\mathfrak{o}'$ is a discrete, rank 1, valuation ring, it is a local (regular) ring of dimension 1. Since $\mathfrak{o}$ is a finite $\mathfrak{o}'$-module, it follows that also $\mathfrak{o}$ is a local ring of dimension 1 (Chevalley [2, Proposition 3, p. 694]). We shall now show that $\mathfrak{o}$ is not regular *and that it cannot be transformed into a regular ring by successive quadratic transformations.*

Let $\tau_1 = (\tau - \xi_1)/x$. We have, by (9), $v(\tau_1) \geqq 0$, whence $v'(\tau_1^p) \geqq 0$. Consequently $\tau_1^p \in \mathfrak{o}'$ and therefore τ_1 is integrally dependent on $\mathfrak{o}$. However $\tau_1 = -\xi_1/x \cdot 1 + 1/x \cdot \tau$, whence, by (8), $\tau_1 \notin \mathfrak{o}$, since $1, \tau, \cdots, \tau^{p-1}$ form an independent basis of Σ over Σ' and since $1/x \notin \mathfrak{o}'$. Consequently $\mathfrak{o}$ is not integrally closed and therefore it cannot be a regular ring.

The maximal ideal $\mathfrak{m}'$ in $\mathfrak{o}'$ is the principal ideal $\mathfrak{o}' \cdot x$, since $v'(x) = +1$. If $\mathfrak{m}$ denotes the maximal ideal in $\mathfrak{o}$, then we have $x \in \mathfrak{m}$, and also $\tau - \xi_1 \in \mathfrak{m}$ (for $(\tau - \xi_1)^p = t - \xi_1^p \in \mathfrak{m}' \subseteq \mathfrak{m}$). From (8) one sees then immediately that x and $\tau - \xi_1$ form a basis for $\mathfrak{m}$.

To apply a "quadratic transformation" to the ring $\mathfrak{o}$ means to pass from $\mathfrak{o}$ to the following ring $\mathfrak{o}_1$:

$$\mathfrak{o}_1 = \mathfrak{o}[\tau_1], \qquad \tau_1 = (\tau - \xi_1)/x.$$

Let $t_1 = (t - \xi_1^p)/x^p$. Then $\Sigma' = \kappa(x, t) = \kappa(x, t_1)$ and, by (7), the valuation v' can be equally well defined by the following analytical branch in the (x, t_1)-plane:

(5) To construct such a power series one may proceed as Schmidt does, that is, take for κ a field of infinite degree of transcendency over the prime field Γ of characteristic p and take for $(\xi_1, \xi_2, \cdots, \xi_n, \cdots)$ a transcendence set in κ/Γ.

$$t_1 = (\xi_2 + \xi_3 x^2 + \cdots)^p.$$

We have $\tau_1 = t_1^{1/p}$ and $\mathfrak{o}_1 = \mathfrak{o}'[\tau_1]$. It follows that the ring $\mathfrak{o}_1$ is in similar relation to $\mathfrak{o}'$ as $\mathfrak{o}$ was. This shows that $\mathfrak{o}_1$ is not a regular ring, and that the successive quadratic transformations will never lead to a regular ring (the next ring $\mathfrak{o}_2$ will be defined as follows: $\mathfrak{o}_2 = \mathfrak{o}_1[\tau_2]$, where $\tau_2 = (\tau_1 - \xi_2)/x$, and so on).

It is easy to verify that the integral closure of $\mathfrak{o}$ in Σ (that is, the valuation ring of v) *is not a finite* $\mathfrak{o}$*-module* (Schmidt [9, p. 447]). This fact is the real reason why it is not possible to transform $\mathfrak{o}$ into a regular ring by successive "quadratic transformations." For it can be easily proved that "any primary integral domain $\mathfrak{o}$ can be transformed into a regular ring by successive quadratic transformations, provided the integral closure of $\mathfrak{o}$ is a finite $\mathfrak{o}$-module."

PART II. JACOBIAN CRITERIA FOR SIMPLE LOCI

6. The vector space $\mathcal{D}(W)$ of local differentials.

6.1. *The local W-differentials in* S_n. Let $x_1, x_2, \cdots, x_n$ be coördinates in a linear S_n and let $(\eta_1, \eta_2, \cdots, \eta_n)$ be the general point of an irreducible ρ-dimensional variety W in S_n. If u is any element of the maximal ideal $\mathfrak{m}$ of the quotient ring $\mathfrak{o}$ of W ($\mathfrak{o} = Q(W/S_n)$), then $u = f(x)/g(x)$, where $f(x)$ and $g(x)$ are in $\kappa[x_1, x_2, \cdots, x_n]$, $f(\eta) = 0$ and $g(\eta) \neq 0$. From $g(\eta) \neq 0$ follows that the partial derivatives $\partial u/\partial x_i$ are elements of $\mathfrak{o}$. The $\mathfrak{m}$-residues $(\partial u/\partial x_i)_{x=\eta}$ of these partial derivatives are therefore elements of the field $\Delta = \mathcal{J}(W) = \kappa(\eta_1, \eta_2, \cdots, \eta_n)$. We call the ordered n-tuple

$$(\partial u/\partial x_1, \partial u/\partial x_2, \cdots, \partial u/\partial x_n)_{x=\eta}$$

of these residues the *local differential of u at W*, or the *local W-differential of u*, and we denote it by $d_W u$. We emphasize our assumption that u is an element of $\mathfrak{m}$, that is, $u = 0$ on W.

We regard $d_W u$ as a vector in the n-dimensional vector space, over Δ as field of scalars, formed by all the ordered n-tuples of elements of Δ. The set of all local differentials $d_W u$ (W is fixed, u varies in $\mathfrak{m}$) is a linear subspace of that n-dimensional vector space. For in the first place, the relation $d_W u \pm d_W v = d_W(u \pm v)$ is obvious ($u, v \in \mathfrak{m}$ and consequently also $u \pm v \in \mathfrak{m}$). In the second place, if $\bar{\delta} \in \Delta$, we have $\bar{\delta} = \phi(\eta)/\psi(\eta)$, where $\phi(x)$, $\psi(x) \in \kappa[x_1, x_2, \cdots, x_n]$ and $\psi(\eta) \neq 0$. If $u \in \mathfrak{m}$ and if we set $\delta = \phi(x)/\psi(x)$, then $\delta u \in \mathfrak{m}$ and we have $d_W(\delta u) = \bar{\delta} \cdot d_W u$, for $(u)_{x=\eta} = 0$.

We shall denote the vector space, over Δ, formed by the local W-differentials $d_W u$, $u \in \mathfrak{m}$, by $\mathcal{D}(W)$.

6.2. *The linear transformation* $\mathcal{M}(W/S_n) \rightarrow \mathcal{D}(W)$. The given variety W in S_n defines also its local vector space $\mathcal{M}(W/S_n)$ over Δ (§2.1), and we have the mapping τ of $\mathfrak{m}$ onto $\mathcal{M}(W/S_n)$ defined by (1). If for any u in $\mathfrak{m}$ we let correspond to the vector τu the local W-differential of u, we obtain a transformation

(10) $$\tau u \to d_W u, \qquad u \in \mathfrak{m},$$

of $\mathfrak{M}(W/S_n)$ onto $\mathfrak{D}(W)$. The linearity of this transformation is obvious. Therefore to prove that (10) is single-valued it is only necessary to show that if $\tau u = 0$ then also $d_W u = 0$. Now if $\tau u = 0$, then $u \in \mathfrak{m}^2$, $u = \sum_{j=1}^{\nu} u_j v_j$, where $u_j, v_j \in \mathfrak{m}$, $j = 1, 2, \cdots, \nu$. Hence $(u_j)_{x=\eta} = (v_j)_{x=\eta} = 0$, and therefore $(\partial u/\partial x_i)_{x=\eta} = 0$, $i = 1, 2, \cdots, n$, that is, $d_W u = 0$.

Hence (10) defines a linear transformation of $\mathfrak{M}(W/S_n)$ onto $\mathfrak{D}(W)$. We know that W is simple for S_n (§4), and that therefore $\mathfrak{M}(W/S_n)$ is of dimension $n-\rho$. Consequently we can state the following result:

LEMMA 5. *The dimension of the space $\mathfrak{D}(W)$ of local W-differentials is at most equal to $n-\rho$, where ρ is the dimension of W, and the dimension is exactly $n-\rho$ if and only if the linear transformation* (10) *is nonsingular.*

7. Jacobian criterion for simple points: the separable case.

7.1. *Criterion for uniformizing parameters.* Let $P(\alpha_1, \alpha_2, \cdots, \alpha_n)$ be a point of S_n. We have constructed in [13, p. 541] a particular set of n polynomials which form not only a minimal local basis (§3.3) of the ideal $\mathfrak{p}(P)$ but even a basis of this ideal. It is a set of polynomials in $\kappa[x_1, x_2, \cdots, x_n]$,

(11) $$f_1(x_1), f_2(x_1, x_2), \cdots, f_n(x_1, x_2, \cdots, x_n),$$

defined and uniquely determined by the following conditions:

a. f_i depends only on $x_1, x_2, \cdots, x_i$ and is monic in x_i;

b. $f_i(\alpha_1, \alpha_2, \cdots, \alpha_i) = 0$, that is, $f_i(x) = 0$ at P;

c. $f_i(\alpha_1, \alpha_2, \cdots, \alpha_{i-1}, x_i)$ is irreducible over the field $\kappa(\alpha_1, \alpha_2, \cdots, \alpha_{i-1})$;

d. If $f_i(x)$ is of degree ν_i in x_i then for $j > i$ the degree of $f_j(x)$ in x_i is less than ν_i.

These n polynomials $f_i(x)$ depend only on the order in which the indeterminates $x_1, x_2, \cdots, x_n$ are labeled. They shall be referred to as the *canonical uniformizing parameters of P.*

By Lemma 5 the dimension of the vector space $\mathfrak{D}(P)$ of local P-differentials is at most equal to n. Making use of the particular form of the canonical parameters $f_i(x)$, we can easily prove the following theorem:

THEOREM 6. *In order that the dimension of the vector space $\mathfrak{D}(P)$ of local P-differentials have its maximum value n, it is necessary and sufficient that the coördinates $\alpha_1, \alpha_2, \cdots, \alpha_n$ of P be separable quantities over κ.*

Proof. By Lemma 5 the dimension of $\mathfrak{D}(P)$ is equal to n if and only if the transformation (10) (with P instead of W) is nonsingular. Since $\tau f_1, \tau f_2, \cdots, \tau f_n$ form an independent basis of $\mathfrak{M}(P/S_n)$, it follows that $\mathfrak{D}(P)$ has dimension n if and only if the local P-differentials $d_P f_1, d_P f_2, \cdots, d_P f_n$ are independent vectors. These vectors are independent if and only if the determinant of their components, that is, the Jacobian determinant $|\partial(f_1, f_2, \cdots, f_n)/\partial(x_1, x_2, \cdots, x_n)|_{x=\alpha}$, is different from zero. Since $f_i(x)$ is independent of x_j, $j>i$, this Jacobian determinant is given by the following product:

$$\prod_{i=1}^{n} (\partial f_i/\partial x_i)_{x=\alpha}.$$

The first factor $(\partial f_1/\partial x_1)_{x=\alpha}$ is not equal to 0 if and only if α_1 is separable over κ. The second factor $(\partial f_2/\partial x_2)_{x_1=\alpha_1, x_2=\alpha_2}$ is not equal to 0 if and only if α_2 is separable over $\kappa(\alpha_1)$. Hence *both* factors are not equal to 0 if and only if α_1 *and* α_2 are separable over κ. In the same fashion it follows that the separability of *all* the coördinates $\alpha_1, \alpha_2, \cdots, \alpha_n$ is necessary and sufficient for the nonvanishing of the Jacobian determinant, q.e.d.

As an immediate corollary of Theorem 6, we obtain the following:

CRITERION FOR UNIFORMIZING PARAMETERS. *Let the coördinates* $\alpha_1, \alpha_2, \cdots, \alpha_n$ *of* P *be separable over* κ *and let* $u_1, u_2, \cdots, u_n$ *be elements of* $\mathfrak{m}(P/S_n)$. *A necessary and sufficient condition that the u's be uniformizing parameters of P is that the Jacobian determinant* $|\partial(u_1, u_2, \cdots, u_n)/\partial(x_1, x_2, \cdots, x_n)|_{x=\alpha}$ *be different from zero.*

To see this it is only necessary to observe that if the α's are separable, then by the preceding theorem and by Lemma 5 of the preceding section the linear transformation (10) (with P instead of W) is nonsingular.

We add the obvious remark that if for a given set of elements $u_1, u_2, \cdots, u_n$ the above Jacobian determinant is different from 0, then the differentials $d_P u_i$ are independent vectors of $\mathcal{D}(P)$, this vector space has then dimension n, and therefore the α's are separable (Theorem 6), and the u's are uniformizing parameters of P.

We shall use the notation $J(v; x)$ for the Jacobian matrix of a set $v_1, v_2, \cdots, v_\nu$ of rational functions of $x_1, x_2, \cdots, x_n$.

In the sequel we shall have occasion to use the following lemma:

LEMMA 6. *Let* $\mathfrak{A}$ *be an ideal in* $\kappa[x_1, x_2, \cdots, x_n]$ *and let* $(F_1(x), F_2(x), \cdots, F_\nu(x))$ *be a basis of* $\mathfrak{A}$. *If* $(\xi_1, \xi_2, \cdots, \xi_n)$ *is a zero of* $\mathfrak{A}$ *in some extension field of* κ *and if the Jacobian matrix* $J(F; x)$ *has rank n at* $x=\xi$, *then the* ξ*'s are separable algebraic over* κ, *and* (ξ) *is a simple zero of* $\mathfrak{A}$. *Moreover, n of the polynomials* $F_i(x)$ *are uniformizing parameters of the point* $P(\xi)$.

Proof. Let V be the irreducible algebraic variety in S_n having (ξ) as general point. Since $J(F; x)$ has rank n on V, it has still rank n at some point $P(\alpha_1, \alpha_2, \cdots, \alpha_n)$ of V. We may assume that $|J(F_1, F_2, \cdots, F_n; x_1, x_2, \cdots, x_n)|_{x=\alpha} \neq 0$. By the preceding remark, the α's are separable over κ, and the polynomials $F_1(x), F_2(x), \cdots, F_n(x)$ are uniformizing parameters of P. This implies that (α) is an isolated zero of the ideal $(F_1(x), F_2(x), \cdots, F_n(x))$, that is, it is not a specialization of a zero of a higher dimension. But since (α) is a specialization of (ξ) and (ξ) is a zero of $\mathfrak{A}$, it follows that necessarily

$(\xi)=(\alpha)$. Moreover, since $F_1(x), F_2(x), \cdots, F_n(x)$ are uniformizing parameters of P, they form a local P-basis of $\mathfrak{A}$. Hence (ξ) is a simple zero of $\mathfrak{A}$ (Definition 2, §3.4).

COROLLARY. *If the matrix $J(F; x)$ of Lemma 6 has rank $n-\rho$ at $x=\xi$, say if $J(F; x_{\rho+1}, x_{\rho+2}, \cdots, x_n)$ has rank $n-\rho$ at $x=\xi$ $(0\leqq\rho\leqq n)$, then $\xi_{\rho+1}, \xi_{\rho+2}, \cdots, \xi_n$ are separable algebraic over the field $\kappa(\xi_1, \xi_2, \cdots, \xi_\rho)$ (and hence the dimension of the "point" (ξ) is not greater than ρ).*

7.2. *Criterion for simple points.* The classical criterion for simple points of an algebraic variety (in terms of Jacobian matrices) can now be readily derived.

We shall take a slightly more general point of view and shall derive a criterion for a given point P to be a simple zero of a given ideal $\mathfrak{A}$ in $\kappa[x_1, x_2, \cdots, x_n]$. If $\mathfrak{A}$ happens to be the ideal $I(V)$ of a given variety V (irreducible or not), then we get the criterion for P to be a simple point of V.

Let $V=\mathfrak{V}(\mathfrak{A})$ be the zero manifold of $\mathfrak{A}$ and let $P(\alpha)$ be a point of an irreducible *r-dimensional* component V_1 of V. Let $(F_1(x), F_2(x), \cdots, F_\nu(x))$ be a basis of $\mathfrak{A}$.

THEOREM 7 (CLASSICAL CRITERION FOR SIMPLE POINTS). *In order that $P(\alpha)$ be a simple zero of the ideal $\mathfrak{A}$ it is sufficient that the Jacobian matrix $J(F; x)$ be of rank $n-r$ at $x=\alpha$. If the coördinates $\alpha_1, \alpha_2, \cdots, \alpha_n$ are separable over κ, then this condition is also necessary.*

Proof. By Theorem 2′ (§3.4), (α) is a simple zero of $\mathfrak{A}$ if and only if exactly $n-r$ of the vectors $\tau F_i(x)$ are independent. If $\alpha_1, \alpha_2, \cdots, \alpha_n$ are separable over κ, the linear transformation (10) (with P instead of W) is nonsingular. Hence, in this separable case, (α) is a simple zero of $\mathfrak{A}$ if and only if exactly $n-r$ of the ν local P-differentials $d_PF_i(x)$ are linearly independent. Hence in the separable case the above condition on the Jacobian matrix $J(F; x)$ is both necessary and sufficient.

In the general case, given that $J(F; x)$ has rank $n-r$ at $x=\alpha$, it follows that $n-r$ of the local P-differentials $d_PF_i(x)$ are linearly independent. Since in the linear transformation (10) we have $\tau F_i(x)\rightarrow d_PF_i(x)$, it follows a fortiori that $n-r$ of the vectors $\tau F_i(x)$ are independent. Hence, by Theorem 2′, (α) is a simple zero of $\mathfrak{A}$.

COROLLARY. *If κ is of characteristic 0 or a perfect field of characteristic $p\neq 0$, then the criterion of Theorem 7 is both necessary and sufficient.*

8. **Continuation of the separable case: generalization to higher varieties in S_n.**

8.1. *Extension of the preceding results.* The results of the preceding section can be easily generalized to the case in which instead of a point P we are dealing with an arbitrary ρ-dimensional irreducible variety W in S_n. Let

$(\eta_1, \eta_2, \cdots, \eta_n)$ be the general point of W and let $t_1, t_2, \cdots, t_{n-\rho}$ be a given set of uniformizing parameters of W. Since the vectors $\tau t_1, \tau t_2, \cdots, \tau t_{n-\rho}$ form an independent basis of the vector space $\mathfrak{M}(W/S_n)$, it follows that *the linear transformation* (10) *is nonsingular if and only if the Jacobian matrix* $J(t; x)$ *has rank* $n-\rho$ at $x=\eta$. If this condition is satisfied for the given set of uniformizing parameters t_i, then the matrix $J(t'; x)$ will be of rank $n-\rho$ at $x=\eta$ for any other set of uniformizing parameters $t_1', t_2', \cdots, t_{n-\rho}'$. It is clear that in all cases the above rank condition for the matrix $J(t'; x)$ is *sufficient* in order that given $n-\rho$ elements t_i' of $\mathfrak{m}(W/S_n)$ be uniformizing parameters of W.

By analogy with Theorem 6 of the preceding section, the condition that $J(t; x)$ have rank $n-\rho$ at $x=\eta$ can be translated into the equivalent condition that *the field* $\mathcal{J}(W)$ $(=\kappa(\eta_1, \eta_2, \cdots, \eta_n))$ *be separably generated over* κ. The proof of this is as follows:

Suppose that $J(t; x)$ is of rank $n-\rho$ at $x=\eta$. We can write $t_i=\phi_i(x)/g(x)$, where $\phi_i(x), g(x) \in \kappa[x_1, x_2, \cdots, x_n]$ and $g(\eta)\neq 0$. Then it is clear that also $J(\phi; x)$ has rank $n-\rho$ at $x=\eta$. We may assume that

$$(12) \qquad \left| J(\phi_1, \phi_2, \cdots, \phi_{n-\rho}; x_{\rho+1}, x_{\rho+2}, \cdots, x_n) \right|_{x=\eta} \neq 0.$$

We take the field $\kappa^*=\kappa(\eta_1, \eta_2, \cdots, \eta_\rho)$ as new ground field and we set $\phi_i^*=\phi_i(\eta_1, \eta_2, \cdots, \eta_\rho, x_{\rho+1}, \cdots, x_n)$. The $n-\rho$ polynomials ϕ_i^* in $\kappa^*[x_{\rho+1}, x_{\rho+2}, \cdots, x_n]$, $i=1, 2, \cdots, n-\rho$, have the common zero $(\eta_{\rho+1}, \eta_{\rho+2}, \cdots, \eta_n)$, and we have by (12) that $\left| J(\phi_1^*, \phi_2^*, \cdots, \phi_{n-\rho}^*; x_{\rho+1}, x_{\rho+2}, \cdots, x_n)\right|_{x=\eta}\neq 0$. Hence by Lemma 6 (where n and ν should be replaced by $n-\rho$) we can conclude that $\eta_{\rho+1}, \eta_{\rho+2}, \cdots, \eta_n$ are separable algebraic over $\kappa(\eta_1, \eta_2, \cdots, \eta_\rho)$. Since $\mathcal{J}(W)$ is of degree of transcendency ρ over κ, it follows that $(\eta_1, \eta_2, \cdots, \eta_\rho)$ is a separating transcendence basis of $\mathcal{J}(W)/\kappa$.

Conversely, if $(\eta_1, \eta_2, \cdots, \eta_\rho)$ is a separating transcendence basis of $\mathcal{J}(W)/\kappa$, then the adjunction of $\eta_1, \eta_2, \cdots, \eta_\rho$ to κ achieves a reduction of W to a *point* in $S_{n-\rho}^{\kappa^*}$ having *separable* coördinates. Hence by Theorem 6 it follows that if $t_1, t_2, \cdots, t_{n-\rho}$ are uniformizing parameters of W then the determinant $|J(t_1, t_2, \cdots, t_{n-\rho}; x_{\rho+1}, x_{\rho+2}, \cdots, x_n)|$ is different from zero at $x=\eta$.

We have thus shown that *if* $t_1, t_2, \cdots, t_{n-\rho}$ *are elements of* $\mathfrak{m}(W/V)$, *then the condition that the determinant* $|J(t_1, t_2, \cdots, t_{n-\rho}; x_{\rho+1}, x_{\rho+2}, \cdots, x_n)|$ *be different from zero at* $x=\eta$ *is equivalent to the condition that the set* $(\eta_1, \eta_2, \cdots, \eta_\rho)$ *be a separating transcendence basis of* $\kappa(\eta_1, \eta_2, \cdots, \eta_n)/\kappa$ *and that* $t_1, t_2, \cdots, t_{n-\rho}$ *be uniformizing parameters of* W. Since by a theorem due to MacLane [8, Lemma 2, p. 380 or Theorem 15, p. 384] the set $(\eta_1, \eta_2, \cdots, \eta_n)$ always contains a separating transcendence basis of the field $\kappa(\eta_1, \eta_2, \cdots, \eta_n)/\kappa$ if this field is separably generated over κ, our assertion is proved.

8.2. *A proof of the theorem of MacLane.* It is of some interest that the italicized statement which we have just proved can be used in order to derive the quoted theorem of MacLane. We proceed as follows:

Let $\{\zeta_1, \zeta_2, \cdots, \zeta_\rho\}$ be a separating transcendence basis of $\kappa(\eta_1, \eta_2, \cdots, \eta_n)/\kappa$ and let

$$\zeta_j = \psi_j(\eta)/\psi_0(\eta), \qquad j = 1, 2, \cdots, \rho.$$

We consider in a linear space $S_{n+\rho}$ the irreducible ρ-dimensional variety W' defined by the general point $(\eta_1, \eta_2, \cdots, \eta_n, \zeta_1, \zeta_2, \cdots, \zeta_\rho)$ (note that $\mathfrak{F}(W)=\mathfrak{F}(W')$, whence W and W' are birationally equivalent varieties). Let

$$(13) \qquad \phi_1(x), \phi_2(x), \cdots, \phi_{n-\rho}(x)$$

be $n-\rho$ polynomials in $\kappa[x_1, x_2, \cdots, x_n]$ which are uniformizing parameters of W. Let moreover

$$(14) \qquad \phi_{n-\rho+j}(x) = x_{n+j}\psi_0(x) - \psi_j(x), \qquad j = 1, 2, \cdots, \rho.$$

The n polynomials $\phi_1(x), \phi_2(x), \cdots, \phi_n(x)$ in $x_1, x_2, \cdots, x_{n+\rho}$ vanish on W', that is, they vanish for $x_i=\eta_i$, $i=1, 2, \cdots, n$, and $x_{n+j}=\zeta_j$, $j=1, 2, \cdots, \rho$. *We assert that these polynomials are uniformizing parameters of W'.* To prove this we have to show the following: if $F(x_i, x_{n+j})\in\kappa[x_1, x_2, \cdots, x_{n+\rho}]$ and $F(\eta_i, \zeta_j)=0$, then there exists a polynomial $A(x_i, x_{n+j})$ such that $A(\eta_i, \zeta_j)\neq 0$ and $A\cdot F$ belongs to the ideal generated by $\phi_1(x), \phi_2(x), \cdots, \phi_n(x)$ in $\kappa[x_1, x_2, \cdots, x_{n+\rho}]$. In view of (14) we can write for an arbitrary polynomial $F(x_i, x_{n+j})$ an identity of the form:

$$[\psi_0(x)]^\nu\cdot F(x) = \sum_{j=1}^{\rho} B_j(x)\phi_{n-\rho+j}(x) + G(x_1, x_2, \cdots, x_n),$$

where $B_j(x)\in\kappa[x_1, x_2, \cdots, x_{n+\rho}]$ and ν is a suitable integer. Now if $F(\eta_i, \zeta_j)=0$ then also $G(\eta)=0$. Since $\phi_1(x), \phi_2(x), \cdots, \phi_{n-\rho}(x)$ are uniformizing parameters of W, there must exist a polynomial $h(x)=h(x_1, x_2, \cdots, x_n)$ such that $h(\eta)\neq 0$ and $h(x)G(x)\equiv 0(\phi_1(x), \phi_2(x), \cdots, \phi_{n-\rho}(x))$. Hence $[\psi_0(x)]^\nu h(x)F(x)\equiv 0(\phi_1(x), \phi_2(x), \cdots, \phi_n(x))$, and this proves our assertion, since $\psi_0(\eta)\neq 0$ and $h(\eta)\neq 0$.

By hypothesis, the set (η_i, ζ_j) contains the separating transcendence basis $(\zeta_1, \zeta_2, \cdots, \zeta_\rho)$ of $\mathfrak{F}(W')$ $(=\mathfrak{F}(W))$, and we have just proved that $\phi_1, \phi_2, \cdots, \phi_n$ are uniformizing parameters of W'. Hence by the result proved above (applied to W' instead of to W), the determinant $|J(\phi_1, \phi_2, \cdots, \phi_n; x_1, x_2, \cdots, x_n)|$ must be different from zero on W' (that is, for $x_i=\eta_i$ and $x_{n+j}=\zeta_j$). This implies that the matrix $J(\phi_1, \phi_2, \cdots, \phi_{n-\rho}; x_1, x_2, \cdots, x_n)$ is of rank $n-\rho$ at $x=\eta$, and therefore, again by the same result proved above (and applied to W itself), the set $(\eta_1, \eta_2, \cdots, \eta_n)$ must contain a separating transcendence basis of $\kappa(\eta_1, \eta_2, \cdots, \eta_n)/\kappa$.

8.3. *The singular manifold of an algebraic variety.* As an immediate consequence of the preceding results we find that *if κ is a field of characteristic zero* *or is a perfect field of characteristic $p \neq 0$, then the singular manifold of an irreducible algebraic variety V in S_n is a proper algebraic subvariety of V.* For let $(\xi_1, \xi_2, \cdots, \xi_n)$ be the general point of V and let $(F_1(x), F_2(x), \cdots, F_\nu(x))$ be a basis of $\mathfrak{p}(V)$. If r is the dimension of V, then under our assumption concerning the field κ, the Jacobian matrix $(J(F; x))_{x=\xi}$ (which can never be of rank greater than $n-r$ since $\mathfrak{D}(V)$ is at most of dimension $n-r$; see Lemma 5, §6.2) must be exactly of rank $n-r$. Hence the points of V where the above matrix is of rank less than $n-r$ form a proper algebraic subvariety of V, and by Theorem 7 this subvariety coincides with the singular manifold of V. The above result will be extended later on to arbitrary ground fields.

We observe that if we had assumed *only* that the field $\mathcal{J}(V)$ is separably generated over κ, then on the basis of Theorem 7 we could only assert the following: *the singular points of V belong to some proper algebraic subvariety of V.* But we could not assert immediately that the singular manifold is algebraic, because Theorem 7 gives only a sufficient condition for simple points.

To conclude this section we shall state the analogue of Theorem 7 for the case in which instead of a point $P(\alpha)$ we are dealing with an irreducible variety W in S_n having $(\eta_1, \eta_2, \cdots, \eta_n)$ as general point. We assume, as in Theorem 7, that W lies on some r-dimensional irreducible component of the zero manifold $\mathfrak{V}(\mathfrak{A})$ of the given ideal $\mathfrak{A}$.

THEOREM 7′ (CLASSICAL CRITERION FOR SIMPLE SUBVARIETIES). *In order that $(\eta_1, \eta_2, \cdots, \eta_n)$ be a simple zero of the ideal $\mathfrak{A}$ it is sufficient that the Jacobian matrix $J(F; x)$ be of rank $n-r$ at $x=\eta$. If the field $\kappa(\eta_1, \eta_2, \cdots, \eta_n)$ is separably generated over κ then the above condition is also necessary.*

The proof is the same as that of Theorem 7.

We observe that if κ is a field of characteristic zero or a perfect field of characteristic $p \neq 0$, then Theorem 3 (§4.2) follows from Theorems 7 and 7′.

9. **The nonseparable case.**

9.1. *The dimension of $\mathfrak{D}(W)$ as a numerical character of the field $\mathcal{J}(W)$.* We say that an irreducible variety W in S_n presents the nonseparable case if the field $\mathcal{J}(W)$ is not separably generated over κ. In particular, if W is a point P, the nonseparable case is characterized by the condition that at least one of the coördinates of P is inseparable over κ.

Whether we are dealing with the separable or the nonseparable case, it is always true that if $(\eta_1, \eta_2, \cdots, \eta_n)$ is the general point of W and if $(u_1, u_2, \cdots, u_\nu)$ is a local W-basis of $\mathfrak{p}(W)$, that is, a basis of $\mathfrak{m}(W/S_n)$, the rank of the Jacobian matrix $J(u; x)$ at $x=\eta$ is independent of the local basis, for this rank gives the dimension of the vector space $\mathfrak{D}(W)$ of local W-differentials. We know that the rank is always less than or equal to $n-\rho$, where ρ is the dimension of W, and that the equality sign holds if and only if W presents the separable case. So at least in the separable case it is true that the difference $n-(n-\rho)$ between the dimension of the ambient space S_n and the dimension of $\mathfrak{D}(W)$ is an intrinsic character of the field $\mathfrak{J}(W)$, namely it is equal to the degree of transcendency of $\mathfrak{J}(W)/\kappa$. We shall now show that also in the nonseparable case, if we denote by $n-\sigma$ the dimension of $\mathfrak{D}(W)$ $(\sigma>\rho)$, then σ is an intrinsic character of the field $\mathfrak{J}(W)$.

THEOREM 8. *If* $\dim \mathfrak{D}(W)=n-\sigma<n-\rho$, *then* $\mathfrak{J}(W)$ *is a* σ*-fold extension of* κ *(that is,* $\mathfrak{J}(W)$ *can be obtained by adjoining to* κ *a suitable set of* σ *elements of* $\mathfrak{J}(W)$*) but is not a* $(\sigma-1)$*-fold extension of* κ[6].

Proof. We select for uniformizing parameters of W as set of polynomials

$$\phi_1(x),\ \phi_2(x),\ \cdots,\ \phi_{n-\rho}(x). \tag{15}$$

We have, by hypothesis, that the Jacobian matrix $J(\phi; x)$ has rank $n-\sigma$ at $x=\eta$. Let, say, $J(\phi; x_{\sigma+1}, x_{\sigma+2}, \cdots, x_n)$ be of rank $n-\sigma$ at $x=\eta$. Then by the corollary to Lemma 6 (§7.1) the field $\mathfrak{J}(W)$ is a separable algebraic extension of the field $\kappa(\eta_1, \eta_2, \cdots, \eta_\sigma)$. Without loss of generality we may assume that $\eta_1, \eta_2, \cdots, \eta_\rho$ are algebraically independent over κ. Since $\sigma>\rho$, the field $\mathfrak{J}(W)$ is an algebraic extension of the field $\Delta_1=\kappa(\eta_1, \eta_2, \cdots, \eta_{\sigma-1})$. Let Δ_1' be the greatest subfield of $\mathfrak{J}(W)$ which is separable algebraic over Δ_1. Every element of $\mathfrak{J}(W)$ is either in Δ_1' or is purely inseparable over Δ_1'. A fortiori every element of $\mathfrak{J}(W)$ is either in $\Delta_1'(\eta_\sigma)$ or is purely inseparable over $\Delta_1'(\eta_\sigma)$. From this we conclude that $\mathfrak{J}(W)=\Delta_1'(\eta_\sigma)$ since $\kappa(\eta_1, \eta_2, \cdots, \eta_\sigma)\subseteq\Delta_1'(\eta_\sigma)$ and since therefore $\mathfrak{J}(W)$ is a separable algebraic extension of $\Delta_1'(\eta_\sigma)$. Now Δ_1' as a separable algebraic extension of Δ_1 is a simple extension of Δ_1: $\Delta_1'=\Delta_1(\theta)$. Hence $\mathfrak{J}(W)=\Delta_1(\theta, \eta_\sigma)$. Of the two elements θ, η_σ one, namely θ, is separable algebraic over Δ_1. Hence $\mathfrak{J}(W)$ is a simple extension of Δ_1: $\mathfrak{J}(W)=\Delta_1(\alpha)=\kappa(\eta_1, \eta_2, \cdots, \eta_{\sigma-1}, \alpha)$, that is, $\mathfrak{J}(W)$ *is a* σ*-fold extension of* κ.

Now let $(\zeta_1, \zeta_2, \cdots, \zeta_m)$ be any set of generators of $\mathfrak{J}(W)$ over κ: $\mathfrak{J}(W)=\kappa(\zeta_1, \zeta_2, \cdots, \zeta_m)$. To complete the proof of the theorem we have to show that

$$m \geqq \sigma. \tag{16}$$

Since $\mathfrak{J}(W)=\kappa(\eta_1, \eta_2, \cdots, \eta_n)$, we can write

$$\zeta_j = \psi_j(\eta)/\psi_0(\eta), \qquad j = 1, 2, \cdots, m,$$

where $\psi_0(x)$, $\psi_j(x)\in\kappa[x_1, x_2, \cdots, x_n]$ and $\psi_0(\eta)\neq 0$. We denote by W' the algebraic variety in S_{n+m} whose general point is $(\eta_1, \eta_2, \cdots, \eta_n, \zeta_1, \zeta_2, \cdots, \zeta_m)$. Let moreover W_1 be the algebraic variety of S_m whose general point is $(\zeta)=(\zeta_1, \zeta_2, \cdots, \zeta_m)$. The three varieties W, W_1, W' are birationally equivalent, and W' is symmetrically related to W and W_1, namely W' is the join of W and W_1.

(6) The theorem is not true in the separable case. If $\sigma=\rho=$degree of transcendency of $\mathfrak{J}(W)/\kappa$, then $\mathfrak{J}(W)$ can always be regarded as a $(\rho+1)$-fold extension of κ (a pure transcendental ρ-fold extension, followed by a simple separable extension), but not always as a ρ-fold extension of κ.

If we set

$$\phi_{n-\rho+j}(x) = x_{n+j}\psi_0(x) - \psi_j(x), \qquad j = 1, 2, \cdots, m,$$

then it follows as in our proof of MacLane's theorem (§8.2) that the $n-\rho$ polynomials (15) together with the m polynomials $\phi_{n-\rho+j}$ just introduced form a set of uniformizing parameters of W' in S_{n+m}. The Jacobian matrix $J(\phi_1, \phi_2, \cdots, \phi_{n+m-\rho}; x_1, x_2, \cdots, x_{n+m})$ has the form

$$\left\| \begin{matrix} J(\phi_1, \phi_2, \cdots, \phi_{n-\rho}; x_1, x_2, \cdots, x_n) & 0 \\ * & \psi_0(x)E_m \end{matrix} \right\|,$$

where E_m is the m-rowed unit matrix. Since the matrix at the upper left corner has rank $n-\sigma$ at $x_i=\eta_i$, $i=1, 2, \cdots, n$, it follows that the above $(n+m-\rho)$-rowed matrix has rank $n+m-\sigma$ at $x_i=\eta_i$ and $x_{n+j}=\zeta_j$ $(j=1, 2, \cdots, m)$. Therefore

$$\dim \mathfrak{D}(W') = n + m - \sigma.$$

If $\dim \mathfrak{D}(W_1) = m-\sigma_1$, then interchanging the roles of W and W_1 we get

$$\dim \mathfrak{D}(W') = m + n - \sigma_1.$$

Consequently $\sigma=\sigma_1$. Since $m-\sigma_1 \geqq 0$ we have the inequality (16), and this completes the proof of the theorem.

9.2. *Abstract differentiations over* κ^p. We shall assume in this section that the ground field κ is of characteristic $p \neq 0$. Our main purpose at this stage is to derive *general* criteria for uniformizing parameters and simple loci, valid whether we are dealing with the separable or the nonseparable case. First we recall a few well known facts about abstract differentiation. Given a field K, a differenitation in K is an operator D in K with the following properties: (1) $Dy \in$ K is defined and is single-valued for all elements y in K; (2) if $y, z \in$ K then $D(y-z)=Dy-Dz$ and $D(yz)=yDz+zDy$. The set of all differentiations in K can be regarded as a vector space over K, if we define: (1) $(D_1+D_2)y = D_1y+D_2y$, (2) $(cD)y=c\cdot Dy$ for any c in K. The zero vector is the trivial differentiation D_0:$D_0y=0$ for all y in K.

If P is a subfield of K, we say that a differentiation D in K *is over* P if $Da=0$ for all a in P. The differentiations in K over P also form a vector space over K (and hence also over P).

Suppose that K is of characteristic $p \neq 0$ and that $K^p \subseteq P$. For any element y of K it is true then that either $y \in$ P or y^p is the least power of y which belongs to P. Let $Z=\{z_i\}$ be a *minimal* (finite or infinite) set of generators of K/P (Z = *a relatively p-independent basis of* K/P; see MacLane [8, p. 376]). The minimality of Z is equivalent to the property that if $\{z_1, z_2, \cdots, z_s\}$ is any finite set of s elements in Z, then the relative degree $[P(z_1, z_2, \cdots, z_s):P]$ is p^s. This property in its turn is equivalent to the condition that each element y of K can be written in *one and only one* form as a polynomial in the z_i, with coefficients in P, of degree not greater than $p-1$ in each of the arguments z_i.

It is clear that an abstract differentiation D over P is uniquely determined if the values $a_i = Dz_i$, $z_i \in Z$, are known, for if $y = \phi(\{z_i\})$ is any element of K, then $Dy = \sum_i (\partial\phi/\partial z_i)\cdot a_i$, only a finite number of partial derivatives $\partial\phi/\partial z_i$ being different from zero. On the other hand, if the elements a_i are preassigned in an arbitrary fashion, then the above expression of Dy defines a differentiation in K over P. The only thing that ought to be verified is that D is then single-valued, and for that it is sufficient to show that if $y=0$ then $Dy=0$. Now if we have a relation $\phi(\{z_i\})=0$ over P, then in view of the p-independence of the elements z_i the polynomial $\phi(\{X_i\})$ in the *indeterminates* X_i must belong to the ideal generated in $P[\{X_i\}]$ by the elements $X_i^p - c_i$, where c_i is the element z_i^p of P. That implies that the partial derivatives $\partial\phi/\partial X_i$ also belong to that ideal, whence $\partial\phi/\partial z_i = 0$.

In particular we have for each z_i in Z the differentiation $D_i = \partial/\partial z_i$ defined by the conditions $D_i z_i = 1$, $D_i z_j = 0$ if $j \neq i$. If the relative degree of K/P is finite, these differentiations form a vector basis for the set of all differentiations in K over P.

We now identify K with the field $\kappa(x_1, x_2, \cdots, x_n)$, where $x_1, x_2, \cdots, x_n$ are indeterminates, *and* P *with the field* κ^p. If $Z = \{z_j\}$ is a p-independent basis of κ/κ^p, then we consider the following differentiations in $\kappa(x_1, x_2, \cdots, x_n)$ over κ^p:

$$D_i = \partial/\partial x_i, \qquad D_j^* = \partial/\partial z_j. \tag{17}$$

The differentiations D_i are actually over κ, while the D_j^* are extensions of differentiations in κ/κ^p.

Let W be an irreducible algebraic variety in S_n, and let $(\eta_1, \eta_2, \cdots, \eta_n)$ be the general point of W. The differentiations (17) leave invariant the quotient ring $\mathfrak{o} = Q(W/S_n)$. For any element y in $\mathfrak{o}$ we set

$$D_i^W y = (D_i y)_{x=\eta}, \qquad D_j^{*W} y = (D_j^* y)_{x=\eta}. \tag{18}$$

The D_i^W and D_j^{*W} are operators D^W satisfying the following conditions: (1) D^W is a single-valued mapping of $\mathfrak{o}$ into the field $\mathcal{J}(W) = \kappa(\eta_1, \eta_2, \cdots, \eta_n)$; (2) $D^W(y-z) = D^W y - D^W z$; (3) $D^W(yz) = \bar{y}D^W z + \bar{z}D^W y$, where $\bar{y}$ and $\bar{z}$ are the W-residues of y and z; (4) if $c \in \kappa^p$ then $D^W c = 0$.

Any operator D^W satisfying the first three of the above 4 conditions may be called a *local W-differentiation*. Condition (4) signifies that the differentiation is over κ^p. The local W-differentiations over κ^p form in an obvious fashion a vector space over $\mathcal{J}(W)$. We denote this vector space by $\mathfrak{K}(W)$.

9.3. *A lemma on the canonical uniformizing parameters.* Our object now is to prove an important auxiliary lemma. Let $P(\alpha_1, \alpha_2, \cdots, \alpha_n)$ be a point in S_n and let $f_1(x_1), f_2(x_1, x_2), \cdots, f_n(x_1, x_2, \cdots, x_n)$ be the canonical uniformizing parameters of P (see §7.1). Let κ_1 be any subfield of κ which is a finite extension of κ^p and which contains all the coefficients of the n polynomials $f_i(x_1, x_2, \cdots, x_i)$. Let $(z_1, z_2, \cdots, z_\nu)$ be a p-independent basis of κ_1/κ^p. The partial derivatives $D_j^* f_i = \partial f_i/\partial z_j$, $i=1, 2, \cdots, n$; $j=1, 2, \cdots, \nu$, are well defined, and it is indifferent whether D_j^* is regarded as a differentiation in $\kappa_1(x_1, x_2, \cdots, x_n)$, or as a differentiation in $\kappa(x_1, x_2, \cdots, x_n)$ defined by choosing a p-independent basis Z of κ/κ^p *containing* the elements $z_1, z_2, \cdots, z_\nu$.

LEMMA 7. *If each polynomial $f_i(x)$, $i=1, 2, \cdots, n$, depends only on $x_1^p, x_2^p, \cdots, x_i^p$, then the Jacobian matrix $J(f; z) = \partial(f_1, f_2, \cdots, f_n)/\partial(z_1, z_2, \cdots, z_\nu)$ has rank n at $x=\alpha$.*

Proof. We shall proceed by induction with respect to the dimension n of the space S_n. For $n=1$ the lemma asserts that the derivatives $\partial f_1/\partial z_j$, $j=1, 2, \cdots, \nu$, are not all zero at $x_1=\alpha_1$. Let us assume that this assertion is false. Our assumption is then that $x=\alpha_1$ is a root of the polynomial $\partial f_1(x_1)/\partial z_j$, $j=1, 2, \cdots, \nu$. Since $f_1(x_1)$ is an irreducible *monic* polynomial over κ which vanishes at $x_1=\alpha_1$, it follows $\partial f_1(x_1)/\partial z_j$ is identically zero, that is, the coefficients of f_1 behave as "constants" under each of the differentiations $\partial/\partial z_j$ in κ_1 over κ^p. Therefore their derivatives are zero under *all* differentiations in κ_1/κ^p, *and consequently all the coefficients of* $f_1(x_1)$ are in κ^p. This implies that $f_1(x_1)$, which by hypothesis is a polynomial in x_1^p, is the pth power of a polynomial in $\kappa[x_1]$, in contradiction with the irreducibility of $f_1(x_1)$.

We assume now that the lemma is true for linear spaces S_{n-1}. Setting $\bar{\kappa} = \kappa(\alpha_1)$ we apply the lemma to $S_{n-1}^{\bar{\kappa}}$ and to the point $\bar{P}(\alpha_2, \alpha_3, \cdots, \alpha_n)$ in this space. The canonical uniformizing parameters of $\bar{P}$ are the polynomials $\Phi_i(x_2, x_3, \cdots, x_i) = f_i(\alpha_1, x_2, x_3, \cdots, x_i)$, $i=2, 3, \cdots, n$, and these depend only on $x_2^p, x_3^p, \cdots, x_n^p$. As field analogous to κ_1 we take the following field $\bar{\kappa}_1$:

$$\bar{\kappa}_1 = \bar{\kappa}^p(z_1, z_2, \cdots, z_\nu) = \kappa_1(\alpha_1^p).$$

This field $\bar{\kappa}_1$ satisfies all the necessary requirements: (1) it is a subfield of $\bar{\kappa}$; (2) it is a finite extension of $\bar{\kappa}^p$; (3) it contains all the coefficients of the polynomials $\Phi_2, \Phi_3, \cdots, \Phi_n$ since by hypothesis only the powers of α_1^p occur explicitly in $f_i(\alpha_1, x_2, \cdots, x_i)$. Thus all the conditions of the lemma are satisfied. *However, this time the elements $z_1, z_2, \cdots, z_\nu$ are not relatively p-independent over $\bar{\kappa}^p$.* While the relative degree $[\kappa_1:\kappa^p]$ was p^ν, *we shall now show that*

$$[\bar{\kappa}_1:\bar{\kappa}^p] = p^{\nu-1}, \tag{19}$$

and consequently $\nu-1$ of the elements z_j form a p-independent basis of $\bar{\kappa}_1/\bar{\kappa}^p$. To see this we observe that $[\bar{\kappa}_1:\kappa^p] = [\bar{\kappa}_1:\bar{\kappa}^p]\cdot[\bar{\kappa}^p:\kappa^p] = [\bar{\kappa}_1:\bar{\kappa}^p]\cdot[\bar{\kappa}:\kappa]$, and that on the other hand $[\bar{\kappa}_1:\kappa^p] = [\bar{\kappa}_1:\kappa_1]\cdot[\kappa_1:\kappa^p]$. Hence

$$[\bar{\kappa}_1:\bar{\kappa}^p]\cdot[\bar{\kappa}:\kappa] = [\kappa_1:\kappa^p]\cdot[\bar{\kappa}_1:\kappa_1]. \tag{20}$$

Since $\bar{\kappa} = \kappa(\alpha_1)$ and α_1 is a root of the irreducible polynomial $f_1(x_1)$ in $\kappa[x_1]$, we have $[\bar{\kappa}:\kappa]=\nu_1=$degree of $f_1(x_1)$. On the other hand, by hypothesis we have $f_1(x_1)=\phi_1(x_1^p)$, ϕ_1 irreducible over κ, and $X_1=\alpha_1^p$ is a root of $\phi_1(X_1)$. Then $\phi_1(X_1)$ is also irreducible over the subfield κ_1 of κ, and since $\bar{\kappa}_1=\kappa_1(\alpha_1^p)$ it follows that $[\bar{\kappa}_1:\kappa_1]=$degree of $\phi_1(X_1)=\nu_1/p$. Hence $[\bar{\kappa}:\kappa]=p\cdot[\bar{\kappa}_1:\kappa_1]$, and therefore, by (20), $[\kappa_1:\kappa^p]=p\cdot[\bar{\kappa}_1:\bar{\kappa}^p]$, as asserted in (19).

We may then assume that $(z_2, z_3, \cdots, z_\nu)$ is a p-independent basis of $\bar{\kappa}_1/\bar{\kappa}^p$. We shall denote differentiation in $\bar{\kappa}_1$ over $\bar{\kappa}^p$ by $\bar{\partial}/\partial z_j$, $j=2, 3, \cdots, \nu$. The differentiation $\bar{\partial}/\partial z_j$, *if applied to elements of the subfield* κ_1 *of* $\bar{\kappa}_1$, is expressible in terms of the old derivatives $\partial/\partial z_j$, $j=1, 2, \cdots, \nu$, by the usual rule of composite differentiation:

$$\bar{\partial}/\partial z_j = \partial/\partial z_j + \partial/\partial z_1\cdot\bar{\partial}z_1/\partial z_j, \qquad j = 2, 3, \cdots, \nu. \tag{21}$$

Here $\bar{\partial}z_1/\partial z_j$ is to be computed by making use of the relation

$$[\partial f_1/\partial z_j + \partial f_1/\partial z_1\cdot\bar{\partial}z_1/\partial z_j]_{x_1=\alpha_1} = 0. \tag{22}$$

From (21) and (22) we conclude that the matrix $(J(f; z))_{x=\alpha}$ is equivalent to the matrix

$$\left\| \begin{matrix} \partial f_1/\partial z_1, & O \\ * & \bar{J}(\bar{\phi}; z) \end{matrix} \right\|_{x=\alpha},$$

where $\bar{J}(\bar{\phi}; z)=\bar{\partial}(\bar{\phi}_2, \bar{\phi}_3, \cdots, \bar{\phi}_n)/\partial(z_2, z_3, \cdots, z_\nu)$.

By our induction, the matrix $\bar{J}(\bar{\phi}; z)$ is of rank $n-1$ at $x=\alpha$. By the case $n=1$ and by (22) we have $\partial f_1/\partial z_1\neq 0$ at $x_1=\alpha_1$. Hence $J(f; z)$ is of rank n at $x=\alpha$, and this completes the proof of the lemma.

9.4. *The vector space* $\mathfrak{D}^*(W)$ *of the mixed local W-differentials.* We go back to the vectors space $\mathfrak{K}(W)$ of the local W-differentiations (§9.2). For a given element ω in $\mathfrak{o}$ we have that $D^W\omega$ is a linear function of the variable element D^W in $\mathfrak{K}(W)$. This function we call the *mixed local W-differential of* ω, and we denote it by $d_W^*\omega$, or simply by $d^*\omega$. The following relations are obvious: $d^*(\omega_1-\omega_2)=d^*\omega_1-d^*\omega_2$; $d^*(\omega_1\omega_2)=\bar{\omega}_1d^*\omega_2+\bar{\omega}_2d^*\omega_1$, where $\bar{\omega}_1$ and $\bar{\omega}_2$ are the $\mathfrak{m}$-residues of ω_1 and ω_2 ($\mathfrak{m}=$the maximal ideal in $\mathfrak{o}$).

We are primarily interested in the differentials of elements in $\mathfrak{m}$. For ω in $\mathfrak{o}$ and u in $\mathfrak{m}$ we have $d^*(\omega u)=\bar{\omega}d^*u$, where $\bar{\omega}$ is the $\mathfrak{m}$-residue of ω, since the $\mathfrak{m}$-residue of u is zero. It follows that the product $\bar{\omega}d^*u$ is again a differential of an element of $\mathfrak{m}$. Therefore, *for variable u in* $\mathfrak{m}$ *the differentials* d^*u *form a vector space over* $\mathfrak{J}(W)$. We shall call this vector space *the space of mixed local W-differentials*, and we shall denote it by $\mathfrak{D}^*(W)$.

From $d^*(\omega_1\omega_2)=\bar{\omega}_1d^*\omega_2+\bar{\omega}_2d^*\omega_1$ follows that if $u\in\mathfrak{m}^2$ then $d^*u=0$. From this we conclude, as in the case of the ordinary local differentials (§6.2) that the mapping

$$\tau u \to d^*u \tag{23}$$

is a linear transformation of the local vector space $\mathfrak{M}(W/S_n)$ onto the vector space $\mathfrak{D}^*(W)$ of mixed local differentials at W.

9.5. *The nonsingular character of the linear transformation* $\mathfrak{M}(W/S_n) \to \mathfrak{D}^*(W)$. Our main result for ground fields κ of characteristic $p \neq 0$ is the following theorem:

THEOREM 9. *The linear transformation* (23) *of* $\mathfrak{M}(W/S_n)$ *onto* $\mathfrak{D}^*(W)$ *is nonsingular (in other words:* $\mathfrak{D}^*(W)$ *is of dimension* $n-\rho$, *where* $\rho=$ *dimension of* W).

Proof. We shall first prove Theorem 9 in the case in which W is a point $P(\alpha)$. Let $n-\sigma=\dim \mathfrak{D}(P)$, $\sigma \geqq 0$. Then for any set of uniformizing parameters $t_1, t_2, \cdots, t_n$ of P the matrix $J(t; x)$ is of rank $n-\sigma$ at P. We take a fixed set of uniformizing parameters t_i and we label the coördinates x_i in S_n in such a fashion that $J(t_1, t_2, \cdots, t_n; x_{\sigma+1}, x_{\sigma+2}, \cdots, x_n)$ be of rank $n-\sigma$ at $x=\alpha$. If $\tau_1, \tau_2, \cdots, \tau_n$ is any other set of uniformizing parameters, then the matrix $J(\tau_1, \tau_2, \cdots, \tau_n; x_{\sigma+1}, x_{\sigma+2}, \cdots, x_n)$ will also be of rank $n-\sigma$ at $x=\alpha$, for the τ's are linear forms in the t's with coefficients in $\mathfrak{o}=Q(P/S_n)$, and the determinant of these coefficients is not 0 at $x=\alpha$. In particular we have then that if $f_i(x_1, x_2, \cdots, x_i)$, $i=1, 2, \cdots, n$, are the canonical uniformizing parameters of P (relative to the particular order in which we have labeled the coördinates x_i), then $J(f_1, f_2, \cdots, f_n; x_{\sigma+1}, x_{\sigma+2}, \cdots, x_n)$ is of rank $n-\sigma$ at $x=\alpha$. Since $f_1, f_2, \cdots, f_\sigma$ are independent of $x_{\sigma+1}, x_{\sigma+2}, \cdots, x_n$, it follows that

$$\left| J(f_{\sigma+1}, f_{\sigma+2}, \cdots, f_n; x_{\sigma+1}, x_{\sigma+2}, \cdots, x_n) \right|_{x=\alpha} \neq 0. \tag{24}$$

We have

$$\begin{aligned} &J(f_1, f_2, \cdots, f_n; x_1, x_2, \cdots, x_n) \\ &= \left\| \begin{matrix} J(f_1, f_2, \cdots, f_\sigma; x_1, x_2, \cdots, x_\sigma) & O \\ * & J(f_{\sigma+1}, f_{\sigma+2}, \cdots, f_n; x_{\sigma+1}, x_{\sigma+2}, \cdots, x_n) \end{matrix} \right\|, \end{aligned} \tag{25}$$

and this matrix must be of rank $n-\sigma$ at $x=\alpha$. Hence, by (24), it follows that $(J(f_1, f_2, \cdots, f_\sigma; x_1, x_2, \cdots, x_\sigma)_{x=\alpha}$ is the zero matrix, that is, we must have: $\partial f_i/\partial x_1 = \partial f_i/\partial x_2 = \cdots = \partial f_i/\partial x_i = 0$, at $x=\alpha$, $i=1, 2, \cdots, \sigma$. This implies, in view of the defining properties of the canonical uniformizing parameters (§7.1, especially properties c and d), that for $i=1, 2, \cdots, \sigma$ the polynomial f_i is a polynomial in $x_1^p, x_2^p, \cdots, x_i^p$. These polynomials are canonical uniformizing parameters of the point $P_1(\alpha_1, \alpha_2, \cdots, \alpha_\sigma)$ in S_σ. We are therefore in position to apply the preceding Lemma 7 (with n replaced by σ), and we may assert that the matrix $J(f_1, f_2, \cdots, f_\sigma; z_1, z_2, \cdots, z_\nu)$ is of rank σ at $x=\alpha$. Here $(z_1, z_2, \cdots, z_\nu)$ is a p-independent basis of κ_1/κ^p, where κ_1 is any field between κ^p and κ which is a finite extension of κ^p and which contains all the coefficients of $f_1, f_2, \cdots, f_\sigma$. If we take κ_1 large enough so as to include in this field also the coefficients of $f_{\sigma+1}, f_{\sigma+2}, \cdots, f_n$, we can introduce the *mixed Jacobian matrix*:

$$J(f;\, x,\, z) = J(f_1, f_2, \cdots, f_n;\, x_1, x_2, \cdots, x_n;\, z_1, z_2, \cdots, z_\nu).$$

By (24) and (25), and in view of the fact that $J(f_1, f_2, \cdots, f_\sigma; z_1, z_2, \cdots, z_\nu)$ is of rank σ at $x=\alpha$, *we conclude that the matrix $J(f; x, z)$ is of rank n at $x=\alpha$*. This implies that *the n mixed local P-differentials d^*f_h are linearly independent vectors of $\mathfrak{D}^*(P)$*, for d^*f_h is, by definition, a linear function on $\mathfrak{K}(P)$ whose value at D_i^P and at D_j^{*P} is respectively $(\partial f_h/\partial x_i)_{x=\alpha}$ and $(\partial f_h/\partial z_j)_{x=\alpha}$ $(i, h=1, 2, \cdots, n; j=1, 2, \cdots, \nu)$. Hence $\mathfrak{D}^*(P)$ is of dimension n, and this completes the proof of Theorem 9 when W is a point.

In the general case we carry out our usual reduction to the zero-dimensional case. However, we must exercise caution in the course of this reduction, for there is one element in our definition of local W-differentiation which depends on the given ground field κ (and not only on the quotient ring of W): it is the requirement that $D^W c=0$ if $c \in \kappa^p$. If $(\eta_1, \eta_2, \cdots, \eta_n)$ is the general point of W and if we take as our new ground field the field $\bar{\kappa}=\kappa(x_1, x_2, \cdots, x_\rho)$ (where we assume that $\eta_1, \eta_2, \cdots, \eta_\rho$ are algebraically independent over κ and hence are identifiable with $x_1, x_2, \cdots, x_\rho$), we shall be dealing with the point $\bar{P}(\eta_{\rho+1}, \eta_{\rho+2}, \cdots, \eta_n)$ in $S_{n-\rho}^{\bar{\kappa}}$. The local $\bar{P}$-differentiations $D^{\bar{P}}$ in $\mathfrak{o}$ $(\mathfrak{o}=Q(W/S_n)=Q(\bar{P}/S_{n-\rho}^{\bar{\kappa}}))$ are then the local W-differentiations D^W which satisfy the stronger requirement: $D^W\omega=0$ for ω in $\bar{\kappa}^p$. Therefore the vector space $\mathfrak{K}(\bar{P})$ is a (proper) subspace of $\mathfrak{K}(W)$. The elements d^*u of $\mathfrak{D}^*(W)$ are the linear functions on $\mathfrak{K}(W)$ defined by the various elements u in the maximal ideal $\mathfrak{m}$ of $\mathfrak{o}$, while the elements $\bar{d}^*u$ of $\mathfrak{D}^*(\bar{P})$ are the linear functions on $\mathfrak{K}(\bar{P})$ defined by the same elements u. Hence $\bar{d}^*u$ is the linear function on $\mathfrak{K}(\bar{P})$ *induced* by d^*u, and *therefore $\mathfrak{D}^*(\bar{P})$ is a projection of $\mathfrak{D}^*(W)$*. The main point that has to be brought out, and this will complete the proof of Theorem 9, is the following: *the two vector spaces $\mathfrak{D}^*(W)$ and $\mathfrak{D}^*(\bar{P})$ have the same dimension*. We have only to show that if $\bar{d}^*u=0$ then also $d^*u=0$. By Theorem 9, which we have already proved in the case of points, it follows that if $\bar{d}^*u=0$ then $\tau u=0$. But then u is necessarily an element in $\mathfrak{m}^2$, and that, of course, implies that also d^*u is equal to zero. Since by Theorem 9 (applied again to the point $\bar{P}$ in $S_{n-\rho}^{\bar{\kappa}}$) the space $\mathfrak{D}^*(\bar{P})$ has dimension $n-\rho$, the proof of Theorem 9 is now complete.

9.6. *General Jacobian criteria for uniformizing parameters and for simple loci*. From Theorem 9 we now can derive readily the desired consequences.

THEOREM 10 (CRITERION FOR UNIFORMIZING PARAMETERS). *A necessary and sufficient condition that given elements $u_1, u_2, \cdots, u_{n-\rho}$ in $\mathfrak{m}(W/S_n)$ be uniformizing parameters of W is that the mixed local W-differentials $d^*u_1, d^*u_2, \cdots, d^*u_{n-\rho}$ be independent vectors of $\mathfrak{D}^*(W)$; or equivalently: if κ_1 is the field obtained by adjoining to κ^p all the coefficients of the rational functions u_i and if $(z_1, z_2, \cdots, z_\nu)$ is a p-independent basis of κ_1/κ^p, then the matrix $J(u; x, z)$ should be of rank $n-\rho$ on W.*

The proof is obvious.

Let $\mathfrak{A}$ be a given ideal in $\kappa[x_1, x_2, \cdots, x_n]$ and let us assume that W lies on an r-dimensional irreducible component of the zero manifold $\mathcal{V}(\mathfrak{A})$ of $\mathfrak{A}$. Let $(F_1(x), F_2(x), \cdots, F_h(x))$ be a basis of the ideal $\mathfrak{A}$, and let $(z_1, z_2, \cdots, z_\nu)$ be a p-independent basis of κ_1/κ^p, where κ_1 is the field obtained by adjoining to κ^p all the coefficients of the polynomials $F_j(x)$. Let moreover $(\eta) = (\eta_1, \eta_2, \cdots, \eta_n)$ be the general point of W.

THEOREM 11 (CRITERION FOR SIMPLE ZEROS). *In order that (η) be a simple zero of the ideal $\mathfrak{A}$ it is necessary and sufficient that the mixed Jacobian matrix $J(F; x, z)$ be of rank $n-r$ on W (that is, for $x=\eta$).*

The proof is the same as that of Theorem 7 (§7.2), with all references to separability to be omitted.

COROLLARY. *If V is an irreducible algebraic r-dimensional variety in S_n and if the ideal $\mathfrak{A}$ of Theorem 11 is identified with the prime ideal $\mathfrak{p}(V)$, then the singular manifold of V consists of those points of V at which the mixed Jacobian matrix $J(F; x, z)$ has rank less than $n-r$. This manifold is a proper algebraic subvariety of V.*

The last assertion in the corollary follows from the fact that, in view of Theorem 11, the rank of $J(F; x, z)$ on V itself must be exactly $n-r$.

10. **Absolutely simple loci.**

10.1. *Algebraic ground field extensions.* Let $R=\kappa[x_1, x_2, \cdots, x_n]$ be a polynomial ring in n indeterminates x_i, over the ground field κ, and let κ' be an algebraic extension of κ. We denote by R' the polynomial ring $\kappa'[x_1, x_2, \cdots, x_n]$.

If $\mathfrak{A}$ is an ideal in R and if ω' is an element of the extended ideal $R'\cdot\mathfrak{A}$, then ω' can be expressed in the form $\omega'=\omega_1u_1'+\omega_2u_2'+\cdots+\omega_su_s'$, where $\omega_i\in\mathfrak{A}$, $u_1'=1$ and $u_1', u_2', \cdots, u_s'$ are elements of κ' which are lineary independent over κ. From this it follows that if $\omega'\in R'\cdot\mathfrak{A}\cap R$, then necessarily $\omega'=\omega_1$ and $\omega_2=\omega_3=\cdots=\omega_s=0$. Hence the relation

$$R'\cdot\mathfrak{A}\cap R = \mathfrak{A} \tag{26}$$

holds for any ideal $\mathfrak{A}$ in R.

LEMMA 8. *If $v_1', v_2', \cdots, v_g'$ are elements of κ' whcih are linearly independent over κ, then an element of R' which is of the form $\omega_1v_1'+\omega_2v_2'+\cdots+\omega_gv_g'$, $\omega_i\in R$, can belong to $R'\cdot\mathfrak{A}$ only if $\omega_i\in\mathfrak{A}$, $i=1, 2, \cdots, g$.*

Proof. Let the element $\sum_{i=1}^{g}\omega_iv_i'$ be denoted by ω' and let κ_1 denote the field $\kappa(v_1', v_2', \cdots, v_g')$. The set $\{v_i'\}$ can be extended to an independent basis $\{v_1', v_2', \cdots, v_g', v_{g+1}', v_{g+2}', \cdots, v_h'\}$ of κ_1/κ. Let $R_1=\kappa_1R=\kappa_1[x_1, x_2, \cdots, x_n]$ and let $\mathfrak{A}_1=R_1\cdot\mathfrak{A}$. We have $R'\cdot\mathfrak{A}=R'\cdot\mathfrak{A}_1$ and $\omega'\in R_1$. Hence $\omega'\in R'\cdot\mathfrak{A}_1\cap R_1$, and therefore, by (26), $\omega'\in\mathfrak{A}_1$. Since $R_1=\sum_{i=1}^{h}R\cdot v_i'$, $\mathfrak{A}_1$

$=\sum_{i=1}^{h}\mathfrak{A}\cdot v_i'$ and since $\omega'\in\mathfrak{A}_1$, we can write ω' in the form $\omega'=\sum_{i=1}^{h}a_iv_i'$, where the a_i are in $\mathfrak{A}$. Since the h elements v_i' are also linearly independent over the field $\kappa(x_1, x_2, \cdots, x_n)$, it follows that we must have $a_i=\omega_i$, $i=1, 2, \cdots, g$, and $a_j=0$ for $j=g+1, g+2, \cdots, h$. Hence the ω_i are in $\mathfrak{A}$, as asserted.

LEMMA 9. *If $\mathfrak{p}$ is a prime ideal in R and if $\mathfrak{p}'$ is a prime ideal of $R'\cdot\mathfrak{p}$, then $\mathfrak{p}'\cap R=\mathfrak{p}$.*

Proof. It is only necessary to show that $\mathfrak{p}'\cap R\subseteq\mathfrak{p}$. Let ω be an element of $\mathfrak{p}'\cap R$. Since $\omega\in\mathfrak{p}'$ we must have $R'\cdot\mathfrak{p}:R'\cdot\omega\neq R'\cdot\mathfrak{p}$. Let ω' be an element of $R'\cdot\mathfrak{p}:R'\cdot\omega$ which is not in $R'\cdot\mathfrak{p}$ and let us write ω' in the form $\omega'=\sum_{i=1}^{g}\omega_iv_i'$, where $\omega_i\in R$, $v_i'\in\kappa'$, $v_1'=1$ and $v_1', v_2', \cdots, v_g'$ are linearly independent over κ. We have $\omega'\omega=\sum_{i=1}^{g}\omega_i\omega v_i'\in R'\cdot\mathfrak{p}$, and hence, by Lemma 8, $\omega_i\omega\in\mathfrak{p}$, $i=1, 2, \cdots, g$. On the other hand not all the g elements ω_i can belong to $\mathfrak{p}$, since $\omega'\notin R'\cdot\mathfrak{p}$. Hence $\omega\in\mathfrak{p}$, $\mathfrak{p}'\cap R\subseteq\mathfrak{p}$, and this establishes the lemma.

THEOREM 12. a. *If κ' is a separable extension of κ then the extension $R'\cdot\mathfrak{p}$ of any prime ideal $\mathfrak{p}$ in R is the intersection of prime ideals in R'.*

b. *If the quotient field of $R/\mathfrak{p}$ is separably generated over κ, then* a *holds also for inseparable extensions κ'/κ.*

c. *Under the assumption made in* b, *the ideal $R'\cdot\mathfrak{p}$ is prime if κ' is a pure inseparable extension of κ.*

Proof. In parts a and b of the theorem we have to show that if some power of an element ω' in R' belongs to $R'\cdot\mathfrak{p}$, then ω' itself belongs to $R'\cdot\mathfrak{p}$. Now if $\omega'^p\in R'\cdot\mathfrak{p}$, then $\omega'^p=\sum_{i=1}^{s}\omega_iu_i'$, $\omega_i\in\mathfrak{p}$, $u_i'\in\kappa'$, so that $\omega'^p\in R_1\cdot\mathfrak{p}$, where $R_1=\kappa_1[x_1, x_2, \cdots, x_n]$ and $\kappa_1=\kappa(u_1', u_2', \cdots, u_s')=$a finite extension of κ. To prove our assertion it is sufficient to prove that $\omega'\in R_1\cdot\mathfrak{p}$. Hence for the proof of the theorem it is permissible to assume that κ' is a finite extension of κ (a similar argument covers part c of the theorem).

Let $u_1, u_2, \cdots, u_\nu$ be an independent basis of κ' over κ, whence $R'=R\cdot u_1+R\cdot u_2+\cdots+R\cdot u_\nu$. By (26) we have $R'\cdot\mathfrak{p}\cap R=\mathfrak{p}$, and hence the integral domain $\bar{R}=R/\mathfrak{p}$ is a subring of $R'/R'\cdot\mathfrak{p}$. Consequently we can write:

$$R'/R'\cdot\mathfrak{p} = \bar{R}\cdot u_1 + \bar{R}\cdot u_2 + \cdots + \bar{R}u_\nu = \kappa'\cdot\bar{R}. \tag{27}$$

By Lemma 8, the elements $u_1, u_2, \cdots, u_\nu$ are linearly independent with respect to $\bar{R}$. By Lemma 9 no element of $\bar{R}$, different from zero, is a zero divisor in $R'/R'\cdot\mathfrak{p}$. Hence if $\bar{\Sigma}$ denotes the quotient field of $\bar{R}$, the ring $R'/R'\mathfrak{p}$ can be embedded in the hypercomplex system

$$\kappa_{\bar{\Sigma}}' = \bar{\Sigma}\cdot u_1 + \bar{\Sigma}\cdot u_2 + \cdots + \bar{\Sigma}\cdot u_\nu,$$

which is an extension of the field κ'/κ regarded as a hypercomplex system $\kappa_\kappa'=\kappa\cdot u_1+\kappa\cdot u_2+\cdots+\kappa\cdot u_\nu$. It is well known that if κ' is a separable extension of κ or if $\bar{\Sigma}$ is separably generated over κ, then $\kappa_{\bar{\Sigma}}'$ is semi-simple (van der Waerden [10, p. 174]). If κ'_{Σ} is semi-simple then $R'/R'\cdot\mathfrak{p}$ (which is a subring of κ'_{Σ}) has no proper nilpotent elements. This establishes parts a and b of the theorem.

As to part c it is sufficient to consider the case $\kappa'=\kappa(a^{1/p})$, where $a\in\kappa$ and $p\neq 0$ is the characteristic of κ. In this case we have by (27): $R'/R'\cdot\mathfrak{p} =\overline{R}[z]/z^p-a$, and hence $\kappa'_{\Sigma}=\overline{\Sigma}[z]/z^p-a$, where z is an indeterminate. Since $\overline{\Sigma}$ is separably generated over κ, z^p-a remains irreducible over $\overline{\Sigma}$. Hence κ'_{Σ} is a field and therefore $R'/R'\cdot\mathfrak{p}$ is an integral domain. This completes the proof of the theorem.

Remark. The ring $\overline{R}[z]/z^p-a$ is at any rate a primary ring. Hence if $R/\mathfrak{p}$ is not separably generated over κ, we can still assert that *if κ' is purely inseparable over κ then $R'\cdot\mathfrak{p}$ is a primary ideal.*

10.2. *Jacobian criterion for absolutely simple loci.* Let V be an irreducible algebraic variety in S_n and let $\mathfrak{p}=\boldsymbol{p}(V)$ be the corresponding prime ideal. We have seen that with one exception (κ' inseparable over κ and $\mathcal{J}(V)$ is not separably generated over κ) the ideal $R'\cdot\mathfrak{p}$ is the intersection of prime ideals in R':

$$R'\cdot\mathfrak{p} = \bigcap_{i=1}^{h} \mathfrak{p}'_i.$$

Correspondingly, the variety V/κ may become reducible over κ'; it splits into h irreducible varieties:

$$V/\kappa \to V/\kappa' = \bigcup_{i=1}^{h} V'_i/\kappa', \tag{28}$$

where $V'_i = \mathcal{V}(\mathfrak{p}'_i)$.

The varieties V'_i are all of the same dimension as V, for $\mathfrak{p}'_i\cap R=\mathfrak{p}$ and R' is integrally dependent over R. In the exceptional case noted above, we have $h=1$ but $R'\cdot\mathfrak{p}$ is, in general, not prime and hence does not coincide with $\boldsymbol{p}(V')$.

If W is an irreducible subvariety of V, let

$$W/\kappa \to W/\kappa' = \bigcup_{i=1}^{g} W'_i/\kappa' \tag{29}$$

describe the splitting of W/κ into irreducible varieties upon the field extension $\kappa\to\kappa'$. Let $\mathfrak{q}=\boldsymbol{p}(W)$, $\mathfrak{q}'_j=\boldsymbol{p}(W'_j)$. Since R' is integrally dependent on R, it is well known that any $\mathfrak{p}'_i$ is contained in at least one $\mathfrak{q}'_j$ [6, Theorem 3; 4, Theorem 3]. Moreover, since R is integrally closed, it is also well known that any $\mathfrak{q}'_j$ contains at least one $\mathfrak{p}'_i$ [6, Theorem 6; 4, Theorem 5]. Hence, each V'_i contains at least one of the varieties W'_j and each W'_j belongs to at least one of the varieties V'_i.

We shall say that W splits into simple subvarieties (under the given ground field extension $\kappa \to \kappa'$) if *each* W_j' is simple for *any* of the varieties V_i' which contain W_j'.

DEFINITION 3. *An irreducible simple subvariety* W/κ *of* V/κ *is absolutely simple for* V/κ *if it splits into simple subvarieties under any algebraic extension of the ground field* κ, *and if* $\mathfrak{J}(V)$ *is separably generated over* κ.

Let $(f_1(x), f_2(x), \cdots, f_\nu(x))$ be a basis of the ideal $\mathfrak{p} = \mathfrak{p}(V)$.

THEOREM 13. *A necessary and sufficient condition that an irreducible subvariety* W *of an irreducible r-dimensional variety* V *in* S_n *be absolutely simple for* V *is that the Jacobian matrix* $J(f; x)$ *be of rank* $n-r$ *on* W.

Proof. Assume that W is absolutely simple for V. We pass from κ to the algebraically perfect field κ' determined by κ. By assumption, $\mathfrak{J}(V)$ is separably generated over κ. Hence, by Theorem 12, part c, *the ideal* $\mathfrak{p}$ *remains prime in* R', so that V remains irreducible over κ' and $(f_1(x), f_2(x), \cdots, f_\nu(x))$ remains a basis of the prime ideal of V over κ'. Also W remains irreducible (see the remark at the end of the proof of Theorem 12). By assumption W/κ' is simple for V/κ', and since κ' is a perfect field, it follows (§8.3) that $J(f; x)$ must be of rank $n-r$ on W.

Conversely, assume that $J(f; x)$ is of rank $n-r$ on W. The matrix $J(f; x)$ is then a fortiori of rank $n-r$ on V, and hence (§8.1) $\mathfrak{J}(V)$ is separately generated over κ. Let κ' be an arbitrary algebraic extension of κ, and let (28) and (29) be the corresponding decompositions of V and W over κ'. If a given W_j' belongs to a given V_i', then we have that $f_\mu(x) = 0$ on V_i', $\mu = 1, 2, \cdots, \nu$, and that $J(f; x)$ has rank $n-r$ on W_j'. Consequently (Theorem 7', §8.3) W_j' is simple for V_i'.

COROLLARY 1. *If* W/κ *is absolutely simple for* V/κ, *then for any algebraic extension* κ' *of the ground field* κ *it is true that each of the irreducible varieties* W_j'/κ' *into which* W/κ *splits belongs to only one of the varieties* V_i'/κ' *into which* V/κ *splits. In other words: each* W_j'/κ' *is simple for the composite variety* V/κ' *in* (28).

For the composite variety V/κ' is the zero manifold of the ideal $R' \cdot (f_1(x), f_2(x), \cdots, f_\nu(x))$, and the fact that $J(f; x)$ is of rank $n-r$ on each W_j' implies that each W_j' is simple for this composite variety (Theorem 7', §8.3)(7).

COROLLARY 2. *If it is known that the condition "$\mathfrak{J}(V)$ is separably generated over* κ*" is already satisfied, then either one of the following two conditions is suffi-*

(7) It is clear from the proof that when we say "W_j'/κ' is simple for the composite variety V/κ'" we actually mean more precisely the following: the general point of W_j'/κ' is a simple zero of the ideal $R' \cdot \mathfrak{p}$, where $\mathfrak{p}$ is the prime ideal of V/κ. This condition can replace in Definition 3 the condition that $\mathfrak{J}(V)$ be separably generated over κ. For if $\mathfrak{J}(V)$ is not separably generated over κ then for a suitable extension κ' of κ (for instance, for κ' = the algebraic closure of κ) the ideal $R' \cdot \mathfrak{p}$ will be primary and therefore nowhere locally prime.

cient in order that W be absolutely simple for V:

(1) *W/κ remains simple under the extension $\kappa \to \kappa'$ = the algebraically perfect field determined by κ.*

(2) *W/κ remains simple under any finite purely inseparable extension of κ*([8]).

The sufficiency of condition (1) has already been established in the course of the proof of Theorem 13. The sufficiency of condition (2) can be seen as follows. Let κ' be the field obtained by adjoining to κ the pth roots of the coefficients of the polynomials $f_1(x), f_2(x), \cdots, f_\nu(x)$ of a basis of $\mathfrak{p}(V)$. The varieties V/κ' and W/κ' remain irreducible (Theorem 12, part c). To these varieties we apply Theorem 11 (§9.6), κ' being our new ground field. Since the coefficients of the polynomials $f_\mu(x)$, $\mu = 1, 2, \cdots, \nu$, already belong to κ'^p and since these polynomials still form a base for the ideal $\mathfrak{p}(V/\kappa')$ (Theorem 12, part c), it follows that the parameters z_j are missing and that the matrix $J(f; x, z)$ coincides with the matrix $J(f; x)$. If W/κ' is simple for V/κ', then it follows that $J(f; x)$ has rank $n-r$ on W, and hence W/κ is absolutely simple for V/κ.

COROLLARY 3. *Any simple irreducible subvariety W of V such that $\mathfrak{J}(W)$ is separably generated over κ is absolutely simple for V.*

This is an immediate consequence of Theorem 7′ (§8.3) and Theorem 13.

11. Intrinsic characterizations of absolutely simple loci.

11.1. *Differential characterization.* Using Theorem 13 it is not difficult to show that the concept of an absolutely simple subvariety W of V belongs to the local geometry of V at W; in other words: whether W is or is not absolutely simple for V depends entirely on the structure of the quotient ring $\mathfrak{o} = Q(W/V)$. The easiest way to see this is to use local W-differentiations *on the variety V*. We mean by such a differentiation a mapping D of the above quotient ring $\mathfrak{o}$ into the field $\mathfrak{J}(W)$ such that the following conditions are satisfied: (1) $Dc = 0$ if $c \in \kappa$; (2) $D(\omega_1 - \omega_2) = D\omega_1 - D\omega_2$; (3) $D(\omega_1\omega_2) = \bar{\omega}_1 D\omega_2 + \bar{\omega}_1 D\omega_2$, where $\bar{\omega}$ denotes the W-residue of ω. These differentiations form a vector space over $\mathfrak{J}(W)$. To prove the local character of the concept of an absolutely simple subvariety, we shall prove in this section the following stronger result:

LEMMA 10. *If $(f_1(x), f_2(x), \cdots, f_\nu(x))$ is a basis of the ideal $\mathfrak{p}(V)$ and if the Jacobian matrix $J(f; x)$ has rank $n-\sigma$ at W, then there exist on V exactly σ local W-differentiations which are linearly independent over $\mathfrak{J}(W)$.*

In view of this lemma, the number of linearly independent local W-differentiations on V is always greater than or equal to r, $r = \dim V$, and (by Theorem 13) *this number is equal to r if and only if W is absolutely simple for V.*

([8]) The proof of (2) exhibits one specific finite, purely inseparable extension of κ, such that W/κ is absolutely simple if and only if it remains simple under that particular extension of κ.

Proof of the lemma. Let $(\xi)=(\xi_1, \xi_2, \cdots, \xi_n)$ and $(\eta)=(\eta_1, \eta_2, \cdots, \eta_n)$ be respectively the general point of V and the general point of W. If $\omega=A(\xi)/B(\xi)$ is an element of $\mathfrak{o}$ (whence $B(\eta)\neq 0$) and if D is a local W-differentiation on V, then

$$(30)\qquad [B(\eta)]^2\cdot D\omega = \sum_{i=1}^{n} [B(\eta)\partial A/\partial\eta_i - A(\eta)\partial B/\partial\eta_i]D\xi_i.$$

Hence the values of the n derivatives $D\xi_i$ determine the differentiation D uniquely. However these derivatives cannot be arbitrary elements of the field $\mathcal{J}(W)$, for they must satisfy the ν relations:

$$(31)\qquad G_j = \sum_{i=1}^{n} \partial f_j/\partial\eta_i\cdot D\xi_i = 0, \qquad i = 1, 2, \cdots, \nu.$$

On the other hand, these relations are sufficient for the existence of a corresponding differentiation D, for then $D\omega$ can be defined by (30) for any element ω in $\mathfrak{o}$, and the only point to check is the following: if $F(x)=F(x_1, x_2, \cdots, x_n)$ is a polynomial with coefficients in κ such that $F(\xi)=0$ then

$$\sum_{i=1}^{n} \partial F/\partial\eta_i\cdot D\xi_i = 0.$$

But this is obvious since $F(x)$ can be expressed in the form $\sum_{j=1}^{\nu} A_j(x)f_j(x)$, whence

$$\sum_{i=1}^{n} \partial F/\partial\eta_i\cdot D\xi_i = \sum_{j=1}^{\nu} A_j(\eta)G_j.$$

If $J(f; x)$ has rank $n-\sigma$ at W, that is, for $x=\eta$, the equations (31) have exactly σ independent solutions $(D^\mu\xi_1, D^\mu\xi_2, \cdots, D^\mu\xi_n)$, $\mu=1, 2, \cdots, \sigma$, and this completes the proof of the lemma.

11.2. *Direct verification of the local character of the definition.* The local character of the concept of absolutely simple loci can also be deduced directly from the definition of these loci (Definition 3), without the use of Theorem 13. We shall only indicate the steps of the proof; the details present no difficulty whatsoever.

It will be sufficient to consider only finite algebraic extensions κ' of κ. Let $\kappa'=\kappa\cdot u_1+\kappa\cdot u_2+\cdots+\kappa\cdot u_g$, where $(u_1, u_2, \cdots, u_g)$ is an independent basis of κ'/κ. We extend the domain of coefficients in the hypercomplex system $\sum_{i=1}^{g}\kappa\cdot u_i$ from κ to the quotient ring $\mathfrak{o}$ $(=Q(W/V))$, that is, we consider the ring

$$\mathfrak{o}' = \mathfrak{o}\cdot u_1 + \mathfrak{o}\cdot u_2 + \cdots + \mathfrak{o}\cdot u_g.$$

We use the notation of §10.1. If $R'\cdot\mathfrak{p}=\bigcap_{i=1}^{h}\mathfrak{p}_i'$, then $\mathfrak{o}'$ is a direct sum of h integral domains:

$$\mathfrak{o}' = \mathfrak{o}_1' \oplus \mathfrak{o}_2' \oplus \cdots \oplus \mathfrak{o}_h'.$$

If V_i' contains s_i of the varieties W_j', then $\mathfrak{o}_i'$ is a semi-local ring having exactly s_i maximal prime ideals, and the s_i quotient rings of these prime ideals in $\mathfrak{o}_i'$ coincide with the s_i quotient rings $Q(W_j'/V_i')$ (i fixed, $W_j' \subseteq V_i'$). Thus everything is described in terms of the quotient ring $\mathfrak{o}$, so that it is entirely a local matter whether or not each W_j' is simple for its carrier V_i'.

One could also proceed as follows. Let $\Sigma = \mathcal{J}(V)$ and let Σ'/κ be a composite extension of Σ/κ and κ'/κ (see Chevalley [1]). Let $\mathfrak{o}' = \kappa'\mathfrak{o}$ be the least subring of Σ'/κ which contains both $\mathfrak{o}$ and κ'. Then $\mathfrak{o}'$ is a local ring and is in fact isomorphic to a quotient ring $Q(W_j'/V_i')$ $(W_j' \subseteq V_i')$. There is only a finite number of nonequivalent or non-isomorphic composite extensions of Σ/κ and κ'/κ, and in this fashion we get the quotient rings $Q(W_j'/V_i')$ for all pairs of indices i and j such that $W_j' \subseteq V_i'$.

11.3. *André Weil's criterion.* Another intrinsic characterization of absolutely simple loci has been communicated to me by André Weil. Given a simple irreducible subvariety W of V and given r elements $\omega_1, \omega_2, \cdots, \omega_r$ ($r = \dim V$) in the quotient ring $\mathfrak{o}$ $(=Q(W/V))$, we say that the ω's are *uniformizing coördinates of* W (*on* V) if the following two conditions are satisfied: (1) the ring $\kappa[\omega_1, \omega_2, \cdots, \omega_r]$ contains a set of uniformizing parameters of W; (2) if $\zeta_1, \zeta_2, \cdots, \zeta_r$ denote the W-residues of the ω's then $\mathcal{J}(W)$ is an algebraic extension of the field $\kappa(\zeta_1, \zeta_2, \cdots, \zeta_r)$(9).

Any simple W possesses uniformizing coördinates. For we can take, for instance, for $\omega_1, \omega_2, \cdots, \omega_\rho$ ($\rho = \dim W$) any set of ρ elements in $\mathfrak{o}$ whose W-residues are algebraically independent over κ, and for $\omega_{\rho+1}, \omega_{\rho+2}, \cdots, \omega_r$ a set of $r-\rho$ uniformizing parameters of W. For later purpose we prove the following lemma:

LEMMA 11. *Any r uniformizing coördinates of W are algebraically independent over κ.*

Proof. If $\rho = \dim W$ we may assume, in view of condition (2), that $\zeta_1, \zeta_2, \cdots, \zeta_\rho$ are algebraically independent over κ. Let $t_1, t_2, \cdots, t_{r-\rho}$ be elements of $\kappa[\omega_1, \omega_2, \cdots, \omega_r]$ which are uniformizing parameters of W. We shall prove the lemma by showing that the r elements $\omega_1, \omega_2, \cdots, \omega_\rho, t_1, t_2, \cdots, t_{r-\rho}$ of $\kappa[\omega_1, \omega_2, \cdots, \omega_r]$ are algebraically independent over κ. Let $F(X_1, X_2, \cdots, X_r)$ be a nonzero polynomial in $\kappa[X_1, X_2, \cdots, X_r]$. We show that $F(t_1, t_2, \cdots, t_{r-\rho}, \omega_1, \omega_2, \cdots, \omega_\rho) \neq 0$. We write $F(X)$ in the following form:

$$F(X) = \sum_{i=s}^{\sigma} F_i(X),$$

where F_i is homogeneous, of degree i, in $X_1, X_2, \cdots, X_{r-\rho}$ and $F_s(X) \neq 0$. We

(9) We find it convenient to include condition (2) in the definition of uniformizing coördinates. When W is a point this condition is, of course, vacuous.

have $F_i(t_1, t_2, \cdots, t_{r-\rho}; \omega_1, \omega_2, \cdots, \omega_\rho) \in \mathfrak{m}^i$, where $\mathfrak{m} = \boldsymbol{m}(W/V)$. On the other hand, since $F_s(X) \neq 0$ and since $\zeta_1, \zeta_2, \cdots, \zeta_\rho$ are algebraically independent over κ, the form $F_s(X_1, X_2, \cdots, X_{r-\rho}; \zeta_1, \zeta_2, \cdots, \zeta_\rho)$ (having coefficients in the residue field $\mathcal{J}(W)$ of $\mathfrak{o}$) is not zero. Since $\mathfrak{o}$ is a regular ring it follows (§5.1) that $F_s(t; \omega) \notin \mathfrak{m}^{s+1}$. Consequently $F(t; \omega) \neq 0$, q.e.d.

Weil's criterion can be stated as follows:

THEOREM 14. *A necessary and sufficient condition that a simple subvariety W of V be absolutely simple is that there exist uniformizing coördinates $\omega_1, \omega_2, \cdots, \omega_r$ of W such that the W-residue $\zeta_1, \zeta_2, \cdots, \zeta_r$ of the ω's generate over κ the entire field $\mathcal{J}(W)$ (that is, $\mathcal{J}(W) = \kappa(\zeta_1, \zeta_2, \cdots, \zeta_r)$).*

Proof. We first show that the condition is sufficient. Let $(\xi) = (\xi_1, \xi_2, \cdots, \xi_n)$ and $(\eta) = (\eta_1, \eta_2, \cdots, \eta_n)$ be respectively the general point of V and of W. We consider the variety V' in S_{n+r} having $(\xi_1, \xi_2, \cdots, \xi_n, \omega_1, \omega_2, \cdots, \omega_r)$ as general point. This variety V' is birationally equivalent to V and it carries a subvariety W' with general point $(\eta_1, \eta_2, \cdots, \eta_n, \zeta_1, \zeta_2, \cdots, \zeta_r)$. Since the ω's belong to the quotient ring $Q(W/V)$, it follows that $Q(W/V) = Q(W'/V')$ (that is, the birational correspondence between V and V' is regular at W). Hence it will be sufficient to show that W' is absolutely simple for V'. Now if we deal with V' and W' we have a particular situation whereby the r uniformizing coördinates ω_i of W' are among the coördinates of the general point of V'. Hence we may assume that this situation prevailed already in the case of W, that is, we assume that the ω's are in the set $(\xi_1, \xi_2, \cdots, \xi_n)$, say $\omega_i = \xi_i$, $i = 1, 2, \cdots, r$.

By hypothesis, the ring $\kappa[\xi_1, \xi_2, \cdots, \xi_r]$ contains a set of uniformizing parameters of W, say $f_j(\xi_1, \xi_2, \cdots, \xi_r)$, $j = 1, 2, \cdots, r-\rho$. Let $f_{r-\rho+\mu}(x_1, x_2, \cdots, x_n)$, $\mu = 1, 2, \cdots, n-r$, be uniformizing parameters of V in S_n. Then it follows from Lemma 2, corollary (§2.3) that the $n-\rho$ polynomials $f_j(x_1, x_2, \cdots, x_r)$, $f_{r-\rho+\mu}(x_1, x_2, \cdots, x_n)$ are uniformizing parameters of W in S_n (that is, of W regarded as subvariety of S_n). By hypothesis, the field $\mathcal{J}(W)$ $(= \kappa(\eta_1, \eta_2, \cdots, \eta_n))$ coincides with the field $\kappa(\eta_1, \eta_2, \cdots, \eta_r)$. Actually, for the purposes of the proof, we need only a weaker hypothesis, namely *that $\mathcal{J}(W)$ is a separable extension of $\kappa(\eta_1, \eta_2, \cdots, \eta_r)$*([10]). For then there exists for $\mu = 1, 2, \cdots, n-r$ a polynomial $F_\mu(x_1, x_2, \cdots, x_r, x_{r+\mu})$ such that $F(\eta_1, \eta_2, \cdots, \eta_r, \eta_{r+\mu}) = 0$ and $(\partial F_\mu/\partial x_{r+\mu})_{x=\eta} \neq 0$. Therefore the Jacobian determinant $|J(F_1, F_2, \cdots, F_{n-r}; x_{r+1}, x_{r+2}, \cdots, x_n)|$ is not 0 at $x = \eta$. Since the $n-\rho$ polynomials $f_j, f_{n-\rho+\mu}$ are uniformizing parameters of W, it follows that a fortiori the Jacobian matrix $J(f_1, f_2, \cdots, f_{n-\rho}; x_{r+1}, x_{r+2}, \cdots, x_n)$ is of rank not less than $n-r$ at $x = \eta$. Since $f_1, f_2, \cdots, f_{r-\rho}$ are independent of $x_{r+1}, x_{r+2}, \cdots, x_n$, it follows that the determinant $|J(f_{r-\rho+1}, f_{r-\rho+2}, \cdots, f_{n-\rho};$

([10]) Hence we can weaken the sufficiency condition and assert that W is absolutely simple for V if there exist uniformizing coördinates $\omega_1, \omega_2, \cdots, \omega_r$ of W such that $\mathcal{J}(W)$ is a separable algebraic extension of the field $\kappa(\zeta_1, \zeta_2, \cdots, \zeta_r)$, where $\zeta_i = W$-residue of ω_i.

$x_{r+1}, x_{r+2}, \cdots, x_n)|$ is not 0 at $x=\eta$. Since $f_{r-\rho+\mu}(x)$ is zero on V ($\mu=1, 2, \cdots, n-r$) we conclude by Theorem 13 that W is absolutely simple for V.

We now prove that the condition is necessary. If W is absolutely simple for V then we may assume that $|J(f_1, f_2, \cdots, f_{n-r}; x_{r+1}, x_{r+2}, \cdots, x_n)| \neq 0$ on W, where $f_1(x), f_2(x), \cdots, f_{n-r}(x)$ are suitable polynomials which vanish on V. In that case $\eta_{r+1}, \eta_{r+2}, \cdots, \eta_n$ are (separable) algebraic over $\kappa(\eta_1, \eta_2, \cdots, \eta_r)$ (Lemma 6, corollary, §7.1), and ρ of the first r η's, say $\eta_1, \eta_2, \cdots, \eta_\rho$, are algebraically independent over κ. We can therefore make the identification $x_i=\eta_i$, $i=1, 2, \cdots, \rho$, and carry out a reduction to the zero-dimensional case. Instead of S_n, V, and W we shall then have an $S_{n-\rho}^{\kappa^*}$, V^*, and W^*, where $\kappa^*=\kappa(x_1, x_2, \cdots, x_\rho)$, dim $V^*/\kappa^*=r-\rho$, and dim $W^*/\kappa^*=0$. Moreover, the general points of V^* and W^* are respectively $(\xi_{\rho+1}, \xi_{\rho+2}, \cdots, \xi_n)$ and $(\eta_{\rho+1}, \eta_{\rho+2}, \cdots, \eta_n)$, while the coördinates in $S_{n-\rho}^{\kappa^*}$ are $x_{\rho+1}, x_{\rho+2}, \cdots, x_n$. From the above inequality, $|J(f_1, f_2, \cdots, f_{n-r}; x_{r+1}, x_{r+2}, \cdots, x_n)| \neq 0$ at W, follows that also W^* is an absolutely simple point of V^*. If we assume that the theorem is true for points, there will exist $r-\rho$ elements $\omega_{\rho+1}, \omega_{\rho+2}, \cdots, \omega_r$ in the quotient ring $\mathfrak{o}$ $(=Q(W/V)=Q(W^*/V^*))$ such that $\kappa[\omega_{\rho+1}, \omega_{\rho+2}, \cdots, \omega_r]$ contains a set of uniformizing parameters of W^* (on V^*) and such that the W^*-residues of $\omega_{\rho+1}, \omega_{\rho+2}, \cdots, \omega_r$ generate *over* κ^* the entire field $\mathcal{J}(W^*)$ $(=\mathcal{J}(W))$. If we set $\omega_i=\xi_i$, $i=1, 2, \cdots, \rho$, the r elements $\omega_1, \omega_2, \cdots, \omega_r$ will satisfy all the conditions of the theorem. Hence it is sufficient to prove the theorem for points of V.

We assume therefore that a given point $P(\alpha_1, \alpha_2, \cdots, \alpha_n)$ of V is absolutely simple for V. Let, say,

$$(32) \qquad |J(f_{r+1}, f_{r+2}, \cdots, f_n; x_{r+1}, x_{r+2}, \cdots, x_n)| \neq 0 \quad \text{at} \quad P,$$

where $f_{r+1}, f_{r+2}, \cdots, f_n$ are suitable polynomials which vanish on V. Let $f_i(x_1, x_2, \cdots, x_i)$, $i=1, 2, \cdots, r$, be the canonical uniformizing parameters of the point $(\alpha_1, \alpha_2, \cdots, \alpha_r)$ in S_r. From (32) it follows that the $n-r$ polynomials $f_{r+j}(\alpha_1, \alpha_2, \cdots, \alpha_r, x_{r+1}, \cdots, x_n)$ are uniformizing parameters of the point $(\alpha_{r+1}, \alpha_{r+2}, \cdots, \alpha_n)$ in $S_{n-r}^{\kappa^*}$, where $\kappa^*=\kappa(\alpha_1, \alpha_2, \cdots, \alpha_r)$. From this one concludes by a simple calculation that the n polynomials $f_1(x), f_2(x), \cdots, f_n(x)$ are uniformizing parameters of the point P *in* S_n. But since the last $n-r$ polynomials vanish on V, it follows that $f_1(\xi_1), f_2(\xi_1, \xi_2), \cdots, f_r(\xi_1, \xi_2, \cdots, \xi_r)$ are uniformizing parameters of the point P *on* V. We have thus shown that $\xi_1, \xi_2, \cdots, \xi_r$ *are uniformizing coördinates of* P *on* V.

Again by (32), we have that the field $\mathcal{J}(P)$ $(=\kappa(\alpha_1, \alpha_2, \cdots, \alpha_n))$ is a separable extension of $\kappa(\alpha_1, \alpha_2, \cdots, \alpha_r)$. From this it follows as in the proof of Theorem 8 (§9.1) that $\mathcal{J}(P)$ is a simple extension of $\kappa(\alpha_1, \alpha_2, \cdots, \alpha_{r-1})$. We set

$$\Delta_1 = \kappa(\alpha_1, \alpha_2, \cdots, \alpha_{r-1}), \qquad \Delta = \mathcal{J}(P) = \Delta_1(\alpha_0).$$

We proceed to show the existence of an element ξ_0 in $Q(P/V)$ whose P-residue is α_0 and such that $\xi_0, \xi_1, \cdots, \xi_{r-1}$ are uniformizing coördinates of the point P on V. This will complete the proof of the theorem.

We consider two cases, according as α_r is or is not separable over Δ_1.

First case: *α_r separable over Δ_1*. In this case also α_0 is separable over Δ_1, since α_0 is separable over $\Delta_1(\alpha_r)$. Let $\phi(z; \alpha_1, \alpha_2, \cdots, \alpha_{r-1})$ be the monic irreducible polynomial in $\Delta_1[z]$ having $z=\alpha_0$ as a root. We may assume that ϕ is a polynomial in $\alpha_1, \alpha_2, \cdots, \alpha_{r-1}$ and z, with coefficients in κ. Moreover, we have

$$\phi'_{\alpha_0}(\alpha_0, \alpha_1, \cdots, \alpha_{r-1}) \neq 0. \tag{33}$$

Let ζ_0 be a fixed element in $Q(P/V)$ whose P-residue is α_0, and let π be a variable element of $\mathfrak{m}(P/V)$. We set:

$$\begin{aligned} \xi_0 &= \zeta_0 + \pi, \\ \tau_0 &= \phi(\zeta_0, \xi_1, \xi_2, \cdots, \xi_{r-1}), \\ t_0 &= \phi(\xi_0, \xi_1, \xi_2, \cdots, \xi_{r-1}), \\ t_i &= f_i(\xi_1, \xi_2, \cdots, \xi_i), \qquad i = 1, 2, \cdots, r, \end{aligned}$$

so that α_0 is also the P-residue of ξ_0, and $\tau_0, t_0 \in \mathfrak{m}$, where $\mathfrak{m} = \mathfrak{m}(P/V)$. We have

$$t_0 \equiv \tau_0 + \partial\phi(\zeta_0, \xi_1, \cdots, \xi_{r-1})/\partial\zeta_0 \cdot \pi \pmod{\mathfrak{m}^2}. \tag{34}$$

By (33) the partial derivative $\partial\phi/\partial\zeta_0$ in (34) does not belong to $\mathfrak{m}$ and is therefore a unit in $Q(P/V)$. We can therefore select π in such a fashion as to have $\tau_0 + \partial\phi/\partial\zeta_0 \cdot \pi = t_r$. For such a choice of π we shall have $t_0 \equiv t_r \pmod{\mathfrak{m}^2}$, and consequently $t_1, t_2, \cdots, t_{r-1}, t_0$ are uniformizing parameters of P on V. From the above definition of the t_i's, $i=0, 1, \cdots, r-1$, it follows then that $\xi_0, \xi_1, \cdots, \xi_{r-1}$ are uniformizing coördinates of P on V.

Second case: *α_r inseparable over Δ_1*. In this case we have that $\partial f_r/\partial x_r$ vanishes at $x=\alpha$, whence

$$\partial f_r(\xi_1, \xi_2, \cdots, \xi_r)/\partial\xi_r \in \mathfrak{m}. \tag{35}$$

Since $\alpha_r \in \kappa(\alpha_0, \alpha_1, \cdots, \alpha_{r-1})$, we can write

$$\alpha_r = g(\alpha_0, \alpha_1, \cdots, \alpha_{r-1}), \tag{36}$$

where g is a polynomial with coefficients in κ. We now take for ξ_0 an *arbitrary* element of $Q(P/V)$ whose P-residue is α_0. Let

$$f_0(x_0, x_1, \cdots, x_{r-1}) = f_r(x_1, x_2, \cdots, x_{r-1}, g(x_0, x_1, \cdots, x_{r-1}))$$

and let $t_0 = f_0(\xi_0, \xi_1, \cdots, \xi_{r-1})$. We have

$$\begin{aligned} t_0 \equiv\ & f_r(\xi_1, \xi_2, \cdots, \xi_{r-1}, \xi_r) \\ &+ \partial f_r(\xi_1, \xi_2, \cdots, \xi_r)/\partial\xi_r \cdot [g(\xi_0, \xi_1, \cdots, \xi_{r-1}) - \xi_r] \pmod{\mathfrak{m}^2}, \end{aligned}$$

since, by (36), $g(\xi_0, \xi_1, \cdots, \xi_{r-1}) - \xi_r \in \mathfrak{m}$. For this same reason and in view of (35) we conclude that $t_0 \equiv f_r(\xi_1, \xi_2, \cdots, \xi_r)$ (mod $\mathfrak{m}^2$), that is, $t_0 \equiv t_r$ (mod $\mathfrak{m}^2$), and therefore, as in the preceding case, we find that $\xi_0, \xi_1, \cdots, \xi_{r-1}$ are uniformizing coördinates of P on V.

11.4. *A criterion of analytical equivalence.* Closely connected with Theorem 14 is another characterization of absolutely simple loci which is very suggestive and which we proceed to derive. For that purpose we introduce first of all the concept of *analytically equivalent varieties.* Let V and V' be two irreducible algebraic varieties over κ, and let W and W' be irreducible subvarieties respectively of V and of V'. Nothing is said about the dimensions of the ambient linear spaces S_n, $S_{n'}$ in which V and V' are immersed; these dimensions may very well be distinct. A priori we do not even assume that the dimensions of V and V', or of W and W', are the same. Let $\mathfrak{o} = Q(W/V)$, $\mathfrak{o}' = Q(W'/V')$; $\mathfrak{m} = \boldsymbol{m}(W/V)$, $\mathfrak{m}' = \boldsymbol{m}(W'/V')$. Let moreover $\mathfrak{o}^*$ be the completion of $\mathfrak{o}$ with respect to the powers of $\mathfrak{m}$, and let similarly $\mathfrak{o}'^*$ be the completion of $\mathfrak{o}'$ with respect to the powers of $\mathfrak{m}'$.

DEFINITION 4. *The varieties V and V' are said to be analytically equivalent at (or in the neighborhood of) W and W' respectively, if the rings $\mathfrak{o}^*$ and $\mathfrak{o}'^*$ are κ-isomorphic.*

Let r and ρ be the dimensions of V and W respectively; similarly let r' and ρ' be the dimensions of V' and W'. The dimensions of the local rings $\mathfrak{o}^*$ and $\mathfrak{o}'^*$ are then $r-\rho$ and $r'-\rho'$ respectively. Consequently, if V and V' are analytically equivalent at W and W', then we must have $r-\rho = r'-\rho'$. Moreover, the residue field of $\mathfrak{o}^*$ is $\mathcal{J}(W)$ and the residue field of $\mathfrak{o}'^*$ is $\mathcal{J}(W')$. Consequently the assumption of analytical equivalence implies that the field $\mathcal{J}(W)$ and $\mathcal{J}(W')$ are κ-isomorphic, that is, *W and W' must be birationally equivalent. Hence $\rho = \rho'$ and therefore also $r = r'$.* Thus analytically equivalent varieties V and V' must have the same dimension (but they need not be birationally equivalent).

Having established the concept of analytical equivalence, we now can state the following theorem:

THEOREM 15. *A necessary and sufficient condition that an irreducible subvariety W of V be absolutely simple for V is that V be, locally at W, analytically equivalent to the linear space S_r, where $r = \dim V$.*

Proof. Assume that W is absolutely simple for V, and let $\omega_1, \omega_2, \cdots, \omega_r$ be elements of the quotient ring $\mathfrak{o}$ of W ($\mathfrak{o} = Q(W/V)$) which have the properties stated in Theorem 14. Let W' denote the irreducible variety in S_r whose general point is $(\zeta_1, \zeta_2, \cdots, \zeta_r)$, where $\zeta_i = W$-residue of ω_i, and let $\mathfrak{o}' = Q(W'/S_r)$. As usual we denote by $\mathfrak{m}$ and $\mathfrak{m}'$ the maximal ideals of $\mathfrak{o}$ and $\mathfrak{o}'$ respectively.

By Lemma 11, the uniformizing coördinates $\omega_1, \omega_2, \cdots, \omega_r$ are algebraically independent over κ. Hence we can identify the ring $\kappa[\omega_1, \omega_2, \cdots, \omega_r]$

with the coördinate ring (the polynomial ring) $\kappa[x_1, x_2, \cdots, x_r]$ of S_r. The elements of $\mathfrak{o}'$ are then of the form $\phi(\omega)/\psi(\omega)$, $\psi(\zeta)\neq 0$. Hence:

(a) $\mathfrak{o}'$ *is a subring of* $\mathfrak{o}$.

The ring $\mathfrak{o}$, topologized by the powers of $\mathfrak{m}$, shall be referred to as the $\mathfrak{m}$-*adic ring* $\mathfrak{o}$. We prove now that:

(b) *The ring* $\mathfrak{o}'$ *is everywhere dense in the* $\mathfrak{m}$-*adic ring* $\mathfrak{o}$.

We have to show that if u is any element of $\mathfrak{o}$ and if i is any positive integer, then there exists a rational function $\phi_i(\omega)/\psi_i(\omega)$ such that $\psi_i(\zeta)\neq 0$ and $u\equiv\phi_i(\omega)/\psi_i(\omega) \pmod{\mathfrak{m}^{i+1}}$. If $i=0$ we make use of the fact that $\mathcal{J}(W) =\kappa(\zeta_1, \zeta_2, \cdots, \zeta_r)$. Because of this we can write: W-residue of $u=\phi_0(\zeta)/\psi_0(\zeta)$, and hence $u=\phi_0(\omega)/\psi_0(\omega) \pmod{\mathfrak{m}}$. Assuming that $\phi_i(\omega)/\psi_i(\omega)$ has already been shown to exist for all $i<s$, we make use of the fact that the ω's are uniformizing coördinates. Let $t_1, t_2, \cdots, t_{r-\rho}$ be elements of $\kappa[\omega_1, \omega_2, \cdots, \omega_r]$ which are uniformizing parameters of W. We have $u-\phi_{s-1}(\omega)/\psi_{s-1}(\omega)\in\mathfrak{m}^s$, hence we can write $u-\phi_{s-1}(\omega)/\psi_{s-1}(\omega)$ as a form in $t_1, t_2, \cdots, t_{r-\rho}$, of degree s, with coefficients in $\mathfrak{o}$. By the case $i=0$, each of these coefficients is congruent mod $\mathfrak{m}$ to elements of $\mathfrak{o}'$. If these elements—rational functions in the ω's—are substituted for the coefficients, then we get a rational function $\alpha(\omega)/\beta(\omega)$ such that $\beta(\zeta)\neq 0$ and such that $u\equiv\phi_{s-1}(\omega)/\psi_{s-1}(\omega)+\alpha/\beta \pmod{\mathfrak{m}^{s+1}}$. This yields the desired rational function $\phi_s(\omega)/\psi_s(\omega)$ and establishes the assertion (b).

An element $\phi(\omega)/\psi(\omega)$ of $\mathfrak{o}'$ belongs to $\mathfrak{m}'$ if and only if $\phi(\zeta)=0$, that is, if and only if $\phi(\omega)/\psi(\omega)\in\mathfrak{m}$. Hence

$$\mathfrak{m}\cap 0 = \mathfrak{m}'. \tag{37}$$

If $u\in\mathfrak{m}$ and if we write, by (b), $u\equiv\phi_1(\omega)/\psi_1(\omega) \pmod{\mathfrak{m}^2}$, then $\phi_1(\omega)/\psi_1(\omega)\in\mathfrak{m}\cap\mathfrak{o}'$, that is, $\phi_1(\omega)/\psi_1(\omega)\in\mathfrak{m}'$, by (37). Consequently $\mathfrak{m}=\mathfrak{o}\cdot\mathfrak{m}'+\mathfrak{m}^2$, and from this we conclude as in §2.1 that

$$\mathfrak{m} = \mathfrak{o}\cdot\mathfrak{m}'. \tag{38}$$

By (38), any basis of $\mathfrak{m}'$ is also a basis of $\mathfrak{m}$. Since W' is of dimension ρ and is immersed in a linear S_r, it follows that a minimal basis of $\mathfrak{m}'$ consists exactly of $r-\rho$ elements. Let $\tau_1, \tau_2, \cdots, \tau_{r-\rho}$ be any minimal basis of $\mathfrak{m}'$, that is, the τ's are uniformizing parameters of W' in S_r'. Then the τ's form also a minimal base of $\mathfrak{m}$, and therefore they are, as well as t's, uniformizing parameters of W on V. Since the t's are in $\mathfrak{m}'$ we have relations of the form: $t_i=\sum_{j=1}^{r-\rho}\alpha_{ij}\tau_j$, $\alpha_{ij}\in\mathfrak{o}'$, $i=1, 2, \cdots, r-\rho$. Since both the t's and the τ's are uniformizing parameters of W on V and since the α_{ij} are also in $\mathfrak{o}$, it follows that the determinant $|\alpha_{ij}|$ is a unit in $\mathfrak{o}$, that is, it is not an element of $\mathfrak{m}$. But since $|\alpha_{ij}|\in\mathfrak{o}'$ it follows by (37) that $|\alpha_{ij}|\notin\mathfrak{m}'$, and therefore $|\alpha_{ij}|$ is also a unit in $\mathfrak{o}'$. Consequently $t_1, t_2, \cdots, t_{r-\rho}$ are also uniformizing parameters of W'. What we have shown is that not only is every minimal basis of $\mathfrak{m}'$ a minimal basis of $\mathfrak{m}$, but also that any minimal basis of $\mathfrak{m}$ whose elements belong to $\mathfrak{o}'$ is necessarily a minimal basis of $\mathfrak{m}'$.

If u is an element of $\mathfrak{m}'^i$ but is not in $\mathfrak{m}'^{i+1}$, then $u=\phi_i(t_1, t_2, \cdots, t_{r-\rho})$, where ϕ_i is a form of degree i with coefficients in $\mathfrak{o}'$, but not all in $\mathfrak{m}'$. Therefore these coefficients are not all in $\mathfrak{m}$, and consequently $u\in\mathfrak{m}^i$, $u\notin\mathfrak{m}^{i+1}$. Hence

$$\mathfrak{m}^i \cap \mathfrak{o}' = \mathfrak{m}'^i, \qquad i = 1, 2, \cdots. \tag{39}$$

The relations (38) and (39) imply that:

(c) *The* $\mathfrak{m}'$*-adic ring* $\mathfrak{o}'$ *is a subspace of the* $\mathfrak{m}$*-adic ring* $\mathfrak{o}$.

From (a), (b) and (c) we conclude that the complete rings $\mathfrak{o}^*$ and $\mathfrak{o}'^*$ coincide (see Chevalley [2, Theorem 1, p. 698]), and therefore V and S_r have analytically equivalent neighborhoods at W and W' respectively.

Conversely, assume that there exists in S_r a ρ-dimensional irreducible variety W' such that V and S_r are analytically equivalent in the neighborhoods of W and W' respectively. Let $x_1, x_2, \cdots, x_r$ be variable coördinates in S_r, and let $(\zeta_1', \zeta_2', \cdots, \zeta_r')$ be the general point of W'. Let $\mathfrak{o}'=Q(W'/S_r)$, $\mathfrak{m}'=\boldsymbol{m}(W'/S_r)$. We denote by $\mathfrak{o}^*$ and $\mathfrak{o}'^*$ the complete rings determined respectively by the $\mathfrak{m}$-adic ring $\mathfrak{o}$ and the $\mathfrak{m}'$-adic ring $\mathfrak{o}'$; and we denote by $\mathfrak{m}^*$ and $\mathfrak{m}'^*$ the maximal ideals in these rings. By hypothesis, $\mathfrak{o}'^*$ and $\mathfrak{o}^*$ are κ-isomorphic. Let f denote a fixed κ-isomorphism of $\mathfrak{o}'^*$ onto $\mathfrak{o}^*$.

We can find a minimal basis of $\mathfrak{m}'$ consisting of polynomials in $\kappa[x_1, x_2, \cdots, x_r]$. Let $(t_1', t_2', \cdots, t_{r-\rho}')$, $t_i'=\phi_i(x_1, x_2, \cdots, x_r)$, be such a basis. Let

$$fx_i = \omega_i^* \in \mathfrak{o}^*. \tag{40}$$

Since $\mathfrak{o}^*$ is the completion of the $\mathfrak{m}$-adic ring $\mathfrak{o}$, we can find elements $\omega_1, \omega_2, \cdots, \omega_r$ in $\mathfrak{o}$ such that

$$\omega_i \equiv \omega_i^* \pmod{\mathfrak{m}^{*2}}. \tag{41}$$

Let ζ_i be the $\mathfrak{m}$-residue of ω_i. By (41), ζ_i is also the $\mathfrak{m}^*$-residue of ω_i^*. Since $\mathfrak{o}'^*/\mathfrak{m}'^*=\mathfrak{o}'/\mathfrak{m}'=\mathcal{J}(W')=\kappa(\zeta_1', \zeta_2', \cdots, \zeta_r')$, it follows from (40) that $\mathfrak{o}^*/\mathfrak{m}^*=\kappa(\zeta_1, \zeta_2, \cdots, \zeta_r)$, that is (since $\mathfrak{o}^*/\mathfrak{m}^*=\mathfrak{o}/\mathfrak{m}=\mathcal{J}(W)$):

(a) *The W-residues of* $\omega_1, \omega_2, \cdots, \omega_r$ *generate over* κ *the entire field* $\mathcal{J}(W)$.

Let

$$t_i^* = ft_i' = \phi_i(\omega_1^*, \omega_2^*, \cdots, \omega_r^*), \tag{42}$$

$$t_i = \phi_i(\omega_1, \omega_2, \cdots, \omega_r). \tag{42'}$$

Since $t_1', t_2', \cdots, t_{r-\rho}'$ is a basis of $\mathfrak{m}'^*$, it follows from (42) that $t_1^*, t_2^*, \cdots, t_{r-\rho}^*$ is a basis of $\mathfrak{m}^*$. In view of (42′) and (41) we have $t_i^*=t_i \pmod{\mathfrak{m}^{*2}}$. Hence $t_1, t_2, \cdots, t_{r-\rho}$ also form a basis of $\mathfrak{m}^*$. Since the t's belong to $\mathfrak{o}$, they also form a basis of $\mathfrak{m}$. We have then that:

(b) *The polynomial ring* $\kappa[\omega_1, \omega_2, \cdots, \omega_r]$ *contains a set of uniformizing parameters of W on V.*

From (a) and (b) it follows, by Theorem 14, that W is absolutely simple for V. This completes the proof of the theorem.

References

1. C. Chevalley, *On the composition of fields*, Bull. Amer. Math. Soc. vol. 48 (1942) pp. 482–487.

2. ———, *On the theory of local rings*, Ann. of Math. vol. 44 (1943) pp. 690–708.

3. I. S. Cohen, *On the structure and ideal theory of complete local rings*, Trans. Amer. Math. Soc. vol. 59 (1946) pp. 54–106.

4. I. S. Cohen and A. Seidenberg, *Prime ideals and integral dependence*, Bull. Amer. Math. Soc. vol. 52 (1946) pp. 252–261.

5. W. Krull, *Idealtheorie*, Ergebnisse der Mathematik und ihrer Grenzgebiete, vol. 4, no. 3, Berlin, 1935.

6. ———, *Zum Dimensionsbegriff der Idealtheorie*, Math. Zeit. vol. 42 (1937) pp. 745–766.

7. ———, *Dimensionstheorie in Stellenringen*, J. Reine Angew. Math. vol. 179 (1938) pp. 204–226.

8. S. MacLane, *Modular fields*, I, *Separating transcendence bases*, Duke Math. J. vol. 5 (1939) pp. 372–393.

9. F. K. Schmidt, *Über die Erhaltung der Kettensätze der Idealtheorie bei beliebigen endlichen Körpererweiterungen*, Math. Zeit. vol. 41 (1936) pp. 443–450.

10. B. L. van der Waerden, *Moderne Algebra*, vol. 2, Berlin, 1931.

11. O. Zariski, *Some results in the arithmetic theory of algebraic varieties*, Amer. J. Math. vol. 61 (1939) pp. 249–294.

12. ———, *Algebraic varieties over ground fields of characteristic zero*, Amer. J. Math. vol. 62 (1940) pp. 187–221.

13. ———, *Foundations of a general theory of birational correspondences*, Trans. Amer. Math. Soc. vol. 53 (1943) pp. 490–542.

14. ———, *The theorem of Bertini on the variable singular points of a linear system of varieties*, Trans. Amer. Math. Soc. vol. 56 (1944) pp. 130–140.

15. ———, *A new proof of Hilbert's Nullstellensatz*, Bull. Amer. Math. Soc. vol. 53 (1947) pp. 362–368.

University of Illinois,
Urbana, Ill.

Part II

Resolution of Singularities

Introduction by H. Hironaka

The most important originality in Zariski's approach toward the problem of resolution of singularities, and for that matter toward algebraic geometry as a whole, is to use as much available power of modern algebra as possible, not only as techniques in each step of solving a specific problem but also in reformulating the problem from its foundation. For instance, in his way of setting the resolution problem (in sharp contrast to earlier works on the problem by Italian geometers and by J. Walker,[19] who gave a first rigorous proof of resolution of singularities for surfaces), we see as a self-evident matter the similarity between the singularity phenomena at the generic point of a subvariety of codimension s in any variety, and those of an s-dimensional variety at a point. We also easily find exactly where his assumption of characteristic zero is used. By this type of fundamental approach (not to mention specific techniques he invented to overcome specific difficulties in the problem), he made it much easier for other mathematicians in later works to follow the track and make further progress, especially Abhyankar's works in the case of positive characteristics and Hironaka's work in the higher dimensions in characteristic zero.

Zariski writes, in the introduction of his first algebraic proof of resolution of singularities of an algebraic surface [39],

> We should say that the requirement of rigor is to be regarded as trivially satisfied in the present proof. More significant, however, is the clarification of the problem brought about by the use of the methods of modern algebra. . . . What is gained concretely thereby is the center of gravity of the proof is shifted from minute details to underlying concepts.

Singularities of algebraic curves are eliminated by integral closure, a conceptually very simple one-step process, as is done in Zariski [38, with Muhly]. This also makes very clear why the singularities of an algebraic curve can be resolved by a *finite* succession of quadratic (or Cremona) transformations, which had been studied earlier in minute details by Italian geometers. There the key is the *finiteness* of integral closure as is made clear in [38]. The problem of resolution of singularities becomes radically more intricate for algebraic surfaces. In [39], Zariski proves that the singularities of an algebraic surface can be eliminated by repeating alternately and a finite number of times the integral closure and the quadratic transformations (applied to all isolated singular points). This is proven roughly in three steps: (a) To classify the valuations of the function field of the given surface by the isomorphism type of their value groups and prove the local uniformization at every valuation by giving an explicit process depending upon the type of the valuation; (b) Using the local uniformization, to prove that a finite repetition of *integral closure and quadratic transformations* transforms the given surface into a surface which locally everywhere birationally dominates at least one nonsingular surface (not necessarily complete); (c) To analyze the factorization properties of *complete* (or integrally closed in a certain sense) ideals in the local rings of a normal surface, locally birationally dominating a nonsingular surface, and to find an explicit recipe to eliminate singularities. In fact, after (b) is accomplished, Zariski proves that the surface continues to be normal after quadratic transformations, and the factorization analysis of the complete ideals involved tells us exactly how many times the quadratic transformations should be repeated. Zariski proves that, after (b), the quadratic transformations give in fact the *minimal* resolution in the sense of birational dominations among those dominating the given surface. The minimal resolution exists for any algebraic surface,[16] but it is often different from the one obtained by the above process of Zariski even if the surface has only rational singularities.[16] The point is that the singularities of a normal surface after (b) are special kinds of rational singularities. (For the classification and the ideal theory of rational singularities, see Du Val,[17] Artin,[9] and Lipman.[16]

The local uniformization was then generalized to all dimensions (for an arbitrary base field of characteristic zero) by Zariski [41]. Here again the proof is based upon the classification of valuations, and the algorithm of repeated Cremona transformations is prescribed by the type of the value group of a given valuation. The result is very local and far from giving a resolution of singularities even for any small neighborhood of a given singular point, but the process is much more canonical and explicit than one can expect in the case of global resolution of singularities. More significant is the fact that the local uniformization is obtained by repeating Cremona transformations of the ambient space. As a by-product of the method of this proof, it is proven that, for every *rational* variety X and every valuation v of the function field of X, there exists a birational transformation $\mathbf{P} \rightarrow X$ that is well-defined at the center of v in $\mathbf{P}$, where $\mathbf{P}$ denotes the projective space of the same dimension as X.

In [44], Zariski gives another proof of resolution of singularities of an algebraic surface (this time over an arbitrary field of characteristic zero). He shows that, for surfaces, the resolution is a very easy consequence of the local uniformization. First of all, he notes a simple fact that the space of all valuations of a function field K/k is in a natural way a *compact* topological space. (It is the inverse limit of all projective models of K/k with respect to the canonical birational dominations.) This is incidentally proven in all dimensions in [46]. Thus the reduction of global resolution to local uniformization is done if the following fact is proven: If X_1 and X_2 are birationally related nonsingular algebraic varieties (not necessarily complete), then there exists a third nonsingular variety $\widetilde{X}$, birationally related to X_1 and hence to X_2, such that, for each i, there exists an open subset U_i of $\widetilde{X}$ for which the birational transformations $U_i \to X_i$ are both *proper* morphisms. Zariski notes that, to find such $\widetilde{X}$, the following is enough:

(*) There exists a finite succession of monoidal (or quadratic) transformations with nonsingular centers, say f_i: $X'_i \to X_i$, for each i, such that if U'_i is the set of those points of X'_i having some corresponding points in X'_j, where $j \neq i$, then,

(i) The U'_i are open and the birational correspondence from U'_1 to U'_2 is a *proper* morphism, and

(ii) The fundamental locus (= the set of indeterminacy points) of its inverse: $U'_2 \to U'_1$ in U'_2 is *complete* (that is, does not extend to the boundary of U'_2).

Note that (ii) is satisfied, for instance, if the fundamental locus of $U'_2 \to U'_1$ consists of only a finite number of points. Hence, in the case of surfaces, (ii) follows from (i), and (i) is immediate from Zariski's lemma that the union of any infinite sequence of local rings obtained by quadratic transformations and integral closure is a valuation ring.

The deduction of global resolution from local uniformization becomes substantially harder in dimension 3, as is seen in [47]. It is done in the following steps:

(1) Prove the local uniformization of an algebraic surface (in a three-dimensional nonsingular variety) only by repeating quadratic and monoidal transformations with nonsingular centers contained in the singular locus. (The most important property of such a process is that it naturally extends to the ambient variety without creating singularities.) Once this type of local uniformization is proven, there follows a constructive global resolution of the surface in the following form, which Zariski called the *theorem of Beppo Levi*[15]: Look at the set S of those points at which the surface X has the maximal multiplicity. If S contains an irreducible curve with singular points, then apply the quadratic transformations to the surface, whose centers are these points. If all the curves in S are nonsingular, then apply the monoidal transformation with one of these curves as center. If S contains no curves,

then apply the quadratic transformation with any one of the points in S as center. Zariski proves this process eliminates all the singularities of the surface X after a finite number of steps.

(2) To prove (*) for three-dimensional nonsingular varieties X_1 and X_2, birationally related. Zariski does this by taking a linear system on X_1 which gives the birational transformation from X_1 to X_2 and then by applying the global resolution of (1) to the generic member of the linear system. The point is that it is easy to eliminate any base point of a linear system of surfaces if it is a smooth point of the generic member. (The intersection of two independent generic members of the linear system consists of curves, whose singularities can be easily eliminated.) What Zariski gets in this manner (which incidentally works in all dimensions and in all characteristics with little modifications) is the deduction of *the elimination of fundamental locus* for a birational transformation, that is, (*) with (i), without requiring (ii), from the global resolution of embedded subvarieties. To get (*) with both (i) and (ii), Zariski proves a more exact form of (*) for the case of surfaces. Namely, he solves the *factorization problem* for a birational transformation of surfaces, that is, when X_1 and X_2 are surfaces, f_1 and f_2 of (*) can be found in such a way that the birational correspondence from U_1' to U_2' of (i) is an *isomorphism.*

The proofs of resolution of algebraic surfaces in [39] and [44] are written in such a way that they extend to the case of arbitrary characteristics as soon as the local uniformization in dimension two is obtained. This essential difficulty in positive characteristic was later overcome by Abhyankar,[1] who first reduced the problem to the case of multiplicity p (p = characteristic), by the techniques of Galois theory and ramification theory, and then established an algorithm for polynomials of degree p whose coefficients are power series in two variables. This algorithm has been simplified and generalized (not to the case of more variables but to more abstract situations) by Abhyankar himself.[2-5,7] Abhyankar thus obtained the local uniformization, and hence the resolution in the form of Zariski's [39] and [44], for two-dimensional schemes of finite type over "good" Dedekind domains.[4] A portion of Abhyankar's algorithm[2] has been conceptualized in terms of the Newton polygon (or Newton polyhedron), so as to give a simpler proof of resolution for an arbitrary excellent surface.[12]

Zariski's work in dimension 3 [47] has been extended to almost all positive characteristic cases by Abhyankar.[6] (Characteristics 2, 3, and 5 are excluded in this work of Abhyankar.) Abhyankar's book[6] follows Zariski's [47] almost faithfully except for two essential points: *local unifor-*

mization in dimension 3 and *local uniformization of an embedded surface by means of quadratic and monoidal transformations.* (See (1) above.) By a comparatively simple global geometric method, due to Albanese[8] and recently brought up to attention by Du Val[18] and, especially from the point of view of arbitrary characteristics, by M. Artin (unpublished note) one can transform an algebraic variety of dimension 3 so that the singularities of multiplicities $> 3!$ are all eliminated. After this is done, if the characteristic $p > 3!$, then the algorithm with respect to at least one transversal parameter becomes completely similar to the case of characteristic zero. In such cases, the uniformization problem is reduced to the same of one less dimension, here to the case of surfaces for which one needs the full strength of Abhyankar's algorithms.[5,7] With this and Zariski's ideas in [47], the resolution of singularities for three-dimensional variety in characteristic > 5 is reduced to the resolution of an embedded algebraic surface by quadratic and monoidal transformations with nonsingular centers, due to Abhyankar.[5,7]

Zariski's resolution papers are written in such a way that, whenever possible, the proof in a particular step is done for all characteristics. A typical example of this is Lemma 7.1 of [47], that asserts: If X is a surface embedded in a three-dimensional nonsingular variety, and if $f\colon X' \to X$ is the quadratic transformation with center $x \in X$, then every point of $f^{-1}(x)$ at which the multiplicity of X' is equal to that of X at x, belongs to the *projective line* in $f^{-1}(x)$ corresponding to the directrix of the tangential cone $C_{X,x}$. Here the directrix means the vector space consisting of those translations that map the cone $C_{X,x}$ into itself. This idea and the result were generalized by Hironaka.[13,14]

In [62], Zariski proposes a possible inductive proof of resolution of singularities in all dimensions. The suggested idea is to look at a generic projection of a given variety to a projective space of the same dimension and then to apply the induction assumption to the branch locus of this finite projection. When the branch locus consists of a finite number of nonsingular hypersurfaces having *normal crossings* everywhere, the covering can be rather easily analyzed, as was done by Jung in the two-dimensional case, and therefore the resolution of singularities of the given variety finds its recipe. This paper is only a partial result as far as the inductive proof of resolution is concerned. Several years later Zariski [78] and [84] took up this approach once again and showed how it can be successfully applied to the case of a surface imbedded in a nonsingular three-dimensional variety. Where the branch locus is nonsingular, the surface is equisingular along the singular locus. This is the case in which the singularity of the surface can be resolved by a finite number of monoidal transformations having the singular loci as their centers, which are nonsingular by themselves. Moreover this process induces parallel resolution of singularities of all the transversal sections of the surface, and in fact it is completely prescribed by the resolution process of any one transversal section curve. In the general case, there are at most a finite number of *exceptional* singular points where the branch locus is singular. Then Zariski shows that after a finite number of quadratic transformations with exceptional singular points as centers, all the remaining exceptional singular points become *quasi-ordinary,* which means that the branch locus is a normal crossing (of two smooth analytic curves). At a quasi-ordinary singular point, the surface as a covering is given by a fractional power series of suitable two variables (similar to Puiseux series in one variable). From this, he extracts a fractional exponent (r_1, r_2) (analogous to the first characteristic exponent of a Puiseux series) describing the exact process which is a combination of monoidal and quadratic transformations and which reduces the multiplicity of the surface. Incidentally this last step is intrinsically contained as a special case in Lemma (D,1), Chapter IV, of Hironaka[11] (viewed as a step in deducing Theorem I from Theorem II in Hironaka's net of inductions).

Hironaka[11] proposed a different induction by means of the Weierstrass preparation theorem and of the language of general schemes (due to Grothendieck) and proved the resolution of singularities in all dimensions (but strictly in the characteristic zero case). The essentially new features of Hironaka's approach are the techniques developed to deal with the ideal bases of subvarieties of codimensions > 1, especially (a) the notion of *normal flatness* (which generalizes equimultiplicity in the case of hypersurfaces, exclusively considered by Zariski) and (b) the numerical characters ν^* of singularities. In Hironaka,[11] the problems about the *total transform into normal crossings* and the *elimination of fundamental locus* are also solved simultaneously with the resolution by a single net of inductions. Bennett[10] studied the *Hilbert-Samuel function* of a scheme at a singular point and showed that this replaces ν^* and has better properties than ν^* in many instances with regard to the effects of monoidal transformations and specialization of singularities. Some of Bennett's theorems are given simpler proofs in Hironaka.[13]

References

1. S. Abhyankar, *Local uniformization on algebraic surfaces over ground fields of characteristic $p \neq 0$*, Ann. of Math., vol. 63 (1956) pp. 491-526.

2. S. Abhyankar, *Reduction to multiplicity less than p in a p-cyclic extension of a two dimensional regular local ring,* Math. Ann., vol. 154 (1964) pp. 28-55.

3. S. Abhyankar, *Uniformization of Jungian local domains,* Math. Ann., vol. 159 (1965) pp. 1-43.

4. S. Abhyankar, *Uniformization in a p-cyclic extension of a two dimensional regular local domain of residue field of characteristic p,* Wissenschaftliche Abh. des Landes Nordrhein-Westfalen, vol. 33 (1966) pp. 243-317.

5. S. Abhyankar, *An algorithm on polynomials in one indeterminate with coefficients in a two dimensional regular local domain,* Ann. Mat. Pura Appl., vol. 71 (1966) pp. 25-60.

6. S. Abhyankar, *Resolution of Singularities of Embedded Algebraic Surfaces,* Academic Press, New York and London, 1966.

7. S. Abhyankar, *Nonsplitting of valuations in extensions of two dimensional regular local domains,* Math. Ann., vol. 170 (1967) pp. 87-144.

8. G. Albanese, *Transformazione birazionale di una superficie algebrica qualunque in un'altra priva di punti multipli,* Rend. Circolo Matem. Palermo, vol. 48 (1924) pp. 321-332.

9. M. Artin, *On isolated rational singularities of surfaces,* Amer. J. Math., vol. 88 (1966) pp. 129-136.

10. B. Bennett, *On the characteristic functions of a local ring,* Ann. Math. vol. 91 (1970) pp. 25-87.

11. H. Hironaka, *Resolution of singularities of an algebraic variety over a field of characteristic zero,* Ann. Math. vol. 79 (1964) pp. 109-326.

12. H. Hironaka, *Characteristic polyhedra of singularities,* J. Math. Kyoto Univ., vol. 7 (1968) pp. 251-293.

13. H. Hironaka, *Certain numerical characters of singularities,* J. Math. Kyoto Univ., vol. 10 (1970) pp. 151-187.

14. H. Hironaka, *Additive groups associated with points of a projective space,* Ann. Math., vol. 92 (1970) pp. 327-334.

15. B. Levi, *Resoluzione delle singolarita puntuali delle superficie algebriche,* Atti. Accad. Sci. Torino, vol. 33 (1897) pp. 66-86.

16. J. Lipman, *Rational singularities, with applications to algebraic surfaces and unique factorization,* Inst. Des Hautes Études Sci., Pub. Math., no. 36 (1969) pp. 195-279.

17. P. Du Val, *On isolated singularities of surfaces which do not affect conditions of adjunction,* I., Proc. Camb. Phil. Soc., vol. 30 (1933-1934) pp. 483-491.

18. P. Du Val, *Removal of singular points from an algebraic surface,* Universite d'Istanbul, Fac. des Sci., Memoires (1948) pp. 21-25.

19. R.J. Walker, *Reduction of singularities of an algebraic surface,* Ann. of Math., vol. 36 (1935) pp. 336-365.

Reprints of Papers

THE RESOLUTION OF SINGULARITIES OF AN ALGEBRAIC CURVE.*

By H. T. MUHLY and O. ZARISKI.

Introduction. The object of this note is to give new and brief proofs of the following well known theorems concerning the resolution of the singularities of an algebraic curve:

1. *Every algebraic curve is birationally equivalent to an hyperspace curve which is free from singularities, both at finite distance and at infinity.*

2. *Any curve free from singularities in an hyperspace can be projected into a curve in the projective 3-space which is also free from singularities.*

3. *Every algebraic curve can be birationally transformed into a curve in the projective plane whose only singularities are ordinary double points.*

We assume that the underlying field of constants, K, is algebraically closed. From the classical theory of algebraic functions of one variable (be it function-theoretic or arithmetic) we use the notion of the Riemann surface of a field of such functions, the existence of a uniformizing parameter at a point of the Riemann surface, and the notion of the order of a function at a point (place).[1]

1. Let Σ be a field of algebraic functions of one variable. Any set of elements $\omega_1, \omega_2, \cdots, \omega_m$ in the field having the property that any element in Σ can be expressed as a rational function of them, defines a curve C in the projective m-dimensional space, S_m; $(\omega_1, \omega_2, \cdots, \omega_m)$ is the generic point of C. Clearly, Σ is the field of rational functions on C. We shall denote the curve C by the symbol $\{\omega_1, \omega_2, \cdots, \omega_m\}$.

A point $P(a_1, a_2, \cdots, a_m)$ of a curve C will be called a *simple point* of C if (a) P corresponds to only one point (place) of the Riemann surface Σ, and if (b) at least one of the elements $\omega_i - a_i$ vainshes to the order one at P (i. e., at the corresponding place). If P is at infinity, and if, for instance, the ratios $\omega_2/\omega_1, \omega_3/\omega_1, \cdots, \omega_m/\omega_1$ are finite and equal to $c_2, c_3, \cdots, c_m$ respectively at P (in which case ω_1 is necessarily infinite at P), then instead of (b) we require that one of the elements $1/\omega_1, \omega_i/\omega_2 - c_i$ vanish to the order one at P.

---

* Received September 19, 1938.

[1] R. Dedekind and H. Weber, "Theorie der algebraischen Funktionen einer Veränderlichen," *Journal für reine und angewandte Mathematik*, vol. 92 (1882).

It is evident that this definition of a simple point is invariant with respect to projective transformations in S_m.

If $\mathfrak{O}$ is the integral closure of the ring $K[\omega_1, \omega_2, \cdots, \omega_m]$ then an equivalent definition of a simple point at finite distance may be formulated as follows: P is a simple point of C if and only if the ideal $\mathfrak{P} = \mathfrak{O} \cdot (\omega_1 - a_1, \omega_2 - a_2, \cdots, \omega_m - a_m)$ is prime. Namely, if there existed two places P_1, P_2 on the Riemann surface of Σ at which $\omega_i = a_i$, $i = 1, 2, \cdots, m$, then there would exist in $\mathfrak{O}$ two prime ideals $\mathfrak{P}_1$, $\mathfrak{P}_2$ such that $\omega_i \equiv a_i(\mathfrak{P}_1)$ and $\omega_i \equiv a_i(\mathfrak{P}_2)$, in contradiction with the hypothesis that $\mathfrak{O} \cdot (\omega_1 - a_1, \cdots, \omega_m - a_m)$ is prime. Moreover, for at least one of the elements $\omega_i - a_i$, say for $\omega_1 - a_1$, we must have $\omega_1 - a_1 \not\equiv 0(\mathfrak{P}^2)$, and hence $\omega_1 - a_1$ vanishes to the order one at the corresponding place.

By a theorem of F. K. Schmidt [2] an independent variable x may be chosen in such a manner that Σ is a separable extension of $K(x)$. Let $\mathfrak{O}$ be the ring of integral functions of x in Σ, and let $\omega_1, \omega_2, \cdots, \omega_n$ be an integral base over $K[x]$ of the ring $\mathfrak{O}$, where n is the relative degree $[\Sigma : k(x)]$. *The curve* $\Gamma \equiv \{x, \omega_1, \omega_2, \cdots, \omega_n\}$ *in* S_{n+1} *is free from singularities at finite distances.* In view of the above ideal theoretic definition of a simple point, our assertion follows from the fact that since $\omega_1, \omega_2, \cdots, \omega_n$ is an integral base of $\mathfrak{O}$, the ring $K[x, \omega_1, \cdots, \omega_n]$ coincides with $\mathfrak{O}$ and hence is integrally closed.

We may assume that x becomes infinite to the order one at n distinct points of the Riemann surface of Σ. (Otherwise we use $x' = 1/(x - c)$ as the independent variable, where c is a constant such that $x - c$ vanishes to the order one at n distinct points of the Riemann surface). Let $P_1, P_2, \cdots, P_n$ be the points of the Riemann surface where x becomes infinite, and let ρ_i be a function in Σ which becomes infinite to the order one at P_i, and is finite at P_j, $j \neq i$. Every element in Σ can be put in the form $\xi/h(x)$, where $\xi \subset \mathfrak{O}$ and $h(x)$ is a polynomial in x. Hence we can find a polynomial $b(x)$ such that $b\rho_i$ is an integral function of x, $i = 1, 2, \cdots, n$. We consider the functions $\eta_1, \eta_2, \cdots, \eta_n$ defined as follows:

$$\eta_j = x^\lambda b(x)\rho_j \qquad (j = 1, 2, \cdots, n),$$

where λ is an integer which is taken so large that each η_j becomes infinite at every point P to an order higher than that of any of the functions ω_i. By construction, the function η_i is infinite to the order say m at P_j, $j \neq i$, and is infinite to the order $m + 1$ at P_i (m independent of i). Since η_i is an integral function of x, it is finite elsewhere.

[2] F. K. Schmidt, "Analytische Zahlentheorie in Körpern der Charakteristik p," *Mathematische Zeitschrift*, vol. 33 (1931).

The curve $\Gamma' \equiv \{x, \omega_1, \cdots, \omega_n; \eta_1, \cdots, \eta_n\}$ *in* S_{2n+1} *is free from singularities both at finite distance and at infinity.* Since the curve Γ was free from singularities at finite distance, the curve Γ' necessarily shares this property with Γ. The points at infinity on Γ' correspond to the points $P_1, P_2, \cdots, P_n$ of the Riemann surface. At P_i the quotients $1/\eta_i$, x/η_i, ω_j/η_i, and η_l/η_i, $l \neq i$, are all zero. Thus P_i gives rise to the point at infinity on the axis η_i. Moreover, at P_i η_j/η_i vanishes to the order one if $j \neq i$. Thus we have proved Theorem 1.

2. Let $C \equiv \{\omega_1, \omega_2, \cdots, \omega_m\}$ be a curve in the projective space $S_m(y_0, y_1, \cdots, y_m)$. On C we will have

$$y_0 : y_1 : \cdots : y_m = 1 : \omega_1 : \omega_2 : \cdots : \omega_m,$$

or introducing a factor of proportionality,

$$y_0 : y_1 : \cdots : y_m = \lambda_0 : \lambda_1 : \cdots : \lambda_m,$$

where $\lambda_i/\lambda_0 = \omega_i \subset \Sigma$. We shall also use the symbol $\{\lambda_0, \lambda_1, \cdots, \lambda_m\}$ to denote C.

Let $\omega'_i = \sum_{j=1}^{m} u_{ij}\omega_j$, $i = 1, 2$, where the u_{ij} are indeterminates. The elements ω'_1 and ω'_2 are connected by an irreducible equation

$$F(\omega'_1, \omega'_2; u_{ij}) = 0.$$

By formal partial differentiation, we obtain the relations

$$\begin{aligned} \omega_j F'_{\omega'_1} + F'_{u_{1j}} &= 0, \\ \omega_j F'_{\omega'_2} + F'_{u_{2j}} &= 0, \end{aligned} \qquad (j = 1, \cdots, n)$$

from which we conclude that if either $F'_{\omega'_1}$ or $F'_{\omega'_2}$ is different from zero, then any element of Σ can be expressed as a rational function of ω'_1 and ω'_2. The functions $F'_{\omega'_1}$, $F'_{\omega'_2}$ cannot both be zero, for if they were then also $F'_{u_{1j}} = F'_{u_{2j}} = 0$, $j = 1, \cdots, n$, and therefore we would have

$$F(\omega'_1, \omega'_2; u_{ij}) = [\phi(\omega'_1, \omega'_2; u_{ij})]^p,$$

where p is the characteristic of K.[3] Moreover, if the u's are selected so that say $F'_{\omega'_2} \neq 0$, then the polynomial $F(\omega'_1, \omega'_2; u_{ij})$ is not a polynomial in ω'^p_2, so that Σ is a separable extension of $K(\omega'_1)$.

Let $A_1, A_2, \cdots, A_k$ be the points of the Riemann surface at which not all ω_i are finite. If

[3] See for example B. L. van der Waerden, *Moderne Algebra*, vol. I, Chapter V.

$$\xi_i = \frac{c_{i0} + c_{i1}\omega_1 + \cdots + c_{im}\omega_m}{c_{00} + c_{01}\omega_1 + \cdots + c_{0m}\omega_m}, \qquad (i = 1, 2, \cdots, m)$$

then by a sufficiently general choice of the constants c we may be certain that each ξ_i is finite at all of the points A_j. Moreover, if we choose c_{00} so that the function $c_{00} + c_{01}\omega_1 + \cdots + c_{0m}\omega_m$ vanishes to the order one at say ν distinct points of the Riemann surface, and c_{i0} so that $c_{i0} + c_{i1}\omega_1 + \cdots + c_{im}\omega_m$ is different from zero at these points, it follows that the functions ξ_i all become infinite to the order one at the same points of the Riemann surface.

In S_m we make the coördinate transformation $T(c_{ij})$:

$$y'_i = \sum_{j=0}^{m} c_{ij} y_j, \qquad (i = 0, 1, \cdots, m).$$

The curve C is represented in the y' coördinate system by $\{\lambda'_0, \lambda'_1, \cdots, \lambda'_m\}$, where

$$\lambda'_i = \sum_{j=0}^{m} c_{ij}\lambda_j$$

and $\lambda'_i/\lambda'_0 = \xi_i \subset \Sigma$.

Obviously, the constants c_{ij} may be chosen so that ξ_2 is a primitive element of Σ over $K(\xi_1)$. Thus we can always obtain a representation $\{\xi_1, \xi_2, \cdots, \xi_m\}$ of a given curve C satisfying the three conditions: (1) *Σ is a separable extension of $K(\xi_1)$*,[4] (2) *ξ_2 is a primitive element of $\Sigma/K(\xi_1)$*, (3) *all of the ξ become infinite to the order one at the same ν points of the Riemann surface, and are finite elsewhere, ν being the relative degree $[\Sigma : K(\xi_1)]$.*

3. Let $\Gamma \equiv \{\xi_1, \cdots, \xi_m\}$ be any curve in S_m free from singularities. We assume that $\xi_1, \cdots, \xi_m$ satisfy conditions (1), (2), (3) of the preceding section. A point P of the Riemann surface of Σ will be called a branch point for the function ξ, if when $\xi = a$ at P, $\xi - a$ vanishes to an order larger than one; or if ξ has a pole of order greater than one at P. Let the set $\mathfrak{A} = (A_1, A_2, \cdots, A_\lambda)$ be the set of the common branch points for ξ_1 and ξ_2. A branch point for ξ_1 gives rise to a zero of $\Delta(\xi_1)$ the field discriminant of $\Sigma/K(\xi_1)$. Since Σ is a separable extension of $K(\xi_1)$, $\Delta(\xi_1)$ has only a finite number of zeros, so that the set $\mathfrak{A}$ contains only a finite number of points.

Let the set of points $\mathfrak{B} = (B_1, B_2, \cdots, B_\mu)$ be defined as follows: A point P of the Riemann surface of Σ is in $\mathfrak{B}$ if and only if there exists a second point $P' \neq P$ such that both ξ_1 and ξ_2 have the same finite values at P' as at P. Since ξ_2 is a primitive element of Σ, the equation

[4] A somewhat stronger result than condition (1) has recently been proved by B. L. van der Waerden, "Zur algebraischen Geometrie," *Mathematische Annalen*, vol. 115 (1938).

$$f(\xi_1, \xi_2) = 0,$$

of degree ν, connecting ξ_1 and ξ_2 is irreducible, so that its discriminant $D(\xi_1)$ is not identically zero. Any point of $\mathfrak{B}$ gives rise to a zero of $D(\xi_1)$, and hence $\mathfrak{B}$ can contain only a finite number of points.

Let $\mathfrak{C} = (C_1, C_2, \cdots, C_\nu)$ be the set of points on the Riemann surface of Σ at which ξ_1 becomes infinite.

Since a point A_i corresponds to a simple point on Γ it follows that if τ_i is a uniformizing parameter on A_i and if

$$\xi_j = c_{ji} + d_{ji}\tau_i + \cdots$$

is the expansion of ξ_j at A_i, then for a given i not all of the constants d_{ji} are zero. Since by hypothesis, $d_{1i} = d_{2i} = 0$, it follows that the linear form in $u_3, \cdots, u_m$

$$\psi_i(u) = \sum_{i=3}^{m} d_{ji}u_j \qquad (i = 1, 2, \cdots, \lambda)$$

is not identically zero.

Similarly, if k_{ji} is the value of ξ_j at B_i, and if B_a and $B_{a'}$ are a pair of points in $\mathfrak{B}$ such that $\xi_1 = a_1$, $\xi_2 = a_2$ at both B_a and $B_{a'}$, then the linear form

$$\phi_{aa'}(u) = \sum_{j=3}^{m} (k_{ja} - k_{ja'})u_j$$

is not identically zero.

Since all ξ_i are infinite to the order one at C_j, $1/\xi_1$ is a uniformizing parameter at C_j, and the ratios ξ_i/ξ_1 are all finite at C_j. Let $\xi_i/\xi_1 = l_{ij}$ at C_j, and let

$$\theta_{a\beta}(u) = \sum_{j=3}^{m} (l_{ja} - l_{j\beta})\, u_j.$$

Then if C_a and C_β are a pair of points in $\mathfrak{C}$ at which $l_{2a} = l_{2\beta}$ the corresponding $\theta_{a\beta}(u)$ is not identically zero, since C_a and C_β correspond to distinct points on Γ.

Choose a set of constants, $c_3, c_4, \cdots, c_m$ such that $\psi_i(c) \neq 0$, $\phi_{aa'}(c) \neq 0$ and $\theta_{a\beta}(c) \neq 0$ when $l_{2a} = l_{2\beta}$. Let

$$\xi = \sum_{j=3}^{m} c_i\xi_i.$$

We assert that *the curve* $L \equiv \{\xi_1, \xi_2, \xi\}$ *in* S_3 *is free from singularities both at finite distance and at infinity.* All points of L which do not correspond to points of the sets $\mathfrak{A}$, $\mathfrak{B}$, $\mathfrak{C}$ are necessarily simple points. Moreover, if $\xi = d_i$ at A_i, then $\xi - d_i$ vanishes to the order one at A_i since $\psi_i(c) \neq 0$. The function ξ assumes distinct values at B_a and $B_{a'}$ since $\phi_{aa'}(c) \neq 0$. At any pair of points C_α, C_β at which the values of ξ_2/ξ_1 are the same, the values of ξ/ξ_1 are distinct since $\theta_{\alpha\beta}(c) \neq 0$.

Finally, at any point C_α the quotients $1/\xi_1, \xi_2/\xi_1, \xi/\xi_1$ are finite and $1/\xi_1$ vanishes to the order one. Thus the points of L which correspond to the points in $\mathfrak{A}$, $\mathfrak{B}$ and $\mathfrak{C}$ are also simple points. This establishes Theorem 2.

4. Let the curve $C \equiv \{\lambda_0, \lambda_1, \cdots, \lambda_r\}$ be free from singularities in the projective space S_r: $(x_0, x_1, \cdots, x_r)$. By placing $x'_{ij} = x_i x_j$, $i, j = 0, 1, 2, \cdots, r$, we refer the hyperquadrics of S_r to the hyperplanes of a new space S_N in which the (x'_{ij}) are homogeneous coördinates. On placing $\lambda'_{ij} = \lambda_i\lambda_j$ we have defined in S_N a curve $C' \equiv \{\lambda'_{ij}\}$. Between C and C' there is a $1:1$ birational correspondence. A straightforward application of our definition of a simple point shows that C' is also free from singularities.

If P'_i, $i = 1, 2, 3$, are any three points of C' and P_1, P_2, P_3 the corresponding points of C, it is possible to find a hyperquadric in S_r passing through P_1 and P_2 but not through P_3, so that the corresponding hyperplane in S_N passes through P'_1 and P'_2 but not through P'_3. Thus no three points of C' are collinear. Similarly, any tangent line of C' has only its point of contact in common with C'.

We repeat this process, applying it on the hyperquadrics of S_N and the curve C'. We obtain thereby a curve C'' in $1:1$ birational correspondence with C' and hence also with C. Since no three points of C' are collinear, there exists a hyperquadric through any three given points of C' which does not pass through a given fourth point. Hence *no four points of C'' are coplanar.* In the same way we conclude from the above-mentioned property of the tangents to C' that *through any two points of C'' there passes a hyperplane which is tangent to C'' at one of them but not at the other.*

Let $\xi_1, \xi_2, \cdots, \xi_M$ be a set of elements in Σ which define the curve C''. (We have $M = \frac{1}{2}(N+2)(N+1) - 1$, the dimension of the space of hyperquadrics over S_N. Likewise, $N = \frac{1}{2}(r+2)(r+1) - 1$.) We assume that the elements $\xi_1, \cdots, \xi_M$ satisfy the conditions (1), (2) and (3) of Paragraph 2.

Let $u_1, u_2, \cdots, u_M$ be indeterminates and let $y = \sum_{i=1}^{M} u_i\xi_i$. The function y is connected with ξ_1 by an irreducible equation

$$f(\xi_1, y; u) = 0$$

of degree ν, where ν is the degree of Σ over $K(\xi_1)$. If $D(\xi_1, u)$ is the discriminant of f, then

$$D(\xi_1, u) = P^2(\xi_1, u)\Delta(\xi_1),$$

where P is a polynomial in ξ_1 and u, and Δ is the field discriminant of Σ, so that Δ is independent of u. Assume for the moment that the u's have special values u^0 and let $y^0 = \Sigma u_i^0 \xi_i$, $P^0(\xi_i) = P(\xi_1, u^0)$. Let $\xi_1 = c$ be a root of $P^0(\xi_1)$. By known properties of the discriminant of f,[5] it follows that (1) either there must exist at least two distinct points of the Riemann surface of Σ at which $\xi_1 = c$ and at which also the values of y^0 are equal to each other; or (2) $y^0 = d$ at one of the points, say at P, at which $\xi_1 = c$, and $y^0 - d$ vanishes at P to an order larger than one. Since C'' is free from singularities, it is true that given any constant c, a linear combination of the ξ_i's can be found such that neither (1) nor (2) take place. For the corresponding values u^0 of the u_i, c will not be a root of $P(\xi_1, u_0)$. It follows that $P(\xi_1, u)$ as a polynomial in ξ_1, does not have roots independent of the u's.

For general values of the u_i, P has no multiple roots in ξ_1. Assuming the contrary, let $P(\xi_1, u)$ have the factor $Q(\xi_1, u)$ all of whose roots are multiple. Then if a is any constant, a sufficient condition that a be a multiple root of P is that the u's satisfy the equation $Q(a, u) = 0$. Hence our assertion will follow if we prove that two independent conditions (in fact linear conditions) are imposed on the indeterminates u by requiring that $\xi_1 = a$ be a multiple root of P.

If $(\xi_1 - a) = \mathfrak{P}_1 \cdot \mathfrak{P}_2 \cdots \mathfrak{P}_\nu$ is the prime decomposition of the ideal $(\xi_1 - a)$ in the ring of integral functions of ξ_1 in Σ, (we may assume $\Delta(a) \neq 0$ so that each factor occurs only to the first power), then one of the following three conditions is necessary in order that $\xi_1 = a$ be a multiple root of P:

a) For some three of the ideals $\mathfrak{P}_i$, say $\mathfrak{P}_1$, $\mathfrak{P}_2$, $\mathfrak{P}_3$, there exists a constant b such that

$$y - b \equiv 0 \ (\mathfrak{P}_1 \cdot \mathfrak{P}_2 \cdot \mathfrak{P}_3).$$

b) $y - b' \equiv 0 \ (\mathfrak{P}_1 \cdot \mathfrak{P}_2)$, $y - b'' \equiv 0 \ (\mathfrak{P}_3 \mathfrak{P}_4)$ for some constants b' and b'' and some four of the ideals $\mathfrak{P}_i$.

c) $y - b - c(\xi_1 - a) \equiv 0 \ (\mathfrak{P}_1^2 \cdot \mathfrak{P}_2^2)$ for some constants, b and c and some two of the ideals $\mathfrak{P}_i$.

If $\xi_i \equiv a_{ij} \ (\mathfrak{P}_j)$ then the matrix

$$A_1 = \begin{pmatrix} a_{11} - a_{12}, a_{21} - a_{22}, \cdots, a_{M1} - a_{M2} \\ a_{11} - a_{13}, a_{21} - a_{23}, \cdots, a_{M1} - a_{M3} \end{pmatrix}$$

is of rank two since otherwise the three points $(a_{1i}, a_{2i}, \cdots, a_{Mi})$, $i = 1, 2, 3$, on C'' would be collinear. Similarly, the matrix

$$A_2 = \begin{pmatrix} a_{11} - a_{12}, a_{21} - a_{22}, \cdots, a_{M1} - a_{M2} \\ a_{13} - a_{14}, a_{23} - a_{24}, \cdots, a_{M3} - a_{M4} \end{pmatrix}$$

[5] See e. g. Dedekind-Weber, "Theorie der algebraischen Funktionen einer Veränderlichen," *loc. cit.*

8

is of rank two since the four points $(a_{1i}, a_{2i}, \cdots, a_{Mi})$, $i = 1, 2, 3, 4$ cannot be coplanar.[6]

Since $\Delta(a) \neq 0$, $(\xi_1 - a)$ is a uniformizing parameter on each of the ideals $\mathfrak{P}_j$. Let $\dfrac{\xi_i - a_{ij}}{\xi_1 - a} = b_{ij}$ at $\mathfrak{P}_j$. The matrix

$$A_3 = \begin{pmatrix} a_{11} - a_{12}, \cdots, a_{M1} - a_{M2} \\ b_{11} - b_{12}, \cdots, b_{M1} - b_{M2} \end{pmatrix}$$

is of rank two. In the contrary case, any hyperplane through the points P_i: $(a_{1i}, a_{2i}, \cdots, a_{Mi})$, $i = 1, 2$ and tangent to C'' at P_1 would also be tangent to C'' at P_2.

In order that condition (a) be fulfilled, the quantities u must satisfy the equations

$$\sum_{j=1}^{M} (a_{j1} - a_{ji}) u_j = 0 \qquad (i = 2, 3).$$

Condition (b) requires that the equations

$$\sum_{j=1}^{M} (a_{j1} - a_{j2}) u_j = 0, \qquad \sum_{j=1}^{M} (a_{j3} - a_{j4}) u_j = 0$$

be satisfied. Finally, the equations

$$\sum_{j=1}^{M} (a_{j1} - a_{j2}) u_j = 0, \qquad \sum_{j=1}^{M} (b_{j1} - b_{j2}) u_j = 0$$

must be satisfied in order that (c) occur. Thus a, b, and c each impose two independent conditions on the indeterminates u. Hence $P(\xi_1, u)$ has no multiple roots for general values of u.

If $P_1, P_2, \cdots, P_\nu$ are the points of the Riemann surface where ξ_1 becomes infinite, and if $\xi_i/\xi_1 = \lambda_{ij}$ at P_j, then (as in Theorem 2),

$$\theta_{\alpha\beta}(u) = \sum_{i=1}^{M} (l_{i\alpha} - l_{i\beta}) u_i$$

is not identically zero if $\alpha \neq \beta$. Thus there exists a set of constants $c_1, c_2, \cdots, c_M$ such that $P(\xi_1, c)$ has only simple roots, none of which coincides with a root of $\Delta(\xi_1)$ and such that $\theta_{\alpha\beta}(c) \neq 0$. If

$$y = c_1\xi_1 + c_2\xi_2 + \cdots + c_M\xi_M,$$

then the curve $\{\xi_1, y\}$ has no singularities at infinity (since $\theta_{\alpha\beta}(c) \neq 0$) and has ordinary double points at finite distance corresponding to the roots of $P(\xi_1, c) = 0$. This proves Theorem 3.

THE JOHNS HOPKINS UNIVERSITY.

[6] Actually the condition imposed by us that no four points be coplanar is stronger than the one we actually use now concerning the ranks of the matrices A_1 and A_2. Since we are projecting our curve in S_M from a particular S_{M-2} at infinity we need only require that if four points on our curve are coplanar then lines joining pairs of them do not meet on this S_{M-2}.

Annals of Mathematics
Vol. 40, No. 3, July, 1939

THE REDUCTION OF THE SINGULARITIES OF AN ALGEBRAIC SURFACE

By Oscar Zariski

(Received February 15, 1939)

Contents

Introduction

The reduction of the singularities of an algebraic surface has been studied by several authors, in particular by B. Levi [5], Chisini [2], Albanese [1] and R. J. Walker [9].[1] All these proofs, except the one by Walker, are essentially geometric. The proof by Walker, based in part on a paper by Jung [3], is analytic and is entirely rigorous. We give in this paper an *arithmetic* proof of the reduction theorem, based on the theory of valuations and the theory of ideals in algebraic function fields. From the standpoint of rigor, the arithmetic proof leaves nothing to be desired. We should say that the requirement of rigor is to be regarded as trivially satisfied in the present proof. More significant, however, is the clarification of the problem brought about by the use of the methods of modern algebra. The formal apparatus of ideal theory and the concepts of valuation theory enable one to gain full control of the intricate aspects of the problem, and to see clearly through the logical necessity of the various steps of the reduction process. What is gained concretely thereby is that the center of gravity of the proof is shifted from minute details to underlying concepts. As for these, we insist above all on clear-cut definitions, in order that at no time should it be less than perfectly clear what are the objects we are dealing with.

In a paper [Z_2], which appeared in the April issue of the American Journal of Mathematics, we used the operation of "integral closure" as a type of birational transformation for surfaces, and, more generally, for varieties. The operation of integral closure leads to the notion of a *normal variety*. Since

[1] For further references and for a critical review of the various proofs, see our Ergebnisse monograph *Algebraic surfaces*, p. 13–23. We may mention here also an unpublished proof by Du Val, based on a careful fusion of procedures outlined by Severi and by Jung.

these concepts play an essential rôle in the proof, we give here the definition of normal varieties and we state briefly some of their most important properties.

Let $\xi_1, \cdots, \xi_n$ be the coördinates of the general point of an algebraic irreducible r-dimensional variety V_r in an affine space S_n. *We assume throughout that the underlying field of coefficients, denoted by* K, *is algebraically closed and of characteristic zero.* The field $\mathsf{K}(\xi_1, \cdots, \xi_n)$, of degree of transcendency r over K, shall be denoted by Σ. We consider a simple transcendental extension Σ' of Σ: $\Sigma' = \Sigma(\omega_0)$. By the *homogeneous coördinates* of the general point of V_r in the projective space P_n, we mean the elements $\omega_0, \omega_1 = \omega_0\xi_1, \cdots, \omega_n = \omega_0\xi_n$.

Let $\mathfrak{o} = \mathsf{K}[\xi_1, \cdots, \xi_n]$ be the ring of polynomials in the ξ's. Similarly, let $\mathfrak{o}' = \mathsf{K}[\omega_0, \omega_1, \cdots, \omega_n]$. We say that V_r *is normal in the affine space* S_n, if $\mathfrak{o}$ is integrally closed in its quotient field Σ. *The variety* V_r *is said to be normal in the projective space* P_n, if $\mathfrak{o}'$ is integrally closed in Σ'. If V_r is normal in P_n, then it is *a fortiori* normal in the affine space S_n, *for any choice of the hyperplane at infinity* (Z_2, section 18; compare also with Theorem 13, loc. cit.). From now on we shall mean by a normal variety one which is normal in P_n.

The importance of normal varieties is due to the following theorem, proved in Z_2 (theorem 11′): *the locus of singular points of a normal* V_r *is an algebraic subvariety of dimension* $\leqq r - 2$. In particular, *a normal surface possesses only a finite number of singular points.*[2]

We associate with a given V_r a *class* of normal varieties which are birationally equivalent to V_r. Let $\bar{\mathfrak{o}}'$ be the integral closure of $\mathfrak{o}'$ in Σ'. The transformation $\omega_i \rightarrow t\omega_i$, $i = 0, 1, \cdots, n$, $t \subset \mathsf{K}$, defines an automorphism τ of $\bar{\mathfrak{o}}'$. An element ζ of $\bar{\mathfrak{o}}'$ is *homogeneous* of *degree* ρ, if $\tau\zeta = t^\rho\zeta$, t-arbitrary in K. The homogeneous elements in $\bar{\mathfrak{o}}'$, of a given degree ρ, have a finite linearly independent base over K. Let $\eta_0, \eta_1, \cdots, \eta_m$ be such a base. It is possible to prove that *if* ρ *is sufficiently high, then the ring* $\mathsf{K}[\eta_0, \eta_1, \cdots, \eta_m]$ *is integrally closed in its quotient field.* This quotient field (a subfield of Σ') is again a simple transcendental extension of Σ, and the quotients η_i/η_0 are elements of Σ (see Z_2, section 20). Hence we may regard the η's as the homogeneous coördinates of the general point of a *normal* V_r', birationally equivalent to V_r. We call V_r' a *derived normal variety* for V_r, relative to the *character of homogeneity* ρ.

Two derived normal varieties relative to the same character of homogeneity are *projectively equivalent.* More generally, the points of two derived normal varieties (for one and the same V_r) are in (1,1) correspondence, *without exceptions*, and *the quotient rings of two corresponding points coincide.*[3] From the

[2] The definition and characteristic properties of a simple point, given in Z_2 (sections 1-9), will be found re-stated in the course of the consideration developed in III, 11.

[3] If P is a point on a V_r and if the homogeneous coördinates of P are $a_0, a_1, \cdots, a_n$, we define the quotient ring $Q(P)$ of P as the set of all quotients $f(\omega_0, \omega_1, \cdots, \omega_n)/g(\omega_0, \omega_1, \cdots, \omega_n)$, where f and g are forms of like degree and $g(a_0, a_1, \cdots, a_n) \neq 0$ (compare with III, 10). The hypothesis that f and g are forms of like degree implies that the quotient f/g is an element of Σ.

standpoint of the analysis of singularities any two derived normal varieties of a given V_r are entirely equivalent. This is so, because to us the structure of the quotient ring $Q(P)$ at a point P is the controlling feature of the type of singularity at the point. Thus, for instance, the following property of $Q(P)$ is characteristic for simple points: *if* $\mathfrak{p}_0$ *denotes the prime zero-dimensional ideal of* $Q(P)$, *then* P *is simple if and only if the ring of residual classes* $\mathfrak{p}_0/\mathfrak{p}_0^2$ *is a* K*-module of rank* r (Z_2, theorem 3.2).

The geometric significance of a normal V_r is the following: *if* V_r *is normal, then the system of its hyperplane sections is complete* (Z_2, Theorem 14). The converse is not true. The geometry of the relationship between a given V_r and a derived normal V_r', whose character of homogeneity is ρ, is simply this: *the system of hyperplane sections of* V_r' *is the complete system,* which is the ρ*-fold of the system of hyperplane sections of* V_r.

The passage from a V_r to a derived normal variety V_r' is one step of our reduction process. Its effect, for surfaces, is the elimination of all the multiple curves of the surface. This elimination, so troublesome in most of the previous proofs, is here achieved in a relatively simple fashion.

We are now in position to state explicitly the exact procedure of reduction. We start with a normal surface F. It has a finite number of singularities. We take one of them, say P, as the fundamental point of a suitable quadratic transformation. Namely, we consider those quadratic forms in the homogeneous coördinates of the general point of F, which vanish at P. We take a linearly independent base for these forms over K and we regard the elements of the base as the homogeneous coördinates of the general point of a new surface F', birationally equivalent to F. Or, in geometric terms: we refer the sections of F by the hyperquadrics on P to the hyperplane sections of a new surface F'.

The surface F' may not be normal. We pass to a derived normal surface of F', say F_1. If F_1 still has singularities (and necessarily only a finite number), we apply the same procedure to the new normal surface F_1, and so we continue. *Our reduction theorem asserts that after a finite number of steps a normal surface is obtained which is free from singularities* (VI, 21).

We wish now to review briefly the salient points of the proof. We shall not follow necessarily the expository order of the paper, but rather we shall attempt to make clear the inner reasons of the various steps. Certain formalities will have to be dispensed with, since we only wish to bring out the conceptual content of the proof.

By the combined transformation from F to F_1 (quadratic transformation and integral closure), the point P is resolved into a fundamental curve on which there may be at most a finite number of singular points of F_1. Let P_1 be one of them. We take P_1 as the fundamental point of our quadratic transformation, getting F_1' and a new derived normal surface F_2. Continuing in this fashion we get an infinite sequence of points: $P, P_1, P_2, \cdots$, belonging to the normal surfaces F, F_1, F_2, $\cdots$ respectively. The question on which depends the solutions of the problem is this: *it should be proved that for some sufficiently high integer m, the point P_m (and consequently also the points P_i, $i > m$) will be simple.* This is essentially our "local reduction" theorem (V, 20). That this theorem is true at all, is an indication of the importance of the "integral closure" transformations. Under quadratic transformation alone infinite sequences of successive singular points can and do occur, as was pointed out by B. Levi [5].

In algebraic geometry[4] one associates intuitively with a sequence of points, such as the one above, a sequence of *successive infinitely near points* on the original surface F, i.e. essentially a branch on F, of origin P. This branch need not be algebraic, nor even analytic.[5] Now the arithmetic analog of a branch is a *zero-dimensional valuation* of the field Σ of rational functions on F. Accordingly, we consider the quotient rings $Q(P_i)$, we observe that they form a strictly ascending sequence of integrally closed rings, and we prove that *the limit of the rings $Q(P_i)$ (i.e. their union) is the valuation ring of a zero-dimensional valuation* (see V, 19, where we prove in fact a more general lemma). In this fashion we can get any given zero-dimensional valuation B. We have only to take, at each step, the fundamental point P_i on F_i at *the center* of B, i.e. at the origin of the branch which B defines on F_i. Should then our "local reduction" theorem be true, it is at least necessary (but not at all sufficient) that the following be true: *given any zero-dimensional valuation of Σ, there exists a projective model F' of Σ on which the center of the valuation is at a simple point.* This is our fundamental lemma, proved in Part II. On the basis of this lemma, the valuation B defined by the sequence $\{P_i\}$ can equally well be defined (and in infinitely many ways) by a sequence $\{A_i\}$ of *simple points* (belonging to suitable projective models of Σ), in the sense that the valuation ring of B is also the limit of the quotient ring $Q(A_i)$. Hence every $Q(P_i)$ is a subring of some $Q(A_j)$, and conversely. The proof is completed by showing that *if i is sufficiently high, then $Q(P_i)$ coincides with some $Q(A_j)$.*

The length of the paper is not entirely due to the lengthiness of the proof. It is due in part to the fact that we treat the entire problem *ab initio*. We also wished to avoid, as much as possible, references to scattered sources, especially when we were not certain whether the reference would exactly cover the topic under consideration. Since the elements of valuation theory are not generally known, we also thought it advisable to devote the first part of the paper to a review and classification of the valuations in fields of algebraic functions of two variables.

---

[4] See for instance C. Segre: *Sulla scomposizione dei punti singolari delle superficie algebriche.* Ann. Mat. pura appl. II. s. Vol. 25 (1897).

[5] In the example of an infinite sequence of successive multiple points given by B. Levi, the corresponding branch is in fact transcendental.

I. Classification of Valuations in Fields of Algebraic Functions of Two Variables

1.

Let Σ be a field of algebraic functions of two variables, over an algebraically closed field K of characteristic zero. The elements of K shall be called *constants*. A *valuation* B of the field Σ is an homomorphic mapping v of the multiplicative group Σ (the element zero excluded) upon an ordered abelian additive group Γ, satisfying the following conditions: if $a \neq 0$ is any element in Σ and if $v(a)$ is the corresponding element in Γ, then

(1) $v(a \cdot b) = v(a) + v(b)$;

(2) $v(a + b) \geqq \min. (v(a), v(b))$;

(3) $v(c) = 0$, if $c \neq 0$ is an element of K.

The condition (1) says merely that the mapping $a \rightarrow v(a)$ is an homomorphism. We shall exclude the *trivial* valuation, in which Γ consist of only the element zero. As a consequence, we must also add the following condition: (4) for some $a \neq 0$ in Σ, $v(a) \neq 0$. The element $v(a)$ of Γ is called the *value* of a. We formally assign to the zero element of Σ the value $+\infty$.

Valuations of Σ are classified first of all according to the structure of the *value group* Γ. If the ordered group Γ *is archimedean*, B is said to be *of rank 1*. In the most general case, one considers the *isolated subgroups* of Γ. A subgroup Γ' of Γ is isolated if it has the property that together with any positive element α of Γ it contains also all the positive elements $\beta < \alpha$. The element 0 is to be regarded by itself as constituting an isolated subgroup of Γ. Given two isolated subgroups Γ' and Γ'' of Γ, one is necessarily a subgroup of the other, say $\Gamma' \subset \Gamma''$, and we say that Γ' *precedes* Γ''. In this sense the isolated proper subgroups of Γ form an ordered set M_Γ, having a definite ordinal type. If M_Γ is a finite set, of n elements, we say that Γ is of *rank n*. The valuation B is then also said to be of rank n. It is well known that if r is the degree of transcendency of Σ over K, then any valuation of Σ is of finite rank $\leqq r$ (see K, p. 116). In our case, we can only have valuations of rank $\leqq 2$.

If Γ is of rank 1, it is archimedean and is isomorphic to a subgroup of the additive group of all real numbers. If Γ is of the type of the group of all rational integers, we say that Γ—and also the valuation B—is *discrete*. More generally, if Γ is of rank n, we consider the n isolated subgroups $\Gamma_1, \Gamma_2, \cdots, \Gamma_n$ in descending order ($\Gamma \supset \Gamma_1 \supset \Gamma_2 \supset \cdots \supset \Gamma_n$) and the quotient groups Γ_{i-1}/Γ_i ($i = 1, \cdots, n$, $\Gamma_0 = \Gamma$). Each quotient group Γ_{i-1}/Γ_i is an ordered archimedean group. If these n quotient groups are all discrete, we say that the group Γ (and the valuation B) is *discrete*, of rank n.

The question of determining the possible types of valuations of Σ can be reduced to the same question for purely transcendental fields, i.e. fields of rational functions of two independent variables. To see this, let B be any valuation of Σ and let x and y be two algebraically independent elements of Σ whose values in B are non-negative. Let ω be any element in Σ and let

$$(1) \qquad F(x, y, \omega) = A_0(x, y)\omega^\nu + \cdots + A_\nu(x, y) = 0$$

be the algebraic relation between x, y, ω, where $\nu = [\Sigma: \mathsf{K}(x, y)]$, the relative degree of Σ with respect to $\mathsf{K}(x, y)$. Since $v(F) = +\infty$, there must exist at least two terms, say $A_i\omega^{\nu-i}$, $A_j\omega^{\nu-j}$, $i \neq j$, whose values in B are the same. Hence, assuming $i > j$, we have $v(A_i/A_j) = (i - j)v(\omega)$, $0 < i - j \leqq \nu$. It follows that for *any* element ω in Σ the product $\nu!v(\omega)$ is the value of a rational function of x, y. The value group Γ' of the valuation induced by B in the field $\mathsf{K}(x,y)$ has thus the property of containing the $\nu!$-multiple of every element of Γ. Hence Γ and Γ' have the same rank, and Γ is discrete if and only if Γ' is discrete.

2.

Another classification of the valuations of Σ is based upon the consideration of the so-called *residue field* Σ^* of the valuation. We consider the totality of elements of Σ whose value in B is non-negative. The set of these elements forms a ring $\mathfrak{B}$, the *valuation ring* of B. The elements of $\mathfrak{B}$ whose value is positive form a prime ideal $\mathfrak{P}$ in $\mathfrak{B}$, the *prime ideal of the valuation.* By the residue field Σ^* of the valuation B is meant the field of residual classes of $\mathfrak{B}$ modulo $\mathfrak{P}$: $\Sigma^* \cong \mathfrak{B}/\mathfrak{P}$.[6] Since no constant different from zero has value > 0, it follows that Σ^* contains a subfield K^* simply isomorphic to K. Now, quite generally, let r be the degree of transcendency of Σ, whence also of $\mathfrak{B}$, over K. (Note that Σ is the quotient field of $\mathfrak{B}$.) Since $\mathfrak{P}$ is not the zero ideal, its dimension r^* is $< r$. Σ^* is of degree of transcendency r^* over K^*. We say that B is an *r^*-dimensional valuation.* In our present case $r = 2$ and thus we may have valuations of dimension zero or one.

Let us again consider the field $\mathsf{K}(x, y)$, where x and y are algebraically independent elements in Σ and $v(x) \geqq 0$, $v(y) \geqq 0$. Let B_1 be the valuation of $\mathsf{K}(x, y)$ induced by B and let $\mathfrak{B}_1$, $\mathfrak{P}_1$ be the valuation ring and the prime ideal of the valuation B_1 in $\mathsf{K}(x, y)$. Clearly $\mathfrak{B}_1 \subsetneqq \mathfrak{B}$ and $\mathfrak{P}_1$ consists of all those elements of $\mathfrak{P}$ which are in $\mathfrak{B}_1$. Hence the residue field $\Sigma_1^* = \mathfrak{B}_1/\mathfrak{P}_1$ of the valuation B_1 is a subfield of Σ^*. Let ω^* be any element of Σ^*, and let ω be some element of $\mathfrak{B}$ belonging to the residual class of $\mathfrak{B}/\mathfrak{P}$ determined by ω^*. We write the equation (1) of algebraic dependence between x, y, ω, and we consider among the polynomials $A_0(x, y), \cdots, A_\nu(x, y)$ one which has the smallest possible value in B. Let, say, A_h be such a polynomial. We may assume $h \neq \nu$, because were $v(A_\nu) < v(A_i)$ for $i = 0, 1, \nu - 1$, then, we would have $v(A_\nu(x, y)) < v(A_i(x, y)\omega^{\nu-i})$, $i = 0, 1, \cdots, \nu - 1$, since $v(\omega) \geqq 0$, and this is impossible. The quotients A_i/A_h have non-negative values in B, whence they are elements of $\mathfrak{B}_1$. Let $a_0^*, a_1^*, \cdots, a_{h-1}^*, 1, a_{h+1}^*, \cdots, a_\nu^*$ be the corre-

[6] That $\mathfrak{B}/\mathfrak{P}$ is a field follows from the fact that $\mathfrak{P}$ is divisorless.

sponding residual classes mod $\mathfrak{P}_1$. In view of the homorphism between $\mathfrak{B}$ and Σ^* we have the relation:

$$a_0^* \omega^{*\nu} + \cdots + a_{h-1}^* \omega^{*\nu-h+1} + \omega^{*\nu-h} + a_{h+1}^* \omega^{*\nu-h-1} + \cdots + a_\nu^* = 0, \quad a_i^* \subset \Sigma_1^*.$$

Since $\nu - h > 0$, ω^* is algebraic over Σ_1^*. Thus, we conclude that Σ^* *is an algebraic extension of* Σ_1^*, and that consequently *the dimension of the valuation B equals the dimension of the induced valuation* B_1 of $\mathsf{K}(x, y)$.

We shall now review the types of valuations existing in Σ.

3. 1-dimensional valuations (prime divisors of Σ).

We fix an element $X f(x, y)/g(x, y)$ in $\mathfrak{B}_1$ which is algebraically independent on K, modulo $\mathfrak{P}$. In other words, we assume that in the homomorphism between $\mathfrak{B}_1$ and $\Sigma_1^*(= \mathfrak{B}_1/\mathfrak{P}_1)$ the element X is not mapped upon an element of K^* (note that we have assumed K to be algebraically closed). Either x or y will actually occur in $f(x, y)/g(x, y)$. Let it be, say, x. Then x is algebraically dependent on X, y, and $\mathsf{K}(x, y)$ is an algebraic extension of $\mathsf{K}(X, y)$. Hence also our original field Σ is an algebraic extension of $\mathsf{K}(X, y)$. We may therefore replace the field $\mathsf{K}(x, y)$ by the field $\mathsf{K}(X, y)$. We assume that this has already been done, and that therefore x is not congruent to a constant modulo $\mathfrak{P}$.

We consider the ring of polynomials $R = \mathsf{K}[x, y]$ in x and y. The elements of this ring which have positive value in B_1 form a prime ideal $\mathfrak{p}$, since the entire ring $\mathsf{K}[x, y]$ is contained in the valuation ring $\mathfrak{B}_1$ of B_1. The ideal $\mathfrak{p}$ cannot be zero-dimensional, since a congruence of the form $x \equiv c(\mathfrak{p})$, $c \supset \mathsf{K}$, would imply the congruence $x \equiv c(\mathfrak{P}_1)$, in contradiction with our assumption on x. Hence $\mathfrak{p}$ is a 1-dimensional ideal, i.e. $\mathfrak{p}$ is a principal ideal $R \cdot f(x, y)$, where f is an irreducible polynomial. If then $h(x, y)$ is a polynomial, then $v(h(x, y))$ is positive if $h \equiv 0(f)$, and is zero if $h \not\equiv 0(f)$. It follows immediately that: (1) the valuation ring $\mathfrak{B}_1$ of B_1 consists of all rational functions which, written in lowest terms, are of the form $g(x, y)/h(x, y)$, $h \not\equiv 0(f)$; (2) $\mathfrak{P}_1 = \mathfrak{B}_1 \cdot (f)$; (3) two elements g/h, g_1/h in $\mathfrak{B}_1$ have the same value in B_1, if and only if g and g_1 are exactly divisible by the same proof of f. If then we normalize the value group Γ_1 of B_1 by putting $v(f) = 1$, then the value of any element of $\mathsf{K}(x, y)$ is a rational integer. Namely, if g is exactly divisible by f^m and if $h \not\equiv 0(f)$, then $v(g/h) = m$; if $g \not\equiv 0(f)$, and h is exactly divisible by f^m, then $v(g/h) = -m$. We conclude, that *any one-dimensional valuation of a field of algebraic functions of two variables is discrete, of rank* 1.

It is well known that the description of a one-dimensional valuation B of the field Σ can be given in terms of the ideal theory of any finite integrally closed domain $\mathfrak{o}$ in Σ, such that (1) $\mathfrak{o}$ is contained in the valuation ring of B and (2) that the *prime ideal* $\mathfrak{p}$ *of* B *in* $\mathfrak{o}$ (i.e. the set of all elements of $\mathfrak{o}$ having positive value in B) is 1-dimensional. For instance, we can take for $\mathfrak{o}$ the ring of integral functions of the elements x, y considered above. Every primary ideal in $\mathfrak{o}$ belonging to $\mathfrak{p}$ is a primary component of some power of $\mathfrak{p}$. The primary component of $\mathfrak{p}^m$ is denoted by $\mathfrak{p}^{(m)}$ (symbolic m^{th} power of $\mathfrak{p}$). Every element α in Σ can be put in the form of a quotient $\alpha = a/b$, where a and b are elements of $\mathfrak{o}$, *not both divisible by* $\mathfrak{p}$. The given valuation B is obtained by putting $v(a/b) = +m$, if $a \equiv 0(\mathfrak{p}^{(m)})$, $a \not\equiv 0(\mathfrak{p}^{(m+1)})$, $b \not\equiv 0(\mathfrak{p})$, and $v(a/b) = -m$, if $a \not\equiv 0(\mathfrak{p})$, $b \equiv 0(\mathfrak{p}^{(m)})$, $b \not\equiv 0(\mathfrak{p}^{(m+1)})$ (see K, p. 37 and p. 104–105).

4. Zero-dimensional valuations of rank 2

The order preserving homomorphic mapping of Γ upon the quotient group $\Delta_1 = \Gamma/\Gamma_1$ (Γ_1, the isolated subgroup) defines an homomorphic mapping of Σ upon the ordered group Δ_1, of rank 1, i.e. a valuation B_1 of Σ, of rank 1. Let $\mathfrak{B}_1$ be the valuation ring of B_1, $\mathfrak{P}_1$ the prime ideal of B_1 in $\mathfrak{B}_1$, and let Σ_1^* be the residue field $\mathfrak{B}_1/\mathfrak{P}_1$ of B_1. The ring $\mathfrak{B}_1$ consists of all those elements of Σ whose value in B is either in Γ_1 or positive, whence $\mathfrak{B}_1 \supset \mathfrak{B}$. $\mathfrak{P}_1$ consists of those elements of Σ whose value in B is a positive element of Γ, not in Γ_1, whence $\mathfrak{P}_1 \subset \mathfrak{P}$. In particular, the value in B of every element of $\mathfrak{B}_1$ which is not in $\mathfrak{P}_1$, is an element of Γ_1. There is thus defined an homomorphic mapping of $\Sigma_1^* = \mathfrak{B}_1/\mathfrak{P}_1$ upon the ordered group Γ_1 of rank 1, i.e. a valuation B_1^* of the field Σ_1^*, of rank 1. Consequently Σ_1^* must be of degree of transcendency 1, since if it was an algebraic extension of $K^*(\cong K)$, it could not possess non-trivial valuation. We conclude that B_1 is a 1-dimensional valuation of Σ, and hence Δ_1 is discrete, of rank 1. Since Σ_1^* is of degree of transcendency 1, its valuation B_1^* must be discrete, of rank 1. Hence Γ *is a discrete group*. The given valuation B is thus composed of a 1-dimensional valuation B_1 of Σ (a *divisor* of Σ), followed up by a valuation B_1^* of the residue field Σ_1^* of B_1. This field Σ_1^* is a field of algebraic functions of one variable and B_1^* is given by a "place", or a "prime divisor" of Σ_1^*. *The given valuation B is thus defined by a branch of an algebraic curve lying on a normal model of our field.*

The actual association of values in Γ to elements in Σ is obtained as follows. We fix again an integrally closed finite domain of integrity $\mathfrak{o}$ in Σ, such that the prime ideal $\mathfrak{p}$ of B_1 in $\mathfrak{o}$ be 1-dimensional. In $\mathfrak{o}$ we fix an element ω, such that $\omega \equiv 0(\mathfrak{p})$, $\omega \not\equiv 0(\mathfrak{p}^{(2)})$. Given any element ξ in Σ, we can put it in the form $\xi = \omega^m \cdot \alpha/\beta$, where $\alpha, \beta \subset \mathfrak{o}$ and $\alpha \not\equiv 0(\mathfrak{p})$, $\beta \not\equiv 0(\mathfrak{p})$. Let $(\alpha^*/\beta^*) \neq 0, \infty$ be the element in the residue field Σ_1^* which corresponds to α/β. Let $v_{B_1^*}(\alpha^*/\beta^*) = n$. Then we put $v(\xi) = (m, n)$. The value group Γ is the group of all pairs of integers (m, n) ordered lexicographically: $(m, n) < (m_1, n_1)$, if $m_1 > m$ or if $m_1 = m$, $n_1 > n$. The isolated group Γ_1 consists of the elements $(0, n)$, n-arbitrary. Note that the integer n depends on the choice of the element ω (except when $m = 0$), but that m is independent on ω. If we choose another element ω' instead of ω, and if in the original construction the value of ω' is $(1, \nu)$, ($\nu \gtreqless 0$), then using ω' instead of ω, we will get for the value of the same element ξ the pair of integers (m', n'), where $m = m'$, $n = \nu m' + n'$. This unimodular transformation defines an order preserving automorphism of Γ.

5. Discrete zero-dimensional valuations of rank 1

The value group Γ is the group of all rational integers. Consider an element ξ of Σ such that $v(\xi) = +1$. If η is any element of Σ and if $v(\eta) = \nu$, then $v(\eta/\xi^\nu) = 0$. Since the residue field Σ^* is simply isomorphic to the field of constants, it follows that there exists a uniquely determined element c in K such that $v(\eta/\xi^\nu - c) > 0$, whence

$$\eta = c\xi^\nu + \eta_1, \quad \text{where} \quad v(\eta_1) = \nu_1 > \nu.$$

Applying the same procedure to η_1 and repeating indefinitely this construction, we get a uniquely determined power series expansion for η:

$$\eta \to c\xi^\nu + c_1\xi^{\nu_1} + \cdots, \qquad \nu < \nu_1 < \cdots. \tag{2}$$

This defines an *isomorphic* mapping of Σ upon a subfield of the field $\mathsf{K}(\{\xi\})$ of integral power series of ξ with coefficients in K. Consider a projective model V_2 of the surface, in some S_n $(x_1, \cdots x_n)$, and let $\omega_1, \cdots, \omega_n$ be the coordinates of the general point of V_2; $\omega_1, \cdots, \omega_n \subset \Sigma$. We assume that the ω's are in the valuation ring of B. Let

$$\omega_i \to c_{i0} + c_{i1}\xi + c_{i2}\xi^2 + \cdots = \omega_i(\xi).$$

These equations represent an analytical arc γ on V_2, $x_i = c_{i0} + c_{i1}\xi + \cdots$, whose origin is the point $(c_{10}, \cdots, c_{n0})$. *This arc is not algebraic.* In fact, if $F(\omega_1, \cdots, \omega_n)$ is a polynomial which does not vanish in Σ, i.e., if $F(x_1, \cdots, x_n)$ does not vanish on V_2, then $F(\omega_1, \cdots, \omega_n)$ has a definite value μ in B, and ξ^μ must be the leading term in the power series $F(\omega_1(\xi), \cdots, \omega_n(\xi))$. In other words, $F(\omega_1(\xi), \cdots, \omega_n(\xi)) \neq 0$ in $\mathsf{K}(\{\xi\})$, if $F(\omega_1, \cdots, \omega_n) \neq 0$ in Σ (*isomorphism of the mapping* (2)). Thus γ does not lie on any algebraic variety of S_n which does not pass through the surface V_2. This proves that γ is not an arc of an algebraic curve.

6. Non-discrete zero-dimensional valuations of rank 1

Here one is led to distinguish between two cases: (1) the value group Γ, which may be assumed to be a subgroup of the set of all real numbers, contains at least two incommensurable numbers; (2) any two numbers in Γ are commensurable, whence the elements of Γ may be assumed to be rational numbers

First Case. Let x and y be two elements whose values $v(x)$ and $v(y)$ are positive incommensurable numbers. We may normalize Γ in such a manner that $v(x)$ is equal to 1. Then $v(y) = \tau$, a positive irrational number. Consider any polynomial in x and y, say $F(x, y) = \sum a_{ij}x^iy^j$. We have $v(x^iy^j) = i + j\tau$, and, moreover, two distinct terms $a_{ij}x^iy^j$, $a_{i'j'}x^{i'}y^{j'}$ have distinct values, since $i + j\tau = i' + j'\tau$ implies $i = i'$, $j = j'$. If then $a_{\alpha\beta}x^\alpha y^\beta$ is *the* term in F of smallest possible value, then $v(F) = \alpha + \beta\tau$. This implies in the first place that *x and y are algebraically independent*, whence Σ is an algebraic extension of the field $\Sigma_1 = \mathsf{K}(x, y)$. In the second place, the value group Γ_1 of the valuation B_1 induced by B in Σ_1 *is the group of all real numbers of the form* $\alpha + \beta\tau$, α and β—arbitrary integers.

We know that a fixed multiple of any element γ in Γ is in Γ_1, i.e., we have $\nu\gamma = \alpha + \beta\tau$, ν — *a* fixed integer. By well known theorem on abelian groups with a finite set of generators, it follows that Γ is generated by two elements, say $\delta_1 = r_1 + s_1\tau$, $\delta_2 = r_2 + s_2\tau$, where r_1, s_1, r_2, s_2 are rational numbers. Every element in Γ is of the form $\alpha\delta_1 + \beta\delta_2$, α, β—integers. We may divide each element in Γ by δ_1 and we may thus assert that also Γ *is generated by the two elements* 1, $\tau^* = \delta_2/\delta_1$, *where* τ^* *is clearly an irrational number.* Thus Γ is of the same type as Γ_1.

The induced valuation B_1 in $\mathsf{K}(x, y)$ can be obtained by putting formally $y = x^\tau$. Then for any polynomial $f(x, y) = \sum a_{ij}x^iy^j$ we find: $v(f) = \min\{i + j\tau\}$. The equation $y = x^\tau$ may be regarded as representing a *transcendental branch* in the (x, y)-plane.[7] If we take as a projective model of Σ a surface F in an (x, y, z)-space, then the valuation B is represented on F by one of the transcendental branches whose projection onto the (x, y)-plane is the branch $y = x^\tau$.

Second Case. The elements of Γ are in this case rational numbers. The denominators must be arbitrarily high, since Γ is not discrete. We consider the prime numbers which occur in the denominators of the various elements m/n of Γ, where we assume that each quotient is written in its lowest terms. We distribute these prime numbers into two classes: in the first class, say $\{p\}$, we put those prime numbers $p_1, p_2, \cdots$ which occur in the denominators n only to a bounded power, say p_i occurs to the power ν_i but not to the power $\nu_i + 1$; in the second class, which we shall denote by $\{q\}$, we put the prime numbers $q_1, q_2, \cdots$ which occur in the denominators n to an arbitrarily high power. Any one of the two sets $\{p\}$ $\{q\}$ may contain a finite or infinite set of elements, or may be vacuous. However, if $\{q\}$ is vacuous, then necessarily the set $\{p\}$ must contain an infinity of prime numbers. We may assume that Γ contains the number 1.

We assert then that Γ *consists of all the rational numbers whose denominators are of the form* $p_1^{m_1}p_2^{m_2}\cdots q_1^{n_1}q_2^{n_2}\cdots$, $m_1 \leqq \nu_1$, $m_2 \leqq \nu_2, \cdots, n_1, n_2, \cdots$—*arbitrary integers*. This assertion follows from the following two remarks.

First: if Γ contains a rational number $a = \alpha/\beta$, $(\alpha, \beta) = 1$, then Γ contains

[7] A suggestive geometric interpretation can be given by using the theory of infinitely near points. Let f_i/g_i, $i = 1, 2, \cdots$, be the convergent fractions of the infinite continued fraction for τ. On the *algebraic branch* $y = x^{f_i/g_i}$ we find a set M_i of infinitely near points $A_1, A_2, \cdots, A_{\sigma_i}$, consisting of a set of free points, followed by a set of satellites (see Enriques-Chisin, *Teoria geometrica delle equazioni e delle curve algebriche*, Vol. II). Moreover, it is well known that the set M_i is a subset of M_{i+1}. The branch $y = x^\tau$ may be associated with the infinite sequence $\{A_1, A_2, \cdots, A_m, \cdots\} = \lim M_i$, of infinitely near points.

$1/\beta$ (and hence all its multiples). In fact, let $m\alpha + n\beta = 1$. Then $1/\beta = ma + n$.

Second: if Γ contains $a = \alpha/\beta$ and $a_1 = \alpha_1/\beta_1$, $(\alpha, \beta) = 1$, $(\alpha_1, \beta_1) = 1$, and if $(\beta, \beta_1) = 1$, then Γ contains $1/\beta\beta_1$. In fact, $a + a_1 = (\alpha\beta_1 + \alpha_1\beta)/\beta\beta_1$. Clearly $\alpha\beta_1 + \alpha_1\beta$ and $\beta\beta_1$ are relatively prime, and our assertion follows from the first remark.

For the actual construction of valuations of this type we proceed as follows. We consider two algebraically independent elements x and y in Σ, contained in the prime ideal of the valuation. We may assign to x the value 1. We pass to the complete field Ω relative to the given valuation B, and containing the field Σ. In Ω we have an extended valuation with the same value group Γ as B and with the same residue field K. Moreover, every convergent sequence of elements in Ω has a limit in Ω. It is well known that Ω is simply isomorphic to a subfield of the field $\mathsf{K}(\{x\})_\Gamma$ of all well ordered power series in x with coefficients in K and exponents in Γ. Hence we have for y an expansion of the form:

$$(3)\quad y = c_1 x^{m_1/n_1} + c_2 x^{m_2/n_2} + c_3 x^{m_3/n_3} + \cdots, \qquad \frac{m_1}{n_1} < \frac{m_2}{n_2} < \frac{m_3}{n_3} < \cdots,$$

each m_i/n_i being a rational number in its lowest terms. We have exhibited in (3) only the simple power series (of type ω) with which the well ordered series begins. It is not difficult to see that by substituting the above simple power series for y into any given rational function η of x and y we get a power series in x such that the value of η is the exponent of the leading term of the series.[8] By a simple Puiseux argument (see Schilling,[7] p. 702) it is not difficult to show that any of the denominators n_i in the exponents actually occurs as a denominator in some element of the value group. Since we may assign the expansion (3) arbitrarily, it follows that any subgroup of the group of rationals is the value group of some valuation of Σ.

II. Uniformization of Valuations. The Fundamental Lemma

7.

Let B be a zero-dimensional valuation of our field Σ. Given any set of elements $\xi_1, \cdots, \xi_n$ which generate the field Σ, there is defined a surface F in S_n, whose general point has coördinates $\xi_1, \cdots, \xi_n$ and which is a projective model of Σ. If the ξ's are in the valuation ring of B, and if, say, $\xi_i - c_i$, $c_i \subset \mathsf{K}$, has positive value in B, then the prime ideal of B in the ring $\mathsf{K}[\xi_1, \cdots, \xi_n]$ is the zero-dimensional ideal $(\xi_1 - c_1, \cdots, \xi_n - c_n)$. The valuation B is then defined by a branch on the surface F, algebraic, analytical, or transcendental, through the point $A(c_1, \cdots, c_n)$. We shall call this point *the center of the valuation* on F.

[8] The substitution may actually lead to a well ordered series of type $> \omega$. However, we associate with η the simple power series with which the well ordered series begins.

FUNDAMENTAL LEMMA. *Given any zero-dimensional valuation B of the field Σ, there exists a projective model F of Σ, on which the center of the valuation B is at a simple point of the surface.*

In the proof of this theorem we have to consider separately the various types of valuations as classified in the preceding sections.

A. VALUATIONS OF RANK 2 (ALGEBRAIC BRANCHES)

Let B_1 be the one-dimensional valuation with which B is composed. We choose two independent elements x and y in Σ such that the prime ideal of B_1 in the ring $\mathfrak{o}$ of integral functions of x, y be one-dimensional (B_1—a divisor of the first kind with respect to the ring $\mathfrak{o}$). Let $\mathfrak{p}_1$ be this prime ideal.

Let x_1, y_1 be any other two elements in $\mathfrak{o}$, such that $\mathfrak{o}$ is the integral closure of $K[x_1, y_1]$, and let z be a primitive element of Σ with respect to the field $K(x_1, y_1)$. Let $f(x_1, y_1, z) = 0$ be the irreducible equation for z, over $K(x_1, y_1)$. We have shown in Z_2 that there always exist elements x_1, y_1, z such that $f'_z(x_1, y_1, z) \not\equiv 0(\mathfrak{p}_1)$.[9] We shall assume that the original elements x, y are the x_1 and y_1 respectively. We have then

$$f(x, y, z) = 0 \tag{4}$$

$$f'_z(x, y, z) \not\equiv 0(\mathfrak{p}_1). \tag{4'}$$

The relation (4′) implies that the *value of the element $f'_z(x, y, z)$ in our valuation B is of the form* $(0, n)$. This remark is of decisive importance.

We consider the surface F defined by (4) in $S_3(x, y, z)$. It is a projective model of our field Σ. On F our valuation B gives rise to an algebraic branch γ. We may assume that the origin of this branch is the point O: $x = y = z = 0$. Let us suppose that O is a multiple point, say an r-fold point, $r > 1$. Then $f(x, y, z)$ is of the form

$$f = f_r + f_{r+1} + \cdots + f_m,$$

where $f_i = f_i(x, y, z)$ is a form of degree i. The elements x, y, z have positive values in B. We shall have to arrange x, y, z in an order corresponding to the order of magnitude of their values. To avoid the necessity of considering various cases, we present the relation (4′) in such a form that will make the element z lose its privileged rôle. Namely, let us consider the three partial derivatives f'_x, f'_y, f'_z, and let us write that

$$\min (v(f'_x), v(f'_y), v(f'_z)) = (0, n),$$

without specifying which of the three derivatives has the minimum value $(0,n)$. The point O is a multiple point of F if and only if $n > 0$. Now it is clear that we may assume, without loss of generality:

$$0 < v(x) \leqq v(y) \leqq v(z).$$

[9] This assertion is equivalent with theorem 11 (section 16) of the quoted paper, to the effect that the normal surfaces defined by the integrally closed ring $\mathfrak{o}$ do not possess *multiple curves*. See also theorem 4 (section 5), loc. cit.

Since B is zero-dimensional, there exist uniquely determined constants c, d such that $v(y - cx) > v(x)$, $v(z - cx) > v(x)$. We apply to F the quadratic transformation

$$(5) \qquad x_1 = x, \qquad y_1 = \frac{y - cx}{x}, \qquad z_1 = \frac{z - dx}{x},$$

whose inverse is given by the formulas:

$$x = x_1, \qquad y = x_1(y_1 + c), \qquad z = x_1(z_1 + d).$$

Substituting into the equation $f(x, y, z) = 0$ we get

$$f(x, y, z) = x_1^r f_1(x_1, y_1, z_1),$$

where

$$f_1(x_1, y_1, z_1) = f_r(1, y_1, z_1) + x_1 f_{r+1}(1, y_1, z_1) + \cdots + x_1^{m-r} f_m(1, y_1, z).$$

Denote by F_1 the transformed surface

$$f_1(x_1, y_1, z_1) = 0.$$

By (5), the values of x_1, y_1, z_1 in B are positive. Hence B is given on F_1 by a branch through the origin $x_1 = y_1 = z_1 = 0$. We find immediately the following relations:

$$f'_x = x_1^{r-1}\{x_1 f'_{1x_1} - (y_1 + c)f'_{1y_1} - (z_1 + d)f'_{1z_1}\}$$
$$f'_y = x_1^{r-1} \cdot f'_{1y_1}$$
$$f'_z = x^{r-1} f'_{1z_1}.$$

Hence

$$\min(v(f'_x), v(f'_y), v(f'_z)) \geqq (r - 1)v(x_1) + \min(v(f'_{1x_1}), v(f'_{1y_1}), v(f'_{1z_1})).$$

By assumption O is a multiple point of F, i.e. $r \geqq 2$. Since $v(x_1) > 0$, we conclude that

$$\min(v(f'_{1x_1}), v(f'_{1y_1}), v(f'_{1z_1})) = (0, n_1) < (0, n),$$

whence $0 < n_1 < n$.

If the origin $x_1 = y_1 = z_1 = 0$, the center of the valuation B on F_1, is still a multiple point, we apply to this surface a new quadratic transformation, getting a new surface

$$f_2(x_2, y_2, z_2) = 0,$$

on which the valuation B is represented by a branch through the point $x_2 = y_2 = z_2 = 0$, and we will also have

$$\min(v(f'_{2x_2}), v(f'_{2y_2}), v(f'_{2z_2})) = (0, n_2), \qquad n > n_1 > n_2 > 0.$$

It follows that after a finite number of steps we will get a surface

$$f_i(x_i, y_i, z_i) = 0,$$

such that at least one of the 3 partial derivatives f_{ix_i}, f_{iy_i}, f_{iz_i} has *value zero*, (i.e. (0, 0)), in the given valuation B. That implies, that on the surface $f_i = 0$, the center of our valuation B is a point at which not all the partial derivatives of f_i vanish. This point is therefore a simple point of the surface $f_i = 0$.

B. Discrete Valuations of Rank 1 (Analytical Branches)

The proof in this case is identical with the proof in the preceding case, and even simpler, since now, because of the archimedean character of the value group, the value of any element in Σ, different from zero, is an integer. Hence in the present case it is permissible to begin with an *arbitrary* projective model $f(x, y, z) = 0$ of Σ, whereas in the preceding case we had to be careful with our choice of this model, because we wished to satisfy the minimum value requirement for the partial derivatives (i.e. that this minimum be a pair of integers (m, n) in which $m = 0$).

C. Non-Discrete Valuations of Rank 1 (Transcendental Branches)

8. First case. The value group consists of all elements of the form $m + n\tau$, m, n—integers, τ—an irrational number

We choose two elements x and y in Σ such that $v(x) = 1$, $v(y) = \tau$. Then x and y are algebraically independent. Let z be an element whose value is positive and which is a primitive element of Σ with respect to the field $\mathsf{K}(x, y)$, and let

$$f(x, y, z) = 0$$

be the corresponding projective model F of Σ. Our valuation is then represented on the surface F by a branch through the origin. Our proof will be essentially a suitably modified Puiseux argument.

Since $f(x, y, z)$ is the element zero of Σ, there must exist two or more terms in the polynomial $f(x, y, z)$ which have equal and minimum value. If two distinct terms $x^i y^j z^\nu$, $x^{i_1} y^{j_1} z^{\nu_1}$ have equal value, then necessarily $\nu \neq \nu_1$. In fact, if $\nu = \nu_1$, then $v(x^i y^j) = v(x^{i_1} y^{j_1})$, i.e. $i + j\tau = i_1 + j_1\tau$, whence $i = i_1$, $j = j_1$. We arrange then the minimum value terms according to increasing powers of z and we write:

$$f(x, y, z) = a_\alpha x^{m_\alpha} y^{n_\alpha} z^\alpha + a_\beta x^{m_\beta} y^{n_\beta} z^\beta + \cdots + a_\delta x^{m_\delta} y^{n_\delta} z^\delta + \Sigma' a_i x^{m_i} y^{n_i} z^{l_i},$$

$$\alpha < \beta < \cdots < \delta,$$

where

$$M + N\tau = v(x^{m_\alpha} y^{n_\alpha} z^\alpha) = \cdots = v(x^{m_\beta} y^{n_\beta} z^\beta) = \cdots = v(a_\delta x^{m_\delta} y^{n_\delta} z^\delta),$$

and where the summation Σ' is extended to all remaining terms whose value is $> M + N\tau$. Here

$$M = m_\alpha + s\alpha = m_\beta + s\beta = \cdots = m_\delta + s\delta,$$

$$N = n_\alpha + \sigma\alpha = n_\beta + \sigma\beta = \cdots = n_\delta + \sigma\delta,$$

and

$$v(z) = s + \sigma\tau, \qquad s, \sigma\text{—integers} \gtreqless 0.$$

Moreover,

$$(m_i + l_i s) + (n_i + l_i\sigma)\tau > M + N\tau, \tag{6}$$

for any term in the sum Σ'.

We may assume that $f(0, 0, z)$ is not identically zero in z (i.e. that the z-axis does not lie on the surface F). Let $z = 0$ be an r-fold root of $f(0, 0, z)$. If $r = 1$, the origin—center of the valuation—is a simple point of F. Assume $r > 1$. Then $f(x, y, z)$ contains the terms az^r, $a \subset K$. Since we must have $v(x^{m_\delta}y^{n_\delta}z^\delta) \leqq v(z^r)$, it follows that

$$\delta \leqq r. \tag{7}$$

We expand τ into a continued fraction:

$$\tau = h_1 + \cfrac{1}{h_2 + \cfrac{1}{h_3 + \cdots}}.$$

Let f_j/g_j be the convergent fractions of τ. Since $\lim f_j/g_j = \tau$, we can find an integer p, sufficiently high, such that the inequalities (6) remain true if τ is replaced by f_{p-1}/g_{p-1} and also by f_p/g_p. We will have then

$$(m_i + l_i s) + (n_i + l_i\sigma)\cdot f_j/g_j > M + N\cdot f_j/g_j, \quad j = p - 1, p, \tag{6'}$$

for any term $a_i x^{m_i}y^{n_i}z^{l_i}$ which is not a minimum value term.

We now pass from the elements x, y, z of Σ to new elements x_1, y_1, z_1 by the following Cremona transformation:

$$x = x_1^{g_p}y_1^{g_{p-1}}, \qquad y = x_1^{f_p}y_1^{f_{p-1}}, \qquad z = x^s y^\sigma(z_1 + c). \tag{8}$$

Here c is a constant determined as follows: since $v(z) = s + \sigma\tau = v(x^s y^\sigma)$, it follows that $v(z/x^s y^\sigma) = 0$. There exists then a uniquely determined constant c such

$$v(z_1) = v\left(\frac{z}{x^s y^\sigma} - c\right) > 0, \qquad c \neq 0. \tag{9}$$

By known properties of convergent fractions we have

$$f_{p-1}g_p - f_p g_{p-1} = \epsilon = \pm 1.$$

We also know that ϵ, $-\tau + f_{p-1}/g_{p-1}$ and $\tau - f_p/g_p$ have like signs. From (8) we find

$$x_1^\epsilon = x^{f_{p-1}}/y^{g_{p-1}}, \qquad y_1^\epsilon = y^{g_p}/x^{f_p},$$

whence

$$\epsilon v(x_1) = g_{p-1}\left(\frac{f_{p-1}}{g_{p-1}} - \tau\right), \qquad \epsilon v(y_1) = g_p\left(\tau - \frac{f_p}{g_p}\right).$$

Hence

(9′) $v(x_1) > 0, \quad v(y_1) > 0, \quad \text{and also} \quad v(z_1) > 0$, by (9).

By our transformation (8) every minimum value term in the equation $f(x, y, z) = 0$ acquires *the same factor*

$$x_1^{Mg_p+Nf_p} y_1^{Mg_{p-1}+Nf_{p-1}},$$

while, in view of (6′), every other term acquires a factor $x_1^{\lambda_i} y_1^{\mu_i}$, where $\lambda_i > Mg_p + Nf_p$, $\mu_i > Mg_{p-1} + Nf_{p-1}$. Hence

$$f(x, y, z) = x_1^{Mg_p+Nf_p} y_1^{Mg_{p-1}+Nf_{p-1}} \cdot f_1(x_1, y_1, z_1),$$

where

$$f_1(x_1, y_1, z_1) = 0$$

is the equation of the transformed surface F_1, and f_1 has the form

$$f_1(x_1, y_1, z_1) = (z_1 + c)^\alpha \{a_\alpha + a_\beta(z_1 + c)^{\beta-\alpha} + \cdots a_\delta(z_1 + c)^{\delta-\alpha}\} + x_1 y_1 H(x_1, y_1, z_1).$$

Since the center of our valuation B on the new surface F_1 is at the origin $x_1 = y_1 = z_1 = 0$ (in view of (9′)), we must have $f(0, 0, 0) = 0$, whence $u = c$ must be a root of the polynomial $\phi(u) = a_\alpha + a_\beta u^{\beta-\alpha} + \cdots + a_\delta u^{\delta-\alpha}$. Let c be an r_1-fold of this polynomial. Then $z_1 = 0$ is an r_1-fold root of $f(0, 0, z_1)$. We have

$$r_1 \leqq \delta - \alpha \leqq r, \text{ by (7).}$$

If $r_1 < r$, we have achieved a reduction, since for $r = 1$ we get a simple point. Assume that $r_1 = r$. This implies that $\alpha = 0$, $\delta = r$ and that $\phi(u) = a_0 + a_\beta u^\beta + \cdots + a_r u^r = a_r(u - c)^r$. Hence the minimum value terms of $f(x, y, z)$ have the form

$$a_0 x^{m_0} y^{n_0} + a_1 x^{m_1} y^{n_1} z + \cdots + a_{r-1} x^{m_{r-1}} y^{n_{r-1}} z^{r-1} + a_r z^r,$$

where $a_{r-1} \neq 0$, $a_r \neq 0$. Since $v(z^r) = v(x^{m_{r-1}} y^{n_{r-1}} z^{r-1})$, it follows that

$$v(z) = m_{r-1} + n_{r-1} \cdot \tau.$$

We express this conclusion as follows: *if our transformation from $f(x, y, z) = 0$ to $f_1(x_1, y_1, z_1) = 0$ does not reduce the value of r, then the value of z in B is of the form $m + n\tau$, where m and n are non-negative integers.*

Let then $v(z) = m + n\tau > 0$, $m, n \geqq 0$. There exist then a uniquely determined constant c such that $v(z - cx^m y^n) > v(z)$. Let

$$z = cx^m y^n + z_1, \qquad v(z_1) > v(z),$$

$$f(x, y, z) = f(x, y, cx^m y^n + z_1) = f_1(x, y, z_1) = 0.$$

Clearly $z_1 = 0$ is also an r-fold root of $f_1(0, 0, z_1)$. We can apply a transformation such as (8) to the new surface $f_1 = 0$. If there is again no reduction as far as r is concerned, then also the value of z_1 in B must be of the form

$$v(z_1) = m_1 + n_1\tau,\ m_1,\ n_1 \geqq 0.$$

We then put

$$z_1 = c_1 x^{m_1} y^{n_1} + z_2,$$

where the constant c_1 is determined in such a fashion that $v(z_2) > v(z_1)$. We get a new surface F_2 :

$$f_2(x, y, z_2) = f_1(x, y, c_1 x^{m_1} y^{n_1} + z_2) = 0.$$

Suppose that we get in this fashion a sequence of birational transforms of our surface $f(x, y, z) = 0$:

$$f_i(x, y, z_i) = 0, \qquad i = 0, 1, \cdots, f_0 = f, \qquad z_0 = z,$$

such that $z_i = 0$ is always also an r-fold root of $f_i(0, 0, z_i)$ and such that our Cremona transformation, of type (8), if applied to the surface $f_i = 0$ is not capable of reducing the value of r. Here

$$z_i = c_i x^{m_i} y^{n_i} + z_{i+1},$$

$$v(z_{i+1}) = m_{i+1} + n_{i+1}\tau > m_i + n_i\tau = v(z_i) > 0,$$

and m_i, n_i are non-negative integers.

Now

$$f_{i+1}(x, y, z_{i+1}) = f_i(x, y, c_i x^{m_i} y^{n_i} + z_{i+1}),$$

whence

$$\frac{\partial f_{i+1}}{\partial z_{i+1}} = \frac{\partial f_i}{\partial z_i}. \tag{10}$$

Since in each polynomial f_i the term in z_i^r must be among the minimum value terms, it follows that

$$v\left(\frac{\partial f_i}{\partial z_i}\right) \geqq (r - 1)v(z_i),$$

or, since $r > 1$,

$$v\left(\frac{\partial f_i}{\partial z_i}\right) \geqq v(z_i).$$

In view of (10) it follows that

$$v\left(\frac{\partial f}{\partial z}\right) \geqq m_i + n_i\tau. \tag{11}$$

Now if our sequence of surfaces $f_i = 0$ was infinite, we would have an infinite ascending sequence of real numbers

$$0 < m + n\tau < m_1 + n_1\tau < \cdots < m_i + n_i\tau < \cdots,$$

whose limit is $+\infty$, *since the integers* m_i, n_i *are non-negative.* This is in contradiction with the inequality (11), since $\partial f/\partial z \neq 0$ in Σ (f being irreducible). Thus, after a finite number of steps we must get a surface $f_i(x, y, z_i) = 0$ such that a Cremona transformation of type (8), applied to the surface, diminishes the value of r. Ultimately we get a surface for which $r = 1$, and on that surface the center of our valuation is at a simple point.

9. Second case. The value group consists of rational numbers

We choose arbitrarily two algebraically independent elements x and y in Σ. Let

$$\frac{v(y)}{v(x)} = \frac{\mu}{\nu}, \qquad (\mu, \nu) = 1$$

and let μ', ν' be positive integers such that

$$\mu'\nu - \mu\nu' = +1.$$

We define a Cremona plane transformation C_1 as follows:

$$C_1\colon x = x_1^{\nu}(y_1 + c_1)^{\nu'}, \qquad y = x_1^{\mu}(y_1 + c_1)^{\mu'},$$

where c_1 is a constant still to be determined.

For C_1^{-1} we find the equations:

$$C_1^{-1}\colon x_1 = x^{\mu'}/y^{\nu'}, \qquad (y_1 + c_1) = y^{\nu}/x^{\mu}.$$

From these equations it follows in the first place that

$$v(x_1) = \mu'\cdot v(x) - \nu' v(y) = \frac{v(x)}{\nu}(\mu'\nu - \nu'\mu) = \frac{v(x)}{\nu} > 0.$$

Moreover, since $v(y^{\nu}/x^{\mu}) = 0$, also $v(y_1 + c_1) = 0$. We choose the constant c_1 in such a fashion that $v(y_1) > 0$. We have now two new elements x_1, y_1 whose values are positive.

In a similar fashion we define a plane Cremona transformation C_2 from the (x_1, y_1) plane to a (x_2, y_2) plane, where again $v(x_2) > 0$, $v(y_2) > 0$:

$$C_2\colon x_1 = x_2^{\nu_1}(y_2 + c_2)^{\nu_1'}, \qquad y_1 = x_2^{\mu_1}(y_2 + c_2)^{\mu_1'},$$

where

$$\frac{v(y_1)}{v(x_1)} = \frac{\mu_1}{\nu_1}, \qquad (\mu_1, \nu_1) = 1 \text{ and } \mu_1'\nu_1 - \mu_1\nu_1' = +1.$$

We get a sequence of Cremona transformations:

$$T_1 = C_1, \qquad T_2 = C_1C_2, \cdots, T_i = C_1C_2\cdots C_i, \cdots,$$

where T_i is a transformation leading to elements x_i, y_i. The elements x, y are polynomials in x_i, y_i.

We prove the following

LEMMA. *Given a polynomial $\phi(x, y)$, then for all sufficiently high values of i the expression of $\phi(x, y)$ as a polynomial in x_i, y_i is of the form*

$$\phi(x, y) = x_i^{\lambda_i}\phi^{(i)}(x_i, y_i), \qquad \lambda_i \geqq 0,$$

where $\phi^{(i)}(0, 0) \neq 0$.

The proof of this lemma is essentially the classical Puiseux argument. We put into evidence the minimum value terms in $\phi(x, y)$:

$$\phi(x, y) = a_\alpha x^{m_\alpha} y^\alpha + a_\beta x^{m_\beta} y^\beta + a_\delta x^{m_\delta} y^\delta + \sum{}' a_i x^{m_i} y^{l_i}, \quad \alpha < \beta < \cdots < \delta,$$

where

$$m_\alpha \nu + \alpha\mu = m_\beta \nu + \beta\mu = \cdots = m_\delta \nu + \delta\mu = \rho$$

and

$$m_i \nu + l_i \mu > \rho$$

for all remaining terms.

Applying the transformation C_1 we find

$$\phi(x, y) = x_1^\rho \phi_1(x_1, y_1),$$

where

$$\phi_1(x_1, y_1) = a_\alpha(y_1 + c_1)^{m_\alpha\nu' + \alpha\mu'} + a_\beta(y_1 + c_1)^{m_\beta\nu' + \beta\mu'} + \cdots$$
$$+ a_\delta(y_1 + c_1)^{m_\delta\nu' + \delta\mu'} + x_1 H(x_1, y_1),$$

or

$$\phi_1(x_1, y_1) = (y_1 + c_1)^{m_\alpha\nu' + \alpha\mu'}\{a_\alpha + a_\beta(y_1 + c_1)^{\frac{\beta-\alpha}{\nu}} + \cdots + a_\delta(y_1 + c_1)^{\frac{\delta-\alpha}{\nu}}\}$$
$$+ x_1 H(x_1, y_1).$$

If $u = c_1$ is not a root of the polynomial

$$g(u) = a_\alpha + a_\beta u^{\frac{\beta-\alpha}{\nu}} + \cdots + a_\delta u^{\frac{\delta-\alpha}{\nu}},$$

then $\phi_1(0, 0) \neq 0$, and our lemma is proved.

If $g(c_1) = 0$, then our transformation C_1 has the same effect as the classical Puiseux substitution $x = x_1^\nu$, $y = x_1^\mu(c_1 + y_1)$, used in the determination of the branches of the curve $\phi(x, y) = 0$ passing through the origin $x = y = 0$. The only difference—and advantage—is that our transformation *does not* lead to elements x_1, y_1 outside the field $\mathsf{K}(x, y)$. The remainder of the proof is now clear. It is sufficient to observe that if our lemma was not true for the given polynomial $\phi(x, y)$, then the above reasoning shows that the valuation B would be the one which is determined by a branch of the curve $\phi(x, y) = 0$ through the origin $x = y = 0$. This is impossible, since B corresponds, by hypothesis, to a transcendental branch.

Let now z be a primitive element of Σ over $\mathsf{K}(x, y)$ and let $v(z) > 0$. Let F be the surface given the irreducible equation

$$f(x, y, z) = 0.$$

We write f as a polynomial in z and we put into evidence the terms of minimum value:

$$f(x, y, z) = a_\alpha(x, y)z^\alpha + a_\beta(x, y)z^\beta + \cdots + a_\delta(x, y)z^\delta + \sum{}' a_i(x, y)z^i,$$

where

$$v(a_\alpha z^\alpha) = v(a_\beta z^\beta) = \cdots = v(a_\delta z^\delta) \tag{12}$$

and

$$v(a_i z^i) > v(a_\alpha z^\alpha) \text{ for all remaining terms.}$$

The center of our valuation B on the surface F is the origin $x = y = z = 0$. Let, as in the preceding section, $z = 0$ be an r-fold root of $f(0, 0, z)$. If $r = 1$, the origin is a simple point. Suppose $r > 1$.

We apply first our Cremona transformation T_i to x and y and we choose i sufficiently high, so that the above lemma be applicable to all the polynomials $a_\alpha, a_\beta, \cdots, a_\delta, a_i$ simultaneously (i.e., to their product). Let us denote the new elements x_i, y_i by $\bar{x}, \bar{y}$. Then $f(x, y, z)$ takes the form:

$$f(x, y, z) = \bar{f}(\bar{x}, \bar{y}, \bar{z}) = \bar{x}^{m_\alpha}\bar{a}_\alpha(\bar{x}, \bar{y})z^\alpha + \bar{x}^{m_\beta}\bar{a}_\beta(\bar{x}, \bar{y})z^\beta + \cdots$$
$$+ \bar{x}^{m_\delta}\bar{a}_\delta(\bar{x}, \bar{y})z^\delta + \sum{}' \bar{x}^{m_i}\bar{a}_i(\bar{x}, \bar{y})z^i,$$

where $\bar{a}_\alpha(0, 0), \bar{a}_\beta(0, 0), \bar{a}_\delta(0, 0), \bar{a}_i(0, 0)$ are all $\neq 0$.

Let now $v(z)/v(\bar{x}) = s/\sigma$, $(s, \sigma) = 1$, and let s', σ' be positive integers such that

$$s'\sigma - s\sigma' = +1.$$

We apply to the surface $\bar{f}(\bar{x}, \bar{y}, z) = 0$ the quadratic transformation:

$$\bar{x} = x_1^\sigma(z_1 + c_1)^{\sigma'}, \qquad z = x_1^s(z_1 + c_1)^{s'}, \qquad \bar{y} = y_1,$$

where the constant c_1 is $\neq 0$ and is determined by the condition: $v(z_1) > 0$. Let

$$f_1(x_1, y_1, z_1) = 0$$

be the equation of the transformed surface. We find immediately, in view of (12),

$$f(x, y, z) = \bar{x}^{m_\alpha\sigma + \alpha s}f_1(x_1, y_1, z_1),$$

and

$$f_1(x_1, y_1, z_1) = (z_1 + c_1)^{m_\alpha\sigma' + \alpha s'}\{\bar{a}_\alpha(\bar{x}, \bar{y}) + \bar{a}_\beta(\bar{x}, \bar{y})(z_1 + c_1)^{(\beta - \alpha)/\sigma} + \cdots$$
$$+ \bar{a}_\delta(\bar{x}, \bar{y})(z_1 + c_1)^{(\delta - \alpha)/\sigma}\} + x_1 H(x_1, y_1, z_1).$$

Naturally, in view of our choice of the constant c_1, we must have $f_1(0, 0, 0) = 0$. As in the preceding section we conclude that the multiplicity of the root $z_1 = 0$

for $f_1(0, 0, z_1)$ is $\leqq r$, and that it equals r if and only if $\alpha = 0$, $\delta = r$ and if $\bar{a}_0(0, 0) + \bar{a}_1(0, 0)u + \cdots + \bar{a}_r(0, 0)u^r = (u - c_1)^r$. The minimum value terms in the original polynomial $f(x, y, z)$ are then the following:

$$a_0(x, y) + a\ (x, y)z + \cdots + a_{r-1}(x, y)z^{r-1} + a_r(x, y)z^r,$$

where $a_r(0, 0) \neq 0$ (since $f(x, y, z)$ contains a term in z^r). As a consequence, $v(z) = v(a_{r-1}(x, y))$, i.e. *the value of z is equal to the value of a polynomial in x and y.* This conclusion is altogether parallel to the one reached in the course of the proof given in the preceding section. The only difference is that in the first case the value of z was the value of a power product $x^m y^n$. The remainder of the proof runs along similar lines, as in the first case. If $v(z) = v(a(x, y))$, a—a polynomial, we can find a constant c such that $v(z - c \cdot a(x, y)) > v(z)$. We pass to the element $z_1 = z - c \cdot a(x, y)$ and to the surface $f_1(x, y, z_1) = 0$. Repeating this procedure, we get a sequence of surfaces $f_i(x, y, z_i) = 0$, such that

(1) $z_i = 0$ is an r-fold root of $f_i(0, 0, z_i)$;
(2) $v(z_i) = v(a_i(x, y))$, a_i—a polynomial;
(3) $v(z) < v(z_1) < v(z_2) < \cdots$; $z_{i+1} = z_i - c_i \cdot a_i(x, y)$.

As in the preceding case we conclude with the inequality

$$v\left(\frac{\partial f}{\partial z}\right) \geqq v(a_i),$$

which shows that our sequence of surfaces f_i must be finite, since it is easily seen that *a strictly ascending sequence of real numbers, in which each element is the value of a polynomial, has $+\infty$ as a limit.*[10] This completes the proof of the fundamental lemma.

III. Ideals in the Quotient Ring of a Simple Point

10.

Let F be an algebraic surface in an S_n, a projective model of the field Σ, and let $\xi_1, \cdots, \xi_n$ be the non-homogeneous coördinates of the general point of F. In Z_2 (section 21) we have associated to every point P of F a *quotient ring* $Q(P)$. Namely, assuming (without loss of generality) that P is a point at finite distance, let $a_1, \cdots a_n$ be its non-homogeneous coördinates. Then $Q(P)$ is the set of all quotients of $f(\xi_1, \cdots, \xi_n)/g(\xi_1, \cdots, \xi_n)$ of polynomials in the ξ's, such that the denominator does not vanish at P, i.e. $g(a_1, \cdots, a_n) \neq 0$.

[10] Let ρ be a fixed positive real number and let $0 < \sigma = \min(v(x), v(y))$. Let n_ρ be the smallest integer such that $n_\rho \cdot \sigma > \rho$. Let $g(x, y)$ be a polynomial whose value is not greater than ρ. We separate in g the terms of degree $< n_\rho$ from those of degree $\geqq n_\rho$, i.e. we write $g = g' + g''$, where g' is the polynomial of degree $< n_\rho$. Every form in x, y, of degree m, has value $\geqq m\sigma$. Hence $v(g'') \geqq n_\rho \sigma > \rho$. Since $v(g) \leqq \rho$, it follows that $v(g') = v(g)$. Hence, if a polynomial has a value not greater than ρ, then its value is the value of some polynomial of degree $< n_\rho$. It follows that among the values assumed by polynomials in x, y there is only a finite number of values which are not greater than a given fixed real number ρ. This proves our assertion.

If $\mathfrak{J}$ denotes the ring of polynomials $K[\xi_1, \cdots, \xi_n]$, then (in ideal theoretic language) $Q(P)$ is the quotient ring $\mathfrak{J}^* = \mathfrak{J}_{\mathfrak{p}_0}$ of the prime zero-dimensional ideal $\mathfrak{p}_0 = \mathfrak{J}\cdot(\xi_1 - a_1, \cdots, \xi_n - a_n)$. It is evident that $Q(P)$ is independent of the choice of the non-homogeneous coördinates in S_n.

The relationship between the ideal theory of $\mathfrak{J}$ and that of $Q(P) = \mathfrak{J}^*$ is well-known (K, p. 17–20). If $\mathfrak{A}$ is any ideal in $\mathfrak{J}$, its extended ideal $\mathfrak{A}^* = \mathfrak{J}^*\cdot\mathfrak{A}$ in $\mathfrak{J}^*$ is the unit ideal, if $\mathfrak{A} \not\equiv 0(\mathfrak{p}_0)$, and is different from the unit ideal if $\mathfrak{A} \equiv 0(\mathfrak{p}_0)$. Thus all the ideals of $\mathfrak{J}$ which are not multiples of $\mathfrak{p}_0$ are lost in $\mathfrak{J}^*$. In particular, all the zero-dimensional prime ideals of $\mathfrak{J}$, except $\mathfrak{p}_0$, are lost, whence $\mathfrak{J}^*$ *possesses only one prime* 0-*dimensional ideal*, namely $\mathfrak{p}_0^* = \mathfrak{p}_0\cdot\mathfrak{J}^*$. If $\mathfrak{A}_1$ and $\mathfrak{A}_2$ are two ideals in $\mathfrak{J}$, then $\mathfrak{J}^*\cdot\mathfrak{A}_1$ and $\mathfrak{J}^*\cdot\mathfrak{A}_2$ coincide, $\mathfrak{J}^*\mathfrak{A}_1 = \mathfrak{J}^*\mathfrak{A}_2 = \mathfrak{A}^*$, if and only if $\mathfrak{A}_1$ and $\mathfrak{A}_2$ differ by primary components which are not divisible by $\mathfrak{p}_0$. Hence, omitting in $\mathfrak{A}_1$, or in $\mathfrak{A}_2$, all these last mentioned primary components, an ideal $\mathfrak{A}_0$ is obtained which is the h.c.d. of all ideals $\mathfrak{A}$ in $\mathfrak{J}$ such that $\mathfrak{A}^* = \mathfrak{J}^*\mathfrak{A}$. For this ideal $\mathfrak{A}_0$ it can be proved that not only $\mathfrak{A}^* = \mathfrak{J}^*\mathfrak{A}_0$, but that also $\mathfrak{A}_0 = \mathfrak{A}^* \cap \mathfrak{J}$. Thus there is a (1, 1) correspondence (in fact, an isomorphism with respect to all ideal theoretic operations) between the ideals in $\mathfrak{J}^*$ and those ideals in $\mathfrak{J}$ of which every primary component is divisible by $\mathfrak{p}_0$. In this correspondence primary ideals go into primary ideals. In particular, the 1-dimensional ideals p^* in $\mathfrak{J}^*$ correspond to the prime 1-dimensional ideals $\mathfrak{p}$ in $\mathfrak{J}$ which are divisible by $\mathfrak{p}_0$.

Geometrically speaking, the ideal theory of $Q(P) = \mathfrak{J}^*$ is the local ideal theory of the surface F in the neighborhood of the point P. Of the irreducible algebraic curves which can be traced on F, only those which pass through P are left in the passage from $\mathfrak{J}$ to $Q(P)$.

11

We wish to derive some properties of the ideal theory of $Q(P) = \mathfrak{J}^*$ in the case when P is a *simple point* of the surface F.

We have shown in Z_2 that if P is a simple point, then the corresponding prime 0-dimensional ideal $\mathfrak{p}_0$ does not divide the conductor $\mathfrak{C}$ of $\mathfrak{J}$ with respect to the integral closure $\bar{\mathfrak{J}}$ of $\mathfrak{J}$. We have also shown (Z_2, theorem 8) that if $\mathfrak{C} \not\equiv 0(\mathfrak{p}_0)$, then $\mathfrak{p}_0\cdot\bar{\mathfrak{J}} = \bar{\mathfrak{p}}_0$ is a prime (0-dimensional) ideal in $\bar{\mathfrak{J}}$. From these facts it follows that $\mathfrak{J}_{\mathfrak{p}_0} = \mathfrak{J}^* = \bar{\mathfrak{J}}_{\bar{\mathfrak{p}}_0}$. But $\bar{\mathfrak{J}}_{\bar{\mathfrak{p}}_0}$ is integrally closed in Σ. Hence *the quotient ring $Q(P)$ of a simple point is integrally closed in* Σ.

By our ideal-theoretic definition of a simple point (Z_2, section 2), there exists a pair of elements x, y in $\mathfrak{J}$ such that $\mathfrak{p}_0$ is an isolated component of $\mathfrak{J}\cdot(x, y)$. It follows that

$$\mathfrak{p}_0^* = \mathfrak{J}^*\cdot(x, y). \tag{13}$$

Whenever necessary, we shall assume that x, y have been so chosen that the elements of $\mathfrak{J}$ are integrally dependent on x and y.

Let R denote the ring of polynomials in x, y, $R = K[x, y]$; let $\mathfrak{P}_0 = R\cdot(x, y)$, and let $R^* = R_{\mathfrak{P}_0}$ be the quotient ring of $\mathfrak{P}_0$ in R. We have $R \subset R^* \subsetneq \mathfrak{J}^*$,

$R \subseteq \mathfrak{J}$. Ideals in R will be denoted by capital German letters; ideals in R^* will be denoted by capital German letters with an asterisk. Thus $\mathfrak{P}_0^* = \mathfrak{P}_0 \cdot R^*$. We proceed to investigate the *relationship between the ideals in* $\mathfrak{J}^*$ *and those of* R^*.

The elements x and y are *uniformizing parameters* of the *complete neighborhood of* P (Z_2, section 4). The uniformization is as follows. Given any element ω in $\mathfrak{J}$, there exists, for any integer $m > 0$, a uniquely determined polynomial $\phi_{m-1}(x, y)$, of degree at most $m - 1$ in x and y, such that

$$\omega \equiv \phi_{m-1}(\mathfrak{p}_0^m). \tag{14}$$

Moreover, $\phi_m - \phi_{m-1} = \psi_m$, where ψ_m is a form of degree m in x, y (or identically zero). We thus get for ω a formal power series in x, y:

$$\omega \to \psi_0 + \psi_1 + \cdots + \psi_m + \cdots, \tag{14'}$$

and (14′) maps $\mathfrak{J}$ isomorphically upon a subring of the ring $\Omega = \mathsf{K}\{x, y\}$ of all formal power series of x and y, with coefficients in K.

Here ψ_0 is a constant. If $\omega \not\equiv 0(\mathfrak{p}_0)$, then $\psi_0 \neq 0$, since $\omega - \psi_0 \equiv 0(\mathfrak{p}_0)$. Hence the elements of $\mathfrak{J}$ which are not in $\mathfrak{p}_0$ are mapped upon units of Ω. It follows that *the entire quotient ring* $Q(P)$ *is mapped upon a subring of* Ω. We shall see later that *the elements of* $Q(P)$ *are the only elements of the field* Σ *which are mapped upon elements of* Ω. Any other element of Σ is a *meromorphic* function of x and y, i.e. is a quotient of two formal power series, not reducible to a formal power series in x and y.

The relation (14) can be expressed symbolically as follows:

$$\mathfrak{J} = (R, \mathfrak{p}_0^m),\ m \text{ arbitrary}, \tag{15}$$

i.e. any element in $\mathfrak{J}$ is the sum of an element in R and of an element in $\mathfrak{p}_0^m$. *We prove a similar relation for* $\mathfrak{J}^*$:

$$\mathfrak{J}^* = (R, \mathfrak{p}_0^{*m}),\ m\text{—an arbitrary integer.} \tag{15'}$$

The proof is immediate. Let ξ be any element in $\mathfrak{J}^*$, and let

$$\xi \to \phi_{m-1} + \psi_m + \psi_{m+1} + \cdots.$$

Then $\xi - \phi_{m-1}$ is represented by the power series $\psi_m + \cdots$ in which the terms of the lowest degree are of degree $\geqq m$. The element $\xi - \phi_{m-1}$ is in $\mathfrak{J}^*$, hence we can write $\xi - \phi_{m-1} = \omega/\omega_0$, where $\omega, \omega_0 \subset \mathfrak{J}$ and $\omega_0 \not\equiv 0(\mathfrak{p}_0)$. Since the power series expansion for ω_0 is a unit in Ω, it follows that the power series for ω must begin with terms of degree $\geqq m$. Hence, by the nature of our uniformization we must have $\omega \equiv 0(\mathfrak{p}_0^m)$. Consequently $\xi - \phi_{m-1} = \omega/\omega_0 \equiv 0(\mathfrak{p}_0^{*m})$, i.e. $\xi \equiv \phi_{m-1}(\mathfrak{p}_0^{*m})$, where $\phi_{m-1} \subset R$, and this proves (15′). We point out that in the proof just given we have shown incidentally that if the power series expansion for an element ω/ω_0 in $\mathfrak{J}^*$ begins with terms of degree $\geqq m$, then the element ω/ω_0 is in $\mathfrak{p}_0^{*m}$. The converse is also true, since $\mathfrak{p}_0^{*m} = \mathfrak{J}^* \cdot (x, y)^m$. Hence $\mathfrak{p}_0^{*m}$ *consists of those and only those elements of* $\mathfrak{J}^*$ *whose power series expansions begin with terms of degree* $\geqq m$.

We have $\mathfrak{p}_0^* = \mathfrak{J}^*\mathfrak{P}_0^*$, where $\mathfrak{P}_0^* = R^* \cdot (x, y)$. Hence also $\mathfrak{p}_0^{*m} = \mathfrak{J}^* \cdot \mathfrak{P}_0^{*m}$. We prove that

$$\mathfrak{P}_0^{*m} = \mathfrak{p}_0^{*m} \cap R^*. \tag{15''}$$

The proof is immediate. Let $g(x, y)/h(x, y)$ be an element of $\mathfrak{p}_0^{*m} \cap R^*$, where h and g are polynomials and $h(0, 0) \neq 0$. The power series for g/h must begin with terms of degree $\geqq m$. Since $h(x, y)$ is a unit, the power series for $g(x, y)$ must begin with terms of degree $\geqq m$. But this power series must be the polynomial g itself, since for all m, $0 = g - g \equiv 0(\mathfrak{p}_0^m)$. Hence $g \equiv 0(\mathfrak{P}_0^m)$ and consequently $g/h \equiv 0(\mathfrak{P}_0^{*m})$. Thus $\mathfrak{p}_0^{*m} \cap R^* \subseteqq \mathfrak{P}_0^{*m}$. Since $\mathfrak{P}_0^{*m} \subset \mathfrak{p}_0^{*m}$, (15″) follows.

By means of (15′) and (15″) we prove the following

THEOREM 1. *There is a* (1,1) *correspondence between the* 0-*dimensional ideals in* $\mathfrak{J}^*$ *and those in* R^*. *Two corresponding ideals* $\mathfrak{q}^*$ *and* $\mathfrak{Q}^*$ *are in the relation of extended and contracted ideal respectively*: $\mathfrak{q}^* = \mathfrak{J}^*\mathfrak{Q}^*$, $\mathfrak{Q}^* = \mathfrak{q}^* \cap R^*$.

PROOF. Let $\mathfrak{Q}^* = \mathfrak{q}^* \cap R^*$. To prove that $\mathfrak{q}^* = \mathfrak{J}^*\mathfrak{Q}^*$, it is sufficient to show that $\mathfrak{q}^* \equiv 0(\mathfrak{J}^*\mathfrak{Q}^*)$, since the relation $\mathfrak{J}^*\mathfrak{Q}^* \equiv 0(\mathfrak{q}^*)$ is obviously true. If we choose m sufficiently high so that $\mathfrak{p}_0^{*m} \equiv 0(\mathfrak{q}^*)$ and if we apply (15′) to those elements of $\mathfrak{J}^*$ which are in $\mathfrak{q}^*$, we find

$$\mathfrak{q}^* = (R \cap \mathfrak{q}^*, \mathfrak{p}_0^{*m}),$$

hence *a fortiori*:

$$\mathfrak{q}^* = (R^* \cap \mathfrak{q}^*, \mathfrak{p}_0^{*m}), \text{ for all } m \text{ sufficiently high.} \tag{16}$$

The relation (16) holds true for any 0-dimensional ideal in $\mathfrak{J}^*$. We have then $\mathfrak{q}^* = (\mathfrak{Q}^*, \mathfrak{p}_0^{*m})$. Since $\mathfrak{J}^*\mathfrak{Q}^*$ is also zero-dimensional, we can take m sufficiently high so as to have $\mathfrak{p}_0^{*m} = 0(\mathfrak{J}^*\mathfrak{Q}^*)$. Then it will follow that $\mathfrak{q}^* \equiv 0(\mathfrak{Q}^*, \mathfrak{J}^*\mathfrak{Q}^*)$ i.e. $\mathfrak{q}^* \equiv 0(\mathfrak{J}^*\mathfrak{Q}^*)$, q.e.d.

On the other hand, let us start with a 0-dimensional ideal $\mathfrak{Q}^*$ in R^* and let $\mathfrak{J}^*\mathfrak{Q}^* = \mathfrak{q}^*$. To prove that $\mathfrak{q}^* \cap R^* = \mathfrak{Q}^*$, it is sufficient to show that $\mathfrak{q}^* \cap R^* \equiv 0(\mathfrak{Q}^*)$, since $\mathfrak{Q}^* \equiv 0(\mathfrak{q}^*)$. Now, by (15′), we have

$$\mathfrak{q}^* = \mathfrak{J}^*\mathfrak{Q}^* \subseteqq (\mathfrak{Q}^*, \mathfrak{p}_0^{*m}),$$

and comparing with (16), we find $(R^* \cap \mathfrak{q}^*, \mathfrak{p}_0^{*m}) \subseteqq (\mathfrak{Q}^*, \mathfrak{p}_0^{*m})$, for all m sufficiently high. Consequently

$$R^* \cap \mathfrak{q}^* \subseteqq (\mathfrak{Q}^*, \mathfrak{p}_0^{*m}),$$

or

$$R^* \cap \mathfrak{q}^* \subseteqq (\mathfrak{Q}^*, \mathfrak{P}_0^{*m}),$$

since, in view of (15″), $\mathfrak{P}_0^{*m}$ is the contracted ideal of $\mathfrak{p}_0^{*m}$. If we now take m sufficiently high so as to have $\mathfrak{P}_0^{*m} \equiv 0(\mathfrak{Q}^*)$, it will follow that $R^* \cap \mathfrak{q}^* \equiv 0(\mathfrak{Q}^*)$, q.e.d.

COROLLARY. *There is also a* (1,1) *correspondence between the zero-dimensional* *ideals of* $\mathfrak{J}^*$ *and those of the ring* Ω *of formal power series in* x *and* y, *the correspondence being again that of contracted and extended ideal.*

The corollary follows from Theorem 1 and from the well-known fact that a correspondence of the type just mentioned exists between the zero-dimensional ideals of R^* and Ω.[11]

12

Let B_1 be a zero-dimensional valuation of $K(x, y)$ whose valuation ring contains R^*. The prime ideal of B_1 in R^* must be $\mathfrak{p}_0^*$, hence x and y have positive values in B_1: $v_{B_1}(x) > 0$, $v_{B_1}(y) > 0$. Similarly let B be a zero-dimensional valuation of Σ whose valuation ring contains $\mathfrak{J}^*$. Again (and for a similar reason) we must have $v_B(x) > 0$, $v_B(y) > 0$. The valuation B is said to be an *extended valuation of* B_1, *if* B *induces in* $K(x, y)$ *the valuation* B_1. An equivalent condition is the following: the *valuation ring of* B_1 *is the intersection of* $K(x, y)$ *with the valuation ring of* B. By general theorems on valuations of algebraic extension fields (van der Waerden [8]) it is well known that if B_1 is given, there always exist extended valuations B (always subject to our condition that the valuation rings contain R^* and $\mathfrak{J}^*$ respectively). In the present special case this can also be proved directly in the following fashion. It is well known (and the proof is elementary) that any zero-dimensional valuation B_1 of $K(x, y)$ such that the valuation ring of B_1 contains R^*, can be extended to a valuation in the field of meromorphic functions of x, y whose valuation ring contains Ω. This extended valuation induces a valuation B in Σ, and B is clearly an extended valuation of B_1 of the desired type.

THEOREM 2. *A zero-dimensional valuation* B_1 *of* $K(x, y)$, *whose valuation ring contains* R^*, *has a unique extension* B *in* Σ *such that the valuation ring of* B *contains* $\mathfrak{J}^*$.

PROOF. Any element ξ in Σ can be written in the form $\xi = \omega_1/\omega_2$, where $\omega_1, \omega_2 \subset \mathfrak{J}$. In view of (14) we have: $\omega_1 \equiv \phi_{m-1}^{(1)}(x, y)(\mathfrak{p}_0^m)$, $\omega_2 \equiv \phi_{m-1}^{(2)}(x, y)(\mathfrak{p}_0^m)$. Our proof will consist in showing that if $v_B(\xi) \geqq 0$, then $v_{B_1}(\phi_{m-1}^{(1)}) \geqq v_{B_1}(\phi_{m-1}^{(2)})$, *for all sufficiently high values of* m. This will show, namely, that the valuation ring of B is uniquely determined by the valuation ring of B_1.

First case. B_1 *is of rank* 1. If $\mathfrak{A}$ is an ideal (in R or in $\mathfrak{J}$), we denote by $v(\mathfrak{A})$ the minimum value assumed by the elements of $\mathfrak{A}$. Since $v_B(x) > 0$, $v_B(y) > 0$, it follows that $\lim_{m \to +\infty} v(\mathfrak{p}_0^m) = +\infty$. Hence if m is sufficiently large we will have $v_B(\omega_i) = v_{B_1}(\phi_{m-1}^{(i)})$, $i = 1, 2$, and the theorem is proved.

Second case. B_1 *is of rank* 2. The valuation B_1 is composed of a divisor B_1', a one-dimensional valuation. Two cases are possible. Either the prime ideal

[11] This can be proved in exactly the same fashion in which we have proved Theorem 1. Namely, the proof is based entirely on the relations (15′) and (15″). These relations continue to hold (and are trivial) if we replace in them the ring $\mathfrak{J}^*$ by the ring Ω (and $\mathfrak{p}_0^*$ by the prime ideal $\Omega(x, y)$).

of B_1' in R is the zero-dimensional ideal $\mathfrak{P}_0$(B_1'—a *divisor of second kind with respect to* R), or the prime ideal of B_1' in R is a 1-dimensional ideal (B_1'—a *divisor of first kind with respect* to R). In the first case the divisor B' with which B is composed is also of the second kind with respect to $\mathfrak{J}$ and we will have $v_B(\mathfrak{p}_0) = (m, n)$ where $m > 0$. Hence again $\lim_{m\to+\infty} v(\mathfrak{p}_0^m) = +\infty$, and the remainder of the proof is as in the preceding case of valuations of rank 1. Suppose then that B_1' (and hence also B') is of the first kind with respect to R and $\mathfrak{J}$ respectively. Let $\mathfrak{p}$ be the prime one-dimensional ideal of B' in $\mathfrak{J}$. The element ξ can always be written in the form of a quotient ω_1/ω_2 where the elements ω_1, ω_2 are not both in $\mathfrak{p}$. If $v_B(\xi) \geqq 0$, then necessarily $\omega_2 \not\equiv 0(\mathfrak{p})$, and the value of ω_2 in B will be of the form $(0, n)$. Hence, if m is sufficiently high, we will have: $v_B(\omega_2) < v_B(\mathfrak{p}_0^m)$, and consequently $v_B(\omega_2) = v_{B_1}(\phi_{m-1}^{(2)}) = (0, n)$. If also $\omega_1 \not\equiv 0(\mathfrak{p})$, then $v_B(\omega_1) = v_{B_1}(\phi_{m-1}^{(1)})$ for all m sufficiently high. If $\omega_1 \equiv 0(\mathfrak{p})$, then $v_B(\omega_1) > v_B(\mathfrak{p}_0^m)$, for all m, whence $v_{B_1}(\phi_{m-1}^{(1)}) = v_B(\mathfrak{p}_0^m)$, and consequently $v_{B_1}(\phi_{m-1}^{(1)}) > v_{B_1}(\phi_{m-1}^{(2)})$, for all m sufficiently high. This completes the proof.[12]

13

We come now to the one-dimensional ideals in $\mathfrak{J}^*$ and we wish to prove the following theorem:

THEOREM 3. *Every unmixed one-dimensional ideal in* $\mathfrak{J}^*$ *is a principal ideal.*

PROOF. It is sufficient to give the proof for prime ideals. Let $\mathfrak{p}^*$ be a prime one-dimensional ideal in $\mathfrak{J}^*$ and let $\mathfrak{p} = \mathfrak{p}^* \cap \mathfrak{J}$ be the corresponding prime ideal in $\mathfrak{J}$. Also $\mathfrak{p}$ is one-dimensional, and moreover $\mathfrak{p} \equiv 0(\mathfrak{p}_0)$ and $\mathfrak{J}^*\mathfrak{p} = \mathfrak{p}^*$ (see section III, 10). We begin by making a judicious choice of the uniformizing parameters. Any two elements x_1 and y_1 in $\mathfrak{J}$ can be taken as uniformizing parameters, provided in the expansions $x_1 = ax + by + \cdots$, $y_1 = a'x + b'y + \cdots$, the Jacobian $ab' - a'b$ be different from zero. This, namely, implies that $\mathfrak{p}_0$ is an isolated component of the ideal $\mathfrak{J}\cdot(x', y')$ (see Z_2, theorem 1), a characteristic property of uniformizing parameters. Now let z be some element of $\mathfrak{J}$ which is in $\mathfrak{p}_0$ but is not in any of the other possible prime zero-dimensional components of the *zero-dimensional* ideal $\mathfrak{J}\cdot(x, y, \mathfrak{p})$. This implies the relation $\mathfrak{p}_0 = (x, y, z, \mathfrak{p})$. We put $x_1 = c_1x + c_2y + c_3z$, $y_1 = c_1'x + c_2'y + c_3'z$. We choose "non-special" values for the coefficients c and c', so as to satisfy the following three conditions: (1) x_1, y_1 are uniformizing parameters; (2) the elements of $\mathfrak{J}$ (which, by our initial choice of x and y, are integrally dependent on x, y) are integrally dependent on x_1, y_1; (3) $\mathfrak{J}\cdot(x_1, y_1, \mathfrak{p}) = \mathfrak{p}_0$.

That conditions (1) and (2) can be satisfied is trivial. As for (3) we observe that if c_1, c_2, c_3 are non-special constants, then the ideal $\mathfrak{J}\cdot(x_1, \mathfrak{p})$ is zero-dimensional. In view of our choice of z, it is clear that of the prime ideals of $\mathfrak{J}\cdot(x_1, \mathfrak{p})$, only $\mathfrak{p}_0$ contains all the three elements x, y and z. Hence, if c_1',

[12] A few additional considerations would make it quite clear that the preceding proof contains implicitly the proof of *existence* of an extended valuation B.

c_2', c_3' are non-special constants, also y_1 will belong to $\mathfrak{p}_0$ but not to any of the remaining prime ideals of $\mathfrak{J}\cdot(x_1, \mathfrak{p})$. If in addition x_1, y_1 are uniformizing parameters, $\mathfrak{p}_0$ will be an isolated component of $\mathfrak{J}\cdot(x_1, y_1)$, whence $\mathfrak{p}_0 = \mathfrak{J}\cdot(x_1, y_1, \mathfrak{p})$.

We shall assume that our original uniformizing parameters x and y already satisfy the above three conditions, in particular:

$$\mathfrak{p}_0 = \mathfrak{J}\cdot(x, y, \mathfrak{p}). \tag{17}$$

Let $\omega_1, \cdots, \omega_m$ be elements of $\mathfrak{J}$ which form a base for $\mathfrak{p}$. Let $\mathfrak{p}_0^{(1)}, \cdots, \mathfrak{p}_0^{(h)}$ be the prime ideals belonging to the ideal $\mathfrak{J}\cdot(x, y)$ and different from $\mathfrak{p}_0$. In view of (17) it is clear that for non-special coefficients c_i the element $c_1\omega_1 + \cdots + c_m\omega_m$ will not belong to any of the ideals $\mathfrak{p}_0^{(j)}$, $j = 1, 2, \cdots, h$. Hence, applying if necessary a preliminary linear homogeneous transformation to the elements ω_i, we may assume that $\omega_i \not\equiv 0(\mathfrak{p}_0^{(j)})$, $i = 1, 2, \cdots, m$, $j = 1, 2, \cdots, h$. By Theorem 4 of Z_2 we have the following: let $F_i(x, y, t) = N(t - \omega_i) =$ norm of $t - \omega_i$ with respect to the field $\mathsf{K}(x, y)$; *then* $t = 0$ *is a simple root of* $F_i(0, 0, t)$. We can write then F_i in the form $F_i = t\cdot H_i(x, y, t) - A_i(x, y)$, where $H_i(0, 0, 0) \neq 0$ and A_i is a polynomial. Since $F_i(x, y, \omega_i) = 0$, it follows that $\omega_i\cdot H_i(x, y, \omega_i) = A_i(x, y)$, where $H_i(x, y, \omega_i) \not\equiv 0(\mathfrak{p}_0)$ (since x, y and ω_i are all $\equiv 0 \pmod{\mathfrak{p}_0}$). Now let $g = g(x, y)$ be the *irreducible* polynomial in x, y which belongs to $\mathfrak{p}$; in other words: let $\mathfrak{p} \cap R$ be the principal prime ideal $R.g$. Since $\omega_i \equiv 0(\mathfrak{p})$, also $A_i \equiv 0(\mathfrak{p})$, and since $A_i \subset \mathsf{K}[x, y]$, A_i must be divisible by g. Since $H_i(x, y, \omega_i) \not\equiv 0(\mathfrak{p}_0)$, we conclude that

$$\omega_i \equiv 0(\mathfrak{J}^*g), \qquad i = 1, 2, \cdots, m,$$

or, since $\mathfrak{p}^* = \mathfrak{J}^*\mathfrak{p} = \mathfrak{J}^*(\omega_1, \cdots, \omega_m)$,

$$\mathfrak{p}^* \equiv 0(\mathfrak{J}^*g).$$

Since, on the other hand, $g \equiv 0(\mathfrak{p}) \equiv 0(\mathfrak{p}^*)$, we conclude with the relation $\mathfrak{p}^* = \mathfrak{J}^*g$, and this proves our theorem.

COROLLARY. $\mathfrak{J}^*$ *coincides with the totality of elements of the field* Σ *which are mapped upon formal power series in* x *and* y. We have to show that if α, β are two elements in $\mathfrak{J}^*$ and if α is divisible by β in $\mathsf{K}\{x, y\}(= \Omega)$, *then* α *is also divisible by* β *in* $\mathfrak{J}^*$. The element α belongs to the principal ideal $\Omega\beta$. This ideal is one-dimensional and unmixed in Ω. Hence its contracted ideal in $\mathfrak{J}^*$ is also one-dimensional and unmixed.[13] By the preceding theorem we can write: $\Omega\beta \cap \mathfrak{J}^* = \mathfrak{J}^*\beta'$, $\beta' \subset \mathfrak{J}^*$, whence α is divisible by β' in $\mathfrak{J}^*$. It remains to prove that β' is divisible by β in $\mathfrak{J}^*$. We have $\beta = \epsilon\beta'$, where $\epsilon \subset \mathfrak{J}^*$. On the other hand β' is in $\Omega\beta$, i.e. β' is at any rate divisible by β in Ω. *Hence* ϵ *is necessarily a unit in* Ω. Now the elements of $\mathfrak{J}^*$ which are mapped upon units

[13] If $\tilde{\mathfrak{p}} = \Omega\cdot\xi$ is a prime 1-dimensional ideal in Ω (ξ—a prime element in Ω, not a unit), the contracted ideal of $\tilde{\mathfrak{p}}$ in $\mathfrak{J}^*$ may be the zero-ideal, but clearly is never the zero-dimensional ideal $\mathfrak{p}_0^* = \mathfrak{J}^*(x, y)$. Since β is in $\mathfrak{J}^*$, the contracted ideal of $\Omega\beta$ is not the zero-ideal.

of Ω are those and only those which are not in $\mathfrak{p}_0^*$. Consequently $\epsilon \not\equiv 0(\mathfrak{p}_0^*)$, and hence $\frac{1}{\epsilon}$ is also an element of $\mathfrak{J}^*$. Since $\beta' = \frac{1}{\epsilon}\cdot\beta$, the corollary follows.

IV. Quotient Rings on Normal Surfaces

14

After having fully discussed in Part III the properties of the quotient ring of a simple point, we now wish to discuss more generally the structure of the quotient ring $Q(P)$ of a point P of a *normal surface* F. The hypothesis that F is normal implies that $Q(P)$ is integrally closed in Σ. It is this last local property which matters, rather than the property at large to the effect that F is normal. Namely, were F not normal and P a point of F such that $Q(P)$ is integrally closed in Σ, then we can always replace F by any of the normal surfaces F^* determined by F. To P there corresponds a unique point P^* of F^*, and the quotient ring of P^* coincides with $Q(P)$. It is therefore to us only a matter of convenience, when, in studying the structure of an integrally closed quotient ring $Q(P)$, we think of P as a point of a normal surface.

The class of integrally closed quotient rings $Q(P)$ whose structure we seek to determine will be, however, somewhat special. *We shall assume namely that $Q(P)$ contains as a subring the quotient ring $Q(A)$ of a simple point A.* For the applications which we wish to make in the sequel, the consideration of such special quotient rings is sufficient. On the other hand, we may observe that the condition imposed by us on $Q(P)$ is consistent with highly complicated types of singular points P.

Our study of the structure of $Q(P)$ will be based upon the results of the preceding section, as applied to the quotient ring of the simple point A, and also on the theory of valuation ideals in a ring $\mathsf{K}\{x, y\}$ of formal power series of two variables, developed in Z_1. In view of the isomorphism between the zero-dimensional ideals in $Q(A)$ and in $\mathsf{K}\{x, y\}$ (x, y—uniformizing parameters at A), the results of that paper are directly applicable to $Q(A)$.

First we shall need a few clear-cut definitions and statements concerning the correspondence between the points of two birationally equivalent surfaces. In agreement with our point of view, we base our definitions on valuation theory.

Let F and F' be birationally equivalent surfaces, projective models of the field Σ. The coördinates $\xi_1', \cdots, \xi_m'$ of the general point of F' are rational functions of the coördinates $\xi_1, \cdots, \xi_n$ of the general point of F, and *vice versa*. The birational transformation from F to F' sets up a *correspondence between the points of the surfaces*. For our present purpose, the most convenient definition of this correspondence is the following: *a point P on F and a point P' on F' correspond to each other, if there exists a zero-dimensional valuation B of the field Σ, such that P is the center of B on F and P' is the center of B on F'.*[14] *A point P*

[14] In other words: there exists a branch on F, with origin at P, to which there corresponds a branch on F', with origin at P'.

on F is a fundamental point, if there exist a 1-dimensional valuation B_1 of Σ, such that the center of B_1 on F is the point P, while the center of B_1 on F' is a curve Γ (necessarily irreducible).[15] The curve Γ is a *fundamental curve* and is said to correspond to the point P. Every point P' of F' which is on the curve Γ corresponds to at least one place of the field of rational functions on Γ. Such a place defines a zero-dimensional valuation of Σ (composed with the divisor B_1) whose center on F' is P' and whose center on F is P. Hence P *corresponds to any point of the fundamental curve* Γ.

It is clear that if P is the center of a zero-dimensional valuation B, then the valuation ring of B contains the quotient ring $Q(P)$, and conversely. It follows that *if P and P' are two corresponding points of F and F' respectively and if $Q(P') \subseteqq Q(P)$, then P' is the only point of F' which corresponds to P*, and thus P is not a fundamental point.

THEOREM 4. *The locus of points P' which correspond to a given point P of F is an algebraic variety of dimension $\leqq 1$. Every one-dimensional irreducible component of this variety is a fundamental curve corresponding to P.*

PROOF. It will be sufficient to prove this theorem for the points at finite distance of F and F', since we can choose for either surface the hyperplane at infinity arbitrarily.

We denote the rings $K[\xi_1, \cdots, \xi_n]$ and $K[\xi_1', \cdots, \xi_m']$ by $\mathfrak{J}$ and $\mathfrak{J}'$ respectively, and we consider the join $\tilde{\mathfrak{J}} = (\mathfrak{J}, \mathfrak{J}') = K[\xi; \xi']$ of the two rings (the smallest ring containing both $\mathfrak{J}$ and $\mathfrak{J}'$). The ring $\tilde{\mathfrak{J}}$ defines an algebraic surface $\tilde{F}$ in an affine space S_{n+m}. The coördinates of the general point of F_1 are $\xi_1, \cdots, \xi_n, \xi_1', \cdots, \xi_m'$ and $\tilde{F}$ is birationally equivalent to F and F'. If $\tilde{P}$ is any point at finite distance on $\tilde{F}$, it is given by a prime zero-dimensional ideal $\tilde{\mathfrak{p}}_0$ in $\tilde{\mathfrak{J}}$. The contracted ideals $\mathfrak{p}_0 = \mathfrak{J} \cap \tilde{\mathfrak{p}}_0$, $\mathfrak{p}_0' = \mathfrak{J}' \cap \tilde{\mathfrak{p}}_0$ are prime and zero-dimensional and define a point P and a point P' on F and F' respectively. It is clear that $Q(\tilde{P}) \supseteqq Q(P)$, $Q(\tilde{P}) \supseteqq Q(P')$. Hence any zero-dimensional valuation of center $\tilde{P}$ has P as its center on F and P' as its center on F'. Consequently, *to a point $\tilde{P}$ on $\tilde{F}$ (at finite distance) there corresponds a unique point P on F and a unique point P' on F', and P, P' are corresponding points.* Conversely, let P and P' be two corresponding points (at finite distance) of F and F' respectively. There exists then a zero-dimensional valuation B whose center on F is P and whose center on F' is P'. Since the valuation ring of B contains both rings $\mathfrak{J}$ and $\mathfrak{J}'$, it also contains their join $\tilde{\mathfrak{J}}$. Hence the center of B on $\tilde{F}$ is at a point $\tilde{P}$ at finite distance. The points P, P', $\tilde{P}$ being the centers of the valuation B on F, F', $\tilde{F}$ respectively, the points, P, P', must be *the points* which correspond to the point $\tilde{P}$. We conclude with the following:

[15] We speak of the center of a 1-dimensional valuation in the following sense. Given B_1, we can so choose the non-homogeneous coördinates ξ_i that their values in B_1 be non-negative. Then the valuation ring of B_1 contains the ring $K[\xi_1, \cdots, \xi_m]$. The prime ideal of B_1 in this ring is either zero-dimensional or one-dimensional. In the first case the center of B_1 on F is a point and B_1 is a divisor of the second kind (with respect to the ring $K[\xi]$). In the second case *the center of* B_1 is an irreducible curve on F, and B_1 is a divisor of the first kind.

The points $\bar{P}$ *at finite distance of* $\bar{F}$ *are in* (1, 1) *correspondence, without exceptions, with the pairs* (P, P') *of corresponding points of* F *and* F' (at finite distance).

Let P be an arbitrary point at finite distance on F, $\mathfrak{p}_0$—its prime zero-dimensional ideal in $\mathfrak{J}$. Let $\bar{\mathfrak{J}} \cdot \mathfrak{p}_0 = \bar{\mathfrak{A}}$ and let $[\bar{\mathfrak{A}}, \mathfrak{J}'] = \mathfrak{A}'$. We assert that *the algebraic locus* Γ' *determined on* F' *by the ideal* $\mathfrak{A}'$ *is the locus of points at finite distance which correspond to* P. In fact, let P' be a point of F', at finite distance, which corresponds to P, and let $\mathfrak{p}_0'$ be the prime zero-dimensional ideal in $\mathfrak{J}'$ given by the point P'. Let, moreover, $\bar{P}$ be the point on $\bar{F}$ which corresponds to the point-pair (P, P'), and let $\bar{\mathfrak{p}}_0$ be the corresponding prime zero-dimensional ideal in $\bar{\mathfrak{J}}$. Since $\bar{P} \rightarrow P$, we must have $\mathfrak{p}_0 = \bar{\mathfrak{p}}_0 \cap \mathfrak{J}$, whence $\bar{\mathfrak{J}} \cdot \mathfrak{p}_0 \equiv 0(\bar{\mathfrak{p}}_0)$, i.e. $\bar{\mathfrak{A}} \equiv 0(\bar{\mathfrak{p}}_0)$. Since $\bar{P} \rightarrow P'$ we also have $\mathfrak{p}_0' = \bar{\mathfrak{p}}_0 \cap \mathfrak{J}$, hence $\mathfrak{A}' \equiv 0(\mathfrak{p}_0')$. This proves that P' is on Γ'.

Consider now the irreducible components of Γ'. They may be points or curves and they are given by the isolated prime ideals of $\mathfrak{A}'$. Since $\mathfrak{A}' = [\bar{\mathfrak{A}}, \mathfrak{J}']$, it is clear that each isolated prime ideal $\mathfrak{p}'$ of $\mathfrak{A}'$ is the contracted ideal of some prime ideal $\bar{\mathfrak{p}}$ of $\bar{\mathfrak{A}}$. Moreover, since $\mathfrak{p}_0 \subseteq \bar{\mathfrak{A}}$, also $\mathfrak{p}_0 \subseteq \bar{\mathfrak{p}}$, whence $\bar{\mathfrak{p}} \cap \mathfrak{J} = \mathfrak{p}_0$ and $\bar{\mathfrak{p}} \cap \mathfrak{J}' = \mathfrak{p}'$. There always exists a valuation whose center on $\bar{F}$ is $\bar{\mathfrak{p}}$ (i.e. the corresponding point or curve, according as $\bar{\mathfrak{p}}$ is zero-dimensional or one-dimensional). The center of this valuation on F is P and its center on F' is the irreducible component of Γ' given by the ideal $\mathfrak{p}'$. Hence every irreducible component of Γ' corresponds to P, and this completes the proof of theorem 4.

A consequence of Theorem 4, which we shall have occasion to use in the sequel, is the following

THEOREM 5. *If* $Q(P)$ *is integrally closed in* Σ *and if* P *is not a fundamental point, then to* P *there corresponds a unique point* P' *on* F'.

PROOF. By Theorem 4, our hypothesis that P is not fundamental implies at any rate that to P there corresponds on F' only a finite number of points, say $P_1', P_2', \cdots P_\sigma'$. We may assume these points to be at finite distance. Let $\mathfrak{p}_0^{(1)}, \mathfrak{p}_0^{(2)}, \cdots, \mathfrak{p}_0^{(\sigma)}$ be the corresponding prime zero-dimensional ideals in $\mathfrak{J}'$. We denote by Δ *the intersection of the quotient rings* $Q(P_1'), Q(P_2'), \cdots, Q(P_\sigma')$. If B is any zero-dimensional valuation whose center is P, its center on F' will be one of the points $P_1', \cdots, P_\sigma'$. Hence the valuation ring of B will contain one of the quotient rings $Q(P_1'), \cdots, Q(P_\sigma')$, and hence it will certainly contain Δ. Hence, if *the valuation ring of a zero-dimensional valuation* B *contains* $Q(P)$, *it also contains* Δ. Since $Q(P)$ is integrally closed, it is the intersection of all the valuation rings which contain it (K, p. 111), and, moreover, it is not difficult to see that in the representation of $Q(P)$ as an intersection of valuation rings it is permissible to omit the one-dimensional valuations.[16] We conclude

[16] A one-dimensional valuation B_1 defines an homomorphic mapping of the valuation ring $\mathfrak{B}_1$ of B_1 upon a field Σ^* of algebraic functions of one variable. If $Q(P)$ is contained in $\mathfrak{B}_1$, then $\mathfrak{J}$ is mapped upon a finite integral domain $\mathfrak{J}^*$ contained in Σ^*, and $Q(P)$ is mapped upon the quotient ring $Q(P^*)$ of a prime ideal in $\mathfrak{J}^*$. The ring $Q(P^*)$ is contained in at least one valuation ring (defined by a "place" of the field Σ^*). Hence $Q(P)$ is contained in the valuation ring $\mathfrak{B}$ of the corresponding zero-dimensional valuation B of Σ. Since $\mathfrak{B} \subset \mathfrak{B}_1$, our assertion follows.

that Δ *is contained in* $Q(P)$. Let $\mathfrak{p}_i'$ be the zero-dimensional prime ideal of $Q(P_i')$, $i = 1, 2, \cdots, \sigma$, and let $\mathfrak{p}_i^* = \Delta \cap \mathfrak{p}_i'$. Let us assume that $\sigma > 1$. The σ prime ideals $\mathfrak{p}_1^*, \cdots, \mathfrak{p}_\sigma^*$ in Δ are distinct, since we can always find an element of Δ which is in $\mathfrak{p}_i^*$ and not in $\mathfrak{p}_j^*$, if $i \neq j$ (such an element already exists in $\mathfrak{J}'$). Let ξ' be an element of $\mathfrak{p}_1^*$, not in $\mathfrak{p}_2^*$. There exists a valuation B_1 whose center on F is at P and whose center on F' is P_1'. Since ξ' belongs to the prime ideal of $Q(P_1')$, necessarily $v_{B_1}(\xi') > 0$. Hence ξ' belongs to the prime zero-dimensional ideal of $Q(P)$. But then necessarily ξ' has a positive value in any valuation whose center is the point P. In particular, let B_2 be a zero-dimensional valuation whose center on F is P and whose center on F' is P_2'. Since $v_{B_2}(\xi') > 0$, ξ' must belong to the prime ideal $\mathfrak{p}_2'$ of $Q(P_2')$. Hence $\xi' \subset \mathfrak{p}_2^*$, a contradiction.

Consequently $\sigma = 1$, q.e.d.

15

After these preliminaries, we come back to the question announced at the beginning of the preceding section. We consider a surface F, given in the affine space by a ring $\mathfrak{J} = \mathsf{K}[\xi_1, \cdots, \xi_n]$, and on F we consider a point P such that $Q(P)$ is integrally closed in Σ. We assume that P is a point at finite distance, so that P is given by a prime zero-dimensional ideal in $\mathfrak{J}$, say by the ideal $\mathfrak{J}\cdot(\xi_1, \cdots, \xi_n)$. As was pointed out before, we may assume, without loss of generality, that $\mathfrak{J}$ itself is integrally closed.

We assume that $Q(P)$ contains the quotient ring $Q(A)$ of a *simple* point A of another surface Φ, birationally equivalent to F. We know that $Q(A)$ is integrally closed in Σ. Let Φ be given in an affine space S_m by the ring $\mathfrak{o} = \mathsf{K}[\omega_1, \cdots, \omega_m]$. As in the preceding case of F and P, we assume also here that A is a point at finite distance, that it is given in $\mathfrak{o}$ by the prime ideal $\mathfrak{o}\cdot(\omega_1, \cdots, \omega_m)$ and that $\mathfrak{o}$ is integrally closed in Σ.

The fact that $Q(P)$ contains $Q(A)$ does not necessarily imply that $\mathfrak{J}$ contains $\mathfrak{o}$. However, we may replace $\mathfrak{J}$ by the join $\mathfrak{J}_1 = (\mathfrak{J}, \mathfrak{o})$. The point pair (P, A) is represented by a prime zero-dimensional ideal $\mathfrak{p}_0^{(1)}$ in $\mathfrak{J}_1$, and since $Q(P) \supseteqq Q(A)$, it is immediately seen that $Q(P)$ coincides with the quotient ring of $\mathfrak{p}_0^{(1)}$ in $\mathfrak{J}_1$. Hence we may replace P by the corresponding point P_1 on the surface defined by $\mathfrak{J}_1$ in the affine space. If $\mathfrak{J}_1$ is not integrally closed, we replace $\mathfrak{J}_1$ by its integral closure. *We assume then that* $\mathfrak{J} \supseteqq \mathfrak{o}$. Under this assumption we have $\mathsf{K}[\xi_1, \cdots, \xi_n] = \mathsf{K}[\xi_1, \cdots, \xi_n, \omega_1, \cdots, \omega_m]$. We may then include the ω's among the generators of $\mathfrak{J}$, and therefore it is permissible to assume that

$$\xi_i = \omega_i, \qquad i = 1, 2, \cdots, m;\ n \geqq m. \tag{18}$$

We shall denote by $\mathfrak{J}^*$ and $\mathfrak{o}^*$ the quotient rings $Q(P)$ and $Q(A)$ respectively. Their prime zero-dimensional ideals shall be denoted by $\mathfrak{P}_0$ and $\mathfrak{p}_0$ respectively. Let

$$\xi_i = \frac{\phi_i(\omega_1, \cdots, \omega_m)}{\phi_0(\omega_1, \cdots, \omega_m)}, \qquad i = 1, 2, \cdots, n, \tag{19}$$

where the ϕ's are polynomials, and

$$\phi_i = \omega_i \phi_0, \qquad i = 1, 2, \cdots, m, \tag{19'}$$

in view of (18). We consider the following ideals in $\mathfrak{o}^*$

$$\mathfrak{M} = \mathfrak{o}^* \cdot (\phi_0, \phi_1, \cdots, \phi_n), \tag{20}$$

$$\mathfrak{m} = \mathfrak{o}^* \cdot (\phi_1, \cdots, \phi_n).$$

We may assume that $\mathfrak{M}$ and $\mathfrak{m}$ are zero-dimensional. In fact, if $\mathfrak{M}$ is one-dimensional, then by theorem 3 (II) we can write

$$\epsilon \phi_i = g(\omega_1, \cdots, \omega_m) \psi_i(\omega_1, \cdots, \omega_m), \qquad i = 0, 1, \cdots, n,$$

where ϵ is a unit in $\mathfrak{o}^*$, g and ψ_i are polynomials and $\psi_0, \cdots, \psi_n$ have no common factor in $\mathfrak{o}^*$ (not a unit). We then replace $\phi_0, \cdots, \phi_n$ by $\psi_0, \cdots, \psi_n$ respectively, since $\xi_i = \psi_i / \psi_0$, and then $\mathfrak{M}$ will be zero-dimensional.[17]

By (19′), a common factor of $\phi_1, \cdots, \phi_n$ (in $\mathfrak{o}^*$) has to divide ϕ_0. Hence if $\mathfrak{M}$ is already zero-dimensional, also $\mathfrak{m}$ is zero-dimensional.

We denote by M the totality of all zero-dimensional valuations B of Σ such that the center of B on F is the point P, or, in other words, such that the valuation ring $\mathfrak{B}$ of B contains $Q(P)$. Consider any member B of M. Since $Q(P) \supset Q(A)$, we have $\mathfrak{B} \supset Q(A)$, and we are in position to consider the *v-ideals* (*valuation ideals*) in $Q(A)$ belonging to the valuation B (see Z_1). Let $\mathfrak{q}$ be the v-ideal such that $v(\mathfrak{q}) = v(\phi_0)$. We consider *the intersection* of all these v-ideals $\mathfrak{q}$ as B varies in M. We denote this intersection by $\mathfrak{A}$. In the terminology of our paper Z_1, $\mathfrak{A}$ is a *complete ideal.*

Again for a fixed B in M, we consider the v-ideal $\mathfrak{q}_1$ in $\mathfrak{o}^*$ which follows immediately the above v-ideal $\mathfrak{q}$, i.e. the first v-ideal $\mathfrak{q}_1$ such that $v(\mathfrak{q}_1) > v(\mathfrak{q})$. We denote by $\mathfrak{a}$ the intersection of all ideals $\mathfrak{q}_1$, as B varies in M. Also $\mathfrak{a}$ is a complete ideal.

Since $\mathfrak{J}^* \cdot (\xi_1, \cdots, \xi_n) = \mathfrak{P}_0$, it follows that $v_B(\xi_i) > 0$, for any B in M and for $i = 1, 2, \cdots, n$. Hence $v_B(\phi_i) > v_B(\phi_0)$, $i \neq 0$, and consequently $\phi_i \subset \mathfrak{q}$, $i = 0, 1, \cdots, n$, and $\phi_i \subset \mathfrak{q}_1$, $i = 1, 2, \cdots, n$. *We conclude that* $\mathfrak{M} \subsetneqq \mathfrak{A}$, $\mathfrak{m} \subsetneqq \mathfrak{a}$ *and that*

$$v_B(\mathfrak{A}) = v_B(\phi_0) > 0, \tag{21}$$

$$v_B(\mathfrak{a}) > v_B(\phi_0) > 0, \tag{21'}$$

for any B in M. Since $v_B(\phi_0) > 0$ (ϕ_0 is not a unit in $\mathfrak{o}^*$, see footnote 17), $\mathfrak{A}$ is not the unit ideal. Since $\mathfrak{M}$ and $\mathfrak{m}$ are *zero-dimensional*, it follows that *both* $\mathfrak{A}$ *and* $\mathfrak{a}$ *are zero-dimensional complete ideals.*

[17] Strictly speaking, it may happen that after this reduction $\mathfrak{M}$ is the unit ideal. Since $\xi_1, \cdots, \xi_n$ are in the prime ideal $\mathfrak{P}_0$ of $\mathfrak{o}^*$ and since $\xi_i \phi_0 = \phi_i$, it follows that $\phi_i \equiv 0(\mathfrak{p}_0)$, $i = 1, \cdots, n$. Hence necessarily ϕ_0 must be a unit in $\mathfrak{o}^*$, and therefore each ξ_i is in σ^*. That implies that $Q(P)$ coincides with $Q(A)$. We naturally exclude this trivial case.

THEOREM 6. *The ideal* $\mathfrak{a}$ *is a maximal subideal of* $\mathfrak{A}$. *The ring of residual classes* $\mathfrak{A}/\mathfrak{a}$ *is a* K-*module of rank* 1.

PROOF. Let α be any element in $\mathfrak{A}$. By (21), we have $v_B(\alpha/\phi_0) \geqq 0$ for any B in M. Since $Q(P)$ is integrally closed, it is the intersection of the valuation rings $\mathfrak{B}$, as B varies in M (see footnote 16). Hence α/ϕ_0 is an element of $Q(P)$. There exists then a constant c in K such that $\alpha/\phi_0 \equiv c(\mathfrak{P}_0)$. For such a constant c it is then true that $v_B(\alpha - c\phi_0) > v_B(\phi_0)$, for all B in M. Hence $\alpha - c\phi_0$ is an element of the v-ideal $\mathfrak{q}_1$, for any B in M, i.e. $\alpha - c\phi_0 \equiv 0(\mathfrak{a})$. This shows that $\mathfrak{A}/\mathfrak{a}$ is a K-module of rank 1 (having ϕ_0 as base) and that consequently $\mathfrak{a}$ is a maximal subideal of $\mathfrak{A}$, q.e.d.

In the course of this proof we have shown that if α is any element in $\mathfrak{A}$, then α/ϕ_0 is an element of $Q(P)$. Notice that ϕ_0 is an element of $\mathfrak{A}$, but not of $\mathfrak{a}$ (in view of (21′)). Now, more generally, we can prove the following

THEOREM 7. *If* $\alpha, \beta \subset \mathfrak{A}$, $\beta \not\subset \mathfrak{a}$, *then* $\alpha/\beta \subset Q(P)$; *moreover* $\alpha/\beta \subset \mathfrak{P}_0$, *if and only if* $\alpha \subset \mathfrak{a}$.

PROOF. Since $\beta \not\subset \mathfrak{a}$, there exists a valuation B in M such that $\beta \not\subset \mathfrak{q}_1$, since $\mathfrak{a}$ is the intersection of the v-ideals $\mathfrak{q}_1$. Since $\mathfrak{q}_1$ is the immediate successor of $\mathfrak{q}$, it follows that $v_B(\beta) = v_B(\mathfrak{q}) = v_B(\phi_0)$. This implies that β/ϕ_0, which we know already to be an element of $Q(P)$, is *not in the prime ideal* $\mathfrak{P}_0$ of $Q(P)$. Therefore we must have $v_B(\beta) = v_B(\phi_0)$, *for any valuation* B in M. Consequently, $v_B(\alpha/\beta) \geqq 0$ for any B in M, and this implies $\alpha/\beta \subset Q(P)$.

The second part of the theorem is now a straightforward consequence of the fact that α/β is an element of $\mathfrak{P}_0$, if and only if $v_B(\alpha/\beta) > 0$, $B \subset M$.

Theorem 7 gives the quotient representation only for *certain* elements of $\mathfrak{J}^*$, because naturally, if α/β is a quotient of the indicated form, then α^ρ/β^ρ, ρ—an integer, will not, in general, be representable as a quotient of two elements α', β' of $\mathfrak{A}$ such that β' is not $\mathfrak{a}$. The main result of the present considerations is the following theorem, which gives full information concerning the connection between the complete ideals $\mathfrak{A}$, $\mathfrak{a}$, and the representation of any element in $\mathfrak{J}^*$ as a quotient of elements in $\mathfrak{o}^*$:

THEOREM 8. *If* ρ *is an integer* > 0 *and* α, β *are two elements in* $\mathfrak{o}^*$ *such that* $\alpha, \beta \equiv 0(\mathfrak{A}^\rho)$, $\beta \not\equiv 0(\mathfrak{A}^{\rho-1}\mathfrak{a})$, *then* α/β *is an element of* $\mathfrak{J}^*$, *and in this fashion all the elements of* $\mathfrak{J}^*$ *can be obtained.* The above element α/β *is in the prime zero-dimensional ideal* $\mathfrak{P}_0$ *of* $\mathfrak{J}^*$, *if and only* if $\alpha \equiv 0(\mathfrak{A}^{\rho-1}\mathfrak{a})$.

PROOF. The products $\omega_1\phi_0, \cdots, \omega_m\phi_0$ have values greater than $v(\phi_0)$ in any valuation B in M, hence they are elements of $\mathfrak{a}$ (this also follows directly from theorem 6). Therefore, by Theorem 6, any element γ of $\mathfrak{A}^\rho$ can be written in the form $c\phi_0^\rho + \gamma_1$, where c is a constant and $\gamma \subset \mathfrak{A}^{\rho-1}\mathfrak{a}$. Since $v_B(\mathfrak{A}^{\rho-1}\mathfrak{a}) > \rho v_B(\phi_0)$, for any B in M, it follows that $\mathfrak{A}^{\rho-1}\mathfrak{a}$ *is a proper maximal subideal* of $\mathfrak{A}^\rho$. Now let α, β be two elements of $\mathfrak{o}^*$ such that $\alpha, \beta \equiv 0(\mathfrak{A}^\rho)$, $\beta \not\equiv 0(\mathfrak{A}^{\rho-1}\mathfrak{a})$. Then we can write $\alpha = c\phi_0^\rho + \alpha_1$, $\beta = d\phi_0^\rho + \beta_1$, where $\alpha_1, \beta_1 \subset \mathfrak{A}^{\rho-1}\mathfrak{a}$ and $d \neq 0$. This yields, for any valuation B in M: $v_B(\alpha) \geqq \rho v_B(\phi_0)$, $v_B(\beta) = \rho v_B(\phi_0)$. Hence $v_B(\alpha/\beta) \geqq 0$, for all B in M, and consequently $\alpha/\beta \subset \mathfrak{J}^*$. This proves the first part of our theorem. Moreover, α/β is an element of $\mathfrak{P}_0$, if and only if $v_B(\alpha) > v_B(\phi_0)$, i.e. if and only if $c = 0$, whence $\alpha \subset \mathfrak{A}^{\rho-1}\mathfrak{a}$. This proves the concluding assertion of the theorem.

To complete the proof, let ξ be any element of $Q(P)$. The element ξ can be written in the form of a quotient $f(\xi_1, \cdots, \xi_n)/\psi(\xi_1, \cdots, \xi_n)$, where $\psi(0, \cdots, 0) \neq 0$. By (19), we can then write ξ in the form of a quotient $F(\phi_0, \phi_1, \cdots, \phi_n)/G(\phi_0, \phi_1, \cdots, \phi_n)$, where F and G are forms of the same degree, say ρ, and G actually contains a term in ϕ_0^ρ. Since all the ϕ's are in $\mathfrak{A}$, the elements F and G are in $\mathfrak{A}^\rho$. On the other hand, since all the power products of $\phi_0, \phi_1, \cdots, \phi_n$, of degree ρ, are elements of $\mathfrak{A}^{\rho-1}\mathfrak{a}$, *except* ϕ_0^ρ, which is certainly not an element of $\mathfrak{A}^{\rho-1}\mathfrak{a}$, it follows that $G \not\equiv 0(\mathfrak{A}^{\rho-1}\mathfrak{a})$. This completes the proof of the theorem.

16

We shall conclude this part of the paper with a theorem concerning the factorization of the complete ideals $\mathfrak{A}$ and $\mathfrak{a}$ into simple v-factors (see Z_1). The notion of a simple v-ideal $\mathcal{P}^{(h)}$, of kind h, is the arithmetic analog of the geometric concept of a point $A^{(h)}$ infinitely near A and in the $(h-1)^{\text{th}}$ neighborhood of A. A given simple v-ideal $\mathcal{P}^{(h)}$ determines uniquely a sequence $\mathcal{P}^{(1)} = \mathfrak{p}_0, \mathcal{P}^{(2)}, \cdots, \mathcal{P}^{(h-1)}$ of simple v-ideals of kind $1, 2, \cdots, h-1$ respectively, such that if B is any zero-dimensional valuation of Σ whose valuation ring contains $\mathfrak{o}^*$ and for which $\mathcal{P}^{(h)}$ is a v-ideal, then also the $\mathcal{P}^{(i)}$, $i = 1, 2, \cdots, h-1$, are v-ideals belonging to B. Any of these $h-1$ simple v-ideals is called a *predecessor of* $\mathcal{P}^{(h)}$. While the simple v-ideal $\mathcal{P}^{(1)}$, of kind 1, corresponds to the proper point $A^{(1)} = A$, the other predecessors $\mathcal{P}^{(i)}$ correspond to the points $A^{(2)}, \cdots, A^{(h-1)}$ which follow the point A in the *first, second,* $\cdots$, $(h-2)^{\text{th}}$ neighborhood and which *precede* the point $A^{(h)}$.

A complete ideal can be factored into a product of simple v-ideals, and the factorization is unique (see Z_1, p. 172). Let

$$\mathfrak{A} = \mathcal{P}_1^{\alpha_1}\mathcal{P}_2^{\alpha_2}\cdots\mathcal{P}_\mu^{\alpha_\mu},$$

$$\mathfrak{a} = \bar{\mathcal{P}}_1^{\beta_1}\bar{\mathcal{P}}_2^{\beta_2}\cdots\bar{\mathcal{P}}_\nu^{\beta_\nu},$$

be the factorization of $\mathfrak{A}$ and $\mathfrak{a}$ respectively, into simple v-ideals. The individual simple v-factors in $\mathfrak{A}$ correspond to the effective base points of the ideal $\mathfrak{A}$, i.e. of the corresponding continuous system of curves $f = 0$, defined by the condition: $f \equiv 0(\mathfrak{A})$. The exponents α_i are related to the effective multiplicities of these curves at the base points.

In order to state the theorem which we wish to prove, we make the following remark. Let $\mathfrak{A}$ be exactly divisible by $\mathfrak{p}_0^\rho$ i.e. let $\mathfrak{A} \equiv 0(\mathfrak{p}_0^\rho)$, $\mathfrak{A} \not\equiv 0(\mathfrak{p}_0^{\rho+1})$. Since $\mathfrak{a} \equiv 0(\mathfrak{A})$, also $\mathfrak{a} \equiv 0(\mathfrak{p}_0^\rho)$. On the other hand, since $\mathfrak{A}\mathfrak{p}_0 \equiv 0(\mathfrak{a})$ (by Theorem 6), we have $\mathfrak{a} \not\equiv 0(\mathfrak{p}_0^{\rho+2})$. Hence $\mathfrak{a}$ is exactly divisible either by $\mathfrak{p}_0^\rho$ or by $\mathfrak{p}_0^{\rho+1}$. Our theorem is the following:

THEOREM 9. *If $\mathfrak{A}$ is exactly divisible by $\mathfrak{p}_0^\rho$ and $\mathfrak{a}$ is exactly divisible by $\mathfrak{p}_0^{\rho+1}$, then every simple v-factor $\bar{\mathcal{P}}_i$ of $\mathfrak{a}$ is a predecessor of or coincides with some simple v-factor $\mathcal{P}_j$ of $\mathfrak{A}$.*

PROOF.[18] Let x, y be uniformizing parameters at A. The elements of $\mathfrak{o}^*$ are formal power series in x and y. Every element α of $\mathfrak{a}$ is a power series $F_{\rho+1}(x, y) + F_{\rho+2}(x, y) + \cdots$, beginning with terms of degree $\rho + 1$ or higher. We call $F_{\rho+1}$ the *subform* of α. As α varies in $\mathfrak{a}$, its subform varies in a linear system, of a certain dimension r (a K-module of rank $r + 1$). Since $\mathfrak{p}_0\mathfrak{A} \equiv 0(\mathfrak{a})$, it follows immediately that $r \geqq 1$. This implies that $\mathfrak{p}_0$ must occur among the simple v-factors of $\mathfrak{a}$, since for any simple v-ideal of kind > 1 it is true that its system of subforms is of dimension zero (Z_1, p. 163, Corollary 3.2). We put into evidence the factor $\mathfrak{p}_0$ and we write $\mathfrak{a} = \mathfrak{p}_0\mathfrak{a}_1$. To prove our theorem, it will now be sufficient to prove that every simple v-factor in $\mathfrak{a}_1$ is a predecessor of, or coincides with, some simple v-factor of $\mathfrak{A}$. To prove this we first show that *the ideal* $\mathfrak{a}_1$ *coincides with the quotient* $\mathfrak{A}:\mathfrak{p}_0$.

Let $\mathfrak{A}: \mathfrak{p}_0 = \mathfrak{B}$. Since $\mathfrak{p}_0\mathfrak{a}_1 = \mathfrak{a} \equiv 0(\mathfrak{A})$, we have at any rate the inclusion

$$\mathfrak{a}_1 \subseteq \mathfrak{B}. \tag{22}$$

We assert that $\mathfrak{p}_0$ *does not occur as a factor in* $\mathfrak{A}$.[19] Since $\mathfrak{a}$ is a maximal subideal of $\mathfrak{A}$, it is clear that

$$\mathfrak{a} = [\mathfrak{A}, \mathfrak{p}_0^{\rho+1}]. \tag{23}$$

Let $ax + by$ be a fixed form which does not divide all subforms of $\mathfrak{A}$, and let $\mathcal{P}_2' = (ax + by, \mathfrak{p}_0^2)$. The ideal $\mathcal{P}_2'$ is a simple v-ideal of second kind. We consider the ideal $\Delta = [\mathfrak{A}, \mathfrak{p}_0^{\rho-1}\cdot\mathcal{P}_2']$. Since $\mathfrak{p}_0^2 \equiv 0(\mathcal{P}_2')$, we have, by (23): $\mathfrak{A} \supseteqq \Delta \supseteqq \mathfrak{a}$. By our choice of the form $ax + by$ it follows that $\mathfrak{A} \not\subseteq \mathfrak{p}_0^{\rho-1}\mathcal{P}_2'$. Hence $\Delta \neq \mathfrak{A}$. Hence $\Delta = \mathfrak{a}$, since $\mathfrak{a}$ is a maximal subideal of $\mathfrak{A}$. Now suppose, if possible, that $\mathfrak{p}_0$ is a factor of $\mathfrak{A}$, and let $\mathfrak{A} = \mathfrak{p}_0\mathfrak{A}_1$, where $\mathfrak{A}_1$ is then necessarily exactly divisible by $\mathfrak{p}_0^{\rho-1}$. We have $\mathcal{P}_2'\mathfrak{A}_1 \subseteq \mathcal{P}_2'\mathfrak{p}_0^{\rho-1}$, and also $\mathcal{P}_2'\mathfrak{A}_1 \subseteq \mathfrak{p}_0\mathfrak{A}_1 = \mathfrak{A}$. Hence $\mathcal{P}_2'\mathfrak{A}_1 \subseteq \Delta$, i.e. $\mathcal{P}_2'\mathfrak{A}_1 \subseteq \mathfrak{a}$. This is impossible, since $\mathcal{P}_2'\mathfrak{A}_1$ is exactly divisible by $\mathfrak{p}_0^{\rho}$, while $\mathfrak{a}$ is divisible by $\mathfrak{p}_0^{\rho+1}$.

[18] It is not without interest to point out the intuitive geometric content of Theorem 9. Interpreted geometrically, Theorem 6 signifies that the complete ideal $\mathfrak{a}$ is obtained by imposing on $\mathfrak{A}$ *one* additional *base condition*, expressed by *one* linear relation. This base condition may be of a two-fold nature: (1) either we impose on the curves defined by $\mathfrak{A}$ the condition of passing through a *new base* point A_{q+1}, which *follows* immediately a sequence $A, A_1, \cdots, A_q$ of successive base points of $\mathfrak{A}$; (2) or we impose on these curves (which, in view of $\mathfrak{A} \equiv 0(\mathfrak{p}_0^{\rho})$, already possess at A an effective multiplicity ρ) the condition of possessing at A a $\rho + 1$-fold point. In order that the new base condition be really equivalent to one linear condition, the following is necessary and sufficient: (1) in the first case, the branches of smallest order which pass through $A, A_1, \cdots, A_q$ and through $A, A_1, \cdots, A_{q+1}$ respectively, must have the same order (in the terminology of Enriques' theory: A_{q+1} must be a free point, not a satellite); (2) in the second case, the general curve defined by $\mathfrak{A}$ must have all its ρ tangents at A fixed. In the first case, the new base condition will not lead automatically to an increased base multiplicity at A, and so $\mathfrak{a}$ will still be exactly divisible by $\mathfrak{p}_0^{\rho}$. In the second case $\mathfrak{a}$ will be exactly divisible by $\mathfrak{p}_0^{\rho+1}$, but—and this is the main point—this change in the effective multiplicity at A may imply a lowering of the effective multiplicities at some other infinitely near base points, but *no new effective base points can appear*.

[19] In other words: the curves defined by elements of $\mathfrak{A}$ and having at A exactly a ρ-fold point, have no variable tangent at A. Compare preceding footnote.

We come back to the quotient $\mathfrak{B} = \mathfrak{A}:\mathfrak{p}_0$. We have $\mathfrak{B} \equiv 0(\mathfrak{p}_0^{\rho-1})$. If we suppose that $\mathfrak{B}$ is exactly divisible by $\mathfrak{p}_0^{\rho-1}$, we would have a contradiction, since the relation $\mathfrak{B}\mathfrak{p}_0 \equiv 0(\mathfrak{A})$ would then imply that the system of subforms of $\mathfrak{A}$ is of dimension $\geqq 1$, whence $\mathfrak{p}_0$ would be a factor of $\mathfrak{A}$. Hence $\mathfrak{B}$ *is exactly divisible by* $\mathfrak{p}_0^{\rho}$. Hence $\mathfrak{B}\mathfrak{p}_0 \equiv 0(\mathfrak{p}_0^{\rho+1})$, and since $\mathfrak{B}\mathfrak{p}_0 \equiv 0(\mathfrak{A})$, it follows by (23), that $\mathfrak{p}_0\mathfrak{B} \equiv 0(\mathfrak{a})$, i.e. $\mathfrak{p}_0\mathfrak{B} \equiv 0(\mathfrak{p}_0\mathfrak{a}_1)$. In view of (22), we also have $\mathfrak{p}_0\mathfrak{a}_1 \equiv 0(\mathfrak{p}_0\mathfrak{B})$. Hence $\mathfrak{p}_0\mathfrak{a}_1 = \mathfrak{p}_0\mathfrak{B}$. Now both $\mathfrak{a}_1$ and $\mathfrak{B}$ are complete ideals: $\mathfrak{a}_1$ is a complete ideal, because it is a product of v-ideals (Z_1, p. 193); $\mathfrak{B}$ is a complete ideal because it is a quotient $\mathfrak{A}:\mathfrak{p}_0$ of a complete ideal $\mathfrak{A}$ and another ideal (Z_1, p. 193). In view of the unique factorization of complete ideals into simple v-factors, we conclude from the relation $\mathfrak{p}_0\mathfrak{a}_1 = \mathfrak{p}_0\mathfrak{B}$, that $\mathfrak{a}_1 = \mathfrak{B}$, i.e. $\mathfrak{a}_1 = \mathfrak{A}:\mathfrak{p}_0$. This proves our assertion.

We have pointed out above that in order to prove our theorem, we have only to prove that every simple v-factor of $\mathfrak{a}_1$ is the predecessor of, or coincides with, some simple v-factor of $\mathfrak{A}$. We now also know that $\mathfrak{a}_1 = \mathfrak{A}:\mathfrak{p}_0$. On this basis our theorem becomes a special case of the following general

LEMMA. *If $\mathfrak{A}$ is a zero-dimensional complete ideal in $\mathfrak{o}^*$ and if $\mathfrak{b}$ is any complete ideal, then every simple v-factor of the ideal $\mathfrak{B} = \mathfrak{A}:\mathfrak{b}$ is the predecessor of, or coincides with some simple v-factor of $\mathfrak{A}$.*

PROOF OF THE LEMMA. In view of the isomorphism between the zero-dimensional ideals of $\mathfrak{o}^*$ and those zero-dimensional primary ideals of the polynomial ring $K[x, y]$ which belong to the ideal (x, y), it is sufficient to prove the lemma for polynomial rings. We assume therefore that $\mathfrak{A}$, $\mathfrak{B}$, $\mathfrak{b}$ are ideals in the polynomial ring $\mathfrak{o} = K[x, y]$. Let $\mathfrak{A} = \mathcal{P}_1^{\alpha_1} \cdots \mathcal{P}_\mu^{\alpha_\mu}$, and let $\mathcal{P}_i$ be of kind h_i. Let $h = \max(h_1, \cdots, h_\mu)$. The lemma is trivial for $h = 1$. In fact, if $h = 1$, then $\mathcal{P}_1, \cdots, \mathcal{P}_\mu$ are distinct prime zero-dimensional ideals, and it is obvious that $\mathfrak{A}:\mathfrak{b}$ will be of the form $\mathcal{P}_1^{\beta_1} \cdots \mathcal{P}_\mu^{\beta_\mu}$, $\beta_i \leqq \alpha_i$. We therefore prove the lemma by induction with respect to h. We assume that the lemma is true for $h < g$, and we prove that it is true for $h = g$. First of all, we observe that it is sufficient to prove the lemma for $h = g$, under the hypothesis that the simple v-factors $\mathcal{P}_1, \cdots, \mathcal{P}_\mu$ of $\mathfrak{A}$ all belong to one and the same prime zero-dimensional ideal $\mathfrak{p}_0$, say $\mathfrak{p}_0 = (x, y)$. Namely, in the general case we can write $\mathfrak{A} = \mathfrak{A}_1\mathfrak{A}_2 \cdots \mathfrak{A}_r = [\mathfrak{A}_1, \mathfrak{A}_2, \cdots, \mathfrak{A}_r]$, where the $\mathfrak{A}_i$ are complete ideals belonging to *distinct* prime zero-dimensional ideals. Then we will have $\mathfrak{A}:\mathfrak{b} = [\mathfrak{A}_1:\mathfrak{b}, \cdots, \mathfrak{A}_r:\mathfrak{b}] = (\mathfrak{A}_1:\mathfrak{b}) \cdots (\mathfrak{A}_r:\mathfrak{b})$. If the Lemma is true for each factor $\mathfrak{A}_i:\mathfrak{b}$, it is also true for the product $\mathfrak{A}:\mathfrak{b}$.

We assume then that $\mathcal{P}_1, \cdots, \mathcal{P}_\mu$ belong to the prime ideal $\mathfrak{p}_0 = (x, y)$. Let $\mathfrak{A}$, $\mathfrak{B}$, $\mathfrak{b}$ be exactly divisible by $\mathfrak{p}_0^\rho$, $\mathfrak{p}_0^\sigma$, $\mathfrak{p}_0^\tau$ respectively. Since $\mathfrak{B}\mathfrak{b} \equiv 0(\mathfrak{A})$, we must have $\sigma + \tau \geqq \rho$. Replacing, if necessary, x by a linear form $ax + by$ with non-special coefficients, we may assume that the subform of $\mathcal{P}_i$ ($i = 1, 2, \cdots, \mu$) is not a power of x (i.e. the line $x = 0$ is not the tangent line of the branches of lowest order defined by $\mathcal{P}_i$). Under this condition, the transform of $\mathcal{P}_i$ by the quadratic transformation

$$x' = x, \; y' = y/x$$

is a simple v-ideal $\mathcal{P}_i'$ in $\mathfrak{o}' = \mathsf{K}[x', y']$, of kind $h_i - 1$ (Z_1, p. 169). Let

$$\text{(24)} \qquad \mathfrak{o}' \cdot \mathfrak{A} = x^{\rho}\mathfrak{A}', \qquad \mathfrak{o}'\mathfrak{B} = x^{\sigma}\mathfrak{B}', \qquad \mathfrak{o}'\mathfrak{b} = x^{\tau}\mathfrak{b}',$$

where $\mathfrak{A}'$, $\mathfrak{B}'$, $\mathfrak{b}'$ are again complete ideals in $\mathfrak{o}'$ (Z_1, p. 196), and $\mathfrak{A}'$, $\mathfrak{B}'$ are zero-dimensional. By hypothesis, $g = \max(h_i)$. Hence $\max(h_i - 1) < g$ and, by our induction, the lemma is true for $\mathfrak{A}'$. *We prove that*

$$\text{(25)} \qquad \mathfrak{B}' = \mathfrak{A}' : x^{\sigma+\tau-\rho}\mathfrak{b}'.$$

From this our lemma will follow, since the simple v-factors of $\mathfrak{B}'$ are the transforms of the simple v-factors of $\mathfrak{B}$ different from $\mathfrak{p}_0$.

Since $\mathfrak{B}\mathfrak{b} \equiv 0(\mathfrak{A})$, we have, by (24), $x^{\sigma+\tau-\rho}\mathfrak{B}'\mathfrak{b}' \equiv 0(\mathfrak{A}')$, whence

$$\text{(26)} \qquad \mathfrak{B}' \subseteq \mathfrak{A}' : x^{\sigma+\tau-\rho}\mathfrak{b}'.$$

Let $\mathfrak{A}' : x^{\sigma+\tau-\rho}\mathfrak{b}' = \mathfrak{B}_1'$, and let x^q be the lowest power of x such that $x^q\mathfrak{B}_1'$ is the extended ideal of an ideal in $\mathfrak{o}$. Then $x^q\mathfrak{B}_1'$ is also the extended ideal of its contracted ideal (Z_1, p. 196) in $\mathfrak{o}$. Let $\mathfrak{B}_1$ be the contracted ideal. From the relation $\mathfrak{B}' \subseteq \mathfrak{B}_1'$ follows the inequality $q \leqq \sigma$ (see Z_1, p. 199). We have $\mathfrak{o}'\mathfrak{B}_1 = x^q\mathfrak{B}_1'$, hence $\mathfrak{o}'\mathfrak{p}_0^{\sigma-q}\mathfrak{B}_1 = x^{\sigma}\mathfrak{B}_1'$. Since $\mathfrak{B}$ and $\mathfrak{p}_0^{\sigma-q}\mathfrak{B}_1$ are the contracted ideals of $x^{\sigma}\mathfrak{B}'$ and $x^{\sigma}\mathfrak{B}_1'$ respectively, and since $\mathfrak{B}' \subseteq \mathfrak{B}_1'$ (by (26)), we conclude that

$$\text{(27)} \qquad \mathfrak{B} \subseteq \mathfrak{p}_0^{\sigma-q}\mathfrak{B}_1.$$

On the other hand, we have $x^{\sigma+\tau-\rho}\mathfrak{b}'\mathfrak{B}' \equiv 0(\mathfrak{A}')$, whence $x^{\sigma+\tau}\mathfrak{b}'\mathfrak{B}' \equiv 0(x^{\rho}\mathfrak{A}')$. Now $x^{\sigma+\tau}\mathfrak{b}'\mathfrak{B}_1'$ is the extended ideal of $\mathfrak{p}_0^{\sigma-q}\mathfrak{b}\mathfrak{B}_1$. Hence, passing to the contracted ideals, we find: $\mathfrak{p}_0^{\sigma-q}\mathfrak{b}\mathfrak{B}_1 \subseteq \mathfrak{A}$, i.e.

$$\text{(27')} \qquad \mathfrak{p}_0^{\sigma-q}\mathfrak{B}_1 \subseteq \mathfrak{B}.$$

Comparing (27) and (27'), we conclude that $\mathfrak{p}_0^{\sigma-q}\mathfrak{B}_1 = \mathfrak{B}$. Passing to the extended ideals in $\mathfrak{o}'$, we have $\mathfrak{B}_1' = \mathfrak{B}'$, i.e. (25), q.e.d.

V. Normal Sequences of Integrally Closed Quotient Rings. "Local Reduction" Theorem

17. Tangential ideals

Let F, a projective model of our field Σ, be given in an affine $S_n(x_1, \cdots, x_n)$ by the ring $\mathfrak{J} = \mathsf{K}[\xi_1, \cdots, \xi_n]$. We consider a point P on F and its quotient ring $\mathfrak{J}^* = Q(P)$. We assume that P is at finite distance, say at the origin $x_1 = \cdots = x_n = 0$. Let $\mathfrak{p}_0 = \mathfrak{J} \cdot (\xi_1, \cdots, \xi_n)$, $\mathfrak{p}_0^* = \mathfrak{J}^* \cdot \mathfrak{p}_0$, be the corresponding prime zero-dimensional ideals in $\mathfrak{J}$ and $\mathfrak{J}^*$ respectively.

The surface F possesses at P a *tangent cone*. We ask: what is, from an arithmetic standpoint, a tangent line of F at P? First of all, any line l through P (tangent or not) can be described by the system of hyperplanes which contain it. They are the hyperplanes $u_1x_1 + \cdots + u_nx_n = 0$, where the coefficients u_i satisfy a given linear relation $c_1u_1 + \cdots + c_nu_n = 0$, $c_i \subset \mathsf{K}$. Any element ξ in $\mathfrak{p}_0$ is a polynomial in $\xi_1, \cdots, \xi_n$, in which the constant term is missing: $\xi = a_1\xi_1 + \cdots + a_n\xi_n +$ terms of higher degree. The elements ξ such that the coefficients a_i satisfy the relation $c_1a_1 + \cdots + c_na_n = 0$ form a *primary maximal subideal* $\mathfrak{q}$ of $\mathfrak{p}_0$. We associate the ideal $\mathfrak{q}$ with the line l. The corresponding ideal in $\mathfrak{J}^*$ shall be denoted by $\mathfrak{q}^*$. Conversely, given a maximal primary subideal $\mathfrak{q}$ of $\mathfrak{p}_0$, the ring $\mathfrak{p}_0/\mathfrak{q}$ is a K-module of rank 1. Hence $\mathfrak{q}$ consists of all those elements $\xi = a_1\xi_1 + \cdots + a_n\xi_n + \cdots$ of $\mathfrak{p}_0$, in which the a_i satisfy a linear relation $c_1a_1 + \cdots + c_na_n = 0$. The hyperplanes $u_1x_1 + \cdots + u_nx_n = 0$ whose coördinates u_i satisfy the relation $c_1u_1 + \cdots + c_nu_n = 0$ have in common a line l through P. The line l corresponds to the ideal $\mathfrak{q}$ in $\mathfrak{J}$ (or $\mathfrak{q}^*$ in $\mathfrak{J}^*$).

Now let l be a *tangent line*. We can think of l as the tangent line of some branch γ (for instance, an algebraic branch) which lies on F and has origin at P. The branch γ defines a zero-dimensional valuation B of the field Σ, and P is the center of B on F. The elements $\xi_1, \cdots, \xi_n$ have positive values in B. It is clear that if ξ is any element in $\mathfrak{p}_0$, or in $\mathfrak{p}_0^*$, then $v_B(\xi) \geqq \min \{v_B(\xi_1), \cdots, v_B(\xi_n)\}$, whence $v_B(\mathfrak{p}_0) = v_B(\mathfrak{p}_0^*) = \min \{v_B(\xi_1), \cdots, v_B(\xi_n)\}$. The elements ξ of $\mathfrak{p}_0^*$, such that $v_B(\xi) > v_B(\mathfrak{p}_0)$, form a primary subideal $\mathfrak{q}^*$ of $\mathfrak{p}_0^*$, since $\mathfrak{J}^*$ is contained in the valuation ring of B. Let η be an element in $\mathfrak{p}_0^*$, not in $\mathfrak{q}^*$. If η_1 is any element in $\mathfrak{p}_0^*$, then $v_B(\eta_1/\eta) \geqq 0$. Since B is zero-dimensional, there exists a constant c such that $v_B(\eta_1/\eta - c) > 0$, i.e. $\eta_1 - c\eta \equiv 0(\mathfrak{q}^*)$. This shows that the ring of residual classes $\mathfrak{p}_0^*/\mathfrak{q}^*$ is a K-module of rank 1, and hence $\mathfrak{q}^*$ is a maximal subideal of $\mathfrak{p}_0^*$. The line l associated with $\mathfrak{q}^*$ shall be, by definition, the tangent line of the branch γ.

A maximal primary subideal $\mathfrak{q}^*$ of $\mathfrak{p}_0^*$ (or the corresponding subideal $\mathfrak{q}$ of $\mathfrak{p}_0$) which corresponds to a tangent line, shall be called a *tangential ideal of* $Q(P)$. From the standpoint of valuation theory, a tangential ideal $\mathfrak{q}^*$ is characterized by the following two properties: (1) $\mathfrak{q}^*$ is a valuation ideal belonging to some zero-dimensional valuation of center P; (2) $\mathfrak{q}^*$ is a maximal subideal of $\mathfrak{p}_0^*$.

18. Tangential ideals and quadratic transformations

Let $y_0, y_1, \cdots, y_n$ be homogeneous coördinates in S_n, $x_i = y_i/y_0$, $i = 1, 2, \cdots, n$. We consider the quadratic transformation:

$$(28)\qquad \begin{aligned} \rho y_{0j} &= y_0y_j, & i, j &= 1, 2, \cdots, n, \\ \rho y_{ij} &= y_iy_j, & y_{ij} &= y_{ji}, \end{aligned}$$

where we interpret the $\frac{1}{2}(n+1)(n+2) - 1$ y_{ij}'s as homogeneous coördinates in a projective P_N, $N = \frac{1}{2}(n+1)(n+2) - 2$. The transformation (28) carries the system of hyperquadrics of S_n which pass through the point $P(y_0 = 1, y_1 = \cdots = y_n = 0)$ into the system of hyperplanes of P_N.

Let $\omega_0, \omega_1, \cdots, \omega_n$ be the homogeneous coördinates of the general point of F. Here ω_0 is a transcendental quantity over the field Σ, and $\omega_i/\omega_0 = \xi_i$. The elements

$$(29)\qquad \begin{aligned} \omega_{0j} &= \omega_0\omega_j, \\ \omega_{ij} &= \omega_i\omega_j, \end{aligned} \qquad i, j = 1, 2, \cdots, n,$$

can be regarded as the homogeneous coördinates of the general point of a surface F' birationally equivalent to F and lying in P_N. The surface F' is *the transform of F by the quadratic transformation* (18.) In the system of non-homogeneous coördinates $x_{ij}^{(\nu)} = y_{ij}/y_{0\nu}$, ν-fixed, the non-homogeneous coördinates of the general point of F' are

$$(30)\qquad \begin{aligned} \xi_{0j}^{(\nu)} &= \xi_j/\xi_\nu, & j &= 1, 2, \cdots, \nu - 1, \nu + 1, \cdots, n, \\ \xi_{ij}^{(\nu)} &= \xi_i\xi_j/\xi_\nu, & i, j &= 1, 2, \cdots, n, \end{aligned}$$

and we have

$$(30')\qquad \mathsf{K}[\xi_{ij}^{(\nu)}] = \mathsf{K}[\xi_\nu, \xi_1/\xi_\nu, \cdots, \xi_{\nu-1}/\xi_\nu, \xi_{\nu+1}/\xi_\nu, \cdots, \xi_n/\xi_\nu] = \mathfrak{J}^{(\nu)}.$$

Let A be a point on F different from P, and let $y_0 = b_0$, $y_1 = b_1$, $y_n = b_n$ be the homogeneous coördinates of A, where $b_1, \cdots, b_n$ are not all zero. Let, for instance, $b_n \neq 0$. If we choose as non-homogeneous coördinates in P_n the ratios $\overset{*}{x}_i = y_i/y_n$ $i = 0, 1, \cdots, n - 1$, then the point A will be at finite distance and will have coördinates $\overset{*}{x}_i = a_i = b_i/b_n$. In this system of coördinates, the coördinates of the general point of F are the elements $1/\xi_n$, $\xi_1/\xi_n, \cdots, \xi_{n-1}/\xi_n$, giving rise to the ring

$$\mathfrak{J}_n = \mathsf{K}[1/\xi_n, \xi_1/\xi_n, \cdots, \xi_{n-1}/\xi_n].$$

Let us choose as non-homogeneous coördinates in P_N the ratios $\overset{*}{x}_{ij} = y_{ij}/y_{nn}$. The coördinates $\overset{*}{\xi}_{ij}$ of the general point of F' will then be the elements:

$$\overset{*}{\xi}_{0j} = \frac{1}{\xi_n}\cdot\frac{\xi_j}{\xi_n}, \qquad \overset{*}{\xi}_{ij} = \frac{\xi_i}{\xi_n}\cdot\frac{\xi_j}{\xi_n}, \quad i, j = 1, 2, \cdots, n - 1;$$

$$\overset{*}{\xi}_{0n} = \frac{1}{\xi_n}, \qquad \overset{*}{\xi}_{in} = \frac{\xi_i}{\xi_n}, \qquad i = 1, 2, \cdots, n - 1.$$

The ring $\mathsf{K}[\overset{*}{\xi}_{ij}]$ coincides with $\mathfrak{J}_n$. Hence to every point A of F at which the $\overset{*}{x}_i$ are finite, there corresponds a point A' on F' at which the $\overset{*}{x}_{ij}$ are finite, *and the quotient rings of the two corresponding points coincide.* From the coincidence of the quotient rings $Q(A)$, $Q(A')$ follows that *A' is the only point of F' which corresponds to A, and vice versa* (see IV, 14). *This, and the invariance of the quotient ring $Q(A)$, holds then for every point A of F which is different from P.*

Let now P' be a point of F' which corresponds to P. There must exist then a zero-dimensional valuation B of Σ whose center on F is P and whose center on F' is P'. The elements $\xi_1, \cdots, \xi_n$ have positive values in B. Let ξ_ν be one of the ξ's for which $v_B(\xi_\nu) = \min\ (v_B(\xi_1), \cdots, v_B(\xi_n))$. Then $v_B(\xi_i/\xi_\nu) \geqq 0$, $i \neq \nu$. Hence, in view of (30), the point P', in the system of non-homogeneous coördinates $x_{ij}^{(\nu)}$, is a point at a finite distance. It is therefore given by a prime zero-dimensional ideal $\mathfrak{p}_0^{(\nu)}$ in the ring $\mathfrak{J}^{(\nu)}$ of (30'). This ideal must be a divisor of the principal ideal $\mathfrak{J}^{(\nu)}\cdot\xi_\nu$, since $v_B(\xi_\nu) > 0$. The point P' therefore belongs to the curve $\Gamma^{(\nu)}$ defined on F' by the principal ideal $\mathfrak{J}^{(\nu)} \cdot \xi_\nu$.[20] Conversely, let P' be any point of $\Gamma^{(\nu)}$, and at finite distance with respect to the non-homogeneous coördinates $x_{ij}^{(\nu)}$. If B is a zero-dimensional valuation whose center is P', then $v_B(\xi_\nu) > 0$, $v_B(\xi_i/\xi_\nu) \geqq 0$, $i \neq \nu$. Hence $v_B(\xi_i) > 0$, $i = 1, 2, \cdots, n$, and therefore the center of B on F is the point P. Consequently, *every irreducible component of* $\Gamma^{(\nu)}$ *is a fundamental curve which corresponds to* P (see IV, 14, theorem 4) *and if* P' *is any point of* $\Gamma^{(\nu)}$, P *is the only point which corresponds to* P'.

It is clear that the relationship between the curves $\Gamma^{(1)}, \Gamma^{(2)}, \cdots, \Gamma^{(n)}$ is simply the following: an irreducible component of a $\Gamma^{(i)}$, $i \neq \nu$, which does not belong to $\Gamma^{(\nu)}$, is necessarily, in the system of non-homogeneous coördinates $x_{ij}^{(\nu)}$, a curve at infinity. It is also not difficult to see that the entire fundamental curve $\Gamma = \Gamma^{(1)} + \cdots + \Gamma^{(n)}$ is given in the ring $K[\omega_{ij}]$ by the *homogeneous* 1-dimensional ideal $(\omega_{11}, \omega_{12}, \cdots, \omega_{nn})$.

We go back to the point P' on $\Gamma^{(\nu)}$ considered above and to the corresponding prime zero-dimensional ideal $\mathfrak{p}_0^{(\nu)}$ in $\mathfrak{J}^{(\nu)}$. Let

$$(31) \quad \begin{aligned} \mathfrak{p}_0^{(\nu)} = \mathfrak{J}^{(\nu)} \cdot (\xi_\nu, \xi_1/\xi_\nu - c_1, \cdots, \xi_{\nu-1}/\xi_\nu - c_{\nu-1}, \\ \cdots, \xi_{\nu+1}/\xi_\nu - c_{\nu+1}, \cdots, \xi_n/\xi_\nu - c_n). \end{aligned}$$

If B is any zero-dimensional valuation whose center is P', then

$$v_B(\xi_i - c_i\xi_\nu) > v_B(\xi_\nu), \qquad i \neq \nu.$$

For all such valuations, the corresponding tangential ideal $\mathfrak{q}^*$ in the quotient ring $Q(P)$, is one and the same, namely:

$$(32) \quad \mathfrak{q}^* = \mathfrak{J}^* \cdot (\xi_1 - c_1\xi_\nu, \cdots, \xi_{\nu-1} - c_{\nu-1}\xi_\nu, \xi_{\nu+1} - c_{\nu+1}\xi_\nu, \cdots, \xi_n - c_n\xi_\nu, \mathfrak{p}_0^{*2}).$$

Hence, *every point* P' *which corresponds to* P *determines uniquely a tangential ideal* $\mathfrak{q}^*$ in $Q(P)$. The relationship between P' and $\mathfrak{q}^*$ is such that, if B is any zero-dimensional valuation with center at P', then $\mathfrak{q}^*$ is a valuation ideal of B in $Q(P)$ (namely, $\mathfrak{q}^*$ is the ideal consisting of all those elements of $Q(P)$ whose value is greater than $v_B(\mathfrak{p}_0^*)$). That, conversely, every tangential subideal of $\mathfrak{p}_0^*$ determines a point P' which corresponds to P, is obvious, in view of our definition of a tangential ideal in terms of zero-dimensional valuations. We conclude that *there is a* (1, 1) *correspondence between the points* P' *of the entire fundamental curve* Γ *and the tangential ideals of* $Q(P)$.

We now wish to elucidate the relationship between the quotient ring $\mathfrak{J}^* = Q(P)$ and the quotient ring $\mathfrak{J}'^* = Q(P')$ of a point P' of the fundamental curve. We assume, as above, that P' is on $\Gamma^{(\nu)}$, at finite distance. We observe first of all that the ring $\mathfrak{J} = \mathsf{K}[\xi_i]$ is a subring of $\mathfrak{J}^{(\nu)}$, as is seen immediately from (30′). Hence, also $\mathfrak{J}^*$ is a subring of $\mathfrak{J}'^*$. Moreover,

$$\mathfrak{J}^{(\nu)} \cdot \mathfrak{p}_0 = \mathfrak{J}^{(\nu)} \cdot (\xi_1, \cdots, \xi_n) = \mathfrak{J}^{(\nu)} \cdot \xi_\nu,$$

[20] It is well known that any principal ideal in $\mathfrak{J}^{(\nu)}$, different from the unit ideal, is one-dimensional, and its prime zero-dimensional components are all imbedded.

and, correspondingly,

$$\mathfrak{J}'^{*}\cdot\mathfrak{p}_0 = \mathfrak{J}'^{*}\cdot\xi_\nu\,. \tag{33}$$

We also point out the relation:

$$\mathfrak{J}'^{*}\cdot\mathfrak{q}^{*} = \xi_\nu\cdot\mathfrak{p}_0^{*(\nu)}, \tag{33'}$$

where $\mathfrak{p}_0^{*(\nu)}$ is the prime zero-dimensional ideal of the quotient ring $Q(P')$. This follows from (31) and (32).

Without loss of generality we may assume that the constants $c_1, \cdots, c_{\nu-1}, c_{\nu+1}, \cdots, c_n$ in (31) are all zero. Then (32) becomes

$$\mathfrak{q}^{*} = (\xi_1, \cdots, \xi_{\nu-1}, \xi_{\nu+1}, \cdots, \xi_n, \mathfrak{p}_0^{*2}). \tag{34}$$

If we put $\xi_i/\xi_\nu = \xi_i'$, $i \neq \nu$, $\xi_\nu = \xi_\nu'$, then any element η' of $Q(P')$ is a quotient of two polynomials $f(\xi_1', \cdots, \xi_n')/g(\xi_1', \cdots, \xi_n')$, with the condition $g(0, \cdots, 0) = 0$. Returning to the elements ξ, we see that η' is of the form:

$$\eta' = \frac{F(\xi)}{G(\xi)} = \frac{F_\rho(\xi) + F_{\rho+1}(\xi) + \cdots}{G_\rho(\xi) + G_{\rho+1}(\xi) + \cdots}, \tag{35}$$

where F_i, G_i are forms of degree i in $\xi_1, \cdots, \xi_n$, with the condition:

$$G_\rho = d\cdot\xi_\nu^\rho + \cdots \qquad d \not= 0. \tag{35'}$$

Conversely, every quotient η' such as in (35), for which (35′) holds true, is an element of $Q(P')$, as can be seen immediately if both F and G are divided through by ξ_ν^ρ. It is clear that both F and G are elements of $\mathfrak{p}_0^{*\rho}$. In view of (34), the inequality $d \not= 0$ in (35′) is certainly satisfied if $G \not\equiv 0(\mathfrak{p}_0^{*\rho-1}\mathfrak{q}^{*})$. Conversely, let us assume that $d \not= 0$. Then $G \equiv 0(\mathfrak{p}_0^{*\rho-1}\mathfrak{q}^{*})$ would imply that ξ_ν^ρ is in $\mathfrak{p}_0^{*\rho-1}\cdot\mathfrak{q}^{*}$. This is impossible, since if B is any zero-dimensional valuation with center of P', then $v_B(\mathfrak{q}^{*}) > v_B(\xi_\nu)$.[21]

It follows that $G \not\equiv 0(\mathfrak{p}_0^{*\rho-1}\mathfrak{q}^{*})$ *is equivalent to the condition* (35′). In a similar fashion it follows that η' belongs to the prime zero-dimensional ideal of $Q(P')$, if and only if $F \equiv 0(\mathfrak{p}^{*\rho-1}\mathfrak{q}^{*})$.

For convenience we reassume the results of this section:

The quadratic transformation (28) (or (29)) *carries our surface F into a birationally equivalent surface F'. With the exception of the fundamental point P on F, the correspondence between the points of F and F' is strictly one-to-one, and there is invariance of quotient rings of corresponding points. To the point P there corresponds a fundamental curve Γ, and we have $Q(P) \subset Q(P')$ for every point P' on Γ. The points P' on Γ are in* (1, 1) *correspondence with the tangential ideals*

[21] This reasoning shows that $\mathfrak{p}_0^{*\rho} \neq \mathfrak{p}_0^{*\rho-1}\mathfrak{q}^{*}$, for all ρ. It is not difficult to show that *these inequalities, together with the condition that $\mathfrak{q}^{*}$ be a maximal subideal of $\mathfrak{p}_0^{*}$, characterize tangential ideals.* Namely, assuming that $\mathfrak{q}^{*}$ is given by (34), it is a straightforward matter to show that the above inequalities imply that the ideal $(\xi_\nu, \xi_1/\xi_\nu, \cdots, \xi_{\nu-1}/\xi_\nu, \xi_{\nu+1}/\xi_\nu, \cdots, \xi_n/\xi_\nu)$ in $\mathfrak{J}^{(\nu)}$ is necessarily zero-dimensional (and not the unit ideal). The point P' defined by this ideal is on $\Gamma^{(\nu)}$ and $\mathfrak{q}^{*}$ is the corresponding tangential ideal.

in $Q(P)$. *The correspondence is such that if to* P' *corresponds the tangential ideal* $\mathfrak{q}^*$, *then the quotient ring* $Q(P')$ *consists of all quotients* $\frac{F}{G}$, F, $G \subset Q(P)$, *such that: if* G *is exactly divisible by* $\mathfrak{p}_0^{*\rho}$ (i.e. if $G \equiv 0(\mathfrak{p}_0^{*\rho})$, $G \not\equiv 0(\mathfrak{p}_0^{*\rho+1})$), *then* $G \not\equiv 0(\mathfrak{p}_0^{*\rho-1}\mathfrak{q}^*)$ *and* $F \equiv 0(\mathfrak{p}_0^{*\rho})$. *The prime zero-dimensional ideal of* $Q(P')$ *consists of all the above quotients* $\frac{F}{G}$ *in which the numerator satisfies the additional condition*: $F \equiv 0(\mathfrak{p}_0^{*\rho-1}\mathfrak{q}^*)$.

The above description of the quotient ring $Q(P')$ is very similar to the one given in Theorem 8 in a different connection. The ideals $\mathfrak{p}_0^*$ and $\mathfrak{q}^*$ now play the rôle of $\mathfrak{A}$ and $\mathfrak{a}$ respectively.

19. Normal sequences of quotient rings

We have seen in the preceding section that the quotient ring $Q(P')$ of a point P' on the fundamental curve can be fully described in terms of $Q(P)$ and of a tangential ideal $\mathfrak{q}^*$ in $Q(P)$. The operation consisting in passing from $Q(P)$ to $Q(P')$ is therefore an intrinsic operation on the quotient ring $Q(P)$. The quadratic transformation (28), (or (29)), is only *a projective realization* of this operation. Nevertheless, we shall continue to say that $Q(P')$ *is obtained from* $Q(P)$ *by a quadratic transformation.* There are as many *distinct* quotient rings obtainable from $Q(P)$ by a quadratic transformation, as there are tangential ideals in $Q(P)$.

Let us now assume that $Q(P)(=\mathfrak{J}^*)$ is integrally closed. Then it is quite possible that $\mathfrak{J}' = Q(P')$ is not integrally closed.[22] Let $\bar{\mathfrak{J}}'$ be the integral closure of $\mathfrak{J}'$. The prime zero-dimensional ideal of $\mathfrak{J}'$ may split up in $\bar{\mathfrak{J}}'$ into a finite number of distinct prime zero-dimensional ideals. Let $\mathfrak{p}_0'$ be one of them, and let $\mathfrak{J}_1^* = \bar{\mathfrak{J}}'_{\mathfrak{p}_0'}$ be the corresponding quotient ring. The quotient ring $\mathfrak{J}_1^*$ is again integrally closed and possesses only one prime zero-dimensional ideal. We shall say that $\mathfrak{J}_1^*$ *is obtained from* $\mathfrak{J}^*$ *by a quadratic transformation followed by integral closure.* It is quite clear that $\mathfrak{J}_1^*$ can be actually realized as the quotient ring of a point on some projective model of Σ. Namely, if F' is the transform of F by the quadratic transformation of the preceding section, and if F_1 is one of the normal surfaces determined by F', then $\mathfrak{J}_1^*$ is the quotient ring $Q(P_1)$ of a point P_1 on F_1.

---

[22] Example: Let P be the point $x = y = z = 0$ on the surface $z^2 = x^4 + y^4$. Apply the quadratic transformation $x' = x$, $y' = \frac{y}{x}$, $z' = \frac{z}{x}$ (i.e. pass to the ring $\mathfrak{J}^{(1)}$, given in (30′), where $\xi_1 = x$, $\xi_2 = y$, $\xi_3 = z$). The transformed surface $z'^2 = x'^2(1 + y'^4)$ has a multiple line $x' = z' = 0$ (the fundamental line of the transformation). Hence while the ring $K[x, y, z]$ is integrally closed, the ring $K[x', y', z']$ is not (for instance $\frac{z'}{x'}$ is integrally dependent on $K[x', y']$.) It is clear that if P' is any point on the line $x' = z' = 0$, the quotient ring $Q(P')$ is not integrally closed. More generally, whenever the fundamental curve Γ on F', created by the quadratic transformation, is a multiple curve, the quotient ring $Q(P')$ of any point on Γ will not be integrally closed.

We say that an infinite sequence of quotient rings

$$\mathfrak{J}^* = Q(P), \qquad \mathfrak{J}_1^* = Q(P_1), \qquad \mathfrak{J}_2^* = Q(P_2), \cdots,$$

is a *normal sequence* if: (1) each $Q(P_i)$ is the quotient ring of a point P_i on some normal projective model F_i of our field Σ; (2) and if $Q(P_{i+1})$ is obtained from $Q(P)_i$ by a quadratic transformation followed by integral closure.

Let $\Omega = \lim_{i \to \infty} \mathfrak{J}_i^*$ be the *limit ring* of the rings $\mathfrak{J}_i^*$, i.e. the union of the rings $\mathfrak{J}_i^*$.

THEOREM 10. *The limit ring Ω of a normal sequence of quotient rings is the valuation ring of a zero-dimensional valuation of Σ.*

PROOF. We first show that there *cannot exist a one-dimensional valuation B_1 of Σ with these two properties*: (1) *the valuation ring of B_1 contains Ω*; (2) *B_1 is of second kind with respect to each ring $\mathfrak{J}_i^*$*. In fact, let B_1 be any one-dimensional valuation of Σ. Since B_1 is one-dimensional, we can find an element α in the valuation ring of B_1 such that α is not congruent to a constant modulo the prime ideal of B_1. Let $\alpha = \xi/\eta$, $\xi, \eta \subset Q(P)$, and let us assume that B_1 is of second kind with respect to $\mathfrak{J}^*$, i.e., that the contracted ideal in $\mathfrak{J}^*$ of the prime ideal of B_1 is the prime zero-dimensional $\mathfrak{p}^*$ of $\mathfrak{J}^*$ (we denote, more generally, the prime zero-dimensional ideal of $\mathfrak{J}_i^*$ by $\mathfrak{p}_i^*$). Under this hypothesis, α is not in $Q(P)$, since every element of $Q(P)$ is congruent to a constant mod $\mathfrak{p}^*$. Hence $\eta \equiv 0(\mathfrak{p}^*)$, whence also $\xi \equiv 0(\mathfrak{p}^*)$, since $v_{B_1}(\alpha) \geqq 0$. Now let $\mathfrak{q}^*$ be the tangential ideal in $\mathfrak{J}^*$ determined by the quotient ring $\mathfrak{J}_1^*$ (or rather by the intermediate quotient ring $\mathfrak{J}'$ which was obtained from $\mathfrak{J}$ by a quadratic transformation and from which $\mathfrak{J}_1^*$ is obtained by the operation of integral closure). Let ζ be an element in $\mathfrak{p}^*$, but not in $\mathfrak{q}^*$. Then ξ/ζ and η/ζ are elements of $\mathfrak{J}'$, and hence also of $\mathfrak{J}_1^*$. Let then $\xi = \zeta\xi_1$, $\eta = \zeta\eta_1$, $\xi_1, \eta_1 \subset \mathfrak{J}_1^*$. Since B_1 is of second kind with respect to $\mathfrak{J}^*$, we have $v_{B_1}(\zeta) > 0$, whence $v_{B_1}(\xi) > v_{B_1}(\xi_1)$, $v_{B_1}(\eta) > v_{B_1}(\eta_1)$, and $\alpha = \xi_1/\eta_1$.

Now, in a similar fashion, if we assume that B_1 is of second kind with respect to $\mathfrak{J}_1^*$, then we can write $\alpha = \xi_2/\eta_2$, where $\xi_2, \eta_2 \subset \mathfrak{J}_2^*$ and

$$v_{B_1}(\xi) > v_{B_1}(\xi_1) > 0, \qquad v_{B_1}(\xi_1) > v_{B_1}(\xi_2),$$

$$v_{B_1}(\eta) > v_{B_1}(\eta_1) > 0, \qquad v_{B_1}(\eta_1) > v_{B_1}(\eta_2).$$

More generally, if we assume that B_1 is of second kind with respect to $\mathfrak{J}^*$, $\mathfrak{J}_1^*, \mathfrak{J}_2^*, \cdots, \mathfrak{J}_m^*$, then we can find elements ξ_i, η_i in each $\mathfrak{J}_i$ $(i = 1, 2, \cdots, m)$ such that $\alpha = \xi_i/\eta_i$ and

$$v_{B_1}(\xi) > v_{B_1}(\xi_1) > \cdots > v_{B_1}(\xi_m) > 0$$

$$v_{B_1}(\eta) > v_{B_1}(\eta_1) > \cdots > v_{B_1}(\eta_m) > 0.$$

Since the value group of B_1 is the group of integers, it follows that *m cannot be greater than* min $(v(\xi), v(\eta))$. Hence, for all i sufficiently high, B_1 must be of first kind with respect to $\mathfrak{J}_i^*$, which proves our assertion.

We next show that Ω is a *proper subring* of Σ. In fact, let α, β be elements of $\mathfrak{J}^*$ such that $\alpha \not\equiv 0(\mathfrak{p}^*)$, $\beta \equiv 0(\mathfrak{p}^*)$. By the nature of the transformation which carries from $\mathfrak{J}^*$ to $\mathfrak{J}_1^*$, we have that α/β is not in $\mathfrak{J}_1^*$. Moreover $\alpha \not\equiv 0(\mathfrak{p}_1^*)$, $\beta \equiv 0(\mathfrak{p}_1^*)$, whence α/β is not in $\mathfrak{J}_2^*$. More generally, $\alpha/\beta \not\subset \mathfrak{J}_i^*$, for all i, i.e. $\alpha/\beta \not\subset \Omega$.

On the basis of these results, theorem 10 now appears as a special case of the following general

LEMMA. *Let*

$$\mathfrak{J}^* \subset \mathfrak{J}_1^* \subset \cdots \subset \mathfrak{J}_i^* \subset \cdots$$

be an arbitrary (non necessarily normal) infinite ascending sequence of integrally closed quotient rings: $\mathfrak{J}_i^* = Q(P_i)$, *and let* $\Omega = \lim \mathfrak{J}_i^*$. *If* Ω *is a proper subring of* Σ, *and if every one-dimensional valuation* B_1 *of* Σ, *whose valuation ring contains* Ω, *is necessarily of the first kind with respect to some* $\mathfrak{J}_i^*$ *(and hence also with respect to all* $\mathfrak{J}_j^*$, $j \geqq i$*), then* Ω *is the valuation ring of a zero-dimensional valuation.*[23]

PROOF OF THE LEMMA. The proper subring Ω of Σ is integrally closed, and is therefore, at any rate, the intersection of the valuation rings containing it. We assert, that in the representation of Ω as an intersection of valuation rings, it is permissible to omit all the rings which belong to one-dimensional valuations. To prove this assertion, it is sufficient to show that if B_1 is a one-dimensional valuation, whose ring $\mathfrak{B}_1$ contains Ω, then Ω is already contained in the valuation ring $\mathfrak{B}_0$ of a zero-dimensional valuation, such that $\mathfrak{B}_0 \subset \mathfrak{B}_1$. The valuation B_1 maps Σ upon a field Σ^* of algebraic functions of one variable. By hypothesis, for some $i = m$, it will be true that B_1 is of first kind with respect to all $\mathfrak{J}_i^*$ $i \geqq m$. The prime ideal of B_1 in $\mathfrak{J}_i^*$ is then one dimensional, say $\bar{\mathfrak{p}}_i$. We consider the ring of residual classes $R_i^* = \mathfrak{J}_i^*/\bar{\mathfrak{p}}_i$. R_i^* is a subring of Σ^*, and Σ^* is the quotient field of R_i^*, since $\mathfrak{J}_i^*$ is integrally closed in Σ. Moreover, R_i^* is the quotient ring of a prime ideal of a finite integral domain in Σ^*, because the same is true of $\mathfrak{J}_i^*$ in the field Σ. Hence R_i^* *is contained in at least one and in at most a finite set of valuation rings of* Σ^*. Let H_i be the (finite and non-vacuous) set of "places" (prime divisors) of Σ^* whose valuation rings contain R_i^*. Since $\bar{\mathfrak{p}}_i = \bar{\mathfrak{p}}_{i+1} \cap \mathfrak{J}_i^*$, it follows that $R_i^* \subseteqq R_{i+1}^*$, whence H_{i+1} *is a subset of* H_i. Since no H_i is vacuous and each H_i consists of a finite number of places, it follows that at least one "place" $\mathfrak{P}$ of Σ^* belongs to all sets H_i. This place $\mathfrak{P}$ of Σ^* determines a zero-dimensional valuation B of Σ, composed with

---

[23] The following is an example in which the last condition of the lemma is not satisfied. Let $\Sigma = K(x, y)$, $\mathfrak{o} = K[x, y]$, $\mathfrak{o}_i = K[x, y, y^2/x, \cdots, y^i/x^{i-1}]$. Each ring $\mathfrak{o}_i$ is integrally closed in Σ and $\mathfrak{o}_i \cdot (x, y, y^2/x, \cdots, y^i/x^{i-1})$ is a prime zero-dimensional ideal $\mathfrak{p}_i$ in $\mathfrak{o}_i$. Let $\mathfrak{J}_i^* = \mathfrak{o}_i \mathfrak{p}_i$. The quotient rings $\mathfrak{J}^*, \mathfrak{J}_1^*, \cdots$ are all integrally closed and form a strictly ascending sequence. If we put $x' = x$, $y' = \frac{y}{x}$, then $\mathfrak{o}_i = K[x', x'y', \cdots, x'y'^i]$, and hence the one-dimensional valuation defined by the prime ideal (x') in the ring $K[x', y']$ is of second kind with respect to each $\mathfrak{J}_i^*$. The elements $\frac{y}{x}$ and $\frac{x}{y}$ do not belong to any of the rings $\mathfrak{J}_i^*$, and hence the union of the rings $\mathfrak{J}_i^*$ is not a valuation ring.

B_1, and clearly the valuation ring of B contains each quotient ring $Q(P_i)$ and is contained in $\mathfrak{B}_1$, q.e.d.

We may then assert that Ω is the intersection of valuation rings belonging to zero-dimensional valuations. Hence, in order to prove our lemma, we have only to prove that Ω *is not contained in two distinct zero-dimensional valuation rings.*

Let us assume that there exist two distinct zero-dimensional valuations B and B' whose valuations rings contain Ω. Let F_1 be a projective model (for instance, a normal model) of the field on which the centers of B and B' are two *distinct* points, say A and A' (the existence of such a surface F_1 follows quite trivially from the hypothesis that B and B' are distinct valuations). Let moreover, $F^{(i)}$ be some projective model of Σ (for instance, a normal model) such that $\mathfrak{I}_i^*$ is the quotient ring of a point P_i on $F^{(i)}$. Since the valuation ring of B contains Ω, its center on $F^{(i)}$ is the point P_i. Hence P_i and A are corresponding points in the birational correspondence between F_1 and $F^{(i)}$. Similarly, P_i and A' are corresponding points. Since to P_i there thus correspond two distinct points of F_1 and since $Q(P_i)$ is integrally closed, it follows, by theorem 5, that P_i must be a fundamental point. Let Γ_i be the set of fundamental curves on F_1 which correspond to P_i. Let $\Gamma_{i+1,j}$ be an irreducible component of Γ_{i+1}. There exists a one-dimensional valuation whose center on F_1 is $\Gamma_{i+1,j}$ and whose center on $F^{(i+1)}$ is P_{i+1}. The center of this valuation on $F^{(i)}$ will then be P_i, since $Q(P_i) \subset Q(P_{i+1})$. Hence $\Gamma_{i+1,j}$ is also a fundamental curve for P_i and for the birational correspondence between $F^{(i)}$ and F_1. Thus, *every irreducible component* of Γ_{i+1} *is also a component of* Γ_i. Since the number of irreducible components in a Γ_i is finite, and > 1, it follows that *there exists an irreducible curve* Γ *on* F_1, *which belongs to each curve* Γ_{i+1}. There exists therefore a 1-dimensional valuation (whose center on F_1 is the curve Γ) which is of second kind with respect to each ring $Q(P_i)$, in contradiction with our hypothesis. This proves our lemma and theorem 10.

20

"Local Reduction" Theorem: *If* $\mathfrak{I}^* = Q(P)$, $\mathfrak{I}_1^* = Q(P_1)$, $\cdots$ *is a normal sequence of quotient rings, then, for some integer m, all points* P_i, $i \geqq m$, *are simple points.*

Proof. Let $\Omega = \text{Lim } \mathfrak{I}_i^*$, and let B be the zero-dimensional valuation whose valuation ring $\mathfrak{B}$ coincides with Ω (by theorem 10). We apply our fundamental lemma (proved in Part II). There exist, according to this lemma, a projective model of Σ on which the center of B is a simple point A. We consider the quotient ring $\mathfrak{o}^* = Q(A)$. It is contained in $\mathfrak{B}$, whence the valuation B determines in $Q(A)$ a tangential ideal $\mathfrak{q}^*$ (V, 18). The ideal $\mathfrak{q}^*$ determines a quotient ring $\mathfrak{o}' = Q(A')$, obtainable from $Q(A)$ by a quadratic transformation and evidently contained in $\mathfrak{B}$. The prime zero-dimensional ideal of $\mathfrak{o}'$ splits, in the integral closure $\bar{\mathfrak{o}}'$ of $\mathfrak{o}'$, into a finite number of prime zero-dimensional ideals, one of which, say $\mathfrak{p}$, will be the contracted ideal of the ideal of non-units in $\mathfrak{B}$. We pass to the quotient ring $\mathfrak{o}_1^* = \bar{\mathfrak{o}}_\mathfrak{p}' = Q(A_1)$. This ring is obtained from $\mathfrak{o}^*$ by a quadratic transformation, followed by integral closure. Continuing this process, we get a normal sequence

$$\mathfrak{o}^* = Q(A), \qquad \mathfrak{o}_1^* = Q(A_1), \qquad \mathfrak{o}_2^* = Q(A_2), \cdots .$$

The quotient rings $\mathfrak{o}_i^*$ are all contained in $\mathfrak{B}$. Hence, by theorem 10, Lim $\mathfrak{o}_i^*$ = Lim $\mathfrak{J}_i^* = \mathfrak{B}$.

Let x, y be uniformizing parameters at the simple point A, which we may assume to be the point $x = y = z = 0$ on some surface $f(x, y, z) = 0$. The elements of $\mathfrak{o}^*$ are those elements of Σ which are formal power series in x and y. We may assume $v(y) > v(x)$, and also $v(z) > v(x)$. Our quadratic transformation with fundamental point at A consists then in passing from the ring $K[x, y, z]$ to the ring $K[x_1, y_1, z_1]$, where $x_1 = x$, $y_1 = y/x$, $z_1 = z/x$, and the point A' is the point $x' = y' = z' = 0$. This point is clearly a simple point of the transformed surface $f_1(x_1, y_1, z_1) = 0$, since $z_1 = 0$ is a simple root of $f_1(0, 0, z_1)$. Hence $Q(A')$ itself is already integrally closed; $Q(A') = Q(A_1)$ and $Q(A_1)$ consists of all elements of Σ which are holomorphic functions of the new elements x_1, y_1. In conclusion, the passage from $Q(A)$ to $Q(A_1)$ is simply effected by a quadratic transformation applied to the uniformizing variables. The points A_i are all simple points.

We shall denote by x_i, y_i the uniformizing parameters at A_i. Following the notations of IV, 15, we denote the prime zero-dimensional ideals of $\mathfrak{J}^*$ and $\mathfrak{o}^*$ by $\mathfrak{P}_0$, $\mathfrak{p}_0$ respectively. The prime zero-dimensional ideals of $\mathfrak{J}_i^*$ and $\mathfrak{o}_i^*$ will be denoted by $\mathfrak{P}_i$ and $\mathfrak{p}_i$ respectively.

Since the rings $\mathfrak{J}_i^*$ and $\mathfrak{o}_i^*$ have the same limit ring, namely $\mathfrak{B}$, and since each of the rings $\mathfrak{J}_i^*$, $\mathfrak{o}_i^*$ is a quotient ring in a finite integral domain, it follows that each $\mathfrak{J}_i^*$ is a subring of some $\mathfrak{o}_j^*$, and each $\mathfrak{o}_i^*$ is a subring of some $\mathfrak{J}_j^*$. Dropping, if necessary, a finite number of elements in the normal sequence $\{\mathfrak{J}_i^*\}$, we may assume that already the first ring, $\mathfrak{J}^*$, contains the ring $\mathfrak{o}^*$. If $\mathfrak{J}^*$ also contains $\mathfrak{o}_1^*, \cdots, \mathfrak{o}_m^*$, but does not contain $\mathfrak{o}_{m+1}^*$, we may start with $\mathfrak{o}_m^*$ as the first member of the sequence $\{\mathfrak{o}_i^*\}$. Hence we may assume that $\mathfrak{J}^* \supseteqq \mathfrak{o}^*$, $\mathfrak{J}^* \not\supseteqq \mathfrak{o}_1^*$.

If $\mathfrak{J}^* = \mathfrak{o}^*$, there is nothing to prove: P is then a simple point. Assume then that $\mathfrak{o}^*$ *is a proper subring of* $\mathfrak{J}^*$. We can apply Theorem 8. According to that theorem, the elements of $\mathfrak{J}^*$ are all of the form α/β, where $\alpha, \beta \subset \mathfrak{o}^*$, $\alpha, \beta \equiv 0(\mathfrak{A}^\sigma)$, $\beta \not\equiv 0(\mathfrak{A}^{\sigma-1}\mathfrak{a})$, σ—an integer. We now make use of theorem 9. The hypothesis of that theorem was that $\mathfrak{a}$ is exactly divisible by $\mathfrak{p}_0^{\rho+1}$, if $\mathfrak{A}$ is exactly divisible by $\mathfrak{p}_0^\rho$. We show that this hypothesis is satisfied in the present case and that it is a consequence of our assumption $\mathfrak{J}^* \not\supseteqq \mathfrak{o}_1^*$. In fact, let us assume that $\mathfrak{a}$ is not divisible by $\mathfrak{p}^{\rho+1}$, whence $\mathfrak{a}$ is exactly divisible by $\mathfrak{p}^\rho$. We consider the decomposition of $\mathfrak{A}$ and $\mathfrak{a}$ into simple v-factors. We make use of the fact that the system of subforms of a simple v-ideal $\mathcal{P}$, of kind > 1, is of dimension zero (see Z_1, p. 163, Corollary 3.2). According to this result, there is associated with every simple v-ideal $\mathcal{P}$, of kind > 1, a linear form $ax + by$.

If $\mathcal{P}$ is exactly divisible by $\mathfrak{p}_0^m$, then every element in $\mathcal{P}$ is of the form: $c \cdot (ax + by)^m$ + terms of higher degree. Let $a_i x + b_i y$, $i = 1, 2, \cdots, s$, be the *distinct* linear forms associated with the various simple v-factors of $\mathfrak{A}$ and $\mathfrak{a}$, We include among these forms also the form y, although *a priori* there may be no simple v-factor associated with y. Let, say, $a_1 x + b_1 y = y$ (i.e. $a_1 = 0$. $b_1 = 1$). We can then write the ideals $\mathfrak{A}$ and $\mathfrak{a}$ in the form:

$$\mathfrak{A} = \mathfrak{p}_0^m \mathfrak{A}_1 \mathfrak{A}_2 \cdots \mathfrak{A}_s,$$

$$\mathfrak{a} = \mathfrak{p}_0^n \mathfrak{a}_1 \mathfrak{a}_2 \cdots \mathfrak{a}_s,$$

where $\mathfrak{A}_i$ is the product of the simple v-factors of $\mathfrak{A}$ associated with the form $a_i x + b_i y$. Similarly for $\mathfrak{a}_i$ and $\mathfrak{a}$. If no simple v-factor of $\mathfrak{A}$ is associated with $a_i x + b_i y$, then we put $\mathfrak{A}_i = (1)$. Similarly for $\mathfrak{a}$ and $\mathfrak{a}_i$.

For a given i we consider the quadratic transformation $x' = x$, $y' = \dfrac{a_i x + b_i y}{x}$ (or $y' = y$, $x' = \dfrac{a_i x + b_i y}{y}$, if $b_i = 0$) and we pass to the ring Ω' of formal power series of x' and y'. Let $\mathfrak{A}_j$ be exactly divisible by $\mathfrak{p}_0^{m_j}$, and similarly, let $\mathfrak{a}_j$ be exactly divisible by $\mathfrak{p}_0^{n_j}$. It is clear that if $j \neq i$, then the transform of $\mathfrak{A}_j$ by our quadratic transformation is the unit ideal, i.e. $\Omega' \cdot \mathfrak{A}_j = (x'^{m_j})$. Similarly for $\mathfrak{a}_j$. On the other hand, since $\mathfrak{a} \equiv 0(\mathfrak{A})$, also $\Omega' \cdot \mathfrak{a} \equiv 0(\Omega' \cdot \mathfrak{A})$. Hence the transform of $\mathfrak{a}_i$ must be a multiple of, or coincide with the transform of $\mathfrak{A}_i$. This implies that $\mathfrak{a}_i$ itself is a multiple of $\mathfrak{A}_i$ (see Z_1, section 12). We have therefore $\mathfrak{a}_i = 0(\mathfrak{A}_i)$, for $i = 1, 2, \cdots, s$. Now $\mathfrak{a}$ is a maximal subideal of $\mathfrak{A}$. Hence at most one of the ideals $\mathfrak{a}_i$ can be a proper multiple of the corresponding ideal $\mathfrak{A}_i$.[24] We show that $\mathfrak{a}_1$ is necessarily a proper subideal of $\mathfrak{A}_1$. Namely, assume that $\mathfrak{a}_1 = \mathfrak{A}_1$. Since in our valuation B we have $v(y) > v(x)$, and since the coefficient b_i in the form $a_i x + b_i y$ is different from zero if $i \neq 1$, it follows that $v(\mathfrak{A}_i) = v(\mathfrak{p}_0^{m_i})$, $i \neq 1$. Similarly, $v(\mathfrak{a}_i) = v(\mathfrak{p}_0^{n_i})$, if $i \neq 1$. Hence, if $\mathfrak{a}_1 = \mathfrak{A}_1$, then $v(\mathfrak{A}) = v(\mathfrak{a})$, since $m + m_1 + \cdots + m_s = n + n_1 + \cdots + n_s = \rho$ (see also preceding footnote). This is in contradiction with the definition of the ideals $\mathfrak{A}$ and $\mathfrak{a}$.

We have then $\mathfrak{A}_1 \neq \mathfrak{a}_1$, $\mathfrak{A}_i = \mathfrak{a}_i$, $i = 2, \cdots, s$. Now consider any valuation B' of center P. We must have $v_{B'}(\mathfrak{A}) < v_{B'}(\mathfrak{a})$. Hence $v_{B'}(\mathfrak{A}_1) < v_{B'}(\mathfrak{a}_1)$, or $v_{B'}(\mathfrak{A}_1 \mathfrak{p}_0) < v_{B'}(\mathfrak{a}_1)$, according as $m = n$, or $m = n + 1$ (see preceding footnote). Since the subform of $\mathfrak{a}_1$ is y^{n_1}, it follows that necessarily $v_{B'}(y) > v_{B'}(x)$. Since

---

[24] *Proof.* We observe that the dimension s of the subforms (of degree ρ) of $\mathfrak{A}$ is m, since $\mathfrak{A}$ has the factor $\mathfrak{p}_0^m$. Similarly, the dimension of the subforms of $\mathfrak{a}$ is n. Since $\mathfrak{a}$ is a maximal subideal of $\mathfrak{A}$, necessarily n is either m or $m - 1$. If $n = m$, the assertion in the text is trivial. Suppose that $n = m - 1$. Since $m + m_1 + \cdots + m_s = n + n_1 + \cdots + n_s$ and $n_j \geqq m_j$, for all j, we deduce that $n_i = m_i + 1$, for some i, while $n_j = m_j$, $j \neq i$. Hence $\mathfrak{a}_i \neq \mathfrak{A}_i$ and therefore $\mathfrak{a}_i$ is a proper multiple of $\mathfrak{A}_i$. Since $\mathfrak{a}_i$ is exactly divisible by $\mathfrak{p}_0^{m_i+1}$, while $\mathfrak{A}_i$ is exactly divisible by m_i, it follows (Z_1, section 12), that $\mathfrak{a}_i$ is also a multiple of $\mathfrak{p}_0 \mathfrak{A}_i$. Hence the ideal $\mathfrak{p}_0^n \mathfrak{A}_1 \cdots \mathfrak{A}_{i-1} \mathfrak{a}_i \mathfrak{A}_{i+1} \cdots \mathfrak{a}_s$ is a proper multiple of $\mathfrak{A}$ and divides $\mathfrak{a}$. Consequently it coincides with $\mathfrak{a}$, q.e.d.

this is true for any zero-dimensional valuation of center P, it follows that the uniformizing parameters $x_1 = x$, $y_1 = y/x$ for the point A_1 are in $Q(P)$. Hence $Q(A_1) \subseteqq Q(P)$, and this proves our assertion.

In view of theorem 9 we can therefore assert that the simple v-factors of $\mathfrak{a}$ are predecessors of, or coincide with, the simple v-factors of $\mathfrak{A}$. Let $\mathfrak{o}_h^*$ be the first ring in the sequence $\{\mathfrak{o}_i^*\}$ such that $\mathfrak{o}_h^* \supseteqq \mathfrak{J}^*$ (such a ring exists, since $\operatorname{Lim} \mathfrak{o}_i^* = \mathfrak{B}$). By theorem 8, the elements of $\mathfrak{J}^*$ are all of the form α/β, where $\alpha, \beta \equiv 0(\mathfrak{A}^\sigma)$, $\beta \not\equiv 0(\mathfrak{A}^{\sigma-1}\mathfrak{a})$, σ—an arbitrary integer. All such quotients α/β must be therefore elements of $\mathfrak{o}_h^*$. But this is only possible if and only if the transform of the ideal $\mathfrak{A}$ by the sequence of quadratic transformations, carrying $\mathfrak{o}^*$ into $\mathfrak{o}_h^*$, is the unit ideal. In other words, the extended ideal $\mathfrak{o}_h^*\mathfrak{A}$ must be a principal ideal in $\mathfrak{o}_h^*$:

$$\mathfrak{o}_h^*\mathfrak{A} = \mathfrak{o}_h^* \cdot \xi^{(h)}.$$

Moreover, h must be the lowest integer for which this reduction of $\mathfrak{A}$ to the unit ideal takes place. Or, to be quite explicit: if $\mathfrak{A} = \mathcal{P}_1^{\alpha_1} \cdots \mathcal{P}_\mu^{\alpha_\mu}$ and if P_i is a simple v-ideal of kind h_i, then $h = \text{maximum } (h_1, \cdots, h_\mu)$. By theorem 9, the reduction of the ideal $\mathfrak{a}$ to the unit ideal will also take place at this stage, or even perhaps earlier, (i.e. in some ring $\mathfrak{o}_i^*$, $i < h$). Now, the prime ideal $\mathfrak{P}_0$ of $\mathfrak{J}^*$ consists of all those quotients α/β in which $\alpha \equiv 0(\mathfrak{A}^{\rho-1}\mathfrak{a})$. Hence —*and this is the goal we were aiming for*—we can state the following

LEMMA. *The extended ideal* $\mathfrak{o}_h^*\mathfrak{P}_0$ *is a principal ideal.*

On the basis of this lemma the proof of the local reduction theorem is rapidly completed. Let

$$\mathfrak{o}_h^*\mathfrak{P}_0 = \mathfrak{o}_h^* \cdot \alpha_h, \qquad \alpha_h \subset \mathfrak{o}_h^*. \tag{36}$$

Let $\mathfrak{Q}^*$ be the tangential ideal of $\mathfrak{J}^*$ (relative the valuation B whose valuation ring is the limit of the rings $\mathfrak{J}_i^*$). We have then

$$v_B(\mathfrak{Q}^*) > v_B(\mathfrak{P}_0). \tag{37}$$

By (36), every element η of $\mathfrak{P}_0$ is of the form: $\eta = \xi \cdot \alpha_h$, $\xi \subset \mathfrak{o}_h^*$. By (37), η is an element of $\mathfrak{Q}^*$ *if and only if* ξ *is not a unit in* $\mathfrak{o}_h^*$. It follows that the intermediate quotient ring $\mathfrak{J}'$, between $\mathfrak{J}^*$ and $\mathfrak{J}_1^*$, obtained from $\mathfrak{J}$ by a quadratic transformation, is contained in $\mathfrak{o}_h^*$. Since $\mathfrak{J}_1^*$ is obtained from $\mathfrak{J}'$ by a process of integral closure and since $\mathfrak{o}_h^*$ is integrally closed, it follows that the *entire ring* $\mathfrak{J}_1^*$ *is contained in* $\mathfrak{o}_h^*$.

Let us review in a few words the essence of the preceding results. We have focused our attention on the last ring of the sequence $\{\mathfrak{o}_i^*\}$ which is contained in $\mathfrak{J}^*$. For simplicity we took that ring as the first ring $\mathfrak{o}^*$ of the sequence. We had then $\mathfrak{J}^* \supseteqq \mathfrak{o}^*$, $\mathfrak{J}^* \not\supseteqq \mathfrak{o}_1^*$. We have assumed that $\mathfrak{o}^*$ is a proper subring of $\mathfrak{J}^*$, and on the basis of this assumption we have proved that, if $\mathfrak{o}_h^*$ is the first ring $\mathfrak{o}_i^*$ such that $\mathfrak{J}^* \subseteqq \mathfrak{o}_i^*$, then $\mathfrak{o}_h^*\mathfrak{P}_0$ is a principal ideal. From this alone we have concluded that also $\mathfrak{J}_1^* \subseteqq \mathfrak{o}_h^*$. Hence we can assert the following: *if* $\mathfrak{o}_h^*$ *is the first ring in the sequence* $\{\mathfrak{o}_i^*\}$ *which contains* $\mathfrak{J}^*$ *and if* $\mathfrak{J}^* \neq \mathfrak{o}_h^*$, *then*

$\mathfrak{o}_h^*$ *also contains* $\mathfrak{J}_1^*$. Obviously, $\mathfrak{o}_h^*$ is also the first ring $\mathfrak{o}_i^*$ which contains $\mathfrak{J}_1^*$. Hence if $\mathfrak{J}_1^*$ does not coincide with $\mathfrak{o}_h^*$, we conclude that $\mathfrak{o}_h^*$ contains $\mathfrak{J}_2^*$, and so on. Since the limit of the rings $\mathfrak{J}_i^*$ is the entire valuation $\mathfrak{B}$ of B, it follows that for some integer m we must have $\mathfrak{J}_m^* = \mathfrak{o}_h^*$. The point P_m, and every point P_i, $i \geqq m$, is then necessarily a simple point, q.e.d.

VI. Reduction of the Singularities

21

We start with a *normal* projective model F of our field Σ in a projective space P_m $(y_0, \cdots, y_m)$. Let $\omega_0, \omega_1, \cdots, \omega_m$ be the homogeneous coördinates of the general point of F. The surface F has at most a finite number of singular points. Let P be one of them. Without loss of generality we may assume that P is the point $y_0 = 1$, $y_1 = \cdots = y_m = 0$. We apply the quadratic transformation (28) of V, 18 and we get the transformed surface F'. The birational correspondence between F and F' has been studied in Part V and its properties were reassumed at the end of section 18. The surface F' may not be normal in its ambient projective space. If it is not normal, we apply to F' the birational transformation of integral closure described in the introduction, getting a new well-defined normal projective model F_1 of our field Σ. The resulting correspondence between the points of F and F_1 is again one to one, with the exception of the point P. In fact, if A is a point of F different from P, and if A' is the corresponding points of F', then we know that $Q(A) = Q(A')$. Since F is normal, $Q(A)$ is integrally closed in Σ. Hence also $Q(A')$ is integrally closed in Σ, and consequently, *A' goes into a unique point A_1 of F_1, and there is invariance of quotient rings*:

$$Q(A_1) = Q(A') = Q(A).$$

The points P_1 of F_1 which correspond to the fundamental point P on F are those which correspond to the points P' on the fundamental curve Γ on F'. The relation between $Q(P)$ and $Q(P_1)$ is clearly the outcome of two operations:

(α) *the quotient formation in terms of a tangential ideal in* $Q(P)$; this yields $Q(P')$;

($\bar{\alpha}$) *the splitting of the prime zero-dimensional ideal of* $Q(P')$ *into primary factors* $\bar{\mathfrak{q}}_i$ *in the integral closure* $\overline{Q(P')}$ of $Q(P')$; the quotient ring $Q(P_1)$ is the quotient ring of one of the corresponding prime ideals $\bar{\mathfrak{p}}_i$. These operations are exactly those described in section 19: a quadratic transformation followed by integral closure.

If A is a simple point of F, the corresponding point A_1 of F_1 is also a simple point. This follows from the invariance of quotient rings and from the fact, pointed out in the introduction, that the property of being a simple point is entirely a property of the quotient ring of that point. Since F_1 is normal, it has at most a finite number of singular points. Among these, there are in the first place those which correspond to the singular points of F, different from P—

and here we have no change in the quotient rings of corresponding points. The remaining singular points of F_1 correspond to P, and their quotient rings are obtained from $Q(P)$ by a quadratic transformation followed by integral closure. Thus *P has been resolved into a finite number of singular points P_1 whose quotient rings are more inclusive than $Q(P)$.*

If F_1 still has singular points, we take one of them, say P_1, as fundamental point of a quadratic transformation, getting a new surface F_1'. From F_1' we pass to the corresponding normal surface F_2. We repeat this process indefinitely, getting a *sequence of normal surfaces*:

$$F, F_1, F_2, \cdots. \tag{38}$$

We now prove our main theorem:

THEOREM OF REDUCTION OF SINGULARITIES: *After a finite number of quadratic and integral closure transformations, a normal surface F_i is obtained which is free from singularities.*

This main theorem is an immediate consequence of the "local reduction" theorem proved in part V. We consider the set M of all *normal* sequences $\{Q(P), Q(P_1), \cdots\}$ where the first point P is a *fixed* point of the given surface F. For each such sequence, let h be the smallest possible value of i such that P_i is a simple point. Then of course also all P_i, $i > h$, are simple points. The integer h depends on the sequence. We show however that *as the sequence $\{Q(P_i)\}$ varies in M, the integer h has a finite upper bound.* In fact, assuming the contrary, there would exist, *for any integer ν*, a sequence

$$\{Q(P_i^\nu)\}: Q(P), Q(P_1^{(\nu)}), Q(P_2^{(\nu)}), \cdots, Q(P_i^{(\nu)}), \cdots$$

such that $P_1^{(\nu)}, \cdots, P_\nu^{(\nu)}$ are singular points (on their respective surfaces). Now the points $P_1^{(\nu)}$ can all be realized on the surface F_1 of the sequence (38) of normal surfaces. They are namely the points which correspond to the point P of F. Since F_1 has only a finite number of singular points, one and the same point, say P_1, must occur in infinitely many of the sequences $\{Q(P_i^{(\nu)})\}$. If F_2 is the normal surface obtained from F_1 by taking P_1 as the fundamental point of the quadratic transformation, then the points $P_2^{(\nu)}$, belonging to the sequences $\{Q(P_i^{(\nu)})\}$ in which $P_1^{(\nu)} = P_1$, are points of F_2. Hence, by the same argument, there must exist infinitely many sequences $\{Q(P_i^{(\nu)})\}$ in which $P_1^{(\nu)} = P_1$ *and* $P_2^{(\nu)} = P_2$, where P_2 is a fixed point of F_2. Continuing in this fashion, we get an *infinite sequence* of points $P, P_1, P_2, \cdots$, all singular on their respective surfaces $F, F_1, F_2, \cdots$, and such that the sequence $\{Q(P_i)\}$ is normal. This is in contradiction with the "local reduction" theorem.

Let $g = g(P)$ be the least upper bound of h. The integer $g(P)$ is a numerical character of P. Clearly, *P is a simple point if and only if $g(P) = 0$.* Let $P', \cdots, P^{(\rho)}$ be the singular points of F, and let $\gamma = \max(g(P'), g(P''), \cdots, g(P^\sigma))$. Let $P', P'', \cdots, P^{(\rho)}$, $\rho \leqq \sigma$, be the points for which the character g is exactly equal to γ. We choose P' as fundamental point of our quadratic transformation and we pass from F to the normal surface F_1. If P_1' is a singular point of F_1 which corresponds to P', then it is obvious that $g(P_1') < g(P')$. If $P_1'', \cdots, P_1^{(\sigma)}$ are the points of F_1 which correspond to $P'', \cdots, P^{(\sigma)}$ respectively, then $g(P^{(j)}) = g(P_1^{(j)})$. Hence F_1 contains only $\rho - 1$ singular points for which the character g equals γ, while for any other singular point of F_1 the character g will be less than γ. If we repeat this procedure ρ times, always taking as fundamental point on F_i $(i = 1, \cdots \rho - 1)$ a singular point whose character g is equal to γ, we will get a normal surface F_ρ on which *the character g for each singular point is less than γ.* Ultimately, after a finite number of steps, a surface F_i will be obtained, in which every point is a simple point. This proves our main theorem.

THE JOHNS HOPKINS UNIVERSITY.

REFERENCES

1. ALBANESE, G.: *Transformazione birazionale di una superficie algebrica qualunque in un'altra priva di punti multipli,* Rend. Circ. mat. Palermo, Vol. 48 (1924).
2. CHISINI, O.: *La risoluzione delle singolarità di una superficie mediante transformazioni birazionali dello spazio,* Mem. Accad. Sci. Bologna VII. s. Vol. 8 (1921).
3. JUNG, H. W. E.: *Darstellung der Funktionen eines algebraischen Körpers zweier unabhängigen Veränderlichen x, y in der Umgebung einer Stelle $x = a$, $y = b$.* J. reine angew. Math. Vol. 133 (1908).
4. KRULL, W. *Idealtheorie,* Ergebnisse der Mathematik und ihrer Grenzgebiete, IV, 3. This monograph is quoted as "K".
5. B. LEVI. *Sulla risoluzione delle singolarità puntuali delle superficie algebriche dello spazio ordinario per transformazioni quadratiche,* Ann. Mat. pura appl. II. s. Vol. 26 (1897).
6. B. LEVI. *Risoluzione delle singolarità puntuali delle superficie algebriche,* Atti Accad. Sci. Torino, Vol. 33 (1897).
7. SCHILLING, O. F. G.: *A generalization of local class field theory,* Am. J. of Math., vol. 60 (1938).
8. VAN DER WAERDEN, B.: *Moderne Algebra,* vol. II (second edition).
9. WALKER, R. J.: *Reduction of singularities of an algebraic surface,* Annals of Mathematics, Vol. 36 (1935).
10. ZARISKI, O.: *Polynomial ideals defined by infinitely near base points,* Am. J. of Math., Vol. 60 (1938) (quoted Z_1).
11. ZARISKI, O.: *Some results in the arithmetic theory of algebraic varieties,* Am. J. of Math.. Vol. 61 (1939) (quoted Z_2).

ANNALS OF MATHEMATICS
Vol. 41, No. 4, October, 1940

LOCAL UNIFORMIZATION ON ALGEBRAIC VARIETIES

BY OSCAR ZARISKI*

(Received March 22, 1940)

CONTENTS

A. INTRODUCTION

I. Uniformization of zero-dimensional valuations

1. In [10] (p. 650) we have proved a uniformization theorem for zero-dimensional valuations on an algebraic surface, over an algebraically closed ground field K (of characteristic zero). In the present paper we generalize this theorem to algebraic varieties, and on the basis of this generalization we obtain a solution of the problem of local uniformization in the classical case (i.e. when K is the field of complex numbers). The exact formulation of the generalized theorem, in its strongest form, will be given in A III and A IV. However, to begin with, we state here the following theorem which is literally a repetition of our theorem for surfaces, with the surface replaced by a variety, and which will be included in our final result:

THEOREM U_1. *The Uniformization Theorem in invariantive form: Given a field Σ of algebraic functions of r independent variables, over an algebraically closed ground field* K *of characteristic zero, and given a zero-dimensional valuation B of Σ, there exists a projective model V of Σ on which the center of B is at a simple point P.*

This theorem is in effect entirely invariantive in nature: it refers exclusively to the field Σ and to the valuation B of Σ. It asserts the existence of *uni-*

* Guggenheim Fellow.

formizing parameters for B: they are the uniformizing parameters $t_1, t_2, \cdots, t_r$ at the simple point P (see [9], p. 258, [11], p. 199). Actually, however, the problem of local uniformization concerns not just the field Σ, but a specific projective model V_0 of Σ, given in advance. We must examine the situation more closely.

2. Let $(\xi_1, \cdots, \xi_n)$ be the general point of V_0 (in the affine space), and let P_0 be the center of B on V_0. The valuation B is represented on V_0 by a "branch" of some sort, and $t_1, \cdots, t_r$ are uniformizing parameters "along the branch." Let $(\eta_1, \cdots, \eta_m)$ be the general point of V. We assume that both P and P_0 are points at finite distance on the respective varieties V and V_0. The (non-homogeneous) coördinates η_i are expressible by means of formal power series of $t_1, \cdots, t_r$ ([9], p. 254). If K is the field of complex numbers, these power series (holomorphic functions of $t_1, \cdots, t_r$) will converge in some neighborhood of the initial values $t_1 = \cdots = t_r = 0$ and will yield (on V) a complete neighborhood E of the simple point P. However, the coördinates ξ_i, which are *rational* functions of $\eta_1, \cdots, \eta_m$, will be, in general, only *meromorphic* functions of $t_1, \cdots, t_r$ (i.e. quotients of power series). Following Walker [8], we shall say that the map of E on V_0 is *a wedge W on V_0* if $\xi_1, \cdots, \xi_n$ are holomorphic functions of $t_1, \cdots, t_r$.[1] What we want is a set of wedges on V_0 and—ultimately—a proof that a complete neighborhood of P_0—and hence also V_0 itself (since V_0 is bicompact)—can be covered by a finite set of wedges.[2] Hence the knowledge that there exist uniformizing parameters for B is not sufficient. We ought to be able to exhibit parameters $t_1, \cdots, t_r$ such that the ξ_i be holomorphic functions of the t's. Now let us observe that any polynomial ψ in $\eta_1, \cdots, \eta_m$ is an holomorphic function of $t_1, \cdots, t_r$ and that ψ vanishes for $t_1 = \cdots = t_r = 0$ if and only if ψ vanishes at P. Hence every element of the *quotient ring $Q(P)$* of the point P is an holomorphic function of $t_1, \cdots, t_r$. Therefore, in order that the ξ's be holomorphic functions of the t's, it is sufficient that they belong to the quotient ring $Q(P)$. If that is the case, then not only the ξ's, but also *the entire quotient ring $Q(P_0)$ will belong to $Q(P)$*. Namely, since P_0 is the center of B on V_0, any polynomial φ in $\xi_1, \cdots, \xi_n$ which does not vanish at P_0 has value zero in B, and therefore φ, as an element of $Q(P)$, must be a unit.

The above considerations suggest the following stronger formulation of the uniformization theorem:

THEOREM U_2. (*The uniformization theorem in projective form*): *Given a projective model V_0 of Σ and given a zero-dimensional valuation B of Σ, with center P_0 on V_0, there exists a projective model V of Σ on which the origin of B is at a simple*

[1] Note that the mapping is then single-valued on E but that the inverse is not necessarily single-valued on W, whence W need not be an open set. Note also that P is mapped upon P_0.

[2] A point Q_0 of V_0 is said to be *covered by the wedge W*, if Q_0 is the image of a point Q of the open set E.

point P and which is such that the quotient ring of P_0 is a subring of the quotient ring of $P(Q(P_0) \subsetneqq Q(P))$.

3. So far we have restricted ourselves to: (a) valuations of dimension zero; (b) ground fields which are algebraically closed. An analysis of the theorem U_2 (see A III) will show, however, that both restrictions are quite illusory, in the sense that the proof of Theorem U_2 leads automatically to the necessity of proving an analogous theorem under the most general hypotheses: (a) B is a valuation of any dimension; (b) K is an arbitrary field (of characteristic zero). But before we undertake this analysis, we proceed to show that from Theorem U_2 follows almost immediately the solution of the problem of local uniformization in the classical case.

II. Solution of the classical problem of local uniformization

4. The ground field K being quite arbitrary, any zero-dimensional valuation B of the field[3] Σ defines an homomorphic mapping of Σ upon (K', ∞), where K' is an algebraic extension of K (finite or infinite). The valuation-theoretic value of an element η of Σ shall be denoted by $v(\eta)$; here $v(\eta)$ is an element of an ordered abelian (additive) group Γ, *the value group of B*. The element c of K' upon which η is mapped, or the symbol ∞ if η is mapped on ∞, shall be called the *function-theoretic value of* η, or also the *B-residue of* η, and shall be denoted by $B(\eta)$. The field K' itself is the *residue field* of B. The relationship between $v(\eta)$ and $B(\eta)$ is the following: (a) $B(\eta) = \infty$, if and only if $v(\eta) < 0$; (b) $B(\eta) = 0$, if and only if $v(\eta) > 0$; (c) if $B(\eta) = \vartheta$ and if $f(\vartheta) = 0$ is the irreducible equation for ϑ, with coefficients in K, then $v(f(\eta)) > 0$. The elements η for which $v(\eta) \geqq 0$ form a ring $\mathfrak{B}$, called the *valuation ring* of B. By (b) and (c), the mapping $\eta \to B(\eta)$ is an homomorphic mapping of $\mathfrak{B}$ upon K'. The elements η of $\mathfrak{B}$ such that $B(\eta) = 0$ form a prime maximal ideal $\mathfrak{P}$ in $\mathfrak{B}$, and we have $\mathfrak{B} \sim \mathfrak{B}/\mathfrak{P} \cong \mathsf{K}'$.

Following Dedekind, we define *a place* $\mathfrak{P}$ *of the field* Σ as an homomorphic mapping of Σ upon (K', ∞), where K' is an algebraic extension of K; provided $c \to c$, for any c in K. We have just shown that any zero-dimensional valuation B of Σ defines a place. *Conversely, given a place* $\mathfrak{P}$, *it determines uniquely a zero-dimensional valuation of* Σ. Namely, the elements of Σ which are mapped upon elements of K', form a ring $\mathfrak{B}$ with the following property: if $\eta \not\subset \mathfrak{B}$ then $1/\eta \subset \mathfrak{B}$. Hence $\mathfrak{B}$ *is a valuation ring* ([3], p. 110), and moreover $\mathsf{K} \subset \mathfrak{B}$. The elements of $\mathfrak{B}$ which are mapped by $\mathfrak{P}$ upon the zero element of K', form the prime ideal of non-units in $\mathfrak{B}$. We denote this ideal by the same letter $\mathfrak{P}$. Since $\mathfrak{B}/\mathfrak{P} \cong \mathsf{K}'$ and since K' is an algebraic extension of K, it follows that the valuation B determined by the ring $\mathfrak{B}$ is zero-dimensional, with K' as residue field. This proves the assertion.

Thus the two terms: "*zero-dimensional valuation*" and "*place*," are inter-

[3] We tacitly assume that only such valuations of Σ are considered in which the elements c of K ($c \neq 0$) have value zero.

changeable. We speak of a valuation if we wish to focus our attention on the mapping $\eta \rightarrow v(\eta) \subset \Gamma$; we speak of a place if we wish instead to emphasize the mapping $\eta \rightarrow B(\eta) \subset (\mathsf{K}', \infty)$. In this second case we shall also use the symbol $\mathfrak{P}(\eta)$ to denote the function-theoretic value of η.

5. DEFINITION. The totality of places of Σ is called the *Riemann manifold* of Σ. This manifold shall be denoted by $\boldsymbol{M}$.

If K is algebraically closed, then $\mathsf{K}' = \mathsf{K}$ for every place of Σ. Let now K be the field of complex numbers. Then (K, ∞) is a sphere H, the sphere of the complex variable. $\boldsymbol{M}$ becomes an Hausdorff space if we define *open sets* in $\boldsymbol{M}$ in the following (well-known) fashion:

Let $\eta_1, \cdots, \eta_m$ be any finite set of elements of Σ and let $E_1, E_2, \cdots, E_m$ be open sets in H. *The totality of all places $\mathfrak{P}$, or zero-dimensional valuations B, such that $B(\eta_i) \subset E_i$, is an open set in $\boldsymbol{M}$.* The open sets of $\boldsymbol{M}$ just defined shall form a *basis* for the totality of open sets on $\boldsymbol{M}$. It is obvious that this definition is equivalent with the following: *if $(\eta_1, \cdots, \eta_m)$ is the general point of a projective model V of Σ, then the zero-dimensional valuations B whose centers lie in a given open subset of V form an open set in $\boldsymbol{M}$.*

Given a projective model V_0 of Σ, every "point" $\mathfrak{P}$ of $\boldsymbol{M}$ lies over a unique point P_0 of V_0 : here P_0 is the center of the zero-dimensional valuation B determined by the place $\mathfrak{P}$. The mapping $\mathfrak{P} \rightarrow P_0$ is obviously a continuous mapping of $\boldsymbol{M}$ on V_0, since any open set on V_0 is the map of an open set on $\boldsymbol{M}$.

By Theorem U_2 it follows that *$\boldsymbol{M}$ can be covered by a collection of open sets whose images on V_0 are wedges $W_1, W_2, \cdots$.* We shall denote these open sets by $W_1', W_2', \cdots$. In the notation of Theorem U_2, each W' is the totality of all places which lie over points of a sufficiently small neighborhood E of the simple point P (so small as to insure the convergence of the power series which give the coördinates of the general point of V and of V_0 in terms of $t_1, \cdots, t_r$). By a preceding definition (see footnote[2]) it follows that if a set W' contains a point $\mathfrak{P}$ of $\boldsymbol{M}$, and if $\mathfrak{P}$ lies over the point A_0 of V_0, then the wedge W "covers" A_0. Namely, let A be the point of E which corresponds to the place $\mathfrak{P}$. Since W is a single-valued image of E, the image of A must be a point A_0^* of W such that *every* place which lies over A lies over A_0^*.[4] But the place $\mathfrak{P}$ lies over A and over A_0. Hence $A_0 = A_0^*$, and consequently W covers A_0.

We conclude that in order to show that V_0 can be covered by a *finite number* of wedges W_i, it is sufficient to show that the Riemann manifold $\boldsymbol{M}$ can be covered by a finite number of open sets W_i', and for that it is sufficient to show (in view of Theorem U_2) that *$\boldsymbol{M}$ is a bicompact space.* Our original proof of the bicompactness of $\boldsymbol{M}$ was tolerably complicated. We are, however, indebted to Chevalley and Tukey for the remark that the bicompactness of $\boldsymbol{M}$ follows

[4] Since $Q(P_0) \subseteqq Q(P)$, the ξ's are rational functions g of $\eta_1, \cdots, \eta_m$ in which the denominators do not vanish at P. Hence if E is a sufficiently small neighborhood of P, these denominators will be $\neq 0$ in E. For any place $\mathfrak{P}$ lying over A, the function-theoretic values of the ξ's will then be equal to the values of the rational functions g at A.

immediately from the theorem of Tychonoff ([5], [6]) on the bicompactness of direct products of bicompact spaces. To see this, we consider the Hausdorff space Ω of *all* mappings of Σ upon H.[5] By Tychonoff's theorem, Ω is bicompact, since H is bicompact.

The Riemann manifold $\boldsymbol{M}$ of Σ is a subset of Ω, consisting of the *homomorphic* mappings upon H (i.e. upon (K, ∞)). It is a straightforward matter to show that *$\boldsymbol{M}$ is a closed subset of Ω*.[6] Hence $\boldsymbol{M}$ is bicompact. This settles the problem of local uniformization for algebraic varieties.

6. The following result, though purely negative, is nevertheless of interest: *if $r > 1$, then the Riemann manifold $\boldsymbol{M}$ of Σ is not metrizable.* To show this it is only necessary to show that the open subsets of $\boldsymbol{M}$ do not possess a countable basis (Urysohn's metrization theorem; see, for instance, [1], p. 88). Let us suppose that such a basis exists. Then, given any basis (countable or not) for the open sets on $\boldsymbol{M}$, that basis will necessarily contain a countable subset also forming a basis ([1], p. 78). It follows that there exists a countable basis $\{N_1, N_2, \cdots\}$ consisting of open sets defined in the preceding section. Each set N_i is defined by a finite set of elements in Σ, say $\eta_{i1}, \eta_{i2}, \cdots, \eta_{i\nu_i}$, and by a corresponding set of open sets on H, say $E_{i1}, E_{i2}, \cdots, E_{i\nu_i}$. Altogether we have a countable set of elements η_{ij} in Σ. We assert that this set enjoys the following property: *there cannot exist two distinct places $\mathfrak{P}_1$, $\mathfrak{P}_2$ such that the function-theoretic value of each η_{ij} at $\mathfrak{P}_1$ is the same as the function-theoretic value of the same element η_{ij} at $\mathfrak{P}_2$.* In other words: *the function-theoretic values of the elements η_{ij} at a place $\mathfrak{P}$ determine the place $\mathfrak{P}$ uniquely.* The proof is immediate. Let us suppose that there exist two distinct places $\mathfrak{P}_1$, $\mathfrak{P}_2$ such that $\mathfrak{P}_1(\eta_{ij}) = \mathfrak{P}_2(\eta_{ij})$, $i = 1, 2, \cdots; j = 1, 2, \cdots, \nu_i$. If a set N_i contains $\mathfrak{P}_1$, we must have $E_{ij} \supset \mathfrak{P}_1(\eta_{ij})$, $j = 1, 2, \cdots, \nu_i$, and since $\mathfrak{P}_1(\eta_{ij}) = \mathfrak{P}_2(\eta_{ij})$, that implies that N_i also contains $\mathfrak{P}_2$. This is in contradiction with the hypothesis that $\{N_1, N_2, \cdots\}$ is a basis for the open sets of the *Hausdorff* space $\boldsymbol{M}$.

We complete our proof that $\boldsymbol{M}$ is not metrizable by showing that *if $r > 1$, then given any countable set of elements in Σ, say $\eta_1, \eta_2, \cdots, \eta_n, \cdots$, there always exist two distinct places $\mathfrak{P}_1$, $\mathfrak{P}_2$ such that $\mathfrak{P}_1(\eta_i) = \mathfrak{P}_2(\eta_i)$, for all i.* For the proof, we consider some *normal* projective model V of Σ in an affine space S_n ([9], p. 279). Let $(\xi_1, \cdots, \xi_n)$ be the general point of V. Since V is normal,

[5] It should be understood that open sets in Ω are defined in the same fashion as the open sets of the Riemann manifold were defined above.

[6] Let $\bar{\boldsymbol{M}}$ be the closure of $\boldsymbol{M}$ in Ω and let f_0 be an element of $\bar{\boldsymbol{M}}$. We have to show that f_0 is a homomorphic mapping of Σ on (K, ∞). Let ω_1, ω_2 be any two elements of Σ and let $\omega_3 = \omega_1 + \omega_2$. Let us assume that $f_0(\omega_i) \neq \infty$ $(i = 1, 2, 3)$. Consider the open set N in Ω which consists of all mapping f such that $|f(\omega_i) - f_0(\omega_i)| < \epsilon/3$, $i = 1, 2, 3$, ϵ—an arbitrary positive real number. Since $f_0 \subset \bar{\boldsymbol{M}}$, N contains an element φ which is an homomorphic mapping of Σ on (K, ∞). We will have then $|\varphi(\omega_i) - f_0(\omega_i)| < \epsilon/3$, and also $\varphi(\omega_3) = \varphi(\omega_1) + \varphi(\omega_2)$. Hence $|f_0(\omega_3) - f_0(\omega_1) - f_0(\omega_2)| < \epsilon$, and since ϵ is arbitrary, we conclude that $f_0(\omega_3) = f_0(\omega_1) + f_0(\omega_2)$. In a similar fashion it follows that $f_0(\omega_1 \cdot \omega_2) = f_0(\omega_1) f_0(\omega_2)$. The case in which some of the $f_0(\omega_i)$ is ∞ is treated in a similar fashion.

the ring $\mathfrak{o} = K[\xi_1, \cdots, \xi_n]$ is integrally closed in Σ. In the sense of *quasi-gleicheit* of van der Waerden we have $\eta_i = \mathfrak{A}_i/\mathfrak{B}_i$, where $\mathfrak{A}_i$ and $\mathfrak{B}_i$ are power products of minimal $((r-1)$-dimensional) prime ideals of $\mathfrak{o}$. The prime factors of $\mathfrak{A}_i$ are distinct from those of $\mathfrak{B}_i$. Hence the ideal $(\mathfrak{A}_i, \mathfrak{B}_i)$ defines on V a subvariety W_i of dimension $\leqq r - 2$. The countable set $\{W_1, W_2, \cdots\}$ cannot fill the entire variety V, which is of dimension r. Hence there exists on V a point $P(a_1, \cdots, a_n)$ which is on none of the varieties W_i. For a given i, the prime ideal $\mathfrak{p} = (\xi_1 - a_1, \cdots, \xi_n - a_n)$ in $\mathfrak{o}$ may divide one of the ideals $\mathfrak{A}_i, \mathfrak{B}_i$, but cannot divide both. From this it follows that η_i can be expressed in the form: $\eta_i = \varphi_i(\xi_1, \cdots, \xi_n)/\psi_i(\xi_1, \cdots, \xi_n)$, where φ_i and ψ_i are polynomials and where $\varphi_i(a_1, \cdots, a_n)$, $\psi_i(a_1, \cdots, a_n)$ are not both zero.[7] Now consider any place $\mathfrak{P}$ of the field Σ which lies over the point P. The function-theoretic value of η_i at $\mathfrak{P}$ is then uniquely determined by P. Namely, if $\psi_i(a_1, \cdots, a_n) \neq 0$, then $\mathfrak{P}(\eta_i) = \varphi_i(a_1, \cdots, a_n)/\psi_i(a_1, \cdots, a_n)$. If $\psi_i(a_1, \cdots, a_n) = 0$, then $\mathfrak{P}(\eta_i) = \infty$. *Since, for $r > 1$, there are infinitely many places $\mathfrak{P}$ which lie over P* (for instance, all the algebraic branches through P), our assertion is proved.

III. The general uniformization theorem

7. Let us assume for a moment that Theorem U_2 has been established, and let us consider a valuation B of Σ, *of an arbitrary dimension $s(s < r)$*. Given a projective model V of Σ, we define the *center of B on V* as follows. Let $(\xi_1, \cdots, \xi_n)$ be the non-homogeneous coördinates of the general point of V. Without loss of generality we may assume that the elements ξ_i have non-negative values $v(\xi_i)$ in B. Then the entire ring $\mathfrak{o} = K[\xi_1, \cdots, \xi_n]$ will be contained in the valuation ring $\mathfrak{B}$ of B. Let $\mathfrak{P}$ be the prime ideal of non-units in $\mathfrak{B}$ and let $\mathfrak{p} = \mathfrak{P} \cap \mathfrak{o}$. The ideal $\mathfrak{p}$ is prime and shall be referred to as *the prime ideal of $\mathfrak{B}$ in $\mathfrak{o}$*; $\mathfrak{p}$ consists of those elements ω of $\mathfrak{o}$ for which $v(\omega) > 0$. Let W be the irreducible algebraic subvariety of V defined by the prime ideal $\mathfrak{p}$. Since $\mathfrak{B}/\mathfrak{P}$ is of degree of transcendency s over K and since $\mathfrak{o}/\mathfrak{p}$ is a subring of $\mathfrak{B}/\mathfrak{P}$, it follows that W is of dimension $\leqq s$. This subvariety W is called *the center of B on V*.

The residue field $\mathfrak{B}/\mathfrak{P}$ of B is a field Σ', of degree of transcendency s over K. We consider an arbitrary zero-dimensional valuation B' of Σ'/K. Compounding B with B' we get a zero-dimensional valuation B_0 of Σ. Let V be a projective model of Σ on which the center of B_0 is a simple point P. Let W be the center of B on V. Since B_0 is compounded with B, *the center of B_0 lies on the center of B*,[8] i.e. P is a point of W. Hence W, containing a simple point of V, *must*

[7] The assertion is obvious. If, for instance, $\mathfrak{B}_i \not\equiv 0(\mathfrak{p})$, then we take an element $\psi_i(\xi_1, \cdots, \xi_n)$ in $\mathfrak{o}$ such that $\psi_i \equiv 0(\mathfrak{B}_i)$, $\psi_i \not\equiv 0(\mathfrak{p})$. Then $\psi_i \cdot \eta_i$ is in $\mathfrak{o}$, $\psi_i \eta_i = \varphi_i(\xi_1, \cdots, \xi_n)$.

[8] This follows from the fact that any element of Σ whose value in B is positive has also positive value in B_0.

itself be a simple subvariety of V ([11], p. 221). *Thus we have a projective model V on which the center of B is a simple subvariety W* (of dimension $\leqq s$).

This result is a generalization of Theorem U_1, but actually it is implied by this theorem. However, in our proof, the relationship between the uniformization of B and that of the compounded valuation B_0 will be inverted: *we first uniformize B and we then use this result in order to uniformize B_0.* The reason for this is that the *value group of B is of lower rank than that of B_0* ([3], p. 113) and that our proof proceeds *by induction with respect to the rank.* From this point of view it begins to appear clearly that the center of gravity of our proof will be found, not in the uniformization of valuations *of dimension zero* and *of arbitrary rank,* but rather in the uniformization of valuations *of rank* 1 *and of arbitrary dimension.* The rest of the proof will be a relatively simple induction from rank σ to rank $\sigma + 1$.

8. The above program calls for another reduction and for a simultaneous generalization of the terms of the problem. Let us adjoin to the ground field K s elements $\zeta_1, \cdots, \zeta_s$ of Σ which are mapped upon algebraically independent elements of the residue field Σ' of B. Every element ($\neq 0$) of the field $K^* = K(\zeta_1, \cdots, \zeta_s)$ has value zero in B, and it is therefore permissible to take K^* as new ground field for Σ. *With respect to this new ground field, our valuation B is of dimension zero.* This is the reduction. However, K^* is not an algebraically closed field. It thus appears that the uniformization of valuations of an arbitrary dimension, in the case of an algebraically closed ground field, is in essence nothing else than the uniformization of zero-dimensional valuation in the case of an arbitrary ground field.

These considerations, *and the additional requirement concerning the quotient rings* (Theorem U_2) lead us to state our *general uniformization theorem* as follows:

THEOREM U_3. *Let Σ be a field of algebraic functions of r independent variables, over an arbitrary ground field K of characteristic zero, and let B be an s-dimensional valuation of Σ. Let, moreover, V be a projective model of Σ and let W be the center of B on V. There exists a projective model V' of Σ, on which the center of B is a simple subvariety W' and such that the quotient ring of W is a subring of the quotient ring of W'.*[9]

IV. Reduction to an hypersurface. The main theorem

9. Let us suppose that Theorem U_3 has already been proved in the case when the given projective model V is an hypersurface, i.e. an r-dimensional variety V lying in an $(r + 1)$-dimensional projective space. We propose to show that Theorem U_3 then follows quite generally.

Let $(\xi_1, \cdots, \xi_n)$ be the general point of V in the affine space S_n and let $\mathfrak{o} = K[\xi_1, \cdots, \xi_n]$. We choose in $\mathfrak{o}$ r elements $\eta_1, \cdots, \eta_r$ such that $\mathfrak{o}$ is

[9] Note that from $Q(W) \subseteq Q(W')$ follows incidentally that *dimension* $W \leqq$ *dimension* W'.

integrally dependent on $K[\eta_1, \cdots, \eta_r]$.[10] Then we fix in $\mathfrak{o}$ a primitive element η_{r+1} of the field $\Sigma/K(\eta_1, \cdots, \eta_r)$. Let $f(\eta_1, \cdots, \eta_r, \eta_{r+1}) = 0$ be the irreducible equation, over K, between the $r + 1$ elements $\eta_1, \cdots, \eta_{r+1}$, and let V^* be the hypersurface defined by this equation. Let $\mathfrak{o}^* = K[\eta_1, \cdots, \eta_r, \eta_{r+1}]$ and let W and W^* be the centers of the given s-dimensional valuation B on V and V^* respectively. Since $\mathfrak{o}$ is integrally dependent on $\mathfrak{o}^*$, and since the prime $\mathfrak{o}^*$-ideal determined by W^* is the contracted ideal of the prime $\mathfrak{o}$-ideal for W, it follows that dimension W^* = dimension W.

By our assumption, there exists a projective model V' of Σ on which the center of B is a simple subvariety W' and such that $Q(W')$ (quotient ring of W') contains $Q(W^*)$. Now the quotient ring of a simple subvariety is integrally closed in Σ ([11], p. 220). Hence $Q(W')$, being integrally closed, *is the intersection of all the valuation rings which contain* $Q(W')$ ([3], p. 111). Since $Q(W') \supseteqq Q(W^*)$, every valuation ring containing $Q(W')$ also contains $Q(W^*)$. Since $\mathfrak{o}$ is integrally dependent on $Q(W^*)$, every valuation ring containing $Q(W^*)$ also contains $\mathfrak{o}$. *Hence* $\mathfrak{o} \subseteqq Q(W')$. *Let* $\mathfrak{p}$ *be the prime ideal of* W *in* $\mathfrak{o}$ (i.e. the prime ideal of B in $\mathfrak{o}$). Every element η of $\mathfrak{o}$ which is not in $\mathfrak{p}$ is such that $v(\eta) = 0$ in B. Such an element η, as an element of $Q(W')$, must be a unit, since also W' is the center of B, on V'. This shows that $\mathfrak{o}_\mathfrak{p}$, i.e. $Q(W)$, is contained in $Q(W')$, which proves Theorem U_3.

10. Our main theorem, to the proof of which most of the paper is devoted, concerns hypersurfaces. This theorem asserts, in part, much more than Theorem U_3, and namely it asserts that *for hypersurfaces, the uniformization of any given valuation of rank* 1 *and dimension zero can be effected by Cremona transformations of the ambient space.* This result is naturally of added interest. We believe that the second part of the paper (Part B), where we build up our Cremona transformations, contains the necessary material for a proof that also valuations of rank > 1 can be uniformized by Cremona transformations. However, we do not wish to overload the present paper with considerations which are not absolutely essential for the proof of the general uniformization theorem. On the other hand, our proof of this last theorem (see C IV) is absolutely dependent on the knowledge that valuation of rank 1 can be uniformized by Cremona transformations (of the type described below in the main theorem).

We also wish to point out that from the main theorem it follows directly that a rank 1 valuation B of arbitrary dimension s can also be uniformized by a Cremona transformation, *provided the center of* B *on the hypersurface* $f = 0$ *is a subvariety of dimension exactly* s. Namely, with this proviso, s of the elements $x_1, \cdots, x_{r+1}$ will have algebraically independent B-residues. If say $B(x_1), \cdots, B(x_s)$ are algebraically independent over K, we adjoin $x_1, \cdots, x_s$ to the ground field K. We take $K(x_1, \cdots, x_s)$ as new ground field and we apply our main theorem to the *zero-dimensional* valuation B of $\Sigma/K(x_1, \cdots, x_s)$. However,

[10] By a "normalization" theorem due to Emmy Noether, it is possible to take for $\eta_1, \cdots, \eta_r$ linear forms in $\xi_1, \cdots, \xi_n$, with "non special" coefficients in K.

in this case, the expressions $\varphi_i(X_1, \cdots, X_{r+1})$, $i = 1, 2, \cdots, r+1$, in the equations $x_i = \varphi_i(X_1, \cdots, X_{r+1})$ (given below), will be polynomials only in $X_{s+1}, \cdots, X_{r+1}$, but they will be in general *rational* functions in $X_1, \cdots, X_s$. (Except for $i = 1, 2, \cdots, s$, for then φ_i is simply X_i.)

MAIN THEOREM. *Let*

$$f(x_1, x_2, \cdots, x_{r+1}) = 0$$

be an irreducible hypersurface V in S_{r+1}, over an arbitrary ground field K (of characteristic zero), and let B be a zero-dimensional valuation, of rank 1, of the field Σ (field of rational functions on V), such that the valuation ring $\mathfrak{B}$ of B contains the x_i ($i = 1, 2, \cdots, r+1$). Let P be the center of B on V. There exists a Cremona transformation of the form:

$$x_i = \varphi_i(X_1, X_2, \cdots, X_{r+1}), \qquad i = 1, 2, \cdots, r+1,$$

the φ_i-polynomials, with coefficients in K, such that the new elements X_i also belong to $\mathfrak{B}$ and such that on the transformed hypersurface

$$F(X_1, X_2, \cdots, X_{r+1}) = 0,$$

the center of B is a simple point P'.

The restriction: $x_i \subset \mathfrak{B}$, implies no loss of generality, since we may always perform a preliminary projective transformation on the x_i so as to satisfy this condition. The condition $Q(P') \supseteqq Q(P)$ is here automatically satisfied, since the ring $\mathsf{K}[x_1, \cdots, x_{r+1}]$ is a subring of $\mathsf{K}[X_1, \cdots, X_{r+1}]$ (the φ_i being polynomials).

B. THE ALGORITHM OF PERRON AND THE ASSOCIATED CREMONA TRANSFORMATIONS

It is well known that given a *rank one* valuation B (i.e. a valuation whose value group consists of real numbers) of a field Σ of algebraic functions of r independent variables, the value group Γ of B is of *rational rank* $m \leqq r$ ([3], p. 116). There exist then in Γ m rationally independent real numbers, say $\tau_1, \cdots, \tau_m$, and every element of Γ is rationally dependent on $\tau_1, \cdots, \tau_m$. In the case of surfaces, m is at most 2. If $m = 2$, we have used in our paper [10] the algorithm of continued fractions for the ratio τ_1/τ_2. With each successive step of this algorithm (i.e. for each convergent fraction of τ_1/τ_2) we have associated a Cremona transformation in a field of two independent variables (see [10], p. 653, (8)). In the quoted paper we have assumed throughout that the ground field is algebraically closed.

Our present investigation depends on a two-fold generalization of the above procedure. First, we need an algorithm of simultaneous approximation of the set of real numbers $\tau_1, \cdots, \tau_m$. We use an algorithm due to O. Perron [7]; see also [2], Chapter IV. In the second place, given a set of elements $x_1, \cdots, x_m$, such that $v(x_i) = \tau_i$, we have to provide ourselves with a sequence of Cremona transformations in the field $\mathsf{K}(x_1, \cdots, x_m)$ running parallel to

Perron's algorithm. The main difficulties of this step arise when K is not algebraically closed, or better, when we are dealing with a zero-dimensional valuation whose residue field is a proper algebraic extension of K.

I. The algorithm of Perron

1. Let $\tau_1, \tau_2, \cdots, \tau_m$ be rationally independent positive real numbers. We consider the transformation (referred to in the sequel as the *elementary transformation* of Perron):

$$(1) \qquad \tau_1 = \tau_m^{(1)},\ \tau_2 = \tau_1^{(1)} + a_2^{(0)} \cdot \tau_m^{(1)},\ \cdots,\ \tau_m = \tau_{m-1}^{(1)} + a_m^{(0)} \tau_m^{(1)},$$

where

$$(1') \qquad a_j^{(0)} = [\tau_j/\tau_1], \qquad j = 2, \cdots, m.$$

The real numbers $\tau_1^{(1)}, \cdots, \tau_m^{(1)}$ are again positive and rationally independent. We apply the same transformation to the numbers $\tau_i^{(1)}$, getting a new set of m positive, rationally independent numbers $\tau_1^{(2)}, \cdots, \tau_m^{(2)}$, and so we continue indefinitely. In this fashion we obtain, for each integer h, a set of m positive, rationally independent real numbers: $\tau_1^{(h)}, \cdots, \tau_m^{(h)}$, where:

$$(2) \quad \tau_1^{(h-1)} = \tau_m^{(h)},\ \tau_2^{(h-1)} = \tau_1^{(h)} + a_2^{(h-1)} \tau_m^{(h)},\ \cdots,\ \tau_m^{(h-1)} = \tau_{m-1}^{(h)} + a_m^{(h-1)} \tau_m^{(h)},$$

and

$$(2') \qquad a_j^{(h-1)} = [\tau_j^{(h-1)}/\tau_1^{(h-1)}].$$

The resulting transformation (referred to in the sequel as the *transformation of Perron*) from $\tau_1, \cdots, \tau_m$ to $\tau_1^{(h)}, \cdots, \tau_m^{(h)}$ is of the form:

$$(3) \qquad \tau_i = A_i^{(h)} \tau_1^{(h)} + A_i^{(h+1)} \tau_2^{(h)} + \cdots + A_i^{(h+m-1)} \tau_m^{(h)}, \qquad i = 1, 2, \cdots, m.$$

Here the coefficients $A_i^{(j)}$ are *non-negative integers*

$$(4) \qquad A_i^{(j)} \geqq 0,$$

and, moreover, the transformation (3) is *unimodular*:

$$(4') \qquad \begin{vmatrix} A_1^{(h)}, & \cdots, & A_1^{(h+m-1)} \\ \cdots & \cdots & \cdots \\ A_m^{(h)}, & \cdots, & A_m^{(h+m-1)} \end{vmatrix} = (-1)^{h(m-1)}.$$

(See [2], p. 437, 439, formulas (5), (14).)

Of fundamental importance is the following: *the ratios* $A_i^{(h)}/A_1^{(h)}$ *are convergent fractions for* τ_i/τ_1 i.e. ([2], p. 440, (18)):

$$(5) \qquad \lim_{h\to\infty} \frac{A_i^{(h)}}{A_1^{(h)}} = \frac{\tau_i}{\tau_1}.$$

2. By means of the algorithm of Perron we derive a theorem concerning *rationally dependent* real numbers. Let $\tau_1, \cdots, \tau_m$ be, as before, rationally independent positive real numbers and let τ_{m+1} be a positive real number which depends rationally on $\tau_1, \cdots, \tau_m$.

THEOREM 1. *There exists a unimodular transformation, with integral non-negative coefficients:*

$$\tau_i = \sum_{j=1}^{m+1} \beta_{ij} \overset{*}{\tau}_j, \qquad i = 1, 2, \cdots, m+1, \tag{6}$$

such that

$$\overset{*}{\tau}_{m+1} = 0, \qquad \overset{*}{\tau}_i > 0, \qquad i = 1, 2, \cdots, m. \tag{6'}$$

PROOF: Let

$$\lambda \tau_{m+1} = \lambda_1 \tau_1 + \cdots + \lambda_m \tau_m, \qquad \lambda > 0, \tag{7}$$

be the relation of rational dependence between $\tau_1, \cdots, \tau_m, \tau_{m+1}$. Here $\lambda, \lambda_1, \cdots, \lambda_m$ are integers and $(\lambda, \lambda_1, \cdots, \lambda_m) = 1$. We apply to $\tau_1, \cdots, \tau_m$ the transformation of Perron (3). By substituting into (7) we get

$$\lambda \tau_{m+1} = \lambda_1^{(h)} \tau_1^{(h)} + \cdots + \lambda_m^{(h)} \tau_m^{(h)}, \tag{8}$$

where

$$\lambda_i^{(h)} = \lambda_1 A_1^{(h+i-1)} + \lambda_2 A_2^{(h+i-1)} + \cdots + \lambda_m A_m^{(h+i-1)} \qquad i = 1, 2, \cdots, m. \tag{9}$$

We take h sufficiently high, so as to satisfy the condition $\lambda_i^{(h)} > 0$. This is possible, in view of (5), since $\lambda_1 \tau_1 + \cdots + \lambda_m \tau_m > 0$.[11] Since the transformation (9) is unimodular, the relation $(\lambda, \lambda_1, \cdots, \lambda_m) = 1$ implies that also $(\lambda, \lambda_1^{(h)}, \cdots, \lambda_m^{(h)}) = 1$. Let, say, $\lambda_1^{(h)}$ not be divisible by λ, and let $\lambda_1^{(h)} = \lambda\mu + \lambda'$,

$$0 < \lambda' < \lambda. \tag{10}$$

We consider the transformation:

$$\tau_{m+1} = \tau_1' + \mu \tau_{m+1}', \quad \tau_1^{(h)} = \tau_{m+1}', \quad \tau_i^{(h)} = \tau_i', \qquad i = 2, \cdots, m. \tag{10'}$$

The $m + 1$ real numbers τ_i' are positive, since $\lambda \tau_{m+1} > \lambda_1^{(h)} \tau_1^{(h)}$, by (8), whence $\lambda \tau_{m+1} > \lambda \mu \tau_1^{(h)}$, i.e. $\lambda \tau_1' > 0$. From (8) we find:

$$\lambda' \tau_{m+1}' = \lambda_1' \tau_1' + \cdots + \lambda_m' \tau_m', \tag{7'}$$

where $\lambda_1' = \lambda$, $\lambda_i' = -\lambda_i^{(h)}$, $i = 2, \cdots, m$. In view of (10) and (7) we have achieved a reduction for λ, and this by a transformation (from the τ_i to the τ_i') which is unimodular and non-negative (i.e. with non-negative integral coefficients). Ultimately we will get by a unimodular non-negative transformation a set of positive real numbers $\bar{\tau}_1, \bar{\tau}_2, \cdots, \bar{\tau}_{m+1}$, such that

$$\bar{\tau}_{m+1} = \bar{\lambda}_1 \bar{\tau}_1 + \cdots + \bar{\lambda}_m \bar{\tau}_m. \tag{11}$$

[11] Note also that, in view of (5), $A_1^{(h)} \neq 0$, when h is sufficiently high, whence, by (4), $A_1^{(h)} > 0$.

A Perron transformation (3) applied to the rationally independent numbers $\bar{\tau}_1, \cdots, \bar{\tau}_m$ will lead to $\bar{\tau}_1^{(h)}, \cdots, \bar{\tau}_m^{(h)}$ and to a new relation:

$$(11') \qquad \bar{\tau}_{m+1} = \bar{\lambda}_1^{(h)} \bar{\tau}_1^{(h)} + \cdots + \bar{\lambda}_m^{(h)} \bar{\tau}_m^{(h)},$$

in which the coefficients $\bar{\lambda}_i^{(h)}$ are non-negative, provided h is sufficiently high (the argument is the same as that applied to the coefficients $\lambda_i^{(h)}$ in (9)). We then put:

$$(11'') \qquad \begin{cases} \bar{\tau}_i^{(h)} = \overset{*}{\tau}_i, & i = 1, 2, \cdots, m, \\ \bar{\tau}_{m+1} = \overset{*}{\tau}_{m+1} + \bar{\lambda}_1^{(h)} \overset{*}{\tau}_1 + \cdots + \bar{\lambda}_m^{(h)} \overset{*}{\tau}_m. \end{cases}$$

Then $\overset{*}{\tau}_{m+1} = 0$ and this completes the proof.

II. Rationally independent values. The Cremona transformation $T^{(h)}$

3. We consider a zero-dimensional valuation B, *of rank 1*, of our field Σ. The residue field K' of B is an algebraic extension of the ground field K. To build up our Cremona transformations, we first consider the special case in which *the residue field* K' *coincides with the ground field.*

Let $x_1, \cdots, x_m$ be arbitrary elements of Σ, but such that $v(x_i) \geqq 0$. Since $\mathsf{K}' = \mathsf{K}$, there exist m uniquely determined constants $c_1, \cdots, c_m$ *in* K such that $v(x_i - c_i) > 0$. Let $v(x_i - c_i) = \tau_i$ and *let us assume that* $\tau_1, \cdots, \tau_m$ *are rationally independent.* We then associate with the Perron transformation (3) the following Cremona transformation $T^{(h)}$ in $\mathsf{K}(x_1, \cdots, x_m)$:

$$(12) \qquad T^{(h)}\colon x_i - c_i = X_1^{A_i^{(h)}} X_2^{A_i^{(h+1)}} \cdots X_m^{A_i^{(h+m-1)}}, \quad i = 1, 2, \cdots, m.$$

$T^{(h)}$ is a Cremona transformation, since the determinant (4′) of the exponents $A_i^{(j)}$ is ± 1 (and is $+1$ if h is even). The x_i are polynomials in $X_1, X_2, \cdots, X_m$, since the $A_i^{(j)}$ are non-negative integers, by (4). Moreover, in view of (3) we have

$$(12') \qquad v(X_i) = \tau_i^{(h)}.$$

The fact that the ratios $A_i^{(h)}/A_1^{(h)}$ $(h = 1, 2, \cdots)$ are convergent fractions for τ_i/τ_1, will be used by us for the following purpose: we prove namely the following

THEOREM 2. *Let*

$$\pi_1 = \prod_{i=1}^{m} (x_i - c_i)^{\mu_{i1}}, \cdots, \qquad \pi_N = \prod_{i=1}^{m} (x_i - c_i)^{\mu_{iN}},$$

be N power products of the differences $x_1 - c_1, \cdots, x_m - c_m$, *and let*

$$\pi_1 = \prod_{i=1}^{m} X_i^{M_{i1}}, \cdots, \qquad \pi_N = \prod_{i=1}^{m} X_i^{M_{iN}},$$

be the form which the elements $\pi_1, \cdots, \pi_N$ *assume after the transformation* $T^{(h)}$. *If h is sufficiently high, then* $\prod X_i^{M_{i\alpha}}$ *is a factor of* $\prod X_i^{M_{i\beta}}$ *for all pairs of indices* α *and* β *such that* $v(\pi_\alpha) < v(\pi_\beta)$.

PROOF. We have

(13) $$\mathrm{M}_{ij} = \mu_{1j}A_1^{(h+i-1)} + \mu_{2j}A_2^{(h+i-1)} + \cdots + \mu_{mj}A_m^{(h+i-1)}.$$

If $v(\pi_\alpha) < v(\pi_\beta)$, then

(14) $$(\mu_{1\beta} - \mu_{1\alpha})\tau_1 + \cdots + (\mu_{m\beta} - \mu_{m\alpha})\tau_m > 0.$$

Since the ratios $A_i^{(h)}/A_1^{(h)}$ are convergent fractions for τ_i/τ_1, it is clear that if h is sufficiently high, say for $h \geqq h_0$, then also

(14′) $$(\mu_{1\beta} - \mu_{1\alpha})A_1^{(h)} + \cdots + (\mu_{m\beta} - \mu_{m\alpha})A_m^{(h)} > 0,$$

for all pairs of indices $\alpha, \beta = 1, 2, \cdots, N$, such that (14) holds true. But (14′) implies $\mathrm{M}_{i\beta} > \mathrm{M}_{i\alpha}$, for $i = 1, 2, \cdots, m$, whenever $h \geqq h_0$, and this proves our theorem.

NOTE. Observe that for two *distinct* power products π_α and π_β it is not possible to have $v(\pi_\alpha) = v(\pi_\beta)$, since $\tau_1, \cdots, \tau_m$ are rationally independent.

4. We now consider the general case in which the residue field K' of our valuation B is an arbitrary algebraic extension of K, finite or infinite.[12] Let K^* be the least Galois extension of K which contains the field K'. For convenience we take K^* as our new ground field. By this ground field extension our original field Σ is extended to a field $\Sigma^* = \mathsf{K}^*\Sigma$, of degree of transcendency r over K^* (see [11], p. 189). It is well known that B admits at least one extension in Σ^* ([4], p. 185) and that K^* is the residue field of each extended valuation.[13] We

[12] The case of an infinite algebraic extension occurs already in the classical case, in view of our reduction of the dimensionality of a valuation through a transcendental extension of the ground field (see A III). The following example will illustrate. Let $\Sigma = \mathsf{K}(x, y, z)$, x, y and z—algebraically independent elements. Consider the one-dimensional valuation B obtained by equating z to the following formal power series in y:

$$z = x^{1/2}\cdot y + x^{1/4}y^2 + x^{1/8}y^3 + \cdots$$

Any element η in Σ becomes, by substitution, a formal power series in y, with coefficients which are algebraic functions of x, and by the value of η in B is meant the exponent of the leading term in the power series. Thus, for instance, $v(z) = 1$, $v(x) = 0$, etc. It is clear that the B-residue of z/y is $x^{1/2}$. We have $z^2/y^2 - x = \left(\frac{z}{y} - \sqrt{x}\right)\left(\frac{z}{y} + \sqrt{x}\right) = 2x^{1/2}x^{1/4}y + \cdots$, whence the B-residue of $\frac{z^2 - xy^2}{y^3}$ is $2x^{1/2}x^{1/4}$. Hence the B-residue of $\frac{z^2 - xy^2}{2zy^2}$ is $x^{1/4}$. Continuing in this fashion we can show that the residue field of B contains $x^{1/2}$, $x^{1/4}$, $x^{1/8}$, etc., i.e. the residue field is an infinite algebraic extension of the field $\mathsf{K}(x)$. But it is this last field which we would have to take as ground field if we wished to regard B as a zero-dimensional valuation.

[13] *If K' is an infinite extension of K, then B has necessarily infinitely many extension in Σ^*.* To see this, we first observe that a field of algebraic functions over a ground field K is by definition a finite extension of K. Hence the *relative algebraic closure* $\bar{\mathsf{K}}$ of K in Σ is also a finite extension of K. Therefore, if K' is an infinite extension of K, it is also an infinite extension of $\bar{\mathsf{K}}$. Therefore, for the proof of our assertion, we may assume that $\mathsf{K} = \bar{\mathsf{K}}$, or that K itself is maximally algebraic in Σ. In this case, the extended valuations of B form a *complete set of conjugate valuations* over K ([4], p. 185). In other words, if an

consider *one* of the extended valuations of B and we continue to denote it by B. However, when necessary we shall denote this extended valuation by B_1 in order to distinguish it from the other extended valuations $B_2, B_3, \cdots$.

Let $x_1, \cdots, x_m$ be elements of Σ contained in the valuation ring of B, and let the B-residue of x_i be α_i. We have then a point of coördinates $\alpha_1, \cdots, \alpha_m$ in the affine space of the x's (over the ground field K^*). Let $(\alpha_{1j}, \cdots, \alpha_{mj})$, $j = 1, 2, \cdots, g$, be the conjugate points over K. Here $(\alpha_{11}, \cdots, \alpha_{m1})$ is the original point $(\alpha_1, \cdots, \alpha_m)$.

Let $v(x_i - \alpha_i) = \tau_i$. *We make the following two assumptions*: (a) $\tau_1, \cdots, \tau_m$ are rationally independent; (b) the g conjugates $\alpha_{11} = \alpha_1, \alpha_{12}, \cdots, \alpha_{1g}$ are distinct.

We apply to $\tau_1, \cdots, \tau_m$ the elementary Perron transformation (1). Next, by using the interpolation formula of Lagrange, we construct a polynomial $\psi_i(x_1)$, $i > 1$, with coefficients in K, such that[14]

$$\psi_i(x_1) - \alpha_{ij} \equiv 0(\text{mod } (x_1 - \alpha_{1j})^{a_i^{(0)}}), \qquad j = 1, 2, \cdots, g; \quad i = 2, \cdots, m. \tag{15}$$

Let $f(x_1) = \prod_{j=1}^{g} (x_1 - \alpha_{1j})$, whence $f(x)$ is in $\mathsf{K}[x_1]$ and is irreducible. We put

$$\varphi_i(x_1) = \psi_i(x_1) + A_i(x_1)[f(x_1)]^{a_i^{(0)}}, \tag{16}$$

where $A_i(x_1)$ is a polynomial in $\mathsf{K}[x_1]$ on which we will impose presently some conditions. However, whatever the polynomial $A_i(x_1)$, it is obvious that $\varphi_i(x_1)$ shares with $\psi_i(x_1)$ the property expressed by (15), i.e. we have

$$\varphi_i(x_1) - \alpha_{ij} \equiv 0 \text{ (mod } (x_1 - \alpha_{1j})^{a_i^{(0)}}), \qquad j = 1, 2, \cdots, g. \tag{17}$$

Let $f_j(x_1) = \dfrac{f(x_1)}{x_1 - \alpha_{1j}}$. By (15) we can write:

$$\frac{\psi_i(x_1) - \alpha_{ij}}{[f(x_1)]^{a_i^{(0)}}} = \frac{H_{ij}(x_1)}{[f_j(x_1)]^{a_i^{(0)}}},$$

where $H_{i1}, H_{i2}, \cdots, H_{ig}$ are polynomials in x_1 which are conjugate over K. Hence, by (16) we have

$$\frac{\varphi_i(x_1) - \alpha_{ij}}{[f(x_1)]^{a_i^{(0)}}} = \frac{H_{ij}(x_1) + [f_j(x_1)]^{a_i^{(0)}} \cdot A_i(x_1)}{[f_j(x_1)]^{a_i^{(0)}}}. \tag{17'}$$

Let us assign, for each $i \geqq 2$, an arbitrary set of g *distinct* conjugate elements

---

element θ of K^* is a residue of an element of η in one extended valuation, each conjugate of θ over K is also a residue of η in some extended valuation. Suppose there were only a finite number, say g, of extended valuations. Then *every* element θ of K^* would have at most g conjugates over K, and this is impossible if K^* is an infinite extension of K.

[14] A polynomial $\psi_i(x_1)$ satisfying (15) exists and is uniquely determined if we impose the condition that it be of degree at most $a_i^{(0)}g - 1$.

$\alpha_{i-1,1}^{(1)}, \cdots, \alpha_{i-1,g}^{(1)}$ of the field $\mathsf{K}(\alpha_{11}, \cdots, \alpha_{1g})$. We choose $A_i(x_1)$ (*in* $\mathsf{K}[x_1]$) in such a fashion as to satisfy the congruences

$$(18) \qquad [f_j(x_1)]^{a_i^{(0)}} A_i(x_1) \equiv -H_{ij}(x_1) - \alpha_{i-1,j}^{(1)} \cdot [f_j(x_1)]^{a_i^{(0)}} (\text{mod } (x_1 - \alpha_{1j})^\rho),$$
$$j = 1, 2, \cdots, g,$$

where ρ is a preassigned positive integer, which we can choose as high as we please. For $\rho = 1$, the existence of such a polynomial $A_i(x_1)$ follows directly from the interpolation formula of Lagrange and from the fact that $f_j(\alpha_{1j}) \neq 0$. For $\rho > 1$ the existence of $A_i(x_1)$ follows by an easy induction.[15]

5. Having set this condition on $A_i(x_1)$ and keeping in mind the definition (16) of the polynomial $\varphi_i(x_1)$, we now define our Cremona transformation $T^{(h)}$, *for* $h = 1$, as follows:

$$(19) \qquad T^{(1)}: \begin{cases} x_1 = x_m^{(1)}, \\ x_i = x_{i-1}^{(1)} \cdot [f(x_m^{(1)})]^{a_i^{(0)}} + \varphi_i(x_m^{(1)}), \qquad i = 2, \cdots, m. \end{cases}$$

From (17), (17′) and (18) we deduce that the equations of $T^{(1)}$ can also be written, for each $j = 1, 2, \cdots, g$, in the following form:

$$(19') \quad T^{(1)}: \begin{cases} x_1 - \alpha_{1j} = x_m^{(1)} - \alpha_{mj}^{(1)}, \qquad \alpha_{1j} = \alpha_{mj}^{(1)} \\ x_i - \alpha_{ij} = (x_m^{(1)} - \alpha_{mj}^{(1)})^{a_i^{(0)}} \{[f_j(x_m^{(1)})]^{a_i^{(0)}} (x_{i-1}^{(1)} - \alpha_{i-1,j}^{(1)}) \\ \qquad\qquad + (x_m^{(1)} - \alpha_{mj}^{(1)})^\rho G_{ij}(x_m^{(1)})\} \end{cases}$$

Here the G_{ij} are polynomials with coefficients in $\mathsf{K}(\alpha_{11}, \cdots, \alpha_{1g})$. For optical reasons we re-write the equations (19′), dropping the index j referring to conjugate elements and denoting $f_j(x_m^{(1)})$ by $f^*(x_1)$:

$$x_1 - \alpha_1 = x_m^{(1)} - \alpha_m^{(1)}, \qquad \alpha_1 = \alpha_m^{(1)};$$
$$(19'') \quad x_i - \alpha_i = (x_m^{(1)} - \alpha_m^{(1)})^{a_i^{(0)}} \{[f^*(x_m^{(1)})]^{a_i^{(0)}} (x_{i-1}^{(1)} - \alpha_{i-1}^{(1)})$$
$$+ (x_m^{(1)} - \alpha_m^{(1)})^\rho G_i(x_m^{(1)})\}, \quad i = 2, \cdots, m.$$

We may, if we wish, regard the relations (19″) as referring to the index $j = 1$.

By (19), $x_1, \cdots, x_m$ are *polynomials* in $x_1^{(1)}, \cdots, x_m^{(1)}$, *with coefficients* in K. By (19″), the B-value of the expression in the curly brackets is equal to $\tau_i - a_i^{(0)}\tau_1$, i.e. to $\tau_{i-1}^{(1)}$. If we take ρ sufficiently large, so as to have $\rho\tau_1 > \tau_{i-1}^{(1)}$, then the value of the expression in the curly brackets will be the same as the value of the element $[f^*(x_m^{(1)})]^{a_i^{(0)}} \cdot (x_{i-1}^{(1)} - \alpha_{i-1}^{(1)})$. But $v(f^*(x_m^{(1)})) = v(f^*(x_1)) = 0$, since $f^*(x_1)$ is not divisible by $x_1 - \alpha_1$ (or, if we pay due respect to the index j: $f_j(x_1)$ is not divisible by $x_1 - \alpha_{1j}$). Hence, for such a large value of ρ we will have:

$$v(x_i^{(1)} - \alpha_i^{(1)}) = \tau_i^{(1)}, \qquad i = 1, 2, \cdots, m.$$

[15] The existence of $A_i(x_1)$ also follows directly from a Lemma proved later on (B III 10, Lemma 1).

Thus the transformation $T^{(1)}$ has the desired effect, leading, as it does, to new elements $x_1^{(1)}, \cdots, x_m^{(1)}$, whose B-residues are $\alpha_1^{(1)}, \cdots, \alpha_m^{(1)}$ and such that the B-values of the differences $x_i^{(1)} - \alpha_i^{(1)}$ are the real numbers $\tau_i^{(1)}$ which are obtained at the first step of the algorithm of Perron. We emphasize that the conjugate elements $\alpha_{i1}^{(1)}, \alpha_{i2}^{(1)}, \cdots, \alpha_{ig}^{(1)}$ can be assigned, for each $i = 1, 2, \cdots, m$, arbitrarily; in particular they may be assumed to be distinct.

6. We now proceed to build up a Cremona transformation $T^{(h)}$ associated with the Perron transformation (3). Essentially $T^{(h)}$ will be obtained by a repeated application of the elementary transformation $T^{(1)}$ defined in the preceding section. We first give a full description of $T^{(h)}$ and we shall then prove its existence, by induction with respect to h.

First of all, $T^{(h)}$ will be of the form:

$$T^{(h)}: x_i = P_i(X_1, X_2, \cdots, X_m), \qquad i = 1, 2, \cdots, m, \tag{20}$$

where P_i is a polynomial with coefficients in K. The X_i belong to the valuation ring of B, and the B-residue of X_i *is an arbitrarily assigned element* γ_i of the field $\mathsf{K}(\alpha_{11}, \alpha_{12}, \cdots, \alpha_{1g})$. Here, we recall, $\alpha_{11} = \alpha_1 = B$-residue of x_1 and $\alpha_{12}, \cdots, \alpha_{1g}$ are the conjugates of α_{11}. We also point out that by hypothesis these g conjugates are distinct, and consequently the field $\mathsf{K}(\alpha_{11}, \cdots, \alpha_{1g})$ contains the B-residues of all the x_i, $i = 1, 2, \cdots, m$, and their conjugates over K. It is important to keep in mind the fact that the residues of X_i lie in the fixed field $\mathsf{K}(\alpha_{11}, \cdots, \alpha_{1g})$, of relative degree g over K. That means that, as long as $\tau_1, \cdots, \tau_m$ are assumed to be rationally independent, the field generated by the coördinates of the centers of the extended valuations in the space of the X_i, and by their conjugates over K, remains fixed for any h. This fixed field $\mathsf{K}(\alpha_{11}, \cdots, \alpha_{1g})$ we shall denote by R. If the center of B in the space of $x_1, \cdots, x_m$ splits into g conjugate points over K, also the center of B in the space of $X_1, \cdots, X_m$ will split into g and only g conjugate points over K. We have g = relative degree of R over K.

The above will be the first of a set of properties of $T^{(h)}$. We have then, as property (a), the following:

(a) $\gamma_i = B$-residue of X_i; γ_i—an arbitrary preassigned element of $\mathsf{K}(\alpha_{11}, \cdots, \alpha_{1g})$. In particular, we may assume that γ_{i1} $(= \gamma_1)$ and its conjugates $\gamma_{i2}, \cdots, \gamma_{ig}$ over K are all distinct.

The other properties follow. For the sake of notations, we omit the index j relative to conjugate elements.[16]

[16] In this connection one must bear in mind the following. The extended valuations of B in Σ^* form a complete set of conjugates under the Galois group of Σ^*/Σ ([4], p. 185). This Galois group coincides with the Galois group of K^*/K if and only if K is maximally algebraic in Σ ([11], p. 193). If then K is not maximally algebraic in Σ, there will certainly exist relative automorphisms of K^*/K which cannot be extended to relative automorphisms of Σ^*/Σ. In particular, it may well happen that a given relative automorphism π of R/K cannot be induced by a relative automorphism of Σ^*/Σ. (R may contain elements

(b) $v(X_i - \gamma_i) = \tau_i^{(h)}$.

(c) In the ring $\mathsf{R}[X_1, \cdots, X_m]$ the equations of $T^{(h)}$ can be put in the following form:

$$(20') \qquad T^{(h)}: x_i - \alpha_i = \Delta_1^{A_i^{(h)}} \Delta_2^{A_i^{(h+1)}} \cdots \Delta_m^{A_i^{(h+m-1)}} \cdot L_i(X_1, \cdots, X_m), \qquad i = 1, 2, \cdots, m.$$

Here $\Delta_1, \cdots, \Delta_m, L_i$ are polynomials with coefficients in R. The Δ_i are of the following form:

$$(21) \qquad \begin{aligned} \Delta_m &= X_m - \gamma_m, \\ \Delta_i &= A_i(X_{i+1}, \cdots, X_m) \cdot (X_i - \gamma_i) + (X_m - \gamma_m)^\rho H_i(X_{i+1}, \cdots, X_m), \end{aligned} \qquad i = 1, 2, \cdots, m - 1.$$

Moreover,[17]

(d) $v(L_i) = v(A_i) = 0$ (whence $L_i(\gamma_1, \cdots, \gamma_m), A_i(\gamma_{i+1}, \cdots, \gamma_m) \neq 0$).

(e) $v(\Delta_i) = v(X_i - \gamma_i) = \tau_i^{(h)}$.

(f) ρ is an arbitrarily large integer.

7. Let us assume that a transformation $T^{(h)}$ with the above properties exists, for a given h. We prove that there also exists such a transformation $T^{(h+1)}$. We have, by (2),

$$\tau_1^{(h)} = \tau_m^{(h+1)}, \qquad \tau_i^{(h)} = \tau_{i-1}^{(h+1)} + a_i^{(h)} \tau_m^{(h+1)}, \qquad i = 2, \cdots, m.$$

With this elementary Perron transformation there is associated a Cremona transformation $\overline{T}^{(1)}$ defined in section 5:

$$(22) \qquad \overline{T}_1: \begin{aligned} X_1 &= Y_m \\ X_i &= Y_{i-1}[f(Y_m)]^{a_i^{(h)}} + \varphi_i(Y_m), \end{aligned}$$

where $f(X_1) = \prod_{j=1}^{g} (X_1 - \gamma_{1j})$. The equations (22) are the same as (19), with a slight change of notations. By (19″), the equations of $\overline{T}_1$ assume in $\mathsf{R}[Y_1, \cdots, Y_m]$ the form:

$$(22') \qquad \begin{aligned} X_1 - \gamma_1 &= Y_m - \epsilon_m, \\ X_i - \gamma_i &= (Y_m - \epsilon_m)^{a_i^{(h)}} \{[f^*(Y_m)]^{a_i^{(h)}} \cdot (Y_{i-1} - \epsilon_{i-1}) + (Y_m - \epsilon_m)^\sigma G_i(Y_m)\}, \end{aligned} \qquad i > 1,$$

---

which are not in K but are in Σ.) If this happens, and if, say, π carries the residues $\gamma_1, \cdots, \gamma_m$ $(= \gamma_{11}, \cdots, \gamma_{m1})$ into $\gamma_{12}, \cdots, \gamma_{m2}$, then $\gamma_{12}, \cdots, \gamma_{m2}$ *are not residues* of $X_1, \cdots, X_m$ for some extended valuation. To understand property (b) correctly it is therefore necessary to observe that it refers to our fixed extended valuation B_1 of B. If we apply the relative automorphisms of Σ^*/Σ, we have to write γ_{ij} instead of γ_i, *but* j *will not necessarily assume all the values* 1, 2, $\cdots$, g. At any rate, for all values which j actually assumes under these automorphisms, the symbol $v(X_i - \gamma_{ij})$ denotes the valuation-theoretic value of $X_i - \gamma_{ij}$ in a suitable extended valuation of B. Naturally, this value is always $\tau_i^{(h)}$, being independent of j.

[17] The remark concerning property (b) made in the preceding footnote also applies to properties (d) and (e).

where

(a′) ϵ_i = *B-residue of* Y_i ; $\epsilon_1, \cdots, \epsilon_{m-1}$, *arbitrary preassigned elements of* R, and $\epsilon_m = \gamma_1$. This corresponds to property (a).

Moreover,

$$f^*(X_1) = \frac{f(X_1)}{X_1 - \gamma_1}, \tag{23}$$

and σ is an arbitrarily large integer. We have also:

(b′) $v(Y_i - \epsilon_i) = \tau_i^{(h+1)}$.

We define: $T^{(h+1)} = T^{(h)} \cdot \bar{T}^{(1)}$. The transformation $T^{(h+1)}$, from $x_1, \cdots, x_m$ to $Y_1, \cdots, Y_m$, is such that the x_i are polynomials in $Y_1, \cdots, Y_m$ with coefficients in K, by (20) and (22). The conditions similar to (a) and (b) are satisfied, in view of (a') and (b').

If we substitute the expressions (22′) of the X_i into (21), we find that Δ_m is divisible by $(Y_m - \epsilon_m)^{a_m^{(h)}}$, and that also Δ_i, $1 < i < m$, acquires the factor $(Y_m - \epsilon_m)^{a_i^{(h)}}$, *provided* $\rho a_m^{(h)} - a_i^{(h)} > 0$. Now from (2′) it follows that if $h \geqq 1$, then $\tau_i^{(h)} < \tau_m^{(h)}$, whence $a_i^{(h)} \leqq a_m^{(h)}$, $i = 2, \cdots, m-1$. Hence by taking ρ sufficiently large, we can make $\rho a_m^{(h)} - a_i^{(h)}$ as large as we please. We therefore can put:

$$\Delta_i = (Y_m - \epsilon_m)^{a_i^{(h)}} \Delta'_{i-1}, \qquad i = 2, 3, \cdots, m, \tag{24}$$

and

$$\Delta_1 = (Y_m - \epsilon_m) \cdot M(Y_1, \cdots, Y_m). \tag{24'}$$

We find that $\Delta'_{i-1} - A_i(X_{i+1}, \cdots, X_m)[f^*(Y_m)]^{a_i^{(h)}}(Y_{i-1} - \epsilon_{i-1})$, $i = 2, \cdots, m-1$, is divisible by $(Y_m - \epsilon_m)^{\rho'}$, where $\rho' \geqq$ minimum $(\rho a_m^{(h)} - a_i^{(h)}, \sigma)$. Hence, by taking ρ and σ sufficiently large, we can make ρ' as large as we please. Let then

$$\begin{aligned} \Delta'_m &= Y_m - \epsilon_m \\ \Delta'_i &= A'_i(Y_{i+1}, \cdots, Y_m) \cdot (Y_i - \epsilon_i) + (Y_m - \epsilon_m)^{\rho'} H'_i(Y_{i+1}, \cdots, Y_m), \\ &\qquad i = 1, 2, \cdots, m-1, \end{aligned} \tag{25}$$

where

(f′) ρ' *is an arbitrarily large integer.* Here

$$A'_i(Y_{i+1}, \cdots, Y_m) = A_{i+1}(X_{i+2}, \cdots, X_m)[f^*(Y_m)]^{a_{i+1}^{(h)}} \qquad i = 1, 2, \cdots, m-2, \tag{26}$$

and, by (22′), for $i = m$,

$$A'_{m-1}(Y_m) = [f^*(Y_m)]^{a_m^{(h)}}. \tag{26'}$$

It is clear that $A_i^{(h+m)} = A_i^{(h)} + A_i^{(h+1)} a_2^{(h)} + \cdots + A_i^{(h+m-1)} a_m^{(h)}$. Hence substituting (24) and (24′) into (20′), and taking into account that by (25) $\Delta'_m = Y_m - \epsilon_m$, we get that the equations of $T^{(h+1)}$ in $\mathsf{R}[Y_1, \cdots, Y_m]$ are of the form:

$$x_i - \alpha_i = \Delta_1'^{A_i^{(h+1)}} \Delta_2'^{A_i^{(h+2)}} \cdots \Delta_m'^{A_i^{(h+m)}} \cdot L'_i(Y_1, \cdots, Y_m), \quad i = 1, 2, \cdots, m$$

where, by (24′),

(27) $L_i'(Y_1, \cdots, Y_m) = L_i(X_1, \cdots, X_m)\cdot[M(Y_1, \cdots, Y_m)]^{A_i^{(h)}}$.

These equations of $T^{(h+1)}$, together with the (25), give property (*c*).

From (24′) and from (21), for $i = 1$, we find that in $\mathsf{R}[Y_1, \cdots, Y_m]$ we have the following congruence:

$$M(Y_1, \cdots, Y_m) \equiv A_1(X_2, \cdots, X_m)(\text{mod } (Y_m - \epsilon_m)^{\rho a_m^{(h)}-1}).$$

Since $v(Y_m - \epsilon_m) = v(X_1 - \gamma_1) > 0$, and since $v(A_1) = 0$, it follows that $v(M) = 0$. Since $f^*(Y_m)$ is not divisible by $Y_m - \epsilon_m$ (by (23); note that $\gamma_{11} = \gamma_1 = \epsilon_m$), also $v(f^*(Y_m)) = 0$. Hence, in view of (26), (26′), (27) and of property (*d*), we conclude that $v(L_i') = v(A_i') = 0$, and this expresses property (*d*) for $T^{(h+1)}$. Finally, by (24), we have: $v(\Delta_i') = \tau_i^{(h+1)}$, for $i = 1, 2, \cdots, m - 1$, and also $v(\Delta_m') = \tau_m^{(h+1)}$, since $\Delta_m' = Y_m - \epsilon_m = X_1 - \gamma_1$. Thus also property (*e*) holds true, and this completes the proof.

8. Theorem 2 admits an immediate generalization. If

$$\pi_1 = \prod_{i=1}^{m} (x_i - \alpha_i)^{\mu_{i1}}, \cdots, \qquad \pi_N = \prod_{i=1}^{m} (x_i - \alpha_i)^{\mu_{iN}},$$

are N distinct power products in the differences $x_i - \alpha_i$, after the transformation $T^{(h)}$ they become:

$$\pi_1 = G_1(X_1, \cdots, X_m) \prod_{i=1}^{m} \Delta_i^{\mathrm{M}_{i1}}, \cdots, \pi_N = G_N(X_1, \cdots, X_m) \prod_{i=1}^{m} \Delta_i^{\mathrm{M}_{iN}},$$

where the M_{ij} are given by the relations (13), and where $v(G_1) = \cdots = v(G_N) = 0$. Here $G_1, \cdots, G_N$ are polynomials with coefficients in the field R. *If h is sufficiently large, then we will have $\mathrm{M}_{i\beta} > \mathrm{M}_{i\alpha}$, $i = 1, 2, \cdots, m$, for all α and β such that $v(\pi_\beta) > v(\pi_\alpha)$.* The proof is the same as that of Theorem 2.

We make the following application of this result. Let $f(x_1, \cdots, x_m)$ be a polynomial with coefficients in K or in some algebraic extension field K_1 of K, for instance in K^*. Clearly, f is a linear combination of power products such as π_i, with coefficients in $(\mathsf{K}_1, \mathsf{R})$, say

$$f = \overset{*}{c}_1\pi_1 + \cdots + \overset{*}{c}_N\pi_N, \qquad \overset{*}{c}_i \neq 0.$$

If, for instance, π_1 is the power product of least value in B, then after the transformation $T^{(h)}$, if h is sufficiently large, f will acquire the factor $\Delta_1^{M_{11}} \cdots \Delta_m^{M_{m1}}$. If we divide through by this factor, what remains will be of the form $\overset{*}{c}_1 G_1 +$ a multiple of the product $\Delta_1\Delta_2 \cdots \Delta_m$. Since $v(G_1) = 0$ and $v(\Delta_1 \cdots \Delta_m) > 0$, we conclude with the following

THEOREM 3. *Given a polynomial $f(x_1, \cdots, x_m)$ with coefficients in K^*, if h is sufficiently high, the transform of f by $T^{(h)}$ will have the form*:

(28) $$f = \Delta_1^{M_1}\Delta_2^{M_2} \cdots \Delta_m^{M_m}F(X_1, \cdots, X_m),$$

where F is a polynomial and where $v(F) = 0$ (i.e. $F(\gamma_1, \cdots, \gamma_m) \neq 0$).

III. Rationally dependent values

9. Let $\tau_1, \cdots, \tau_m$ be rationally independent real numbers and let τ_{m+1} be rationally dependent on $\tau_1, \cdots, \tau_m$:

$$\lambda\tau_{m+1} = \lambda_1\tau_1 + \cdots + \lambda_m\tau_m, \qquad \lambda > 0, \qquad (\lambda, \lambda_1, \cdots, \lambda_m) = 1.$$

In that case we have Theorem 1 (I2). *Let us assume that the residue field of a given zero-dimensional valuation B of Σ (of rank 1) coincides with the ground field* K, and let $x_1, \cdots, x_{m+1}$ be elements of Σ such that $v(x_i - c_i) = \tau_i$, where $c_i = B(x_1) \subset \mathsf{K}$. We associate with the transformation (6) of Theorem 1 the following Cremona transformation:

$$x_i - c_i = \prod_{j=1}^{m+1} X_j^{\beta_{ij}}, \qquad i = 1, 2, \cdots, m+1. \tag{29}$$

Then

$$\begin{aligned} v(X_i) &= \overset{*}{\tau_i} > 0, \qquad i = 1, 2, \cdots, m, \\ v(X_{m+1}) &= 0, \end{aligned} \tag{30}$$

where naturally $\overset{*}{\tau_1}, \cdots, \overset{*}{\tau_m}$ are rationally independent. The way in which we will apply the transformation (29) will be as follows. Let

$$\pi = \prod_{i=1}^{m+1} (x_i - c_i)^{\mu_i}, \qquad \pi' = \prod_{i=1}^{m+1} (x_i - c_i)^{\mu_i'},$$

be two power products of the differences $x_i - c_i$, where the μ_i and the μ_i' are non-negative integers, and let us assume that *π and π' have equal values in B*:

$$v(\pi) = v(\pi'). \tag{31}$$

After the transformation (29), the power products π and π' assume the form:

$$\pi = \prod_{i=1}^{m+1} X_i^{\mathrm{M}_i}, \qquad \pi' = \prod_{i=1}^{m+1} X_i^{\mathrm{M}_i'},$$

where

$$\begin{aligned} \mathrm{M}_i &= \mu_1\beta_{1i} + \cdots + \mu_{m+1}\beta_{m+1,i}, \\ \mathrm{M}_i' &= \mu_1'\beta_{1i} + \cdots + \mu_{m+1}'\beta_{m+1,i}. \end{aligned} \tag{32}$$

Since $v(\pi) = \mathrm{M}_1\overset{*}{\tau_1} + \cdots + \mathrm{M}_m\overset{*}{\tau_m}$ and $v(\pi') = \mathrm{M}_1'\overset{*}{\tau_1} + \cdots + \mathrm{M}_m'\overset{*}{\tau_m}$, and since $\overset{*}{\tau_1}, \cdots, \overset{*}{\tau_m}$ are rationally independent, it follows, by (31), that

$$\mathrm{M}_i = \mathrm{M}_i', \qquad i = 1, 2, \cdots, m. \tag{33}$$

In view of the relations (32) we have therefore the following equations:

$$\begin{aligned} \beta_{1i}(\mu_1' - \mu_1) + \cdots + \beta_{m+1,i}(\mu_{m+1}' - \mu_{m+1}) &= 0, \qquad i = 1, 2, \cdots, m \\ \beta_{1,m+1}(\mu_1' - \mu_1) + \cdots + \beta_{m+1,m+1}(\mu_{m+1}' - \mu_{m+1}) &= \mathrm{M}_{m+1}' - \mathrm{M}_{m+1}. \end{aligned} \tag{34}$$

The $(m + 1)$-rowed determinant $|\beta_{ij}|$ is ± 1. Hence, by (34), we have

$$\text{(35)} \qquad M'_{m+1} - M_{m+1} = \frac{\mu'_{m+1} - \mu_{m+1}}{d},$$

where d is, to within a $\pm$ sign, equal to the m-rowed determinant $|\beta_{ij}|$, $i, j = 1, 2, \cdots, m$.

10. In order to generalize the considerations of the preceding section to the general case of an arbitrary residue field, we first prove two lemmas.

Let R be a finite algebraic extension of the ground field K, of relative degree g. Let $P_j(\alpha_{1j}, \cdots, \alpha_{mj})$, $j = 1, 2, \cdots, g$, be g *distinct* conjugate points over K in the affine space $S_m(x_1, \cdots, x_m)(\alpha_{ij} \subset \mathsf{R})$. Let, moreover, $A_1(x_1, \cdots, x_m)$, $B_1(x_1, \cdots, x_m)$ be polynomials with coefficients in R, and let $A_j(x_1, \cdots, x_m)$, $B_j(x_1, \cdots, x_m)$ be the conjugate polynomials over K ($j = 1, 2, \cdots, g$).

LEMMA 1. *If*

$$\text{(36)} \qquad A_j(\alpha_{1j}, \cdots, \alpha_{mj}) \neq 0, \qquad j = 1, 2, \cdots, g,$$

then given any positive integer ρ_0, there exists a polynomial $H(x_1, \cdots, x_m)$ with coefficients in K such that $A_jH + B_j$, written as a polynomial in the differences $x_1 - \alpha_{1j}, \cdots, x_m - \alpha_{mj}$, begins with terms of degree $\geqq \rho_0$.

PROOF. Let $\mathfrak{p}_j$ denote the prime zero-dimensional ideal $(x_1 - \alpha_{1j}, \cdots, x_m - \alpha_{mj})$ in the polynomial ring $\mathsf{R}[x_1, \cdots, x_m]$. Since the ideals $\mathfrak{p}_1, \cdots, \mathfrak{p}_g$ are distinct, any two of them are free from common divisors (*teilfremd*), and the same holds true for the powers $\mathfrak{p}_1^{\rho_0}, \cdots, \mathfrak{p}_g^{\rho_0}$. We therefore can find a polynomial ϕ_i in $\mathsf{R}[x_1, \cdots, x_m]$ such that

$$\text{(37)} \qquad \phi_i \equiv 0(\mathfrak{p}_j^{\rho_0}), j \neq i; \qquad \phi_i \equiv 1(\mathfrak{p}_i^{\rho_0}),$$

and we may assume that $\phi_1, \cdots, \phi_g$ are conjugate polynomials over K.

Since, by hypothesis (36), $A_j \not\equiv 0(\mathfrak{p}_j)$, the ideal $(A_j, \mathfrak{p}_j^{\rho_0})$ is the unit ideal. Hence we can find a polynomial h_j in $\mathsf{R}[x_1, \cdots, x_m]$ such that $h_jA_j + B_j \equiv 0(\mathfrak{p}_j^{\rho_0})$, and again we may assume that $h_1, \cdots, h_g$ are conjugate polynomials over K. We put $H = h_1\phi_1 + \cdots + h_g\phi_g$. Then H is a polynomial with coefficients in K. Moreover, by (37), we have: $H \equiv h_j(\mathfrak{p}_j^{\rho_0})$, whence $HA_j + B_j \equiv h_jA_j + B_j \equiv 0(\mathfrak{p}_j^{\rho_0})$, q.e.d.

11. Let $\lambda_1, \cdots, \lambda_m$ be arbitrary non-negative integers.

LEMMA 2. *Under the hypothesis that $\alpha_{11}, \cdots, \alpha_{1g}$ are distinct conjugates, given a set of conjugate elements $\gamma_1, \cdots, \gamma_g$ of R over K and given a positive integer ρ_0, there exists a polynomial $\varphi(x_1, \cdots, x_m)$ with coefficients in K and, for each $i = 1, 2, \cdots, m$, there exists a set of conjugate polynomials $\delta_{ij}(x_1, \cdots, x_i)$ ($j = 1, 2, \cdots, g$) with coefficients in R, such that:*

$$\text{(38)} \quad \varphi - \gamma_j = \delta_{1j}\delta_{2j} \cdots \delta_{mj};$$

(39) $\delta_{1j} = A_{1j}(x_1)\cdot(x_1 - \alpha_{1j})^{\lambda_1}$,

$$\delta_{ij} = A_{ij}(x_1, \cdots, x_i)\cdot(x_i - \alpha_{ij})^{\lambda_i} + B_{ij}(x_1, \cdots, x_i),$$

$$i = 2, \cdots, m;$$

(40) *each* B_{ij} *begins with terms of degree* $\geqq \rho_0$ *in* $x_1 - \alpha_{1j}, \cdots, x_i - \alpha_{ij}$;
(40′) $A_{ij}(\alpha_{1j}, \cdots, \alpha_{ij}) \neq 0$;
(40″) $\delta_{ij}(\alpha_{1j'}, \cdots, \alpha_{ij'}) \neq 0$, if $j \neq j'$.

PROOF. For $m = 1$ the lemma follows directly from the interpolation formula of Lagrange. We assume the lemma to be true for a given m and for given $\lambda_1, \cdots, \lambda_m(\lambda_m \geqq 0)$, and we prove it for $m, \lambda_1, \cdots, \lambda_{m-1}, \lambda_m + 1$ (if $\lambda_m = 0$, we put $\delta_{mj} = 1$, and our induction is actually from $m - 1$ to m).

Let $d_i = \prod_{j=1}^{g} \delta_{ij}$ and $\delta_{ij}^* = d_i/\delta_{ij}$. Then d_i is a polynomial with coefficients in K. We put

$$\varphi' = \varphi + [cx_m + H(x_1, \cdots, x_{m-1})]\, d_1 d_2 \cdots d_m,$$

where c is some element of K and where H is a polynomial with coefficients in K, both to be determined. From (38) it follows that

$$\varphi' - \gamma_j = \delta_{1j}\delta_{2j} \cdots \delta_{mj}\cdot\delta_j', \tag{41}$$

where

$$\delta_j' = 1 + [cx_m + H(x_1, \cdots, x_{m-1})]\delta_{1j}^* \cdots \delta_{mj}^* . \tag{41′}$$

Let a_j and b_j be respectively the constant term and the coefficient of $x_m - \alpha_{mj}$ in the product $(\delta_{1j}^*\,\delta_{2j}^* \cdots \delta_{mj}^*)$, when this polynomial is written as a polynomial in $x_1 - \alpha_{1j}, \cdots, x_m - \alpha_{mj}$. In view of (40″), we have $a_j \neq 0$. We choose c in K so as to have $-b_j/a_j + ca_j \neq 0$. Having fixed the constant c, we find $H(x_1, \cdots, x_{m-1})$ in such a fashion that after the substitution $x_m = \alpha_{mj}$ the polynomial δ_j', written as a polynomial in $x_1 - \alpha_{1j}, \cdots, x_{m-1} - \alpha_{m-1,j}$, begin with terms of degree $\geqq \rho_0$. This is possible, by Lemma 1, since $a_j \neq 0$. Then δ_j' has the following form:

$$\delta_j' = A_j(x_1, \cdots, x_m)\cdot(x_m - \alpha_{mj}) + B_j(x_1, \cdots, x_{m-1}),$$

where B_j begins with terms of degree $\geqq \rho_0$ in $x_1 - \alpha_{1j}, \cdots, x_{m-1} - \alpha_{m-1,j}$. Moreover, $A_j(\alpha_{1j}, \cdots \alpha_{mj}) \neq 0$, as a consequence of our choice of the constant c. It is clear that if we put $\delta_{mj}' = \delta_{mj}\delta_j'$, then δ_{mj}' will be of the same form as δ_{mj}, with $\lambda_m + 1$, instead of λ_m, and that the condition analagous to (40′) will be also satisfied. As to condition (40″), for δ_{mj}', its validity follows from the definition (41′) of δ_j' : namely, we have obviously

$$(\delta_j')_{x_i = \alpha_{ij'}} = 1, \qquad \text{if } j \neq j'.$$

Since, by (41), $\varphi' - \gamma_j = \delta_{1j}\delta_{2j} \cdots \delta_{m-1,j}\cdot\delta_{mj}'$, the proof is now complete.

12. The significance of Theorem 3, as shown by (28), resides in the conclusion that the transform of a polynomial $f(x_1, \cdots, x_m)$, under $T^{(h)}$, if h is sufficiently high, differs from the transform of a power product of the differences $x_1 - \alpha_1, \cdots, x_m - \alpha_m$ (see (20′)) only by a *trivial factor*, i.e. by a factor whose value in B is zero. If, namely, $v(f) = \mu_1\tau_1 + \cdots + \mu_m\tau_m$, then the power product in question is $(x_1 - \alpha_1)^{\mu_1} \cdots (x_m - \alpha_m)^{\mu_m}$. In particular, if we have m arbitrary polynomials $\delta_1, \delta_2, \cdots, \delta_m$ in $\mathsf{K}^*[x_1, \cdots, x_m]$, such that $v(\delta_i) = \tau_i$, then, to within trivial factors, the δ_i will transform under $T^{(h)}$ like the differences $x_i - \alpha_i$, provided h is sufficiently high, i.e. we will have:

$$T^{(h)}\delta_i = \Delta_1^{A_i^{(h)}}\Delta_2^{A_i^{(h+1)}} \cdots \Delta_m^{A_i^{(h+m-1)}} \cdot L_i(X_1, \cdots, X_m), \tag{42}$$

where $v(L_i) = 0$. These relations are similar to (20′). More generally, a power product $\delta_1^{\mu_1} \cdots \delta_m^{\mu_m}$ will transform, to within trivial factors, as the product $(x_1 - \alpha_1)^{\mu_1} \cdots (x_m - \alpha_m)^{\mu_m}$; i.e., for h sufficiently high we will have:

$$T^{(h)}(\delta_1^{\mu_1} \cdots \delta_m^{\mu_m}) = \Delta_1^{\mathrm{M}_1} \cdots \Delta_m^{\mathrm{M}_m} G(X_1, \cdots, X_m), \tag{42′}$$

where $v(G) = 0$. While the real numbers $\tau_1, \cdots, \tau_m$ undergo the Perron transformation (3), it is clear that the exponents $\mu_1, \cdots, \mu_m$ undergo the *contragredient transformation* (13), i.e.:

$$\mathrm{M}_i = \mu_1 A_1^{(h+i-1)} + \cdots + \mu_m A_m^{(h+i-1)}, \qquad i = 1, 2, \cdots, m. \tag{42″}$$

Note that the polynomials $\Delta_1, \cdots, \Delta_m$ themselves form a particular set of polynomials such as $\delta_1, \cdots, \delta_m$, with respect, however, to the new variables $X_1, \cdots, X_m$. This is so because of property (*e*): $v(\Delta_i) = \tau_i^{(h)}$ (see B II 6).

More generally, suppose we consider a polynomial f which depends on $x_1, \cdots, x_m$ *and on other elements* $x_{m+1}, \cdots, x_n$ contained in the valuation ring $\mathfrak{B}$ of B. We write f as a polynomial in $x_1 - \alpha_1, \cdots, x_m - \alpha_m$, with coefficients which are polynomials in $x_{m+1}, \cdots, x_n$. Let $A(x_{m+1}, \cdots, x_n)(x_1 - \alpha_1)^{\mu_1} \cdots (x_m - \alpha_m)^{\mu_m}$ be the term for which $\mu_1\tau_1 + \cdots + \mu_m\tau_m$ is minimum. We shall say that f *is monovalent in* $x_1, \cdots, x_m$ if $v(A) = 0$. If f is monovalent, then $v(f) = \mu_1\tau_1 + \cdots + \mu_m\tau_m$, and it is obvious that under the transformation $T^{(h)}$, which affects only $x_1, \cdots, x_m$, f will transform like $(x_1 - \alpha_1)^{\mu_1} \cdots (x_m - \alpha_m)^{\mu_m}$, to within a trivial factor; provided h is sufficiently high. Similarly, if $\delta_1, \cdots, \delta_m$ are monovalent in $x_1, \cdots, x_m$ and if $v(\delta_i) = \tau_i$, then they also transform under $T^{(h)}$ according to (42); only the trivial factor L_i now depends on $X_1, \cdots, X_m$ *and* $x_{m+1}, \cdots, x_n$.

Remark. It is important to bear in mind that while the coefficients of the Cremona transformation $T^{(h)}$ are elements of K, the equations (20′) are the equations of $T^{(h)}$ written over the field R. Here R is the field generated by the B-residues of $x_1, \cdots, x_m$ and by their conjugates over K. Thus all the polynomials Δ_i, L_i in (20′) have their coefficients in R. As a consequence, if a given polynomial f in $x_1, \cdots, x_m$, and perhaps in other variables $x_{m+1}, \cdots, x_n$, has its coefficients in R, the coefficients of f, *after the transformation* $T^{(h)}$, *will remain in* R.

13. Let x_{m+1} be an element of Σ whose B-residue α_{m+1} is in the field R, i.e. in the field of the B-residues of $x_1, \cdots, x_m$ and of their conjugates. Let $v(x_{m+1} - \alpha_{m+1}) = \tau_{m+1} > 0$, and let us assume that τ_{m+1} is rationally dependent on $\tau_1, \cdots, \tau_m$:

$$(43) \qquad \lambda\tau_{m+1} = \lambda_1\tau_1 + \cdots + \lambda_m\tau_m, \qquad \lambda > 0.$$

The transformation (6) of Theorem 1 consisted of two steps: the first step served to lower the value of λ [see (7), (10) and (7')], and at each stage of this step the transformations affected only rationally independent numbers. For $\lambda = 1$, we had the second step which consisted of a preliminary Perron transformation (which led to (11')) and of the transformation (11''). At this stage we shall be concerned with the first step. We wish to generalize in some way the considerations of the preceding section.

Let $\delta_1, \delta_2, \cdots, \delta_m$ be polynomials in $\mathsf{R}[x_1, \cdots, x_m]$ of the same form in $x_1, \cdots, x_m$ as the Δ_i's were in the X_i, i.e. (see (21)):

$$\delta_m = x_m - \alpha_m,$$

$$(44) \qquad \delta_i = a_i(x_{i+1}, \cdots, x_m)\cdot(x_i - \alpha_i) + (x_m - \alpha_m)^\sigma h_i(x_{i+1}, \cdots, x_m),$$
$$i = 1, 2, \cdots, m - 1,$$

and with properties similar to (*d*), (*e*) and (*f*) (B II 6). Moreover, let

$$(44') \quad \delta_{m+1} = a_{m+1}(x_1, \cdots, x_m)\cdot(x_{m+1} - \alpha_{m+1}) + (x_m - \alpha_m)^\sigma h_{m+1}(x_1, \cdots, x_m)$$

be another polynomial, of the indicated form, with coefficients in R. Such a set of $m + 1$ polynomials shall be called a *normal set* for $x_1, \cdots, x_m, x_{m+1}$, provided *$\sigma$ is sufficiently large*. How large must σ be will appear in the course of our considerations. *We also assume that, for each $i = 1, 2, \cdots, m$, the conjugates α_{i1} $(= \alpha_i)$, $\alpha_{i2}, \cdots, \alpha_{ig}$ are distinct* (compare with property (*a*) of B II 6).

We fix an integer ρ_0 such that

$$(45) \qquad \rho_0\tau_m > \tau_i, \qquad i = 1, 2, \cdots, m - 1, m + 1.$$

Our *first* condition on σ will be:

$$(45') \qquad \sigma \geqq \rho_0 .$$

We now pass by a Perron transformation from $\tau_1, \cdots, \tau_m$ to $\tau_1^{(h)}, \cdots, \tau_m^{(h)}$, getting (8) as the new relation of rational dependence between $\tau_1^{(h)}, \cdots, \tau_m^{(h)}$ and τ_{m+1}. We take h so high as to satisfy the following conditions:
(a) The coefficients $\lambda_1^{(h)}, \cdots, \lambda_m^{(h)}$ in (8) are all positive.
(b) $\lambda_m^{(h)} > \lambda$ (note that in view of (9), $\lim_{h\to+\infty} \lambda_i^{(h)} = +\infty$);
(c) $\lambda_m^{(h)} \not\equiv 0(\lambda)$. That it is possible to satisfy this condition can be seen as follows. The coefficients $\lambda_m^{(h)}, \lambda_m^{(h+1)}, \cdots, \lambda_m^{(h+m-1)}$, corresponding to m consecutive values of the upper index, are linear forms of $\lambda_1, \cdots, \lambda_m$, and the determinant of the coefficients of these forms is ± 1. Since $(\lambda_1, \cdots, \lambda_m, \lambda) = 1$,

at least one of the integers $\lambda_m^{(h)}, \lambda_m^{(h+1)}, \cdots, \lambda_m^{(h+m-1)}$ is not divisible by λ. If, say, $\lambda_m^{(h+i)} \not\equiv 0(\lambda)$, then we replace h by $h + i$.

(d) $\lambda \rho_0 A_m^{(h+m-1)} > \lambda_1 A_1^{(h+m-1)} + \cdots + \lambda_m A_m^{(h+m-1)}$, i.e. $\lambda \rho_0 A_m^{(h+m-1)} > \lambda_m^{(h)}$.

This inequality, for all high values of h, follows from the inequality (45): $\rho_0 \tau_m > \tau_{m+1}$, and from (43).

(e) $\delta_1, \cdots, \delta_m$ *should transform under* $T^{(h)}$ *as* $x - \alpha_1, \cdots x_m - \alpha_m$ *respectively.* That this is the case, for sufficiently high values of h, follows from (45), for $i = 1, 2, \cdots, m - 1$, and from (45′). Namely, these inequalities tell us that $\delta_1, \cdots, \delta_m$ are monovalent in $x_1, \cdots, x_m$ and that $v(\delta_i) = \tau_i$. How high h must be in order to satisfy the condition (e) depends of course on the value of σ: the larger the value of σ is the sooner we will encounter satisfactory values of h. However, we wish to fix a lower bound for h which is *independent of* σ, since σ has not yet been fixed. For that it is sufficient to subject h to the following condition: h should be taken so high that $\delta_1, \cdots, \delta_m$ transform under $T^{(h)}$ as $x - \alpha_1, \cdots, x_m - \alpha_m$, *also if in* (44) *we replace* σ *by* ρ_0. This condition is stronger than condition (e), in view of (45′).

Now that h has been fixed we apply the transformation $T^{(h)}$ (x_{m+1} is so far not affected). *The integer* ρ *in* (21) *we take so high as to satisfy the condition*:

$$(45'') \qquad \rho \tau_1^{(h)} > \tau_i^{(h)}, \qquad i = 2, \cdots, m - 1.$$

The polynomials $\delta_1, \cdots, \delta_m$ are transformed according to (42). Instead of the set $\delta_1, \cdots, \delta_m, \delta_{m+1}$ we have now to deal with the polynomials $\Delta_1, \cdots, \Delta_m, \delta_{m+1}$. Concerning δ_{m+1} we only wish to observe that the factor $(x_m - \alpha_m)^\sigma$ in the second term of (44′) now takes the form:

$$(46) \qquad (x_m - \alpha_m)^\sigma = \Delta_1^{\sigma A_m^{(h)}} \Delta_2^{\sigma A_m^{(h+1)}} \cdots \Delta_m^{\sigma A_m^{(h+m-1)}} \cdot L,$$

where $\Delta_m = X_m - \gamma_m$ and where L is a trivial factor. Since, by (d) and (45′), $\sigma A_m^{(h+m-1)} > \frac{\lambda_m^{(h)}}{\lambda}$, it follows that if we put

$$(47) \qquad \lambda_m^{(h)} = g\lambda + \lambda', \qquad 0 < \lambda' < \lambda \qquad \text{(by (}c\text{))},$$

then

$$(48) \qquad (x_m - \alpha_m)^\sigma \equiv 0(\text{mod } (X_m - \gamma_m)^g).$$

We now operate on $\tau_m^{(h)}$ *and* τ_{m+1} *only.* We put, namely

$$(47') \qquad \tau_{m+1} = \bar{\tau}_{m+1} + g\tau_m^{(h)}, \qquad \bar{\tau}_{m+1} > 0, \text{ by (47)}.$$

For this elementary transformation involving *two* real numbers we have of course a Cremona transformation of the type $T^{(1)}$, from the variables X_m, x_{m+1} to the variables X_m, X_{m+1} (note that the conjugate residues $\gamma_{m1}, \gamma_{m2}, \cdots, \gamma_{m,g}$ of X_m are distinct, by property (a), B II 6). Written in the field R, the equation of the transformation $T^{(1)}$ is of the following form (compare with (19′)):

$$(49) \quad x_{m+1} - \alpha_{m+1} = (X_m - \gamma_m)^g \{C(X_m) \cdot (X_{m+1} - \gamma_{m+1}) + (X_m - \gamma_m)^{\bar{\rho}} \cdot D(X_m)\}.$$

Here we may assume that $\bar{\rho}$ is as high as we please and that the conjugate B-residues $\gamma_{m+1,1}$ $(= \gamma_{m+1})$, $\gamma_{m+1,2}, \cdots, \gamma_{m+1,g}$ of X_{m+1} are distinct. These residues are in the field R.

By (48) and (49) δ_{m+1} acquires the factor $(X_m - \gamma_m)^g$. Let

(50) $$\delta_{m+1} = (X_m - \gamma_m)^g \Delta_{m+1} .$$

Then, in view of (46), (48), and (49),

(50′) $$\Delta_{m+1} \equiv A(X_1, \cdots, X_m)C(X_m)(X_{m+1} - \gamma_{m+1}) \bmod ((X_m - \gamma_m)^{\bar{\rho}}, \Delta_1^{\sigma A_m^{(h)}}),$$

and

(50″) $$v(AC) = 0.$$

We now have our new variables: $X_1, \cdots, X_m, X_{m+1}$, and our new set of polynomials: $\Delta_1, \cdots, \Delta_m, \Delta_{m+1}$ whose values are $\tau_1^{(h)}, \cdots, \tau_m^{(h)}$ and $\bar{\tau}_{m+1}$ respectively. The transformation (50) from δ_{m+1} to Δ_{m+1} is in an obvious relationship with the transformation (47′) from τ_{m+1} to $\bar{\tau}_{m+1}$.

We assume that $\bar{\rho}$ is so high as to satisfy the inequality:

(51) $$\bar{\rho}\tau_1^{(h)} > \bar{\tau}_{m+1} .$$

We now impose our second condition on the integer σ (in addition to (45′)):

(51′) $$\sigma A_m^{(h)} \tau_1^{(h)} > \bar{\tau}_{m+1} .$$

We know that $\tau_m^{(h)} > \tau_1^{(h)}$ (by (2)). Our final step will be an operation on $\tau_m^{(h)}$ and $\tau_1^{(h)}$; namely, we put:

$$\tau_m^{(h)} = \tau_1^{(h)} + \bar{\tau}_m .$$

At the same time we operate on X_1 and X_m by the corresponding Cremona transformation $T^{(1)}$, getting a new variable $\bar{X}_m$:

(52) $$X_m - \gamma_m = (X_1 - \gamma_1)\bar{\Delta}_m ,$$

where

(52′) $$\bar{\Delta}_m = \bar{C}(X_1)\cdot(\bar{X}_m - \bar{\gamma}_m) + (X_1 - \gamma_1)^{\rho'}\bar{D}(X_1), \qquad v(\bar{C}) = 0.$$

Here again ρ' can be as high as we please, and the conjugate B-residues $\bar{\gamma}_{m1}(= \bar{\gamma}_m), \bar{\gamma}_{m2}, \cdots, \bar{\gamma}_{mg}$ of $\bar{X}_m$ may be assumed to be distinct. They are elements of R, and so are the coefficients of $\bar{C}$ and $\bar{D}$.

From the expression (21) of Δ_1, we find:

53) $$\Delta_1 = (X_1 - \gamma_1)\cdot G(X_1, \cdots, X_{m-1}, \bar{X}_m), \qquad v(G) = 0.$$

We put

(53′) $$X_1 - \gamma_1 = \bar{\Delta}_1 .$$

Again from the expression (21) of the Δ_i, we deduce that each Δ_i, $i = 2, \cdots, m - 1$, *is monovalent in* X_i *and* X_1:

$$(54)\qquad \Delta_i \equiv \bar{A}_i(X_1, X_{i+1}, \cdots, X_{m-1}, \bar{X}_m)\cdot(X_i - \gamma_i) \bmod (X_1 - \gamma_1)^\rho,$$
$$v(\bar{A}_i) = 0, \qquad i = 2, \cdots, m-1.$$

Finally, as far as Δ_{m+1} is concerned, we deduce from (50′), (51), (51′), and (53) that Δ_{m+1} is monovalent in X_{m+1} and X_1. In conclusion, our new set of polynomials is the following: $\bar{\Delta}_1, \Delta_2, \cdots, \Delta_{m-1}, \Delta_{m+1}; \bar{\Delta}_m$. The polynomial Δ_1 differs from $\bar{\Delta}_1$ by a trivial factor $G(X_1, \cdots, X_{m-1}, \bar{X}_m)$ (see (53)), which, however, has this important property:

$$(55)\qquad G(\gamma_1, \cdots, \gamma_{m-1}, \bar{X}_m) = \text{const.}$$

This follows immediately from (53) and from the expression (21) of Δ_1. As a matter of fact, it is immediately seen that $G(\gamma_1, X_2, \cdots, X_{m-1}, \bar{X}_m)$ is a constant. As for $\bar{\Delta}_m$, we have, by (52): $\Delta_m = \bar{\Delta}_1\bar{\Delta}_m$, and this corresponds exactly to the relation $\tau_m^{(h)} = \tau_1^{(h)} + \bar{\tau}_m$. Finally, $\bar{\Delta}_1, \Delta_2, \cdots, \Delta_{m-1}, \Delta_{m+1}$ are monovalent in $X_1, X_2, \cdots, X_{m-1}, X_{m+1}$, and we have: $v(\bar{\Delta}_1) = \tau_1^{(h)}$, $v(\Delta_i) = \tau_i^{(h)}$ $i = 2, \cdots, m-1$; $v(\Delta_{m+1}) = \bar{\tau}_{m+1}$ and $v(\bar{\Delta}_m) = \bar{\tau}_m$. In the relation of rational dependence between $\tau_1^{(h)}, \cdots, \tau_{m-1}^{(h)}, \bar{\tau}_{m+1}$ and $\bar{\tau}_m$, the coefficient of $\bar{\tau}_m$ is λ', where $\lambda' < \lambda$ (see (47)).

14. Our result will become clearer if we change our notation. We denote $\tau_1^{(h)}$ by τ_m'; $\tau_i^{(h)}$ by τ_i' $(i = 2, \cdots, m-1)$; $\bar{\tau}_{m+1}$ by τ_1' and $\bar{\tau}_m$ by τ_{m+1}'. Accordingly we denote $X_1, X_2, \cdots, X_{m-1}, \bar{X}_m, X_{m+1}$ by $x_m', x_2', \cdots, x_{m-1}', x_{m+1}', x_1'$ respectively and $\bar{\Delta}_1, \Delta_2, \cdots, \Delta_{m-1}, \bar{\Delta}_m, \Delta_{m+1}$ respectively by $\delta_m', \delta_2', \cdots, \delta_{m-1}', \delta_{m+1}', \delta_1'$. We denote the B-residue of x_i' by α_i'. We reassume our results by listing the following properties of the Cremona transformation from the x's to the x''s:

a). The variables $x_i (i = 1, 2, \cdots, m+1)$ are polynomials in $x_1', \cdots, x_{m+1}'$ with coefficients in K. These polynomials reduce to constants, namely to $\alpha_1, \cdots, \alpha_{m+1}$, if $x_1', \cdots, x_m'$ are replaced by their B-residues $\alpha_1', \cdots, \alpha_m'$.[18]

b). The conjugate B-residues $\alpha_{i1}'(= \alpha_i'), \alpha_{i2}', \cdots, \alpha_{ig}'$ of x_i' are elements of R and may be assumed to be distinct.

c). If $\tau_i = \sum_{j=1}^{m+1} \beta_{ij}\tau_j'$, $i = 1, 2, \cdots, m+1$, are the equations of transformation of the values $\tau_1, \cdots, \tau_{m+1}$, then $\delta_i = G_i(x_1', \cdots, x_{m+1}')\cdot\delta_1'^{\beta_{i1}} \cdots \delta_{m+1}'^{\beta_{i,m+1}}$.

d). $v(G_i) = 0$ and $G_i(\alpha_1', \cdots, \alpha_m', x_{m+1}')$ is a constant; the coefficients of G_i are in R.

e). The polynomials $\delta_1', \cdots, \delta_m'$ are monovalent in $x_1', \cdots, x_m'$. The polynomial δ_{m+1}' is of the form $C'(x_m')\cdot(x_{m+1}' - \alpha_{m+1}') + (x_m' - \alpha_m)^{\rho'} D'(x_m')$, where $v(C') = 0$ and ρ' is as high as we please. The coefficients of $\delta_1', \cdots, \delta_{m+1}'$ are in the field R.

[18] In other words: all the points of the line $x_1' = \alpha_1', \cdots, x_m' = \alpha_m'$ in the space of $x_1', \cdots, x_m', x_{m+1}'$ correspond to the point $(\alpha_1, \cdots, \alpha_m, \alpha_{m+1})$ in the space of $x_1, \cdots, x_m, x_{m+1}$. Our assertion follows directly, by inspection, from the consecutive steps of the inverse Cremona transformation (from $X_1, \cdots, X_{m-1}, \bar{X}_m, X_{m+1}$ to $x_1, \cdots, x_{m+1}$).

f). $v(x_i' - \alpha_i') = v(\delta_i') = \tau_i'$.

g). We have the relations of rational dependence:
$\lambda\tau_{m+1} = \lambda_1\tau_1 + \cdots + \lambda_m\tau_m$ and $\lambda'\tau_{m+1}' = \lambda_1'\tau_1' + \cdots + \lambda_m'\tau_m'$, and in these relations λ' *is less than* λ.

We now operate on $x_1', \cdots, x_m'$ alone, by a transformation $T^{(h)}$. Under such a transformation, for h sufficiently high, the polynomials $\delta_1', \cdots, \delta_m'$ transform as $x_1' - \alpha_1', \cdots, x_m' - \alpha_m'$. It is clear that none of the above listed properties will be affected (in particular, the coefficient λ' in g) is not changed). The polynomials $\delta_1', \cdots, \delta_m'$ are replaced by new polynomials, which we shall continue to denote by $\delta_1', \cdots, \delta_m'$: they are of the same type in the new variables (which we shall continue to call $x_1', \cdots, x_m'$), as the polynomials $\Delta_1, \cdots, \Delta_m$, given in (21), are in $X_1, \cdots, X_m$. It is therefore clear that, provided we take the exponent ρ in (21) and the exponent ρ' in e) sufficiently high, the polynomials $\delta_1', \cdots, \delta_m', \delta_{m+1}'$ will form a *normal set*. Hence we may replace property e) by the following:

e'). *The polynomials* $\delta_1', \cdots, \delta_m', \delta_{m+1}'$ *form a normal set* (*with coefficients in* R).

Proceeding in this fashion we shall ultimately get a Cremona transformation which enjoys all the above properties and *such that in addition the coefficient* λ' *in* g) *is equal to* 1.

15. Suppose now that the coefficient λ was originally equal to 1:

$$\tau_{m+1} = \lambda_1\tau_1 + \cdots + \lambda_m\tau_m .$$

Let $\sigma_1, \cdots, \sigma_{m+1}$ be a normal set of polynomials in $\mathsf{R}[x_1, \cdots, x_m, x_{m+1}]$. We apply Lemma 2, where we put $\gamma_j = \alpha_{m+1,j}$. Let

$$x_{m+1}' = \frac{x_{m+1} - \varphi}{d_1 d_2 \cdots d_m} + H(x_1, \cdots, x_m), \qquad H \subset \mathsf{K}[x_1, \cdots, x_m], \tag{56}$$

where φ is the polynomial satisfying (38) and where $d_i = \prod_{j=1}^{g} \delta_{ij}$. In view of (38), we can rewrite (56) as follows:

$$x_{m+1}' = \frac{x_{m+1} - \alpha_{m+1,j}}{d_1 d_2 \cdots d_m} - \frac{1}{\delta_{1j}^* \cdots \delta_{mj}^*} + H, \tag{57}$$

where $\delta_{ij}^* = d_i/\delta_{ij}$.

In view of (40″) and of Lemma 1, we can find H in such a fashion that

$$H\delta_{1j}^*, \cdots, \delta_{mj}^* - 1 = B_j(x_1, \cdots, x_m) = \varphi_{\sigma j}(x_1 - \alpha_{1j}, \cdots, x_m - \alpha_{mj}) + \text{terms of higher degree}, \tag{57'}$$

where $\varphi_{\sigma j}$ is a form of arbitrarily high degree σ.

Substituting into (57) we get

$$x_{m+1} - \alpha_{m+1,j} = \delta_{1j}\delta_{2j} \cdots \delta_{mj}(\delta_{1j}^* \cdots \delta_{mj}^* x_{m+1}' - B_j). \tag{57''}$$

If ρ_0 in (40) is sufficiently high, then $v(\delta_{ij}) = \lambda_i \tau_i$, $i = 1, 2, \cdots, m$. Hence, by (57″),

$$v(\delta_{1j}^* \cdots \delta_{mj}^* x'_{m+1} - B_j) = 0,$$

and since $v(B_j) > 0$ (if $\sigma \geqq 1$), it follows that $v(x'_{m+1}) = 0$. By substitution of (57″), the polynomial σ_{m+1} will assume the following form (we denote δ_{i1} by δ_1)

$$\sigma_{m+1} = \delta_1 \delta_2 \cdots \delta_m \sigma'_{m+1} + (x_m - \alpha_m)^\sigma h(x_1, \cdots, x_m, x_{m+1}),$$

where

$$\sigma'_{m+1} = A(x_1, \cdots, x_m) x'_{m+1} + B(x_1, \cdots, x_m)$$

and $v(A) = 0$, $v(\sigma'_{m+1}) = 0$. In this form σ_{m+1} now appears as a monovalent polynomial in $x_1, \cdots, x_m$. Moreover, from (57″) it follows that if we replace $x_1, \cdots, x_m$ by their B-residues $\alpha_1, \cdots, \alpha_m$ (while letting x'_{m+1} be arbitrary), the expression of x_{m+1} as a polynomial in $x_1, \cdots, x_m, x_{m+1}$ reduces to a constant, namely to the B-residue α_{m+1} of x_{m+1}. If we now apply to $x_1, \cdots, x_m$ a Cremona transformation $T^{(h)}$, h sufficiently high, we obtain, in view of the results of the preceding section, the following

THEOREM 4. *Given a normal set of polynomials $\delta_1, \cdots, \delta_m, \delta_{m+1}$, there exists a unimodular integral non-negative transformation:*
$\tau_i = \sum \beta_{ij} \tau_j^*$, *such that* $\tau_{m+1}^* = 0$; *and there exists a corresponding Cremona transformation*

(58) $$x_i = P_i(X_1, \cdots, X_m, X_{m+1}), \qquad P_i \subset \mathsf{K}[X_1, \cdots, X_{m+1}],$$

where

(58′) $$P_i(\gamma_1, \cdots, \gamma_m, X_{m+1}) = \alpha_i, \quad \gamma_i = B \text{ residue of } X_i, \quad i = 1, 2, \cdots, m+1,$$

such that

$$\delta_i = G_i(X_1, \cdots, X_m, X_{m+1}) \Delta_1^{\beta_{i1}} \cdots \Delta_m^{\beta_{im}} \Delta^{\beta_{i,m+1}}, \qquad i = 1, 2, \cdots, m+1,$$

and
(a) $\Delta_1, \cdots, \Delta_m$ *are of the form* (21);
(b) $v(X_i - \gamma_i) = v(\Delta_i) = \tau_i^*$, $i = 1, 2, \cdots, m$;
(c) $\Delta_{m+1} = A(X_1, \cdots, X_m) X_{m+1} + B(X_1, \cdots, X_m)$, $v(\Delta_{m+1}) = v(A) = 0$;
(d) $G_i(\gamma_1, \cdots, \gamma_m, X_{m+1}) = \text{const.}$
(e) *The residues γ_i of the elements $X_i (i = 1, 2, \cdots, m)$ and the coefficients of the Δ_i and of the $G_i (i = 1, 2, \cdots, m+1)$ are in the field* R.

It is important to point out that *the B-residue of X_{m+1} need not be in the field* R. Namely, from (57″) it follows that the B-residue of x'_{m+1} may very well be a proper algebraic quantity over R. Thus, of our various Cremona transformations, it is the last transformation, namely (56), that may lead for the first time to new variables whose B-residues generate a larger field than the one generated by the B-residues of the original variables $x_1, \cdots, x_{m+1}$.

From this theorem we can draw the same consequences as those given in B III 10. Namely, given two power products of $\delta_1, \cdots, \delta_{m+1}$:

$$\pi = \prod_{i=1}^{m+1} \delta_i^{\mu_i}, \qquad \pi' = \prod_{i=1}^{m+1} \delta_i^{\mu_i'},$$

and assuming that $v(\pi) = v(\pi')$, then after the transformation (58) has been performed, π and π' assume the form:

$$\pi = G(X_1, \cdots, X_m, X_{m+1}) \prod_{i=1}^{m+1} \Delta_i^{M_i},$$

$$\pi' = G'(X_1, \cdots, X_m, X_{m+1}) \prod_{i=1}^{m+1} \Delta_i^{M_i'},$$

where $G(\gamma_1, \cdots, \gamma_m, X_{m+1})$ and $G'(\gamma_1, \cdots, \gamma_m, X_{m+1})$ are *constants* different from zero, and where the exponents M_i, M_i' satisfy the relations (33) and (35).

C. Uniformization of Valuations

I. Zero-dimensional valuations of rank 1. A lemma

1. The proof of the main theorem is based on a lemma concerning rank 1 zero-dimensional valuations in a field $K(x_1, \cdots, x_r, x_{r+1})$, where $x_1, \cdots, x_{r+1}$ are *algebraically independent* elements. Let B be such a valuation, with K' as residue field. Let $\alpha_1, \cdots, \alpha_{r+1}$ be the B-residues of $x_1, \cdots, x_{r+1}$ respectively (we assume that $v(x_i) \geqq 0$), and let R be the field generated by these residues and by their conjugates over K. Finally, let K^* be the least Galois extension of K containing K'.

Lemma. *Given a polynomial* $f(x_1, \cdots, x_{r+1})$, *with coefficients in* R, *there exists a Cremona transformation*:

$$x_i = P_i(X_1, \cdots, X_{r+1}), \qquad i = 1, 2, \cdots, r+1,$$

where the P_i *are polynomials with coefficients in* K, *such that the expression of* f *in terms of the new variables* X_i *is of the form*:

$$f = \Delta_1^{\mu_1} \cdots \Delta_m^{\mu_m} F(X_1, \cdots, X_{r+1}), \tag{59}$$

where:

(a) $v(X_i) \geqq 0, i = 1, 2, \cdots, r+1$;

(b) $v(\Delta_i) = v(X_i - \gamma_i)$, *where* γ_i *is the B-residue of* X_i;

(c) $v(\Delta_1), \cdots, v(\Delta_m)$ *are rationally independent numbers*;

(d) $\Delta_1, \cdots, \Delta_m$ *are polynomials in* $X_1, \cdots, X_m$, *of the form* (21); *the coefficients of the* Δ_i *and of* F *are in the field generated by* $\gamma_1, \cdots, \gamma_{r+1}$ *and by their conjugates over* K. *If* $K' = K$, *then* $\Delta_i = X_i$.

(e) $v(F) = 0$.

It should be understood that the values $v(\Delta_i)$, $v(F)$ are taken relatively to some extended valuation of B in $K^*\Sigma$.

Main theorem and Lemma will be proved simultaneously, *under the hypothesis that the lemma is true in the case of r independent variables.*

In the case $r = 1$, the lemma is trivial. In fact, let $\alpha_1 = B(x_1)$. If $v(f(x_1)) = 0$, there is nothing to prove. If $v(f(x_1)) > 0$, then $f(x_1)$ is divisible by $x_1 - \alpha_1$. If $f(x_1)$ is exactly divisible by $(x_1 - \alpha_1)^{\mu_1}$, then the factorization $f(x_1) = (x_1 - \alpha_1)^{\mu_1} F(x_1)$ yields (59).

II. Special case: residue field = ground field

2. We shall first treat the case in which the residue field of B coincides with the ground field K. Let $f(x_1, x_2, \cdots, x_{r+1})$ be the given polynomial with coefficients in K. In the case of the main theorem, f is irreducible and $f = 0$ is the defining equation of our hypersurface. In the case of the lemma, $x_1, \cdots, x_{r+1}$ are algebraically independent elements. Without loss of generality we may assume that $v(x_i) > 0$, $i = 1, 2, \cdots, r + 1$, i.e. that the point $(0, \cdots, 0)$ is the center of the valuation. We write f as a polynomial in x_{r+1}:

$$f = a_0(x_1, \cdots, x_r) + a_1(x_1, \cdots, x_r)x_{r+1} + \cdots + a_\nu(x_1, \cdots, x_r)x_{r+1}^\nu.$$

In the case of the theorem we assume that $f(0, \cdots, 0, x_{r+1}) \neq 0$, i.e. that our hypersurface does not contain the line $x_1 = \cdots = x_r = 0$. In the case of the lemma, if $h(x_1, \cdots, x_r)$ is the h.c.d. of $a_0, \cdots, a_\nu$, and if we put $f/h = \varphi$, we assume that $\varphi(0, \cdots, 0, x_{r+1}) \neq 0$. These assumptions can always be satisfied if we perform a preliminary linear homogeneous transformation on $x_1, \cdots, x_{r+1}$.

Whether we are dealing with the theorem or with the lemma, it is true in both cases that $x_1, \cdots, x_r$ are algebraically independent elements. Hence, by our induction, we can apply the lemma to the product $a_0 a_1 \cdots a_\nu$ and to the valuation induced by B in the field $\mathsf{K}(x_1, \cdots, x_r)$. There exists then a Cremona transformation: $x_i = P_i(X_1, \cdots, X_r)$, $i = 1, 2, \cdots, r$, such that each polynomial a_i assumes the following form:

$$a_i(x_1, \cdots, x_r) = X_1^{\mu_{1i}} \cdots X_m^{\mu_{mi}} b_i(X_1, \cdots, X_r),$$

where $\mu_{1i}, \cdots, \mu_{mi}$ are non-negative integers and $v(b_i) = 0$. Moreover, $v(X_i) \geqq 0$, $i = 1, 2, \cdots, r$, and $v(X_1), \cdots, v(X_m)$ are rationally independent numbers. We assume that this Cremona transformation has already been performed, and we identify the X's with the x's. For convenience, we denote $X_{m+1}, \cdots, X_r$ and X_{r+1} by $x_{m+2}, \cdots, x_{r+1}$ and x_{m+1} respectively. Then f has the following form:

$$f = b_0 x_1^{\mu_{10}} \cdots x_m^{\mu_{m0}} + b_1 x_1^{\mu_{11}} \cdots x_m^{\mu_{m1}} x_{m+1} + \cdots + b_\nu x_1^{\mu_{1\nu}} \cdots x_m^{\mu_{m\nu}} x_{m+1}^\nu, \tag{60}$$

where $b_0, b_1, \cdots, b_\nu$ are polynomials in $x_1, \cdots, x_m, x_{m+2}, \cdots, x_{r+1}$, and $v(b_i) = 0$.

Let $v(x) = \tau_i$, $i = 1, 2, \cdots, m + 1$. We have two possible cases: (1) $\tau_1, \cdots, \tau_{m+1}$ are rationally independent; (2) they are rationally dependent.

In the first case we are dealing necessarily with the lemma, since in the case of the theorem the defining equation $f = 0$ implies, in view of (60), that τ_{m+1} is rationally dependent on $\tau_1, \cdots, \tau_m$ (if $f = 0$, at least two of the $\nu + 1$ terms of f must have the same value in B). In this case the lemma follows directly from Theorem 2 (B II 3). Namely, if we replace in Theorem 2 the integer m by $m + 1$ and if we identify the power product π_i with $x_1^{\mu_{1i}} \cdots x_m^{\mu_{mi}} x_{m+1}^m$, $i = 0, 1, 2, \cdots, \nu$, then under our Cremona transformation $T^{(h)}$, applied to $x_1, \cdots, x_{m+1}$, h sufficiently high, f will assume the form:

$$f = X_1^{M_{1\alpha}} \cdots X_{m+1}^{M_{m+1,\alpha}}[b_\alpha + X_1X_2 \cdots X_{m+1}H(X_1, \cdots, X_{r+1})],$$

$$X_i = x_i, i = m + 2, \cdots, r + 1,$$

where we assume that π_α is the term of lowest value in B. Since $v(b_\alpha) = 0$ and $v(X_i) > 0$, our lemma is proved.

3. We now consider the case in which $\tau_1, \cdots, \tau_{m+1}$ are rationally dependent. In the case of the theorem we have assumed that $f(0, \cdots, 0\ x_{r+1}) \neq 0$ and also that $f(0, \cdots, 0, 0) = 0$, since, by assumption, $v(x_i) > 0$, $i = 1, 2, \cdots, r + 1$. In the case of the lemma we have assumed that $\varphi(0, \cdots, 0, x_{r+1}) \neq 0$, where $f = h(x_1, \cdots, x_r)$. $\varphi(x_1, \cdots, x_{r+1})$, and where h is the h.c.d. of $a_0, \cdots, a_\nu$. Let s be the multiplicity of the root $x_{r+1} = 0$ for the polynomial $f(0, \cdots, 0, x_{r+1})$, in the case of the theorem; for the polynomial $\varphi(0, \cdots, 0, x_{r+1})$, in the case of the lemma. Then $s \geqq 1$, in the first case; $s \geqq 0$ in the second case. In the case of the theorem, if $s = 1$, then the center $(0, \cdots, 0)$ of B is already a simple point of our hypersurface, and there is nothing to prove. In the case of the lemma, if $s = 0$, then $v(\varphi) = 0$ and, by our induction, the proof of the lemma is completed by applying the lemma to the polynomial h *of the* r *variables* $x_1, \cdots, x_r$. We proceed to show that it is always possible to achieve a reduction of the multiplicity s, by Cremona transformations of the desired type, as long as $s > 1$ or $s > 0$, according as we are dealing with the theorem or with the lemma.

4. To the rationally dependent numbers $\tau_1, \cdots, \tau_m, \tau_{m+1}$ we apply the transformation (6) of Theorem 1, and we operate on $x_1, \cdots, x_m, x_{m+1}$ by the corresponding Cremona transformation (29), where we now put $c_i = 0$ (B III 9). Let

$$\pi_i = b_i x_1^{\mu_{1i}} \cdots x_m^{\mu_{mi}} x_{m+1}^i, \qquad i = 0, 1, \cdots, \nu,$$

and let $\pi_\alpha, \pi_\beta \cdots, \pi_\delta, \alpha < \beta < \cdots < \delta$ be the terms of *lowest value*. Then

$$\pi_i = b_i X_1^{\mathrm{M}_{1i}} \cdots X_m^{\mathrm{M}_{mi}} X_{m+1}^{\mathrm{M}_{m+1,i}},$$

where the M_{ji} are given by relations similar to (32), in which, however, $\mu_{m+1,i}$ should be replaced by i. In view of (33), the polynomial f assumes the form:

$$f = X_1^{\mathrm{M}_1} \cdots X_m^{\mathrm{M}_m}[B_\alpha X_{m+1}^{\mathrm{M}_{m+1,\alpha}} + B_\beta X_{m+1}^{\mathrm{M}_{m+1,\beta}} + \cdots + B_\delta X_{m+1}^{\mathrm{M}_{m+1,\delta}}] + \sum{}' B_i X_1^{\mathrm{M}_{1i}} \cdots X_m^{\mathrm{M}_{mi}} X_{m+1}^{\mathrm{M}_{m+1,i}}, \tag{61}$$

where $M_i = M_{i\alpha} = M_{i\beta} = \cdots = M_{i\delta}$ $(i = 1, 2, \cdots, m)$. The summation symbol $\sum'$ is extended to all terms π_i other than the minimum value terms. Moreover, $B_i(X_1, \cdots, X_{m+1}, x_{m+2}, \cdots, x_{r+1}) = b_i(x_1, \cdots, x_{r+1})$. We know that $v(X_{m+1}) = 0$. Hence, by (29), it follows that, for a given $i = 1, 2, \cdots, m + 1$, the exponents $\beta_{i1}, \cdots, \beta_{im}$ cannot be all zero. Hence

$$B_i(0, \cdots, 0, X_{m+1}, 0, \cdots, 0) = b_i(0, \cdots, 0). \tag{61'}$$

By (35) we have

$$\begin{aligned} M_{m+1,\beta} - M_{m+1,\alpha} &= \frac{\beta - \alpha}{d}, \\ &\cdots\cdots\cdots\cdots \\ M_{m+1,\delta} - M_{m+1,\alpha} &= \frac{\delta - \alpha}{d}. \end{aligned} \tag{62}$$

According as $d > 0$ or $d < 0$, the power $X_{m+1}^{M_{m+1,\alpha}}$, or the power $X_{m+1}^{M_{m+1,\delta}}$ will factor out from all the terms in the square brackets in (61), i.e. we shall have either

$$\pi_\alpha + \pi_\beta + \cdots + \pi_\delta = X_1^{M_1} \cdots X_m^{M_m} X_{m+1}^{M_{m+1,\alpha}} [B_\alpha + B_\beta X_{m+1}^{(\beta-\alpha)/d} + \cdots + B_\delta X_{m+1}^{(\delta-\alpha)/d}],$$

or

$$\pi_\alpha + \pi_\beta + \cdots + \pi_\delta = X_1^{M_1} \cdots X_m^{M_m} X_{m+1}^{M_{m+1,\delta}} [B_\alpha X_{m+1}^{(\alpha-\delta)/d} + B_\beta X_{m+1}^{(\beta-\delta)/d} + \cdots + B_\delta],$$

according as $d > 0$ or $d < 0$.

Since $\pi_\alpha, \pi_\beta, \cdots, \pi_\delta$ were the lowest value terms, we have $M_1\overset{*}{\tau_1} + \cdots + M_m\overset{*}{\tau_m} < M_{1i}\overset{*}{\tau_1} + \cdots + M_{mi}\overset{*}{\tau_m}$, for all i such that π_i is not a lowest value term. We now operate on $X_1, \cdots, X_m$ by a Cremona transformation $T^{(h)}$, h sufficiently high, and we apply Theorem 2. Let $y_1, \cdots, y_m$ be the new variables (in Theorem 2 the new variables were denoted by $X_1, X_2, \cdots$). For convenience we denote by $y_{m+1}, y_{m+2}, \cdots, y_{r+1}$ the variables $X_{m+1}, x_{m+2}, \cdots, x_{r+1}$. Then the polynomial f assumes the following form:

$$f = y_1^{\Lambda_1} \cdots y_m^{\Lambda_m} F(y_1, \cdots, y_{r+1}), \tag{63}$$

where

$$F = y_{m+1}^{\Lambda_{m+1}}\psi(y_1, \cdots, y_{r+1}) + y_1y_2 \cdots y_mG(y_1, \cdots, y_{r+1}), \ G, \text{ a polynomial.} \tag{63'}$$

Here

$$\psi = A_\alpha + A_\beta y_{m+1}^{(\beta-\alpha)/d} + \cdots + A_\delta y_{m+1}^{(\delta-\alpha)/d} \text{ and } \Lambda_{m+1} = M_{m+1,\alpha} \text{ if } d > 0, \tag{64}$$

and

$$\psi = A_\alpha y_{m+1}^{(\alpha-\delta)/d} + A_\beta y_{m+1}^{(\beta-\delta)/d} + \cdots + A_\delta \text{ and } \Lambda_{m+1} = M_{m+1,\delta} \text{ if } d < 0. \tag{64'}$$

Moreover

$$A_i(y_1, \cdots, y_{r+1}) = B_i(X_1, \cdots, X_m, X_{m+1}, x_{m+2}, \cdots, x_{r+1}),$$

whence, by (61′),

(65) $$A_i(0, \cdots, 0, y_{m+1}, 0, \cdots, 0) = b_i(0, \cdots, 0) \neq 0.$$

In the case of the theorem, the defining equation of the transformed hypersurface is either $F = 0$ or is $F_1 = 0$, where F_1 is a factor of F.

Let c be the B-residue of y_{m+1}. Then $c \neq 0$, since $v(X_{m+1}) = v(y_{m+1}) = 0$. By (65) it follows that ψ does not vanish identically if we put $y_i = 0$, $i \neq m + 1$. Hence, by (63′), the same holds true for the polynomial F. Let $y_{m+1} = c$ be a root of multiplicity $s'(s' \geqq 0)$ for the polynomial $F(0, \cdots, 0, y_{m+1}, 0, \cdots, 0)$. From (64), (64′) and (65) it follows that $s' \leqq (\delta - \alpha)/|d|$. Let us compare $(\delta - \alpha)/|d|$ with the original multiplicity s. We have $\pi_s = a_s(x_1, \cdots, x_m, x_{m+2}, \cdots, x_{r+1}) \cdot x_{m+1}^s$. In the case of the theorem, we must have $a_s(0, \cdots, 0) \neq 0$, since $x_{m+1} = 0$ is an s-fold root of $f(0, \cdots, 0, x_{m+1}, 0, \cdots, 0)$. In the case of the lemma, $f = h\varphi$ and $a_s = h\bar{a}_s$, where h is the h.c.d. of $a_0, a_1, \cdots, a_\nu$ and again we must have $\bar{a}_s(0, \cdots, 0) \neq 0$. In either case we conclude that $v(\pi_i) > v(\pi_s)$ for all $i > s$, since $v(\pi_i) \geqq v(x_{m+1}^i)$ and $v(\pi_s) = v(x_{m+1}^s)$. Since $\pi_\alpha, \pi_\beta, \cdots, \pi_\delta$ are the lowest value terms, it follows that

(66) $$0 \leqq \alpha < \delta \leqq s.$$

Hence

(66′) $$s' \leqq \frac{|\delta - \alpha|}{|d|} \leqq s.$$

Let us suppose that there is no reduction in the value of s, i.e. that $s' = s$. (Note that y_{m+1} plays now the role of x_{m+1}, or, in our old notations, of x_{r+1}). This implies in the first place $(\delta - \alpha)/|d| = s$, by (66′), whence, by (66), $\alpha = 0$ and $\delta = s$, $|d| = 1$. In the second place, we must have that $\psi(0, \cdots, 0, y_{m+1}, 0, \cdots, 0)$, to within a constant factor, is equal to $(y_{m+1} - c)^s$. This implies that $\alpha = 0, \beta = 1, \cdots, \delta = s$, i.e. originally there must have been in the polynomial f terms π_i of degree i in x_{m+1}, for all $i = 0, 1, \cdots, s$, and these terms were the lowest value terms. In particular, we have the terms $\pi_{s-1} = a_{s-1}x_{m+1}^{s-1}$ and $\pi_s = a_s x_{m+1}^s$, and since $v(\pi_{s-1}) = v(\pi_s)$, it follows that $v(a_{s-1}) = v(a_s) + v(x_{m+1})$. In the case of the theorem we have $a_s(0, \cdots, 0) \neq 0$, whence $v(x_{m+1}) = v(a_{s-1})$. Similarly, in the case of the lemma, we have $v(\bar{a}_{s-1}) = v(\bar{a}_s) + v(x_{m+1})$ and $v(\bar{a}_s) = 0$, whence $v(x_{m+1}) = v(\bar{a}_{s-1})$. In either case we conclude that *if the multiplicity s is not reduced by our Cremona transformation (from the x's to the y's), then the value of x_{m+1} must be equal to the value of a polynomial in the remaining variables $x_1, \cdots, x_m, x_{m+2}, \cdots, x_{r+1}$.*

5. For convenience we go back to our original notations in which x_{r+1} was the variable which latter was denoted by x_{m+1}. We assume that our Cremona transformation failed to achieve a reduction of the multiplicity s, and that consequently

$$v(x_{r+1}) = v(H), \qquad H \subset K[x_1, \cdots, x_r].$$

Since $v\left(\frac{x_{r+1}}{H}\right) = 0$, there exists a constant $c \neq 0$ in K, such that $v\left(\frac{x_{r+1}}{H} - c\right) > 0$. Let

$$x_{r+1}^{(1)} = x_{r+1} - cH,$$

whence $v(x_{r+1}) < v(x_{r+1}^{(1)})$.

Let $f_1(x_1, \cdots, x_r, x_{r+1}^{(1)}) = f(x_1, \cdots, x_r, x_{r+1}^{(1)} + cH)$. We now deal with the polynomial f_1 in the same way as we dealt with f. We note that the multiplicity of the root $x_{r+1}^{(1)} = 0$ is still s, both in the case of the theorem and of the lemma. If our Cremona transformation, applied to $x_1, \cdots, x_r, x_{r+1}^{(1)}$, again fails to achieve a reduction of s, then we must have $v(x_{r+1}^{(1)}) = v(H_1)$, $H_1 \subset K[x_1, \cdots, x_r]$. We then determine a constant c_1 such that $v\left(\frac{x_{r+1}^{(1)}}{H_1} - c_1\right) > 0$, and we put $x_{r+1}^{(2)} = x_{r+1}^{(1)} - c_1 H_1$, whence $v(x_{r+1}^{(1)}) < v(x_{r+1}^{(2)})$, and so we continue indefinitely.

Let us suppose that we get in this fashion a sequence of polynomials $f_i(x_1, \cdots, x_r, x_{r+1}^{(i)})$ for which no immediate reduction of s is possible by the method of section 4. Here

$$x_{r+1}^{(i)} = x_{r+1}^{(i-1)} - c_{i-1}H_{i-1}(x_1, \cdots, x_r), \tag{67}$$

$$f_i(x_1, \cdots, x_r, x_{r+1}^{(i)}) = f_{i-1}(x_1, \cdots, x_r, x_{r+1}^{(i)} + c_{i-1}H_{i-1}), \tag{67'}$$

$$v(x_{r+1}) < v(x_{r+1}^{(1)}) < \cdots < v(x_{r+1}^{(i)}) < \cdots, \tag{67''}$$

and where, in view of (67) and (67″), we have:

$$v(x_{r+1}^{(i)}) = v(H_i). \tag{68}$$

We assert *that such a sequence must be finite, provided* $s \geqq 2$ *or* $s \geqq 1$, *according as we deal with the theorem or with the lemma.* The proof of this assertion will complete the proof of both theorem and lemma.

In the case of the theorem we observe that $v\left(\frac{\partial f}{\partial x_{r+1}}\right) \geqq v(x_{r+1}^{s-1})$, since π_s must be a minimum value term (as a consequence of our hypothesis that our Cremona transformations fail to reduce the multiplicity s). Hence, if $s > 1$, then $v\left(\frac{\partial f}{\partial x_{r+1}}\right) > v(x_{r+1})$. Similarly, we find $v\left(\frac{\partial f_i}{\partial x_{r+1}^{(i)}}\right) > v(x_{r+1}^{(i)})$, and consequently, since $\frac{\partial f}{\partial x_{r+1}} = \frac{\partial f_i}{\partial x_{r+1}^{(i)}}$,

$$v\left(\frac{\partial f}{\partial x_{r+1}}\right) > v(x_{r+1}^{(i)}). \tag{69}$$

Now $v(x_{r+1})$, $v(x_{r+1}^{(1)})$, $\cdots$, $v(x_{r+1}^{(i)})$, $\cdots$, is a strictly ascending sequence of real numbers, and each element of the sequence is, by (68), the value of a polynomial in $x_1, \cdots, x_r$. It is easily seen (see [10], p. 659, footnote) that this last property implies that *if the sequence* $v(x_{r+1}^{(i)})$ *were infinite, then its limit would be* $+\infty$. But then (69) would imply that $\frac{\partial f}{\partial x_{r+1}}$ is necessarily the zero element of the field $\mathsf{K}(x_1, \cdots, x_r, x_{r+1})$ $(= \Sigma)$. This is impossible, since f is irreducible.

In the case of the lemma, we observe that $v(f) \geqq v(x_{r+1}^{s})$. Hence, if $s \geqq 1$, then $v(f) \geqq v(x_{r+1})$. Similarly, $v(f_i) \geqq v(x_{r+1}^{(i)})$, and consequently

$$v(f) \geqq v(x_{r+1}^{(i)}).$$

Again, if the sequence $\{f_i\}$ was infinite, then by the same argument as above the above inequality would imply that $f = 0$, and this is impossible since $x_1, \cdots, x_r, x_{r+1}$ are algebraically independent.

III. The general case

6. Let $\alpha_1, \cdots, \alpha_{r+1}$ be the B-residues of $x_1, \cdots, x_{r+1}$ respectively, and let $\alpha_{i1} (= \alpha_i)$, $\alpha_{i2}, \cdots, \alpha_{ig}$ be the conjugates of α_i over K $(i = 1, 2, \cdots, r+1)$. As before, we denote by R the field $\mathsf{K}(\alpha_{11}, \cdots, \alpha_{1g}, \cdots, \alpha_{r+1,1} \cdots, \alpha_{r+1,g})$. By a linear transformation of the coördinates $x_1, \cdots, x_{r+1}$, with coefficients in K, we may arrange matters so that, for each i, the g conjugate elements $\alpha_{i1}, \cdots, \alpha_{ig}$ be distinct. We write our polynomial f as a polynomial in $x_{r+1} - \alpha_{r+1}$:

$$f = a_0(x_1, \cdots, x_r) + a_1(x_1, \cdots, x_r)(x_{r+1} - \alpha_{r+1}) + \cdots + a_\nu(x_1, \cdots, x_r)(x_{r+1} - \alpha_{r+1})^\nu = \pi_0 + \pi_1 + \cdots + \pi_\nu.$$

By hypothesis, the coefficients of f are in R. Since $\alpha_{r+1} \subset \mathsf{R}$, it follows that the coefficients of the polynomials $a_0, a_1, \cdots, a_\nu$ are also in R, i.e. in the field generated by the B-residues of $x_1, \cdots, x_r$ and by their conjugates. By our induction, we therefore may apply the lemma to the product $a_0 a_1 \cdots a_\nu$. We have therefore a Cremona transformation from $x_1, \cdots, x_r$ to new variables $X_1, \cdots, X_r$, where the x's are polynomials in the X's, with coefficients in K, and such that each polynomial a_i assumes the form:

$$a_i(x_1, \cdots, x_r) = \Delta_1^{\mu_{1i}} \cdots \Delta_m^{\mu_{mi}} b_i(X_1, \cdots, X_r).$$

Here $\Delta_1, \cdots, \Delta_m$ are polynomials of the form (21) and $v(b_i) = 0$. Moreover, if $\gamma_1, \cdots, \gamma_r$ denote the B-residues of $X_1, \cdots, X_r$ and if we denote by Ω the field generated by these residues and by their conjugates (over K), then the polynomials b_i are in $\Omega[X_1, \cdots, X_r]$ (since this is true of $\Delta_1, \cdots, \Delta_m$).

The $\nu + 1$ terms π_i are, to within trivial factors b_i (i.e. factors having value zero in B), power products of $\Delta_1, \cdots, \Delta_m, x_{r+1} - \alpha_{r+1}$. The field Ω contains the residues $\alpha_1, \cdots, \alpha_r$ and their conjugates, since $x_1, \cdots, x_r$ are polynomials in $X_1, \cdots, X_r$, with coefficients in K. Hence this field Ω also contains the B-residue α_{r+1} of x_{r+1} and its g conjugates, since the g conjugates $\alpha_{i1}, \cdots, \alpha_{ig}$

are distinct. Now if we assume that the value of $x_{r+1} - \alpha_{r+1}$ is rationally independent on $\tau_1^*, \cdots, \tau_m^* (\tau_i^* = v(\Delta_i) = v(X_i - \gamma_i))$, then we can apply Theorem 3 and our lemma follows immediately by the same argument as that used in the preceding special case (C II 2). It is only necessary to recall from B III 12 that $\Delta_1, \cdots, \Delta_m, x_{r+1} - \alpha_{r+1}$ transform under $T^{(h)}$ as $X_1 - \gamma_1, \cdots, X_m - \gamma_m, x_{r+1} - \alpha_{r+1}$, provided h is sufficiently high.

7. If the value of $x_{r+1} - \alpha_{r+1}$ is rationally dependent on $\tau_1^*, \cdots, \tau_m^*$, our problem will be to achieve a reduction similar to the reduction achieved in the preceding special case (C II 4, 5). We go back to the original polynomial $f(x_1, \cdots, x_r, x_{r+1})$. We assume first of all, both in the case of the theorem and of the lemma, that $f(\alpha_1, \cdots, \alpha_r, x_{r+1})$ is not identically zero. In the case of the theorem we write f as a polynomial in $x_1 - \alpha_1, \cdots, x_r - \alpha_r, x_{r+1} - \alpha_{r+1}$. Let f begin with terms of lowest degree s in these differences. Then *without loss of generality we may assume that* α_{r+1} *is an s-fold root of* $f(\alpha_1, \cdots, \alpha_r, x_{r+1})$. All these assumptions can be satisfied by applying a preliminary linear transformation, with coefficients in K, to the elements $x_1, \cdots, x_{r+1}$.

In the case of the lemma we decompose f into its *irreducible factors* in $\mathsf{R}[x_1, \cdots, x_{r+1}]$ and for each irreducible factor φ_i we make the same assumptions as we have made above for f in the case of the theorem, namely: if φ_i begins with terms of lowest degree s_i in the differences $x_1 - \alpha_1, \cdots, x_{r+1} - \alpha_{r+1}$, then α_{r+1} is an s_i-fold root of $\varphi_i(\alpha_1, \cdots, \alpha_r, x_{r+1})$. To indicate the fact that φ_i begins with terms of lowest degree s_i, we shall say that *φ_i has an s_i-fold point at the center of the valuation B*. Similarly for f and s. In the case of the theorem, if $s = 1$, and in the case of the lemma, if all s are zero, there is nothing to prove.[19]

For convenience we indicate by φ a typical irreducible factor of f. We write f—or each φ (if we are dealing with the lemma)—as a polynomial in $x_{r+1} - \alpha_{r+1}$:

$$f(\text{or } \varphi) = a_0(x_1, \cdots, x_r) + a_1(x_1, \cdots, x_r)(x_{r+1} - \alpha_{r+1}) + \cdots + a_\nu(x_1, \cdots, x_r)(x_{r+1} - \alpha_{r+1})^\nu.$$

By our induction we may apply the lemma to the product $a_0 a_1 \cdots a_\nu$, or, if we are dealing with the lemma, to the product of *all* the coefficients a_i of *all* the irreducible factors φ_i. Then f—or each irreducible factor φ—assumes the form (we replace x_{r+1} by x_{m+1}):

$$f = \pi_0 + \pi_1 + \cdots + \pi_\nu,$$

where

$$\pi_i = b_i(X_1, \cdots, X_m, X_{m+2}, \cdots, X_{r+1}).\ \Delta_1^{\mu_{1i}} \cdots \Delta_m^{\mu_{mi}} \cdot (x_{r+1} - \alpha_{r+1})^i.$$

Here $\Delta_1, \cdots, \Delta_m$ are polynomials in $X_1, \cdots, X_m$ of type (21), and $v(b_i) = 0$.

---

[19] In the case of the theorem, if $s = 1$, the derivative $\partial f/\partial x_{r+1}$ is $\neq 0$ at the center of B, and hence the center is a simple point ([11], p. 214, Theorem 11).

Moreover, the coefficients of b_i and of the Δ_i are in the field Ω of the B-residues $\gamma_1, \cdots, \gamma_m, \gamma_{m+2}, \cdots, \gamma_{r+1}$ of the X's. Let $v(\Delta_i) = \tau_i$, $v(x_{m+1} - \alpha_{m+1}) = \tau_{m+1}$. We have to assume that τ_{m+1} is rationally dependent on $\tau_1, \cdots, \tau_m$.

The polynomials $\Delta_1, \cdots, \Delta_m, x_{m+1} - \alpha_{m+1}$ form a normal set for the variables $X_1, \cdots, X_m, x_{m+1}$. We apply Theorem 4 (B III 15). We shall continue to denote the new variables of that theorem by letters X. We have:

$$\pi_i = B_i(X_1, \cdots, X_m, X_{m+1}, \cdots, X_{r+1})\Delta_1^{\mathrm{M}_{1i}} \cdots \Delta_m^{\mathrm{M}_{mi}}\Delta_{m+1}^{\mathrm{M}_{m+1,i}},$$

where the M_{ji} are given by the relations (32), with μ_{ji} instead of μ_j. The factors B now depend on all the $r + 1$ variables X_i, but in view of (58′) and of property (d) of Theorem 4, we have

$$(70) \qquad B_i(\gamma_1, \cdots, \gamma_m, X_{m+1}, \gamma_{m+2}, \cdots, \gamma_{r+1}) = \text{const.} \neq 0.$$

Let $\pi_\alpha, \pi_\beta, \cdots \pi_\delta$ be the terms of lowest value. We know that Δ_{m+1} has value zero and that the values of $\Delta_1, \cdots, \Delta_m$ are rationally independent. Recalling what we said in section 12 concerning the behavior of the polynomials $\Delta_1, \cdots, \Delta_m$ under a Cremona transformation of type $T^{(h)}$, and taking into account Theorem 3, we conclude that by a Cremona transformation $T^{(h)}$ applied by $X_1, \cdots, X_m$ only, we can force the exponents $\mathrm{M}_{j\alpha}, \mathrm{M}_{j\beta}, \cdots, \mathrm{M}_{j\delta}$ to become equal, for all $j = 1, 2, \cdots, m$, and to be less than M_{ji} for any i such that π_i is not a lowest value term. We can do that for all irreducible factors φ_i of f simultaneously, in the case of the lemma. This transformation does not affect the exponents $\mathrm{M}_{m+1,i}$ and the validity of (70). Suppose that this has already been done. Then, if we put $\mathrm{M}_{j\alpha} = \mathrm{M}_{j\beta} = \cdots \mathrm{M}_{j\delta} = \mathrm{M}_j$, and if we recall the relation (35), we find that f(or each φ) has the following form:

$$(71) \qquad f(\text{or } \varphi) = \Delta_1^{\mathrm{M}_1} \cdots \Delta_m^{\mathrm{M}_m} F(X_1, \cdots, X_{r+1}),$$

where

$$F = \Delta_{m+1}^{\mathrm{M}_{m+1}}\psi(X_1, \cdots, X_{r+1}) + \Delta_1\Delta_2 \cdots \Delta_m G(X_1, \cdots, X_{r+1}).$$

Here

$$(71') \qquad \psi = B_\alpha + B_\beta\Delta_{m+1}^{(\beta-\alpha)/d} + \cdots + B_\delta\Delta_{m+1}^{(\delta-\alpha)/d} \quad \text{and} \quad \mathrm{M}_{m+1} = \mathrm{M}_{m+1,\alpha},$$

if $d > 0$, and

$$(71'') \qquad \psi = B_\alpha\Delta_{m+1}^{(\alpha-\delta)/d} + B_\beta\Delta_{m+1}^{(\beta-\delta)/d} + \cdots + B_\delta \text{ and } \mathrm{M}_{m+1} = \mathrm{M}_{m+1,\delta},$$

if $d < 0$. We continue to denote by $\gamma_1, \cdots, \gamma_{r+1}$ the B-residues of $X_1, \cdots, X_{r+1}$, and by Ω the field generated by these residues and by their conjugates. The coefficients of the Δ_i and of the polynomial F are in Ω. Let γ_{r+1} be a root of multiplicity s' for the polynomial $F(\gamma_1, \cdots, \gamma_m, X_{m+1}, \gamma_{m+2}, \cdots, \gamma_{r+1})$. *Let us assume that* $s' < s$; we assume that this is so *for each irreducible factor* φ, if we are dealing with the lemma.

We consider first the situation in the case of the theorem. Our Cremona transformation from the x's to the X's has its coefficients in K. Let $F_1(X_1, \cdots, X_{r+1}) = 0$ be the new (irreducible) equation of the hypersurface (over K). The polynomial F has its coefficients in K and must be a factor of $\Delta_1^{M_1}\Delta_2^{M_2} \cdots \Delta_m^{M_m}F$. The polynomials Δ_i depend only on $X_1, \cdots, X_m$, and these m elements of Σ are *algebraically independent*, since the values $v(X_1 - \gamma_1), \cdots, v(X_m - \gamma_m)$ are rationally independent. On the other hand $F_1(X_1, \cdots, X_{r+1}) = 0$ in Σ. Consequently, for any $i = 1, 2, \cdots, m$, F_1 and Δ_i can have no common factors in $\Omega[X_1, \cdots, X_{r+1}]$ (as a matter of fact, F_1 and Δ_i remain relatively prime even if we pass from Ω to the algebraically closed field determined by Ω). *It follows that F_1 is a factor of F.* Since F does not vanish identically in X_{m+1} when we put $X_i = \gamma_i (i = 1, 2, \cdots, m, m + 2, \cdots, r + 1)$, the same holds true also for F_1. On the other hand, since γ_{m+1} is a root of multiplicity s' of $F(\gamma_1, \cdots, \gamma_m, X_{m+1}, \gamma_{m+2}, \cdots, \gamma_{r+1})$, F has at most an s'-fold point at the center of B (in the space of the X's). Hence also F_1 has at most an s'-fold point at the center of B, and since $s' < s$, we have achieved a reduction. Ultimately, we will get an hypersurface for which $s = 1$, and our main theorem will be proved in the case under consideration.[20]

Let us now consider the case of the lemma. We observe that $\Delta_1, \cdots, \Delta_m$ have each a simple point at the center of B; this follows from the expression (21) of the Δ_i. Hence our Cremona transformation has the effect of replacing each irreducible factor φ_i of f, by a factor F_i which has at the center of B (in the space of $X_1, \cdots, X_{r+1}$) a multiplicity s_i' which is less than the multiplicity s_i of φ_i at the center of B (in the space of $x_1, \cdots, x_{r+1}$). In addition, new irreducible factors Δ_i appear, but these have each *a simple point* at the center of $B(s = 1)$. If we now apply the same procedure to f, expressed in terms of the new variables $X_1, \cdots, X_{r+1}$, and if the reduction ($s' < s$) succeeds at each step, then it is clear that ultimately f will assume the form (71) and in this F *will have value zero,* which establishes the lemma.

8. There remains the case in which $s' = s$ for f, or—in the case of the lemma—for at least one irreducible factor φ of f. The reasoning given in the special case (C II 4) can be repeated here. The equality $s' = s$ implies $\alpha = 0$,

[20] We add a remark which will be useful to us in the sequel. We know that in the ring $\mathsf{K}[X_1, \cdots, X_{r+1}]$ the polynomial $f(x_1, \cdots, x_{r+1})$ (expressed in terms of the X's) is divisible by $F_1(X_1, \cdots, X_{r+1})$. *We assert that f is not divisible by F_1^2* (the x's as well as the X's are now regarded temporarily as indeterminates). To see this, we observe that the prime principal ideal (F_1) in the ring $\mathsf{K}[X_1, \cdots, X_{r+1}]$ defines in the field $\mathsf{K}(X_1, \cdots, X_{r+1})$ a valuation of rank 1 and of dimension r. In this valuation the value of F_1 is 1. The prime ideal of this valuation in the smaller ring $\mathsf{K}[x_1, \cdots, x_{r+1}]$ is (f) (since $f \equiv 0(F_1)$ in $\mathsf{K}[X_1, \cdots, X_{r+1}]$). Hence the same valuation is defined by the principal ideal (f) in $\mathsf{K}[x_1, \cdots, x_{r+1}]$. Therefore also the value of f is 1, and this proves our assertion. Thus we may write the following identity (the x's and the X's being indeterminates related by our Cremona transformation): $f(x_1, \cdots, x_{r+1}) = F_1(X_1, \cdots, X_{r+1}) \cdot H(X_1, \cdots, X_{r+1})$, where $H \not\equiv 0(F_1)$.

$\beta = 1, \cdots, \delta = s$, and we conclude, as before, that $v(x_{r+1} - \alpha_{r+1}) = v(a_{s-1})$.[21] But at this stage there is a new complication. We can of course find a constant c in K^* such that

$$(72) \qquad v\left(\frac{x_{r+1} - \alpha_{r+1}}{a_{s-1}} - c\right) > 0.$$

The transformation used in C II 5 was the following: $x_{r+1} - ca_{s-1} = x_{r+1}^{(1)}$, from which it would now follow that $v(x_{r+1}^{(1)} - \alpha_{r+1}) > v(x_{r+1} - \alpha_{r+1})$. *But now this transformation cannot be used, since its coefficients are not in* K. To overcome this difficulty, we proceed as follows.

We first observe that the coefficients of a_{s-1} are in the field R. *We assert that also the constant c is in* R. This is a consequence of our hypothesis $s' = s$, and is shown as follows.

We divide the polynomial f (or φ) by $a_{s-1}^s a_s$. Since $v(a_{s-1}) = v(x_{r+1} - \alpha_{r+1})$, we have: $v(a_{s-1}^s a_s) = v(a_s \cdot (x_{r+1} - \alpha_{r+1})^s) = v(\pi_s)$, whence $v(a_{s-1}^s a_s) = v(\pi_s) = v(\pi_{s-1}) = \cdots = v(\pi_0)$. Hence the quotients $\pi_0/a_{s-1}^s a_s, \cdots, \pi_s/a_{s-1}^s a_s$ have value zero, while the quotients $\pi_i/a_{s-1}^s a_s$, $i > s$, have positive values. On the other hand, the B-residue of $f/a_{s-1}^s a_s$, in the case of the theorem, and the B-residue of $\varphi/a_{s-1}^s a_s$, in the case of the lemma, in zero.[22] Therefore, if we replace in $f/a_{s-1}^s a_s$ (or in $\varphi/a_{s-1}^s a_s$) each term $\pi_i/a_{s-1}^s a_s$ by its B-residue, and if we observe that $B(\pi_i/a_{s-1}^s a_s) = B\left(\frac{a_i}{a_{s-1}^{s-i} a_s} \cdot \left(\frac{x_{r+1} - \alpha_{r+1}}{a_{s-1}}\right)^i\right) = c^i \cdot B\left(\frac{a_i}{a_{s-1}^{s-i} a_s}\right)$, where c is the B-residue of $\frac{x_{r+1} - \alpha_{r+1}}{a_{s-1}}$ according to (72), we conclude that c satisfies the equation:

$$(73) \qquad d_0 + d_1 c + \cdots + d_{s-1} c^{s-1} + c^s = 0,$$

where

$$(74) \qquad d_i = B(a_i/a_{s-1}^{s-i} a_s), \qquad i = 0, 1, \cdots, s - 1.$$

We have

$$\frac{\pi_i}{\pi_s} = \frac{a_i}{(x_{r+1} - \alpha_{r+1})^{s-i} a_s} = \frac{a_i}{a_{s-1}^{s-i} a_s} \cdot \left(\frac{a_{s-1}}{x_{r+1} - \alpha_{r+1}}\right)^{s-i},$$

whence, if we put

---

[21] We return to the original notation: x_{r+1}, instead of x_{m+1}. The polynomials $a_1, \cdots, a_\nu$ are the coefficients of f or—in the case of the lemma—of that irreducible factor φ for which there is no reduction of s.

[22] This is obvious in the case of the theorem, since in that case $f = 0$. In the case of the lemma, our assertion follows from the hypothesis that there was no reduction of the multiplicity s. This hypothesis implies at any rate that the polynomial F in (71) has positive value in B, i.e. $v(F) > 0$. Since $v(\Delta_1^{M_1} \cdots \Delta_m^{M_m}) = v(\pi_\alpha) = v(\pi_\beta) = \cdots = v(\pi_\delta)$, it follows that $v(\varphi/\pi_\alpha) > 0$. In the present case we have $\pi_\alpha = \pi_0$ and $v(\pi_0) = v(a_{s-1}^s a_s)$. Hence $v(\varphi/a_{s-1}^s a_s) > 0$, i.e. $B(\varphi/a_{s-1}^s a_s) = 0$.

(75) $$d_i' = B(\pi_i/\pi_s), \qquad i = 0, 1, \cdots, s,$$

then

(75′) $$d_i = d_i' \cdot c^{s-i}, \qquad i = 0, 1, \cdots, s; d_s = 1.$$

On the other hand, from our hypothesis that the B-residue γ_{r+1} of X_{r+1} is an s-fold root of the polynomial $\psi(\gamma_1, \cdots, \gamma_r, X_{r+1})$ (see (71′) and (71″)), follows that if t is an indeterminate, then the polynomial $d_0' + d_1't + \cdots + d_{s-1}'t^{s-1} + t^s$ is a perfect s^{th} power. Namely,

$$d_0' + d_1't + \cdots + d_{s-1}'t^{s-1} + t^s = \left(t - \frac{1}{\gamma_{r+1}}\right)^s.$$

Hence, by (75′), also the polynomial $d_0 + d_1t + \cdots + d_{s-1}t^{s-1} + t^s$ is a perfect s^{th} power, i.e., by (73),

$$d_0 + d_1t + \cdots + d_{s-1}t^{s-1} + t^s = (t - c)^s.$$

Consequently,

$$c = -\frac{d_{s-1}}{s}, \quad \text{i.e., by (74), for } i = s - 1,$$

$$c = -\frac{1}{s} \cdot B\left(\frac{1}{a_s}\right),$$

and since $v(a_s) = 0$, whence $a_s(\alpha_1, \cdots, \alpha_r) \neq 0$, we conclude that $c = -\dfrac{1}{s \cdot a_s(\alpha_1, \cdots, \alpha_r)}$, and therefore $c \subset \mathsf{R}$. This proves our assertion.

Now that we have recognized that the coefficients of the polynomial $c \cdot a_{s-1}(x_1, \cdots, x_r)$ are in the field R, we may apply Lemma 1 (B III 10), where we put $A_j = 1$ and where we take for $B_1, \cdots, B_g$ the polynomial $c \cdot a_{s-1}$ and its conjugates over K. There exists then a polynomial $H(x_1, \cdots, x_r)$ in $\mathsf{K}[x_1, \cdots, x_r]$, such that $H - c \cdot a_{s-1}$, written as a polynomial in $x_1 - \alpha_1$, $x_r - \alpha_r$, begins with terms of degree $\geqq \rho_0$. If we now take ρ_0 sufficiently high, then we will have: $v(H - ca_{s-1}) > v(x_{r+1} - \alpha_{r+1} - ca_{s-1})$. If we put

(76) $$x_{r+1} = x_{r+1}^{(1)} + H(x_1, \cdots, x_r),$$

then $v(x_{r+1}^{(1)} - \alpha_{r+1}) = v(x_{r+1} - H - \alpha_{r+1}) \geqq \min. \{v(x_{r+1} - \alpha_{r+1} - ca_{s-1}), v(H - c \cdot a_{s-1})\} = v(x_{r+1} - \alpha_{r+1} - ca_{s-1}) > v(x_{r+1} - \alpha_{r+1})$, i.e.

(76′) $$v(x_{r+1}^{(1)} + \alpha_{r+1}) > v(x_{r+1} - \alpha_{r+1}) = v(H).$$

Thus we have the transformation (76) *whose coefficients are in* K and, in view of (76′) this transformation plays the same role as the transformation (67) in the special case. Moreover, by (76′), the B-residue of $x_{r+1}^{(1)}$ is the same as that of x_{r+1}, whence the field R remains the same. Finally, the integer s is obviously unaltered by the transformation (76). The rest of the proof is now the same in C II 5.

IV. Valuation of arbitrary rank and dimension

9. We assume that the uniformization theorem U_3 has already been proved for zero-dimensional valuations of rank σ. We proceed to prove the theorem, first for valuations of rank σ and of any dimension, and then for zero-dimensional valuations of rank $\sigma + 1$. This will complete the proof of the general uniformization theorem U_3.

Let B be an s-dimensional valuation of rank σ, and let V be a given projective model of Σ. Let $(\xi_1, \cdots, \xi_n)$ be the general point of V and let $\mathfrak{o} = K[\xi_1, \cdots, \xi_n]$. We may assume that $\mathfrak{o}$ belongs to the valuation ring $\mathfrak{B}$ of B. The center of B on V is a subvariety W of dimension $s' \leqq s$, given by a prime s'-dimensional ideal $\mathfrak{p}$ in $\mathfrak{o}$. Of the elements $\xi_1, \cdots, \xi_n$, s' and only s', say $\xi_1, \cdots, \xi_{s'}$, are algebraically independent mod $\mathfrak{p}$, and their B-residues are algebraically independent elements of the residue field of B. Suppose that $s' < s$. We can choose in Σ $s - s'$ elements $\eta_1, \cdots, \eta_{s-s'}$, such that the B-residues of $\xi_1, \cdots, \xi_{s'}, \eta_1, \cdots, \eta_{s-s'}$ be algebraically independent (over K). Let V' be the projective model of Σ whose general point is $(\xi_1, \cdots, \xi_n, \eta_1, \cdots, \eta_{s-s'})$, and let $\mathfrak{o}' = K[\xi_1, \cdots, \xi_n, \eta_1, \cdots, \eta_{s-s'}]$. The ring $\mathfrak{o}'$ is contained in the valuation ring of B. The center W' of B on V' is exactly of dimension s, since $\mathfrak{o}'$ contains s elements whose B-residues are algebraically independent. The quotient ring $Q(W')$ contains $Q(W)$, since $\mathfrak{o}$ is a subring of $\mathfrak{o}'$. Hence it is sufficient to prove the uniformization theorem for B and V'. We may therefore assume that the center W of B on the original variety V is of dimension s.

Let the s ξ's whose B-residues are algebraically independent over K be $\xi_1, \cdots, \xi_s$. We take as new ground field the field $\Delta = K(\xi_1, \cdots, \xi_s)$. Over this new ground field the general point $(\xi_1, \cdots, \xi_n)$ defines an $(r - s)$-dimensional variety V^*. The valuation B is now zero-dimensional, of rank σ. Its origin on V^* is a point P^*, and we have: $Q(P^*) = Q(W)$ ([11], p. 219). By our induction, there exists a projective model V_1^* of the field Σ/Δ on which the origin of B is a simple point P_1^*, such that $Q(P_1^*) \supseteqq Q(P^*)$. Let $(\eta_1, \cdots, \eta_m)$ be the general point of V_1^*. We may assume that the η's belong to the valuation ring of B. Let V_1 be the projective model of Σ/K whose general point is $(\xi_1, \cdots, \xi_s, \eta_1, \cdots, \eta_m)$, and let W_1 be the center of B on V_1. We have: $Q(W_1) = Q(P_1^*)$, and W_1 is a simple subvariety of V_1 ([11], p. 219–220). Since $Q(W_1) \supseteqq Q(P^*) = Q(W)$, the proof in the present case is complete.

10. Let now B be a zero-dimensional valuation of rank $\sigma + 1$, and let V be the given projective model of Σ/K. The valuation B is composite with a valuation B_1, of rank σ and of a certain dimension s. Let P and W be the centers on V of B and B_1 respectively. Let $\xi_1, \cdots, \xi_n$ be the general point of V and let $\mathfrak{o} = K[\xi_1, \cdots, \xi_n]$, where we assume that the ξ's belong to the valuation ring of B.

Let, by the preceding proof, V' be a projective model of Σ/K on which the center of B_1 is a simple subvariety W' such that $Q(W') \supseteqq Q(W)$, and let

$\eta_1, \cdots, \eta_m$ be the general point of V'. Again we may assume that the η's belong to the valuation ring of B (note that the quotient ring $Q(W')$ is invariant under projective transformations of the coördinates η_i). *A fortiori*, the η_i will belong to the valuation ring of B_1. We consider the projective model $\bar{V}$ of Σ/K whose general point is $(\xi_1, \cdots, \xi_n, \eta_1, \cdots, \eta_m)$. The ring $\bar{\mathfrak{o}} = \mathsf{K}[\xi_1, \cdots, \xi_n, \eta_1, \cdots, \eta_m]$ is the join of the two rings $\mathfrak{o} = \mathsf{K}[\xi_1, \cdots, \xi_n]$, $\mathfrak{o}' = \mathsf{K}[\eta_1, \cdots, \eta_m]$. Since $\mathfrak{o} \subseteq Q(W) \subseteq Q(W')$ and since $\mathfrak{o}' \subseteq Q(W')$, it follows that $\bar{\mathfrak{o}} \subseteq Q(W')$. On the other hand $\mathfrak{o}' \subseteq \bar{\mathfrak{o}}$. From this it follows that if $\bar{W}$ is the center of B_1 on $\bar{V}$, then $Q(\bar{W}) = Q(W')$, (and hence $\bar{W}$ is a simple subvariety of $\bar{V}$). Since $\mathfrak{o} \subseteq \bar{\mathfrak{o}}$, and since the center $\bar{P}$ of B on $\bar{W}$ is a point at finite distance, it follows that $Q(P) \subseteq Q(\bar{P})$. *It is therefore permissible to replace in the proof the variety V by the variety $\bar{V}$. We may therefore suppose that the center W of B_1, on the original variety V, is simple.*

11. Let ρ be the dimension of W. We choose in $\mathfrak{o}(= \mathsf{K}[\xi_1, \cdots, \xi_n])$ a set of $r + 1$ elements $\eta_1, \cdots, \eta_{r+1}$, such that: (a) $\mathfrak{o}$ is integrally dependent on $\mathsf{K}[\eta_1, \cdots, \eta_r]$; (b) η_{r+1} is a primitive element of Σ with respect to $\mathsf{K}(\eta_1, \cdots, \eta_r)$; (c) $\eta_{\rho+1}, \cdots, \eta_r$ are uniformizing parameters of the simple subvariety W; (d) if $\mathfrak{p}$ is the prime ideal of W in $\mathfrak{o}$ and if $F(\eta_1, \cdots, \eta_{r+1}) = 0$ is the irreducible relation between $\eta_1, \cdots, \eta_{r+1}$, then $F'_{\eta_{r+1}} \not\equiv 0(\mathfrak{p})$ ([11], p. 214). Let V' be the hypersurface $F(\eta_1, \cdots, \eta_{r+1}) = 0$, and let W' and P' be the centers of B_1, B respectively, on V'. Exactly as in A IV it is shown that it is sufficient to prove the uniformization theorem for V' and B.

The prime ideal $\mathfrak{p}'$ of W' in $\mathsf{K}[\eta_1, \cdots, \eta_{r+1}]$ is such that $\mathfrak{p} \cap \mathsf{K}[\eta_1, \cdots, \eta_{r+1}] = \mathfrak{p}'$. From (d) it follows that $F'_{\eta_{r+1}} \not\equiv 0(\mathfrak{p}')$, whence W' is a simple subvariety of V'. *Hence we may assume that our original V is an hypersurface*

$$f(\xi_1, \cdots, \xi_r, \xi_{r+1}) = 0,$$

that $\xi_{\rho+1}, \cdots, \xi_r$ are uniformizing parameters of W, and that $f'_{\xi_{r+1}} \not\equiv 0(\mathfrak{p})$, where $\mathfrak{p}$ is the prime (ρ-dimensional) ideal of W in $\mathsf{K}[\xi_1, \cdots, \xi_{r+1}]$. Moreover, ξ_{r+1} is integrally dependent on $\xi_1, \cdots, \xi_r$.

12. The ideal $\mathfrak{p}$ is ρ-dimensional. Since $\xi_i \equiv 0(\mathfrak{p})$, $i = \rho + 1, \cdots, r$, and since ξ_{r+1} is integrally dependent on $\mathsf{K}[\xi_1, \cdots, \xi_r]$, it follows that $\xi_1, \cdots, \xi_\rho$ are algebraically independent mod $\mathfrak{p}$. Let $g(\xi_1, \cdots, \xi_\rho, \xi_{r+1}) \equiv 0(\mathfrak{p})$ be the irreducible congruence which $\xi_1, \cdots, \xi_\rho, \xi_{r+1}$ satisfy mod $\mathfrak{p}$. Since $f(\xi_1, \cdots, \xi_{r+1}) \equiv f(\xi_1, \cdots, \xi_\rho, 0, \cdots, 0, \xi_{r+1})$ (mod $\mathfrak{p}$), the polynomial $f(\xi_1, \cdots, \xi_\rho, 0, \cdots, 0, \xi_{r+1})$ is divisible by $g(\xi_1, \cdots, \xi_\rho, \xi_{r+1})$. It is *not* divisible by g^2, since $f'_{\xi_{r+1}} \not\equiv 0(\mathfrak{p})$. Let therefore

$$\begin{aligned} &f(\xi_1, \cdots, \xi_\rho, 0, \cdots, 0, \xi_{r+1}) \\ &\qquad = g(\xi_1, \cdots, \xi_\rho, \xi_{r+1}) \cdot \varphi(\xi_1, \cdots, \xi_\rho, \xi_{r+1}), \qquad \varphi \not\equiv 0(g). \end{aligned} \tag{77}$$

If then $x_1, \cdots, x_{r+1}$ are indeterminates, then we have identically:

$$(78)\quad f(x_1, \cdots, x_{r+1}) = g(x_1, \cdots, x_\rho, x_{r+1})\cdot\varphi(x_1, \cdots, x_\rho, x_{r+1}) + a_1x_{\rho+1} + \cdots + a_{r-\rho}x_r,$$

where $a_i \subset \mathsf{K}[x_1, \cdots, x_{r+1}]$.

The valuation B_1, with which B is composite, is of rank σ. Hence B defines *in the residue field of* B_1 a zero-dimensional valuation of *rank* 1. Since $\mathfrak{p}$ is the prime ideal of B_1 in $\mathsf{K}[\xi_1, \cdots, \xi_{r+1}]$, the B-residues of $\xi_1, \cdots, \xi_\rho, \xi_{r+1}$ are the residues of these elements mod $\mathfrak{p}$. Let $\bar\xi_1, \cdots, \bar\xi_\rho, \bar\xi_{r+1}$ be these residues. We consider the ring $\bar{\mathfrak{o}} = \mathsf{K}[\bar\xi_1, \cdots, \bar\xi_\rho, \bar\xi_{r+1}]$ and its quotient field $\bar\Sigma$. This field is of degree of transcendency ρ over K, and the hypersurface

$$(79)\quad g(x_1, \cdots, x_\rho, x_{r+1}) = 0.$$

is a projective model of $\bar\Sigma$. The general point of this hypersurface is $(\bar\xi_1, \cdots, \bar\xi_\rho, \bar\xi_{r+1})$. The field $\bar\Sigma$ is contained in the residue field of B_1. Hence the original zero-dimensional valuation B of Σ induces a zero-dimensional, *rank* 1, valuation $\bar B$ of $\bar\Sigma$. By our main theorem, there exists a Cremona transformation

$$(80)\quad x_i = P_i(y_1, \cdots, y_\rho, y_{r+1}), \qquad i = 1, 2, \cdots, \rho, r+1,$$

where $P_i \subset \mathsf{K}[y_1, \cdots, y_\rho, y_{r+1}]$, by which the hypersurface (79) is transformed into an hypersurface

$$(81)\quad G(y_1, \cdots, y_\rho, y_{r+1}) = 0,$$

on which the center of $\bar B$ is at a simple point. Moreover, the coördinates $\bar\eta_1, \cdots, \bar\eta_\rho, \bar\eta_{r+1}$ of the general point of (81) are in the valuation ring of $\bar B$. The $\bar B$-residues $\alpha_1, \cdots, \alpha_\rho, \alpha_{r+1}$ of $\bar\eta_1, \cdots, \bar\eta_\rho, \bar\eta_{r+1}$ are the coördinates of the center of B on the hypersurface (81). Since the center is a simple point, we may assume that

$$(82)\quad G'_{y_{r+1}}(\alpha_1, \cdots, \alpha_\rho, \alpha_{r+1}) \neq 0.$$

The polynomial $g(x_1, \cdots, x_\rho, x_{r+1})$, expressed in terms of $y_1, \cdots, y_\rho, y_{r+1}$, assumes the form

$$g = G(y_1, \cdots, y_\rho, y_{r+1})\cdot H(y_1, \cdots, y_\rho, y_{r+1}),$$

where (see footnote[20]),

$$(83)\quad H \not\equiv 0(G).$$

If we now put $\varphi(x_1, \cdots, x_\rho, x_{r+1}) = \psi(y_1, \cdots, y_\rho, y_{r+1})$ and $a_i(x_1, \cdots, x_{r+1}) = A_i(y_1, \cdots, y_\rho, x_{\rho+1}, \cdots, x_r, y_{r+1})$, the identity (78) assumes the following form:

$$(84)\quad f(x_1, \cdots, x_{r+1}) = GH\psi + A_1x_{\rho+1} + \cdots + A_{r-\rho}x_r.$$

If we replace in (80) the indeterminates $x_1, \cdots, x_\rho, x_{r+1}$ by the elements $\xi_1, \cdots, \xi_\rho, \xi_{r+1}$ of the field Σ, we have to replace the indeterminates $y_1, \cdots, y_\rho, y_{r+1}$ by well-defined elements $\eta_1, \cdots, \eta_\rho, \eta_{r+1}$ of the field Σ, such that:

$\xi_i = P_i(\eta_1, \cdots, \eta_\rho, \eta_{r+1})$. The B_1-residues of the η's are the coördinates $\bar{\eta}_1, \cdots, \bar{\eta}_\rho, \bar{\eta}_{r+1}$ of the general point of the hypersurface (81). In view of (83), it follows that the B_1-residue of $H(\eta_1, \cdots, \eta_\rho, \eta_{r+1})$ is $\neq 0$, whence $v_{B_1}(H(\eta_1, \cdots, \eta_\rho, \eta_{r+1})) = 0$. Similarly, since $\varphi \not\equiv 0(g)$ (see (77)), also $v_{B_1}(\varphi(\xi_1, \cdots, \xi_\rho, \xi_{r+1})) = v_{B_1}(\psi(\eta_1, \cdots, \eta_\rho, \eta_{r+1}) = 0$. Hence $v_{B_1}(H\psi) = 0$. Since $\xi_{\rho+1}, \cdots, \xi_r$ are the uniformizing parameters of the center W of B_1 on the hypersurface $f = 0$, their values in B_1 are positive. Since B is composite with B_1, it follows that if we put

$$(85) \qquad \eta_i = \xi_i/H\psi, \qquad i = \rho + 1, \cdots, r,$$

then $\eta_{\rho+1}, \cdots, \eta_r$ *have positive values in* B. We have now $r + 1$ elements of Σ: $\eta_1, \cdots, \eta_{r+1}$. The elements $\xi_1, \cdots, \xi_{r+1}$ are polynomials in the η's, with coefficients in K. We have namely: $\xi_i = P_i(\eta_1, \cdots, \eta_\rho, \eta_{r+1})$, $i = 1, 2, \cdots, \rho, r + 1$, and $\xi_i = H(\eta_1, \cdots, \eta_\rho, \eta_{r+1}) \cdot \psi(\eta_1, \cdots, \eta_\rho, \eta_{r+1}) \cdot \eta_i$, $i = \rho + 1, \cdots, r$. The hypersurface $f(\xi_1, \cdots, \xi_{r+1}) = 0$ is now transformed into an hypersurface $F(\eta_1, \cdots, \eta_{r+1}) = 0$, where, in view of (84) and (85):

$$(86) \qquad F = G(\eta_1, \cdots, \eta_\rho, \eta_{r+1}) + B_1\eta_{\rho+1} + \cdots + B_{r-\rho}\eta_r .$$

The B-residues of the η's are: $B(\eta_i) = \bar{B}(\bar{\eta}_i) = \alpha_i$, $i = 1, 2, \cdots, \rho, r + 1$; $B(\eta_i) = 0$, $i = \rho + 1, \cdots, r$. The point P_1^* $(\alpha_1, \cdots, \alpha_\rho, 0, \cdots, 0, \alpha_{r+1})$ is the center P^* of B on the hypersurface $F = 0$ (or, more precisely: the center P^* consists of that point P_1^* and of its conjugates over K). From the form (86) of F it follows immediately, in view of (82), that P^* is a simple point.

If P is the center of B on the original hypersurface $f = 0$, the quotient ring of P is a subring of the quotient ring of P^*, since $\xi_1, \cdots, \xi_{r+1}$ are polynomials in $\eta_1, \cdots, \eta_{r+1}$. This completes the proof of the uniformization theorem.

THE JOHNS HOPKINS UNIVERSITY AND
INSTITUTE FOR ADVANCED STUDY

REFERENCES

[1] P. ALEXANDROFF AND H. HOPF: *Topologie.*
[2] P. BACHMANN: *Zahlentheorie*, vol. 4.
[3] W. KRULL: *Idealtheorie*, Ergebnisse der Mathematik und ihrer Grenzgebiete, IV, 3.
[4] W. KRULL: *Allgemeine Bewertungstheorie*, J. reine angew. Math., vol. 167 (1932).
[5] A. TYCHONOFF: *Über die topologische Erweiterung von Räumen*, Math. Ann., vol. 102 (1929).
[6] A. TYCHONOFF: *Über einen Funktionenraum*, Math. Ann., vol. 111 (1935).
[7] O. PERRON: *Grundlagen für eine Theorie des Jacobischen Kettenbruchalgorithmen.* Habilitationsschrift, 1906.
[8] R. J. WALKER: *Reduction of singularities of an algebraic surface*, Annals of Mathematics, vol. 36 (1935).
[9] O. ZARISKI: *Some results in the arithmetic theory of algebraic varieties*, Am. J. Math., vol. 61 (1939).
[10] O. ZARISKI: *The reduction of the singularities of an algebraic surface*, Annals of Mathematics, vol. 40 (1939).
[11] O. ZARISKI: *Algebraic varieties over ground fields of characteristic zero*, Am. J. of Math., vol. 62 (1940).

ANNALS OF MATHEMATICS
Vol. 43, No. 3, July, 1942

A SIMPLIFIED PROOF FOR THE RESOLUTION OF SINGULARITIES OF AN ALGEBRAIC SURFACE

BY OSCAR ZARISKI

(Received March 19, 1942)

1. Introduction

We presuppose the theorem of local uniformization, as proved elsewhere.[1] By this theorem, any valuation of a field Σ/K of algebraic functions of r independent variables, over a given ground field K of characteristic zero, can be "uniformized" on a suitable projective model V of Σ, i.e. the center of the valuation on a suitable model V will be a simple subvariety of V.[2] If the ground field K (the field of coefficients) is the field of complex numbers (the classical case), then the above result, in conjunction with the bicompactness of the Riemann manifold **M** of Σ/K,[3] implies the existence of a *finite* set of models of Σ, say

$$V_1, V_2, \cdots, V_n,$$

such that any zero-dimensional valuation of Σ/K is uniformized on at least one of the models V_i of the set.[4] Any finite set of projective models with the above property shall be called a *resolving system* of the Riemann manifold **M**.

In the case of abstract varieties, where we cannot fall back on topology, it is necessary to give an algebraic proof of the existence of resolving systems of **M**. In the special case of algebraic surfaces the algebraic proof of the existence of resolving systems is strikingly simple (see section 6 of this paper). The general case of algebraic varieties will be treated in a subsequent paper.

The theorem of the resolution of singularities of an algebraic variety can be formulated in terms of resolving systems, as follows: *There exists a resolving system of the Riemann manifold* **M** *which consists of only one model.* In view of the existence of resolving systems, this theorem will be established if it can be

---

[1] For the case of algebraic surfaces, see our paper "The reduction of singularities of an algebraic surface", Annals of Mathematics, vol. 40 (July, 1939), pp. 649–659. For the general case of varieties of any dimension, see our paper "Local uniformization on algebraic varieties", Annals of Mathematics, vol. 41 (October, 1940), pp. 852–896. These two papers will be referred to respectively as "Reduction" and "Uniformization".

[2] For the definition of the center of a valuation see "Uniformization", p. 857.

[3] **M** is the totality of all *zero-dimensional* valuations (or *places*) of Σ/K; see "Uniformization", p. 855.

[4] It follows then necessarily that *any* valuation (whether of dimension zero or greater than zero) will be uniformized on at least one of the models V_i. To see this, it is only necessary to observe that: a) If B is any valuation of dimension > 0, then there exist zero-dimensional valuations compounded with B; b) if B_0 is such a zero-dimensional valuation then on every model V of Σ the center of B_0 will lie on the center of B; c) if an irreducible subvariety W of V contains a simple point, then W itself is simple.

proved that *the existence of a resolving system of* **M** *consisting of n models* ($n > 1$) *implies the existence of a resolving system of* **M** *consisting of* $n - 1$ *models.* To carry out this induction from n to $n - 1$, it is sufficient to prove the *fundamental theorem* which we are now going to state.

Let N be an arbitrary subset of **M**, i.e. let N be an arbitrary set of places of our field Σ/K. In the same fashion as we have defined resolving system of **M**, we can define resolving systems of N. A resolving system of N will be therefore any finite set of models of Σ such that any valuation in N has a simple center on at least one of the models in the set. In particular, if a resolving system of N consists of only one model, this model shall be called a *resolving model for* N.

Fundamental Theorem. *If N is an arbitrary subset of* **M** *and if there exists a resolving system of N consisting of two models, then there also exists a resolving model for N.*

From the fundamental theorem, the theorem on the resolution of singularities follows immediately. For, let $V_1, \cdots, V_n$ be a resolving system of **M**. Let N be the subset of **M** consisting of those valuations which have a singular center on *each* of the models $V_1, \cdots, V_{n-2}$. It is clear that the pair V_{n-1}, V_n constitutes a resolving system for N. If we assume the fundamental theorem, then there exists a resolving model V'_{n-1} for N. The $n - 1$ models $V_1, \cdots, V_{n-2}, V'_{n-1}$ constitute a resolving system for **M**, and our induction from n to $n - 1$ is complete.

The aim of this paper is to give a proof of the fundamental theorem in the case of algebraic surfaces. Our present proof for the resolution of singularities is much simpler than our earlier proof (see "Reduction") and is also more general, since at present we do not assume that the ground field K is algebraically closed. In the course of the proof we have to make use of certain properties of fundamental loci of birational correspondences. This preliminary material, strictly confined to the precise needs of our proof, is presented in the next section. In a forthcoming paper dealing with the general theory of birational correspondences we study these properties systematically. A brief account of this general theory will be found in my address "Normal varieties and birational correspondences" (delivered before the annual meeting of the Society in Bethlehem in 1941) which has appeared in the Bulletin of the American Mathematical Society, Vol. 48, No. 6 (June 1942).

I. Preliminary Concepts

2. Birational correspondences

Let V and V' be two models of our field Σ/K.

Definition. Two points, P, P' of V and V' respectively are corresponding points if there exists a valuation of Σ/K such that its center on V is the point P and its center on V' is the point P'.[5]

[5] Compare with p. 666 of "Reduction". Our study of the general theory of birational correspondences is based on this valuation-theoretic definition.

We say that P is a *fundamental point* of the birational correspondence between V and V', if there exists a corresponding point P' on V' such that $Q(P) \not\supseteq Q(P')$ (here $Q(P)$ denotes the quotient ring of P).

Suppose that there exists a point P' which corresponds to P and which is such that $Q(P') \subseteq Q(P)$. If v is a valuation of centers P and P' on V and V' respectively and if $\mathfrak{P}_v$ denotes the prime ideal of v, then $\mathfrak{P}_v \cap Q(P) = \mathfrak{p}$ and $\mathfrak{P}_v \cap Q(P') = \mathfrak{p}'$, where $\mathfrak{p}(\mathfrak{p}')$ is the ideal of non units in $Q(P)(Q(P'))$. Hence $\mathfrak{p} \cap Q(P') = \mathfrak{p}'$, and from this it follows that *any* valuation whose center on V is P has P' as center on V', i.e. *P' is the only point of V' which corresponds* to P. Therefore, if P is a point of V which is not fundamental, then to P there corresponds a unique point P' on V', and we have $Q(P') \subseteq Q(P)$. On the other hand, if P is fundamental, then it follows that $Q(P) \not\supseteq Q(P')$ for any point P' which corresponds to P.

In the case of surfaces we have proved elsewhere that if V and V' are *normal* surfaces and if to P there corresponds a finite number of points on V', then P is not fundamental.[6] Therefore, if P is fundamental, the locus of corresponding points P' is a variety of dimension 1.[7]

If neither V nor V' carry fundamental points, then the birational correspondence between V and V' shall be called *regular*.

In any case, the number of fundamental points on either surface is always finite.

Let $\xi_0^*, \xi_1^*, \cdots, \xi_n^*$ be the homogeneous coördinates of the general point of V, and let $\eta_0^*, \eta_1^*, \cdots, \eta_m^*$ be the homogeneous coördinates of the general point[8] of V'. The $(n + 1)(m + 1)$ products

$$\omega_{ij}^* = \xi_i^* \eta_j^*$$

can be regarded as homogeneous coördinates of the general point of a variety V^* which is birationally equivalent to V and to V': for the quotient of any two ω^*'s, say $\omega_{ij}^*/\omega_{\alpha\beta}^*$, is equal to $\xi_i^*/\xi_\alpha^* \cdot \eta_j^*/\eta_\beta^*$, and is therefore an element of the field Σ. Moreover, the non-homogeneous coördinates of the general point of V^*, i.e. the quotients of the ω^*'s by a fixed ω^*, say by ω_{00}^*, generate the field Σ/K, since the quotients ξ_i^*/ξ_0^* (and also the quotients η_j^*/η_0^*) are among them. The variety V^* is called the *join* of V and V'. It is clear that the ring

[6] See "Reduction", p. 688, Theorem 5, but the proof is much simpler and is as follows. If $P_1', \cdots, P_h'$ are the points on V' which correspond to P, then for any valuation v whose center is P it must be true that the valuation ring of v contains one of h quotient rings $Q(P_i')$, whence it also contains the intersection $\mathfrak{J}$ of these quotient rings. Since $Q(P)$ is integrally closed, it is the intersection of the above mentioned valuation rings. Hence $Q(P) \supseteq \mathfrak{J}$. If $\mathfrak{p} \cap \mathfrak{J} = \bar{\mathfrak{p}}$, then $\bar{\mathfrak{p}}$ is one of the h prime maximal ideals of $\mathfrak{J}$, say the one relative to the point P_1', and it is clear that any valuation of center P will have center P_1' on V', i.e. $h = 1$. [For the definition of normal varieties, see our paper "Some results in the arithmetic theory of algebraic varieties", p. 282, American Journal of Mathematics, vol. 61 (April 1939). This paper will be referred to as "Results".]

[7] See "Reduction", p. 667.

[8] For the concept of homogeneous coördinates of the general point of an irreducible algebraic variety, see "Results" p. 284.

$K[\omega_{10}^*/\omega_{00}^*, \cdots, \omega_{nm}^*/\omega_{00}^*]$ of polynomials in these non-homogeneous coördinates —briefly: the ring of non-homogeneous coördinates—coincides with the ring $K[\xi_1^*/\xi_0^*, \cdots, \xi_n^*/\xi_0^*, \eta_1^*/\eta_0^*, \cdots, \eta_m^*/\eta_0^*]$ and is therefore the join of the corresponding rings of non-homogeneous coördinates of the general points of V and V'. From this one concludes immediately that the birational correspondence between V^* and either one of the two given models V, V' has no fundamental points on V^*.

It follows that each point P^* of V^* represents a unique pair (P, P') of *corresponding* points of V and V'. Conversely, each such pair (P, P') is represented by at least one point P^* of the join V^*. For this reason the join V^* is often referred to in the literature as the variety of pairs of corresponding points of V and V'. However, this last term may be misleading when we deal with a ground field K which is not algebraically closed. For in this case (and only in this case) it may happen that one and the same pair (P, P') of corresponding points of V and V' is represented by (or, we may say, splits into) more than one point of V^*.

The join of *any finite number* of models $V_1, \cdots, V_h$ can be defined as the join of V_h and of the join of $V_1, \cdots, V_{h-1}$. It is seen immediately that this definition is symmetric in $V_1, \cdots, V_h$.

3. Quadratic transformations

Let P be a point of V and let $\mathfrak{p}^*$ be the corresponding prime homogeneous ideal (of projective dimension zero) in the ring $K[\xi_0^*, \xi_1^*, \cdots, \xi_n^*]$. Let η_0^*, $\eta_1^*, \cdots, \eta_m^*$ be a set of forms in $\xi_0^*, \xi_1^*, \cdots, \xi_n^*$, *of like degree*, such that the ideal $(\eta_0^*, \eta_1^*, \cdots, \eta_m^*)$ differs from $\mathfrak{p}^*$ only by an irrelevant primary component i.e. a component belonging to the irrelevant prime ideal $\mathfrak{p}_0 = (\xi_0^*, \xi_1^*, \cdots, \xi_n^*)$.[9] By a *quadratic transformation of center* P we mean the birational transformation which carries V into the variety $\bar{V}$ whose general point has the following homogeneous coördinates:

$$\omega_{ij}^* = \xi_i^*\eta_j^*, \qquad i = 0, 1, \cdots, n; j = 0, 1, \cdots, m. \tag{1}$$

This transformation depends of course on the choice of the forms η_j^*, but in a non-essential fashion. Namely, if $\bar{V}_1$ is the transform of V by a quadratic transformation, relative to another set of forms, *then* $\bar{V}$ *and* $\bar{V}_1$ *are in regular birational correspondence.*

PROOF. Let $\zeta_0^*, \zeta_1^*, \cdots, \zeta_\mu^*$ be the set of forms which defines $\bar{V}_1$, and let $\bar{\omega}_{ij}^* = \xi_i^*\zeta_j^*$. Let: α = degree η_j^*, β = degree ζ_j^*, $\mathfrak{A} = (\eta_0^*, \cdots, \eta_m^*)$, $\mathfrak{B} = (\zeta_0^*, \cdots, \zeta_\mu^*)$. Since the ideals $\mathfrak{A}$ and $\mathfrak{B}$ differ only in their primary irrelevant components, we will have for sufficiently high integers a and b: $\mathfrak{A}\cdot\mathfrak{p}_0^a \equiv 0(\mathfrak{B})$, $\mathfrak{B}\cdot\mathfrak{p}_0^b \equiv 0(\mathfrak{A})$. Select a and b so as to have $\alpha + a = \beta + b$.

[9] Let $\mathfrak{p}_0^* = (\phi_0, \phi_1, \cdots, \phi_s)$, where the ϕ_i are forms in the ξ^*'s and let ν_i = degree of ϕ_i, $\nu = \max.(\nu_0, \nu_1, \cdots, \nu_s)$. Then the forms $\xi_j^{*\nu-\nu_i}\phi_i$, $j = 0, 1, \cdots, n$, $i = 0, 1, \cdots, s$, satisfy the desired condition.

The elements ω_{ij}^* constitute a linear base for the forms of degree $\alpha + 1$ which belong to $\mathfrak{A}$. If we multiply the ω_{ij}^* by $\xi_0^*, \cdots, \xi_n^*$, then the products give a base for the forms of degree $\alpha + 2$ in $\mathfrak{A}$. These products are the homogeneous coördinates of the general point of the join V^* of V and $\bar{V}$. Since, as will be pointed out later on, the quadratic transformation has no fundamental points on $\bar{V}$, it follows that $\bar{V}$ and V^* are in regular birational correspondence. We now multiply the homogeneous coördinates of the general point of V^* by $\xi_0^*, \cdots, \xi_n^*$, getting the join of V^* and V. Proceeding in this fashion we construct a model V^*, such that: a) the homogeneous coördinates of the general point of V^* constitute a linear base for the forms of degree $\alpha + a$ in $\mathfrak{A}$; i.e., they are elements of $\mathfrak{A}\mathfrak{p}_0^a$, and consequently also of $\mathfrak{B}$; b) V^* and $\bar{V}$ are in regular birational correspondence. In a similar fashion we construct a model V_1^* in regard to $\mathfrak{B}$, $\bar{V}_1$ and the integer $\beta + b$. Since $\alpha + a = \beta + b$, it follows immediately that the homogeneous coördinates of the general point of V^* are linearly dependent on the homogeneous coördinates of the general point of V_1^*, and *vice versa*. Consequently, V^* and V_1^* are related projectively to each other, q.e.d.

If K is algebraically closed, then it is permissible to assume that the coördinates of P (elements of K) are $1, 0, 0, \cdots, 0$. Then $\mathfrak{p}^* = (\xi_1^*, \cdots, \xi_n^*)$, and our quadratic transformation is given by the equations: $\omega_{ij} = \xi_i\xi_j$, $i = 0, 1, \cdots, n$; $j = 1, 2, \cdots, n$. This is the ordinary quadratic transformation defined by the system of hyperquadrics passing through the point P (see "Reduction," p. 676).

Since P is the only point of V at which all the forms $\xi_i^*\eta_j^*$ vanish simultaneously, P is the only fundamental point of the quadratic transformation. To any other point A of V there corresponds a unique point $\bar{A}$ on $\bar{V}$, and moreover $Q(\bar{A})$ coincides with $Q(A)$. The transformation has no fundamental points on $\bar{V}$. The proofs of all these assertions are straightforward and do not differ essentially from the proofs in the case of an algebraically closed ground field (see "Reduction," p. 676–679).

To see what happens to the point P, we pass to non-homogeneous coördinates: $\xi_i = \xi_i^*/\xi_0^*$ $(i \neq 0)$, and $\omega_{ij} = \omega_{ij}^*/\omega_{00}^*$ (i, j not both zero). Let η_i denote the non-homogeneous polynomial obtained from the form η_i^* by the usual process (replace ξ_0^* by 1 and ξ_i^* by ξ_i, $i \neq 0$). The ring of non-homogeneous coördinates for V, resp. $\bar{V}$, will be: $\mathfrak{o} = \mathsf{K}[\xi_1, \cdots, \xi_n]$ and

$$\bar{\mathfrak{o}} = \mathsf{K}[\xi_1, \cdots, \xi_n, \eta_1/\eta_0, \cdots, \eta_m/\eta_0]$$

respectively. The point P will be given by the prime ideal $\mathfrak{p} = (\eta_0, \cdots, \eta_m)$ in $\mathfrak{o}$. Since the ideal $\bar{\mathfrak{o}}\mathfrak{p}$ is the principal ideal $\bar{\mathfrak{o}} \cdot \eta_0$, we conclude immediately that *the transform $T(P)$ of the point P is a pure $(r - 1)$-dimensional subvariety of $\bar{V}$* (not necessarily irreducible).[10]

[10] If Γ is an irreducible subvariety of $\bar{V}$, at finite distance with respect to the non-homogeneous coördinates ω_{ij}, Γ is given by a prime ideal $\bar{\mathfrak{p}}$ in the ring $\bar{\mathfrak{o}}$. If Γ corresponds to the

4. Quadratic transformations with simple center. The $\mathfrak{p}$-adic divisor

We are now especially interested in the case in which P is a simple point. Let $\mathfrak{J}$ be the quotient ring of P and let $\mathfrak{P}$ denote the ideal of non-units in $\mathfrak{J}$. Let $t_1, \cdots, t_r$, $t_i \in \mathfrak{J}$, be uniformizing parameters of P, i.e. a set of r elements in $\mathfrak{J}$ with the property $\mathfrak{J}\cdot(t_1, \cdots, t_r) = \mathfrak{P}$.[11]

A significant property of the uniformizing parameters is the following: if $\phi(t_1, \cdots, t_r) = 0$, where ϕ is a polynomial in $t_1, \cdots, t_r$ with coefficients in $\mathfrak{J}$, and if $\phi_\rho(t_1, \cdots, t_r)$ is the sum of terms of ϕ of lowest degree ρ (ϕ_ρ = the *leading form* of ϕ), then the coefficients of ϕ_ρ must all be divisible by $\mathfrak{P}$.[12] This property enables us to construct a valuation of Σ in the following fashion. If $\alpha \in \mathfrak{J}$ and if α is exactly divisible by $\mathfrak{P}^\rho$, then $\alpha = \phi_\rho(t_1, \cdots, t_r)$, where ϕ_ρ is a form of degree ρ with coefficients in $\mathfrak{J}$ but not all in $\mathfrak{P}$. Let β be another element of $\mathfrak{J}$, exactly divisible by $\mathfrak{P}^\sigma$, so that $\beta = \psi_\sigma(t_1, \cdots, t_r)$. We have $\alpha\beta \equiv 0(\mathfrak{P}^{\rho+\sigma})$, and since the coefficients of the form $\phi_\rho\psi_\sigma$ are not all in $\mathfrak{P}$, we conclude, by the property of the uniformizing parameters stated above, that $\alpha\beta$ is not divisible by $\mathfrak{P}^{\rho+\sigma+1}$. This shows that if we put $v(\alpha) = \rho$, we get a *discrete valuation* B of Σ, *of rank* 1. We shall call B the *$\mathfrak{p}$-adic divisor of center* P (P — *a* simple point!).

It is not difficult to see that B *is of dimension* $r - 1$, i.e. B is a divisor. For, we have $v(t_i) = 1$, $v(t_i/t_1) = 0$. Were the B-residues of $t_2/t_1, \cdots, t_r/t_1$ algebraically dependent over K, there would exist a form $\phi_\rho(t_1, \cdots, t_m)$, with coefficients in K, not all zero, such that $v(\phi_\rho) > v(t_1^\rho)$. This would imply $\phi_\rho \equiv 0(\mathfrak{P}^{\rho+1})$, which is impossible.

The following theorem puts into evidence the effect of a quadratic transformation of center P in regard to our $\mathfrak{p}$-adic divisor:

THEOREM 1. *If $\bar{V}$ is the transform of V under a quadratic transformation T of simple center P, then the $\mathfrak{p}$-adic divisor of center P is of the first kind with respect to $\bar{V}$, i.e. the center of the divisor on $\bar{V}$ is $(r - 1)$-dimensional. This center coincides with the transform $T(P)$ (whence $T(P)$ is irreducible).*

PROOF. We use the notations of the preceding section. We assume that the variety $T(P)$ is not entirely at infinity, i.e. that the principal ideal $\bar{\mathfrak{o}}\cdot\eta_0$ is not unit ideal (see footnote 10).

---

point P, then we must have $\bar{\mathfrak{p}} \cap \mathfrak{o} = \mathfrak{p}$ (see footnote 7). Hence $\bar{\mathfrak{o}}\mathfrak{p} \equiv \mathfrak{o}(\mathfrak{p})$, i.e. Γ lies on the subvariety $\bar{W}$ of $\bar{V}$ defined by the principal ideal $\bar{\mathfrak{o}}\cdot\eta_0$. Conversely, if Γ is on $\bar{W}$, then $\bar{\mathfrak{o}}\cdot\mathfrak{p} \equiv 0(\bar{\mathfrak{p}})$, whence $\bar{\mathfrak{p}} \cap \mathfrak{o} = 0$ (since $\mathfrak{p}$ is a maximal ideal of $\mathfrak{o}$), and therefore Γ corresponds to P. This shows that $T(P)$ consists of $\bar{W}$ and perhaps of other irreducible components at infinity. Since any irreducible component of $T(P)$ can be assumed to be at finite distance, for a proper choice of the non-homogeneous coördinates (if $\omega^*_{\alpha\beta}$ is different from zero on the given component, we use the quotients $\omega^*_{ij}/\omega^*_{\alpha\beta}$) and since if the principal ideal $\bar{\mathfrak{o}}\cdot\eta_0$ is not the unit ideal, all its isolated prime ideals are $(r - 1)$-dimensional, our assertion follows.

[11] See our paper "Algebraic varieties over ground fields of characteristic zero", American Journal of Mathematics, vol. 62 (January, 1940), p. 199.

[12] See loc. cit. in footnote 11, p. 202 and p. 207–208. As a consequence of this property, the quotient ring of a simple point (and also, more generally, the quotient ring of a simple subvariety) is a "p-Reihenring" in the sense of Krull (see Krull, Dimensionstheorie in Stellenringen, Crelle's Journal, vol. 179 (1938)).

To prove the theorem it will be sufficient to prove that the ideal $\bar{\mathfrak{J}}\cdot\eta_0$ is prime, where $\bar{\mathfrak{J}} = \mathfrak{J}\cdot\bar{\mathfrak{o}}$ ($\mathfrak{J}$ = quotient ring of P),[13] and that the irreducible subvariety of $\bar{V}$ defined by this prime ideal is the center of B. Let B' be an arbitrary valuation of center P, whose center on $\bar{V}$ is at finite distance (such valuations exist, since we have assumed that $T(P)$ is not entirely at infinity).

Among the uniformizing parameters $t_1, \cdots, t_r$ of P we take one which has least value in B'. Let it be t_1. Since $\mathfrak{P} = \mathfrak{J}\cdot(\eta_0, \eta_1, \cdots, \eta_m)$, it follows that $t_i \in \mathfrak{J}\cdot(\eta_0, \eta_1, \cdots, \eta_m)$. Consequently $t_i/\eta_0 \in \bar{\mathfrak{J}}$, and, in particular, $t_1/\eta_0 \in \bar{\mathfrak{J}}$. Since the center of B' is at finite distance we conclude that $v_{B'}(t_1) \geqq v_{B'}(\eta_0)$. *This inequality implies that* $\eta_0 \not\equiv 0(\mathfrak{P}^2)$. For, assume that $\eta_0 \equiv 0(\mathfrak{P}^2)$. Then $\eta_0 = \phi_2(t_1, \cdots, t_r)$, where ϕ_2 is a quadratic form in the t's, with coefficients in $\mathfrak{J}$. We can write $\eta_0/t_1 = t_1\phi_2(1, t_2/t_1, \cdots, t_r/t_1)$. By hypothesis, $v_{B'}(t_i/t_1) \geqq 0$. We also have $v_{B'}(t_1) > 0$ and $v_{B'}(\alpha) \geqq 0$, for any element α in $\mathfrak{J}$ (since P is the center of B'). Consequently, $v_{B'}(\eta_0/t_1) > 0$, a contradiction.

Since $\eta_0 \not\equiv 0(\mathfrak{P}^2)$, we have $v_B(\eta_0) = 1$, and since $v_B(\eta_i) \geqq 1$, we conclude that the quotients η_i/η_0 belong to the valuation ring of B. Consequently *the entire ring* $\bar{\mathfrak{J}}$ *is contained in the valuation ring of* B. Therefore the center of B on V is a (irreducible) subvariety $\bar{W}$ of V at finite distance, where $\bar{W}$ will be given by the prime ideal $\bar{\mathfrak{P}}$ in $\bar{\mathfrak{J}}$ consisting of the elements of positive value in B. Now it is clear that $\bar{\mathfrak{J}}\cdot\eta_0 \equiv 0(\bar{\mathfrak{P}})$. On the other hand, let $\bar{\alpha} \in \bar{\mathfrak{J}}$, $v(\bar{\alpha}) > 0$. We can write $\bar{\alpha}$ in the form: $\bar{\alpha} = \phi_\rho(\eta_0, \eta_1, \cdots, \eta_m)/\eta_0^\rho$, where ϕ_ρ is a form of degree ρ, with coefficients in $\mathfrak{J}$. Since $v_B(\bar{\alpha}) > 0$ and $v_B(\eta_0^\rho) = \rho$, we must have $v_B(\phi_\rho) \geqq \rho + 1$. Hence $\phi_\rho = \psi_{\rho+1}(t_1, \cdots, t_r)$, where $\psi_{\rho+1}$ is a form of degree $\rho + 1$ in $t_1, \cdots, t_r$, with coefficients in $\mathfrak{J}$. Since $t_i \equiv 0(\eta_0, \eta_1, \cdots, \eta_m)$, we will also have $\phi_\rho = g_{\rho+1}(\eta_0, \cdots, \eta_m)$, where $g_{\rho+1}$ is again a form of degree $\rho + 1$ in $\eta_0, \cdots, \eta_m$, with coefficients in $\mathfrak{J}$. Consequently, $\bar{\alpha} = g_{\rho+1}/\eta_0^\rho = \eta_0\cdot g_{\rho+1}(1, \eta_1/\eta_0, \cdots, \eta_m/\eta_0)$, whence $\bar{\alpha} \equiv 0(\bar{\mathfrak{J}}\cdot\eta_0)$. We conclude that $\bar{\mathfrak{J}}\cdot\eta_0 = \bar{\mathfrak{P}}$, and this completes the proof.

From the fact that the *irreducible* variety $T(P)$ is defined by a *principal* ideal, namely by the ideal $\bar{\mathfrak{o}}\cdot\eta_0$, it follows $T(P)$ is a *simple* subvariety of $\bar{V}$ (of dimension $r - 1$, and having η_0 as uniformizing parameter). Therefore $T(P)$ contains points which are simple for $\bar{V}$. We shall need, however, the following stronger result:

THEOREM 2. *Every point of* $T(P)$ *is a simple point of* $\bar{V}$.

PROOF. Let $\bar{P}$ be a point of $T(P)$, which we may assume to be a point at finite distance. We consider an arbitrary but fixed valuation B' with centers P and $\bar{P}$ (on V and $\bar{V}$ respectively). Assuming, as we did before, that $v_{B'}(t_1) \leqq v_{B'}(t_i)$, $i = 1, 2, \cdots, r$, we will have $t_i/\eta_0 \in \bar{\mathfrak{J}}$, and since $\mathfrak{J} \subset Q(\bar{P})$, it follows that

$$\frac{t_i}{\eta_0} \in Q(\bar{P}), \qquad i = 1, 2, \cdots, r. \tag{2}$$

[13] Every prime ideal $\bar{\mathfrak{p}}$ of $\bar{\mathfrak{o}}\cdot\eta_0$ contracts to $\mathfrak{p}$ (see footnote 11), and hence $\mathfrak{J} = \mathfrak{o}_p \subseteqq \bar{\mathfrak{o}}_{\bar{p}}$. From this it follows immediately the ideals $\bar{\mathfrak{o}}\cdot\eta_0$ and $\bar{\mathfrak{J}}\cdot\eta_0$ have like decompositions into maximal primary components.

We consider the ring $\mathfrak{o}^* = \mathsf{K}[\xi_1, \cdots, \xi_n, t_2/t_1, \cdots, t_r/t_1]$. Let V^* be the model whose general point (in non-homogeneous coördinates) is $(\xi_1, \cdots, \xi_n, t_2/t_1, \cdots, t_r/t_1)$. Since $v_{B'}(t_i/t_1) \geqq 0$, the center P^* of B' on V^* is a point at finite distance. We will have also the following relations, similar to (2):

$$\frac{\eta_i}{t_1} \in Q(P^*), \qquad i = 0, 1, \cdots, m. \tag{3}$$

From (2) and (3) it follows that t_1/η_0 is an element of $Q(\bar{P})$ and that its reciprocal has non-negative value in the valuation B'. Since $\bar{P}$ is the center of B' on $\bar{V}$, we conclude *that* t_1/η_0 *is a unit in* $Q(\bar{P})$. Similarly, t_1/η_0 is a unit in $Q(P^*)$. But then the quotients $t_i/t_1(= t_i/\eta_0 \cdot \eta_0/t_1)$ are in $Q(\bar{P})$, and the quotients $\eta_i/\eta_0(= \eta_i/t_1 \cdot t_1/\eta_0)$ are in $Q(P^*)$. Therefore $\bar{\mathfrak{o}} \subset Q(P^*)$ and $\mathfrak{o}^* \subset Q(\bar{P})$, *and this implies*: $Q(P^*) = Q(\bar{P})$. Thus, we have only to show that P^* is a simple point of V^*. For that we have to exhibit uniformizing parameters at P^*.

Let Δ be the residue class field of the point P, i.e. let $\Delta = \mathfrak{o}/\mathfrak{p}$. Here Δ is an algebraic extension of K. Similarly, let Δ^* be the residue class field of P^*. We have $\Delta^* \supseteq \Delta$, since $\mathfrak{o}^* \supseteq \mathfrak{o}$ and since the prime $\mathfrak{o}^*$-ideal $\mathfrak{p}^*$ of the point P^* contracts in $\mathfrak{o}$ to $\mathfrak{p}$. Let $c_1, \cdots, c_n$ be the P-residues of $\xi_1, \cdots, \xi_n$ $(c_i \in \Delta)$, and let $d_2, \cdots, d_r$ be the P^*-residues of $t_2/t_1, \cdots, t_r/t_1$ $(d_i \in \Delta^*)$. The residue d_i will be the root an irreducible polynomial $f_i(z)$ with coefficients in Δ. Replace each coefficient of $f_i(z)$ by an arbitrary but fixed element of $\mathfrak{o}$ of which it is the residue. We get a certain polynomial $F_i(z)$ with coefficients in $\mathfrak{o}$. We assert that *the r elements*

$$t_1' = t_1, \qquad t_2' = F_2\left(\frac{t_2}{t_1}\right), \quad \cdots, \qquad t_r' = F_r\left(\frac{t_r}{t_1}\right)$$

are uniformizing parameters at P^*, i.e. $t_1', t_2', \cdots, t_r'$ generate in $Q(P^*)$ the ideal of non-units. It is sufficient to show that *the ideal* $\mathfrak{o}^*(t_1', t_2', \cdots, t_r')$ *is the intersection of prime zero-dimensional ideals* (one of these ideals will have to be the ideal defined by the point P^*, since $t_i' = 0$ at P^*). Now this ideal is contained in the prime $(r - 1)$-dimensional ideal $\mathfrak{o}^*t_1(= \mathfrak{o}^*\mathfrak{p}$; the center of the $\mathfrak{p}$-adic divisor on V^*). Modulo this prime ideal the elements $t_2/t_1, \cdots, t_r/t_1$ are algebraically independent, while the residues of $\xi_1, \cdots, \xi_n$ are $c_1, \cdots, c_n$ respectively. Therefore, if we pass to the ring $\mathfrak{o}^*/\mathfrak{o}^*t_1$, we see immediately that the above assertion is equivalent to the following statement: if $z_2, \cdots, z_r$ are indeterminates over Δ, then the polynomials $f_2(z_2), \cdots, f_r(z_r)$ generate in the polynomial ring $\Delta[z_2, \cdots, z_r]$ an ideal which is the intersection of prime zero-dimensional ideals. Since the polynomials $f_i(z_i)$ are irreducible over Δ and *since* Δ *is of characteristic zero*, this statement is true and its proof is straightforward.

II. Resolution of the Singularities of an Algebraic Surface

5. Two lemmas

If W is an arbitrary collection of points on a given irreducible algebraic variety V (for instance, if W is an algebraic subvariety of V), we denote by

$N(W)$ the set of all zero-dimensional valuations B (of the field Σ of rational functions on V) such that the center of B on V is in W.

LEMMA 1. *If W is an algebraic subvariety of V and if for any point P of W there exists a resolving system for $N(P)$, then there also exists a resolving system of $N(W)$.*

PROOF. Let $s =$ dimension W, i.e. s is the highest dimension of the irreducible components of W. For $s = 0$ the lemma is trivial, because W consists then of a finite number of points. We therefore assume that the lemma is true for $s = \rho - 1$ and we prove that it is also true if $\dim W = \rho$.

We fix a point P_i on *each* irreducible ρ-dimensional component of W and we consider a resolving system $V_1, \cdots, V_h$ of $N(P_1, P_2, \cdots)$. Let V^* be the join of $V, V_1, \cdots, V_h$. The points of V^* to which there correspond singular points of V_i form an algebraic subvariety W_i^* of V^* $(i = 1, 2, \cdots, h)$. Let W^* be the intersection of $W_1^*, \cdots, W_h^*$. The points of W which correspond to points of W^* form an algebraic subvariety W_1 of W, *of dimension* $< \rho$, since $P_i \notin W_1$. It is clear that $(V_1, \cdots, V_h)$ is also a resolving system of $N(W - W_1)$. Hence, if there exists a finite resolving system for $N(W_1)$, this system, together with $V_1, \cdots, V_h$, will give a resolving system for $N(W)$. Since $\dim W_1 < \rho$, our induction is complete.

The second lemma deals with algebraic surfaces. Let F be a normal surface, and let P be a point of F. We apply to F a quadratic transformation of center P, getting a surface F_1'. If F_1' is not normal, we pass from F_1' to a corresponding derived normal surface F_1 (see "Results," p. 290). The birational transformation which consists in passing from a given variety to a corresponding derived normal variety shall be referred to in the sequel as a *normal transformation*. We know from section 3 that the transform of P on F_1' is a pure $(r - 1)$-dimensional subvariety of F_1'. We take an arbitrary point P_1' of this subvariety. To P_1' there will correspond on F_1 at most a finite number of points. Let P_1 be one of these points. We now repeat the above procedure, starting with the normal surface F_1 and with the point P_1. We will get first a quadratic transform F_2' of F_1 (with P_1 as center of the quadratic transformation) and then a derived normal variety F_2 of F_2'. On F_2' we select at random a point P_2' which corresponds to P_1, and then we let P_2 be one of the points of F_2 which corresponds to P_2'. In this fashion we proceed indefinitely, getting an infinite sequence of models F; F_1', F_1; F_2', F_2; $\cdots$; F_i', F_i; $\cdots$, and of points P; P_1', P_1; P_2', P_2; $\cdots$; P_i', P_i; $\cdots$, where: a) $P_i' \in F_i'$, $P_i \in F_i$; b) the quotient ring $Q(P_i)$ is integrally closed and c) $Q(P) \subset Q(P_1') \subseteq Q(P_1) \subset Q(P_2') \subseteq Q(P_2) \cdots$.

LEMMA 2. *The union of the quotient rings $Q(P_i)$ (or $\lim Q(P_i)$) is the valuation ring of a zero-dimensional valuation of Σ.*

The proof is exactly the same as the one we gave in the case of algebraically closed ground fields ("Reduction," p. 681, Theorem 10).[14]

[14] In that proof we selected an element ζ in $Q(P)$ which is a non-unit of $Q(P)$ but is not in the "tangential" ideal. The consideration of the tangential ideal can be omitted. We know that if $\mathfrak{P}$ is the ideal of non-units in $Q(P)$, $\mathfrak{P} = (\eta_0, \eta_1, \cdots, \eta_m)$, then the ex-

6. The existence of resolving systems

Let F be a normal surface. We shall prove that the hypothesis that the field Σ of rational functions on F does not possess a resolving system leads to a contradiction. Under this hypothesis it follows, in view of Lemma 1, that there must exist a point P on F such that $N(P)$ does not possess a resolving system. By a quadratic transformation of center P, we transform F into a surface F_1' and into the derived normal surface F_1 of F_1'. Let W_1 be the subvariety of F_1 whose points correspond to P. It is clear that $N(P) = N(W_1)$, and hence there must exist a point P_1 on W_1 such that $N(P_1)$ does not possess a resolving system. By repeated application of this argument, we get an infinite sequence of points $P, P_1, \cdots, P_i, \cdots$ of the type considered in the preceding section, such that $N(P_i)$ does not possess a resolving system, $i = 1, 2, \cdots$. Let B be the zero-dimensional valuation whose valuation ring B is the union of the rings $Q(P_i)$. By the local uniformization theorem, let F^* be a model on which the center P^* of B is a simple point. We have $Q(P^*) \subset B$, hence $Q(P^*) \subset Q(P_i)$, for i sufficiently high, say $i \geqq m$. *Every* valuation of center P_i, $i \geqq m$, will have P^* as center on F^*, i.e. $N(P_i) \subseteq N(P^*)$. Therefore F^* *is a resolving surface for* $N(P_i)$, if $i \geqq m$, a contradiction.

7. Proof of the fundamental theorem

Let F, F' be the pair of surfaces which constitute a resolving system for a given set N of zero-dimensional valuations of our field Σ. The birational correspondence between F and F' may have fundamental points on either surface. Our first step consists in the elimination of the fundamental points of one of the two surfaces, say of F, by applying to F a sequence of successive quadratic and normal transformations, as described in the preceding section. As center of the quadratic transformation we take a fundamental point of F, *one at a time*, until we have exhausted all the fundamental points of F. As a result we get some new normal surface, say F_1. If the birational correspondence between F_1 and F' still has fundamental points on F_1, these points must be among the points which correspond to the fundamental points of F. In that case we proceed with F_1 in the same fashion. We assert that *after a finite number of steps we will get a surface* $F_i = \bar{F}$ *such that the birational correspondence between* $\bar{F}$ *and* F' *has no fundamental points on* $\bar{F}$. For, otherwise there would exist an infinite sequence of points $P, P_1, P_2, \cdots, P_i \in F_i$, such that each point P_i is fundamental and such that $Q(P) \subset Q(P_1) \subset \cdots$. Let $B = \lim Q(P_i)$, and let B be the corresponding zero-dimensional valuation (Lemma 2). Let P' be the center of B on F'. Then if i is sufficiently high we will have $Q(P_i) \supseteq Q(P')$, and this is in contradiction with our hypothesis that P_i is a fundamental point of the birational correspondence between F_i and F' (see section 3).

Since under quadratic transformations simple points go into simple points

---

tended ideal of $\mathfrak{P}$ in $Q(P_1)$ is a principal ideal generated by one of the elements η_i, say by η_0. Take as ζ the element η_0.

(section 4) and since under normal transformations simple points are not affected at all, it follows that the pair of surfaces $(\bar{F}, F')$ is also a resolving system of N.

Our next step is similar to the step just executed, namely we now eliminate by quadratic and normal transformations the fundamental points of F' in the birational correspondence between $\bar{F}$, F'. However, *this time we only eliminate those fundamental points of F' which are simple points of F'.* No singular point of F' will be affected, even if it is a fundamental point. Let $\bar{F}'$ be the transform of F' thus obtained.

We assert that the join F^* of $\bar{F}$ and $\bar{F}'$ is a resolving surface for N.

The proof is straightforward. For, let B be any valuation in the set N. Let $\bar{P}$, P', $\bar{P}'$ and P^* be the centers of B on $\bar{F}$, F', $\bar{F}'$ and F^* respectively. We have to show that P^* is a simple point.

FIRST CASE: *P' is a singular point.* Then $\bar{P}$ is simple (since $(\bar{F}, F')$ is a resolving system for N). We have $Q(\bar{P}) \supseteq Q(P')$ (since the birational correspondence between $\bar{F}$ and F' has no fundamental points on $\bar{F}$) and also $Q(P') = Q(\bar{P}')$ (since the singular points of F' have not been affected in the passage from F' to $\bar{F}'$). Therefore $Q(\bar{P}) \supseteq Q(\bar{P}')$ and consequently[15] $Q(\bar{P}) = Q(P^*)$. Since $\bar{P}$ is a simple point, it follows that also P^* is a simple point.

SECOND CASE: *P' is a simple point.* If P' is not a fundamental point of the birational correspondence between $\bar{F}$ and F', then $Q(P') = Q(\bar{P})$ (since $\bar{P}$ is also not a fundamental point), and moreover $Q(P') = Q(\bar{P}')$ (since the points of F' which are not fundamental in the above birational correspondence have not been affected in the passage from F' to $\bar{F}'$). Hence $Q(\bar{P}) = Q(\bar{P}') = Q(P^*)$, and therefore P^* is a simple point. If, on the other hand, P' *is* a fundamental point in the birational correspondence between $\bar{F}$ and F', then $\bar{P}'$ is *not* fundamental for the birational correspondence between $\bar{F}$ and $\bar{F}'$ (because the *simple* fundamental points of F' are eliminated in the course of the passage from F' to $\bar{F}'$). Hence $Q(\bar{P}') \supseteq Q(\bar{P})$, $Q(P^*) = Q(\bar{P}')$. Since P' is simple, also $\bar{P}'$ is simple, and consequently P^* is a simple point, q.e.d.

THE JOHNS HOPKINS UNIVERSITY

[15] We make use of the following property of the join V^* of two varieties V and V': if P^*, P and P' are corresponding points of V^*, V and V' respectively and if $Q(P) \supseteq Q(P')$, then $Q(P) = Q(P^*)$. The proof is straightforward.

THE COMPACTNESS OF THE RIEMANN MANIFOLD OF AN ABSTRACT FIELD OF ALGEBRAIC FUNCTIONS

OSCAR ZARISKI

1. The existence of finite resolving systems. In an earlier paper[1] we have announced the result that the existence of a resolving system of the Riemann manifold of an abstract field of algebraic functions (in any number of variables) or—what is the same—the local uniformization theorem[2] implies the existence of *finite* resolving systems of the Riemann manifold. We have proved this result for algebraic surfaces by arithmetic considerations.[1] The proof for the general case of varieties, which at that time was in our possession,[3] and which we have promised to publish in a subsequent paper, was of similar nature, that is, it was based upon considerations involving the structure of certain infinite sequences of quotient rings. However, we have succeeded lately in finding a much simpler proof which is based on topological considerations.

Let Σ be a field of algebraic functions of several variables, over an arbitrary ground field k. By the Riemann manifold $\boldsymbol{M}$ of Σ we mean the totality of places of Σ, that is, the totality of zero-dimensional valuations v of Σ/k. If V is a projective model of Σ/k, and if H is any subset of V, we denote by $N(H)$ the subset of $\boldsymbol{M}$ consisting of those valuations $\mathfrak{v}$ which have center in H. By a resolving system of $\boldsymbol{M}$ we mean a collection $\mathfrak{B} = \{V_a\}$ of projective models (finite or infinite in number) with the property that for any v in $\boldsymbol{M}$ there exists a V_a in $\mathfrak{B}$ such that the center of v on V_a is a simple point.

The topology which we introduce in $\boldsymbol{M}$ is simply this: *we choose as a basis for the closed sets of* $\boldsymbol{M}$ *the sets* $N(W)$, *where* W *is any algebraic subvariety of any projective model of* Σ. We prove that if topologized in this fashion, *the set* $\boldsymbol{M}$ *is a compact*[4] *topological space*. From this the result announced above follows immediately. For if $\{V_a\}$ is a resolving system, and if we denote by S_a the singular locus of V_a, then $N(V_a - S_a)$ is an open set and $\{N(V_a - S_a)\}$ is an open covering of $\boldsymbol{M}$.

Received by the editors April 10, 1944.

[1] *A simplified proof for the resolution of singularities of an algebraic surface*, Ann. of Math. vol. 43 (1942) p. 583.

[2] See loc. cit. footnote 1.

[3] That proof was presented by us at a seminar in algebraic geometry at Johns Hopkins in 1942.

[4] We use the term compact in the same sense as it is used by S. Lefschetz in his *Algebraic topology* (Amer. Math. Soc. Colloquium Publications, vol. 27, 1942). The old term is bicompact.

Hence this covering contains a finite subcovering $\{N(V_i - S_i)\}$, $i = 1, 2, \cdots, m$, and this means that $\{V_1, V_2, \cdots, V_m\}$ is a finite resolving system of M.

The proof of compactness of M given in the next section is based in part on some simple algebro-geometric considerations, and in part on a theorem of Steenrod[5] on the compactness of the limit space of an inverse system of compact T_1-spaces.

2. **The Riemann manifold as the limit space of an inverse system.** Let $\mathfrak{B} = \{V_a\}$ be the collection of *all* projective models of Σ/k. By a point of V_a we mean a zero-dimensional prime ideal in a suitable coördinate ring of V_a, or, in other terms, a point is a prime one-dimensional *homogeneous* ideal in the ring of homogeneous coördinates of the general point of V_a. This defines V_a set-theoretically as a set of points. We topologize V_a by choosing as closed sets the algebraic subarieties of V_a. It is obvious that V_a then becomes a compact topological space in which points are closed sets (whence V_a is a T_1-space; however, it is not a Hausdorff space).

If V_a and V_b are two projective models of Σ/k, we denote by π_a^b the transformation of V_b onto V_a defined by the birational correspondence between V_a and V_b. We define a partial ordering $<$ of the collection $\mathfrak{B}$ as follows: $V_a < V_b$ if whenever P_a and P_b are corresponding points of V_a and V_b under π_a^b then $Q(P_a) \subseteq Q(P_b)$. Here $Q(P)$ denotes the quotient ring of P. It is clear that if $V_a < V_b$ then π_a^b is a single-valued continuous and *closed* mapping. Moreover, if V_a and V_b are arbitrary projective models of Σ/k, and if V_c denotes the join[6] of V_a and V_b, then $V_a < V_c$ and $V_b < V_c$. Hence we have here an inverse system $\{V_a; \pi_a^b | V_a < V_b\}$ of compact T_1-spaces. Let M be the limit space of the system. By Steenrod's theorem M is compact. Every point P^* of M represents an infinite collection of points $\{P_a\}$, $P_a \in V_a$, $V_a \in \mathfrak{B}$, with the property that if $V_a < V_b$ then $Q(P_a) \subseteq Q(P_b)$. We shall denote by π_a^* the mapping $P^* \to P_a$ of M into V_a. If V_a is any projective model of Σ/k and if W is any algebraic subvariety of V_a, then $(\pi_a^*)^{-1} W$ is a closed subset of M, and the closed sets obtained in this fashion form a basis for the closed subsets of M.

The compactness of the Riemann manifold of Σ/k and the implications stated in the preceding section are immediate consequences of the following theorem.

THEOREM. *There is a* (1, 1) *correspondence between the points* P^*

---

[5] N. E. Steenrod, *Universal homology groups*, Amer. J. Math. vol. 58 (1936) p. 666.

[6] See our paper *Foundations of a general theory of birational correspondences*, Trans. Amer. Math. Soc. vol. 53 (1943) p. 516.

of M and the zero-dimensional valuations v of the field Σ/k. If P^* and v are corresponding elements, and if V_a is any projective model of Σ/k, then $\pi_a^*P^*$ is the center of v on V_a.

PROOF. Let v be a zero-dimensional valuation of Σ/k and let $P_{a,v}$ be the center of v on any given projective model V_a of Σ/k. For any two projective models V_a, V_b it is then true that $P_{a,v}$ and $P_{b,v}$ are corresponding points in the birational correspondence π_a^b. Hence $P_v^* = \{P_{a,v}\}$ is a point of M. *Thus every zero-dimensional valuation v determines uniquely a point P_v^* of M.*

If v_1 and v_2 are two distinct zero-dimensional valuations, then there exists at least one projective model V_a such that $P_{a,v_1} \neq P_{a,v_2}$. *Hence if $v_1 \neq v_2$ than $P_{v_1}^* \neq P_{v_2}^*$.*

Now let P^* be an arbitrary point of M, $P^* = \{P_a\}$. We denote by $\mathfrak{B}$ the least ring containing the quotient rings $Q(P_a)$. Let V_b be a fixed projective model of Σ/k and let $P_b = \pi_b^*P^*$. *We assert that if ω is a non-unit in $Q(P_b)$ then* $1/\omega \notin \mathfrak{B}$. For assume that $1/\omega \in \mathfrak{B}$. Then $1/\omega$ will belong to the ring generated by a finite number of quotient rings $Q(P_a)$, say $Q(P_{a_1})$, $Q(P_{a_2})$, $\cdots$, $Q(P_{a_m})$. Let V_c be the join of the varieties V_b, V_{a_1}, V_{a_2}, $\cdots$, V_{a_m} and let $P_c = \pi_c^*P^*$. Since $\pi_{a_i}^*P^* = P_{a_i}$ and $\pi_b^*P^* = P_b$, we have $\pi_{a_i}^c P_c = P_{a_i}$ and $\pi_b^c P_c = P_b$, and hence $Q(P_{a_i}) \subseteq Q(P_c)$, $Q(P_b) \subseteq Q(P_c)$. Therefore $1/\omega \in Q(P_c)$. This is a contradiction since any non-unit of $Q(P_b)$ is obviously also a non-unit in $Q(P_c)$.

We have therefore shown that $\mathfrak{B}$ is a proper ring (not a field). We now show that $\mathfrak{B}$ *is a valuation ring*. For this it is sufficient to show that if ξ is any element of Σ then either $\xi \in \mathfrak{B}$ or $1/\xi \in \mathfrak{B}$. We consider again a fixed projective model V_b of Σ/k. We select a system of nonhomogeneous coördinates x_1, x_2, $\cdots$, x_n of the general point of V_b in such a fashion that the point P_b $(= \pi_b^*P^*)$ is at finite distance with respect to these coördinates. Let V_d be the projective model whose general point has as nonhomogeneous coördinates the elements x_1, x_2, $\cdots$, x_n, ξ. If the point P_d $(= \pi_d^*P^*)$ is at finite distance with respect to these coördinates, then $\xi \in Q(P_d) \subseteq \mathfrak{B}$. If P_d is a point at infinity, we observe first of all that in the above proof of our assertion $1/\omega \notin \mathfrak{B}$ we have shown incidentally the following: if V_a and V_b are *any* two projective models of Σ/k and if $\pi_a^*P^* = P_a$ and $\pi_b^*P^* = P_b$, then P_a and P_b are *corresponding* points of the birational correspondence between V_a and V_b. For on the join V_c of V_a and V_b we have the point $P_c = \pi_c^*P^*$ and the relations $Q(P_c) \supseteq Q(P_a)$, $Q(P_c) \supseteq Q(P_b)$. These relations show that if v is any zero-dimensional valuation whose center on V_c is the point P_c, then the center of v on V_a is P_a and its center on V_b is P_b. Hence P_a and P_b are indeed corresponding points.[7] With this observation in mind, let v be a zero-dimensional valuation whose center on V_b is the point P_b and whose center on V_d is the point P_d. Since P_b is at finite distance, we have $v(x_i) \geqq 0$, $i=1, 2, \cdots, n$. Since P_d is at infinity, we must have $v(\xi)<0$. Hence $v(1/\xi)>0$, $v(x_i/\xi)>0$, and this shows that if we take $1/\xi, x_1/\xi, \cdots, x_n/\xi$ as nonhomogeneous coördinates of the general point of V_d, then P_d is at finite distance. Hence $1/\xi \in Q(P_d) \subseteq \mathfrak{B}$. This completes the proof of our assertion that $\mathfrak{B}$ is a valuation ring.

Let v be the valuation defined by the valuation ring $\mathfrak{B}$. We assert that *v is zero-dimensional*. For let v be of dimension s. We can find a projective model V_b on which the center of v is an s-dimensional variety W. If $P_b = \pi_b^* P^*$, then $Q(P_b) \subseteq \mathfrak{B}$ and this implies that $P_b \in W$.[8] If $s>0$, then we can find a non-unit ω in $Q(P_b)$ such that $\omega \neq 0$ on W, whence $1/\omega \in Q(W) \subseteq \mathfrak{B}$, a contradiction. Hence $s=0$, as asserted.

The above relation $P_b \in W$ implies now $P_b = W$. This is true for any projective model V_b, that is, the center of v on any projective model V_b is the point $P_b = \pi_b^* P^*$. This completes the proof of the theorem.

3. **A generalization. Infinite direct products of projective lines.** The idea of topologizing an algebraic variety V by choosing as closed sets the algebraic subvarieties of V can be used with good effect in order to topologize the set M^* of all homomorphic mappings of any abstract field A into another abstract field K. In this general case we are dealing essentially with a generalization of the concept of the Riemann manifold of a field of algebraic functions (see the Remark at the end of the paper). We begin with some topological preliminaries.

Let $\{R_a\}$ be a system of compact topological spaces indexed by a set $A = \{a\}$. We assume that each R_a is a T_1-space; that is, that the points of R_a are closed sets. Elements of A shall be denoted by small Latin letters, $a, b, c, \cdots$; subsets of A shall be denoted by small Greek letters, $\alpha, \beta, \gamma, \cdots$. If α is a subset of A we shall denote by R_α the direct product $P_{a \in \alpha} R_a$. If $\alpha \subset \beta$ we denote by π_α^β *the projection* of R_β onto R_α. Finally, elements of R_a and R_α shall be denoted by $x_a, y_a, z_a, \cdots$ and by $\mathfrak{x}_\alpha, \mathfrak{y}_\alpha, \mathfrak{z}_\alpha, \cdots$ respectively. If $a \in \alpha$ and if $\pi_a^\alpha \mathfrak{x}_\alpha = x_a$, then x_a shall be referred to as the a-component of $\mathfrak{x}_\alpha$.

We assume that for each *finite* subset α of A a topology has been assigned to R_α and that the following three conditions are satisfied: (1) *R_α is a compact topological space*; (2) *if $\alpha \subset \beta$ then π_α^β is a closed*

[7] See our definition of corresponding points of a birational transformation, loc. cit. footnote 6, p. 505.

[8] See loc. cit. footnote 6, Theorem 3, p. 497.

mapping (mapping = single-valued continuous transformation); (3) *if α is a set with one element a then the topology assigned to R_α is exactly the topology of R_a.* It is clear that in virtue of these two conditions R_α is a T_1-space. For if x_a is the a-component of x_α, then $(\pi_a^\alpha)^{-1}x_a$ is closed and x_α is the intersection of the closed sets $(\pi_a^\alpha)^{-1}x_a$, $a \in \alpha$.

If we consider only finite subsets α of A and if we define a partial ordering in the collection $\{R_\alpha\}$ by setting $R_\alpha < R_\beta$ if $\alpha \subset \beta$, then we have an inverse system $\{R_\alpha; \pi_\alpha^\beta\}$. It is clear that set-theoretically the limit space R^* of the system coincides with the direct product $R^* = \mathbf{P}_{a \in A} R_a$. However, the topology in R^* is not necessarily the usual topology of the product space, for our topology in R^* depends not only on the topology of each factor R_a but also on the topology which has been assigned to each R_α, for α any finite subset of A.

Our space R^* is compact, by Steenrod's theorem. We are dealing here with a special case of Steenrod's theorem, and the proof of the compactness of R^* can be somewhat simplified. For this reason, and also for the convenience of the reader, we shall include here a proof of the compactness of R^*.

We have to show that if a family of closed sets in R^* has the finite intersection property (that is, if every finite subfamily has a non-empty intersection), then the intersection of the entire family is non-empty. It will be sufficient to prove this for families of basic closed sets F_α^*, $F_\alpha^* = \pi_\alpha^{-1} F_\alpha$, where F_α denotes a closed set in R_α and where π_α is the projection of R^* onto R_α. Let then $\{F_\alpha^*\}$ be a family $\mathfrak{F}$ of basic closed sets which has the finite intersection property. By Zorn's lemma the family $\{F_\alpha^*\}$ is contained in a maximal family $\{G_\alpha^*\}$ of basic closed sets which has the finite intersection property. It will be sufficient to show that $\bigcap G_\alpha^*$ is non-empty. We shall therefore assume that our original family $\{F_\alpha^*\}$ is not contained properly in another family of basic closed sets which has the finite intersection property.[9]

We first observe that *the intersection of any finite collection of basic closed sets is itself a basic closed set.* For let $\{\pi_{\alpha_i}^{-1} F_{\alpha_i}\}$ be a finite collection of basic closed sets. We put $\alpha = \bigcup \alpha_i$, $F_\alpha = \bigcap (\pi_{\alpha_i}^\alpha)^{-1} F_{\alpha_i}$. Then it is clear that $\bigcap \pi_{\alpha_i}^{-1} F_{\alpha_i} = \pi_\alpha^{-1} F_\alpha$.

In virtue of this remark and in virtue of the maximality property

[9] The idea of passing to a maximal family is taken from the proof of Tychonoff's theorem as given in Lefschetz, *Algebraic topology*, p. 19. There is only this difference: the maximal family in Lefschetz is not a family of closed sets, while ours is. This modification of the proof succeeds because we restrict ourselves to families of basic closed sets and because in our case the mappings π_α^β are closed.

of the given family $\mathfrak{F}$, it follows that every finite intersection of sets in $\mathfrak{F}$ is again in the family $\mathfrak{F}$.

For any element a in A and for any member F_α^* in $\mathfrak{F}$ let $\pi_a F_\alpha^* = F_{\alpha,a}$. If $a \notin \alpha$ then it is clear that $F_{\alpha,a} = R_a$, for then the a-component of the points of F_α^* is not restricted. If $a \in \alpha$ and if $F_\alpha^* = \pi_\alpha^{-1} F_\alpha$, then $F_{\alpha,a} = \pi_a^\alpha F_\alpha$. In either case $F_{\alpha,a}$ is a closed set in R_a, for we have assumed that π_α^β is closed whenever $\alpha \subset \beta$. For a given a the family $\mathfrak{F}_a$ of closed sets $\{F_{\alpha,a}\}$ has the finite intersection property. Since R_a is compact, the intersection $\bigcap_\alpha F_{\alpha,a}$ is non-empty. Let x_a be a point common to all the sets in $\mathfrak{F}_a$. Then $\pi_a^{-1} x_a$ is a basic closed set (since R_a is a T_1-space) which meets every set F_α^* in $\mathfrak{F}$. Consequently $\pi_a^{-1} x_a \in \mathfrak{F}$, $x_a \in \mathfrak{F}_a$, and the intersection $\bigcap_\alpha F_{\alpha,a}$ consists only of the point x_a.

Let then $x = \{x_a\}$, where $x_a = \bigcap_\alpha F_{\alpha,a}$. We show that x is a common point of the sets F_α^* in $\mathfrak{F}$. Since $\pi_a^{-1} x_a \in \mathfrak{F}$, for any a, it follows that $\bigcap_{a\in\alpha} \pi_a^{-1} x_a \in \mathfrak{F}$, that is, $\pi_\alpha^{-1} x_\alpha \in \mathfrak{F}$, where $x_\alpha = \pi_\alpha x$. Therefore $\pi_\alpha^{-1} x_\alpha$ meets F_α^*, that is, $\pi_\alpha^{-1} F_\alpha$; hence $x_\alpha \in F_\alpha$ and $x \in \pi_\alpha^{-1} x_\alpha \subset \pi_\alpha^{-1} F_\alpha = F_\alpha^*$, q.e.d.

Now let K be a fixed abstract field and let the sets R_a be projective lines over K, so that the points of each set R_a are in (1, 1) correspondence with the elements of K together with the symbol ∞. We topologize R_a by choosing as closed sets the finite subsets of R_a. Then each R_a becomes a compact topological T_1-space.

We still have to topologize each set R_α, for α a finite subset of A. For this purpose we introduce on each line R_a a pair of homogeneous coördinates x_{a1}, x_{a2} and we define an algebraic variety V_α by the following parametric equations (in which the $X_{(\epsilon)}^{(\alpha)}$ denote the homogeneous coördinates of the general point of V_α):

$$\rho \cdot X_{\epsilon_1\epsilon_2\cdots\epsilon_n}^{(\alpha)} = x_{a_1\epsilon_1} x_{a_2\epsilon_2} \cdots x_{a_n\epsilon_n}, \tag{1}$$

where $\alpha = \{a_1, a_2, \cdots, a_r\}$ and where each ϵ_j can take the values 1 or 2. It is well known that V_α is a Segre variety, of dimension n, immersed in a projective space of dimension $2^n - 1$. The points of V_α are in (1, 1) correspondence with n-tuples of ratios $\{x_{a2}/x_{a1}\}$, $a \in \alpha$, that is, with the points of the direct product $R_\alpha = R_{a_1} \times R_{a_2} \times \cdots \times R_{a_n}$. It should be noted that here we only consider points X^α whose homogeneous coördinates are in K. *We topologize V_α by choosing as closed sets the algebraic subvarieties of V_α.* Then it is clear that each V_α becomes a compact topological T_1-space.

If $\alpha = \{a_1, a_2, \cdots, a_n\}$ and if β is a subset of α, say if $\beta = \{a_1, a_2, \cdots, a_m\}$, $m < n$, then the projection π_β^α of V_α onto V_β is given by the equations:

$$X_{\epsilon_1\epsilon_2\cdots\epsilon_m}^{(\beta)} : X_{\delta_1\delta_2\cdots\delta_m}^{(\beta)} = X_{\epsilon_1\epsilon_2\cdots\epsilon_m\gamma_{m+1}\cdots\gamma_n}^{(\alpha)} : X_{\delta_1\delta_2\cdots\delta_m\gamma_{m+1}\cdots\gamma_n}^{(\alpha)},$$

where each ϵ, δ and γ can take *independently* the values 1 or 2. Thus π^{α}_{β} is a single-valued *rational* transformation of V_α onto V_β, and therefore π^{α}_{β} is closed and $(\pi^{\alpha}_{\beta})^{-1}$ is open. It is clear that the closed sets in the infinite direct product R^*, as defined above, are the sets defined by (finite or infinite) systems of homogeneous equations, each equation involving the variables $X^{(\alpha)}$ relative to some finite subset α of A.

4. The space of homomorphic mappings of one abstract field into another. We now further specialize our application by assuming that the set A is a field. The space R^* is then the space of all single-valued transformations $x^*: a \to x_a = x_{a1}/x_{a2}$, of the field A into the set consisting of the elements of the field K and of the symbol ∞. We shall now express in an appropriate homogeneous form the conditions that a given mapping x^* be a homomorphism. Let α be a subset of A consisting of three elements, $\alpha = \{a_1, a_2, a_3\}$. On the corresponding variety V_α let F_{a_1, a_2, a_3} be the algebraic subvariety obtained by imposing on the 6 parameters x_{i1}, x_{i2}, $i = a_1, a_2, a_3$, the following condition:

$$x_{a_1 1} x_{a_2 2} x_{a_3 2} + x_{a_1 2} x_{a_2 1} x_{a_3 2} = x_{a_1 2} x_{a_2 2} x_{a_3 1}. \tag{2}$$

Similarly we define another algebraic subvariety G_{a_1, a_2, a_3} of V_α by the equation

$$x_{a_1 1} x_{a_2 1} x_{a_3 2} = x_{a_1 2} x_{a_2 2} x_{a_3 1}. \tag{3}$$

Let $x_{a_j 1}/x_{a_j 2} = x_j$, $j = 1, 2, 3$, where x_j may be ∞. Suppose that equation (2) holds true. Then if x_1 and x_2 are both different from ∞ we find $x_3 = x_1 + x_2$. If $x_1 = \infty$ and $x_2 \neq \infty$, then $x_{a_1 2} = 0$, $x_{a_1 1} \cdot x_{a_2 2} \neq 0$, whence (2) yields $x_{a_3 2} = 0$, that is, $x_3 = \infty$. Assume now that equation (3) holds. Again we find that $x_3 = x_1 x_2$, if both x_1 and x_2 are different from ∞. If $x_1 = \infty$ and $x_2 \neq 0$, then $x_{a_1 2} = 0$, $x_{a_1 1} \cdot x_{a_2 1} \neq 0$, and (3) yields $x_{a_3 2} = 0$, that is, $x_3 = \infty$. Thus the equations (2) and (3) are the homogeneous counterparts of the equations $x_3 = x_1 + x_2$ and $x_3 = x_1 x_2$ respectively, and they include the conventions which are usually made for the symbol ∞. We can therefore assert that x^* represents a homomorphic mapping of A into (K, ∞) if and only if the following conditions are satisfied: for any three elements a_1, a_2, a_3 of A such that respectively $a_3 = a_1 + a_2$ or $a_3 = a_1 a_2$, the projection $\pi_\alpha^{-1} x^*$ (where $\alpha = \{a_1, a_2, a_3\}$) must lie respectively on F_{a_1, a_2, a_3} or on G_{a_1, a_2, a_3}. Therefore, if we denote by M the set of all *homomorphic* mappings of A into (K, ∞), we see that

$$M = \bigcap_{\alpha} \overset{-1}{\pi_\alpha} F_{a_1 a_2 a_3} \bigcap_{\beta} \overset{-1}{\pi_\beta} G_{b_1 b_2 b_3},$$

where the index α ranges over all sets $\alpha=\{a_1, a_2, a_3\}$ such that $a_3=a_1+a_2$, and the index β ranges over the sets $\beta=\{b_1, b_2, b_3\}$ such that $b_3=b_1b_2$. We see thus that $\boldsymbol{M}$ is an intersection of basic closed subsets of R^*. Hence $\boldsymbol{M}$ is closed, and since R^* is compact $\boldsymbol{M}$ is also compact.

The case which is of special interest to us is that in which K is a subfield of A. In this case we are interested in the relative homomorphisms of A into (K, ∞), that is, in the homomorphisms x^* which leave each element of K invariant. If $\boldsymbol{M}^*$ is the set of all these relative homomorphisms, then it is clear that $\boldsymbol{M}^*$ is the intersection of $\boldsymbol{M}$ with the closed set $\bigcap_{a\in K}\pi_a^{-1}a$. Here, according to our notations, $\pi_a^{-1}a$ denotes that subset of R^* which consists of the points x^* whose a-component x_a is a itself $(a\in K)$. Hence also $\boldsymbol{M}^*$ is a compact space.

It is convenient to describe in algebro-geometric terms the relative topology induced in $\boldsymbol{M}^*$ by the topology of $\boldsymbol{M}$. Let $x_1, x_2, \cdots, x_n$ be a finite set of elements of A. For each x_i we introduce a pair of homogeneous parameters x_{i1}, x_{i2} such that $x_{i1}/x_{i2}=x_i$. We consider the algebraic variety Z over K whose general point has as homogeneous coördinates the quantities $X_{(\epsilon)}$ defined by the parametric equations

$$\rho X_{\epsilon_1\epsilon_2\cdots\epsilon_n} = x_{1\epsilon_1}x_{2\epsilon_2}\cdots x_{n\epsilon_n}, \tag{4}$$

where each ϵ_j can take the values 1 or 2. If the quantities x_i are algebraically independent, then the variety Z coincides with the variety V_α defined by the equations (1), α being the subset $\{x_1, x_2, \cdots, x_n\}$ of A. But in general Z is a subvariety of V_α. If $x^*\in\boldsymbol{M}^*$, then the mapping x^* of A into (K, ∞) must preserve all the algebraic relations between $x_1, x_2, \cdots, x_n$ over K, since x^* is a homomorphism. It follows that the point $\pi_\alpha x^*$ of V_α must lie on Z. Now we observe that the homomorphism x^* defines a unique valuation of A/K whose residue field is K itself and whose center on Z is the point $\pi_\alpha x^*$. Conversely, every valuation of A/K whose residue field is K defines a relative homomorphic mapping of A/K onto (K, ∞). We conclude that *if W is any algebraic subvariety of Z, then the set of all valuations of A/K having K as residue field and having center on W is a closed subset of $\boldsymbol{M}^*$.* By taking different finite subsets $\{x_1, x_2, \cdots, x_n\}$ of A and different subvarieties W of Z we obtain a family of closed subsets of $\boldsymbol{M}^*$ which form a basis for the closed subsets of $\boldsymbol{M}^*$.

REMARK. Suppose that A is a field of algebraic functions in any number of variables, over a given ground field k. We identify the field K with the algebraically closed field determined by k. The Riemann manifold $\boldsymbol{M}$ of A is the set of all zero-dimensional valuations v

of A. By the ground field extension $k \rightarrow K$ we can embed A in a field $A' = KA$. Every relative homomorphic mapping of A' onto (K, ∞) determines uniquely a zero-dimensional valuation of A'/K, and vice versa. Every zero-dimensional valuation of A'/K induces a unique zero-dimensional valuation of A/K, but a given zero-dimensional valuation of A/K may be extendable in more than one way to a zero-dimensional valuation of A'/K. It follows that the Riemann manifold M' of A'/K coincides with the space M^* of all relative homomorphic mappings of A' onto (K, ∞) and is therefore a compact space. The Riemann manifold M of A/K is obtainable from M' by topological identification and therefore can also be converted into a compact topological space. That is precisely what we have proved in §2.

The Johns Hopkins University

ANNALS OF MATHEMATICS
Vol. 45, No. 3, July, 1944

REDUCTION OF THE SINGULARITIES OF ALGEBRAIC THREE DIMENSIONAL VARIETIES

BY OSCAR ZARISKI

(Received January 1, 1944)

CONTENTS

INTRODUCTION

1. In the present paper we give a solution of the problem of reduction of singularities of algebraic three-dimensional varieties.* In contrast with the

* The solution was first announced by the author in a conference report at the 1941 Summer Meeting of the American Mathematical Society in Chicago. See Bull. Amer. Math. Soc., vol. 48, n. 3 (1942), pp. 172–173.

two-dimensional case which has been studied by many authors, the three-dimensional problem is totally unexplored territory. As a challenge to the investigator it is second only to the general problem of the reduction of singularities of varieties of arbitrary dimension. How much more difficult is the general problem is of course impossible to say with certainty and precision at the present moment. We are inclined to conjecture that the difficulties in the general case and in the three-dimensional case are of comparable order of magnitude. However that may be, the three-dimensional problem offers an excellent testing ground for ideas and methods in relation to the general case, and for this reason its solution is presented here as a possible contribution to the general problem of reduction of singularities. We shall now endeavor to outline the main ideas of our solution.

2. By the theorem of local uniformization (see [6], p. 858), given any zero-dimensional valuation v of a field $\sum$ of algebraic functions of any number of variables (over a ground field of characteristic zero), there exists a projective model V of $\sum$ on which the center of v is a simple point. We shall show in a separate paper** that this theorem implies the existence of *a finite resolving system* for $\sum$, i.e. of a finite number of projective models of $\sum$ such that any zero-dimensional valuation v of $\sum$ has a simple center on at least one of the models (for the classical case, see our proof in [6], p. 854; compare also with footnote[15], section 25). We have shown in [7] that, in virtue of the existence of finite resolving systems, the solution of the problem of the reduction of singularities hinges entirely on the proof of the "fundamental theorem" stated in Part IV, section 25. Now the assumption of that theorem is that two given projective models V and V' constitute a resolving system for a given set N of zero-dimensional valuations. One is naturally led to consider tentatively the join V^* of V and V' ([8], p. 516). In general, V^* will not be a resolving model for N. However, the following condition is sufficient in order that V^* be a resolving model for N:

If v is any valuation in N and if P and P' are the centers of v on V and V' respectively, then one of the points P, P' is simple and at the same time is not fundamental for the birational correspondence T between V and V' (i.e. its quotient ring contains the quotient ring of the second point).

The gist of the considerations developed in the last section of this paper (section 25) is that in the three-dimensional case it is possible to transform V and V', by successive quadratic and monoidal transformations, into another pair of varieties for which the above condition is satisfied. The centers of these transformations are always selected as follows: if we are dealing with a simple point P of V which is an *isolated* fundamental point of T, then we apply to V a quadratic transformation of center P; if P lies on a simple fundamental curve Δ of T then we use a monoidal transformation of center Δ. In the second case it is, however, necessary to assume that *P is a simple point of Δ*, for otherwise the monoidal

** "The compactness of the Riemann manifold of an abstract field of algebraic functions", in course of publication in Bull. Amer. Math. Soc.

transformation will not necessarily satisfy the requirement of transforming the simple point P into simple points (section 5, a)). For this reason it is necessary first to resolve the singularities of Δ before a monoidal transformation is applied (section 20). Similar considerations apply to V' and to the successive transforms of V and V'. Given any valuation v in N, with centers P and P' on V and V' respectively, and if, say, P is the simple center, the intended effect of the successive transformations applied to V is *to augment the quotient ring of* P, so that ultimately P is replaced by a simple point whose quotient ring contains the quotient ring of P'. In this procedure we come up against the following complication (see case C, section 25): when we apply a monoidal transformation whose center is a *simple* curve Δ, we automatically augment the quotient ring of *each* point of Δ, hence also of such points of Δ which are *singular* for V and occur as centers of valuations in N. A lemma proved in section 24 enables us to overcome this difficulty.

3. When all incidental questions are discounted, there remains in the proof of the fundamental theorem a central idea, and that is the elimination, by successive quadratic and monoidal transformations, of the simple fundamental curves and of the isolated simple fundamental points of a given birational transformatión T of a three-dimensional variety V into another variety V'. That this elimination is always possible is precisely what is asserted in Theorem 7 (section 19), or in the essentially equivalent Theorem 7′ (section 19). In this last theorem the elimination of the simple fundamental locus of T is presented as a process of elimination of the simple base locus (base curves and isolated base points outside the singular locus of V) of the linear system $|F|$ of surfaces on V which corresponds to the system of hyperplane sections of V'. *The exigencies of the proof of Theorem 7′ determine to a considerable extent the contents of the rest of the paper.* The scope of this theorem will be best understood if we analyze one implication of the theorem.

By a well-known theorem of Bertini, the singular points of the variable surface in the linear system $|F|$, outside the singular locus of V, lie necessarily on the base manifold of $|F|$. This theorem of Bertini is re-stated in algebraic terms, more appropriate for our purpose, in section 23, and we prove it for abstract varieties in a separate paper [9]. In virtue of this theorem of Bertini, any birational transformation of V which brings about the elimination of the simple base locus of the linear system $|F|$ *automatically reduces the singularities of the variable surface of the system* (not counting the singularities which fall at singular points of V). We render the term "variable surface" more precise by interpreting the linear system of *surfaces* $|F|$ as *one* surface F/K, over a suitable extended ground field K, embedded in the variety V/K. This way of looking at the linear system $|F|$ is part of our formulation of the theorem of Bertini. We see therefore that *Theorem 7′ implies a reduction theorem for algebraic surfaces,* namely *that it is possible to reduce the singularities of a surface F/K embedded in a three-dimensional variety V/K by successive quadratic and monoidal transformations of the ambient variety V/K* (Theorem 6, section 17). A geometric proof of this theorem in the classical case, for surfaces embedded in a linear space, was presented by Beppo Levi in [2]. In Part III we prove this theorem for abstract surfaces. Here we make use of a local reduction theorem (Theorem 5, section 10) proved in Part II.

With the aid of the above reduction theorem, the proof of Theorem 7′ is carried out smoothly in two steps. First the theorem is proved directly for linear systems $| F |$ which, outside of the singular manifold of V, are free from singular base points (i.e., base points which are singular for each surface in the system). This is done in sections 20, 21, and 22. For arbitrary linear systems $| F |$ the proof is completed in section 23 by reducing the base singularities of $| F |$.

4. Our solution of the three-dimensional problem is complete in the case of ground fields of characteristic zero. However, we can point out with precision what has still to be done in order that the proof be complete also for fields of characteristic $p \neq 0$. Throughout the paper we have used the assumption that the field is of characteristic zero only in two instances: a) in Part IV, where we presuppose the local uniformization theorem for three-dimensional varieties; b) in sections 13, 14 and 15, where we prove the local reduction theorem for valuations of rank 1. These then are the only two gaps which would have to be filled in in order that our solution be valid for arbitrary ground fields of characteristic p. We must state explicitly that these gaps remain *even if the ground field is perfect.*

When we speak of a possible generalization of our solution of the three-dimensional problem to arbitrary fields of characteristic $p \neq 0$, we have in mind the following definition of a simple point: *a point P of an r-dimensional algebraic variety is simple if the ideal of non-units in the quotient ring of P has a base of r elements.* The elements of such a base are called *uniformizing parameters of P.* A simple point, according to this definition, is therefore characterized by the intrinsic condition that its quotient ring be a p-series ring in the sense of Krull [1]. It was pointed out to me by Chevalley and André Weil that this definition is in general not equivalent with the following classical definition: *if $(f_1, f_2, \cdots, f_m)$ is a base for the prime polynomial ideal in $k[x_1, x_2, \cdots, x_n]$ which defines the variety V, then a point P of V is simple if and only if the Jacobian matrix $\| \partial f_i/\partial x_j \|$ is of rank $n - r$ at P.* Simple points, according to this classical definition, remain simple upon an arbitrary extension of the ground field. For such points the term "absolutely simple" has therefore been proposed by Weil. An absolutely simple point P is necessarily a simple point, but if the residue field of P is not separable over the ground field the converse is not necessarily true. In a separate paper we shall analyse the relationship between simple and absolutely simple points, and we shall also characterize simple points by an algebraic condition involving the rank of a certain Jacobian matrix.

It is clear that if we mean by a non-singular variety one whose points are all absolutely simple, then not every field of algebraic functions has a non-singular model. For instance, for the existence of such a model it is at least necessary (but not sufficient) that the function field be separably generated over the ground field. On the other hand, we conjecture that *every* field of algebraic functions (in any number of variables) possesses a model which is non-singular in the sense of our definition of simple points. It is in this sense that we speak, for instance, of the solution of the three-dimensional reduction problem for arbitrary ground fields of characteristic p.

One more remark. We make frequent use of the theorem that *every minimal prime ideal in the quotient ring of a simple point is a principal ideal.* In the case of characteristic zero and more generally for absolutely simple points, this theorem will be proved in our paper to be printed elsewhere (for algebraic surface, see our proof in [5], p. 664, Theorem 3). However, this theorem follows in all its generality, i.e. for arbitrary ground fields, from an unpublished result concerning p-series rings, due to I. Cohen. This theorem of Cohen states that *any p-series ring, whose characteristic coincides with the characteristic of its residue field K, can be embedded in a full power series ring over this field K* [see abstract of paper "Some theorems on local rings", by I. S. Cohen, Bull. Amer. Math. Soc., vol. 49, n. 1 (1943), p. 37].

PART I

QUADRATIC AND MONOIDAL TRANSFORMATIONS OF SURFACES EMBEDDED IN A V_3

1. Multiple points of embedded surfaces

The object of our considerations in Parts I, II and III shall be an irreducible 3-dimensional algebraic variety V and a fixed irreducible algebraic surface F on V. The variety V may have singularities. *However, we stipulate that from now on whenever points or curves of V are considered it will be tacitly understood that these points or curves are simple for V.* To this we add that in the case of a simple curve of V it is not excluded that some special points of the curve may be singular for V, but if a point of the curve is mentioned it shall be understood that it is a simple point of V.

On the other hand, no such restrictions shall be imposed on the points and curves of the surface F. On the contrary, we shall be primarily interested in those points or curves of F which are singular for F (but are simple for V).

Quite generally, if H and G are irreducible algebraic varieties and if $H \subset G$, we shall denote by $Q_G(H)$ the quotient ring of H, regarded as a subvariety of G.

If H is simple for G and if say H and G are of dimension h and g respectively, then it is known that the ideal of non units in $Q_G(H)$ has a (minimal) base of $g - h$ elements. The elements $t_1, t_2, \cdots, t_{g-h}$ of any such minimal base shall be referred to as *uniformizing parameters* of $H(G)$. The notations $H(G)$ and $Q_G(H)$ are dictated by the necessity of avoiding confusion in cases when H occurs as a subvariety of more than one variety G. In all other cases the letter G will be omitted.

Let H be a point P or a curve Δ on F, and let $\mathfrak{m}$ be the ideal of non units in $\mathfrak{J} = Q_V(H)$. The surface F is defined in $\mathfrak{J}$ by a minimal prime ideal which is a principal ideal, say $\mathfrak{J} \cdot \omega$, for H is simple for V (see Introduction).

DEFINITION 1. *H is ν-fold for the surface F if $\omega \equiv 0(\mathfrak{m}^\nu)$, $\omega \not\equiv 0(\mathfrak{m}^{\nu+1})$.*

This definition implies that ω can be expressed as a form f of degree ν in the uniformizing parameters t_i of $H(V)$, whose coefficients are in $\mathfrak{J}$ and do not all belong to $\mathfrak{m}$. Thus, if H is a point P, then

$$\omega = f(t_1, t_2, t_3), \tag{1}$$

and if H is a curve Δ, then

$$\omega = f(t_1, t_2). \tag{1'}$$

We shall also denote the multiplicity ν by $m_F(H)$.

LEMMA 1.1 *If $P \subset \Delta \subset F$, then $m_F(P) \geqq m_F(\Delta)$.* This is well known in the classical case. If the ground field is of characteristic zero or is a perfect field of characteristic p, the lemma can be proved by methods of formal differentiation (see Part II, section 13). However, since we shall give elsewhere (compare with Introduction, 4), a proof of Lemma 1.1 in the most general case, we shall omit the proof here.

2. Quadratic transformations

We apply to V a quadratic transformation T_1 whose center is a point P on F, and we denote by V' and F' respectively the transforms of V and of F under T_1. Here F' is an irreducible surface on V' which is birationally equivalent to F. We denote by T the quadratic transformation from F to F' which is induced by T_1. We shall now state several salient features of T which we shall need in the sequel and which we have established elsewhere (see [8], p. 532).

By T_1 the point P is transformed into a surface $\Phi' = T_1[P]$.

a) *The surface Φ' is irreducible and each point of Φ' is simple for V'.*

Let $\mathfrak{J} = Q_V(P)$ and let $\mathfrak{m}$ be the ideal of non units in $\mathfrak{J}$. We denote by K the residue field of P, i.e. the field $\mathfrak{J}/\mathfrak{m}$. Let $S = S_2^K$ be the projective plane over K.

b) *The surface Φ' is in regular[1] birational correspondence with S.*

Let t_1, t_2, t_3 be a fixed set of uniformizing parameters of $P(V)$. If $f(z_1, z_2, z_3)$ is a form with coefficients in $\mathfrak{J}$, where z_1, z_2, z_3 are indeterminates, then we denote by $F(z_1, z_2, z_3)$ the form in $K[z_1, z_2, z_3]$ whose coefficients are the P-residues (i.e. the residues mod. $\mathfrak{m}$) of the coefficients of f. If $\xi \in \mathfrak{J}$, $\xi \equiv 0(\mathfrak{m}^\rho)$, $\xi \not\equiv 0(\mathfrak{m}^{\rho+1})$, we can write $\xi = f(t_1, t_2, t_3)$, where f is a form of degree ρ and where $F(z_1, z_2, z_3) \neq 0$. In this case we shall call $F(z_1, z_2, z_3)$ the *leading form* of ξ (see Krull [1], p. 207); it is independent of the particular way of representing ξ as a form of degree ρ in t_1, t_2, t_3. We identify the indeterminates z_1, z_2, z_3 with the homogeneous coördinates in S.

Let now H' be a point or an irreducible curve on Φ', or Φ' itself, and let $\bar{H}$ be the corresponding locus in S [by b)].

c) *The quotient ring $\mathfrak{J}' = Q_{V'}(H')$ consists of all quotients $f(t_1, t_2, t_3)/g(t_1, t_2, t_3)$,*

[1] For the definition of a regular correspondence see our paper [8], p. 513.

where $f(z_1, z_2, z_3)$, $g(z_1, z_2, z_3)$ *are forms of like degree, with coefficients in* $\mathfrak{J}$, *and where* $G(z_1, z_2, z_3) \neq 0$ *on* $\bar{H}$. *In other words,* $f(t)/g(t) \in Q_{V'}(H')$ *if and only if* $F(z)/G(z) \in Q_S(\bar{H})$.

From c), in the particular case $H' = \Phi'$, it follows that $Q_{V'}(\Phi')$ consists of all quotients $f(t)/g(t)$ such that $G(z) \neq 0$. In other words: if ξ and η are elements of $\mathfrak{J}$, then $\xi/\eta \in Q_{V'}(\Phi')$ if and only if the degree of the leading form of η is not greater than the degree of the leading form of ξ. Let μ be the degree of the leading form of η. If ξ/η is a non unit in $Q_{V'}(\Phi')$, then the degree of the leading form of ξ must be $\geqq \mu + 1$, and conversely. In this case we have: $\xi/t_1^{\mu+1} \in Q_{V'}(\Phi')$, $t_1^{\mu}/\eta \in Q_{V'}(\Phi')$, *whence* ξ/η *is divisible by* t_1 *in* $Q_{V'}(\Phi')$. Conversely, if this condition is satisfied, then ξ/η is a non unit in $Q_{V'}(\Phi')$. From this we conclude, on account of c), as follows:

d) *If* $z_1 \neq 0$ *on* $\bar{H}$, *the surface* Φ' *is defined in* $\mathfrak{J}'$ *by the principal ideal* $\mathfrak{J}' \cdot t_1$, *and we have*:

$$Q_S(\bar{H}) = \mathfrak{J}'/\mathfrak{J}' \cdot t_1; \qquad \mathfrak{J}' \sim Q_S(\bar{H}). \tag{2}$$

Let $\{\bar{t}_2, \bar{t}_3\}$ or $\{\bar{t}_2\}$ be a set of uniformizing parameters of $\bar{H}(S)$ according as H' (and therefore also $\bar{H}$) is a point or a curve. Let t_i' be an element of $\mathfrak{J}'$ which is mapped into $\bar{t}_i$ by the homomorphism (2). We put $t_1' = t_1$.

e) *According as* H' *is a point or a curve, the elements* t_1', t_2', t_3' *or the elements* t_1', t_2' *are uniformizing parameters of* $H'(V')$.

3. The effect of a quadratic transformation on the multiple points of F

We now turn our attention on the surfaces F and F' and on the quadratic transformation T, of center P, which T_1 induces between these two surfaces. To the fundamental point P there will correspond on F' an algebraic curve $T[P]$ which may be reducible. It is clear that this curve is the intersection of F' and Φ' and that [see (1)]

$$F(z_1, z_2, z_3) = 0 \tag{3}$$

is the equation of the plane curve in S which corresponds to $T[P]$ in the regular birational correspondence between Φ' and S. This plane curve is of order ν, where $\nu = m_F(P)$.

We take for H' [see c)] a point P' of $T[P]$ and we denote by $\bar{P}$ the point of S which corresponds to P'. We put $\omega' = \omega/t_1^{\nu}$.

Lemma 3.1. *If* $z_1 \neq 0$ *at* $\bar{P}$, *then the principal ideal* $\mathfrak{J}' \cdot \omega'$ *is prime and defines the irreducible surface* F'.[2]

Proof. We have to show that: 1) ω' is zero on F' and 2) that if an element ζ' of $\mathfrak{J}'$ is zero on F' then $\zeta' \in \mathfrak{J}' \cdot \omega'$. The first assertion is obvious, since $\omega = 0$ on F and $t_1 \neq 0$ on F (in view of our assumption that $z_1 \neq 0$ at $\bar{P}$). To prove 2), let $\zeta' = \xi/\eta$, where $\xi, \eta \in \mathfrak{J}$. Since $\zeta' = 0$ on F, ξ must be zero on F,

[2] Since P' is a point of F', the surface F' is given in $\mathfrak{J}'$ $(= Q_{V'}(P'))$ by a prime minimal ideal. When we say that $\mathfrak{J}' \cdot \omega'$ defines the surface F' we mean that this prime ideal coincides with $\mathfrak{J}' \cdot \omega'$.

whence $\xi = \omega \cdot \xi_1$, $\xi_1 \in \mathfrak{J}$. If ρ is the degree of the leading form of η, then the degree of the leading form of ξ is at least ρ [c)] and hence the degree of the leading form of ξ_1 is at least $\rho - \nu$. By c) we may assume that the leading form of η does not vanish at $\bar{P}$. Under this assumption the quotient $\xi_1 t_1'^{\nu}/\eta$ is in $\mathfrak{J}'$, whence $\zeta'/\omega' = \xi_1 t_1'^{\nu}/\eta \in \mathfrak{J}'$. This completes the proof.

LEMMA 3.2. *If P' is a point of $T[P]$, then $m_{F'}(P') \leqq m_F(P)$.*

PROOF. Let $\mathfrak{m}'$ and $\bar{\mathfrak{m}}$ denote the ideals of non-units in $\mathfrak{J}'$ and in $\bar{\mathfrak{J}} = Q_S(\bar{P})$ respectively, so that [see e)] $\mathfrak{m}' = \mathfrak{J}' \cdot (t_1', t_2', t_3')$ and $\bar{\mathfrak{m}} = \bar{\mathfrak{J}} \cdot (\bar{t}_2, \bar{t}_3)$. In the homomorphism $\mathfrak{J}' \sim \bar{\mathfrak{J}} = \mathfrak{J}'/\mathfrak{J}' \cdot t_1$ [see (2)], the ideal $\mathfrak{m}'$ is mapped into the ideal $\bar{\mathfrak{m}}$. The element ω' is mapped into the element $\bar{\omega} = F(z_1, z_2, z_3)/z_1^{\nu}$. If $\nu' = m_{F'}(P')$, then we must have, by Lemma 3.1, $\omega' \equiv 0(\mathfrak{m}'^{\nu'})$. Hence $\bar{\omega} \equiv 0(\bar{\mathfrak{m}}^{\nu'})$, i.e. *the point $\bar{P}$ must be a ν'-fold point of the curve* (3). Our lemma is therefore equivalent to the assertion that the plane curve (3), *which is of order* ν, cannot possess points of multiplicity $> \nu$. This assertion is trivial if the ground field k is algebraically closed. On the other hand, it is obvious from our definition 1 that the multiplicity of P for the curve $F(z) = 0$ does not diminish if we extend the ground field k. This establishes the lemma.

THEOREM 1. *If $T[P]$ possesses an irreducible component which is ν-fold for F' then $T[P]$ itself is irreducible and is a rational curve free from singularities.*

PROOF. Let Γ' be an irreducible component of $T[P]$ and let us assume that Γ' is ν-fold for F'. Let $\bar{\Gamma}$ be the irreducible plane curve in S which corresponds to Γ' in the regular birational correspondence between Φ' and S. The proof of Lemma 3.2 shows that $\bar{\Gamma}$ must be a ν-fold component of the curve (3). Since this curve is itself of order ν, it follows that the form $F(z_1, z_2, z_3)$ is necessarily the νth power of a linear form, i.e. the plane curve in S into which $T[P]$ is transformed must be a straight line, q.e.d.

4. Auxiliary considerations concerning ruled surfaces

Let Δ be an irreducible algebraic curve in an n-dimensional projective space S_n^k and let $\eta_0, \eta_1, \cdots, \eta_n$ be the homogeneous coördinates of the general point of Δ. If z_1, z_2 are algebraically independent over the field $k(\eta_0, \eta_1, \cdots, \eta_n)$, we consider the ruled surface R over k whose general point has the following homogeneous coördinates:

$$\rho\eta_{ij} = \eta_i z_j, \qquad i = 0, 1, 2, \cdots, n; \; j = 1, 2,$$

where ρ is a factor of proportionality. This surface lies in a projective space of dimension $2n + 1$. Assuming that $\eta_0 \neq 0$ let $\xi_i = \eta_i/\eta_0$, $i = 1, 2, \cdots, n$, and let K be the field of rational functions on Δ: $K = k(\xi_1, \xi_2, \cdots, \xi_n)$. For a suitable choice of the factor ρ we can write: $\rho\eta_{0j} = z_j$, $\rho\eta_{ij} = \xi_i z_j$, and this shows that if K is taken as ground field then R is a straight line L and that the field of rational functions on R (over k) is the field $K(x)$, where $x = z_1/z_2$. Let $P(\alpha_0, \alpha_1, \cdots, \alpha_n)$ be a point of Δ, where the ratios of the α's are algebraic over k, and let z_1, z_2 be still regarded as algebraically independent over the residue field k^* of P. The irreducible curve, over k, which is defined by the general point

(η^*_{ij}), where $\eta^*_{ij} = \alpha_i z_j$, is a generator of the ruled surface R. We shall denote this generator by g_P. It is clear that g_P is a straight line l^* over k^*.

LEMMA 4.1. *The generator g_P is simple for R if and only if the point P is simple for Δ. If t is a uniformizing parameter of $P(\Delta)$ then t is also a uniformizing parameter of $g_P(R)$.*

PROOF. Let $\mathfrak{o} = Q_\Delta(P)$, $\mathfrak{o}_1 = Q_R(g_P)$. Any element of $\mathfrak{o}_1$ is of the form $f(\eta_{ij})/g(\eta_{ij})$, where f and g are forms of like degree with coefficient in k and where $g(\eta^*_{ij}) \neq 0$. If for any polynomial $F(x)$ in $\mathfrak{o}[x]$ we agree to denote by $F^*[x]$ the polynomial in $k^*[x]$ whose coefficients are the P-residues of the coefficients of $F(x)$, then we can say that $\mathfrak{o}_1$ consists of all quotients $F(x)/G(x)$, $F(x)$ and $G(x)$ in $\mathfrak{o}[x]$ and $G^*(x) \neq 0$.

Suppose that P is a simple point of Δ, with uniformizing parameter t. If $\omega_1 = F(x)/G(x) \in \mathfrak{o}_1$, then ω_1 is a non unit in $\mathfrak{o}_1$ if and only if $F^*(x) = 0$, i.e. if and only if each coefficient of $F(x)$ is a multiple of t (in $\mathfrak{o}$). Hence each non unit of $\mathfrak{o}_1$ is divisible by t (in $\mathfrak{o}_1$), and since t is itself obviously a non unit in $\mathfrak{o}_1$ it follows that g_P is simple for R and that t is a uniformizing parameter of $g_P(R)$.

Conversely, assume that g_P is simple for R and let t be a uniformizing parameter of $g_P(R)$. Without loss of generality we may assume that t is a polynomial in $\mathfrak{o}[x]$, say

$$t = t_0 x^m + t_1 x^{m-1} + \cdots + t_m, \qquad t_i \in \mathfrak{o}.$$

Since $t = 0$ on g_P, it follows $t_i = 0$ at P, whence the coefficients t_i are non units in $\mathfrak{o}$. We can therefore write:

$$(4) \qquad t_i \cdot G(x) = tF_i(x), \qquad G^*(x) \neq 0, \qquad i = 0, 1, \cdots, m.$$

Let now ω be an arbitrary non unit in $\mathfrak{o}$. Since ω is also a non unit in $\mathfrak{o}_1$, we have: $\omega \cdot B(x) = t \cdot A(x)$, $B(x)$ and $A(x)$ in $\mathfrak{o}[x]$ and $B^*(x) \neq 0$. If therefore b is any coefficient of $B(x)$, the product ωb belongs to the ideal $\mathfrak{o}\cdot(t_0, t_1, \cdots, t_m)$. In particular, if b is one of the coefficients of $B(x)$ which is not zero at P, then b is a unit in $\mathfrak{o}$, and we conclude that $\omega \in \mathfrak{o}\cdot(t_0, t_1, \cdots, t_m)$. Hence $\mathfrak{o}\cdot(t_0, t_1, \cdots, t_m)$ is the ideal $\mathfrak{m}$ of non units in $\mathfrak{o}$. Since $\mathfrak{m} \neq \mathfrak{m}^2$ it follows from (4) that we cannot have $F^*_i(x) = 0$ for all i. Let, say, $F^*_\nu(x) \neq 0$, and let c be a coefficient of $F_\nu(x)$ which is not zero at P and which is therefore a unit in $\mathfrak{o}$. We have $t_i F_\nu(x) = t_\nu F_i(x)$ and consequently $ct_i \equiv 0(\mathfrak{o}\cdot t_\nu)$. This shows that $\mathfrak{m} = \mathfrak{o}\cdot t_\nu$, q.e.d.

LEMMA 4.2. *If P is a simple point of Δ, then every point of g_P is simple for R.*

PROOF. The general point of g_P is (η^*_{ij}), where $\eta^*_{ij} = \alpha_i z_j$. Any point P^* of g_P has as homogeneous coördinates the products $\alpha_i\beta_j$, where the ratio β_1/β_2 is algebraic over k. The quotient ring $Q_R(P^*)$ consists of all quotients $F(z_1, z_2)/G(z_1, z_2)$, where F, G are forms of like degree with coefficients in $\mathfrak{o}$ and where the form $G^*(z_1, z_2)$ in $k^*[z_1, z_2]$ is such that $G^*(\beta_1, \beta_2) \neq 0$. The quotient ring $Q_{g_P}(P^*)$ consists of the quotients $F^*(z_1, z_2)/G^*(z_1, z_2)$. This shows in the first place that P^* is a simple point of g_P, and this is true even if P is singular for Δ. This is in agreement with a preceding remark to the effect that g_P is a straight line. Let $\tau^* = F^*(z_1, z_2)/G^*(z_1, z_2)$ be a uniformizing parameter of $P^*(g_P)$ and—assuming that P is simple for Δ—let t be a uniformizing parameter of $P(\Delta)$. Then it is immediately seen that if we put $\tau = F(z_1, z_2)/G(z_1, z_2)$, then t and τ are uniformizing parameters of $P^*(R)$.

5. Monoidal transformations

We apply to V a monoidal transformation T_1 whose center is an irreducible curve Δ on F and we denote by V' and F' the transforms of V and F respectively. We collect in this section some fundamental properties of T_1. For the proofs see our paper [8].

Under T_1 the curve Δ is transformed into a surface, which may be reducible. However, only one irreducible component of this surface corresponds to Δ itself. We shall denote this component by Φ'. The remaining components correspond to special points of Δ, points which are either singular for V or for Δ, and do not interest us at present.

a) *The points of the surface Φ' which correspond to simple points*[3] *of Δ are simple for V' and for Φ'.*

We denote by $\mathfrak{J}$ the quotient ring $Q_V(\Delta)$ and by $\mathfrak{m}$ the ideal of non units in $\mathfrak{J}$. Let $\eta_0, \eta_1, \cdots, \eta_n$ be the homogeneous coördinates of the general point of Δ, and let $K = \mathfrak{J}/\mathfrak{m}$ be the field of rational functions on Δ. We introduce the ruled surface R relative to the curve Δ (see preceding section) and we denote by N' the subvariety of Φ' whose points correspond to the points of Δ which are singular either for Δ or for V.

b) *The surfaces Φ' and R are birationally equivalent, and corresponding points P' and P^* ($P' \in \Phi'$, $P^* \in R$) are such that if $T_1(P) = P'$ then $P^* \in g_P$. This birational correspondence is regular at each point of $\Phi' - N'$.*

We shall denote by g'_P the transform of P on Φ'. By a), b) and by Lemmas 4.1 and 4.2 it follows that:

c) *If P is simple for Δ (and also for V), then g'_P is in regular birational correspondence with the generator g_P and is therefore an irreducible rational curve free from singularities; moreover, g'_P and each point of g'_P is simple for V' and for Φ'.*

In addition to the ground field k and to the field K of rational functions on Δ we shall have occasion to deal with the residue field of P. This last field shall be denoted by k^*. We shall use small Latin letters f, g, h, etc. to denote polynomials (in any number of variables) with coefficients in $\mathfrak{J}$. If the coefficients of such polynomials are reduced mod Δ, the resulting polynomials will have coefficients in K and shall be denoted by the corresponding capital Latin letters F, G, $H \cdots$. If the coefficients of f, g, $h \cdots$ are not only in $\mathfrak{J}[= Q_V(\Delta)]$ but also in $Q_V(P)$, then their coefficients can be reduced mod P, and the resulting polynomials, with coefficients in k^*, shall be denoted by f^*, g^*, h^*, $\cdots$ respectively.

Let t_1, t_2 be a fixed pair of uniformizing parameters of $\Delta(V)$ and let z_1, z_2

[3] But these points must also be simple for V (see section 1).

be indeterminates. If $\mathfrak{m}$ denotes the ideal of non units in $\mathfrak{J}$ and if ξ is an element of $\mathfrak{J}$ which is exactly divisible by $\mathfrak{m}^\rho$, then $\xi = f(t_1, t_2)$, where f is a form of degree ρ, and $F(z_1, z_2)$—the leading form of ξ—is not zero.

In the sequel only points and curves of Φ' are considered which do not belong to N'.

Let H' be an irreducible curve or a point on Φ', or Φ' itself. If H' is a curve, not in the set $\{g'_P\}$, or if $H' = \Phi'$, let H^* be the corresponding locus on R. Since H^* is not a generator g_P, no element of K, different from zero, is zero on H^*. Hence if R is regarded as a straight line over K, then H^* is a point of that line or the line itself. Consequently it makes sense to say that a form $F(z_1, z_2)$ does or does not vanish on H^*.

On the other hand, if H' is either a point of Φ' or if $H' = g'_P$, then H' corresponds to a point P of Δ. Then H^* is either a point P^* or a generator g_P. We select uniformizing parameters t_1, t_2, t_3 of $P(V)$ in such a fashion that t_1, t_2 be uniformizing parameters of $\Delta(V)$.

In both cases let H denote the corresponding locus in Δ ($H = \Delta$ in the first case, $H = P$ in the second case).

d) *The quotient ring* $\mathfrak{J}' = Q_{V'}(H')$ *consists of all quotients* $f(t_1, t_2)/g(t_1, t_2)$ *of forms of like degree, with coefficients in* $Q_V(H)$, *such that* $F(z_1, z_2)/G(z_1, z_2) \in Q_R(H^*)$.

In particular, if $H' = \Phi'$, then it follows that $Q_{V'}(\Phi')$ consists of the quotients ξ/η, ξ and η in $\mathfrak{J}$, such that the degree of the leading form of η is not greater than the degree of the leading form of ξ. From this it follows as in section 2 that:

e) *If* $z_1 \neq 0$ *on* H^* (*i.e. if* $\eta_{01}, \eta_{11}, \cdots, \eta_{n1}$ *are not all zero on* H^*), *then the surface* Φ' *is defined in* $\mathfrak{J}'$ *by the prime principal ideal* $\mathfrak{J}' \cdot t_1$, *and we have*

$$Q_R(H^*) \cong \mathfrak{J}'/\mathfrak{J}' \cdot t_1. \tag{5}$$

A similar result holds if $z_2 \neq 0$ *on* H^*, *in which case* t_1 *is to be replaced by* t_2.

f) *The elements* t_1, t_3 *and also the elements* t_2, t_3 *are uniformizing parameters of* $g'_P(V')$. *If* $H' = P'$ *is a point of* g'_P *with* $H^* = P^*$ *as the corresponding point on* R, *and if* $z_1 \neq 0$ *at* P^*, *then* (t_1, t'_2, t_3) *are uniformizing parameters of* $P'(V')$, *where* t'_2 *is any element of* $\mathfrak{J}'$ *which in the homomorphism*[4] $\mathfrak{J}' \sim Q_{g_P}(P^*)$ *is mapped onto a uniformizing parameter of* $P^*(g_P)$.

6. The effect of a monoidal transformation on the multiple points of F

We now study the birational transformation T from F to F' induced by the monoidal transformation T_1. To the curve Δ there will correspond on F' an algebraic curve $T[\Delta]$, which may be reducible. The total intersection of F' and Φ' may contain, in addition to $T[\Delta]$, also a finite number of curves g'_P and a finite number of isolated points P'. These isolated points necessarily are

[4] We note that $\mathfrak{J}' \sim Q_R(H^*)$, by (5), and that if H^* is a point P^* on g_P then $Q_R(P^*) \sim Q_{g_P}(P^*)$.

singular for V' (and hence belong to N'), for at any common point of F' and Φ' which is simple for V' the intersection of F' and Φ' is locally pure 1-dimensional. As to the curves g'_P which may possibly occur in the total intersection of F' and Φ', we shall see later on that they necessarily correspond to certain special points P of Δ (see Lemmas 6.2 and 6.5).

Let H' be an irreducible component or a point of $T[\Delta]$ and let H and H^* denote respectively the locus on F and R which corresponds to H'. We denote by $\mathfrak{J}'$ the quotient ring $Q_{V'}(H')$. We put $\omega' = \omega/t_1^\nu$, where $\omega \in Q_V(H)$ [see (1) and (1')].

LEMMA 6.1. *If* $z_1 \neq 0$ *on* H^*, *then the principal ideal* $\mathfrak{J}' \cdot \omega'$ *is prime and defines the irreducible surface* F'.

The proof is the same as that of Lemma 3.1.

LEMMA 6.2. *If P is a point of Δ which is simple for Δ (and also for V) and if the curve g'_P lies on F', then necessarily* $m_F(P) > m_F(\Delta)$.

PROOF. If $\nu = m_F(\Delta)$ and if t_1, t_2, t_3 are uniformizing parameters of $P(V)$ such that t_1, t_2 are uniformizing parameters of $\Delta(V)$, then ω [$= f(t_1, t_2, t_3)$] can be expressed as a form $g(t_1, t_2)$, of degree ν, with coefficients in the ring $Q_V(P)$. Assuming that $z_1 \neq 0$ on the generator g_P of $\dot{R}$, it follows from d), section 5, that $g(t_1, t_2)/t_1^\nu \in Q_{V'}(g'_P)$. If we now assume that $g'_P \in F'$, then the element $g(t_1, t_2)/t_1^\nu$ must be a non-unit in $Q_{V'}(g'_P)$. That implies $g^*(z_1, z_2) = 0$, where the coefficients of g^* are the P-residues of the coefficients of g. Hence the coefficients of $g(t_1, t_2)$ are non units in $Q_V(P)$, whence ω can be written as a form in t_1, t_2, t_3, with coefficients in $Q_V(P)$, *of degree greater than* ν. Hence $m_F(P) > \nu$, q.e.d.

LEMMA 6.3. *If P is a simple point of Δ such that $m_F(P) = m_F(\Delta)$ and if P' is a point of F' which corresponds to P, then* $m_{F'}(P') \leqq m_F(P)$.

PROOF. Let $\nu = m_F(P) = m_F(\Delta)$, $\nu' = m_{F'}(P')$, and let P^* be the point of R which corresponds to P' [see b), section 5]. Let $\mathfrak{J}$, $\mathfrak{J}'$ and $\bar{\mathfrak{J}}$ denote respectively the quotient rings $Q_V(P)$, $Q_{V'}(P')$ and $Q_R(P^*)$, and let $\mathfrak{m}$, $\mathfrak{m}'$ and $\bar{\mathfrak{m}}$ denote the ideal of non units in these rings. Assuming that $z_1 \neq 0$ at P^*, we have:

$$\omega' = \omega/t_1^\nu = g(t_1, t_2)/t_1^\nu \equiv 0(\text{mod } \mathfrak{m}'^{\nu'}).$$

By the homomorphism $\mathfrak{J}' \sim \bar{\mathfrak{J}}$ [see (5)] the element ω' is mapped into the element $\bar{\omega} = G(z_1, z_2)/z_1^\nu$, and we have

$$\bar{\omega} = G(z_1, z_2)/z_1^\nu \equiv 0(\bar{\mathfrak{m}}^{\nu'}). \tag{6}$$

Let now $\mathfrak{J}^* = Q_{g_P}(P^*)$ and let $\mathfrak{m}^*$ be the ideal of non units in $\mathfrak{J}^*$. From (6) follows (see footnote[4]):

$$\omega^* = g^*(z_1, z_2)/z_1^\nu \equiv 0(\mathfrak{m}^{*\nu'}). \tag{7}$$

Since $m_F(P) = m_F(\Delta)$, $g^*(z_1, z_2)$ is not identically zero and is a form of degree ν with coefficients in k^* (k^* = residue field of P). Hence if we put $z_2/z_1 = x$, then ω^* is a polynomial in $k^*[x]$, of degree $\leqq \nu$, while $\mathfrak{J}^*$ is the quotient ring of a prime ideal in $k^*[x]$. Consequently, by (7), we conclude that $\nu' \leqq \nu$, q.e.d.

Lemma 6.4. *If H' is an irreducible component of $T[\Delta]$, then $m_{F'}(H') \leqq m_F(\Delta)$.*

Proof. In this case, if K is taken as ground field, then H^* is a *point* of R (over K) and the lemma follows, as in the preceding lemma, from the fact that $G(z_1, z_2)$ is a form of degree ν.

Lemma 6.5. *If $\nu = m_F(\Delta)$, all but a finite number of points of Δ are exactly ν-fold for F.*

Proof. The surface F is defined in $Q_V(\Delta)$ by a principal ideal (ω), where ω is a form $g(t_1, t_2)$ of degree ν with coefficients in $Q_V(\Delta)$. Here t_1, t_2 are uniformizing parameters of $\Delta(V)$, and we may suppose that they are polynomials in the non-homogeneous coördinates ξ_i of the general point of V. If $\mathfrak{o}$ denotes the ring $k[\xi_1, \xi_2, \cdots, \xi_n]$ (we assume that V is immersed in an S_n^k), we consider the ideal $\mathfrak{o}\cdot(t_1, t_2)$ and the points P of Δ satisfying the following conditions:

1) P is at finite distance (with respect to the above non-homogeneous coördinates ξ_i) and is simple for V and for Δ.

2) The coefficients of $g(t_1, t_2)$ are in $Q_V(P)$ and are not zero at P.

It is clear that the number of points P on Δ which do not satisfy both conditions 1) and 2), is finite. Now we show that if conditions 1) and 2) are satisfied, then $m_F(P) = \nu$.

Since P is simple for Δ, there exists an element t_3 in $Q_V(P)$ such that t_1, t_2, t_3 are uniformizing parameters of $P(V)$. But then, in view of condition 2, $g(t_1, t_2)$ is a form *in the uniformizing parameters* of P (although t_3 does not actually occur in the form), with coefficients in $Q_V(P)$ which are not all zero at P. Our lemma will therefore follow if we show that ω remains prime in $Q_V(P)$. Suppose that ω factors in $Q_V(P)$, say $\omega = \omega_1\omega_2$. Since ω is prime in $Q_V(\Delta)$, one of the factors, say ω_1, must be a unit in $Q_V(\Delta)$, and the other factor ω_2 must be divisble by the ideal $(t_1, t_2)^\nu$, in $Q_V(P)$. But then ω_1 must also be a unit in $Q_V(P)$, for otherwise all the coefficients of $g(t_1, t_2)$ would be zero at P.

Theorem 2. *If $T[\Delta]$ possesses an irreducible component which is ν-fold for F', then $T[\Delta]$ itself is an irreducible curve, birationally equivalent to Δ, and the birational correspondence between Δ and $T[\Delta]$ is regular at each simple point P of Δ, provided $m_F(P) = m_F(\Delta)$.*

Proof. Let Δ' be an irreducible component of $T[\Delta]$ which is ν-fold for F' and let Δ^* be the corresponding curve on R. The curves on R which correspond to the various components of $T[\Delta]$ (in the birational correspondence between Φ' and R), *regarded as points over* K, are given by the zeros of the form $G(z_1, z_2)$. This form in $K[z_1, z_2]$ is of degree ν. If Δ' is ν-fold for F', then Δ^* must correspond to a ν-fold root of $G(1, x)$, where $x = z_2/z_1$. Hence $G(z_1, z_2)$ must be the νth power of a linear form in $K[z_1, z_2]$, and this shows that $T[\Delta]$ itself is irreducible, $T[\Delta] = \Delta'$, and that the field of rational functions on Δ' coincides with K, i.e. Δ' is birationally equivalent to Δ.

Let now P and P' be corresponding points of Δ and of Δ', where we assume that $m_F(P) = m_F(\Delta)$ and that P is simple for Δ. To complete the proof of the theorem we have to show that $Q_\Delta(P) = Q_{\Delta'}(P')$. Since P is simple for Δ we may assume that the uniformizing parameters t_1, t_2 of $\Delta(V)$ together with some element t_3 form a set of uniformizing parameters of $P(V)$. Then the coefficients of $g(t_1, t_2)$ are in $Q_V(P)$ and are not all zero at P, since $m_F(P) = m_F(\Delta)$ (see proof of Lemma 6.5). Therefore the coefficients of $G(z_1, z_2)$ are in $Q_\Delta(P)$ and are not all zero at P. By the first part of the proof we have:

$$G(z_1, z_2) = (a_2 z_1 - a_1 z_2)^\nu, \qquad a_1, a_2 \in K,$$

and now we know that a_1 and a_2 are both in $Q_\Delta(P)$ and are not both zero at P. Let, say, $a_2 \neq 0$ at P and let a_1^*, a_2^* be the P-residues of a and b. In the regular birational correspondence between Φ' and R let Δ^* be the irreducible curve on R which corresponds to Δ' and let P^* be the point of Δ^* which corresponds to P'. Then it is clear that $a_2 z_1 - a_1 z_2 = 0$ on Δ^* and that $z_1 : z_2 = a_1^* : a_2^*$ at P^*. The quotient ring $Q_R(P^*)$ consists of all quotients $A(z_1, z_2)/B(z_1, z_2)$ of forms of like degree, with coefficients in $Q_\Delta(P)$, such that $B(a_1^*, a_2^*) \neq 0$. Since $z_1 : z_2 = a_1 : a_2$ on Δ^*, it follows that $Q_{\Delta^*}(P^*)$ consists of all quotients $A(a_1, a_2)/B(a_1, a_2)$, such that $B(a_1^*, a_2^*) \neq 0$. That shows that $Q_\Delta(P) = Q_{\Delta^*}(P^*)$, and since the birational correspondence between Φ' and R is regular at P^*, we also have $Q_{\Delta^*}(P^*) = Q_{\Delta'}(P')$. This completes the proof.

COROLLARY. *If T transforms the ν-fold curve Δ of F into a ν-fold curve Δ' of F', then simple points of Δ which are ν-fold for F go into simple points of Δ'. In particular, if Δ is free from singularities and if no point of Δ is of higher multiplicity than ν for F, then Δ' is free from singularities.*

7. Further remarks concerning singular points of F whose multiplicity is not lowered by a quadratic or by a monoidal transformation

We have seen in the preceding section (Theorem 2) that if a ν-fold curve Δ of F is transformed by the monoidal transformation T of center Δ into a ν-fold curve Δ' of F', then the two curves Δ and Δ' are birationally equivalent. We can express this result by saying that:

a) *residue field of Δ = residue field of Δ'.*

Now consider the case of a ν-fold point P of F belonging to the ν-fold curve Δ, and let us assume that a point P' of F' which corresponds to P is also ν-fold for F'. From the proof of Lemma 6.3 we deduce that the form $g^*(z_1, z_2)$ must be the νth power of a linear form in $k^*[z_1, z_2]$, where k^* is the residue field of P. Hence if P^* denotes the point of R which corresponds to P', then the P^*-residue of the ratio z_2/z_1 is an element of k^*, since $g^*(z_1, z_2) = 0$ at P^*. This shows that the residue field of P^*, and hence also the residue field of P', coincides with the residue field of P. Moreover this also shows that P' is the only point of $T[\Delta]$ which corresponds to P. Hence

b) *residue field of P = residue field of P', and $T[P] = P'$.*

We now go back to the case of a quadratic transformation T, studied in section 3. We suppose that $T[P]$ carries a point P' which is ν-fold for F', where $\nu = m_F(P)$. This point P' must then correspond to a ν-fold point P of the curve (3) (see proof of Lemma 3.2). The conditions under which a plane curve of order ν can possess a ν-fold point are elicited in the following lemma:

Lemma 7.1. *If $F(z_1, z_2, z_3)$ is a form of degree ν in $K[z_1, z_2, z_3]$, different from zero, and if the plane curve $F(z_1, z_2, z_3) = 0$ has a ν-fold point P, then the curve consists of ν (not necessarily distinct) lines through P.*

Proof. We assume that $z_3 \neq 0$ at P and we use the non-homogeneous coordinates $x = z_1/z_3$, $y = z_2/z_3$. Let $F(x, y, 1) = f(x, y)$. Our lemma asserts that if the point P is ν-fold for the curve $f(x, y) = 0$, then the polynomial $f(x, y)$ is either the νth power of a linear polynomial in $K[x, y]$ or is a form of degree ν in two such linear polynomials which both vanish at P. Thus the lemma implies at any rate that there exists a linear polynomial which is zero at P. On the other hand, if we can show that there exists a linear polynomial which is zero at P, then the proof of the lemma can be readily completed as follows. Without loss of generality we may assume that $x = 0$ at P. Let S denote the (x, y)-plane and let l denote the line $x = 0$. We put $\mathfrak{o} = Q_S(P)$, $\mathfrak{o}_1 = Q_l(P)$ and we denote by $\mathfrak{m}$ and $\mathfrak{m}_1$ respectively the ideals of non-units in $\mathfrak{o}$ and $\mathfrak{o}_1$. We have by hypothesis: $f(x, y) \equiv 0(\mathfrak{m}^\nu)$. Hence if we put $f_0(y) = f(0, y)$, we also have: $f_0(y) \equiv 0(\mathfrak{m}_1^\nu)$. If $f_0(y) \neq 0$, then the congruence $f_0(y) \equiv 0(\mathfrak{m}_1^\nu)$ is possible only if $f_0(y) = (c_0 y + c_1)^\nu$, $c_0, c_1 \in K$, where $c_0 y + c_1 = 0$ at P. Without loss of generality we may assume that $f_0(y) = y^\nu$. The point P is then the origin $x = y = 0$ and $f(x, y)$ is necessarily a form of degree ν in x and y. Suppose now that $f_0(y) = 0$. Then $f(x, y) = xf_1(x, y)$, and since $x \not\equiv 0(\mathfrak{m}^2)$, it follows that $f_1(x, y) \equiv 0(\mathfrak{m}^{\nu-1})$. Since f_1 is of degree $\nu - 1$, the proof is completed by induction with respect to ν.

In order to prove the existence of a linear polynomial in $K[x, y]$ which vanishes at P, we use formal differentiation. In order to include fields of arbitrary characteristic, we shall use the well-known operator:

$$D_{ij}^{(m)} = \frac{1}{i!\,j!} \cdot \frac{\partial^{(m)}}{\partial x^i \partial y^j}, \quad i + j = m.$$

It is immediately seen that if $f(x, y) \equiv 0(\mathfrak{m}^\nu)$, then $D_{ij}^{(m)} f(x, y) \equiv 0(\mathfrak{m}^{\nu-m})$. Since $D_{ij}^{(m)} f(x, y)$ is of degree $\leqq \nu - m$, the existence of at least one linear polynomial which vanishes at P is assured (by induction on ν), provided not all the partial derivatives $D_{ij}^{(m)}$ of $f(x, y)$, $m = 1, 2, \cdots, \nu - 1$, vanish identically. In the contrary case we must have $\nu = p^e$ and f must be a polynomial of the form

$$f(x, y) = c_0 + c_1 x^{p^e} + c_2 y^{p^e}, \tag{8}$$

where we may assume that $c_2 \neq 0$.

Let $\bar{x}$, $\bar{y}$ be the coördinates of P in a suitable algebraic extension field of K. Let $g(x)$, $h(x, y)$ be polynomials in $K[x, y]$ defined by the following conditions: 1) the leading coefficient of $g(x)$ is 1 and also the leading coefficient of $h(x, y)$, regarded as polynomial in y with coefficients in $k[x]$, is 1; 2) $g(x)$ is irreducible in $K[x]$ and $g(\bar{x}) = 0$; 3) the degree of $h(x, y)$ in x is less than the degree of $g(x)$; 4) $h(\bar{x}, \bar{y}) = 0$ and $h(\bar{x}, y)$ is irreducible in $K(\bar{x})[y]$. Then it is known ([8], p. 541) that $g(x)$ and $h(x, y)$ are not only uniformizing parameters of P, but also form a base for the prime ideal of P in $K[x, y]$. If $g(x)$ is of degree 1, then there is nothing to prove. We therefore assume that $g(x)$ is of degree greater than 1.

From our hypothesis that $f(x, y) \equiv 0(\mathfrak{m}^{\nu})$ follows immediately that $f(\bar{x}, y)$ must be divisible by $h(\bar{x}, y)^{\nu}$ (in $k(\bar{x})[y]$). Since $f(\bar{x}, y) \neq 0$, it follows that $h(x, y)$ must be of degree 1 in y, say

$$h(x, y) = y + A(x).$$

Consider the following polynomial in $K[x]$:

$$B(x) = f(x, y) - c_2[y + A(x)]^{p^e}.$$

Since the polynomial on the right belongs to the ideal $\mathfrak{m}^{\nu}$ and since $B(x)$ depends only on x, we conclude that $B(x)$ is divisible by $[g(x)]^{p^e}$. Since the degree of $A(x)$ is less than the degree of $g(x)$ and since by hypothesis the degree of $g(x)$ is greater than 1, comparison of the degrees of $B(x)$ and of $[g(x)]^{p^e}$ shows that $B(x)$ must be identically zero. Hence

$$c_0 + c_1x^{p^e} + c_2y^{p^e} = c_2y^{p^e} + c_2[A(x)]^{p^e}.$$

From this we conclude that $A(x)$ is linear, say $A(x) = d_0x + d_1$. Hence, $h(x, y)$ is a linear polynomial which vanishes at P. This completes the proof of the lemma.

In applying this lemma to the ν-fold point P of the curve (3) we observe, in the case of the lemma, that if there exist two independent linear polynomials in $K[x, y]$ which vanish at P, then the coordinates $\bar{x}$, $\bar{y}$ of P are in K. Since the coefficients of the $F(z_1, z_2, z_3)$ in (3) are in the residue field K of P, we conclude with the following theorem:

THEOREM 3. *If P is a ν-fold point of F and if among the points of F' which correspond to P in the quadratic transformation of center P there is one, say P', which is also ν-fold for F', then either $T[P]$ is an irreducible rational curve free from singularities (and—in fact—is equivalent to a straight line over K under a regular birational transformation) or $T[P]$ consists of several such curves through P', and then the residue field of P' coincides with the residue field of P.*

In the sequel we shall find it convenient to use the following phraseology: we shall say that the multiplicity of the center H of a quadratic or of a monoidal transformation $T(H \subset F)$ is *uniformly reduced* by T, if for any two corresponding points P and P' of F and F' respectively, such that P is a simple point of H and $m_F(P) = m_F(H)$, it is true that $m_{F'}(P') < m_F(P)$. (In the case of a quadratic transformation P is H itself and the condition $m_F(P) = m_F(H)$ is vacuous). In the problem of the reduction of the singularities of F it is true that the reduction process suffers a setback whenever at some step of the process a quadratic or monoidal transformation T fails to reduce uniformly the multiplicity of its own center. However, the results a), b) and Theorem 3 of this section show that, with one exception (see Theorem 3), the setback is always compensated, to a certain degree, by *the invariance of the residue field* of the point or of the curve under consideration. This compensation is brought about by the fact that when the residue field is invariant the transformation T operates *locally* very much in the same fashion as it would if the ground field k was algebraically closed. We proceed to develop this idea.

1) Let first T be a monoidal transformation of center Δ, and let us suppose that some irreducible component Δ' of $T[\Delta]$ has the same residue field as Δ (i.e. that Δ and Δ' are birationally equivalent). Let t_1 and t_2 be uniformizing parameters of $\Delta(V)$. By d), section 5, either t_2/t_1 or t_1/t_2 belongs to $Q_{V'}(\Delta')$. Let, say, $t_2/t_1 \in Q_{V'}(\Delta')$. Since Δ and Δ' have the same residue field, there exists an element ω in $Q_V(\Delta)$ which has the same Δ'-residue as t_2/t_1, and hence $(t_2 - t_1\omega)/t_1$ is a non unit in $Q_{V'}(\Delta')$. It is clear that also t_1 and $t_2 - t_1\omega$ are uniformizing parameters of $\Delta(V)$. Hence we may replace t_2 by $t_2 - t_1\omega$ and we may therefore assume that t_2/t_1 is a non unit in $Q_{V'}(\Delta')$. Using the results of section 5, especially d), we now conclude as follows:

If we put

$$t_1' = t_1, \qquad t_2' = t_2/t_1, \tag{9}$$

then t_1' and t_2' are uniformizing parameters of $\Delta'(V')$, and the quotient ring $Q_{V'}(\Delta')$ consists of all quotients $\phi(t_1', t_2')/\psi(t_1', t_2')$, where ϕ and ψ are polynomials with coefficients in $Q_V(\Delta)$ and where the "constant term" of ψ is a unit in $Q_V(\Delta)$.

The equations (9) can be regarded *locally* (i.e. in regard to the pair of corresponding curves Δ and Δ') as the equations of T. The form of these equations is the same as that of a quadratic transformation between two planes (t_1, t_2) and (t_1', t_2'), over an algebraically closed ground field.

2) The transformation T still being monoidal, with center Δ, let P be a simple point of Δ and let us suppose that a point P' of F' which corresponds to P is such that its residue field is the same as that of P. Let t_1, t_2, t_3 be uniformizing parameters of $P(V)$ such that t_1, t_2 are uniformizing parameters of $\Delta(V)$. By d), section 5, we may assume that $t_2/t_1 \in Q_{V'}(P')$, and then we can conclude as in the preceding case that for a suitable choice of t_2 the quotient t_2/t_1 is a non unit in $Q_{V'}(P')$. Using the results of section 5, especially d) and f), we conclude as follows:

If we put

$$t_1' = t_1, \qquad t_2' = t_2/t_1, \qquad t_3' = t_3, \tag{10}$$

then t_1', t_2' and t_3' are uniformizing parameters of $P'(V')$ and the quotient ring $Q_{V'}(P')$ consists of all quotients $\phi(t_1', t_2', t_3')/\psi(t_1', t_2', t_3')$, where ϕ and ψ are polynomials with coefficients in $Q_V(P)$ and where the "constant term" of ψ is a unit in $Q_V(P)$.

Here again, the equations (10) can be regarded locally (i.e. in regard to the pair of corresponding points P and P') as the equations of T. The form of these equations is the same as that of a monoidal transformation between two affine three-dimensional spaces (t_1, t_2, t_3) and (t_1', t_2', t_3'), over an algebraically closed field, the center being the line $t_1 = t_2 = 0$. The points P and P' play the role of the origins $(t) = 0$, $(t') = 0$.

3) We finally consider the case of a quadratic transformation T with center P, and we assume that $T[P]$ contains a point P' with the same residue field as P. By the same reasoning as in the preceding two cases and making use of the results of section 2, we conclude as follows:

For a suitable choice of the uniformizing parameters t_1, t_2, t_3 *of* $P(V)$, *the elements*

$$(11) \qquad t_1' = t_1, \qquad t_2' = t_2/t_1, \qquad t_3' = t_3/t_1$$

are uniformizing parameters of $P'(V')$. *The quotient ring* $Q_{V'}(P')$ *consists of all quotients* $\phi(t_1', t_2', t_3')/\psi(t_1', t_2', t_3')$, *where* ϕ *and* ψ *are polynomials with coefficients in* $Q_V(P)$ *and where the "constant term" of* ψ *is a unit in* $Q_V(P)$.

The form of the equations (11) is typical for a quadratic transformation between two three-dimensional affine spaces, over an algebraically closed field, the center of the transformation being the origin $t_1 = t_2 = t_3 = 0$.

8. Normal crossings

Let Δ be an irreducible curve on V and let P be a simple point of Δ. The prime ideal of Δ in $Q_V(P)$ has then a base consisting of two elements ω_1 and ω_2 whose leading forms are linear and linearly independent. That means that if t_1, t_2, t_3 are uniformizing parameters of $P(V)$, and if k^* denotes the residue field of P, then $\omega_i = f_i(t_1, t_2, t_3)$, where $f_i(z_1, z_2, z_3)$ is a linear form with coefficients in $Q_V(P)$ and where $f_1^*(z_1, z_2, z_3)$ and $f_2^*(z_1, z_2, z_3)$ are linearly independent over k^*. Here $f_i^*(z_1, z_2, z_3)$ denotes, as usual, the form in $k^*[z_1, z_2, z_3]$ whose coefficients are the P-residues of the coefficients of $f_i(z_1, z_2, z_3)$. The two forms f_1^*, f_2^* define a point in the projective plane $S_2^{k^*}$, or also a direction through the origin $z_1 = z_2 = z_3 = 0$ in an affine three-dimensional space over k^*. It is the *tangential direction of* Δ *at the point* P.

If P is a common point of two irreducible curves Δ and $\bar{\Delta}$ we shall say that P is a *normal crossing* of Δ and $\bar{\Delta}$ if P is simple point of both curves and if the tangential directions of Δ and $\bar{\Delta}$ at P are distinct.

If Δ and $\bar{\Delta}$ are two ν-fold curves on F we shall say that P is a normal crossing *on* F if P is a normal crossing and *if it is exactly* ν*-fold for* F.

LEMMA 8.1. *If* P *is a normal crossing of two irreducible curves* Δ *and* $\bar{\Delta}$, *then there exist uniformizing parameters* t_1, t_2, t_3 *of* $P(V)$ *such that* (t_1, t_2) *is a basis for the prime ideal of* Δ *in* $Q_V(P)$ *and* (t_2, t_3) *is a basis for the prime ideal of* $\bar{\Delta}$ *in* $Q_V(P)$.

PROOF. Let τ_1, τ_2 and τ_3, τ_4 be bases respectively for the ideals of Δ and $\bar{\Delta}$ in $Q_V(P)$. Since P is a normal crossing, three of the leading forms of the 4 elements τ_i must be linearly independent. Assuming that the leading forms of τ_1, τ_2, τ_3 are linearly independent, we will then have that τ_1, τ_2, τ_3 are uniformizing parameters of $P(V)$. We therefore can write $\tau_4 = a_1\tau_1 + a_2\tau_2 + a_3\tau_3$, $a_i \in Q_V(P)$. The elements a_1 and a_2 cannot be both zero at P, for otherwise the leading forms of τ_3 and τ_4 would be linearly dependent, contrary to the assumption that P is a simple point of $\bar{\Delta}$. Let, say, $a_2 \neq 0$ at P, and let us put $t_1 = \tau_1$, $t_2 = a_1\tau_1 + a_2\tau_2$, $t_3 = \tau_3$. Since a_2 is a unit in $Q_V(P)$, the elements

(t_1, t_2) form a basis for the prime ideal of Δ in $Q_V(P)$. On the other hand we have $t_2 = \tau_4 - a_3\tau_3$, and hence t_2 and t_3 form a basis for the prime ideal of $\bar{\Delta}$ in $Q_V(P)$, q.e.d.

Let now Δ and $\bar{\Delta}$ be two irreducible ν-fold curves of the surface F and let F' be the transform of F be the monoidal transformation of center Δ.

LEMMA 8.2. *If T carries Δ into a ν-fold curve Δ' of F' and if P is a normal crossing of Δ and $\bar{\Delta}$ on F, then the corresponding point P' of $\bar{\Delta}' = T[\bar{\Delta}]$ is a normal crossing of Δ' and $\bar{\Delta}'$ on F'.*

PROOF. We select the uniformizing parameters of $P(V)$ as in Lemma 8.1. By Theorem 2 of section 6 the transform $T[\Delta]$ is the irreducible curve Δ', and by b), section 7, there corresponds to P a unique point P' of Δ'. Since P' is ν-fold for F', the residue fields of P and P' coincide. We are therefore in position to apply the results of the preceding section. In order to decide which one of the two quotients t_1/t_2 and t_2/t_1 certainly belongs to $Q_{V'}(P')$, let us consider the principal ideal (ω) in $Q_V(P)$ which defines the surface F. Since both curves Δ and $\bar{\Delta}$ are ν-fold for F, ω must be expressable in the following form:

$$\omega = \omega_0 t_2^\nu + \omega_1 t_2^{\nu-1} t_1 t_3 + \cdots + \omega_\nu (t_1 t_3)^\nu, \tag{12}$$

where the ω_i are in $Q_V(P)$ and where ω_0 *is not zero at* P, since $m_F(P) = \nu$. Thus the leading form of ω is $\overset{*}{\omega}_0 z_2^\nu$, where $\overset{*}{\omega}_0$ is the P-residue of ω_0. Consequently [see f), section 5], $t_2/t_1 \in Q_{V'}(P')$, but t_1/t_2 is definitely not in $Q_{V'}(P')$. Therefore t_2/t_1 is a non unit in $Q_{V'}(P')$, and the equations (10) of the preceding section are valid.

The surface F' is defined in $Q_{V'}(P')$ by the principal ideal (ω'), where

$$\omega' = \omega/t_1^\nu = \omega_0 t_2'^\nu + \omega_1 t_2'^{\nu-1} t_3' + \cdots + \omega_\nu t_3'^\nu. \tag{12'}$$

The curve $\bar{\Delta}'$ is defined in $Q_{V'}(P')$ by the prime ideal (t_2', t_3') and is naturally ν-fold for F', as is also shown clearly by (12′). Therefore P' is a simple point of $\bar{\Delta}'$. As to Δ', we know that it lies on the surface Φ' $(= T_1[\Delta])$, and therefore [e), section 5], t_1' is one of the two generators of the prime ideal of Δ' in $Q_{V'}(P')$. Hence P' is a normal crossing of Δ' and $\bar{\Delta}'$ on F', as was asserted.

In the sequel we shall have occasion to follow up the monoidal transformation T_1, of center Δ, by a monoidal transformation T_1', of center $\bar{\Delta}'$. Now we could have first applied a monoidal transformation of center Δ' and then a monoidal transformation whose center is the transform of Δ. It turns out that interchanging the order in which the curves Δ and $\bar{\Delta}$ are used as centers of a monoidal transformation does not affect essentially the resulting composite transformation of the surface F, i.e. *to within a regular birational correspondence.* We proceed to state precisely the result.

We suppose then that Δ and $\bar{\Delta}$ are irreducible ν-fold curves of F and we apply to V a monoidal transformation T_1 of center Δ. We denote, as usual, by V' and F' the transforms of V and of F respectively, and by T—the transformation from F to F' induced by T_1. We suppose that $T[\Delta]$ is an irreducible ν-fold curve Δ' of F', and we denote the transform $T[\bar{\Delta}]$ by $\bar{\Delta}'$. We next apply to V'

a monoidal transformation $\bar{T}'_1$, of center $\bar{\Delta}'$, and we denote by V'' and F'' the transforms of V' and F' respectively, and by $\bar{T}'$—the transformation from F' to F'' which is induced by $\bar{T}'_1$. We suppose that $\bar{T}'[\bar{\Delta}']$ is irreducible and ν-fold for F''.

We now interchange the order in which the two curves Δ and $\bar{\Delta}$ have been used. We first apply to V a monoidal transformation $\bar{T}_1$ of center $\bar{\Delta}$, getting entities V^*, F^* and $\bar{T}$ whose meaning is similar to that of V', F' and T. Since T does not affect the quotient ring of $\bar{\Delta}$ we have $Q_F(\bar{\Delta}) = Q_{F'}(\bar{\Delta}')$, and since we have assumed that $\bar{T}'[\bar{\Delta}']$ is irreducible and ν-fold for F'', it follows therefore that also $\bar{T}[\bar{\Delta}]$ is irreducible and ν-fold for F^*. Let $\bar{T}[\bar{\Delta}]$ and $\bar{T}[\Delta]$ be denoted respectively by $\bar{\Delta}^*$ and Δ^*. We next apply to V^* a monoidal transformation T_1^* of center Δ^*, getting V^{**}, F^{**} and T^*. Since $\bar{T}$ does not affect the quotient ring of Δ, we have $Q_F(\Delta) = Q_{F^*}(\Delta^*)$, and since we have assumed that $T[\Delta]$ is irreducible and ν-fold for F', it follows that also $T^*[\Delta^*]$ is irreducible and ν-fold for F^{**}.

Lemma 8.3. *If P'' and P^{**} are two corresponding points of F'' and F^{**}, i.e. if P'' and P^{**} correspond to one and the same point P of F, then $Q_{F''}(P'') = Q_{F^{**}}(P^{**})$, except when P is a common point of Δ and $\bar{\Delta}$ which is not a normal crossing on F.*

Proof. Let P' and P^* be the points of F' and F^* respectively which correspond respectively to P'' and P^{**}. We shall first suppose that P is not a common point of Δ and $\bar{\Delta}$. If P does not lie on either curve, then all the transformations used above do not affect the quotient ring $Q_V(P)$ at all, and hence $Q_{V''}(P'') = Q_{V^{**}}(P^{**}) = Q_V(P)$. Assume now that $P \in \Delta$, $P \notin \bar{\Delta}$. Since $P \notin \bar{\Delta}$, we have $Q_F(P) = Q_{F^*}(P^*)$ and $P^* \in \Delta^*$. Therefore the monoidal transformations T and T^* are locally identical at P and P^* respectively, whence $Q_{F'}(P') = Q_{F^{**}}(P^{**})$. On the other hand, since $P \notin \bar{\Delta}$, we also have $P' \notin \bar{\Delta}'$, and therefore $Q_{F'}(P') = Q_{F''}(P'')$, and the assertion of the lemma is proved in this case.

Suppose now that P is a normal crossing, on F, of Δ and $\bar{\Delta}$. Locally, for the pair of corresponding points P and P', the equations of T_1 are given by (10), i.e.

$$t'_1 = t_1, \qquad t'_2 = t_2/t_1, \qquad t'_3 = t_3. \tag{13'}$$

We know already that in $Q_{V'}(P')$ the curve $\bar{\Delta}'$ is given by the prime ideal (t'_2, t'_3). From the form (12′) of the element ω' and from the fact that $\omega_0 \neq 0$ at P' (since $\omega_0 \neq 0$ at P), we conclude that $t'_2/t'_3 \in Q_{V''}(P'')$. Since the residue fields of P and P'' are the same, we can find an element a in $Q_V(P)$ such that $t'_2/t'_3 - a$ be a non unit in $Q_{V''}(P'')$. Now note that $(t_1, t_2) = (t_1, t_2 - at_1t_3)$ and $(t_2, t_3) = (t_2 - at_1t_3, t_3)$, all ideals being in $Q_V(P)$. Hence we may replace t_2 by $t_2 - at_1t_3$. With this choice of t_2, t'_2/t'_3 will be a non unit in $Q_{V''}(P'')$ since $t'_2 - at'_3 = (t_2 - at_1t_3)/t_1$. Hence the equations of $\bar{T}'_1$ are as follows:

$$t''_1 = t'_1, \qquad t''_2 = t'_2/t'_3, \qquad t''_3 = t'_3, \tag{13''}$$

where t''_1, t''_2, t''_3 are uniformizing parameters of $P''(F'')$.

In a similar fashion we find that the equations of $\overline{T}_1$ are

$$(13^*) \qquad t_1^* = t_1, \qquad t_2^* = t_2/t_3, \qquad t_3^* = t_3,$$

where t_1^*, t_2^*, t_3^* are the uniformizing parameters of $P^*(V^*)$, and that the equations of T_1^* are as follows:

$$(13^{**}) \qquad t_1^{**} = t_1^*, \qquad t_2^{**} = t_2^*/t_1^*, \qquad t_3^{**} = t_3^*,$$

where t_1^{**}, t_2^{**}, t_3^{**} are uniformizing parameters of $P^{**}(V^{**})$. From (13′), (13″), (13*) and (13**) we conclude that

$$(14) \qquad t_1'' = t_1^{**} = t_1, \qquad t_2'' = t_2^{**} = t_2/t_1t_3, \qquad t_3'' = t_3^{**} = t_3,$$

and this shows the complete identity of the composite transformations $T_1\overline{T}_1'$ and $\overline{T}_1T_1^*$ locally, at (P, P'') and (P, P^{**}). This completes the proof of the Lemma.

Lemma 8.4. *Let Δ be an irreducible ν-fold curve of F and let P be a simple point of Δ which is exactly ν-fold for F. If a quadratic transformation T, of center P, transforms P into a ν-fold curve Γ' of F', then the transform Δ' of Δ intersects Γ' in one point only, and that point is a normal crossing of Δ' and Γ' on F'.*

Proof. We know from b), section 2, that the points of Φ' are in (1, 1) correspondence with the directions about P (since each direction about P is represented by a point of the projective plane S). Since P is simple for Δ, it follows that Δ' intersects Φ', and hence also Γ', in only one point. Let P' be the common point of Δ' and Γ'. If t_1, t_2, t_3 are uniformizing parameters of $P(V)$ such that t_2, t_3 are uniformizing parameters of $\Delta(V)$, then the equations of T_1 are given by (11), section 7, where t_1', t_2', t_3' are uniformizing parameters of $P'(V')$. The curve Δ' is given in $Q_{V'}(P')$ by the ideal (t_2', t_3'), while the surface Φ' is defined by the ideal (t_1'). Since $\Gamma' \subset \Phi'$ and Γ' has a simple point at P' (Theorem 1, section 3), and since moreover P' is exactly ν-fold for F', it follows that P' is a normal crossing of Δ' and Γ', as asserted.

Part II

Local Reduction Theorem for Surfaces Embedded in a V_3

9. Reduction of the singularities of an algebraic curve by quadratic transformations

In the sequel we shall have to make use of the theorem that the singularities of any higher space curve can be resolved by quadratic transformations of the ambient space. This theorem is well-known in the classical case. In this section we shall prove this theorem for arbitrary ground fields. It is clear that all depends on proving the following theorem:

Theorem 4. *If*

$$\Gamma_1, \Gamma_2, \cdots, \Gamma_i, \cdots$$

is an infinite sequence of irreducible algebraic curves (over a common ground field k) and if P_1, P_2, $\cdots$, P_i, $\cdots$ is a corresponding sequence of points such that: 1) P_i is a point of Γ_i; 2) Γ_{i+1} is obtained from Γ_i by a quadratic transformation T_i of center P_i; 3) the point P_{i+1} is one of the points of Γ_{i+1} which correspond to P_i

(there is only a finite number of such points), then if i is sufficiently high, P_i is a simple point of Γ_i.

PROOF. Let $\mathfrak{J}_i$ be the quotient ring $Q_{\Gamma_i}(P_i)$, so that we have

$$\mathfrak{J}_1 \subseteqq \mathfrak{J}_2 \subseteqq \cdots \subseteqq \mathfrak{J}_i \subseteqq \cdots,$$

and let Ω be the union of the rings $\mathfrak{J}_i$. Any non unit in $\mathfrak{J}_1$ is also a non unit in $\mathfrak{J}_2, \mathfrak{J}_3, \cdots$, and consequently Ω is a proper ring (i.e., not a field). Therefore Ω is contained in the valuation ring R_v of at least one valuation v of the field of algebraic functions of which the curves Γ_i are projective models.

LEMMA 9.1. *If $\Omega \subseteqq R_v$, then $\Omega = R_v$.*

PROOF. Let $\mathfrak{P}$ be the ideal of non units of R_v and let $\mathfrak{m}_i$ be the ideal of non units of $\mathfrak{J}_i$, so that $\mathfrak{P} \cap \mathfrak{J}_i = \mathfrak{m}_i$.

Let ξ_1/η_1 be an arbitrary element of R_v, where $\xi_1, \eta_1 \in \mathfrak{J}_1$. If $\eta_1 \not\equiv 0(\mathfrak{m}_1)$, then $\xi_1/\eta_1 \in \mathfrak{J}_1$. Suppose that $\eta_1 \equiv 0(\mathfrak{m}_1)$. Then necessarily also $\xi_1 \equiv 0(\mathfrak{m}_1)$. Considerations similar to those developed in section 2 show immediately that the extended ideal $\mathfrak{J}_2 \cdot \mathfrak{m}_1$ is a principal ideal, say $\mathfrak{J}_2 \cdot \mathfrak{m}_1 = \mathfrak{J}_2 \cdot \beta_2$, $\beta_2 \in \mathfrak{m}_2$ (see also [5], p. 679, formula (33)). Let $\xi_1 = \xi_2\beta_2$, $\eta_1 = \eta_2\beta_2$, where $\xi_2, \eta_2 \in \mathfrak{J}_2$, whence $\xi_1/\eta_1 = \xi_2/\eta_2$. If $\eta_2 \not\equiv 0(\mathfrak{m}_2)$, then $\xi_2/\eta_2 \in \mathfrak{J}_2$. In the contrary case we will have by a similar argument: $\xi_2 = \xi_3\beta_3$, $\eta_2 = \eta_3\beta_3$, $\beta_3 \equiv 0(\mathfrak{m}_3)$. Since $\beta_2 = 0(\mathfrak{m}_2)$, we have $v(\beta_2) > 0$, whence $v(\xi_1) > v(\xi_2) \geqq 0$ and $v(\eta_1) > v(\eta_2) \geqq 0$. Similarly, $v(\xi_2) > v(\xi_3) \geqq 0$ and $v(\eta_2) > v(\eta_3) \geqq 0$. Since the value group of v is the additive group of integers, it follows that if i is sufficiently high then $v(\eta_i) = 0$, and then we will have $\eta_i \not\equiv 0(\mathfrak{m}_i)$, $\xi_1/\eta_1 = \xi_i/\eta_i \in \mathfrak{J}_i$. Thus every element of R_v is contained in one of the quotient rings $\mathfrak{J}_i$, and this completes the proof of the lemma.

Each point P_i is the center of only a finite number of valuations (P_i is the origin of only a finite number of branches of Γ_i). From the preceding lemma we therefore may conclude as follows:

a) *If i is sufficiently high then R_v is the only valuation ring containing $\mathfrak{J}_i$* (i.e. P_i is the origin of only one branch of Γ_i).

The residue field $R_v/\mathfrak{P}$ of the valuation v is a finite extension of the ground field k, and on the other hand this residue field must be, by Lemma 9.1, the union of the residue fields $\mathfrak{J}_i/\mathfrak{m}_i$. Since these form an ascending sequence of fields, it follows that:

b) *If i is sufficiently high then $\mathfrak{J}_i/\mathfrak{m}_i = \mathfrak{J}_{i+1}/\mathfrak{m}_{i+1} = \cdots =$ residue field of v.*

From Lemma 9.1 it also follows that

c) *If i is sufficiently high then $\mathfrak{m}_i$ contains elements of value 1 in v; in other words: $R_v \cdot \mathfrak{m}_i = \mathfrak{P}$, for i sufficiently high.*

In view of a), b) and c), Theorem 4 is a consequence of the following lemma:

LEMMA 9.2. *A point P of an irreducible algebraic curve Γ is a simple point of Γ if the following conditions are satisfied:*[5]

[5] It is obvious that the conditions a), b) and c) are also necessary, since a point of an algebraic curve is a simple point of the curve if and only if the quotient ring of the point is a valuation ring.

a) *The point P is the center of only one valuation of the field of rational functions on* Γ.

b) *If v is the valuation of center P, then the residue field of P coincides with the residue field of the valuation v.*

c) *The quotient ring $Q_\Gamma(P)$ contains elements of value* 1.

Proof. Let $\mathfrak{J} = Q_\Gamma(P)$ and let $\bar{\mathfrak{J}}$ denote the integral closure of $\mathfrak{J}$ in its quotient field. Since $\bar{\mathfrak{J}}$ is the intersection of valuation rings, it follows by a) that $\bar{\mathfrak{J}} = R_v$. Since $\mathfrak{J}$ is the quotient ring of a prime ideal of a finite integral domain (namely of the ring generated over k by the non-homogeneous coördinates of the general point of Γ), it is known that $\bar{\mathfrak{J}}$ is a finite $\mathfrak{J}$-module (see F. K. Schmidt [3]). Hence the conductor $\mathfrak{C}$ of $\bar{\mathfrak{J}}$ with respect to $\mathfrak{J}$ is not the zero ideal. The conductor $\mathfrak{C}$ is therefore a primary ideal in $\mathfrak{J}$, whose associated prime ideal is the ideal $\mathfrak{m}$ of non-units in $\mathfrak{J}$; unless $\mathfrak{C} = \mathfrak{J}$, in which case $\mathfrak{J} = \bar{\mathfrak{J}} = R_v$, and P is then indeed a simple point of Γ. In either case, since $\mathfrak{J}$ is a chain-theorem ring, we will have $\mathfrak{m}^\rho \equiv 0(\mathfrak{C})$, for some integer ρ. Since $\mathfrak{C}$ is also an ideal in $\bar{\mathfrak{J}}$, the congruence $\mathfrak{m}^\rho \equiv 0(\mathfrak{C})$ implies, by c), that *every element of R_v whose value is $\geqq \rho$ is necessarily an element of* $\mathfrak{J}$. Let now t be an element of $\mathfrak{J}$ such that $v(t) = 1$, in view of c), and let ξ be an arbitrary non unit in $\mathfrak{J}$, whence $v(\xi) = m > 0$. By b), there exists an element α in $\mathfrak{J}$ whose P-residue coincides with the residue of ξ/t^m in the valuation v. Therefore, if we put

$$\xi = \alpha t^m + \xi_1,$$

then we will have $v(\xi_1) = m_1 > m$. In a similar fashion we can find an element α_1 in $\mathfrak{J}$ such that $v(\xi_1 - \alpha_1 t^{m_1}) = m_2 > m_1$, whence

$$\xi = \alpha t^m + \alpha_1 t^{m_1} + \xi_2, \qquad m < m_1 < m_2 = v(\xi_2).$$

Let

$$\xi = \alpha t^m + \alpha_1 t^{m_1} + \alpha_2 t^{m_2} + \cdots + \alpha_h t^{m_h} + \xi_{h+1}, \tag{15}$$

where

$$1 \leqq m < m_1 < m_2 < \cdots < m_h < m_{h+1} = v(\xi_{h+1})$$

and $\alpha, \alpha_1, \alpha_2, \cdots, \alpha_h, \xi_{h+1} \in \mathfrak{J}$. Let us take h so high that $m_{h+1} \geqq \rho + 2$. Then $v(\xi_{h+1}/t^2) \geqq \rho$, whence $\xi_{h+1}/t^2 \in \mathfrak{J}$ and consequently $\xi_{h+1} \equiv 0(\mathfrak{J}\cdot t^2)$. From (15) we therefore conclude that $\xi \equiv 0(\mathfrak{J}\cdot t)$. Since this holds for every element of $\mathfrak{m}$, it follows that $\mathfrak{m} = \mathfrak{J}\cdot t$, whence P is a simple point, with t as uniformizing parameter, q.e.d.

Let Γ be an irreducible ν-fold curve of a surface F, $\nu > 1$, and let $F^{(1)}$ be a transform of F by a monoidal transformation of center Γ. To Γ there will correspond on $F^{(1)}$ a finite number of irreducible curves. Suppose that one of these curves, say $\Gamma^{(1)}$, is still ν-fold for $F^{(1)}$. We apply to $F^{(1)}$ a monoidal transformation of center $\Gamma^{(1)}$, getting a surface $F^{(2)}$. If $F^{(2)}$ still contains a ν-fold curve $\Gamma^{(2)}$ which corresponds to $\Gamma^{(1)}$, we repeat the process. Suppose that we get in this fashion a sequence of consecutive ν-fold curves

$$\Gamma, \Gamma^{(1)}, \Gamma^{(2)}, \cdots, \Gamma^{(i)}, \cdots$$

all ν-fold for their respective carriers $F, F^{(1)}, F^{(2)}, \cdots, F^{(i)}, \cdots$, where $F^{(i+1)}$ is obtained from $F^{(i)}$ by a monoidal transformation of center $\Gamma^{(i)}$ and where $\Gamma^{(i)}$ corresponds to $\Gamma^{(i+1)}$ in that monoidal transformation.

THEOREM 4′. *The sequence* $\Gamma, \Gamma^{(1)}, \Gamma^{(2)}, \cdots$ *is necessarily finite.*

PROOF. Let ξ be an element of the field of rational functions on F such that the Γ-residue of ξ is a transcendental with respect to the ground field k. Over the extended ground field $k(\xi)$ the surface F becomes a curve, Γ becomes a point, and our successive *monoidal* transformations become quadratic. It is thus seen that Theorem 4′ is nothing but Theorem 4 in a different "dimension terminology."

10. A local reduction theorem for algebraic surfaces

Let F be an irreducible algebraic surface embedded in a V_3, and let v be a zero-dimensional valuation of the field of rational functions on F such that the center of v on F is a point P which is simple for V_3. Let P be a ν-fold point of F. We make the following assumption: *If P is not an isolated ν-fold point of F then it is either a simple point or a normal crossing of the total (possibly reducible) ν-fold curve of F.* By a *permissible transformation* of V_3 we shall mean either a quadratic transformation of center P (regardless whether P is or is not an isolated ν-fold point) or a monoidal transformation whose center is a ν-fold curve through P.

Let T_1 be a permissible transformation of V_3 and let V_3' and F' be the transforms of V_3 and F respectively, under T_1. Let T be the birational transformation between F and F' induced by T_1. *We suppose that the center P' of v on F' is still ν-fold* and we consider various possible cases.

a) *P is an isolated ν-fold point of F.* In this case T_1 is necessarily quadratic and any ν-fold curve through P' can only be a new ν-fold curve created by the transformation T, i.e. it must belong to $T[P]$. It then follows from Theorem 1, section 3, that if P' is not an isolated ν-fold point it is a simple point of the ν-fold curve of F'.

b) *P is a simple point of the ν-fold curve of F.* Let Δ be the irreducible component of the ν-fold curve which passes through P. If T_1 is quadratic, then only the following cases are possible: 1) P' is an isolated ν-fold point P'; 2) $T[\Delta]$ is the only irreducible ν-fold curve of F' which passes through P'; P' is a simple point of $T[\Delta]$ since P is simple for Δ; 3) $T[P]$ is ν-fold for F' and is the only irreducible ν-fold curve of F' passing through P'; 4) $T[P]$ is ν-fold for F', and P' belongs to both curves $T[P]$ and $T[\Delta]$; in this case P' is a normal crossing, by Lemma 8.4.

Now suppose that T_1 is a monoidal transformation of center Δ. If $T[\Delta]$ is not ν-fold for F', then P' is an isolated ν-fold point. If $T[\Delta]$ is ν-fold, then it is irreducible (Theorem 2, section 6), and P' must be a simple point of $T[\Delta]$ (Corollary to Theorem 2), and there are no other ν-fold curves through P'.

c) *P is a normal crossing of two ν-fold curves Δ and $\bar{\Delta}$.* If T_1 is a quadratic transformation, then P' cannot belong to both curves $T[\Delta]$ and $T[\bar{\Delta}]$ since Δ

and $\bar{\Delta}$ have at P distinct tangent lines. This case does not differ from case b) considered above. If T_1 is a monoidal transformation of center Δ, then in addition to the possibilities which have arisen in the preceding case b) we have also the possibility that P' is a normal crossing of the two ν-fold curves $T[\Delta]$ and $T[\bar{\Delta}]$ (see Lemma 8.2).

We conclude that in all cases the center P' of the valuation v on F' is again either an isolated ν-fold point, or a simple point of the ν-fold curve of F' or a normal crossing of the ν-fold curve of F'. All this provided that P satisfy these same conditions, that P' be a ν-fold point and that T_1 be a permissible transformation.

We next apply to V_3' and F' a permissible transformation T_1', relative to the point P', getting V_3'', F'' and P''. If P'' is still ν-fold for F'', we apply a permissible transformation T_1'' to V_3'', relative to the point P'', and so we continue indefinitely as long as the center of the valuation v remains a ν-fold point. In this fashion we get a sequence of surfaces F, F', F'', $\cdots$, $F^{(i)}$, $\cdots$, where $F^{(i+1)}$ is obtained from $F^{(i)}$ by a permissible transformation $T_1^{(i)}$ relative to the center $P^{(i)}$ of v on $F^{(i)}$ where it is supposed that $P^{(i)}$ is still ν-fold for $F^{(i)}$. We observe that whenever $P^{(i)}$ is not an isolated ν-fold point of its carrier $F^{(i)}$, then, in view of our definition of a permissible transformation, $T_1^{(i)}$ may be either quadratic or monoidal. We have therefore, in general, a certain degree of freedom in building up the sequence of successive transforms $F^{(i)}$

The main purpose of this second part of the paper is to establish the following *local reduction theorem*:

THEOREM 5. *By a suitably chosen sequence of permissible transformations it is possible to obtain a birational transform F^* of F on which the center of the valuation v is a point of multiplicity less than ν.*

This theorem naturally implies the theorem of uniformization of valuations which we have first proved in [5] for surfaces over an algebraically closed ground field of characteristic zero and later on, in [6], for higher varieties over arbitrary ground fields of characteristic zero. However, our present local reduction theorem is stronger than the pure existence theorem ("there exists a birational transform of F on which the center of the given valuation is a simple point") proved in the quoted papers, since it calls for a proof that the valuation can be uniformized by birational transformations of a *specified type*. Another difference is this: in the proof of the general uniformization theorem for surfaces, it was sufficient to carry out explicitly the uniformization for surfaces embedded in a linear 3-space. In the present case our surface is embedded in an arbitrary V_3 and it is essential that the uniformization be carried out by birational transformations of the ambient V_3.

The broad lines of the proof of Theorem 5 are essentially the same as those of our old proof of the uniformization of valuations. However, their adaptation to the new set-up calls for a technique in which the local character of the problem comes to the fore, for the controlling ring is now not a ring of polynomials in three variables but the quotient ring of a simple point P of V_3. The resulting reorganization of the proof actually implies an improvement on our old proof, especially when the ground field k is not algebraically closed.

In the following sections we take up the various types of zero-dimensional valuations and prove Theorem 5 for each type separately. As was pointed out in the Introduction, our proof of Theorem 5 for valuations of rank 1 applies only to fields of characteristic zero. For valuations of rank 2, our proof applies to arbitrary fields.

11. Some preliminary lemmas

Let V be an irreducible algebraic variety and let v be a zero-dimensional valuation of the field of rational functions on V. The valuation v determines an infinite sequence of successive birational transforms of V, say

$$V, V_1, V_2, \cdots, V_i, \cdots$$

where V_{i+1} is obtained by applying to V_i a quadratic transformation T_i whose center is the center P_i of v on V_i. Let Γ be an irreducible algebraic curve on V passing through P and let $T[\Gamma] = \Gamma_1$, $T_1[\Gamma_1] = \Gamma_2, \cdots, T_i[\Gamma_i] = \Gamma_{i+1}, \cdots$.

Lemma 11.1. *If $P_i \in \Gamma_i$ for all i, then v is compounded with a valuation whose center on V is the curve Γ.*

Proof. Let $\mathfrak{J}_i = Q_{V_i}(P_i)$, $\bar{\mathfrak{J}}_i = Q_{\Gamma_i}(P_i)$ and let $\mathfrak{m}_i$ and $\bar{\mathfrak{m}}_i$ be respectively the ideal of non units in $\mathfrak{J}_i$ and $\bar{\mathfrak{J}}_i$. The ideal $\mathfrak{J}_{i+1} \cdot \mathfrak{m}_i$ is principal [cf. section 2; also [5], p. 679, formula (33)], say $\mathfrak{J}_{i+1} \cdot \mathfrak{m}_i = \mathfrak{J}_{i+1} \cdot \beta_{i+1}$. If $\bar{\beta}_{i+1}$ is the Γ_{i+1}-residue of β_{i+1}, then $\bar{\beta}_{i+1} \neq 0$ and $\bar{\mathfrak{J}}_{i+1} \cdot \bar{\mathfrak{m}}_i = \bar{\mathfrak{J}}_{i+1} \cdot \bar{\beta}_{i+1}$. Let now ξ and η be elements of $\mathfrak{J}$ and let us assume that ξ is zero on Γ and $\eta \neq 0$ on Γ. We have then $\xi = \xi_1\beta_1$, $\xi_1 \in \mathfrak{J}_1$, $\xi_1 = 0$ on Γ_1. If η is a non unit in $\mathfrak{J}$ we can also write $\eta = \eta_1\beta_1$, $\eta_1 \in \mathfrak{J}_1$. If η_1 is a non unit in $\mathfrak{J}_1$, we define the element η_2 in $\mathfrak{J}_2$ by the relation $\eta_1 = \eta_2\beta_2$, and we also put $\xi_1 = \xi_2\beta_2$. More generally, if the elements $\eta_1, \eta_2, \cdots, \eta_s$ are definable by the above construction, and if η_s is a non unit in $\mathfrak{J}_s$, we put $\xi_s = \xi_{s+1}\beta_{s+1}$, $\eta_s = \eta_{s+1}\beta_{s+1}$. If $\bar{\xi}_i$, $\bar{\eta}_i$ denote the Γ_i-residues of ξ_i and η_i respectively, then $\bar{\xi}_i = 0$ and $\bar{\eta}_i = \bar{\eta}_{i+1}\bar{\beta}_{i+1}$ ($i = 1, 2, \cdots, s$). Since $\bar{\mathfrak{J}}_{i+1} \cdot \bar{\mathfrak{m}}_i = \bar{\mathfrak{J}}_{i+1} \cdot \bar{\beta}_{i+1}$, the proof of Lemma 9.1 shows that for some integers s we must get an element $\bar{\eta}_{s+1}$ which is a unit in $\bar{\mathfrak{J}}_{s+1}$. Then η_{s+1} is a unit in $\mathfrak{J}_{s+1}$, and consequently $v(\xi_{s+1}) > v(\eta_{s+1})$, since ξ_{s+1} is zero at P_{s-1}. Since $\xi_{s+1}/\eta_{s+1} = \xi/\eta$, we conclude that $v(\xi) > v(\eta)$. We have thus proved the following: *if* $\xi, \eta \in \mathfrak{J}$, $\xi = 0$ *on* Γ *and* $\eta \neq 0$ *on* Γ, *then* $v(\xi) > v(\eta)$. This shows that *the valuation v is of rank greater than* 1, for if η is such that $\eta = 0$ at P, $\eta \neq 0$ on Γ then $v(\eta) > 0$ and $v(\xi) > v(\eta^n)$, for any positive integer n.

Let m be the rank of v and let

$$\mathfrak{P} \supset \mathfrak{P}_1 \supset \cdots \supset \mathfrak{P}_{m-1} \neq (0)$$

be the strictly descending chain of the prime ideal of the valuation ring R_v, where $\mathfrak{P}$ is the ideal of non units in R_v. Since P is the center of v, we must have: $\mathfrak{P} \cap \mathfrak{J} = \mathfrak{m}$. On the other hand, $\mathfrak{P}_{m-1} \cap \mathfrak{J} \neq \mathfrak{m}$, since, as we have just seen, there exist pairs of elements ξ and η in $\mathfrak{m}$ such that $v(\xi)$ is greater than any positive multiple of $v(\eta)$; for such a pair of elements it is true that η cannot belong to $\mathfrak{P}_{m-1}$. Let therefore s be an integer, $1 \leqq s \leqq m - 1$, with the property:

$$\mathfrak{P}_{s-1} \cap \mathfrak{J} = \mathfrak{m}, \qquad \mathfrak{P}_s \cap \mathfrak{J} = \mathfrak{p} \neq \mathfrak{m}.$$

From the preceding considerations it follows that every element ξ of $\mathfrak{J}$ which vanishes on Γ must belong to $\mathfrak{P}_s$, hence also to $\mathfrak{p}$. Hence the prime ideal $\mathfrak{p}$ is at most one-dimensional, and since $\mathfrak{p} \neq \mathfrak{m}$ we conclude that $\mathfrak{p}$ *is the prime ideal of the curve* Γ. If we now consider the valuation v_1 of the field of rational functions on V whose valuation ring is the quotient ring of $\mathfrak{P}_s$ in R_v, then v is compounded with v_1 and from the fact that $\mathfrak{p}$ is the prime ideal of Γ it follows that Γ is the center of v_1, q.e.d.

The following is an application of the preceding lemma. Let us suppose that P is a simple point of V, whence also P_i is a simple point of V_i. Let $t_1, t_2, \cdots, t_r$ be uniformizing parameters of $P(V)$. The point P_1 corresponds to a definite tangential direction of V at P, i.e. to a definite point $\bar{P}$ of a projective $(r - 1)$-dimensional space S over the residue field of P (cf. section 8 for the case $r = 3$; the general case is treated exactly in the same fashion). If $z_1, z_2, \cdots, z_r$ are homogeneous coordinates in S and if, say, $z_1 \neq 0$ at $\bar{P}$, then $\mathfrak{J}_1 \cdot \mathfrak{m} = \mathfrak{J}_1 \cdot t_1$ [cf. section 2, d)]. Let ω be an element of $\mathfrak{J}$ and let us suppose that ω is exactly divisible by $\mathfrak{m}^\nu$. Then ω is divisible by t_1^ν in $\mathfrak{J}_1$, say $\omega = t_1^\nu \cdot \omega_1$. We call ω_1 the *transform of* ω *by the quadratic transformation* T, or briefly: *the* T*-transform of* ω.

If $\nu \geqq 1$ then the ideal $\mathfrak{J} \cdot \omega$ defines a pure $(r - 1)$-dimensional subvariety W of V which passes through P (W may be reducible and may possess multiple components). The ideal $\mathfrak{J}_1 \cdot \omega_1$ defines that part of the variety $T[W]$ which passes through P_1 (see Lemma 3.1).

We now suppose that V is a surface $(r = 2)$. Let $\omega_1, \omega_2, \cdots, \omega_i, \cdots$ be the successive transforms of ω by $T, T_1, \cdots$. Let $\Gamma^{(1)}, \Gamma^{(2)}, \cdots, \Gamma^{(h)}$ be the irreducible components of the algebraic curve defined on V by the ideal $\mathfrak{J} \cdot \omega$. Suppose that each ω_i is a non unit in the corresponding ring $\mathfrak{J}_i$, i.e. $\omega_i = 0$ at P_i. Then for at least one of the curves $\Gamma^{(i)}$, say for $\Gamma^{(1)}$, it must be true that $T_i[\Gamma^{(1)}]$ passes through P_i. But then, by Lemma 11.1, we conclude that the valuation v must be of rank 2 and must be compounded with a divisor whose center is the curve $\Gamma^{(1)}$. We therefore can state the following

LEMMA 11.2. *Let v be a zero-dimensional valuation of the field of rational functions on a given surface F, and let*

$$F_1, F_2, \cdots, F_i, \cdots$$

be the sequence of quadratic transforms of F determined by the valuation v [i.e. let F_{i+1} $(i \geqq 0)$ be the transform of F_i by a quadratic transformation whose center P_i on F_i is the center of v]. We assume that the center P of v on F is a simple point of F. Under the hypothesis that v is of rank 1, it is true that if ω is any element of $Q_F(P)$ and if $\omega_1, \omega_2, \cdots, \omega_i, \cdots$ are the successive transforms of ω by $T, T_1, T_2, \cdots$ (i.e., if ω_{i+1} is the T_i transform of ω_i), then for i sufficiently high the element ω_i is a unit in $Q_{F_i}(P_i)$.

In the next lemma, r is again arbitrary, although we shall actually use the lemma only for $r = 2, 3$.

Lemma 11.3. *If $\omega \in Q_V(P)$, where P is a simple point of V, and if $\omega_1, \omega_2, \cdots$ are the successive quadratic transforms of ω by $T, T_1, T_2, \cdots$, then*

$$\omega = \epsilon_i t_{1i}^{\rho_{1i}} t_{2i}^{\rho_{2i}} \cdots t_{ri}^{\rho_{ri}} \omega_i, \tag{16}$$

where ϵ_i is a unit in $Q_{V_i}(P_i)$ and where $t_{1i}, t_{2i}, \cdots, t_{ri}$ form a set of uniformizing parameters of $P_i(V_i)$, while the exponents ρ_{ji} are non-negative rational integers.

Proof. The lemma is true for $i = 1$, since if $\omega = t_1^{\nu}\omega_1$ then t_1 is one of a set of r uniformizing parameters of $P_1(V_1)$ [see section 2, e)]. We assume that the lemma is true for $i = n$ and we prove it for $i = n + 1$. Without loss of generality we may assume that

$$\omega_i = t_{1i}^{\nu_i}\omega_{i+1}, \tag{17}$$

and that t_{1i} is one of a set of r uniformizing parameters of $P_{i+1}(V_{i+1})$. We know from section 2 that the quotients t_{ji}/t_{1i}, $j = 2, 3, \cdots, r$, belong to $Q_{V_{i+1}}(P_{i+1})$. Without loss of generality we may assume that the first $s - 1$ of these quotients ($1 \leqq s \leqq r$) are non units, while the remaining quotients are units in $Q_{V_{i+1}}(P_{i+1})$. From the considerations developed in section 2 it follows immediately that the elements $t_{1i}, t_{2i}/t_{1i}, \cdots, t_{si}/t_{1i}$ form a subset of a set of uniformizing parameters of $P_{i+1}(V_{i+1})$. Hence we may put

$$t_{1,i+1} = t_{1i}, \qquad t_{2,i+1} = t_{2i}/t_{1i}, \cdots, t_{s,i+1} = t_{si}/t_{1i},$$

and we get from (16) and (17)

$$\omega = \epsilon_{i+1} t_{1,i+1}^{\rho_{1,i+1}} t_{2,i+1}^{\rho_{2,i+1}} \cdots t_{s,i+1}^{\rho_{s,i+1}} \omega_{i+1},$$

where

$$\rho_{1,i+1} = \rho_{1i} + \rho_{2i} + \cdots + \rho_{ri} + \nu_i,$$

$$\rho_{j,i+1} = \rho_{ji}, \qquad j = 2, 3, \cdots, s,$$

and where

$$\epsilon_{i+1} = \epsilon_i \prod_{j=s+1}^{r} (t_{ji}/t_{1i})^{\rho_{ji}},$$

whence ϵ_{i+1} is a unit in $Q_{V_{i+1}}(P_{i+1})$, q.e.d.

The next—and last—lemma of this section refers to a sequence of permissible birational transformations *all elements of which are quadratic transformations* (see section 10). Using the notations of section 10, let $\mathfrak{J} = Q_V(P)$, $\mathfrak{J}_i = Q_{V^{(i)}}(P^{(i)})$, and let the surface F be defined in $\mathfrak{J}$ to be the prime principal ideal $\mathfrak{J}\cdot\omega$.

Lemma 11.4. *If the points $P, P^{(1)}, P^{(2)}, \cdots$ are all ν-fold for their respective carriers $F, F^{(1)}, F^{(2)}, \cdots$ (all the successive transformations $T, T^{(1)}, T^{(2)}, \cdots$ being quadratic) and if the leading form of ω is not a νth power (of a linear form), then the* *valuation v is discrete (and is therefore defined by an algebraic or analytical branch through P).*

PROOF. Let t_1, t_2, t_3 be uniformizing parameters of $P(V)$ and let $F(z_1, z_2, z_3)$ be the leading form of ω; here F is a form of degree ν whose coefficients are in the residue field K of P. Since $P^{(1)}$ is a ν-fold point of $F^{(1)}$, and since, by hypothesis, F is not the νth power of a linear form, it follows from Lemma 7.1 and Theorem 3 (section 7), that F is a binary form of two linear forms in $K[z_1, z_2, z_3]$. Without loss of generality we may assume that $F = G(z_2 - a_1z_1, z_3 - b_1z_1)$, where where $a_1, b_1 \in K$ and G is a form with coefficients in K. The plane curve $F(z_1, z_2, z_3) = 0$ consists of straight lines through the point $z_1:z_2:z_3 = 1:a_1:b_1$, and since this curve is, by hypothesis, not a ν-fold line, the above point is the only ν-fold point of the curve and therefore represents the tangential direction of V at P which corresponds to the ν-fold point $P^{(1)}$ It follows that the local equations of the quadratic transformation T are as follows; [cf. equations (11), section 7]:

$$(18) \qquad t_1^{(1)} = t_1, \qquad t_2^{(1)} = \frac{t_2 - \alpha_1 t_1}{t_1} \qquad t_3^{(1)} = \frac{t_3 - \beta_1 t_1}{t_1},$$

where α_1 and β_1 are any two elements of $Q_\nu(P)$ whose P-residues are respectively a_1 and b_1 and where $t_1^{(1)}, t_2^{(1)}, t_3^{(1)}$ are uniformizing parameters of $P^{(1)}(V^{(1)})$. Let $g(u_1, u_2)$ be a form with coefficients in $\mathfrak{J}$ such that the reduced form mod P is $G(u_1, u_2)$. Then it is clear that ω and $g(t_2 - \alpha_1t_1, t_3 - \beta_1t_1)$ have the same leading form. Consequently

$$\omega = g(t_2 - \alpha_1t_1, t_3 - \beta_1t_1) + \zeta, \qquad \zeta \equiv 0(\text{mod } \mathfrak{m}^{\nu+1}),$$

where $\mathfrak{m}$ is the ideal of non units in $\mathfrak{J}$.

The surface $F^{(1)}$ is defined by the principal ideal $\mathfrak{J}_1\cdot\omega_1$, where ω_1 is the T-transform of ω. Using (18) we find

$$(19) \qquad \omega_1 = g(t_2^{(1)}, t_3^{(1)}) + t_1^{(1)}\zeta_1, \qquad \zeta_1 \in \mathfrak{J}_1.$$

Let $F_1(z_1^{(1)}, z_2^{(1)}, z_3^{(1)})$ denote the leading form of ω_1. By hypothesis, F_1 is of degree ν, since $P^{(1)}$ is ν-fold for $F^{(1)}$. Hence it follows from (19) that

$$(20) \qquad F_1(0, z_2^{(1)}, z_3^{(1)}) = G(z_2^{(1)}, z_3^{(1)}),$$

whence $F_1(z_1^{(1)}, z_2^{(1)}, z_3^{(1)})$ is not the νth power of a linear form. Therefore we have now at $P^{(1)}$ the same situation as at P. Since $P^{(2)}$ is also ν-fold for $F^{(2)}$, we conclude in view of (20) and recalling that the residue field of $P^{(1)}$ is the same as that of $P^{(1)}$ (Theorem 3, section 7), that $F_1(z_1^{(1)}, z_2^{(1)}, z_3^{(1)})$ must be a binary form of two linear forms in the $z_i^{(1)}$, all the coefficients being in K. In view of (20) these two linear forms can be assumed to be $z_2^{(1)} - a_2z_1^{(1)}, z_3^{(1)} - b_2z_1^{(1)}$, $a_2, b_2 \in K$. If α_2, β_2 are elements of $\mathfrak{J}$ whose residues are a_2 and b_2 respectively, then we have relations similar to (18):

$$(21) \qquad t_1^{(2)} = t_1^{(1)}, \qquad t_2^{(2)} = \frac{t_2^{(1)} - \alpha_2 t_1^{(1)}}{t_1^{(1)}}, \qquad t_3^{(2)} = \frac{t_3^{(1)} - \beta_2 t_1^{(1)}}{t_1^{(1)}},$$

where $t_1^{(2)}$, $t_2^{(2)}$, $t_3^{(2)}$ are uniformizing parameters of $P^{(2)}(V^{(2)})$. More generally, we find:

(22) $$t_1^{(i+1)} = t_1^{(i)}, \qquad t_2^{(i+1)} = \frac{t_2^{(i)} - \alpha_{i+1}t_1^{(i)}}{t_1^{(i)}}, \qquad t_3^{(i+1)} = \frac{t_3^{(i)} - \beta_{i+1}t_1^{(i)}}{t_1^{(i)}},$$

where $\alpha_{i+1}, \beta_{i+1} \in \mathfrak{J}$.

From (18), (21) and (22) we obtain the following relations:

(23) $$\begin{aligned} t_2 &= \alpha_1 t_1 + \alpha_2 t_1^2 + \cdots + \alpha_i t_1^i + (\alpha_{i+1} + t_2^{(i+1)})t_1^{i+1}, \\ t_3 &= \beta_1 t_1 + \beta_2 t_1^2 + \cdots + \beta_i t_1^i + (\beta_{i+1} + t_3^{(i+1)})t_1^{i+1}. \end{aligned}$$

The relations (23) point obviously to the fact that the successive centers $P, P^{(1)}, P^{(2)}, \cdots$ of the valuation v arise from a sequence of "infinitely near" points on an algebraic or analytical branch γ on F, given by the expansions: $t_2 = \alpha_1 t_1 + \alpha_2 t_1^2 + \cdots$, $t_3 = \beta_1 t_1 + \beta_2 t_1^2 + \cdots$, and that the valuation v is therefore either algebraic (i.e. has rank 2) or analytical (i.e. has rank 1 and is discrete). The formal proof is as follows:

For simplicity of notation we agree to use one and the same symbol for an element of $\mathfrak{J}$ and for the F-residue of that element. Thus we shall treat t_1, t_2, t_3, if necessary, as elements of the field of rational functions on F, although in reality they are elements of $Q_V(P)$, i.e. elements of the field of rational functions on V. This convention permits us to speak of the values $v(t_1)$, $v(t_2)$, $v(t_2 - \alpha_1 t_1)$, etc. We shall adhere to this convention throughout this part of the paper.

Let

(24) $$\begin{aligned} g_i &= t_2 - \alpha_1 t_1 - \alpha_2 t_1^2 - \cdots - \alpha_i t_1^i, \\ h_i &= t_3 - \beta_1 t_1 - \beta_2 t_1^2 - \cdots - \beta_i t_1^i, \end{aligned}$$

whence, by (23),

(25) $$v(g_i) \geqq v(t_1^{i+1}), \qquad v(h_i) \geqq v(t_1^{i+1}).$$

Let ζ be an arbitrary element of $Q_F(P)$, and let $n_i(\zeta)$ be an integer defined as follows: a) if ζ, regarded as an element of $\mathfrak{J}$ belongs to the ideal $\mathfrak{J}\cdot(g_i, h_i)$, then $n_i(\zeta) = i + 1$; b) if ζ does not belong to the ideal $\mathfrak{J}\cdot(g_i, h_i)$, there exists an integer j (depending on ζ and i) such that

(26) $$\zeta \equiv \epsilon\cdot t_1^j \pmod{\mathfrak{J}(g_i, h_i)},$$

where ϵ is a unit in $\mathfrak{J}$[6]. In this case we put

(27) $$n_i(\zeta) = \min.\,(j, i + 1).$$

[6] The leading forms of g_i and h_i are respectively $z_2 - a_1 z_1$, $z_3 - b_1 z_1$. Hence t_1, g_i and h_i are uniformizing parameters of $P(V)$, and therefore the ideal $\mathfrak{J}\cdot(g_i, h_i)$ defines an algebraic curve Γ_i on V which has at P a simple point. If $\bar{\zeta}$ and $\bar{t}_1$ are the Γ_i-residues of ζ and t_1 respectively, then $\bar{\zeta} \neq 0$ and $\bar{t}_1$ is a uniformizing parameter of $P(\Gamma_i)$. Therefore we can write $\bar{\zeta} = \bar{\epsilon}\bar{t}_1^j$, where $\bar{\epsilon}$ is a unit in $Q_{\Gamma_i}(P)$. From this, congruence (26) follows immediately (ϵ = any element of $\mathfrak{J}$ whose Γ_i-residue is $\bar{\epsilon}$).

We have from (25) and (26): $v(\zeta) \geqq n_i(\zeta)v(t_1)$. If $n_i(\zeta) \to +\infty$ as $i \to +\infty$, then the valuation v is of rank 2 and is composed of a divisor whose center is a component of the curve defined on F by the principal ideal (ζ). Suppose now that v is not of rank 2. Then $n_i(\zeta)$ is bounded for any non zero element ζ in $Q_F(P)$. We conclude from (25), (26) and (27) that if i is sufficiently high we must have

$$v(\zeta) = n_i(\zeta)v(t_1),$$

whence *the value of any non zero element* ζ *in* $Q_F(P)$ *is an integer*. Therefore the valuation v is of rank 1. This completes the proof of the lemma.

12. Valuations of rank 2

The valuation v is composite with a divisor (a 1-dimensional valuation). This divisor may be of second kind with respect to F, i.e., its center may be the point P. It is known, however, that a finite number of quadratic transformations will necessarily lead to a surface $F^{(i)}$ with respect to which the divisor is of the first kind (see [5], p. 681; the reasoning is identical with that employed in the proof of Lemma 9.1). Hence it is permissible to assume that the center of the divisor is an irreducible curve Γ on F, through P. If Γ is a ν-fold curve of F (where ν is the multiplicity of the point P), then P is a simple point of Γ (see section 10). Hence successive monoidal transformations operating on Γ and on its transforms are permissible. By Theorem 4′ (section 9), a finite number of such transformations will eliminate Γ *as a* ν*-fold curve*. Hence we may assume that the multiplicity of Γ for F is less than ν (we assume of course that $\nu > 1$). *From now on we shall employ only quadratic transformations.*

First of all we may eliminate by quadratic transformations the singularity which Γ may possibly possess at P. Note that the successive centers $P^{(1)}$, $P^{(2)}$, $\cdots$ of the valuation v will always lie on successive transforms of the curve Γ, for the valuation v is composite with a divisor whose center is Γ. Hence it is permissible to assume that P *is a simple point of* Γ. We may then select uniformizing parameters x, y, z of $P(V)$ in such a fashion that the curve Γ be given by the equations $y = z = 0$ [i.e. Γ is given in $Q_V(P)$ by the ideal (y, z)]. Since $v(y) > nv(x)$ and $v(z) > nv(x)$ for any integer n, it follows that the local equations of our successive quadratic transformations $T^{(i)}$ will all be of the form:

$$x_{i+1} = x, \qquad y_{i+1} = \frac{y_i}{x}, \qquad z_{i+1} = \frac{z_i}{x},$$

where x, y_i, z_i are uniformizing parameters of $P^{(i)}(V^{(i)})$. On each surface $F^{(i)}$ the center of the divisor is the curve $y_i = z_i = 0$.

Let (ω_i) be the principal ideal in $\mathfrak{J}_i$ [$= Q_{V^{(i)}}(P^{(i)})$] which defines the surface $F^{(i)}$. Here ω_{i+1} is the $T^{(i)}$-transform of ω_i (see section 11). As long as P_i is ν-fold for $F^{(i)}$ we will have

$$\omega_i = x^{\nu}\omega_{i+1}. \tag{28}$$

In particular, F is defined by the principal ideal $\mathfrak{J}\cdot\omega$.

Let $\mathfrak{A}$ denote the ideal $\mathfrak{J}\cdot(y, z)$, and let $\mathfrak{m}_i$ be the ideal of non units in $\mathfrak{J}_i$.

LEMMA 12.1. *Let ξ be an element of $\mathfrak{J}$ which is exactly divisible by $\mathfrak{m}^\rho$ and let ξ_1 be the T-transform of ξ, i.e. let $\xi = x^\rho \xi_1$. A necessary condition that ξ_1 be divisible by $\mathfrak{m}_1^\sigma$ is that the following congruence be satisfied:*

$$\xi \equiv 0(\mathfrak{A}^\sigma \mathfrak{m}^{\rho-\sigma}, \mathfrak{m}^{\rho+1}). \tag{29}$$

PROOF. We shall use induction with respect to σ, since for $\sigma = 0$ the lemma is trivial. Assume that the lemma is true for $\sigma = s$ and let ξ_1 be divisible by $\mathfrak{m}_1^{s+1}$. Then ξ_1 is also divisible by $\mathfrak{m}_1^s$, (29) holds (with σ replaced by s), and we can write:

$$\xi = x^{\rho-s}\varphi_s(y, z) + \varphi_{s+1}(y, z) + \psi_{\rho+1}(x, y, z), \tag{30}$$

where φ_s is a form with coefficients in $\mathfrak{J}$, φ_{s+1} is a form with coefficients in $\mathfrak{m}^{\rho-s-1}$ and $\psi_{\rho+1}$ is a form with coefficients in $\mathfrak{J}$, the degrees of these forms being s, $s + 1$ and $\rho + 1$ respectively. We have:

$$\xi_1 = \varphi_s(y_1, z_1) + \overset{*}{\varphi}_{s+1}(y_1, z_1) + x\psi_{\rho+1}(1, y_1, z_1),$$

where $\overset{*}{\varphi}_{s+1}$ is a form, of degree $s + 1$, with coefficients in $\mathfrak{J}$. From this expression of ξ_1 it follows immediately that if $\xi_1 \equiv 0(\mathfrak{m}_1^{s+1})$ then the coefficients of φ_s must all belong to $\mathfrak{m}$. Hence $\varphi_s(y, z) \equiv 0(\mathfrak{m}^{s+1})$, and consequently, by (30), $\xi \equiv 0(\mathfrak{A}^{s+1}\mathfrak{m}^{\rho-s-1}, \mathfrak{m}^{\rho+1})$, q.e.d.

LEMMA 12.2. *If ρ, s and σ are non-negative integers, $s \geqq \sigma$, then*

$$(\mathfrak{A}^\rho, \mathfrak{m}^s):x^\sigma = (\mathfrak{A}^\rho, \mathfrak{m}^{s-\sigma}). \tag{31}$$

PROOF. It is sufficient to prove the lemma for $\sigma = 1$. Letting $\sigma = 1$, we shall use induction with respect to s, since if $s \leqq \rho$ then the lemma is trivial. Let ξ be any element of $(\mathfrak{A}^\rho, \mathfrak{m}^{s+1}):x$. Then by our induction, ξ is an element in $(\mathfrak{A}^\rho, \mathfrak{m}^{s-1})$, and we can write $\xi = \alpha_\rho + \pi$, when $\alpha_\rho \equiv 0(\mathfrak{A}^\rho)$ and $\pi \equiv 0(\mathfrak{m}^{s-1})$. We must have $\pi x \equiv 0(\mathfrak{A}^\rho, \mathfrak{m}^{s+1})$, i.e.

$$\pi x = \phi_\rho(y, z) + \pi', \tag{32}$$

where ϕ_ρ is a form, of degree ρ, with coefficients in $\mathfrak{J}$ and where $\pi' \equiv 0(\mathfrak{m}^{s+1})$. If $\pi \equiv 0(\mathfrak{m}^s)$, then $\xi \equiv 0(\mathfrak{A}^\rho, \mathfrak{m}^s)$, and there is nothing to prove. Suppose then that $\pi \not\equiv 0(\mathfrak{m}^s)$. In this case πx is exactly divisible by $\mathfrak{m}^s$, and the relation (32) is possible if and only if also $\phi_\rho(y, z)$ is in $\mathfrak{m}^s$ and if πx and $\phi_\rho(yz)$ have the same leading form, for $\pi' \equiv 0(\mathfrak{m}^{s+1})$. Now no term of the leading form of $\phi_\rho(y, z)$ is of degree less than ρ in y and z. Hence the same must be true of the leading form of π. This implies that $\pi \equiv 0(\mathfrak{A}^\rho, \mathfrak{m}^s)$, and consequently also $\xi \equiv 0(\mathfrak{A}^\rho, \mathfrak{m}^s)$, q.e.d.

By means of the preceding two lemmas the proof of the local reduction theorem for the valuation v is readily completed. If $P^{(1)}$ is a ν-fold point of $F^{(1)}$, the leading form of ω must be a binary form in y, z, with coefficients in the residue field of P. Hence $\omega \equiv 0(\mathfrak{A}^\nu, \mathfrak{m}^{\nu+1})$. On the other hand, the curve Γ $(y = z = 0)$ is not ν-fold for the surface F, i.e. $\omega \not\equiv 0(\mathfrak{A}^\nu)$. There exists therefore an integer $n \geqq 1$ such that[7]

$$\omega \equiv 0(\mathfrak{A}^\nu, \mathfrak{m}^{n+\nu}), \qquad \omega \not\equiv 0(\mathfrak{A}^\nu, \mathfrak{m}^{n+\nu+1}),$$

and we can write: $\omega = \phi_\nu(y, z) + \xi$, where $\xi \equiv 0(\mathfrak{m}^{n+\nu})$ and ϕ_ν is a form of degree ν, with coefficients in $\mathfrak{J}$. By (28) we have: $\omega_1 = \phi_\nu(y_1, z_1) + x^n\xi_1$, $\xi_1 \epsilon \mathfrak{J}_1$. *We assert that* $\omega_1 \not\equiv 0(\mathfrak{A}_1^\nu, \mathfrak{m}_1^{n+\nu})$, where $\mathfrak{A}_1 = \mathfrak{J}_1 \cdot (y_1, z_1)$. For in the contrary case we would have: $x^n\xi_1 \equiv 0(\mathfrak{A}_1^\nu, \mathfrak{m}_1^{n+\nu})$, and hence, by Lemma 12.2 (applied to $\mathfrak{A}_1$ and $\mathfrak{m}_1$), $\xi_1 \equiv 0(\mathfrak{A}_1^\nu, \mathfrak{m}_1^\nu)$, i.e. $\xi_1 \equiv 0(\mathfrak{m}_1^\nu)$. It then follows from Lemma 12.1 (in which we now have $\rho = n + \nu$, $\sigma = \nu$) that $\xi \equiv 0(\mathfrak{A}^\nu\mathfrak{m}^n, \mathfrak{m}^{n+\nu+1})$, and consequently $\omega \equiv 0(\mathfrak{A}^\nu, \mathfrak{m}^{n+\nu+1})$. This is in contradiction with our definition of the integer n.

We thus conclude that if $P^{(1)}$ is still ν-fold for $F^{(1)}$, and if n_1 is the integer with the property

$$\omega_1 \equiv 0(\mathfrak{A}_1^\nu, \mathfrak{m}_1^{n_1+\nu}), \qquad \omega_1 \not\equiv 0(\mathfrak{A}_1^\nu, \mathfrak{m}_1^{n_1+\nu+1}),$$

then

$$n > n_1 \geqq 0.$$

More generally, if $P^{(1)}, P^{(2)}, \cdots, P^{(i)}$ are ν-fold points of $F^{(1)}, F^{(2)}, \cdots, F^{(i)}$, and if n_i is the integer with the property

$$\omega_i \equiv 0(\mathfrak{A}_i^\nu, m_i^{n_i+\nu}), \qquad \omega_i \not\equiv 0(\mathfrak{A}_i^\nu, m_i^{n_i+\nu+1}),$$

where $\mathfrak{A}_i = \mathfrak{J}_i \cdot (y_i, z_i)$, then

$$n > n_1 > n_2 > \cdots > n_i \geqq 0.$$

This shows that the number of successive ν-fold points $P^{(i)}$ is necessarily finite. This completes the proof of the local reduction theorem for valuations of rank 2.

13. Differentiation with respect to uniformizing parameters

In this and in the remaining sections of Part II we assume that the ground field k is of characteristic zero. Let P be a simple point of an r-dimensional variety V and let $t_1, t_2, \cdots, t_r$ be uniformizing parameters of $P(V)$. If $f(z_1, z_2, \cdots, z_r)$ is a non-zero polynomial in the indeterminates z_i, with coefficients in k, and if $f_\rho(z_1, z_2, \cdots, z_r)$ is the leading form of f, i.e. the sum of terms of lowest degree ρ, then $f_\rho(z_1, z_2, \cdots, z_r)$ is also the leading form of $f(t_1, t_2, \cdots, t_r)$. Hence $f(t_1, t_2, \cdots, t_r)$ is different from zero since it has a leading form which is different from zero. We conclude that *the uniformizing parameters* $t_1, t_2, \cdots, t_r$ *are algebraically independent over* k. Hence every element ω of the field of rational functions on V is algebraic over the field $k(t_1, t_2, \cdots, t_r)$, and the partial derivatives $\partial\omega/\partial t_\alpha$, $\alpha = 1, 2, \cdots, r$, are defined and are elements of the field.

LEMMA 13.1. *If* $\omega \epsilon Q_V(P)$, *then also* $\dfrac{\partial\omega}{\partial t_\alpha} \epsilon Q_V(P)$, $\alpha = 1, 2, \cdots, r$.

[7] See, for instance, Krull [1], p. 207.

PROOF. Let $\xi_1, \xi_2, \cdots, \xi_n$ be a set of non-homogeneous coördinates of V such that P is at finite distance with respect to this system of coördinates. Since every element in $Q_V(P)$ is a quotient of polynomials in $k[\xi_1, \xi_2, \cdots, \xi_n]$ in which the denominator is $\neq 0$ at P, to prove the lemma it is sufficient to prove that the derivatives $\partial\xi_i/\partial t_\alpha$ belong to $Q_V(P)$.

Let $g_i(u)$ be an irreducible polynomial in $k[u]$ such that $g_i(\xi_i) = 0$ at P. Then $g_i(\xi_i)$ is a non-unit in $Q_V(P)$, and therefore, by the very definition of uniformizing parameters, we must have relations of the form:

$$H_i(\xi, t) = A(\xi)g_i(\xi_i) + \sum_{\alpha=1}^{r} B_{i\alpha}(\xi)t_\alpha = 0, \tag{33}$$

where ξ and t stand for the sets of elements $\xi_1, \xi_2, \cdots, \xi_n$ and $t_1, t_2, \cdots, t_r$ respectively. Here A and the $B_{i\alpha}$ are polynomials with coefficients in k and $A(\xi) \neq 0$ at P. Since $H_i(\xi, t) = 0$, we have

$$\frac{\partial H_i}{\partial t_\alpha} + \sum_{j=1}^{n} \frac{\partial H_i}{\partial \xi_j} \cdot \frac{\partial \xi_j}{\partial t_\alpha} = 0, \qquad \begin{array}{l} i = 1, 2, \cdots, n \\ \alpha = 1, 2, \cdots, r. \end{array} \tag{33'}$$

On the other hand, since $g_i'(\xi_i) \neq 0$ at P, we see from (33) that

$$\frac{\partial H_i}{\partial \xi_j} = 0 \quad \text{at } P, \text{ if } i \neq j,$$

$$\frac{\partial H_i}{\partial \xi_i} \neq 0 \quad \text{at } P.$$

Hence the determinant $\left|\frac{\partial H_i}{\partial \xi_j}\right|$ is different from zero at P, and consequently $\frac{\partial \xi_i}{\partial t_\alpha} \in Q_V(P)$, in view of (33′), q.e.d.

The following is an immediate consequence of the lemma. Let ω be an element of $Q_V(P)$ and let $F(z)$ be the leading form of ω, where F is of degree ν in $z_1, z_2, \cdots, z_r$. *Then* $\frac{\partial F}{\partial z_i}$ *is the leading form of* $\frac{\partial \omega}{\partial t_i}$, *unless* $\frac{\partial F}{\partial z_i} = 0$, *in which case* $\frac{\partial \omega}{\partial t_i} \equiv 0(\mathfrak{m}^\nu)$, *where* $\mathfrak{m}$ *is the ideal of non-units in* $Q_V(P)$. For if $F(z_1, z_2, z_3) = \sum_{(i)} a_{(i)} z_1^{i_1} z_2^{i_2} \cdots z_r^{i_r}$, then we can write ω in the form: $\omega = \sum_{(i)} \alpha_{(i)} t_1^{i_1} t_2^{i_2} \cdots t_r^{i_r}$, where $\alpha_{(i)}$ is some element of $Q_V(P)$ whose P-residue is $a_{(i)}$. Since $\partial\alpha_{(i)}/\partial t_j \in Q_V(P)$, it follows that $\frac{\partial \omega}{\partial t_1} \equiv \sum i_1 \alpha_{(i)} t_1^{i_1 - 1} t_2^{i_2} \cdots t_r^{i_r} \pmod{\mathfrak{m}^\nu}$, and from this our assertion follows. As a corollary we have the following familiar result: *if* ω *is exactly divisible by* $\mathfrak{m}^\nu$, *then all partial derivatives of* ω, *of order* $1, 2, \cdots, \nu - 1$, *with respect to the uniformizing parameters, are zero at* P, *but at least one partial derivative of order* ν *is different from zero at* P; *and conversely.*

In the following considerations we shall denote by z one of the uniformizing parameters of $P(V)$, say the parameter t_r. Thus in the case $r = 3$, which is of particular interest to us, the uniformizing parameters will be denoted by x, y, z, as in the preceding section.

We introduce in $\mathfrak{J}$ [$= Q_V(P)$] an operator Δ_n, n — a positive integer, defined as follows: if ξ is any element of $Q_V(P)$, then

$$(34) \qquad \Delta_n(\xi) = \xi - z\frac{\partial \xi}{\partial z} + \frac{z^2}{2!}\cdot\frac{\partial^2 \xi}{\partial z^2} - \cdots + (-1)^n \frac{z^n}{n!}\cdot\frac{\partial^n \xi}{\partial z^n}.$$

We have

$$(35) \qquad \frac{\partial \Delta_n(\xi)}{\partial z} = (-1)^n \frac{z^n}{n!}\cdot\frac{\partial^{n+1}\xi}{\partial z^{n+1}} \equiv 0(\mathfrak{J}\cdot z^n).$$

Given any element ω in $\mathfrak{J}$ and given any positive integer ν, it is always possible to write ω in the form

$$(36) \qquad \omega = \alpha_0 z^\nu + \alpha_1 z^{\nu-1} + \cdots + \alpha_{\nu-1} z + \alpha_\nu,$$

where $\alpha_i \in \mathfrak{J}$. Moreover, if $\omega \equiv 0(\mathfrak{m}^\nu)$, the α's can be so selected that they satisfy the congruences:

$$(37) \qquad \alpha_i \equiv 0(\mathfrak{m}^i), \qquad i = 0, 1, \cdots, \nu.$$

Definition 1. The expression (36) of ω is *normal* if the elements α_i satisfy the congruences

$$(38) \qquad \frac{\partial \alpha_i}{\partial z} \equiv 0(z^i), \qquad i = 0, 1, \cdots, \nu.$$

Definition 2. If $\omega \equiv 0(\mathfrak{m}^\nu)$, then the expression (36) of ω is *strongly normal* if the elements α_i satisfy the following condition: if α_i is exactly divisible by $\mathfrak{m}^{\nu_i}$, then $\nu_i \geqq i$ and

$$(39) \qquad \frac{\partial \alpha_i}{\partial z} \equiv 0(z^{\nu_i}), \qquad i = 0, 1, \cdots, \nu.$$

We shall now prove the following lemma:

Lemma 13.2. *Any element ω of $\mathfrak{J}$ can be expressed in a normal form (for any integer ν). If $\omega \equiv 0(\mathfrak{m}^\nu)$ then the elements α_i which occur in a normal expression of ω satisfy the congruences* (37). *Any element ω in $\mathfrak{m}^\nu$ can be expressed in a strongly normal form.*

Proof. We start from an arbitrary expression of ω, of the form (36). By (34) we have $\Delta_n(\alpha_\nu) \equiv \alpha_\nu (\text{mod } z)$. Hence, if we put $\alpha'_\nu = \Delta_n(\alpha_\nu) = \alpha_\nu + \beta_{\nu-1} z$, then

$$\omega = \alpha_0 z^\nu + \alpha_1 z^{\nu-1} + \cdots + \alpha_{\nu-2} z^2 + (\alpha_{\nu-1} - \beta_{\nu-1}) z + \alpha'_\nu.$$

This expression of ω is similar to (36), and we have $\dfrac{\partial \alpha'_\nu}{\partial z} \equiv 0(z^n)$, by (35). Hence if $n \geqq \nu$, then (38) is satisfied for $i = \nu$. Suppose that the congruences (38) are satisfied for $i = s + 1, s + 2, \cdots \nu$. Then we put $\alpha_s' = \Delta_n(\alpha_s) = \alpha_s + \beta_{s-1}z$, and we replace the expression (36) of ω by the following:

$$\omega = \alpha_0 z^\nu + \alpha_1 z^{\nu-1} + \cdots + (\alpha_{s-1} - \beta_{s-1})z^{\nu-s+1} + \alpha_s' z^{\nu-s} + \alpha_{s+1} z^{\nu-s-1} + \cdots + \alpha_\nu ,$$

and for this new expression of ω the congruences (38) are satisfied for $i = s$, $s + 1, \cdots, \nu$. Ultimately we get in this fashion a normal expression of ω.

Let us suppose now that (36) is a normal expression of ω, and let $\omega \equiv 0(\mathfrak{m}^\nu)$. All the partial derivatives of ω with respect to the uniformizing parameters $x, y, \cdots, z$, of order $\leqq \nu - 1$, vanish at P. Therefore all the partial derivatives of α_ν, of order $\leqq \nu - 1$, with respect to the parameters $x, y, \cdots$, other than z, must be zero at P. Since $\dfrac{\partial \alpha_\nu}{\partial z} \equiv 0(z^\nu)$, and since α_ν is certainly zero at P, it follows that $\alpha_\nu \equiv 0(\mathfrak{m}^\nu)$. Consequently

$$\omega' = \alpha_0 z^{\nu-1} + \alpha_1 z^{\nu-2} + \cdots + \alpha_{\nu-1} \equiv 0(\mathfrak{m}^{\nu-1}),$$

and since this is obviously a normal expression of ω', we conclude in a similar fashion that $\alpha_{\nu-1} \equiv 0(\mathfrak{m}^{\nu-1})$. Thus the congruences (37) follow by induction with respect to ν.

To derive a strongly normal expression of the element ω, we start with an arbitrary normal expression (36) of ω, and we assume that α_i is exactly divisible by $\mathfrak{m}^{\nu_i}$. From the preceding part of the proof we know that the inequalities $\nu_i \geqq i$ are automatically satisfied, since, by hypothesis, $\omega \equiv 0(\mathfrak{m}^\nu)$. Let us assume that the congruences (39) are already satisfied for $i = s + 1$, $s + 2, \cdots, \nu$, $s \leqq \nu$. We consider the element α_s. If the leading form of α_s is divisible by z, then α_s can be written in the form: $\alpha_s = \beta_{s-1}z + \bar{\alpha}_s$, where $\beta_{s-1} \in \mathfrak{J}$ and $\bar{\alpha}_s \equiv 0(\mathfrak{m}^{\nu_s+1})$, whence

$$\alpha_s \equiv 0(z, \bar{\alpha}_s). \tag{40}$$

This leads to another expression of ω which differs from (36) in that α_{s-1} is replaced by $\alpha_{s-1} + \beta_{s-1}$ and α_s is replaced by $\bar{\alpha}_s$. This expression may not be normal, but can be rendered normal according to the procedure of the first part of the proof. In this procedure the coefficients $\alpha_{s+1}, \alpha_{s+2}, \cdots \alpha_\nu$ are not affected, and $\bar{\alpha}_s$ is replaced by $\alpha_s' = \Delta_n(\bar{\alpha}_s)$, where $n \geqq s$. By the definition of the operator Δ_n it follows that

$$\alpha_s' \equiv \bar{\alpha}_s(\mathfrak{J}\cdot z) \quad \text{and} \quad \alpha_s' \equiv 0(\mathfrak{m}^{\nu_s+1}),$$

whence, by (40),

$$\alpha_s \equiv 0(z, \alpha_s') \equiv 0(z, \mathfrak{m}^{\nu_s+1}).$$

If the leading form of α_s' is still divisible by z, the above procedure leads to a new normal form of ω:

$$\omega = \alpha_0'' z^\nu + \alpha_1'' z^{\nu-1} + \cdots + \alpha_s'' z^{\nu-s} + \alpha_{s+1} z^{\nu-s+1} + \cdots + \alpha_\nu ,$$

and we will have

$$\alpha_s' \equiv 0(z, \alpha_s''), \qquad \alpha_s'' \equiv 0(\mathfrak{m}^{\nu_s+2}),$$

whence

$$\alpha_s \equiv 0(z, \alpha_s'') \equiv 0(z, \mathfrak{m}^{\nu_s+2}).$$

If that process continues indefinitely, so that the leading form of $\alpha_s^{(i)}$ is always divisible by z, then $\alpha_s \equiv 0(z, \mathfrak{m}^{\nu_s+i})$, for all i, whence $\alpha_s \equiv 0(z)$, say $\alpha_s = z\beta_{s-1}$. In that case we get a normal form of ω:

$$\omega = \alpha_0 z^\nu + \alpha_1 z^{\nu-1} + \cdots + (\alpha_{s-1} + \beta_s)z^{\nu-s+1} + \alpha_{s+1}z^{\nu-s-1} + \cdots + \alpha_\nu$$

in which the coefficient of $z^{\nu-s}$ is zero. But if $\alpha_s = 0$, then the congruences (39) are satisfied for $i = s, s+1, \cdots, \nu$.

We may therefore assume that the leading form of α_s is not divisible by z. As was pointed out above, we may replace the original normal expression (36) of ω by another normal expression, in which α_s is replaced by $\Delta_n(\alpha_s)$, without affecting the coefficients $\alpha_{s+1}, \alpha_{s+2}, \cdots, \alpha_n$. Since $\alpha_s \equiv 0(\mathfrak{m}^{\nu_s})$, each of the $n + 1$ terms in the expression of $\Delta_n(\alpha_s)$, except the first (i.e. α_s), is in $\mathfrak{m}^{\nu_s}$, and its leading form *is* divisible by z, while α_s is *exactly* divisible by $\mathfrak{m}^{\nu_s}$ and its leading form is *not* divisible by z. Hence $\Delta_n(\alpha_s)$ *is still exactly divisible by* $\mathfrak{m}^{\nu_s}$. It is therefore sufficient to take $n \geqq \nu_s$ in order to satisfy the congruence $\partial\alpha_s/\partial z \equiv 0(z^{\nu_s})$ [see (35)]. This completes the proof of the lemma.

We conclude this section with two lemmas concerning the effect of a quadratic transformation on a normal form of an element ω. Let T be a quadratic transformation of V, of center P, and let P' be one of the points of the transform V' of V which correspond to V. *We assume that the quotients* $t_2/t_1, t_3/t_1, \cdots, t_r/t_1$ *are finite at* P' *and that, in particular,* t_r/t_1, *i.e.* z/t_1, *is zero at* P'. Then $z' = z/t_1$ is one of the uniformizing parameters of $P'(V')$. We consider a normal expression (36) of an element ω in $\mathfrak{J}$ and we assume that ω is exactly divisible by $\mathfrak{m}^\nu$. Then $\omega = t_1^\nu\omega'$, where ω' is the T-transform of ω. Moreover, by (37), we can write: $\alpha_i = t_1^i\alpha_i'$, $\alpha_i' \in \mathfrak{J}' = Q_{V'}(P')$, and therefore:

$$\omega' = \alpha_0' z'^\nu + \alpha_1' z'^{\nu-1} + \cdots + \alpha_{\nu-1}' z' + \alpha_\nu'. \tag{41}$$

LEMMA 13.3. *If* (36) *is a normal expression of* ω, *then* (41) *is likewise a normal expression of* ω'.

PROOF. We have, by hypothesis, $\dfrac{\partial\alpha_i}{\partial z} \equiv 0(z^i)$. We also have:

$$t_1^i \frac{\partial\alpha_i'}{\partial z'} = \frac{\partial\alpha_i}{\partial z'} = \frac{\partial\alpha_i}{\partial z}\cdot\frac{\partial z}{\partial z'} = t_1\frac{\partial\alpha_i}{\partial z} \equiv 0(z^i) \equiv 0(t_1^i z'^i).$$

Hence $\partial\alpha_i'/\partial z' \equiv 0(z'^i)$, q.e.d.

The principal ideal $\mathfrak{J}\cdot z$ defines an irreducible $(r-1)$-dimensional subvariety of V which has at P a simple point. The uniformizing parameters of $P(W)$ are $t_1, t_2, \cdots, t_{r-1}$ (more precisely: the residues of $t_1, t_2, \cdots, t_{r-1}$ mod $\mathfrak{J}\cdot t_r$). Under our assumption that the quotients t_i/t_1 are finite at P' and that z/t_1

is zero at P', it is clear that the T-transform of W is a variety W' passing through P', and that W' is defined in $\mathfrak{J}'$ [$= Q_{V'}(P')$] by the principal ideal $\mathfrak{J}' \cdot z'$. The transformation $\bar{T}$ from W to W' which is induced by T is quadratic with center at P. Let $\mathfrak{m}$ denote the ideal of non units in $Q_W(P)$ and let $\mathfrak{m}'$ denote the ideals of non units in $Q_{V'}(P')$.

LEMMA 13.4. *If α is an element of $\mathfrak{J}$ such that the congruence $\alpha \equiv 0(\mathfrak{m}^\rho)$ always implies the congruence $\partial\alpha/\partial z \equiv 0(\mathfrak{J} \cdot z^\rho)$, then the T-transform α' of α satisfies the same condition (with respect to $\mathfrak{m}'$, $\mathfrak{J}'$ and z'). Moreover, if α is exactly divisible by $\mathfrak{m}^\nu$, and if $\bar{\alpha}$ is the W-residue of α, then also $\bar{\alpha}$ is exactly divisible by $\bar{\mathfrak{m}}^\nu$, and the $\bar{T}$-transform of $\bar{\alpha}$ is the W'-residue of the T-transform of α.*

PROOF. If α is exactly divisible by $\mathfrak{m}^\nu$, then α' is at most divisible by $\mathfrak{m}'^\nu$, (Lemma 3.2), while from the proof of the preceding lemma it follows that the congruence $\partial\alpha/\partial z \equiv 0(z^\nu)$ implies the congruence $\partial\alpha'/\partial z' \equiv 0(z'^\nu)$.

Since α is exactly divisible by $\mathfrak{m}^\nu$ and since $\partial\alpha/\partial z \equiv 0(z^\nu) \equiv 0(\mathfrak{m}^\nu)$, it follows that the leading form of α must be independent of z. Hence the leading form of α is not divisible by z and consequently $\alpha \not\equiv 0(z, \mathfrak{m}^{\nu+1})$. This implies that $\bar{\alpha} \not\equiv 0(\bar{\mathfrak{m}}^{\nu+1})$. Hence $\bar{\alpha}$ is exactly divisible by $\bar{\mathfrak{m}}^\nu$.

We have therefore: $\alpha = t_1^\nu \alpha'$, $\bar{\alpha} = \bar{t}_1^\nu \bar{\alpha}'$, where $\bar{t}_1$ is the W'-residue of t_1 and where $\bar{\alpha}'$ is the $\bar{T}$-transform of $\bar{\alpha}$. From this it follows that $\bar{\alpha}'$ is the W'-residue of α', and this completes the proof of the lemma.

14. Valuations of rank 1

We now proceed to prove the local reduction theorem for any zero-dimensional valuation v of rank 1. We use the notations of section 10. If it is possible to lower the multiplicity ν of P by using quadratic transformations only, then there is nothing to prove. *We shall therefore assume that it is not possible to lower the multiplicity of P by using only quadratic transformations of the ambient variety V.* If we then denote by

$$F, F^{(1)}, F^{(2)}, \cdots, F^{(i)}, \cdots$$

the successive quadratic transforms of F which are *determined* by the valuation v, then our assumption implies that the centers P, $P^{(1)}$, $P^{(2)}$, $\cdots$ of v on the surfaces F, $F^{(1)}$, $F^{(2)}$, $\cdots$ are all ν-fold points. Let

$$\mathfrak{J}_i = Q_{V^{(i)}}(P^{(i)}), \qquad \mathfrak{m}^{(i)} = \text{ideal of non units in } \mathfrak{J}_i,$$

and let $\mathfrak{J}_i \cdot \omega_i$ denote the principal ideal which defines the surface $F^{(i)}$. Here ω_{i+1} is the $T^{(i)}$-transform of ω_i.

We begin by observing that *the leading form of ω_i is necessarily the νth power of a linear form, for all i.* For suppose that the contrary is true. Then we may assume that the leading form of ω itself ($i = 0$) is not a νth power. In that case the valuation v is discrete, by Lemma 11.4. Moreover, the leading form of no ω_i is a νth power [see (20) and the statement which immediately follows that formula], and the residue fields of the consecutive centers P, $P^{(1)}$, $P^{(2)}$, $\cdots$, all coincide (Theorem 3, section 7). Under these conditions, our assumption that all points $P^{(i)}$ are ν-fold, $\nu > 1$, is impossible. This has been proved by us in [5], p. 651–652, under the hypothesis that the ground field is algebraically closed. But it is clear from the proof that what really matters is that the points $P^{(i)}$ all have the same residue field, and this condition is satisfied in the present case.

Let x, y, z be uniformizing parameters of $P(V)$. We assume that z actually occurs in the leading form of ω. If z' is any other element of $\mathfrak{J}$ such that also x, y, z' are uniformizing parameters of $P(V)$, then it is clear that also z' will actually occur in the leading form of ω. We shall select the element z so as to satisfy a certain condition.

LEMMA 14.1. *The parameters x and y being fixed, there exists a third uniformizing parameter z such that the following condition is satisfied: if z' is any element of $\mathfrak{J}$ such that x, y, z' are also uniformizing parameters of $P(V)$, then the relations:*

$$v(z') > v(z), \tag{42}$$

$$v\left(\frac{\partial\omega}{\partial z}\right) \geqq v(z) \tag{42'}$$

are never satisfied simultaneously.

PROOF. Suppose that the parameter z does not satisfy the required condition. Then (42′) holds and there exists another parameter z' for which (42) holds. If we write z' in the form $z' = \alpha x + \beta y + \gamma z$, where α, β, γ are elements of $\mathfrak{J}$, then the hypothesis that also x, y and z' are uniformizing parameters of $P(V)$ implies that γ is a unit in $\mathfrak{J}$, whence $v(\gamma) = 0$. Since

$$\partial z'/\partial z = \gamma + \frac{\partial\alpha}{\partial z}x + \frac{\partial\beta}{\partial z}y + \frac{\partial\gamma}{\partial z}z,$$

it follows that also $\partial z'/\partial z$ is a unit in $\mathfrak{J}$. Consequently

$$v\left(\frac{\partial\omega}{\partial z}\right) = v\left(\frac{\partial\omega}{\partial z'}\cdot\frac{\partial z'}{\partial z}\right) = v\left(\frac{\partial\omega}{\partial z'}\right).$$

If also z' does not satisfy the required condition, then

$$v\left(\frac{\partial\omega}{\partial z}\right) = v\left(\frac{\partial\omega}{\partial z'}\right) \geqq v(z'),$$

and there will exist another parameter z'' such that

$$v(z'') > v(z').$$

If this continues indefinitely, then we get a sequence of elements z, z', z'', $\cdots$, $z^{(i)}$, $\cdots$ such that

$$v(z) < v(z') < v(z'') < \cdots < v(z^i) < \cdots \tag{43}$$

$$v\left(\frac{\partial\omega}{\partial z}\right) = v\left(\frac{\partial\omega}{\partial z^i}\right) \geqq v(z^{(i)}). \tag{43'}$$

From the fact that the valuation v is of rank 1 and that every ideal in $\mathfrak{J}$ has a finite base, follows immediately that the values assumed by the elements of $\mathfrak{J}$, if arranged in order of magnitude, form a simple sequence whose limit is $+\infty$[8]. Consequently, by (43), we have $\lim. v(z^{(i)}) = +\infty$, whence, by (43′), $\partial\omega/\partial z = 0$ on F. Hence $\partial\omega/\partial z$ must be divisible by ω (in $\mathfrak{J}$). This is impossible, since the leading form of ω actually involves z and therefore $\partial\omega/\partial z$ is exactly divisible by $\mathfrak{m}^{\nu-1}$, while ω is divisible by $\mathfrak{m}^{\nu}$. This contradiction completes the proof of the lemma.

From now on we assume that z satisfies the condition of Lemma 14.1.

We denote by $v(\mathfrak{m})$ the least value assumed by elements of $\mathfrak{m}$. It is clear that $v(\mathfrak{m}) = \text{minimum } [v(x), v(y), v(z)]$. Since the leading form of ω actually involves z, either the ratios y/x, z/x, or the ratios x/y, z/y are finite at $P^{(1)}$; in other words: either $v(x) = v(\mathfrak{m})$ or $v(y) = v(\mathfrak{m})$. We fix our notation so that y/x and z/x are finite at $P^{(1)}$, whence

$$(44) \qquad v(x) = v(\mathfrak{m}).$$

We shall prove in a moment that $v(z) > v(\mathfrak{m})$ (Lemma 14.2). Assuming this, we see that z/x is zero at $P^{(1)}$, whence x and $z_1 = z/x$ can be taken as two of a set of three uniformizing parameters of $P^{(1)}(V^{(1)})$, while as third uniformizing parameter we can take an element of the form $f(y/x)$, where f is a polynomial with coefficients in $\mathfrak{J}$. The polynomial $f(t)$ must satisfy the condition that if its coefficients are reduced modulo $\mathfrak{m}$, the resulting polynomial $F(t)$ (with coefficients in the residue field of P) is irreducible and vanishes for $t = t_0 = P^{(1)}$-residue of y/x (all this follows from section 2). Thus the uniformizing parameters of $P^{(1)}(V^{(1)})$ are x, $f(y/x)$ and z_1. Let the pair of parameters $(x, f(y/x))$ be denoted by (x_1, y_1), where we fix our notation so that $v(x_1) \leqq v(y_1)$ (so that x_1 is not necessarily the element x). Since $v(z_1) \geqq v(\mathfrak{m}^{(1)})$, and since $z = xz_1$, it follows from (44) that $v(z) \geqq v(\mathfrak{m}) + v(\mathfrak{m}^{(1)})$. We have thus shown that if $v(z) > v(\mathfrak{m})$, then $v(z) \geqq v(\mathfrak{m}) + v(\mathfrak{m}^{(1)})$.

Since the leading form of ω involves z, the leading form of ω_1 will actually involve z_1 [see formula (19), section 11]. Hence we may assert again that y_1/x_1 and z_1/x_1 are finite at $P^{(2)}$, i.e. that $v(x_1) = v(\mathfrak{m}^{(1)})$. *The Lemma* 14.2 *will show that again we must have* $v(z_1) > v(x_1)$, i.e. $v(z_1) > v(\mathfrak{m}^{(1)})$, and consequently we find, as before, uniformizing parameters x_2, y_2, z_2 of $P^{(2)}(V^{(2)})$, where $z_2 = z_1/x_1$. Here $v(x_2) \leqq v(y_2)$ and one of the two elements x_2, y_2 coincides with x_1. The other element is of the form $f_1(y_1/x_1)$, where f_1 is a polynomial with coefficients in $\mathfrak{J}_1$. The relation $v(z_1) > v(x_1)$ is equivalent to the relation $v(z) > v(m) + v(\mathfrak{m}^{(1)})$, and again we point out that this relation implies the following: $v(z) \geqq v(\mathfrak{m}) + v(\mathfrak{m}^{(1)}) + v(\mathfrak{m}^{(2)})$ (since $v(z_2) \geqq v(\mathfrak{m}^{(2)})$).

---

[8] Let α be an arbitrary element of $\mathfrak{J}$ and let $\mathfrak{A}$ be the ideal in $\mathfrak{J}$ whose elements have value $\geqq v(\alpha)$. Since all elements of $\mathfrak{m}$ have positive value and since v is of rank 1, we have $\mathfrak{m}^{\rho} \equiv 0(\mathfrak{A})$ for ρ sufficiently high. Since $\mathfrak{m}^{\rho}$ has finite length it follows that if the values $v(\alpha)$, $\alpha \in \mathfrak{J}$, are arranged in order of magnitude, then each $v(\alpha)$ is preceded by a finite number of values $v(\beta)$, $\beta \in \mathfrak{J}$. Since $\lim_{\rho \to +\infty} v(\mathfrak{m}^{\rho}) = +\infty$, our assertion follows.

The possibility of continuing the above procedure indefinitely depends at each stage on the validity of the inequality: $v(z_i) > v(x_i) = v(\mathfrak{m}^{(i)})$, or of the equivalent inequality $v(z) > v(\mathfrak{m}) + v(\mathfrak{m}^{(1)}) + \cdots + v(\mathfrak{m}^{(i)})$. Hence we shall prove the following lemma:

LEMMA 14.2. *The inequality*

$$v(z) > v(\mathfrak{m}) + v(\mathfrak{m}^{(1)}) + \cdots + v(\mathfrak{m}^{(i)}) \tag{45}$$

holds for all values of i.

PROOF. Let us assume that the lemma is false and let us denote by $n - 2(n \geqq 1)$ the greatest value of i for which (45) holds true. At each of the points $P, P^{(1)}, P^{(2)}, \cdots, P^{(n-1)}$ we will have then uniformizing parameters (x_i, y_i, z_i), such that: 1) $z_i = z_{i-1}/x_{i-1}$; 2) $v(x_i) = v(\mathfrak{m}^{(i)})$; 3) one of the two elements x_i, y_i coincides with x_{i-1}, while the second element is of the form $f_{i-1}(y_{i-1}/x_{i-1})$, where f_{i-1} is a polynomial with coefficients in $\mathfrak{J}_{i-1}$. But while $v(z_i/x_i) > 0$, for $i = 0, 1, \cdots, n - 2$ $[(x_0, y_0, z_0) = (x, y, z)]$, we must have, by hypothesis, $v(z_{n-1}/x_{n-1}) = 0$, *whence* z_{n-1}/x_{n-1} *is a unit in* $Q_{V^{(n)}}(P^{(n)})$.

Let

$$\omega = \alpha_0 z^{\nu} + \alpha_1 z^{\nu-1} + \cdots + \alpha_{\nu} \tag{46}$$

be a normal expression for ω (see Definition 1, section 13). Then

$$\omega_1 = \alpha_0^{(1)} z_1^{\nu} + \alpha_1^{(1)} z_1^{\nu-1} + \cdots + \alpha_{\nu}^{(1)}, \tag{46.1}$$

where

$$\omega = x^{\nu}\omega_1, \qquad \alpha_j = x^j \alpha_j^{(1)}, \qquad \alpha_j^{(1)} \in \mathfrak{J}_1, \tag{47.1}$$

and the expression (46.1) for ω_1 is normal (Lemma 13.3). Continuing in this fashion we get a normal expression for each of the elements $\omega_1, \omega_2, \cdots, \omega_{n-1}$:

$$\omega_i = \alpha_0^{(i)} z_i^{\nu} + \alpha_1^{(i)} z_i^{\nu-1} + \cdots + \alpha_{\nu}^{(i)}, \qquad i = 1, 2, \cdots, n - 1, \tag{46.i}$$

where

$$\omega_{i-1} = x_{i-1}^{\nu}\omega_i, \qquad \alpha_j^{(i-1)} = x_{i-1}^j \alpha_j^{(i)}, \qquad \alpha_j^{(i)} \in \mathfrak{J}_i. \tag{47.i}$$

We also have a similar expression for ω_n:

$$\omega_n = \alpha_0^{(n)} z_n^{\nu} + \alpha_1^{(n)} z_n^{\nu-1} + \cdots + \alpha_{\nu}^{(n)}, \tag{46.n}$$

where $z_n = z_{n-1}/x_{n-1}$ and

$$\omega_{n-1} = x_{n-1}^{\nu}\omega_n, \qquad \alpha_j^{(n-1)} = x_{n-1}^j \alpha_j^{(n)}, \qquad \alpha_j^{(n)} \in \mathfrak{J}_n, \tag{47.n}$$

but this time z_n *is a unit in* $\mathfrak{J}_n$.

Let c_j and d be the P_n-residues of $\alpha_j^{(n)}$ and of z_n respectively $(d \neq 0)$. Since $\omega_n \equiv 0(\mathfrak{m}^{(n)\nu})$, we must have, by (46.n):

$$c_0 u^{\nu} + c_1 u^{\nu-1} + \cdots + c_{\nu} = c_0(u - d)^{\nu}.$$

From this we deduce the following consequences:

a) All c_j are different from zero. Consequently the $\nu + 1$ terms $\alpha_j^{(n)} z_n^{\nu-j}$ have value zero in the valuation v. But since each of these terms differs from the corresponding term $\alpha_j z^{\nu-j}$ in (46) by the factor $x^\nu x_1^\nu \cdots x_{n-1}^\nu$, it follows that

$$(48) \qquad v(\alpha_0 z^\nu) = v(\alpha_1 z^{\nu-1}) = \cdots = v(\alpha_\nu).$$

b) The residue d of z_n coincides with the residue $-c_1/\nu c_0$ of $-\alpha_1^{(n)}/\nu\alpha_0^{(n)}$, or—what is the same—the residue of $\alpha_1^{(n)}/\nu\alpha_0^{(n)} z_n$ equals -1. Since

$$\alpha_1^{(n)}/\alpha_0^{(n)} z_n = \alpha_1^{(n)} z_n^{\nu-1}/\alpha_0^{(n)} z_n^\nu = \alpha_1 z^{\nu-1}/\alpha_0 z^\nu = \alpha_1/\alpha_0 z,$$

we conclude that the residue of $\alpha_1/\nu\alpha_0 z$ in our valuation v is equal to -1. Taking into account that α_0 is a unit ($\alpha_0 = \alpha_0^{(n)}$ and $c_0 \neq 0$), whence $v(\alpha_0) = 0$, we conclude that

$$v\left(z + \frac{\alpha_1}{\nu\alpha_0}\right) > v(z).$$

Since (46) is a normal expression of ω, we have $\alpha_1 \equiv 0(\mathfrak{m})$ and $\dfrac{\partial \alpha_1}{\partial z} \equiv 0(\mathfrak{m})$. Hence either $\alpha_1 \equiv 0(\mathfrak{m}^2)$ or the leading form of α_1 is independent of z. Consequently, if we put

$$z' = z + \frac{\alpha_1}{\nu\alpha_0},$$

then x, y, z' are uniformizing parameters of $P(V)$, and

$$(49) \qquad v(z') > v(z).$$

We consider $\partial\omega/\partial z$:

$$\frac{\partial \omega}{\partial z} = [\nu\alpha_0 z^{\nu-1} + (\nu - 1)\alpha_1 z^{\nu-2} + \cdots + \alpha_{\nu-1}]$$
$$+ \left[\frac{\partial \alpha_0}{\partial z} z^\nu + \frac{\partial \alpha_1}{\partial z} z^{\nu-1} + \cdots + \frac{\partial \alpha_\nu}{\partial z}\right].$$

All terms in the first parenthesis have the same value, equal to $v(z^{\nu-1})$. All terms in the second parenthesis are divisible by z^ν, since $\dfrac{\partial \alpha_i}{\partial z} \equiv 0(z^i)$. Hence $v(\partial\omega/\partial z) \geqq v(z^{\nu-1}) \geqq v(z)$, since $\nu > 1$. This equality and the inequality (49) are in contradiction with our hypothesis that z satisfies the condition of Lemma 14.1. This completes the proof of Lemma 14.2.

In view of Lemma 14.2, the parameters x_i, y_i, z_i as described above for $i = 0, 1, \cdots, n - 1$, are actually defined for all values of i, and so is the normal expression (46.i) of ω_i. For each point $P^{(i)}$ we consider the surface $W^{(i)}$ through $P^{(i)}$ defined on $V^{(i)}$ by the principal ideal $\mathfrak{J}_i \cdot z_i$. As was pointed out in the preceding section in connection with Lemma 13.4, $W^{(i+1)}$ is the transform of $W^{(i)}$ by our quadratic transformation $T^{(i)}$, and x_i, y_i are uniformizing parameters of $P^{(i)}(W^{(i)})$. If we now assume that the expression (46) of ω is not only normal but also strongly normal (Lemma 13.2) and if we take into account Lemmas 11.2, 11.3 and 13.4 (these lemmas have to be applied to $\alpha_0, \alpha_1, \cdots, \alpha_\nu$), we conclude that for a sufficiently high value s of i, the expression (46.i) of ω_i will take the following form:

$$(50) \quad \omega_s = \epsilon_0^{(s)} z_s^\nu + \epsilon_1^{(s)} x_s^{m_1} y_s^{n_1} z_s^{\nu-1} + \cdots + \epsilon_{\nu-1}^{(s)} x_s^{m_{\nu-1}} y_s^{n_{\nu-1}} z_s + \epsilon_\nu^{(s)} x_s^{m_\nu} y_s^{n_\nu},$$

where m_j, n_j are non-negative integers and where each $\epsilon_j^{(s)}$ is either a unit in $\mathfrak{J}_s$ or is zero (but $\epsilon_0^{(s)} = \alpha_0^{(s)} = \alpha_0 \neq 0$).

At this stage of the proof the following two cases must be considered separately:

FIRST CASE. *$v(x_s)$ and $v(y_s)$ are rationally dependent.*

SECOND CASE. *$v(x_s)$ and $v(y_s)$ are rationally independent.* In the second case we are definitely dealing with a valuation of rational rank 2. This case will be considered in the next section. Here we shall deal with the first case.

If $v(x_s)$ and $v(y_s)$ are rationally dependent, there will exist a least integer σ ($\sigma \geqq s$) such that $v(x_\sigma) = v(y_\sigma)$, while for any integer i, $s < i \leqq \sigma$, we will have either $x_i = x_{i-1}$, $y_i = y_{i-1}/x_{i-1}$, or $x_i = y_{i-1}/x_{i-1}$ and $y_i = x_{i-1}$. It follows therefore from (50), that ω_σ will have the following form:

$$(51) \qquad \omega_\sigma = \epsilon_0 z_\sigma^\nu + \epsilon_1 x_\sigma^{\lambda_1} z_\sigma^{\nu-1} + \cdots + \epsilon_{\nu-1} x_\sigma^{\lambda_{\nu-1}} z_\sigma + \epsilon_\nu x_\sigma^{\lambda_\nu},$$

where each ϵ_j is either a unit in $\mathfrak{J}_\sigma$ or is zero.

Since $P^{(\sigma)}$ is a ν-fold point of $F^{(\sigma)}$, and since the ϵ's which are not zero are units in $\mathfrak{J}_\sigma$, we must have

$$\lambda_j - j \geqq 0$$

for all j such that $\epsilon_j \neq 0$. Hence the curve Δ defined by the ideal $\mathfrak{J}_\sigma(x_\sigma, z_\sigma)$ is a ν-fold curve of $F^{(\sigma)}$. We shall now apply to $F^{(\sigma)}$ a monoidal transformation M of center Δ. This transformation is permissible since $P^{(\sigma)}$ is a simple point of Δ [besides, we know in advance from the discussion in section 10 that the quadratic transformations which led from F to $F^{(\sigma)}$ could not possibly lead to ν-fold points other than those of types a), b) and c) described in section 10]. Let F_1 and V_1 be respectively the M-transforms of $F^{(\sigma)}$ and of $V^{(\sigma)}$, and let P_1 be the center of the valuation v on F_1.

We put $z_{\sigma 1} = z_\sigma/x_\sigma$ and

$$(51.1) \quad \omega_{\sigma 1} = \epsilon_0 z_{\sigma 1}^\nu + \epsilon_1 x_\sigma^{\lambda_1 - 1} z_{\sigma 1}^{\nu-1} + \cdots + \epsilon_{\nu-1} x_\sigma^{\lambda_{\nu-1}-1-\nu+1} z_{\sigma 1} + \epsilon_\nu x_\sigma^{\lambda_\nu - \nu}.$$

Since ϵ_0 is a unit, the leading form of ω_σ involves z_σ, whence $z_{\sigma 1}$ is finite at P_1. The surface F_1 is defined in $Q_{V_1}(P_1)$ by the principal ideal $(\omega_{\sigma 1})$. *If $z_{\sigma 1}$ is zero at P_1*, then x_σ, y_σ, $z_{\sigma 1}$ are uniformizing parameters of $P_1(V_1)$ [see section 7, equations (10)]. *Note that $z_{\sigma 1}$ is certainly zero at P_1 if* $\lambda_j - j > 0$, for all j such that $\epsilon_j \neq 0$.

Under the hypothesis that $z_{\sigma 1}$ is zero at P_1, the expression (51.1) of $\omega_{\sigma 1}$ is similar to that of ω_σ (51). If P_1 is still ν-fold for F_1, then we must have $\lambda_j - 2j \geqq 0$, and the curve Δ_1 defined by the ideal $(x_\sigma, z_{\sigma 1})$ is ν-fold for F_1. We then apply a second monoidal transformation M_1 of center Δ_1, getting new transforms F_2 and V_2. We then put

$$z_{\sigma 2} = z_{\sigma 1}/x_\sigma = z_\sigma/x_\sigma^2,$$

$$\omega_{\sigma 2} = \epsilon_0 z_{\sigma 2}^{\nu} + \epsilon_1 x_\sigma^{\lambda_1-2} z_{\sigma 2}^{\nu-1} + \cdots + \epsilon_{\nu-1} x_\sigma^{\lambda_{\nu-1}-2(\nu-1)} z_{\sigma 2} + \epsilon_\nu x_\sigma^{\lambda_\nu - 2\nu},$$

where again $z_{\sigma 2}$ is certainly finite at the center P_2 of v on F_2 and where F_2 is defined in $Q_{V_2}(P_2)$ by the principal ideal $(\omega_{\sigma 2})$. If $\lambda_j - 3j \geqq 0$, then the curve Δ_2 defined by the ideal $(x_\sigma, z_{\sigma 2})$ is ν-fold, and we can again apply a monoidal transformation.

Let h be the greatest integer such that $\lambda_j - hj \geqq 0$ for all j such that $\epsilon_j \neq 0$. Then the preceding considerations show that after h monoidal transformations with centers at successive ν-fold curves we will get a surface F_h on a V_h, on which the center of v is a point P_h, and which is defined in $Q_{V_h}(P_h)$ by the principal ideal $(\omega_{\sigma h})$, where

$$\omega_{\sigma h} = \epsilon_0 z_{\sigma h}^{\nu} + \epsilon_1 x_\sigma^{\lambda_1'} z_{\sigma h}^{\nu-1} + \cdots + \epsilon_{\nu-1} x_\sigma^{\lambda_{\nu-1}'} z_{\sigma h} + \epsilon_\nu x_\sigma^{\lambda_\nu'}. \tag{52}$$

Here

$$\lambda_j' = \lambda_j - hj,$$

and

$$z_{\sigma h} = z_\sigma/x_\sigma^h.$$

If $z_{\sigma h}$ is zero at P_h, then x_σ, y_σ, $z_{\sigma h}$ are uniformizing parameters of $P_h(V_h)$. By the definition of the integer h, we cannot have $\lambda_j' - j \geqq 0$ for all j such that $\epsilon_j \neq 0$. Hence in this case P_h is of multiplicity less than ν for F_h, and the proof is complete.

Suppose, however, that $z_{\sigma h} \neq 0$ at P_h, and let d be the residue of $z_{\sigma h}$ at P_h, $d \neq 0$. Then if $c_0, c_1, \cdots c_\nu$ are the P_h-residues of ϵ_0, $\epsilon_1 x_\sigma^{\lambda_1'}$, $\cdots$, $\epsilon_\nu x_\sigma^{\lambda_\nu'}$, *and if suppose that P_h is ν-fold for F_h*, then $u = d$ must be a ν-fold root of the polynomial

$$c_0 u^\nu + c_1 u^{\nu-1} + \cdots + c_\nu.$$

This implies that $c_j \neq 0$, for $j = 0, 1, 2, \cdots, \nu$, whence $\lambda_j' = 0$ and no ϵ_j is zero. Thus (52) becomes:

$$\omega_{\sigma h} = \epsilon_0 z_{\sigma h}^{\nu} + \epsilon_1 z_{\sigma h}^{\nu-1} + \cdots + \epsilon_{\nu-1} z_{\sigma h} + \epsilon_\nu,$$

and each of the $\nu + 1$ terms $\epsilon_j z_{\sigma h}^{\nu-j}$ is a unit, and therefore has value zero in the valuation v. Since these $v + 1$ terms are proportional to the $\nu + 1$ terms $\alpha_j z^{\nu-j}$ in (46), we find again the equations (48). Moreover, since $d = -c_1/\nu c_0 = P_h$-residue of $-\epsilon_1/\nu\epsilon_0$, it follows that the residue of z (in the valuation v) equals the residue of $-\alpha_1/\nu\alpha_0$. As in the proof of Lemma 14.2, so also here these conclusions lead to a contradiction with our original hypothesis that z satisfies the condition of Lemma 14.1. Therefore P_h cannot be a ν-fold point of F_h. This completes the proof of the local reduction theorem under the hypothesis that $v(x_s)$ and $v(y_s)$ are rationally dependent.

15. Valuations of rational rank 2

We now consider the case where $v(x_s)$ and $v(y_s)$ are rationally independent. Then also $v(x_i)$ and $v(y_i)$ will be rationally independent for all $i > s$ and we will have either

$$x_{i+1} = x_i, \qquad y_{i+1} = y_i/x_i, \qquad z_{i+1} = z_i/x_i,$$

if $v(x_i) < v(y_i/x_i)$, or

$$x_{i+1} = y_i/x_i, \qquad y_{i+1} = x_i, \qquad z_{i+1} = z_i/x_i,$$

if $v(y_i/x_i) < v(x_i)$, for all $i \geqq s$. Let $s < j_1 < j_2 \cdots$ be the sequence of integers defined as follows: $v(x_i) > v(y_i/x_i)$, if $i = j_\alpha - 1$, $i \geqq s$, and $v(x_i) < v(y_i/x_i)$ for all other values of i, $i \geqq s$. We combine the successive quadratic transformations $T^{(s)}, T^{(s+1)}, \cdots, T^{(j_\alpha-1)}$ into one transformation, which we shall denote by C_α, and we put

$$X = x_s, \qquad Y = y_s, \qquad X_\alpha = x_{j_\alpha}, \qquad Y_\alpha = y_{j_\alpha}, \qquad \bar{z}_\alpha = z_{j_\alpha}.$$

If we put $v(Y)/v(X) = \tau$, then it is immediately seen ([5], p. 653) that

$$(53) \qquad X = X_\alpha^{g_{\alpha-1}} Y_\alpha^{g_\alpha}, \qquad Y = X_\alpha^{f_{\alpha-1}} Y_\alpha^{f_\alpha},$$

where f_α/g_α are the convergent fractions of the irrational number τ ($f_0 = 1$, $g_0 = 0$).

For the element ω_{j_α} we will have an expression similar to (50):

$$(54) \quad \bar{\omega}_\alpha = \bar{\epsilon}_0^{(\alpha)} \bar{z}_\alpha^\nu + \bar{\epsilon}_1^{(\alpha)} X_\alpha^{\bar{m}_1} Y_\alpha^{\bar{n}_1} \bar{z}_\alpha^{\nu-1} + \cdots + \bar{\epsilon}_{\nu-1}^{(\alpha)} X_\alpha^{\bar{m}_{\nu-1}} Y_\alpha^{\bar{n}_{\nu-1}} \bar{z}_\alpha + \bar{\epsilon}_\nu^{(\alpha)} X_\alpha^{\bar{m}_\nu} Y_\alpha^{\bar{n}_\nu},$$

where we have put $\bar{\omega}_\alpha = \omega_{j_\alpha}$, $\epsilon^{(j_\alpha)} = \bar{\epsilon}^{(\alpha)}$, and where the integers $\bar{m}_j$, $\bar{n}_j$ depend on j and α. We denote the surface $F^{(j_\alpha)}$ and the point $P^{(j_\alpha)}$ respectively by $\bar{F}_\alpha$ and $\bar{P}_\alpha$.

If $\bar{m}_j \geqq j$, for $j = 1, 2, \cdots, \nu$, then the curve $X_\alpha = \bar{z}_\alpha = 0$ is a ν-fold curve of $\bar{F}_\alpha$, and we can apply to $\bar{F}_\alpha$ a monoidal transformation with center at that curve. Similarly, if $\bar{n}_j \geqq j$, $j = 1, 2, \cdots, \nu$, we apply a monoidal transformation with center at the curve $Y_\alpha = \bar{z}_\alpha = 0$. If both curves are ν-fold for $\bar{F}_\alpha$, then $\bar{P}_\alpha$ is a normal crossing of these curves, and we can apply both monoidal transformations, the order in which these transformations are applied being immaterial (Lemma 8.3).

More generally, if s is the greatest integer such that $\bar{m}_j - sj \geqq 0$, for all j, and if t is the greatest integer such that $\bar{n}_j - tj \geqq 0$, for all j, then we can apply to $\bar{F}_\alpha$ s consecutive monoidal transformations, with center at the ν-fold curve $X_\alpha = \bar{z}_\alpha = 0$ and at the $s - 1$ successive transforms of this curve (all these $s - 1$ transforms will be ν-fold curves), and t consecutive monoidal transformations with center at the ν-fold curve $Y_\alpha = \bar{z}_\alpha = 0$ and at the successive transforms of this curve. Let the new surface thus obtained be denoted by F_α, and let P_α be the center of v on F_α. If we put

$$\bar{z}_\alpha = X_\alpha^s Y_\alpha^t Z_\alpha,$$

then the surface F_α will be defined in $Q(P_\alpha)$ by the principal ideal (Ω_α), where

(55) $\Omega_\alpha = \bar{\epsilon}_0^{(\alpha)} Z_\alpha^\nu + \bar{\epsilon}_1^{(\alpha)} X_\alpha^{\bar{M}_1} Y_\alpha^{\bar{N}_1} Z_\alpha^{\nu-1} + \cdots + \bar{\epsilon}_{\nu-1}^{(\alpha)} X_\alpha^{\bar{M}_{\nu-1}} Y_\alpha^{\bar{N}_{\nu-1}} Z_\alpha + \bar{\epsilon}_\nu^{(\alpha)} X_\alpha^{\bar{M}_\nu} Y_\alpha^{\bar{N}_\nu}$,

where the integers $\bar{M}_j$, $\bar{N}_j$ (which depend on j and α) will satisfy the inequalities:

(56) $\text{minimum } (\bar{M}_j - j) < 0, \qquad \text{minimum } (\bar{N}_j - j) < 0.$

The case where $Z_\alpha \neq 0$ at P_α is settled by means of considerations similar to those developed in the preceding section in connection with the element $z_{\sigma h}$ occurring in (52). It follows namely in this case that the hypothesis that P_α is a ν-fold point of F_α is in contradiction with the hypothesis that our original element z satisfies the condition of Lemma 14.1. Hence we may assume that $Z_\alpha = 0$ at P_α, for all α, whence X_α, Y_α and Z_α are uniformizing parameters of $P_\alpha(V_\alpha)$.

We have now transformed the surface $F^{(s)}$ into the surface F_α by a sequence of permissible transformations. If we put $z_s = Z$, then the uniformizing parameters of $P^{(s)}(V^{(s)})$ are X, Y, Z, and $F^{(s)}$ is given in $Q_{V^{(s)}}(P^{(s)})$ by the principal ideal (ω_s), where [see (50)],

(57) $\omega_s = \epsilon_0^{(s)} Z^\nu + \epsilon_1^{(s)} X^{m_1} Y^{n_1} Z^{\nu-1} + \cdots + \epsilon_{\nu-1}^{(s)} X^{m_{\nu-1}} Y^{n_{\nu-1}} Z + \epsilon_\nu^{(s)} X^{m_\nu} Y^{n_\nu}$.

The equations of the transformation from $F^{(s)}$ to F_α are the following [see (53)]:

(58) $X = X_\alpha^{g_{\alpha-1}} Y_\alpha^{g_\alpha}, \qquad Y = X_\alpha^{f_{\alpha-1}} Y_\alpha^{f_\alpha}, \qquad Z = X_\alpha^p Y_\alpha^q Z_\alpha.$

Since by the transformation (58) the element ω_s is transformed into Ω_α (after a suitable power of X_α^ν and of Y_α^ν has been deleted), it follows, in view of the inequalities (56), that if we put

(59)
$$M_j = m_j g_{\alpha-1} + n_j f_{\alpha-1},$$
$$N_j = m_j g_\alpha + n_j f_\alpha,$$

then

(60) $p = \text{minimum } [M_j/j], \qquad q = \text{minimum } [N_j/j],$

and

(61) $\bar{M}_j = M_j - pj, \qquad \bar{N}_j = N_j - qj.$

Using (59) we write M_j and N_j in the following form

$$M_j = g_{\alpha-1}[m_j + n_j\tau + n_j(f_{\alpha-1}/g_{\alpha-1} - \tau)],$$
$$N_j = g_\alpha[m_j + n_j\tau + n_j(f_\alpha/g_\alpha - \tau)].$$

Since $\left|\dfrac{f_{\alpha-1}}{g_{\alpha-1}} - \tau\right| < \dfrac{1}{g_{\alpha-1} g_\alpha}$ and $\left|\dfrac{f_\alpha}{g_\alpha} - \tau\right| < \dfrac{1}{g_{\alpha-1} g_\alpha}$, it follows that

$$g_{\alpha-1}(m_j + n_j\tau) - M_j \to 0$$
$$g_\alpha(m_j + n_j\tau) - N_j \to 0,$$

as $\alpha \to \infty$. Hence if α is sufficiently high, then

$$[M_j/j] - 1 \leqq \left[\frac{(m_j + n_j\tau)g_{\alpha-1}}{j}\right] \leqq [M_j/j],$$

$$[N_j/j] - 1 \leqq \left[\frac{(m_j + n_j\tau)g_{\alpha}}{j}\right] \leqq [N_j/j],$$

for $j = 1, 2, \cdots, \nu$ [or better: for all j such that $\epsilon_j^{(s)} \neq 0$]. From these inequalities and from the fact that $g_\alpha \to \infty$, we draw the following consequence: if k is one of the integers $1, 2, \cdots, \nu$ such that

$$\frac{m_k + n_k\tau}{k} \leqq \frac{m_j + n_j\tau}{j}, \quad \text{all } j, \tag{62}$$

then, for α sufficiently high,

$$\begin{cases} [M_j/j] - [M_k/k] \to +\infty, \\ [N_j/j] - [N_k/k] \to +\infty, \end{cases} \tag{63}$$

if in (62) the sign $<$ holds. On the other hand, if the equality sign holds in (62), then $m_k/k = m_j/j$ and $n_k/k = n_j/j$, since τ is irrational, and hence, by (59),

$$\begin{aligned} [M_j/j] &= [M_k/k], \\ [N_j/j] &= [N_k/k]. \end{aligned} \tag{63'}$$

From (63), (63′), (60) and (61) we conclude that if α is sufficiently high, then

$$p = [M_k/k], \qquad q = [N_k/k],$$

while

$$\bar{M}_j \to +\infty, \qquad \bar{N}_j \to \infty$$

if (63) holds, and

$$\bar{M}_j < j, \qquad \bar{N}_j < j,$$

if (63′) holds. In particular, we have

$$\bar{M}^k < k, \qquad \bar{N}^k < k. \tag{64}$$

If minimum $(\bar{M}_j + \bar{N}_j - j) < 0$, then P_α is of multiplicity less than ν for F_α. In the contrary case we apply to F_α a quadratic transformation of center P_α. Interchanging, if necessary, X_α and Y_α, we may assume that the equations of the quadratic transformations are of the form:

$$X_{\alpha 1} = X_\alpha, \qquad Y_{\alpha 1} = Y_\alpha/X_\alpha, \qquad Z_{\alpha 1} = Z_\alpha/X_\alpha.$$

Let $F_{\alpha 1}$ be the transform of the surface F_α and let $P_{\alpha 1}$ be the center of v on $F_{\alpha 1}$. The surface $F_{\alpha 1}$ will be defined in the quotient ring of $P_{\alpha 1}$ by the principal ideal $(\Omega_{\alpha 1})$, where [see (55)]

$$\Omega_{\alpha 1} = \bar{\epsilon}_0^{(\alpha)} Z_{\alpha 1}^{\nu} + \bar{\epsilon}_1^{(\alpha)} X_{\alpha 1}^{\bar{M}_{11}} Y_{\alpha 1}^{\bar{N}_{11}} Z_{\alpha 1}^{\nu-1} + \cdots + \bar{\epsilon}_{\nu-1}^{(\alpha)} X_{\alpha 1}^{\bar{M}_{\nu-1,1}} Y_{\alpha 1}^{\bar{N}_{\nu-1,1}} Z_{\alpha 1} + \bar{\epsilon}_{\nu}^{(\alpha)} X_{\alpha 1}^{\bar{M}_{\nu 1}} Y_{\alpha 1}^{\bar{N}_{\nu 1}},$$

and where

$$\bar{M}_{j1} = \bar{M}_j + \bar{N}_j - j, \qquad \bar{N}_{j1} = N_j,$$

whence, by (64),

$$\bar{M}_{k1} < M_k. \tag{65}$$

If $Z_{\alpha 1} \neq 0$ at $P_{\alpha 1}$, we conclude by the usual argument based on the condition of Lemma 14.1, that $P_{\alpha 1}$ is of multiplicity less than ν for $F_{\alpha 1}$. If $Z_{\alpha 1} = 0$ at $P_{\alpha 1}$, then $X_{\alpha 1}$, $Y_{\alpha 1}$ and $Z_{\alpha 1}$ are uniformizing parameters of $P_{\alpha 1}$. The expression of $\Omega_{\alpha 1}$ is similar to that of Ω_{α} in (55). Moreover, by (65), we still have $\bar{M}_{k1} < k$, while N_k has not been affected. But since, by (65), the exponent M_k has been replaced by a lower exponent, we conclude that after a finite number of quadratic transformations applied successively to F_{α}, $F_{\alpha 1}$, $F_{\alpha 2}$, $\cdots$ we must ultimately get a surface F_{α} on which the center of the valuation v is of multiplicity less than ν. This surface *may* be reached before the sum $\bar{M}_{kl} + \bar{N}_{kl}$ becomes less than k, but at any rate the inequalities

$$\bar{M}_k + \bar{N}_k > \bar{M}_{k1} + \bar{N}_{k1} > \bar{M}_{k2} + \bar{N}_{k2} > \cdots$$

guarantee that it will be reached after a finite number of steps. This concludes the proof of the local reduction theorem.

Part III

Reduction of the singularities of an embedded surface by quadratic and monoidal transformations of the ambient V_3

16. Preparation of the singular locus of the surface F

The singular locus of F—i.e., the set of singular points of F—is a proper algebraic subvariety of F. This statement, whose proof is trivial in the case of ground fields of characteristic zero or of perfect fields of characteristic p, will be proved in all generality in our paper to be printed elsewhere. Anticipating this result, we have therefore that the singular locus of F consists of a finite number of singular curves and of isolated points. A singular curve which is s-fold for F contains at most a finite number of points whose multiplicity for F is greater than s (Lemma 6.5). Hence the multiplicities of the singular points of F have a maximum, say ν, and the singular points of highest multiplicity ν form again a proper algebraic subvariety of F. Let $\Delta_1, \Delta_2, \cdots, \Delta_h$ be the irreducible ν-fold curves of F. All points of each Δ_i are exactly ν-fold for F. Each Δ_i may have a finite number of singular points, and these, together with the intersections of pairs of curves Δ_i, constitute the singular locus of the total ν-fold curve $\Delta_1 + \Delta_2 + \cdots + \Delta_h$ of F.

The singularities of each curve Δ_i can be resolved by quadratic transformation of the ambient space V (Theorem 4, section 9). However, a quadratic transformation will generally introduce new singular curves on the transform of F. We must analyze the situation more closely.

Let T be a quadratic transformation of the ambient V, with center at a singular point P of one of the curves Δ_i, say of Δ_1. Let V' and F' be the T-transforms of V and of F respectively. The quotient ring of any point of F, different from P, is not affected by T, nor is the quotient ring of any irreducible curve on F affected by T. Hence if A' is any point of F' and if A is the corresponding point of F, then $m_{F'}(A') = m_F(A) \leqq \nu$, if $A \neq P$. If $A = P$, then we know (Lemma 3.2) that $m_{F'}(A') \leqq m_F(A)$. Hence ν is still the maximum multiplicity of the singular points of F', and to each curve Δ_i $(i = 1, 2, \cdots, h)$ there will correspond on F' an irreducible curve Δ_i' of multiplicity ν. If F' possesses another ν-fold curve Δ', different from each Δ_i', such a curve can only correspond to the point P. But in that case that ν-fold curve Δ' must be irreducible and *free from singularities* (Theorem 1, section 3), and naturally will remain free from singularities under all further successive quadratic transformations which we may have to apply.

We therefore conclude that after a finite number of successive quadratic transformations we shall get a variety V^* and a birational transform F^* of F, on V^*, such that *the irreducible singular curves of F^*, of highest multiplicity ν, are all free from singularities.* We shall assume therefore that the original surface F already satisfies this condition, i.e. we have now that each curve Δ_i is free from singularities, $i = 1, 2, \cdots, h$.

The total ν-fold curve $\Delta_1 + \Delta_2 + \cdots + \Delta_h$ may still have singularities at points which are common to two or more curves Δ_i. To these points we apply quadratic transformations. Let then P be a singular point of the total ν-fold curve of F, and let V' and F' be respectively the transforms of V and of F by the quadratic transformation of center P. If Δ_i goes through P and if $T[\Delta_i] = \Delta_i'$, then Δ_i' will carry exacty one point, say P_i', which corresponds to P, since P is a simple point of Δ_i. If Δ_j is another ν-fold curve through P and if P_j' is the point of Δ_j' $(= T[\Delta_j])$ which corresponds to P, then $P_i' = P_j'$ if and only if the two curves Δ_i and Δ_j have the same tangential direction at P (see section 8). If T creates a new ν-fold curve, say Γ', then, by Lemma 8.4, the tangential direction of Δ_i' at P_i' is different from the tangential direction of Γ' at P_i'. Hence if no two of the curves $\Delta_1', \Delta_2', \cdots, \Delta_h'$ have a common point, then the only singularities of the total ν-fold curve of F' are normal crossings.

In the contrary case we repeat the procedure, i.e. we apply to F' a quadratic transformation T' whose center is a common point of two curves Δ_i' and Δ_j', getting V'', F'' and ν-fold curves $\Delta_1'', \Delta_2'', \cdots, \Delta_h''$. We observe that if T has created a new ν-fold curve Γ', then the ν-fold Γ'' $(= T[\Gamma'])$ of F'' will not intersect any of the ν-fold curves $\Delta_1'', \Delta_2'', \cdots, \Delta_h''$, by the remark just made, while its intersection with the ν-fold curve of F'' which possibly has been created by T' will be a normal crossing. We assert that in this fashion we will get, after a finite number of quadratic transformations, models $V^{(s)}$ and $F^{(s)}$ such that no two of the transforms $\Delta_1^{(s)}, \Delta_2^{(s)}, \cdots, \Delta_h^{(s)}$ of $\Delta_1, \Delta_2, \cdots, \Delta_h$ have common points, and *that consequently the only singularities of the total ν-fold curve of $F^{(s)}$ are normal crossings.* To prove our assertion, let us suppose that the assertion is false. We will have an infinite sequence of models $V, F; V', F'; \cdots ; V^{(i)}, F^{(i)}; \cdots$ and an infinite sequence of points $P, P^{(1)}, P^{(2)}, \cdots, P^{(i)}, \cdots$ such that: 1) $F^{(i+1)}$ is the transform of $F^{(i)}$ by a quadratic transformation of center $P^{(i)}$; 2) P is a common point of two curves Δ_i, say of Δ_1 and Δ_2, and the successive transforms $\Delta_1', \Delta_1'', \cdots ; \Delta_2', \Delta_2'', \cdots$ are such that $P^{(i)}$ is a common point of $\Delta_1^{(i)}$ and $\Delta_2^{(i)}$. Let Ω denote the union of the quotient rings $Q_{F^{(i)}}(P^{(i)})$. This ring is a proper ring, (i.e. not a field). Hence Ω is contained in the valuation ring of at least one zero-dimensional valuation v. The center of v on $F^{(i)}$ is the point $P^{(i)}$.

We reach the absurd conclusion, in view of Lemma 11.1, that the valuation v must be composed with a divisor whose center on F is at the same time the curve Δ_1 *and* the curve Δ_2. Our assertion is therefore established.

From now on we shall assume that the above preparation of the singular ν-fold locus of F has already been accomplished. Therefore the ν-fold curves $\Delta_1, \Delta_2, \cdots, \Delta_h$ of F are now not only free from singularities, but their mutual intersections are normal crossings and no intersection is common to more than two curves Δ_i. At this stage it may be well to recall our stipulation made in the beginning of section 1, according to which all our considerations apply only to *simple points of the ambient space* V. All points of F, in particular all singular points of F, which fall at singular points of V are excluded.

17. Reduction of the singularities of F (Theorem of Beppo Levi)

In section 10 we have defined permissible transformations, and we have observed that, in general, there is a certain degree of freedom in the selection of a sequence of permissible transformations. This was due to the fact that if a ν-fold point P lies on a ν-fold curve Δ, then a quadratic transformation of center P *and* a monoidal transformation of center Δ are both permissible. We shall now restrict the type of permissible transformations to be used, by stipulating that *a quadratic transformation of center P shall be used only if P is an isolated ν-fold point.* A sequence of permissible transformations in which each transformation is restricted according to the above stipulation shall be called *a normal sequence of permissible transformations.* If then

$$F, F_1, F_2, \cdots, F_i, \cdots$$

is a sequence of birational transforms of F by successive permissible quadratic and monoidal transformations $T, T_1, T_2, \cdots, T_i, \cdots$ which form a normal sequence, then the following conditions are satisfied:

1) either T_i is a monoidal transformation whose center is a ν-fold curve of F_i,

2) or T_i is a quadratic transformation whose center is an isolated ν-fold point of F_i.

It is clear that the theorem of reduction of singularities of the algebraic surface F is established if it can be proved that by a suitable birational transformation it is possible to lower the maximum of the multiplicities of the singular points of F. This has been established by Beppo Levi [2] in the classical case. We state the theorem of Beppo Levi in the following form:

THEOREM 6. *Let ν be the maximum multiplicity of the singular points of F and let the total ν-fold curve of F have only normal crossings as singularities. Under these conditions, if*

$$F, F_1, F_2, \cdots, F_i, \cdots \tag{66}$$

is a sequence of surfaces obtained from F by permissible transformations forming a normal sequence (the center of each transformation being a ν-fold point or a ν-fold curve), then the sequence (66) *is necessarily finite.*

Before we proceed with the proof of Theorem 6, we make a few observations concerning normal sequences of permissible transformations which will clarify the reduction process called for by this theorem. In building up a normal sequence of permissible transformations we have a certain degree of freedom, for at each step we may select any ν-fold curve or any isolated ν-fold point as center of the transformation. But granted that each *maximal* normal sequence is finite, it is not difficult to see that *any two maximal normal sequences of transformations lead to one and the same surface.* This can be seen as follows. First, if $P^{(1)}$ and $P^{(2)}$ are two isolated ν-fold points on F, and if we apply to F a quadratic transformation of center $P^{(1)}$, getting a surface F_1, and then apply to F_1 a quadratic transformation whose center is the transform of the point $P^{(2)}$, getting a surface F_2, then it is clear that the same surface F_2 would be obtained if the points $P^{(1)}$, $P^{(2)}$ were interchanged. Similarly, if P and Δ are respectively a ν-fold point and a ν-fold curve of F, then P is a permissible center of a quadratic transformation only if P is isolated, hence at any rate not on Δ. But then it is clear that also in this case a quadratic transformation of center P followed by a monoidal transformation whose center is the transform of Δ, leads to the same surface as does a monoidal transformation of center Δ followed by a quadratic transformation whose center is the transform of P. This possibility of interchanging two consecutive transformations of a normal sequence of transformations extends also to the case where both transformations are monoidal whose centers are represented by ν-fold curves of F. This is obvious if the two ν-fold curves do not meet. If they do meet, then their intersections are normal crossings, and our assertion follows from Lemma 8.3.

It follows that the context of Theorem 6 is not narrowed down to a more special procedure if we, for instance, proceed as follows. If the surface F possesses ν-fold curves, we apply first only monoidal transformations. Then Theorem 6 says, in part, that after a finite number of steps all ν-fold curves will be eliminated[9]. *This, however, we know already from Theorem 4′* (section 9). To the new surface, free from ν-fold curves, which is thus obtained, we now

[9] We again recall that our considerations are limited only to the points or curves on F which are not singular for the ambient V_3.

apply successive quadratic transformations whose centers are isolated ν-fold points. This second stage of the process terminates whenever a quadratic transformation creates a new ν-fold curve. This curve must then first be resolved by monoidal transformations before quadratic transformations come again into play. Theorem 6 asserts that by this procedure, where quadratic and monoidal transformations alternate according to a well-defined pattern, we must obtain after a finite number of steps a birational transform of F whose singular points are all of multiplicity less than ν.

Let $F, F_1, F_2, \cdots$, be a sequence of birational transforms of F, and let $P, P_1, P_2, \cdots$ be a corresponding sequence of points, $P_i \in F_i$. If T_i denotes the birational transformation from F_i to F_{i+1}, then we shall say that $\{P_i\}$ is a *normal sequence of ν-fold points*, if the following conditions are satisfied: 1) each point P_i is a ν-fold point of its carrier F_i, and is either an isolated ν-fold point or lies on only one ν-fold curve Δ of which it is a simple point, or is a normal crossing of two ν-fold curves; 2) if P_i is an isolated ν-fold point, then T_i is a quadratic transformation with center P_i; 3) if P_i is not isolated, then T_i is a monoidal transformation whose center is a ν-fold curve through P_i; 4) P_i and P_{i+1} are corresponding points under T_i.

It is clear that Theorem 6 is equivalent to the assertion that *a normal sequence of consecutive ν-fold points is necessarily finite.* Let us suppose for a moment that Theorem 6 is false. Then some ν-fold point P of F will give rise to an infinite normal sequence of consecutive ν-fold points

$$P, P_1, P_2, \cdots, P_i, \cdots. \tag{67}$$

The quotient rings $Q_{F_i}(P_i)$ form a strictly ascending chain of rings, and their union will be contained in the valuation ring of at least one zero-dimensional valuation. Let v be one such valuation. Now *using this valuation v* we apply to F an *arbitrary sequence of permissible transformations* as described in section 10 in connection with the local reduction theorem 5. Let $F, F_1', F_2', \cdots, F_i'$, $\cdots$ be the corresponding sequence of consecutive transforms of F, and let P_i' be the center of v on F_i'. By the local reduction theorem there exists *some sequence* of permissible transformations such that only a finite number of the points P_i' are ν-fold. We shall prove, however,—and this will establish Theorem 6—the following lemma:

LEMMA. *If the normal sequence* (67) *of consecutive ν-fold points is infinite, then any sequence of consecutive centers*

$$P, P_1', P_2', \cdots, P_i', \cdots \tag{68}$$

of the valuation v, obtained by using an arbitrary sequence of permissible transformations, consists entirely of ν-fold points; *or, in other words*; *if a normal sequence of permissible transformations is not capable of lowering the center of a given valuation* v (the first center being the point P) *then no sequence of permissible transformations will lower the multiplicity of the center of v.*

This lemma expresses, so to speak, the dominant character of a normal sequence of transformations in regard to the reduction process. The proof of this lemma is given in the next section.

18. Proof of the lemma

If two points A and B are members of a *normal* sequence of consecutive ν-fold points, and if A precedes B in that sequence, we shall express this by the notation: $A \prec B$. Thus, assuming—as we do—that (67) is a normal sequence, we have $P_i \prec P_j$ if $i < j$ $(i = 0, 1, 2, \cdots, P_0 = P)$.

We observe that any point P_i in the sequence (67), which is not an isolated ν-fold point, can be followed *immediately* at most by a finite number of ν-fold points which are not isolated. This follows from the fact that any ν-fold curve is ultimately eliminated by monoidal transformations. Let therefore $P_{i_1} = A_1$ and $P_{i_2} = A_2$ be the first two isolated ν-fold points *after* P in the sequence (67). We shall prove that

$$P_1' \prec A_2. \tag{69}$$

The proof of the relation (69) *will establish the lemma.* For the relation (69) implies in the first place that P_1' is a ν-fold point of F_1' (this is part of the definition of the symbol $\prec$). In the second place, it implies that the (uniquely determined) normal sequence of consecutive centers of the valuation v which begins with the ν-fold point P_1' contains the point A_2. Hence that sequence also contains the points $P_{i_2+1}, P_{i_2+2}, \cdots$, since

$$P_{i_2}, P_{i_2+1}, P_{i_2+2}, \cdots$$

is *the* normal sequence of ν-fold points which begins with P_{i_2}. Therefore, under the assumption of the lemma, also the normal sequence of consecutive centers of v which begins with P_1' is infinite. Thus the assumption is true not only for the point P, but also for P_1'. But then, by the same argument, also P_2' is a ν-fold point of F_2', and the normal sequence of consecutive centers of v which begins with P_2' is also infinite, and so on. Consequently all the points P_i' are ν-fold as asserted.

We shall now proceed to prove the relation (69). Let T and T' denote the birational transformation respectively from F to F_1 and from F to F_1'. *If P is an isolated ν-fold point,* then a quadratic transformation of center P is the only permissible transformation. Hence in this case $T = T'$, $F_1 = F_1'$, $P_1' = P_1$, and there is nothing to prove (since $P_1 \prec A_2$). We therefore may assume that P lies on a ν-fold curve Δ. In that case T is necessarily monoidal. As to T', it may be either monoidal or quadratic. If T' is monoidal, then its center is either Δ or another ν-fold curve Γ through P (this last case may arise if P is a normal crossing). In the first case $T = T'$, and again there is nothing to prove. In the second case we have essentially the same situation, since we may interchange the order in which the curves Δ and Γ are treated without affecting the *normal* sequence (67) (only the order of the *non isolated* ν-fold points, i.e. the points between P and A_1 and between A_1 and A_2, will be affected).

Hence it is sufficient to prove relation (69) under the following assumptions: *T is a monoidal transformation whose center is a ν-fold curve Δ through P, and T' is a quadratic transformation of center P.*

We select uniformizing parameters t_1, t_2, t_3 of $P(V)$ in such a fashion that t_1 and t_2 are uniformizing parameters of $\Delta(V)$. Let V_1 and V_1' denote respectively the transforms of V under T and T', and let F be defined in $Q(P)$ by the principal ideal (ω). Since P_1 is a ν-fold point, we have [section 7, b)] that T may be assumed to be given by the equations (10) of that section, i.e.:

$$(70) \qquad t_{11} = t_1, \qquad t_{12} = t_2/t_1, \qquad t_{13} = t_3,$$

where t_{11}, t_{12} and t_{13} are uniformizing parameters of $P_1(V_1)$; and that ω has the form (see the proof of Lemma 6.3):

$$(71) \qquad \omega = t_2^{\nu} + \sum_{i=0}^{\nu} \alpha_i t_1^i t_2^{\nu-i},$$

where the α_i are non units in $Q_V(P)$. As to the equations of the quadratic transformation T' we must consider separately two cases according as t_3/t_1 is or is not finite at P_1'.

First case. *t_3/t_1 is finite at P_1'.* Let $\tau_3 = t_3/t_1$ and let a be the residue of τ_3 at P_1'. Denoting by k^* the residue field of P, let a be a root of an irreducible polynomial $h^*(u)$ with coefficients in k^* and let $h(\tau_3)$ be a polynomial with coefficients in $Q_V(P)$ such that $h(u)$ reduces to $h^*(u)$ if the coefficients of $h(u)$ are replaced by their P-residues. Then t_1, t_2/t_1, and $h(\tau_3)$ are uniformizing parameters of $P_1'(V_1')$, so that we can write the equations of T' as follows:

$$(72) \qquad t_1' = t_1, \qquad t_2' = t_2/t_1 = t_{12}, \qquad t_3' = h(\tau_3) = h(t_3/t_1).$$

The quotient ring $Q_{V_1}(P_1)$ consists of all quotients $f(t_2')/g(t_2')$, where f and g are polynomials with coefficients in $Q_V(P)$ and where $g^*(0) \neq 0$ [see section 5, d)][10].

Similarly [see section 2, c)], the quotient ring $Q_{V_1'}(P_1')$ consists of all quotients $f(t_2', \tau_3)/g(t_2', \tau_3)$, where f and g are polynomials with coefficients in $Q_V(P)$ and where $g^*(0, a) \neq 0$. This shows that

$$(73) \qquad Q_{V_1}(P_1) \subsetneqq Q_{V_1'}(P_1').$$

Both surfaces F_1 and F_1' are defined in the respective quotient rings $Q_{V_1}(P_1)$ and $Q_{V_1'}(P_1')$ by the principal ideal (ω_1), where $\omega_1 = \omega/t_1^{\nu}$, i.e., by (71),

$$(74) \qquad \omega_1 = t_{12}^{\nu} + \sum_{i=0}^{\nu} \alpha_i t_{12}^{\nu-i}.$$

Since P_1 is a ν-fold point of F_1, we must have $\omega_1 \equiv 0(t_{11}, t_{12}, t_{13})^{\nu}$ in $Q_{V_1}(P_1)$. Hence $\omega_1 \equiv 0(t_1', t_2', t_1' \cdot \tau_3)^{\nu} \equiv 0(t_1', t_2')^{\nu}$. This shows that the curve L' defined·

[10] Here and throughout this section, an asterisk affixed to a polynomial with coefficients in $Q_V(P)$ indicates that the coefficients of the polynomial have been replaced by their P-residues.

by the ideal (t'_1, t'_2) is a ν-fold curve of F'_1. This ν-fold curve is obviously one which has been created by the quadratic transformation T'.

We now apply to V'_1 a monoidal transformation M with center L'. Let V^*_2 and F^*_2 be the transforms of V'_1 and of F'_1 respectively, under M, and let P^*_2 be the center of the valuation v on F^*_2. It has been pointed out above that the elements α_i in (71) are non units in $Q_V(P)$. Hence all α_i belong to the principal ideal (t'_1) *in the ring* $Q_{V'_1}(P'_1)$. Therefore, by (74), $\omega_1 \equiv t_2'^{\nu}(t'_1)$. Since $\omega_1 \equiv 0(t'_1, t'_2)^{\nu}$, and since $\omega_1 = 0$ on F, it follows immediately that $v(t'_2) \geqq v(t'_1)$[11], and *consequently* t'_2/t'_1 *is finite* at P^*_2. Therefore if we put $\tau^* = t'_2/t'_1$ and if we denote by b the P^*_2-residue of τ^* (b = v-residue of τ^*), then the quotient ring $Q_{V^*_2}(P^*_2)$ consists of all quotients $f'(\tau^*)/g'(\tau^*)$, where f' and g' are polynomials with coefficients in $Q_{V'_1}(P'_1)$ and where $g'^*(b) \neq 0$; here g'^* is the polynomial obtained from g' by replacing the coefficients of g' by their P'-residues.

On the other hand, let us apply to F_1 a quadratic transformation of center P_1 and let us denote by V^{**}_2 and F^{**}_2 the transforms of V_1 and F_1 respectively. Let P^{**}_2 be the center of v on F^{**}_2. We know already that the quotients t_{13}/t_{11} (= $t_3/t_1 = \tau_3$) and t_{12}/t_{11} (= $t'_2/t'_1 = \tau^*$) have finite residue in the valuation v, hence are finite at P^{**}_2. Let us compare the two quotient rings $\mathfrak{J}^* = Q_{F^*_2}(P^*_2)$ and $\mathfrak{J}^{**} = Q_{F^{**}_2}(P^{**}_2)$. We have: $Q_V(P) \subset Q_{V_1}(P_1) \subset \mathfrak{J}^{**}$, and also $t'_2 \in \mathfrak{J}^{**}$, $\tau_3 \in \mathfrak{J}^{**}$. From this it follows that $Q_{V'_1}(P'_1) \subseteqq \mathfrak{J}^{**}$. Since also τ^* is contained in $\mathfrak{J}^{**}$ we conclude that

$$\mathfrak{J}^* \subseteqq \mathfrak{J}^{**}.$$

On the other hand, since $\tau^* \in \mathfrak{J}^*$ we conclude from (73) that

$$\mathfrak{J}^{**} \subseteqq \mathfrak{J}^*.$$

Therefore $\mathfrak{J}^* = \mathfrak{J}^{**}$, i.e. *the point* P^*_2 *can be obtained directly from the point* P_1 *by applying to* F_1 *a quadratic transformation.* Before we draw conclusions from this result, we consider the second case.

SECOND CASE. $t_1/t_3 = 0$ at P'_1. The uniformizing parameters of $Q_{V'_1}(P'_1)$ are now:

$$(75) \qquad t'_1 = t_1/t_3, \quad t'_2 = t_2/t_3, \quad t'_3 = t_3,$$

and $Q_{V'_1}(P'_1)$ consists of all quotients $f(t'_1, t'_2)/g(t'_1, t'_2)$ of polynomials in t'_1, t'_2 with coefficients in $Q_V(P)$, such that $g(0, 0) \neq 0$. The surface F'_1 is now defined in $Q_{V'_1}(P'_1)$ by the principal ideal (ω'_1), where [see (71)],

$$\omega'_1 = \omega/t_3^{\nu} = t_2'^{\nu} + \sum_{i=0}^{\nu} \alpha_i t_1'^{i} t_2'^{\nu-i}.$$

The curve Δ'_1 defined by the ideal (t'_1, t'_2) is ν-fold for F'_1. This curve is merely the transform Δ' of the ν-fold curve Δ under T'.[12] We shall show now that

[11] The elements t'_1 and t'_2 are now thought of as elements of the field of rational functions on the surface F'_1.

[12] This shows incidentally that the second case under consideration arises when the point P'_1 corresponds to the tangential direction of Δ at P.

there is another ν-fold curve of F_1' through P_1', namely the curve L' defined by the ideal (t_2', t_3'). To see this, we prove that ω can be written as a finite sum of the form

$$\omega = \sum_i \beta_{(i)} t_1^{i_1} t_2^{i_2} t_3^{i_3}, \qquad \beta_{(i)} \in Q_V(P), \tag{76}$$

where the exponents i_1, i_2, i_3 in each term satisfy the inequality:

$$i_1 + 2i_2 + i_3 \geqq 2\nu. \tag{77}$$

The proof is immediate. We observe that if we have a monomial $\beta t_1^{i_1} t_2^{i_2} t_3^{i_3}$ in which β is a non-unit in $Q_V(P)$, then $\beta = \beta_1 t_1 + \beta_2 t_2 + \beta_3 t_3$, $\beta_i \in Q_V(P)$, and consequently the monomial can be written as a sum of three similar monomials with higher exponents. Hence we can certainly write ω as a finite sum of the form:

$$\omega = \sum_j \epsilon_{(j)} t_1^{j_1} t_2^{j_2} t_3^{j_3} + \sum_i \beta_{(i)} t_1^{i_1} t_2^{i_2} t_3^{i_3},$$

where the $\epsilon_{(j)}$ are units in $Q_V(P)$, $j_1 + 2j_2 + j_3 < 2\nu$ and $i_1 + 2i_2 + i_3 \geqq 2\nu$. Since Δ is ν-fold for F, we also must have $j_1 + j_2 \geqq \nu$, $i_1 + i_2 \geqq \nu$. We find then [see (74)]:

$$\omega_1 = \sum_j \epsilon_{(j)} t_{11}^{j_1+j_2-\nu} t_{12}^{j_2} t_{13}^{j_3} + \sum_i \beta_{(i)} t_{11}^{i_1+i_2-\nu} t_{12}^{i_2} t_{13}^{i_3}.$$

The point P_1 being ν-fold for F_1, we must have $\omega_1 \equiv 0(t_{11}, t_{12}, t_{13})^\nu$. Each term in the second sum belongs to the ideal $(t_{11}, t_{12}, t_{13})^\nu$, in view of the inequality $i_1 + 2i_2 + i_3 \geqq 2\nu$. On the other hand, no term of the first sum belongs to the ideal $(t_{11}, t_{12}, t_{13})^\nu$, since $j_1 + 2j_2 + j_3 < 2\nu$ and since $\epsilon_{(j)}$ are units in $Q_V(P)$. Since no two terms in the first sum have the same exponents, we have a contradiction, unless the first sum is not present at all. This proves that ω is of the form (76) where the exponents satisfy (77). From (76) we obtain the following expression of ω_1':

$$\omega_1' = \sum_i \beta_{(i)} t_1'^{i_1} t_2'^{i_2} t_3'^{i_1+i_2+i_3-\nu},$$

and since in each term the sum of the exponents of t_2' and t_3' is not less than ν, in view of (77), the curve L' is indeed ν-fold for F_1', as asserted.

The point P_1' is therefore a normal crossing of the two ν-fold curves Δ_1' and L'.

We now apply to F_1' a birational transformation M' combined of two monoidal transformations of centers Δ_1' and L' respectively. The order in which these two monoidal transformations are applied is immaterial (see Lemma 8.3). Let V_2^* and F_2^* be the transforms of V_1' and F_1' respectively, and let P_2^* be the center of v on F_2^*. From the expression (74) of ω_1 we find that t_{12}/t_{13} has finite v-residue [i.e. $v(t_{12}/t_{13}) \geqq 0$], since $v(t_{11}) > v(t_{13})$ and we cannot have simultaneously $v(t_{12}) < v(t_{11})$, $v(t_{12}) < v(t_{13})$. Now $t_{12}/t_{13} = t_2/t_1t_3 = t_2'/t_1't_3'$. Hence, taking into account the equations (14) of section 8, we conclude that the equations of M' are of the form

$$t_1^* = t_1', \qquad \tau_2^* = t_2'/t_1't_3', \qquad t_3^* = t_3',$$

where t_1^*, t_3^* are two of the uniformizing parameters of $P_2^*(V_2^*)$, while τ_2^* is not necessarily zero at P_2^*, but is at any rate finite at P_2^*. From these equations of M' and from (75) and (70) we find

$$t_1^* = t_{11}/t_{13}, \qquad \tau_2^* = t_{12}/t_{13}, \qquad t_3^* = t_{13},$$

and from these equations we conclude as in the preceding case that *the point P_2^* can be obtained directly from the point P_1 by a quadratic transformation of center P_1.*

In either case we have the following situation: a) $P_1' < P_2^*$, since F_2^* is obtained from F_1' by one or two monoidal transformations with center at ν-fold curves; b) P_2^* is obtained from P_1 by a quadratic transformation of center P_1. If $i_1 = 1$, i.e. if P_1 is isolated, then $P_2^* = P_2$ and (69) follows, since $P_1' < P_2^* = P_2$ and either P_2 is A_2 or $P_2 < A_2$. If $i_1 > 1$, then P_1 and P_2^* are in the same relation as P and P_1', and there will therefore exist a point P_3^* such that: a) $P_2^* < P_3^*$; b) P_3^* is obtainable directly from P_2 by a quadratic transformation. If then $i_1 = 2$, then $P_3^* = P_3$, and since $P_1' < P_3^*$ the relation (69) follows. If $i_1 > 2$, we repeat the same procedures. Ultimately, after i_1 steps we will be able to conclude that $P_1' < A_2$. This completes the proof of the lemma and also of Theorem 6.

PART IV

REDUCTION OF THE SINGULARITIES OF THREE-DIMENSIONAL VARIETIES

19. Elimination of the simple fundamental locus of a birational transformation of a three-dimensional variety

Let V and V' be two birationally equivalent three-dimensional irreducible algebraic varieties. We denote by T the birational transformation of V into V'. We mean by the "simple fundamental locus" of T the set of fundamental curves and of isolated fundamental points of T which are *simple* for V.

Suppose that by another birational transformation T^* we have succeeded in transforming V into a variety V^* in such a fashion that the following two conditions are satisfied: a) T^{*-1} has no fundamental points on V^*; b) if a point P^* of V^* corresponds to a simple point of V then P^* is not fundamental for the birational correspondence between V^* and V'. Under these conditions we shall say that *the birational transformation T^* has eliminated the simple fundamental locus of T.*

A preliminary but essential step in our reduction of singularities of three-dimensional varieties, a step to which most of this last part of our investigation has to be dedicated, consists in proving the following theorem:

THEOREM 7. *Given a birational transformation T of a three-dimensional variety V into another variety V', it is possible to eliminate the simple fundamental locus of*

T by successive quadratic and monoidal transformations and to carry out this elimination in such a fashion that simple points of V are transformed into simple points by the successive transformations.

In the proof of this theorem we must therefore show that there exists a sequence of quadratic and monoidal transformations to be applied to V and to the successive transforms of V, such that if T^* denotes the product of these transformations and if V^* denotes the transform of V by T^*, then the following conditions are satisfied:

a) If a point P^* of V^* corresponds to a simple point of V, then P^* is not fundamental for the birational correspondence between V^* and V'.

b) If a point P^* of V^* corresponds to a simple point of V, then P^* itself is a simple point of V^*.

c) T^{*-1} has no fundamental points (on V^*).

Since the only transformations to be used are quadratic and monoidal, not only is condition c) automatically satisfied, but we also have that to every point of V^* there will correspond a unique point of V. Moreover, the quadratic and monoidal transformations which will actually be used will always have their centers on the simple fundamental locus of T and on the successive transforms of that locus. Hence T^* will not affect any point of V which is not fundamental for T. Or, more precisely: if P^* is any point of V^* and if the corresponding point P of V is not fundamental for T, then $Q_V(P) = Q_{V^*}(P^*)$ [whereas for an arbitrary point P^* of V^* we can only assert that $Q_V(P) \subseteqq Q_{V^*}(P^*)$].

For the purpose of the proof of Theorem 7 we find it necessary to re-formulate that theorem in terms of *linear systems of surfaces* on V and of the *base loci* of such systems. Quite generally, let Σ be a field of algebraic functions of r independent variables, and let $\omega_0, \omega_1, \cdots, \omega_n$ be elements of some extension field of Σ such that the quotients ω_i/ω_j are elements of Σ. If V is a given model of Σ, the elements ω_i determine a unique set of divisors $\mathfrak{A}_0, \mathfrak{A}_1, \cdots, \mathfrak{A}_n$ of Σ, of the first kind with respect to V, free from common factors and such that

$$\omega_i/\omega_j = \mathfrak{A}_i/\mathfrak{A}_j.$$

Each divisor $\mathfrak{A}_i$ determines on V (and—if V is normal—is defined by) an $(r-1)$-dimensional subvariety F_i whose irreducible components are counted to well-defined multiplicities. If $\lambda_0, \lambda_1, \cdots, \lambda_n$ are arbitrary elements in the ground field k, not all zero, then we can write

$$(\lambda_0\omega_0 + \lambda_1\omega_1 + \cdots + \lambda_n\omega_n)/\omega_0 = \mathfrak{A}(\lambda)/\mathfrak{A}_0,$$

where $\mathfrak{A} = \mathfrak{A}(\lambda)$ is a well-defined divisor of Σ of the first kind with respect to V. Let $F = F(\lambda)$ be the $(r-1)$-dimensional subvariety of V defined by $\mathfrak{A}(\lambda)$. The set $|F|$ of all subvarieties F obtained by letting the λ's vary in k is called a *linear system*; it is the linear system determined by the elements $\omega_0, \omega_1, \cdots, \omega_n$, or rather by the linear (projective) space spanned by the element $\rho(\lambda_0\omega_0 + \lambda_1\omega_1 + \cdots + \lambda_n\omega_n)$, where ρ is an arbitrary factor of proportionality. This system $|F|$ contains of course the particular members $F_0, F_1, \cdots, F_n$, and since the divisors $\mathfrak{A}_0, \mathfrak{A}_1, \cdots, \mathfrak{A}_n$ have no common factor *the system* $|F|$ *is free from fixed components* (a fixed component being an $(r - 1)$-dimensional variety common to all varieties of the system.)

Let $\eta_0, \eta_1, \cdots, \eta_m$ be the homogeneous coördinates of the general point of V, and let $y_0, y_1, \cdots, y_m$ be the corresponding homogeneous coördinates in the ambient projective space of V. Since the quotients ω_i/ω_j belong to Σ, the elements $\omega_0, \omega_1, \cdots, \omega_n$ are proportional to forms $f_0(\eta), f_1(\eta), \cdots, f_n(\eta)$ of like degree in $\eta_0, \eta_1, \cdots, \eta_m$, with coefficients in the ground field k. It is then clear that the linear system $|F|$ is cut out on V, outside of fixed components, by the linear system of hypersurfaces

$$\lambda_0 f_0(y) + \lambda_1 f_1(y) + \cdots + \lambda_n f_n(y) = 0. \tag{78}$$

If $|F'|$ is the linear system determined by $\{\omega_0, \omega_1, \cdots, \omega_n\}$ on another model V' of Σ, then the two linear systems $|F|$ and $|F'|$ are regarded as corresponding linear systems in the birational correspondence between the two models. This fixes without ambiguity the *law of transformation of linear systems* under birational transformations. By stipulating, as we did, that both systems $|F|$ and $|F'|$ be despoiled of fixed components, we have in reality sidetracked certain difficulties which would unavoidably arise upon a more thorough examination of the situation (in the case of algebraic surfaces see our monograph [4], p. 41). However, for our present purpose the transformation law for linear systems as formulated above is entirely adequate.

Let us suppose now that $\omega_0, \omega_1, \cdots, \omega_n$ are the homogeneous coordinates of the general point of a projective model V' of the field Σ[13]. In this case the linear system $|F'|$ determined on V' by $\{\omega_0, \omega_1, \cdots, \omega_n\}$ is merely the system of hyperplane sections of V', according to (78). On any other projective model V the corresponding linear system $|F|$ is therefore the birational image of the system of hyperplane sections of V'. It is thus seen that given an arbitrary linear system $|F|$ on a variety V, and provided that the quotients $\omega_1/\omega_0, \omega_2/\omega_0, \cdots, \omega_n/\omega_0$ generate the field Σ of rational functions on V, then this linear system determines uniquely a birational transformation of V into another variety V' on which the linear system which corresponds to $|F|$ is the system of hyperplane sections, i.e. uniquely to within a projective transformation of V'.

Going back to Theorem 7, let $|F|$ be the linear system of surfaces on V which is the image of the system of hyperplane sections of V'. It is cut out on V by a linear system of hypersurfaces (78). It is known ([8], p. 528) that if H is a point or a curve on V such that the quotient ring $Q_V(H)$ is integrally closed (i.e. if V is *locally normal* at H; see [8], p. 512) then H is fundamental for the birational transformation T if and only if H belongs to the base locus of the

[13] In order that it should be possible to regard the ω's as homogeneous coordinates of the general point of a projective model of Σ, it is necessary and sufficient to have: $\Sigma = k(\omega_1/\omega_0, \omega_2/\omega_0, \cdots, \omega_n/\omega_0)$.

linear system $|F|$, i.e. if H lies on every F of the system. The condition that $Q_V(H)$ be integrally closed is certainly satisfied if H is a simple point or a simple curve of V. Hence *the simple fundamental locus of T coincides with the base locus of* $|F|$, outside those base curves and isolated base points of $|F|$ which are singular for V.

It is now clear how Theorem 7 can be formulated in terms of the linear system $|F|$ and of the base locus of $|F|$. In re-formulating below this theorem, we drop the condition that $|F|$ be the image of the system of hyperplane sections of a variety V' birationally equivalent to V. Thus the theorem which we are going to state presently is somewhat stronger than Theorem 7.

THEOREM 7'. *Given a linear system $|F|$ (free from fixed components) on a three-dimensional variety V, it is possible to transform V by successive quadratic and monoidal transformations into another variety V^* in such a fashion as to satisfy the following two conditions:*

a) *If $|F^*|$ is the linear system (free from fixed components) on V^* which corresponds to $|F|$, then every base point of $|F^*|$ corresponds necessarily to a singular point of V.*

b) *If a point P^* of V^* corresponds to a simple point of V, then P^* is itself a simple point of V^*.*

20. Linear systems free from singular base points: preparation of the base locus of $|F|$

Let the linear system $|F|$ be cut out on V by the system of hypersurfaces (78), and let $F_0, F_1, \cdots, F_n$ be the particular surfaces in the system which are cut out by the hypersurfaces $f_0 = 0, f_1 = 0, \cdots, f_n = 0$ respectively. Let P be a base point of $|F|$ which is simple for V. We say that P is a *singular base point of the linear system* $|F|$ if P is a singular point of *each* surface F in the system. A *simple base point of* $|F|$ is a base point which is non-singular; such a base point must therefore be simple for at least one surface F in the linear system $|F|$.

LEMMA 20.1. *If a simple point P of V is a simple base point of $|F|$, it is necessarily a simple point of at least one of the $n + 1$ surfaces $F_0, F_1, \cdots, F_n$.*

PROOF. Without loss of generality we may assume that $\eta_0 \neq 0$ at P, where—we recall—$\eta_0, \eta_1, \cdots, \eta_m$ denote the homogeneous coördinates of the general point of V. If we put $\omega_i = f_i(1, \eta_1/\eta_0, \cdots, \eta_m/\eta_0)$, then $\omega_0, \omega_1, \cdots, \omega_n$ are elements of $Q_V(P)$. By the theorem of unique factorization in $Q_V(P)$ let δ be the highest common divisor of $\omega_0, \omega_1, \cdots, \omega_n$ and let $\omega_i = \delta\zeta_i, \zeta_i \in Q_V(P)$. It is then clear that each surface $F(\lambda)$ of $|F|$ is defined locally, at P, by the principal ideal $(\lambda_0\zeta_0 + \lambda_1\zeta_1 + \cdots + \lambda_n\zeta_n)$ in $Q_V(P)$, i.e. this principal ideal gives all the components of $F(\lambda)$ which pass through P, and the exponents which occur in the decomposition of the element $\lambda_0\zeta_0 + \lambda_1\zeta_1 + \cdots + \lambda_n\zeta_n$ into prime factors give the multiplicities of the corresponding components of $F(\lambda)$. The point P is simple for $F(\lambda)$ if and only if the leading form of $\lambda_0\zeta_0 + \lambda_1\zeta_1 + \cdots + \lambda_n\zeta_n$ is linear (whence $F(\lambda)$ is necessarily locally irreducible at P). For that, however, it is necessary that the leading form of at least one of the elements ζ_i be linear, and this proves the lemma.

In this and in the next two sections we prove Theorem 7′ in the special case where *the linear system* $|F|$ *has no singular base points* (at simple points of V; *singular points of* V *are left out entirely from our considerations*). In this case the base locus of $|F|$ consists of simple base curves of $|F|$ and of isolated simple base points. Our first step is to eliminated by quadratic transformations the singularities of the total base curve, other than normal crossings. We prove namely the following lemma:

LEMMA 20.2. *By successive quadratic transformations it is possible to transform* V *into a variety* V^* *satisfying condition* b) *of Theorem* 7′ *and such that also the following conditions are satisfied*: 1) *if* $|F^*|$ *denotes the linear system on* V^* *which corresponds to* $|F|$, *then* $|F^*|$ *has no singular base points* (*outside the singular points of* V^*); 2) *each irreducible simple base curve of* $|F^*|$ *is free from singularities* (*outside the singular locus of* V^*); 3) *if two irreducible simple base curves of* $|F^*|$ *have a point* P^* *in common* (P^* — *a simple point of* V^*), *then the tangential directions of the two curves at* P^* *are distinct, and* P^* *does not belong to a third irreducible simple base curve.*

PROOF. Let P be a simple point of V which is a singular point of the total base curve of $|F|$. We apply to V a quadratic transformation T of center P and we denote by V' and $|F'|$ the T-transforms of V and of $|F|$ respectively. Let Φ' denote the surface, free from singularities [see section 2, a) and b)], which corresponds to the point P. What is the base locus of $|F'|$? In the first place, it is clear that if H' is a base curve or an isolated base point of $|F'|$, then H' either corresponds to a base variety of $|F|$, *of the same dimension* as H', or H' corresponds to the point P. Hence the base of locus of $|F'|$ consists 1) of the proper transform of the base locus of $|F|$ and possibly 2) of a certain number of new base curves and of isolated base points which lie on the surface Φ'. We are interested in these new base curves on Φ'. By hypothesis, P is a simple base point of $|F|$, hence, by Lemma 20.1, P is a simple point of one of the surfaces $F_0, F_1, \cdots, F_n$. Let, say, P be a simple point of F_0. If $\mathfrak{P}$ denotes the divisor determined by the surface Φ', then ζ_0 is exactly divisible by $\mathfrak{P}$ [see section 2, d) and Lemma 3.1; in the present case ζ_0 plays the role of ω, and ν is equal to 1], while each ζ_i is at least divisible by the first power of $\mathfrak{P}$. Hence if $F'_0, F'_1, \cdots, F'_n$ denote the members of the system $|F'|$ which correspond respectively to $F_0, F_1, \cdots, F_n$, then we can assert that *the surface* Φ' *is not a component of* F'_0. Hence F'_0 is the proper transform of F_0, i.e. $F'_0 = T[F_0]$. Therefore the intersection of F'_0 with Φ' is *an irreducible curve, free from singularities* (Theorem 1, section 3, where again $\nu = 1$). This is the only curve on Φ' which *may* be a base curve of $|F'|$, and this curve is certainly a simple curve of F'_0. We conclude that our quadratic transformation T may create *at most* one new base curve of the linear system under consideration, and that such a new base curve *is necessarily free from singularities* and is a simple base curve. Moreover all its points are simple base points of $|F'|$, for they are all simple for F'_0. The rest of the proof is achieved by considerations which are identical with those developed in section 16, Part III. We only note that since the quadratic transformations used have centers at simple points, conditions analogous to condition b) of Theorem 7′ and to condition 1) of Lemma 20.2 are certainly satisfied.

From now on we shall therefore suppose that the base locus of our original system $|F|$ satisfies conditions analogous to conditions 2) and 3) of Lemma 20.2.

21. Linear systems free from singular base points: elimination of the base curves by monoidal transformations

We have proved elsewhere ([7], p. 592), for algebraic surfaces, theorems analogous to theorems 7 and 7′. We therefore know that given a linear system of curves on an algebraic surface (free from fixed components) it is possible to eliminate the base points of the system by successive quadratic transformations. Actually this result has been proved again in the present paper: it is an immediate consequence of Lemma 11.1. Only a change of "dimension terminology" is necessary in order to express this result as a statement concerning linear systems $|F|$ of $(r-1)$-dimensional varieties on a V_r and the $(r-2)$-dimensional base varieties of $|F|$. In particular, the above result applies to the base curves of our linear system of surfaces $|F|$ on the three-dimensional variety V, and it can be used as a stepping stone toward the elimination of these base curves by successive monoidal transformations. However, if $|F|$ was an arbitrary linear system, this extrapolation from base points of a linear system of curves on a surface to base curves of a linear system of surfaces on a three-dimensional variety would have met with a serious difficulty. We proceed to present this difficulty and to show why it disappears in the present case where we deal with a linear system $|F|$ free from singular base points.

Let Δ be an irreducible simple base curve of the system $|F|$. If Σ denotes the field of rational functions on V and if we adjoin to the ground field k an element ξ of Σ *whose Δ-residue is transcendental over k*, then over the new ground field $k(\xi)$ the variety V becomes a "surface," the system $|F|$ becomes a linear system of "curves" on the "surface" $V/k(\xi)$ and Δ becomes a base "point" of $|F|$. Let T be a monoidal transformation of V/k, of center Δ, and let V'/k and $|F'|$ be the T-transforms of V/k and of $|F|$ respectively. The transformation T is a quadratic transformation of $V/k(\xi)$ into $V'/k(\xi)$ with center at the "point" $\Delta/k(\xi)$. If the system $|F'|$ on $V'/k(\xi)$ still possesses base "points" which correspond to the point Δ, these points represent base curves of $|F'|$ on V'/k, *which correspond to the curve* Δ under the monoidal transformation T. To these curves we again apply monoidal transformation and so we continue until the base "point" Δ is eliminated.

All this refers to $k(\xi)$ as ground field. Let us see how our monoidal transformation T actually affects the base locus of $|F|$ over our original ground field k. The transform $T[\Delta]$ of Δ is an irreducible surface Φ'. Since Δ is free from singularities (outside the singular locus of V), all points of Φ' which correspond to simple points of V are simple both for Φ' and for V' [section 5, a)]. Thus condition analogous to condition b) of Theorem 7′ is satisfied, since we know (section 5) that any two-dimensional component of the total transform $T\{\Delta\}$ of Δ, other than Φ', must correspond to points of Δ which are singular for V.

Now let P' be a base point of $|F'|$ which corresponds to a simple point P of V. If P is not on Δ, then it has not been affected by T, and hence P', as well as P, is a simple base point. Suppose that P is on Δ. There exists a surface F, say F_0, in $|F|$, which has at P a simple point. If $\mathfrak{P}$ denotes the divisor defined by the irreducible surface Φ', then we know [section 5, e) and Lemma 6.1] that ω_0 is exactly divisible by $\mathfrak{P}$. Since each element ω_i $(i = 0, 1, \cdots, n)$ is divisible at least by the first power of $\mathfrak{P}$, it follows that Φ' is not a component of F_0', where F_0' denotes that member of the linear system $|F'|$ which corresponds to F_0. Hence F_0' is the proper transform of F_0, i.e. $F_0' = T[F_0]$. Since P is a simple point of F_0 and of Δ, it follows that also P' is a simple point of F_0' (Lemma 6.3). Hence the base points of $|F|$, which (outside the singular locus of V) are by hypothesis simple base points, are transformed into simple base points of $|F'|$.

The points of V' which correspond to singular points of V do not interest us. With the understanding that these points are left out altogether from our consideration, we analyse the possible base curves of $|F'|$.

In general, there could be curves on Φ' which are base curves of $|F'|$ and *which correspond to special points*[14] *of* Δ. Such new base curves, created by the monoidal transformation T, are lost when we pass to the new ground field $k(\xi)$, since ξ is not a transcendental on any such curve. The knowledge that the base "point" $\Delta/k(\xi)$ can be eliminated by "quadratic" transformations would have been of little help to us if in the course of eliminating base "points" over $k(\xi)$ we would find again and again new base curves which correspond to special points of the centers of the successive monoidal transformations. Fortunately, *as a consequence of our assumption that* $|F|$ *has no singular base points, this is not the case.* For suppose that $|F'|$ possesses a base curve Γ' which corresponds to a point P of Δ (P-simple for V). Again assuming that P is a simple point of F_0, it would follow by Lemma 6.2 that $m_{F_0}(P) > m_{F_0}(\Delta)$. This is a contradiction, since $m_{F_0}(P) = m_{F_0}(\Delta) = 1$.

We conclude that any base curve of $|F'|$ which lies on Φ' must correspond to Δ itself, and not to a point of Δ. By Theorem 2, section 6, there can only be at most one such curve, say Δ', since Δ is a simple curve for some surface F in the system $|F|$. This curve Δ' is free from singularities, and (Lemma 8.2) has only normal crossings with other base curves of $|F'|$. If we apply the same treatment to Δ' as we have applied to Δ and if we do that for all the base curves of $|F|$, we see that ultimately we will succeed in eliminating all the base curves

[14] We also may have on Φ' a finite number of isolated base points of $|F'|$ which corresponds to special points of Δ. These points do not interest us in this section, but they have to be taken care of in the next section where we carry out the elimination of isolated base points.

of $|F|$, and that, without violating condition b) of Theorem 7′. We therefore assume from now on that the linear system $|F|$ on our original surface has only isolated simple base points (outside of the singular locus of V).

22. Linear systems free from singular base points: elimination of the isolated base points

Let now P be an isolated simple base point of $|F|$. We apply to V a quadratic transformation T of center P and we denote by V' and $|F'|$ the T-transforms of V and of $|F|$ respectively. We know already from section 20 that T will resolve the base point P *either* into a finite number of isolated simple base points of $|F'|$ *or* into a simple base curve Δ' of $|F'|$. In the first case we continue with quadratic transformations, taking as centers the new isolated base points. In the second case we first eliminate the base curve Δ' by monoidal transformation, as in the preceding section. However, after Δ' has been eliminated, isolated base points (all simple) may remain which correspond to special points of Δ' (see footnote 14). To these we apply again quadratic transformations, and so we continue indefinitely. *We assert that this process must terminate after a finite number of steps.* For suppose that the contrary is true. We will have then an infinite sequence of successive transforms of V:

$$V, \ V^{(1)}, \ V^{(2)}, \cdots, \ V^{(i)}, \cdots$$

and of successive simple points $P, P^{(1)}, P^{(2)}, \cdots, P^{(i)}, \cdots$, where $P^{(i)} \epsilon V^{(i)}$, with the following properties:

1. $Q(P) \subset Q(P^{(1)}) \subset Q(P^{(2)}) \subset \cdots$
2. If $|F^{(i)}|$ denotes the linear system on $V^{(i)}$ which corresponds to $|F|$, then $P^{(i)}$ is an isolated base point of $F^{(i)}$.

Let $\bar{V}$ be the variety birationally equivalent to V on which the linear system which corresponds to $|F|$ is the system of hyperplane sections (section 19). Since $P^{(i)}$ is an isolated base point of the linear system $|F^{(i)}|$, it is an isolated fundamental point of the birational transformation which carries $V^{(i)}$ into $\bar{V}$. Since on the other hand $P^{(i)}$ is a simple point of $V^{(i)}$, it follows (see [8], p. 532) that to $P^{(i)}$ there must correspond on $\bar{V}$ a pure two-dimensional variety. Let then $H_i = \{H_{i1}, H_{i2}, \cdots\}$ be the set of irreducible surfaces H_i on $\bar{V}$ which correspond to the point $P^{(i)}$. Since $Q(P_i) \subset Q(P_j)$ if $i < j$, it follows that $H \supseteqq H_1 \supseteqq H_2 \supseteqq \cdots$. Since no set H_i is empty and every H_i is a finite set, it follows that they have an element in common, say G. Then G is an irreducible surface on $\bar{V}$ which corresponds to every point $P^{(i)}$. Let $\mathfrak{P}$ be a divisor of the field Σ whose center on $\bar{V}$ is the surface G. This divisor is then of second kind with respect to each variety $V^{(i)}$, its center on $V^{(i)}$ being the point $P^{(i)}$. If then $\mathfrak{p}^{(i)}$ denotes the ideal of non units in $Q(P^{(i)})$, then $v_{\mathfrak{P}}(\mathfrak{p}^{(i)}) > 0$, where $v_{\mathfrak{P}}(\mathfrak{p}^{(i)})$ denotes the least value assumed by the elements of $\mathfrak{p}^{(i)}$ in the discrete valuation defined by the divisor $\mathfrak{P}$. Now, each $V^{(i)}$ is obtained from $V^{(i-1)}$ by a quadratic transformation of center $P^{(i-1)}$, followed up, perhaps, by a finite number of monoidal transformations. Consequently (see proof of Lemma 9.1)

$$v_{\mathfrak{P}}(\mathfrak{p}^{(1)}) > v_{\mathfrak{P}}(\mathfrak{p}^{(2)}) > \cdots > v_{\mathfrak{P}}(\mathfrak{p}^{(i)}) > \cdots > 0.$$

This yields a contradiction (since each $v_{\mathfrak{P}}(\mathfrak{p}^{(i)})$ is a positive integer) and establishes Theorem 7′ for linear systems free from singular base points.

23. Reduction of the singular base points of a linear system of surfaces

Let now $|F|$ be an arbitrary linear system on V. To complete the proof of Theorem 7′ it is only necessary to show that it is possible to eliminate the singular base points of $|F|$ by quadratic and monoidal transformations, always subject to the additional condition that simple points of V be transformed into simple points. This we proceed to show by using on one hand the reduction theorem for surfaces (Theorem 6, section 17) and on the other hand the theorem of Bertini on the variable singular points of a variety which varies in a linear system. The proof of this theorem of Bertini for abstract varieties is given by us in [9]. For our present application we need some of the concepts which are developed in [9] and the abstract formulation of Bertini's theorem. For simplicity of exposition we shall confine ourselves to three-dimensional varieties.

Denoting as usual the homogeneous coördinates of the general point of V by $\eta_0, \eta_1, \cdots, \eta_m$, let $|F|$ be our linear system (78) on V, free from fixed components, and let $t_0, t_1, \cdots, t_n$ be indeterminates (i.e. we assume that the t's are algebraically independent over the field $k(\eta_0, \eta_1, \cdots, \eta_m)$. We adjoin to the ground field k these indeterminates and we denote by K the extended field $k(t_0, t_1, \cdots, t_n)$. The variety V/k can be regarded as a three-dimensional variety V/K over the new ground field, with the same general point $(\eta_0, \eta_1, \cdots, \eta_m)$ as V.

Every irreducible subvariety W/k of V/k, with general point $(\bar{\eta}_0, \bar{\eta}_1, \cdots, \bar{\eta}_m)$ defines a subvariety W/K of V/K, of the same dimension as W, with the same general point as W and with the property that $t_0, t_1, \cdots, t_n$ are algebraically independent over k $(\bar{\eta}_0, \bar{\eta}_1, \cdots, \bar{\eta}_m)$. The subvariety W/K shall be called *the extension of the subvariety* W/k. Conversely, if an irreducible subvariety W/K of V/K, with general point $(\bar{\eta}_0, \bar{\eta}_1, \cdots, \bar{\eta}_m)$, is such that $t_0, t_1, \cdots, t_n$ are algebraically independent over k $(\bar{\eta}_0, \bar{\eta}_1, \cdots, \bar{\eta}_m)$, then W/K is the extension of a unique subvariety W/k of V/k.

It can be proved that *the singular locus of V/K is the extension of the singular locus of V/k.*

For an *arbitrary* irreducible subvariety W^*/K of V/K, with general point $(\eta_0^*, \eta_1^*, \cdots, \eta_m^*)$, it is true that there exists a unique subvariety W/k of V/k, with general point $(\bar{\eta}_0, \bar{\eta}_1, \cdots, \bar{\eta}_m)$, such that

$$k[\eta_0^*, \eta_1^*, \cdots, \eta_m^*] \cong k(\bar{\eta}_0, \bar{\eta}_1, \cdots, \bar{\eta}_m),$$

where the isomorphism is such that η_i^* and $\bar{\eta}_i$ are corresponding elements. The variety W/k shall be called the *contraction* of W^*/K. We have always:

$$W^*/K \subseteqq \text{extension } W/K \text{ of } W/k,$$

whence dimension of $W^*/K \leqq$ dimension of W/k.

The essence of our formulation of Bertini's theorem consists in regarding the *linear system* of surfaces (78) on V/k as *one irreducible surface* F^*/K on V/K. This surface is defined by a general point $(\zeta_0, \zeta_1, \cdots, \zeta_m)$ satisfying the following conditions:

1) $k[\zeta_0, \zeta_1, \cdots, \zeta_m] \cong k[\eta_0, \eta_1, \cdots, \eta_m]$,
whence the contraction of F^*/K is the entire variety V/k.

2) $t_0 f_0(\zeta) + t_1 f_1(\zeta) + \cdots + t_n f_n(\zeta) = 0$.
It is proved that *if a variety* W/K *is an extension of* W/k *then* W/K *lies on* F^*/K *if and only if* W/k *belongs to the base locus of the linear system* $|F|$, *and that* W/K *is singular for* F^*/K *if and only if* W/k *is a singular base variety of* $|F|$.

After these preliminaries we state the theorem of Bertini on the variable singular points of $|F|$ as follows:

THEOREM OF BERTINI. *If* W_1/K *is an irreducible singular subvariety of* F^*/K, *then the contracted subvariety* W/k *of* V/k *is either singular for* V/k *or belongs to the base locus of the linear system* $|F|$.

From this theorem and on the basis of the preliminary results stated above, we conclude that, outside of varieties W_1/K which correspond to singular subvarieties of V/k, the surface F^*/K may possess only the following singular curves and points:

a) Singular curves which are extensions of singular base curves of $|F|$.

b) Isolated singular points which are extensions of isolated singular base points of $|F|$.

c) Isolated singular points whose contractions are simple base curves of $|F|$ (it can be proved that along these curves the surfaces of the linear system have *variable* singular points, *provided the ground field* k *is of characteristic zero;* see [9].)

If the variety V/k is subjected to a birational transformation which carries V and $|F|$ into V' and $|F'|$ respectively, then V/K and F^*/K undergo a corresponding birational transformation, and the transform F'^*/K of F^*/K has the same relationship to $|F'|$ as F^*/K has to $|F|$. In other words, the surface F^*/K is a birational covariant of the linear system $|F|$. In particular, if we apply to V/k a quadratic or a monoidal transformation with center W/k, then V/K and F^*/K undergo respectively a quadratic or a monoidal transformation whose center is the extension W/K of W/k. If we then apply the reduction Theorem 6 of section 17 to the surface F^*/K and to its ambient variety V/K, we see that by successive monoidal and quadratic transformation of V/k it is possible to resolve the singular curves a) and the singular points b) of the surface F^*/K, i.e. it is possible to resolve the singular base locus of the linear system $|F|$. It should be pointed out that a preparation of the singular curves a) of F^*/K is a preliminary step of the reduction process (section 16). The quadratic transformations involved in this step are actually quadratic transformations over k (i.e. of V/k) since the singular points of the singular curves a) are necessarily extensions of the singular points of the corresponding singular base curves of $|F|$. Finally, we also point out that we do nothing to the isolated singular points c) of F^*/K; they do not correspond to singular base loci of $|F|$ and their reduction is not necessary for our purposes. Besides, since these singular points are not extensions of points over k, the transformations needed for their reduction are not directly expressable as transformations of V/k.

This completes the proof of Theorems 7 and 7′.

24. A lemma on simple points of an algebraic surface

Let F and F' be two birationally equivalent algebraic surfaces.

LEMMA. *If P and P' are corresponding simple points in the birational correspondence between F and F' and if $Q(P) \subseteqq Q(P')$, then either $Q(P) = Q(P')$ or P' can be obtained from P by successive quadratic transformations.*

This lemma implies the following corresponding result for $(r-2)$-dimensional subvarieties of a V_r:

If W and W' are corresponding $(r-2)$-dimensional varieties of a birational correspondence between two r-dimensional varieties V_r and V'_r and if $Q(W) \subseteqq Q(W')$, then either $Q(W) = Q(W')$ or W' can be obtained from W by successive monoidal transformations.

PROOF OF THE LEMMA. Assuming that $Q(P) \neq Q(P')$, we have to show that there exists a finite sequence of birational transforms of F:

$$F, F_1, F_2, \cdots, F_n,$$

and a corresponding sequence of points $P_1, \cdots, P_n, P_i \in F_i$, such that: 1) F_i is the transform of F_{i-1} by a quadratic transformation of center $P_{i-1}(F_0 = F, P_0 = P)$; 2) P_{i-1} and P_i are corresponding points under this quadratic transformation; 3) $Q(P_n) = Q(P')$.

Let x and y be uniformizing parameters of $P(F)$, and let P_1 be a point of F_1 which corresponds to P' (in the birational correspondence between F_1 and F'). Since $Q(P) \subset Q(P')$, every valuation of center P' on F' has its center at P on V. Consequently also P_1 and P are corresponding points. Without loss of generality we may assume that y/x is finite at P_1. Then x can be taken as one of a pair of uniformizing parameters of $Q_{F_1}(P_1)$.

We shall show now that necessarily $Q(P_1) \subseteqq Q(P')$. Assume the contrary. Then P' is a fundamental point of the birational correspondence between F' and F_1. Hence to P' there must correspond on F_1 at least one irreducible curve through P_1. Any such curve must also correspond to P, since $Q(P) \subset Q(P')$. But the irreducible curve Γ_1 defined in $Q(P_1)$ by the principal ideal (x) is the only curve on F_1 which corresponds to P. Hence this is the only curve through P_1 which corresponds to P'.

Let $\mathfrak{P}$ be the divisor defined by the curve Γ_1, and let $\mathfrak{m}$ and $\mathfrak{m}'$ be the ideals of non units in $Q_F(P)$ and $Q_{F'}(P')$ respectively. We have $v_{\mathfrak{P}}(x) = 1$ and $v_{\mathfrak{P}}(m') > 0$. Hence the element x, which is certainly an element of $\mathfrak{m}'$, does not belong to $\mathfrak{m}'^2$. Now P is fundamental for the birational correspondence between F and F', and therefore there must exist on F' an irreducible curve Δ' through P' which corresponds to P. Since both x and y vanish on Δ', this curve must be defined by a prime element of $Q(P')$ which divides *in* $Q(P')$ both x and y. But x itself is a prime element of $Q(P')$, since $x \not\equiv 0(\mathfrak{m}'^2)$. Hence x

divides y, $y/x \in Q(P')$, whence $Q(P_1) \subseteqq Q(P')$. This shows that our assumption that $Q(P_1) \nsubseteqq Q(P')$ is absurd.

If $Q(P_1) = Q(P')$, there is nothing more to prove. If $Q(P_1)$ is a proper subring of $Q(P')$, we pass to the quadratic transform F_2 of F_1 taking P_1 as center, and we find as before that F_2 carries a point P_2 such that $Q(P_1) \subset Q(P_2) \subseteqq Q(P')$. Since the union of the quotient rings of successive corresponding points P, P_1, P_2, $\cdots$, obtained by successive quadratic transformations, is a valuation ring (see [5], p. 681, Theorem 10; also [7], p. 591, Lemma 2, footnote 14) and since $Q(P')$ is not a valuation ring nor can it contain a valuation ring, it follows that after a finite number of quadratic transformations we must get a point P_n such that $Q(P_n) = Q(P')$, q.e.d.

25. Reduction of the singularities of three-dimensional varieties

Let Σ be a field of algebraic functions of three independent variables over an arbitrary ground field k, and let N be an arbitrary set of zero-dimensional valuations of Σ. A finite set of models V_1, V_2, $\cdots$, V_n of Σ shall be called a *resolving system* of N if every valuation in N has a simple center on at least one of the modes V_i. The existence of resolving systems for the set of all valuations of Σ will be established by us in a separate paper, for any field Σ of algebraic functions for which the theorem of local uniformization holds true (hence, in particular, for function fields over ground fields of characteristic zero; see [6])[15]. Assuming the existence of a resolving system for the totality of valuations of Σ, we have shown in [7], p. 584 that the theorem of the reduction of singularities is equivalent to the following

FUNDAMENTAL THEOREM. *If there exists a resolving system of N consisting of two models V and V', then there also exists a resolving model for N (i.e. a model of Σ on which every valuation in N has a simple center).*

Our proof of the fundamental theorem will be based exclusively on Theorem 7 of section 19 and on the lemma of the preceding section.

Let V and V' be a resolving pair of N, and let T denote the birational transformation which carries V into V'. We first apply to V the birational transformation T^* of Theorem 7 which eliminates the simple fundamental locus of T on V. Let V^* denote the T^*-transform of V. Since no singular point of V^* corresponds to a simple point of V, also V^* and V' form a resolving pair of N. Let V_1 denote the join of the two varieties V^* and V' (see [8], p. 516). *I assert that also V_1 and V' form a resolving pair for N.* For let v be any valuation in N and let P, P', P^* and P_1 be the centers of v respectively on V, V', V^*, V_1. If P' is a simple point, there is nothing to prove. Let P' be a singular point. Then P is necessarily a simple point. If P is not a fundamental point of T, then $Q(P) \supseteqq Q(P')$, and on the other hand $Q(P) = Q(P^*)$, since in this case P has not been affected by the quadratic and monoidal transformations which compose the transformation T^* (see section 19). Hence $Q(P^*) \supseteqq Q(P')$ and

---

[15] For the proof of the existence of finite resolving systems in the classical case (k = field of complex numbers) see [6], p. 855, and in the case of abstract surfaces see [7], p. 592. For the general case see the Bulletin note quoted in footnote **, Introduction.

hence ([8], p. 516) $Q(P_1) = Q(P^*) = Q(P)$. Therefore P_1 is a simple point. If, however, P *is* fundamental for T, then it belongs to the *simple* fundamental locus of T, and therefore P^* is not fundamental for the birational transformation between V^* and V'. Hence $Q(P') \subseteqq Q(P^*) = Q(P_1)$, and since P^* is a simple point of V^* (for P is simple for V) we again conclude that P_1 is a simple point. Thus, one of the two points P_1, P' must be simple, and therefore V_1 and V' constitute indeed a resolving pair of N.

The net effect of the preceding considerations is that we have replaced the original resolving pair (V, V') by the resolving pair (V_1, V'), and for this last resolving pair it is true that the birational correspondence between the two models has no fundamental points on one of them (namely on V_1, since V_1 is the join of V' and of another model). We may therefore assume that this condition is already satisfied for our original resolving pair. *Accordingly, we assume that T^{-1} has no fundamental points on V'.* Then if H and H' are any two corresponding loci of V and V', the following relation will always hold true:

$$Q(H) \subseteqq Q(H'). \tag{79}$$

We now again apply Theorem 7, and we eliminate the simple fundamental locus of T on V. Let V^* be the birational transform of V which is obtained as a result of this elimination. Again we see that V^* and V' constitute a resolving pair of N.

Let v be an arbitrary valuation in N and let P, P' and P^* be the centers of v on V, V' and V^* respectively. We analyse the relationship between the quotient rings of these three points and we divide this analysis into several cases. For simplicity we denote by F_s the simple fundamental locus of T on V. *We include in F_s all the points of that locus*, hence also points of a simple base curve which are singular for V.

Case A: $P \notin F_s$. In this case P is either a singular point or is not fundamental for T. In both cases the following is true:

$$P' \textit{ is simple and } Q(P') \supseteqq Q(P^*). \tag{A}$$

For $P \notin F_s$ implies that P has not been affected by the birational transformation from V to V^*. Hence $Q(P) = Q(P^*)$, and therefore $Q(P') \supseteqq Q(P^*)$, by (79). Moreover P' must be a simple point even if P is simple, for if P is simple it is not fundamental for T ($P \notin F_s$), i.e. $Q(P) \supseteqq Q(P')$, and hence $Q(P) = Q(P')$, by (79). Hence assertion (A) is established.

Case B. *P is a simple point and $P \in F_s$.* In this case it is obvious that

$$P^* \textit{ is simple and } Q(P^*) \supseteqq Q(P'). \tag{B}$$

Case C. *P is a singular point and $P \in F_s$.* In this case P' is necessarily a simple point. If P' is not fundamental for the birational correspondence between V' and V^*, then we will have:

$$P' \textit{ is simple and } Q(P') \supseteqq Q(P^*). \tag{C'}$$

In (A), (B) and (C') we have always the same situation, namely of the two points P' and P^* one has the property that it is a simple point and that its quotient ring contains the quotient ring of the other point. If we now pass to the join of V' and V^*, the center of the valuation v will therefore be a simple point. There remains to consider the case C *under the hypothesis that P' is fundamental for the birational correspondence between V' and V^*.* We proceed to study this case.

The point P' is either an isolated fundamental point of the birational transformation between V' and V^* or lies on some fundamental curve of this transformation. We are interested especially in this second possibility. Let Γ' be a *fundamental curve* of the birational transformation between V' and V^* which passes through P'. Since T^{-1} has no fundamental points on V', $T^{-1}(\Gamma')$ is either a unique point or a unique curve on V (note that since P' is simple for V', also Γ' is simple). I *assert that $T^{-1}(\Gamma')$ is necessarily a point.* For suppose the contrary, and let $T^{-1}(\Gamma') = \Gamma$, an irreducible curve on V.

Suppose that $\Gamma \notin F_s$. Then Γ is regular for the birational correspondence between V and V^*. If then Γ^* is the corresponding curve on V^*, then we will have $Q(\Gamma) = Q(\Gamma^*)$. On the other hand we have $Q(\Gamma') \supseteqq Q(\Gamma)$ [by (79)]. Hence $Q(\Gamma') \supseteqq Q(\Gamma^*)$, in contradiction with our assumption that Γ' is fundamental for the birational correspondence between V' and V^*.

Hence Γ must belong to F_s, and is therefore a simple curve of V. We have now the following situation: Γ and Γ' are corresponding *simple* curves of V and V' and, by (79), $Q(\Gamma') \supset Q(\Gamma)$ [not $Q(\Gamma') = Q(\Gamma)$, since $\Gamma \in F_s$]. By the lemma of the preceding section, Γ' can be obtained from Γ by a sequence of monoidal transformations. But in eliminating the simple fundamental curves of T we have used just monoidal transformations. Hence the variety V^* must contain a curve Γ^* such that $Q(\Gamma^*) = Q(\Gamma')$. This is in contradiction with our assumption that Γ' is fundamental for the birational correspondence between V' and V^*.

We have therefore proved that $T^{-1}(\Gamma')$ is necessarily a point. This point is necessarily the point P, since $P' \in \Gamma'$.

Accordingly we consider on V' those isolated fundamental points O' and those fundamental curves Γ' of the birational transformation between V' and V^* *which correspond to singular points of V which lie on F_s and which actually carry centers of valuations in N.* Since (V, V') is a resolving pair of N it is clear that these points O' and these curves Γ' must be simple for V', *and that no point of a Γ' which is singular for V' can occur as center of a valuation in N.* According to Theorem 7 applied to V' and V^* we can eliminate these fundamental points and curves, and by the *italicized* remark just made we do not have to be concerned about the fate of the points of a curve Γ' which are singular for V'. We denote by Ω' the set of the points O' and curves Γ' and we also denote by V_1' the birational transform of V' obtained in the course of the elimination of the fundamental points O' and the fundamental curves Γ'.

Now let again v be any valuation in N and let P, P', P^* and P_1' be the centers of v on V, V', V^* and V_1' respectively. In view of (A), (B), (C′) we find immediately the following:

(A_1) *If $P \notin F_s$, then P_1' is simple and $Q(P_1') \supseteqq (P^*)$.*

For by (A), P' is not fundamental for the birational transformation between V' and V^*. Hence P' is not affected when we pass from V' to V_1', and consequently $Q(P') = Q(P_1')$.

(B_1) *If P is simple and $P \in F_s$, then P^* is simple and $Q(P^*) \supseteqq Q(P_1')$.*

For in this case $P' \notin \Omega'$, whence $Q(P') = Q(P_1')$.

The relation (C') yields (since $P' \notin \Omega'$):

(C_1') *P_1' is simple and $Q(P_1') \supseteqq Q(P^*)$.*

There remains the case C, under the hypothesis that $P' \in \Omega'$. In this case, in view of our elimination of the fundamental locus Ω', we must have $Q(P_1') \supseteqq Q(P^*)$, and moreover P_1' is simple since P' is simple. Hence in this case the statement C_1' still holds true.

We have now in all cases, i.e. for any valuation v in N, the situation whereby of the two centers of v on V^* and V_1' one always has the property that it is a simple point and that its quotient ring contains the quotient ring of the second center. *Therefore the join of V^* and V_1' is a resolving model of N.* This completes the proof of the fundamental theorem.

THE JOHNS HOPKINS UNIVERSITY.

REFERENCES

[1] W. KRULL, *Dimensionstheorie in Stellenringen*, J. Reine Angew. Math., vol. 179 (1938), pp. 204–226.
[2] BEPPO LEVI: *Risoluzione delle singolarità puntuali delle superficie algebriche*, Atti Accad. Sci. Torino, vol. 33 (1897), pp. 66–86.
[3] F. K. SCHMIDT: *Ueber die Erhaltung der Kettensätze der Idealtheorie bei beliebigen endlichen Körpererweiterungen*, Math. Zeit., vol. 41 (1936), pp. 443–450.
[4] O. ZARISKI: *Algebraic surfaces*, Ergebnisse der Mathematik und ihrer Grenzgebiete, vol. 3 (1935).
[5] O. ZARISKI: *The reduction of the singularities of an algebraic surface*, Ann. of Math., vol. 40 (1939), pp. 639–689.
[6] O. ZARISKI: *Local uniformization on algebraic varieties*, Ann. of Math., vol. 41 (1940), pp. 852–896.
[7] O. ZARISKI: *A simplified proof for the resolution of singularities of an algebraic surface*, Ann. of Math., vol. 43 (1942), pp. 583–593.
[8] O. ZARISKI: *Foundations of a general theory of birational correspondences*, Trans. Amer. Math. Soc., vol. 53 (1943), pp. 490–542.
[9] O. ZARISKI: *The theorem of Bertini on the variable singular points of a linear system of varieties*, Trans. Amer. Math. Soc., vol. 56 (1944), pp. 130–140.

LE PROBLÈME DE LA RÉDUCTION DES SINGULARITÉS D'UNE VARIÉTÉ ALGÉBRIQUE [1];

par M. O. ZARISKI.

1. Le problème général de la réduction des singularités a jusqu'ici résisté à tous les efforts de ceux qui ont essayé de le résoudre. Je veux seulement parler des solutions véritablement mathématiques, non des pseudo-solutions. Depuis le temps où j'ai donné la solution pour les variétés de dimension 3 (sur des corps de base de caractéristique nulle), dix ans se sont écoulés et aucun progrès nouveau n'a été réalisé. Ma seule contribution personnelle depuis cette époque prit la forme d'une revue critique dont l'objet principal était de décourager les chercheurs futurs de prendre des libertés injustifiées avec ce difficile et important problème.

Récemment, je revins au problème de la réduction que j'avais laissé provisoirement de côté, depuis dix ans déjà. Quoique mon présent travail sur ce problème n'ait pas encore été poussé à fond, les résultats que j'ai obtenus, et que je désire présenter ici, représentent, je crois, un progrès certain en vue de la solution générale.

Mon présent essai sur le problème de la réduction des singularités est essentiellement un retour à une vieille idée, utilisée pour la première fois, avec un succès complet, par Jung dans sa solution du problème de l'uniformisation locale des surfaces algébriques. La même idée avait été aussi utilisée par Kneser dans son étude, seulement esquissée et par suite incertaine, du problème général de l'uniformisation locale des variétés supérieures.

Pour la simplicité de l'exposition je supposerai que le corps des constantes est le corps des nombres complexes. Tous nos

(1) Cet article est le résumé d'une conférence qui devait avoir lieu à Paris en mai 1953.

résultats sont valables pour des corps de base de caractéristique nulle.

2. Pour uniformiser une surface algébrique F, donnée par

$$f(x, y, z) = 0, \tag{1}$$

Jung considère la courbe de diramation [2] $D(x, y) = 0$ de la fonction algébrique z définie par (1). L'uniformisation en un point (x_0, y_0, z_0) de F ne présente pas de difficultés si le point (x_0, y_0) dans le plan des (x, y) n'appartient pas à D ou est un point simple de D. Si (x_0, y_0) est un point singulier de D, Jung résout la singularité de D en (x_0, y_0) par des transformations localement quadratiques appliquées au voisinage du point $(x_0\, y_0)$. De cette manière, il ramène le problème au cas pour lequel (x_0, y_0) est un point double ordinaire de D, et passe ensuite à la solution du problème de l'uniformisation locale dans ce cas simple.

Nous souhaitons adapter cette méthode de Jung au problème global de la réduction des singularités des variétés supérieures. Notre adaptation conduit à un lemme préliminaire fondamental dont l'énoncé requiert une soigneuse préparation.

Soit V_r une variété non mixte de dimension r [3], non nécessairement irréductible.

Nous ferons les hypothèses suivantes :

(1) V_r est plongée dans une variété F_{r+1} de dimension $r+1$, produit direct $F_r \times S_1$ d'une variété F_r non singulière, de dimension r, et d'une droite projective S_1.

(2) Si T est la restriction à V_r de la projection de F_{r+1} sur F_r, T applique V_r *sur* F_r et T^{-1} transforme *chaque* point de F_r en un nombre *fini* de points de V_r.

On appellera les conditions (1) et (2) les « *conditions d'immersion* ».

Soit n le degré de T^{-1}. Le lieu des points Q de F_r dont le transformé $T^{-1}\{Q\}$ est constitué par moins de n points est

---

(2) En anglais : branch curve.

(3) ou *purement r-dimensionnelle.*

une sous-variété Γ_{r-1} (non mixte de dimension $r-1$) de F_r. On appellera cette variété Γ_{r-1} la « *variété critique associée à* V_r ». Elle est la somme $D \cup \Delta$ de deux variétés non mixtes de dimension $r-1$, D et Δ, où D est le lieu des points de diramation de T^{-1}, tandis que chaque point de Δ est nécessairement la projection d'un point singulier de V_r. La projection du lieu singulier de V_r est contenue dans Γ.

Nous allons appliquer à la variété ambiante F_{r+1} de V_r des transformations birationnelles φ qui sont le produit de « *dilatations monoïdales* » successives, prenant soin que le centre de chaque dilatation soit une variété sans singularités. On sait que chaque transformation φ a les propriétés suivantes :

(*a*) φ transforme F_{r+1} en une variété non singulière F'.

(*b*) φ^{-1} est régulière sur F'.

(*c*) le lieu des points de F' en lesquels φ^{-1} n'est pas birégulière est une variété non mixte W' de dimension r.

Les composantes irréductibles de la variété W' sont les variétés exceptionnelles introduites par φ; elles correspondent à des sous-variétés de F_{r+1}, de dimension $\leq r-1$.

Étant donné une transformation φ satisfaisant aux conditions (*a*)-(*c*), nous désignons par $\varphi\{V_r\}$ la variété comprenant :

1° les transformées propres des composantes irréductibles de V_r et 2° les variétés exceptionnelles qui correspondent à des sous-variétés de V_r. La variété $\varphi\{V_r\}$ est non mixte de dimension r.

Tout cela étant précisé, le problème spécial de réduction auquel nous sommes conduit demande la démonstration du :

Théorème fondamental. — *Si* V_r *satisfait aux conditions d'immersion* (1) *et* (2) *et si toutes les singularités de la variété critique* Γ_{r-1} *associée à* V_r *sont des « croisements normaux »* [4], *il existe une transformation birationnelle* φ *de* F_{r+1} *qui satisfait aux conditions* (*a*)-(*c*) *et est telle que les singularités de* $\varphi\{V_r\}$ *sont des « croisements normaux »*.

[Un point Q d'une variété W est un « croisement normal » si les conditions suivantes sont satisfaites :

---

[4] En anglais : normal crossings.

(1) Q appartient à deux, et deux seulement, composantes irréductibles W_1 et W_2 de W;

(2) Q est un point simple de W_1 et W_2;

(3) les espaces tangents de W_1 et W_2 en Q sont indépendants.

Il s'ensuit que si W a seulement des « croisements normaux » chaque composante irréductible de W est sans singularités].

Nous allons maintenant montrer comment le théorème fondamental conduit au théorème de la réduction des siugularités.

Nous supposerons à nouveau qu'une variété V_r satisfait les conditions d'immersion (1) et (2), mais cette fois nous supposerons que F_r est elle-même le produit de r droites projectives. Nous supposerons aussi que la variété critique $V_{r-1} (= \Gamma)$ de V_r, considérée comme sous-variété de $F_r = F_{r-1} \times S_1$ satisfait aux conditions d'immersion. Nous aurons alors une variété critique V_{r-2} sur F_{r-1}, associée à V_{r-1}. Nous supposerons à nouveau que V_{r-2}, considérée comme sous-variété de $F_{r-1} = F_{r-2} \times S_1$, satisfait aux conditions d'immersion, et ainsi de suite. Notre hypothèse est alors que toutes les variétés critiques successives $V_{r-1} V_{r-2}, \ldots, V_1$ sont définies et satisfont aux conditions d'immersion (1), (2). Nous l'exprimerons en disant que V_r est « *régulièrement immergée* » dans le produit direct $F_{r+1} = S_1 \times S_1 \times \ldots \times S_1$ ($r + 1$ fois). Il est clair que si V_r est régulièrement immergée dans F_{r+1}, chaque variété critique successive est régulièrement immergée dans $F_{r-i+1} (= S_1 \times S_1 \times \ldots \times S_1, r - i + 1$ fois).

Il est alors aisé de prouver le résultat suivant : étant donne une variété non mixte V'_r de dimension r, dans un espace projectif S_{r+1} il existe une transformation birationnelle ψ de S_{r+1} en le produit direct F_{r+1} de $r + 1$ droites projectives telle que la transformée propre V_r de V'_r par ψ est régulièrement immergée dans F_{r+1}.

Il s'ensuit que, dans le problème général de réduction, nous pouvons nous restreindre aux variétés qui sont régulièrement immergées dans F_{r+1}.

Nous pouvons maintenant formuler le théorème général de réduction comme suit :

Théorème de réduction. — *Étant donné une variété non*

mixte V_r *de dimension* r, *régulièrement immergée dans le produit direct* F_{r+1} *de* $r+1$ *droites projectives, il existe une transformation birationnelle* φ *de* F_{r+1} *satisfaisant aux conditions* (a)-(c) *telle que les seules singurités de* $\varphi\{V_r\}$ *soient des croisements normaux.*

La démonstration se fait par induction sur r de la manière suivante.

Soit V_{r-1} la variété critique de V_r : $V_{r-1}\subset F_r=$ le produit direct de r droites projectives. Puisque V_{r-1} est régulièrement immergée dans F_r, il existe, d'après l'hypothèse d'induction, une transformation birationnelle ψ de F_r en une variété F'_r non singulière, telle que $\psi\{V_{r-1}\}$ n'ait que des croisements normaux [la transformation ψ satisfaisant aux conditions (a)—(c)]. Nous prolongeons ψ d'une manière évidente en une transformation birationnelle de $F_{r+1}(=F_r\times S_1)$ en $F'_{r+1}(=F'_r\times S_1$ et nous notons encore ψ cette extension.

On peut établir que :

(1) ψ satisfait aux conditions analogues à (a)—(c);

(2) la transformée $\psi\{V_r\}$ coïncide avec la transformée propre V'_r de V_r par ψ;

(3) V'_r, en tant que sous-variété de $F'_r\times S_1$, satisfait aux conditions d'immersion (1) et (2);

(4) La variété critique V'_{r-1} de V'_r est une sous-variété de $\psi\{V_{r-1}\}$, et par conséquent V'_{r-1} a seulement des croisements normaux.

D'après le théorème fondamental, il existe une transformation birationnelle φ' de F'_{r+1} satisfaisant aux conditions (a)—(c) et telle que $\varphi'\{V'_r\}$ n'ait que des croisements normaux. Alors la transformation $\varphi=\psi\varphi'$ possède toutes les propriétés requises dans le « théorème de réduction ».

3. La solution du problème général de réduction est maintenant ramenée à la démonstration du « théorème fondamental ». Nous avons pu établir ce « théorème fondamental » dans l'hypothèse supplémentaire suivante :

V_r *est une variété irréductible, analytiquement irréductible en chaque point* (cela a lieu en particulier si V_r est normale).

Je pense qu'il sera possible de montrer, compte tenu du résultat précédent, que la solution du problème de la réduction sera complète, si l'on arrive à démontrer le théorème suivant :

Si V_r *est une variété irréductible, immergée dans une variété sans singularités* G_{r+1} *(par exemple dans l'espace projectif* S_{r+1}*) il existe une transformation birationnelle* φ *de* G_{r+1} *[satisfaisant aux conditions usuelles (a)-(c)] telle que la transformée* V'_r *de* V_r *par* φ *soit analytiquement irréductible en chaque point.*

En d'autres termes, le problème de la réduction des singularités est, selon toute vraisemblance, ramené au problème plus simple de la « séparation des branches » par une transformation birationnelle convenable de l'espace ambiant contenant V_r. La normalisation d'une variété V sépare bien les branches, mais elle n'est pas une transformation induite par une transformation birationnelle de l'espace ambiant contenant V_r.

Je vais maintenant esquisser la démonstration du résultat énoncé plus haut. Pour des raisons de simplicité — et aussi pour abréger — je n'esquisserai celle-ci que dans le cas où V est une variété normale.

4. Nous supposons donc que V est normale et que les conditions du « théorème fondamental » sont réalisées. Ces conditions ont un caractère global et ne sont plus satisfaites quand on applique à V les dilatations monoïdales nécessaires. Elles entraînent toutefois qu'en chaque point de V certaines conditions locales sont réalisées et il se trouve que ces conditions locales sont préservées et peuvent être contrôlées après l'application des dilatations monoïdales en question à V. Nous allons maintenant extraire des conditions globales leur contenu local.

Soit P un point quelconque de V, $\mathfrak{o}$ l'anneau local de V en P et $\{x_1, x_2, \ldots, x_r\}$ un système de paramètres de $\mathfrak{o}$ [au sens de Chevalley : l'idéal $\mathfrak{o}.(x_1, x_2, \ldots, x_r)$ est primaire pour l'idéal $\mathfrak{m}$ des non-unités de $\mathfrak{o}$]. Désignons par $\mathfrak{o}^*$ la complétion de $\mathfrak{o}$. On sait que $\mathfrak{o}^*$ est un domaine d'intégrité, que $\mathfrak{o}^*$ est intégralement dépendant sur l'anneau des séries de puissances $k\langle x_1, x_2, \ldots, x_r\rangle$

et que le corps des quotients F^* de $\mathfrak{o}^*$ est une extension algébrique finie du corps $k\{x_1, x_2, \ldots, x_r\}$ des quotients de $k\langle x_1, x_2, \ldots, x_r\rangle$. Soit D_x la variété (locale) de diramation ([5]) de l'extension algébrique $F^*/k\{x\}$; D_x est définie par un idéal principal (d) dans l'anneau $k\langle x\rangle$.

Si D_x est vide $(d=1)$, $\mathfrak{o}^*=k\langle x\rangle$ et P est un point simple de V. On voit aussi aisément que si D_x a un point simple en $(x)=(0)$, alors P est un point simple. Car dans ce cas on peut supposer que $d=x_1$ et que $F^*=k\{x\}(\sqrt[n]{x_1})$, où n est un entier convenable. Alors $\mathfrak{o}^*=k\langle\sqrt[n]{x_1}, x_2, \ldots, x_r\rangle$ et $\mathfrak{o}$ est un anneau local régulier.

Définition. — *Un point singulier* P *de* V *est « faiblement singulier » (weakly singular) si par un choix convenable des paramètres locaux* $x_1, x_2, \ldots, x_r$ *de* $\mathfrak{o}$, *la variété de diramation* D_x *possède un « croisement normal » (normal crossing) en* $(x)=(0)$.

Si P est un point singulier faible de V nous appellerons *rang de* P la plus petite valeur possible du degré relatif $[F^*:k\{x\}]$, pour tous les choix possibles de systèmes de paramètres $\{x_1, x_2, \ldots, x_r\}$ tels que D_x ait un croisement normal en $(x)=(0)$. On désignera le rang de P par $\nu(P)$. On a $\nu(P)\geqq 2$ car $\nu(P)=1$ entraîne que le point P est simple.

Les hypothèses du « théorème fondamental » impliquent que *tous les points singuliers de* V *sont faiblement singuliers*.

Nous pourrons donc supposer que nous avons affaire à une variété V normale avec seulement des points singuliers « faibles ».

Soit P un point singulier de V, soit n le rang de P et soit $\{x_1, x_2, \ldots, x_r\}$ un système de paramètres de $\mathfrak{o}$ tels que $[F^*:k\{x\}]=\nu(P)=n$.

Nous pouvons supposer que D_x est constitué par les deux branches $x_1=0$ et $x_2=0$. Il est bien connu qu'il existe deux entiers positifs q_1, q_2 tels que $F^*\subset\Delta^*=k\{x\}(\sqrt[q_1]{x_1}, \sqrt[q_2]{x_2})$, supposant q_1 et q_2 choisis les plus petits possible. Le groupe de Galois de $\Delta^*/k\{x\}$ est le produit direct $G_1\times G_2$, de deux sous-

([5]) En Anglais : the (local) branch variety.

groupes cycliques d'ordres respectifs q_1 et q_2. Désignons par τ_i la substitution :

$$\sqrt[q_i]{x_i} \to \sqrt[q_i]{x_i}\, e^{\frac{2\pi\sqrt{-1}}{q_i}} \qquad (i = 1, 2);$$

on voit alors que τ_i est un générateur de G_i. Notons H le sous-groupe de $G_1 \times G_2$ laissant les éléments de F^* invariant. Nous allons maintenant obtenir quelques propriétés de H.

1. $H \cap G_i = (1)$ *pour* $i = 1, 2$. — Supposons $H \cap G_1 \neq (1)$ par exemple. Alors $\tau_1^{p_1} \in H$, p_1 désignant un diviseur de q_1 avec $1 \leq p_1 < q_1$. Le corps fixe du groupe cyclique $\{\tau_1^{jp_1}\}$ est le corps $k\{x\}(\sqrt[p_1]{x_1}, \sqrt[q_2]{x_2})$. Puisque ce groupe cyclique est contenu dans H, on a l'inclusion $F^* \subset k\{x\}(\sqrt[p_1]{x_1}, \sqrt[q_2]{x_2})$, ce qui contredit la propriété minimale du couple (q_1, q_2).

2. H *est un groupe cyclique, isomorphe à un sous-groupe de* $G_i (i = 1, 2)$. Cela résulte directement de **1**.

3. $q_1 = q_2$. — Désignons en effet par m le plus grand commun diviseur de q_1 et q_2 en sorte que $q_1 = mp_1$, $q_2 = mp_2$. D'après **2**, H est contenu dans le groupe engendré par $\tau_1^{p_1}$ et $\tau_2^{p_2}$. On en déduit que

$$F^* \supset k\{x\}(\sqrt[p_1]{x_1}, \sqrt[p_2]{x_2}, x_3, \ldots, x_r) = k\{\sqrt[p_1]{x_1}, \sqrt[p_2]{x_2}, x_3, \ldots, x_r\}.$$

Posons $x'_i = \sqrt[p_i]{x_i}$ pour $i = 1, 2$; $x'_j = x_j$ pour $j = 3, 4, \ldots, r$.

Alors $\{x'_1, x'_2, \ldots, x'_r\}$ est un système de paramètres de $\mathfrak{o}^*$ tel que la variété de diramation $D_{x'}$ est donnée par $x'_1 x'_2 = 0$. On en déduit que $D_{x'}$ a un croisement normal au point $(x') = (0)$. Puisque $[F^* : k\{x\}] < [F^* : k\{x'\}] = \nu(P)$, on a une contradiction avec la définition de $\nu(P)$, si l'on a soit $p_1 > 1$, soit $p_2 > 1$. Donc $p_1 = p_2 = 1$ et $q_1 = q_2 = m$.

4. H *est un groupe cyclique d'ordre* m. — En effet soit $\tau = \tau_1^{\alpha_1}\tau_2^{\alpha_2}$ un générateur de H et soit $(m, \alpha_1) = p_1$. Nous trouverons que $F^* \supset k\{\sqrt[p_1]{x_1}, x_2, x_3, \ldots, x_r\}$, et le raisonnement utilisé dans la preuve de **3** montre que l'on doit avoir $p_1 = 1$; de même on a $(m, \alpha_2) = 1$.

En résumé, nous avons prouvé que G_1, G_2 et H sont des groupes cycliques de même ordre m. Puisque F^* est le corps fixe de H il s'ensuit que le *groupe de Galois de* $F^*/k\{x\}$ *est un groupe cyclique d'ordre m. Donc* $\nu(P) = m$.

Fixons dans H un générateur τ de la forme $\tau_1\tau_2^s$ et posons $u_1 = \sqrt[m]{x_1}$, $u_2 = \sqrt[m]{x_2}$. Désignons ensuite par α_i la plus petite solution positive de la congruence $\alpha_i + is \equiv (\bmod m)$ pour $i = 1, 2, \ldots, m-1$. Posons

$$z_0 = u_1^m, \qquad z_i = u_1^{\alpha_i} u_2^i \qquad (i = 1, 2, \ldots, m-1); \qquad z_m = u_2^m.$$

Alors les $m+1$ éléments z_i sont invariants dans tous les automorphismes de H et par conséquent appartiennent à F^*. Comme ce sont des fonctions intégrales de x_1, x_2, ils appartiennent à $\mathfrak{o}^*$. On voit facilement que $\{1, z_1, \ldots, z_{m-1}\}$ est une base intégrale de $\mathfrak{o}^*$ sur $k\langle x\rangle$. Ces résultats nous donnent de la manière la plus complète la structure analytique de V en P : soit V′ la variété algébrique définie par le point général $(x_1, x_2, \ldots, x_r; z_1, z_2, \ldots, z_{m-1})$. et soit O l'origine de l'espace affine de dimension $m+r-1$; dans ces conditions, le voisinage de V en P est analytiquement équivalent au voisinage de V′ en O.

Notons que nous avons introduit plus haut deux caractères numériques de P : son rang $\nu(P) = m$ et l'entier s.

Nous avons $m \geqq 2$ et $1 \leqq s < m$, $(s, m) = 1$.

Au lieu de s nous pouvons considérer l'entier s' défini par $ss' \equiv 1 \ (\bmod m)$, $1 \leqq s' < m$, car $\tau_1^{s'}\tau_2$ est aussi un générateur de H.

Nous n'insisterons pas sur cette analyse et nous nous limiterons à indiquer les résultats et les conclusions finales.

5. Soit V une variété normale avec seulement des points singuliers faibles. On prouve d'abord que le *lieu singulier* S *de* V *est non mixte de dimension* $r-2$; *que chaque composante irréductible de* S *est une variété sans singularités; mais que deux quelconques de ces composantes de* S *n'ont aucun point commun.*

Soit W une composante irréductible de S. On démontre que les *caractères numériques* $m = m(P)$ *et* $s = s(P)$, *définis précédement, sont les mêmes pour tous les points* P *de* W.

Décrivons maintenant l'effet d'une dilatation monoïdale T de centre W sur la variété V.

Soit m, s les deux caractères numériques associés à tous les points de W. Nous développons $\frac{m}{s}$ en fraction continue

$$\frac{m}{s} = h_1 + \cfrac{1}{h_2 + \cdots + \cfrac{1}{h_n}}.$$

On trouve que T transforme V en une variété V′ qui est normale et dont toutes les singularités sont faibles. La composante W du lieu singulier S de V sera remplacée par un certain nombre de composantes irréductibles du lieu singulier S′ de V′. *Ce nombre est égal au nombre des entiers* h_i *tels que* i *est pair*, $i < n$ *et* $h_i > 1$.

A chaque entier h_i *ainsi caractérisé correspond une composante irréductible* W′ *de* S′ *telle que* $\nu(P') = h_i$ et $s(P') = 1$ pour tout $P' \in W'$. Puisque $h_i < m$, nous avons une réduction du rang ν, et ainsi le « théorème fondamental » est établi, au moins pour les variétés normales.

Observons que si $s = 1$, les résultats précédents montrent que la dilatation monoïdale T résout simultanément les singularités de V, aux points de W. Il s'ensuit, dans le cas général, que la réduction des singularités de V requiert au plus deux transformations monoïdales successives (pour chaque composante W de S).

(**Extrait du *Bulletin des Sciences mathématiques*, 2e série, t. LXXVIII, janvier-février 1954.**)

ACCADEMIA NAZIONALE DEI LINCEI
Estratto dai *Rendiconti della Classe di Scienze fisiche, matematiche e naturali*
Serie VIII, vol. XXXI, fasc. 3-4 – Settembre–Ottobre (Ferie)
e dal fasc. 5 – Novembre 1961

Matematica. — *La risoluzione delle singolarità delle superficie algebriche immerse.* Nota (*) I del Socio straniero OSCAR ZARISKI (**).

1. INTRODUZIONE. – Nel mio lavoro *Reduction of singularities of algebraic three-dimensional varieties* («Annals of Mathematics», vol. 45, 1944, pp. 472–542) ho dato una dimostrazione della possibilità di risolvere le singolarità delle varietà algebriche tridimensionali, definite sopra un corpo base algebricamente chiuso e di caratteristica zero. La maggior parte di quella lunga Memoria (e anche la parte più difficile) consisteva della dimostrazione di un teorema ausiliare di riduzione delle singolarità di una superficie algebrica F qualunque, *immersa in una varietà tridimensionale* V *non-singolare* (e definita sopra un corpo base di caratteristica zero). Questo teorema ausiliare afferma che è possibile risolvere le singolarità di F mediante successive dilatazioni monoidali della varietà ambiente V. Ho ottenuto recentemente una nuova dimostrazione di questo teorema ausiliare, la quale è molto più semplice della mia previa dimostrazione, sia concettualmente sia metodologicamente. I dettagli di questa nuova dimostrazione saranno pubblicati altrove. In questa Nota e nella Nota successiva presenterò a grandi linee la dimostrazione e i concetti a cui essa s'ispira.

(*) Pervenuta all'Accademia il 29 agosto 1961.

(**) I risultati di questa Nota sono stati ottenuti dall'autore nel corso di ricerche fatte sotto gli auspici dello United States Air Force Office of Scientific Research, Air Reseacrh and Development Command.

Zariski. *

2. ENUNCIATO DEL TEOREMA. – Sia F una superficie algebrica astratta (*non necessariamente irriducibile*). Dunque F non è necessariamente proiettiva o completa. Si supponga solo che F sia definita sopra un corpo base k algebricamente chiuso e di caratteristica zero. Se P è un punto di F indicheremo con $\mathfrak{o}$ l'anello locale di P su F, con $\mathfrak{m}$ l'ideale massimo di $\mathfrak{o}$, con $\mathfrak{o}^*$ il completamento di $\mathfrak{o}$ e con $\mathfrak{m}^*$ l'ideale massimo dell'anello locale completo $\mathfrak{o}^*$.

Definizione 1. – *F è una superficie* IMMERSA *se per ogni punto* P *di* F *l'ideale massimo* $\mathfrak{m}$ *dell'anello locale corrispondente* $\mathfrak{o}$ *ha una base di tre elementi* (« *ipotesi d'immersione locale* »).

In linguaggio geometrico la nostra ipotesi d'immersione locale significa che un intorno conveniente di ogni punto P di F può essere immerso in uno spazio affine a tre dimensioni. Osserviamo che i punti *semplici* di F sono caratterizzati dalla condizione che l'ideale massimo $\mathfrak{m}$ corrispondente ammette una base di *due* elementi.

In linguaggio di geometria analitica formale, l'ipotesi d'immersione locale significa che per ogni punto P di F l'anello locale completo $\mathfrak{o}^*$ di questo punto è il quoziente $k\,[[X_1, X_2, X_3]]/(f)$ dell'anello $k\,[[X_1, X_2, X_3]]$ delle serie formali in tre variabili X_1, X_2, X_3, con coefficienti in k, rispetto a un ideale principale (f). Dunque, nell'intorno di P la superficie F è definita da una sola equazione,

$$(1) \qquad f(X_1, X_2, X_3) = 0,$$

dove f è una serie formale. Questa serie è definita a meno di una trasformazione analitica biregolare delle variabili X_1, X_2, X_3 (corrispondente a un cambiamento della base $\{x_1, x_2, x_3\}$ di $\mathfrak{m}^*$) e di un fattore unitario arbitrario in $\mathfrak{o}^*$. Se la serie f comincia con termini di grado s ($s \geqq 1$), allora P è un punto s-uplo di F (punto semplice se $s = 1$).

Definizione 2. – *Una dilatazione monoidale* T *di* F *dicesi* TRASFORMAZIONE ELEMENTARE *se il centro di* T *è* 1) *o un punto* P *di* F, *con coordinate in* k (« *trasformazione localmente quadratica, di centro* P »); 2) *o una curva* Γ (*non necessariamente irriducibile*) *su* F, *definita sopra* k, *a condizione però che questa curva abbia le proprietà seguenti*: a) Γ *è chiusa in* F; b) *le singolarità eventuali di* Γ *sono punti doppi ordinari*; c) *tutti i punti di* Γ *hanno la stessa molteplicità per* F.

È ben noto che una trasformazione localmente quadratica $T : F \to F'$, di centro P, è definita intrinsecamente sulla superficie F (a meno di una trasformazione biregolare arbitraria di F'), in termini dell'anello locale $\mathfrak{o}$ di P su F. Similmente, una dilatazione monoidale $T : F \to F'$, il cui centro è una curva irriducibile Γ di F, è definita intrinsecamente sulla F (a meno di una trasformazione biregolare arbitraria di F'), in termini dell'anello locale del punto *generale* di Γ/k. Una trasformazione *elementare* il cui centro è una curva riducibile Γ risulta il prodotto delle trasformazioni elementari i cui centri sono le componenti irriducibili di Γ/k (questo prodotto è indipendente dall'ordine dei fattori).

Una trasformazione localmente quadratica T, di centro P, è biregolare in ogni punto di F, differente da P, mentre la trasformata T (P) del punto P è una curva (che può anche essere riducibile). D'altra parte, una trasformazione elementare $T : F \rightarrow F'$, il cui centro è una curva Γ, è biregolare in ogni punto di F non appartenente a Γ, mentre la trasformata T (P) di ogni punto P di Γ è un insieme *finito* di punti di F' (e quindi T non ha punti fondamentali): questo segue dalla nostra ipotesi che tutti i punti di Γ hanno la stessa molteplicità per F. Ne segue che nel caso di una trasformazione elementare $T : F \rightarrow F'$, il cui centro è una curva Γ, si ha $F \preccurlyeq F' \preccurlyeq \bar{F}$, dove il simbolo $\preccurlyeq$ si riferisce alla relazione di *dominazione* di superficie e dove $\bar{F}$ è la normalizzazione di F.

Si vede subito che *trasformazioni elementari mandano superficie immerse in superficie immerse.*

Ciò premesso, il teorema di risoluzione delle singolarità delle superficie immerse è il seguente:

Teorema di risoluzione. – Ogni superficie immersa F *può essere trasformata in una superficie priva di singolarità per mezzo di trasformazioni elementari successive.*

3. PUNTI SINGOLARI ISOLATI. – Sia P un punto *singolare s*–uplo della nostra superficie immersa F e sia $\mathfrak{m}^*$ l'ideale massimo dell'anello locale completo $\mathfrak{o}^*$ del punto P. Se $\{x_1, x_2, x_3\}$ è una base di $\mathfrak{m}^*$, diremo che gli elementi x_1, x_2, x_3 sono *coordinate locali* nel punto P.

Definizione 3. – Due elementi x_1, x_2 di $\mathfrak{m}^*$ *si diranno* PARAMETRI LOCALI *nel punto* P *se* (1) *essi fanno parte di un sistema di coordinate locali nel punto* P *e* (2) *se l'ideale* $\mathfrak{o}^*(x_1, x_2)$ *è primario, avente* $\mathfrak{m}^*$ *come ideale primo corrispondente.*

Consideriamo l'equazione locale (1) della F, nel punto P, relativa a un dato sistema di coordinate locali x_1, x_2, x_3. Gli elementi x_1, x_2 sono parametri locali nel punto P quando e solo quando la serie $f(O, O, X_3)$ non è identicamente nulla.

Se questa condizione è soddisfatta, allora, per il teorema di preparazione di Weierstrass, si potrà scrivere la f (a meno di un fattore unitario in $\mathfrak{o}^*$) sotto la forma

$$f(X_1, X_2, X_3) = X_3^{\nu} + \Sigma_{i=1}^{\nu} A_i(X_1, X_2) X_3^{\nu-i}, \tag{2}$$

dove le $A_i(X_1, X_2)$ sono serie formali in X_1, X_2 che si annullano per $X_1 = X_2 = 0$. È chiaro che si ha $\nu \geqq s =$ molteplicità di P.

Definizione 4. – Diremo che due elementi x_1, x_2 *di* $\mathfrak{m}^*$ *sono parametri locali* TRASVERSALI *nel punto* P *se essi sono parametri locali nel punto* P *e se, inoltre, la direzione* $X_1 = X_2 = 0$ *non è tangente alla* F *in* P (*oppure, sotto forma equivalente, se nella* (2) *si ha* $\nu = s$).

Riguardando la f in (2) come polinomio in X_3 (con coefficienti in $k[[X_1, X_2]]$), indichiamo con $D(X_1, X_2)$ il suo discriminante (che è una serie formale in X_1 e X_2) e con Δ_{x_1,x_2} la *curva* analitica (non necessariamente irriducibile nel punto (O, O)) definita dall'equazione $D(X_1, X_2) = 0$ (si noti che siccome

$f(O, O, X_3) = X_3^\nu$ e $\nu \geqq s > 1$, si ha $D(O, O) = 0$; si noti anche che intendiamo con Δ_{x_1, x_2} una curva e *non un ciclo*, dimodoché nella definizione di Δ_{x_1, x_2} non si deve tener conto delle molteplicità dei vari fattori irriducibili di $D(X_1, X_2)$). Chiameremo Δ_{x_1, x_2} la *curva critica locale di* F, nel punto P, relativa alla coppia dei parametri locali x_1, x_2. Si vede subito che Δ_{x_1, x_2} non dipende dalla scelta della terza coordinata locale x_3.

Definizione 5. – Un punto singolare P *di* F *dicesi* ISOLATO *se per* OGNI *scelta dei parametri locali* x_1, x_2 *nel punto* P *la curva critica locale corrispondente* Δ_{x_1, x_2} *ha un punto singolare nell'origine* $X_1 = X_2 = 0$.

Si hanno le proposizioni seguenti:

Proposizione 1. – Sia P *un punto singolare di* F *e siano* x_1, x_2 *parametri locali* TRASVERSALI *nel punto* P. *Se* Δ_{x_1, x_2} *ha un punto singolare nell'origine* $X_1 = X_2 = 0$, *allora* P *è un punto singolare isolato.*

Proposizione 2. – L'insieme dei punti singolari isolati di F *è finito, e questo insieme contiene in particolare i punti seguenti*: (a) *i punti singolari di* F *che non appartengono a curve singolari di* F; (b) *i punti singolari della curva singolare di* F; (c) *i punti* P *di* F *che non appartengono a nessuna curva di molteplicità* s, *dove* s *è la molteplicità di* P (*si noti che i punti della categoria* (a) *appartengono alla categoria* (c)).

4. Riduzione delle singolarità non-isolate. – Sia L il luogo delle singolarità non-isolate di F. L'insieme L, se non è vuoto, è una curva algebrica priva di singolarità (Proposizione 2), non necessariamente chiusa in F o irriducibile. Sia P un punto di L e sia $\mathfrak{p}^*$ l'ideale della curva L nell'anello completo locale $\mathfrak{o}^*$ di P su F. Siccome P è punto semplice di L, l'ideale $\mathfrak{p}^*$ è primo, e l'anello completo $\mathfrak{o}^*/\mathfrak{p}^*$ è regolare, di dimensione 1. L'ideale massimo $\mathfrak{m}^*/\mathfrak{p}^*$ di $\mathfrak{o}^*/\mathfrak{p}^*$ è dunque principale. Sia z un elemento di $\mathfrak{m}$ tale che $\mathfrak{m}^*/\mathfrak{p}^* = \mathfrak{o}^*/\mathfrak{p}^* \cdot z^*$, dove z^* è il $\mathfrak{p}^*$–residuo di z.

Proposizione 3. – L'ideale principale $\mathfrak{o}^* z$ *è intersezione di ideali primi.* Il contenuto geometrico di questa proposizione è il seguente: se Φ è una superficie (nello spazio affine ambiente S_3 dell'intorno di P su F) che passa semplicemente per P e non è tangente ivi alla curva singolare L (l'ideale principale $\mathfrak{o}^* z$ definisce precisamente una tale superficie), allora l'intersezione locale di F con Φ (nel punto P) non ha mai componenti multipli.

Indicheremo con $F_{P,z}$ la curva analitica (non necessariamente irriducibile) di origine P, definita dall'ideale principale $\mathfrak{o}^* z$ e diremo che $F_{P,z}$ è sezione di F, *trasversale alla* L nel punto P. La curva $F_{P,z}$ è ovviamente una curva immersa (nell'intorno di P), essendo contenuta nella superficie Φ (passante *semplicemente* per P) definita dall'ideale principale $\mathfrak{o}^* z$. Ora, per curve analitiche immerse si ha il concetto di *singolarità equivalenti*. Questo concetto si può definire, impiegando una trasformazione quadratica, per via induttiva come segue:

Sia Γ una curva analitica immersa, di origine P, sia s la molteplicità di P per Γ e sia q il numero delle tangenti distinte di Γ nel punto P. Applicando a Γ una trasformazione quadratica di centro P si otterranno q curve analitiche $\Gamma'_1, \Gamma'_2, \cdots, \Gamma'_q$, di origini distinte $P'_1, P'_2, \cdots, P'_q$. Sia $\bar{\Gamma}$ un'altra curva analitica immersa, di origine $\bar{P}$, e per questa curva si definiscano in un modo analogo gli interi $\bar{s}, \bar{q}$ e le curve $\bar{\Gamma}'_1, \bar{\Gamma}'_2, \cdots, \bar{\Gamma}'_q$, le cui origini siano $\bar{P}'_1, \bar{P}'_2, \cdots, \bar{P}'_{\bar{q}}$. Allora possiamo dare la definizione seguente:

Definizione 6. – Le due curve analitiche Γ, $\bar{\Gamma}$ *hanno singolarità equivalenti nelle loro origini rispettive* P, $\bar{P}$, *se sono soddisfatte le condizioni seguenti*:

a) $s = \bar{s}$, $q = \bar{q}$.

b) *Per una conveniente enumerazione delle curve* Γ'_i, $\bar{\Gamma}'_i$, *si ha che* Γ'_i e $\bar{\Gamma}'_i$ *hanno singolarità equivalenti nelle loro origini rispettive* P'_i, $\bar{P}'_i$ $(i = 1, 2, \cdots, q)$.

Se si aggiunge poi la stipulazione ovvia che tutte le curve analitiche regolari Γ (cioè curve Γ aventi un punto semplice nella loro origine P) sono equivalenti, ne risulta che la definizione precedente è una definizione effettiva, per induzione rispetto al numero delle trasformazioni quadratiche successive che si debbono impiegare per risolvere la singolarità della curva Γ nel punto P.

Ciò premesso, si dimostra la proposizione seguente:

Proposizione 4. – La classe d'equivalenza a cui appartiene la curva analitica $F_{P,z}$ *immersa non dipende dalla scelta dell'elemento z* (*di cui alla Proposizione 3*), *e questa classe non cambia quando* P *varia su una componente irriducibile della curva* L (L = *luogo delle singolarità non-isolate di* F).

In virtù di questa proposizione, a ogni componente irriducibile *l* di L viene associata una classe d'equivalenza *c* (*l*) di singolarità di curve analitiche immerse: per *ogni* punto P di *l* si ha che *ogni* sezione di F, trasversale a *l* nel punto P, è una curva analitica appartenente alla classe *c* (*l*). Nel campo complesso si dimostra addirittura che, fissato un elemento Γ in *c* (*l*), esiste un intorno N, su Γ, dell'origine O di Γ, tale che in ogni punto P di *l* la superficie F è, localmente, nel punto P, prodotto topologico diretto di N e di un intorno E_P di P su *l*.

Applichiamo adesso alla superficie F una dilatazione monoidale $T: F \to F'$ di centro *l* e applichiamo nello stesso tempo alla curva Γ ($\Gamma \varepsilon c(l)$) una trasformazione localmente quadratica φ, di centro O. Là proposizione seguente dice in sostanza che l'effetto risolutivo della T sulle singolarità che la superficie F possiede lungo la curva irriducibile *l* è lo stesso che φ ha sulla singolarità che Γ possiede nel punto O.

Proposizione 5. – Se $\varphi(\Gamma)$ *si compone di q curve* $\Gamma'_1, \Gamma'_2, \cdots, \Gamma'_q$ *aventi origini distinte* $O'_1, O'_2, \cdots, O'_q$, *allora a ogni parte* P *di l corrispondono q punti distinti* $P'_1, P'_2, \cdots, P'_q$ *sulla superficie* F' (= T (F)).

Al variare di P *su l questi q punti* $P'_1, P'_2, \cdots, P'_q$ *descrivono una curva algebrica l'* (*non necessariamente irriducibile*) *priva di singolarità, e nessun punto di l' è un punto singolare isolato di* F'. *Le sezioni di* F', *trasversali a l' nei punti* $P'_1, P'_2, \cdots, P'_q$, *appartengono alle classi d'equivalenza determinate dalle curve* $\Gamma'_1, \Gamma'_2, \cdots, \Gamma'_q$ *rispettivamente*.

Osserviamo subito che *se la superficie* F *non ha singolarità isolate*, allora la Proposizione 5 ci dice che *la normalizzazione* $\bar{F}$ *di* F *è priva di singolarità e che la* $\bar{F}$ *si può ottenere dalla* F *applicando trasformazioni elementari, prive di punti fondamentali* (*che sono quindi dilatazioni monoidali i cui centri sono curve*).

Nel caso generale rimane però il compito di risolvere le eventuali singolarità *isolate* della superficie F, ed è questo il vero nocciolo del problema. Le singolarità isolate di F possono essere di due specie: (*a*) punti singolari non appartenenti a nessuna curva singolare (quindi isolati come insieme di punti; (*b*) punti singolari « particolari » appartenenti a curve singolari. È chiaro che la riduzione delle singolarità non-isolate non ha nessun effetto sulle singolarità isolate della specie (*a*). Quanto alle singolarità della specie (*b*), consideriamo, per esempio, la curva irriducibile l di cui alla Proposizione 5. Tutti i punti di questa curva sono singolarità non-isolate di F. Detta $\bar{l}$ la chiusura di l in F, sarà $\bar{l}$ una curva irriducibile, e l'insieme (*finito*) $\bar{l} - l$ sarà costituito di punti singolari isolati di F, di specie (*b*). La trasformazione monoidale T, di centro l, si estende (in modo unico) a una trasformazione monoidale $\bar{T}$ di centro $\bar{l}$. Mentre T è una trasformazione elementare (della superficie aperta ottenuta dalla F omettendo i punti singolari isolati di F), $\bar{T}$ non è necessariamente una trasformazione elementare della F, giacché (1) i punti di $\bar{l} - l$ possono essere punti singolari di $\bar{l}$, e anche (2) le molteplicità di questi punti per F può essere superiore alla molteplicità della curva l per F. Nel caso (1) la superficie trasformata $\bar{T}$ (F) non è più necessariamente una superficie immersa. Nel caso (2) i punti di $\bar{l} - l$ sono necessariamente punti fondamentali di $\bar{T}$. Ad ogni modo, il comportamento di $\bar{T}$ nei punti di $\bar{l} - l$ può essere assai complicato e difficilmente controllabile.

Perció *non è consigliabile* di cominciare il procedimento della riduzione delle singolarità di F con la riduzione delle singolarità non-isolate (indicata nella Proposizione 5). Anzi, giova cominciare *con l'eliminazione delle singolarità isolate di* F. Nella prossima Nota II ci proponiamo infatti di mostrare (a grandi linee) *che con trasformazioni elementari successive si può trasformare la* F *in una superficie priva di singolarità isolate*. Questo completerà la dimostrazione del nostro teorema della riduzione delle singolarità delle superficie algebriche immerse.

Matematica. — *La risoluzione delle singolarità delle superficie algebriche immerse.* Nota[(*)] II del Socio straniero OSCAR ZARISKI[(**)].

1. SINGOLARITÀ ISOLATE QUASI-ORDINARIE. – Il primo passo del nostro procedimento d'eliminazione delle singolarità isolate della superficie immersa F consiste nel trasformare F in un'altra superficie dotata di singolarità isolate di natura speciale e relativamente semplice.

Definizione 1. – *Un punto singolare isolato* P *di* F *dicesi* QUASI-ORDINARIO *se esistono parametri locali* TRASVERSALI x_1, x_2 *nel punto* P *tali che la curva critica locale corrispondente* Δ_{x_1,x_2} *ha nell'origine* $X_1 = X_2 = 0$ *un punto doppio ordinario.*

Sia P un punto singolare s-tuplo della superficie F e sia $T : F \to F'$ la trasformazione localmente quadratica di F, di centro P. Al punto P corrisponderà su F' una curva algebrica E', e si vede subito che ogni punto di questa curva è di molteplicità $\leq s$ per F'. D'altronde, v'è soltanto un numero finito di punti di E' che sono singolarità isolate di F' (Nota I, Proposizione 2).

Proposizione 1. – *Siano* x_1, x_2 *parametri locali trasversali nel punto* P. *Se un punto* P' *di* E' *è ancora s-tuplo per* F', *allora esiste una costante c tale*

(*) Pervenuta all'Accademia il 29 agosto 1961.

(**) I risultati di questa Nota sono stati ottenuti dall'autore nel corso di ricerche fatte sotto gli auspici dello United States Air Force Office of Scientific Research, Air Research and Development Command.

che gli elementi $x'_1 = x_1$, $x'_2 = \frac{x_2}{x_1} - c$ *(oppure* $x'_1 = \frac{x_1}{x_2} - c$, $x'_2 = x_2$*) sono parametri locali trasversali nel punto* P'. *Inoltre, la curva critica locale* $\Delta_{x'_1, x'_2}$ *di* F' *nel punto* P' *è costituita da una delle q trasformate quadratiche locali* $\varphi(\Delta_{x_1,x_2})$ *della curva critica locale* Δ_{x_1,x_2} *di* F *nel punto* P (q = *numero delle tangenti alla curva* Δ_{x_1,x_2} *nell'origine* $X_1 = X_2 = 0$) *e – eventualmente – dalla curva fondamentale della trasformazione quadratica* φ.

Da questa proposizione (la cui dimostrazione è elementarissima) si ottiene come conseguenza immediata il primo risultato di riduzione a cui abbiamo accennato al principio di questa Nota:

Proposizione 2. – («*Preparazione dei punti singolari isolati*»). *Con successive trasformazioni localmente quadratiche, accentrate in punti singolari isolati, è possibile trasformare la superficie* F *in una superficie* F' *tale che ogni punto singolare isolato di* F' *è un punto singolare quasi-ordinario.*

Un'altra conseguenza della Proposizione 1 è la seguente:

Proposizione 3. – *Se* P *è un punto singolare quasi ordinario di* F, *di molteplicità s, e se un punto* P' *di* E'(=T (P), T : F → F' *è la trasformazione localmente quadratica di* F, *di centro* P) *ha ancora molteplicità s (per* F' = T (F)) *ed è punto singolare isolato di* F', *allora anche* P' *è un punto singolare quasi-ordinario di* F'.

2. Alcune proprietà di punti singolari quasi-ordinari. – Sia P un punto singolare quasi-ordinario di F, di molteplicità s. Siano x_1, x_2 parametri trasversali nel punto P, tali che la curva critica locale Δ_{x_1,x_2} ha un punto doppio ordinario nell'origine $X_1 = X_2 = 0$. Possiamo allora supporre che Δ_{x_1,x_2} sia data localmente dall'equazione $x_1 x_2 = 0$. Completiamo la coppia $\{x_1, x_2\}$ in un sistema $\{x_1, x_2, z\}$ di coordinate locali nel punto P. Allora z è una funzione algebroide di x_1, x_2, a s valori $z_1, z_2, \cdots, z_s$. La funzione z può non essere diramata nell'origine $x_1 = x_2 = 0$, ma se lo è allora la curva di diramazione (locale) dev'essere ad ogni modo contenuta nella curva critica locale $x_1 x_2 = 0$. Ne consegue (per risultati ben noti) che *ogni ramo* z_α *della funzione z è una serie di potenze frazionarie* di x_1, x_2 (a esponenti che sono numeri razionali non-negativi aventi un commun denominatore). Ora il fatto che Δ_{x_1,x_2} è la curva $x_1 x_2 = 0$ significa che le differenze $z_\alpha - z_\beta$ ($\alpha, \beta = 1, 2, \cdots, s$; $\alpha \neq \beta$) non si annullano, nell'intorno di $x_1 = x_2 = 0$, in nessun punto dove le coordinate x_1 e x_2 sono tutte e due differenti da zero. L'analisi di questa proprietà delle serie frazionarie $z_\alpha - z_\beta$ porta alla proposizione seguente:

Proposizione 4. – *Sostituendo eventualmente la funzione z con* $z - g(x_1, x_2)$, *dove* $g(x_1, x_2)$ *è una certa serie di potenze* INTERE *di* x_1, x_2, *che si annulla nell'origine* $x_1 = x_2 = 0$, *e ordinando i rami* $z_1, z_2, \cdots, z_s$ *di z in un modo conveniente, si può supporre che le serie* z_α *abbiano la forma seguente:*

$$z_\alpha = x_1^{\lambda_1} x_2^{\lambda_2} G_\alpha(x_1, x_2), \qquad \alpha = 1, 2, \cdots, s, \tag{1}$$

dove λ_1, λ_2 *sono numeri razionali non-negativi, tali che* $\lambda_1 + \lambda_2 \geqq 1$, *e dove le* $G_\alpha(x_1, x_2)$ *sono serie di potenze frazionarie. Inoltre* $G_1(0, 0) \neq 0$.

Definizione 2. – Un punto singolare isolato P, *di molteplicità* s, dicesi STRETTAMENTE ISOLATO *se* P *non appartiene a nessuna curva* s*–tupla di* F. *Nel caso contrario si dirà che* P *è* DEBOLMENTE ISOLATO.

Con riferimento alla Proposizione 4 e alle notazioni ivi introdotte, si trova facilmente il risultato seguente:

Proposizione 5. – Sia P *un punto singolare quasi–ordinario, di molteplicità* s. Allora:

a) P *è strettamente isolato quando, e solo quando,* $\lambda_1 < 1$ *e* $\lambda_2 < 1$.

b) *Se* P *è debolmente isolato e se indichiamo con* C_s *la curva* s*–tupla di* F, *allora i rami irriducibili della* C_s *nel puno* P *appartengono alla coppia* (γ_1, γ_2), *dove* γ_i *è la curva analitica definita dalle equazioni* $x_i = z = 0$, *e precisamente* γ_i *è un ramo di* C_s *quando e solo quando* $\lambda_i \geqq 1$. (*Ne segue che* C_s *ha in* P *o un punto semplice o un punto doppio ordinario*).

3. RISULTATI LOCALI DI RIDUZIONE DI SINGOLARITÀ QUASI–ORDINARIE. – Indichiamo con $\lambda(P)$ la somma $\lambda_1+\lambda_2$ degli esponenti che figurano nell'equazione (1) (P è un punto singolare isolato, *quasi–ordinario*, di molteplicità s).

Consideriamo prima il caso in cui P è strettamente isolato. Sia $T : F \rightarrow F'$ la trasformazione localmente quadratica, di centro P.

Proposizione 6.

a) *Se* $\lambda(P) = 1$ *allora ogni punto di* F′ *che corrisponde a* P *è di molteplicità minore di* s.

b) *Se* $\lambda(P) > 1$ *allora vi sono al più due punti* P′ *di* F′ *corrispondenti a* P *e aventi molteplicità* s, *e per ognuno di questi punti* (*che sono ancora quasi–ordinari, per la Proposizione 3, e ovviamente ancora strettamente isolati*) *si ha* $\lambda(P') < \lambda(P)$.

Occupiamoci adesso del caso in cui P è debolmente isolato. Per la Proposizione 5, parte *b*), la dilatazione monoidale $T : F \rightarrow F'$ di centro C_s è una trasformazione elementare *nell'intorno* di P (per il momento il comportamento di T fuori di P non c'interessa). Orbene, si dimostra facilmente la proposizione seguente:

Proposizione 7. – Se un punto P′ *dell'insieme* (*finito*) T (P) *è ancora* s*–tuplo per* F′ *ed è punto singolare isolato, allora* P′ *è ancora quasi–ordinario*.

4. LO SCHEMA FINALE DELLA RISOLUZIONE DELLE SINGOLARITÀ ISOLATE DELLE SUPERFICIE IMMERSE. – Basandoci sui risultati della Nota I e sui risultati precedenti della Nota presente possiamo adesso indicare con precisione le tappe successive del nostro metodo di risoluzione delle singolarità isolate della superficie immersa F.

A) Indichiamo con $\sigma(F)$ il massimo delle molteplicità dei punti singolari *isolati* di F, e sia $s = \sigma(F) > 1$. L'intero $\sigma(F)$ non viene aumentato se si applicano a F trasformazioni elementari. Per la Proposizione 2, noi possiamo trasformare la F con successive trasformazioni localmente quadratiche, accentrate in punti singolari isolati della massima molteplicità s, in una superficie F′ tale che si verifichi uno dei due casi seguenti: (*a*) $\sigma(F') < \sigma(F)$;

(*b*) $\sigma(F') = \sigma(F)$ *e tutti i punti singolari isolati di* F′, *di molteplicità massima* $\sigma'(F)$ $(= s)$, *sono quasi-ordinari.*

Nel caso (*a*) abbiamo già una effettiva riduzione dei punti singolari isolati di F. Occupiamoci adesso del caso (*b*).

B) Nel caso (*b*), il nostro prossimo compito è quello di eliminare i punti singolari isolati di F′ che hanno la massima molteplicità *s e che sono debolmente isolati* (vedi Definizione 2). Più precisamente, faremo vedere che *con trasformazioni elementari successive aventi come centro curve s-tuple, e quindi prive di punti fondamentali, si può trasformare* F′ *in una superficie* F″ *tale che* $\sigma(F'') \leqq \sigma(F')$ *e – nell'ipotesi che* $\sigma(F'') = \sigma(F')$ $(= s)$ *– che i punti singolari isolati di* F′, *di molteplicità s, siano quasi-ordinari* E STRETTAMENTE ISOLATI. A questo scopo, consideriamo la curva *s*-tupla C'_s di F′. Siccome $s = \sigma(F')$, *ogni* punto di C'_s è *s*-tuplo. I punti di C'_s che *non* sono singolarità isolate sono punti semplici di C'_s. (Nota I, Proposizione 2, parte (*b*)). I punti di C'_s che sono singolarità isolate sono quasi ordinari e quindi sono o punti semplici o punti doppi ordinarii di C'_s (Proposizione 5, parte (*b*)). Ne segue che la dilatazione monoidale $T' : F' \rightarrow \bar{F}'$ accentrata in C'_s è una trasformazione elementare (Nota I, Definizione 2). Sia $\bar{P}'$ un punto singolare isolato di $\bar{F}'$, *di molteplicità s*, e sia P′ il punto corrispondente di F′. Se non è $P' \varepsilon C'_s$, allora T′ è biregolare in P′, e $\bar{P}'$ è dunque quasi-ordinario e *strettamente isolato.* Se $P' \varepsilon C'_s$, allora P′ dev'essere una singolarità *isolata* (Nota I, Proposisizione 5) e quindi quasi-ordinaria (debolmente isolata). Ne segue, per la Proposizione 7, che anche $\bar{P}'$ è un punto singolare *quasi-ordinario.* Siccome T′ è priva di punti fondamentali, ogni curva irriducibile *s*-tupla di $\bar{F}'$ può corrispondere solo a una curva irriducibile *s*-tupla di F′. Dunque, se esistono ancora su $\bar{F}'$ punti singolari isolati $\bar{P}'$ di molteplicità *s*, che sono *debolmente isolati*, allora almeno una componente irriducibile della curva $T'(C'_s)$ deve ancora essere *s*-tupla per $\bar{F}'$. Ripetendo lo stesso procedimento a partire di $\bar{F}'$ e ricordando quanto si è detto nella Nota I sull'argomento di riduzione di singolarità non-isolate (Nota I, Proposizione 5), si conclude che si arriverà finalmente a una superficie F″ priva di curve *s*-tuple, e questa superficie soddisferà a tutte le condizioni dell'enunciato.

C) Se $\sigma(F'') < \sigma(F')$ $(= \sigma(F) = s)$ allora abbiamo effettuato una riduzione. Se $\sigma(F'') = s$, allora tutti i punti singolari isolati di F″, di molteplicità *s*, sono quasi-ordinarii e *strettamente isolati*. La prossima – e ultima – tappa del nostro procedimento consiste nell'applicare a F″ trasformazioni successive localmente quadratiche, accentrate in punti singolari isolati di molteplicità *s*. Per la Proposizione 6 della presente Nota si otterrà, dopo un numero finito di tali trasformazioni, una superficie (immersa) F‴ tale che $\sigma(F''') < s$.

Come conseguenza delle tappe *A*), *B*) e *C*) si otterrà finalmente una superficie immersa $\tilde{F}$ priva di punti singolari isolati. A questa superficie possiamo applicare la conclusione finale della Nota I (vedi Nota I, l'enunciato che segue immediatamente la Proposizione 5).

Questo completa la risoluzione delle singolarità di F per mezzo di trasformazioni elementari.

ROMA, 1962 — Dott. G. Bardi, Tipografo dell'Accademia Nazionale dei Lincei

ACCADEMIA NAZIONALE DEI LINCEI
Estratto dai *Rendiconti della Classe di Scienze fisiche, matematiche e naturali*
Serie VIII, vol. XLIII, fasc. 3–4 – Settembre–Ottobre – Ferie 1967

Matematica. — *Exceptional singularities of an algebroid surface and their reduction.* Nota (*) del Socio Straniero OSCAR ZARISKI (**).

RIASSUNTO. — In lavori precedenti l'Autore ha definito il concetto di singolarità eccezionale di una superficie algebrica o algebroide F e ha dato un procedimento canonico per la risoluzione delle singolarità di F nel caso in cui F è priva di singolarità eccezionali. In questa Nota il processo dello scioglimento delle singolarità della F viene completato. L'Autore dà, cioè, un procedimento canonico per la riduzione del massimo $e(F)$ delle molteplicità dei punti singolari eccezionali della F (i quali sono sempre in numero finito).

INTRODUCTION. We deal with a (not necessarily irreducible) algebraic or algebroid surface F, defined over an algebraically closed ground field k of characteristic zero and having the property that locally, at each of its closed points, F can be embedded in the affine 3-space over k (we shall often refer to this property of F by saying that F is an " embedded surface "). If F is algebroid, we are dealing only with the local case, i.e., we assume that F is the spectrum of a complete equidimensional local ring $\mathfrak{o}$, of Krull dimension 2, having k as field of representatives and free from nilpotent elements (other than zero). In the algebroid case, therefore, F has only one closed point (represented by the maximal ideal $\mathfrak{m}$ of $\mathfrak{o}$); this point will be referred to as the center of F.

In [5] we have defined the concept of *equisingularity* on F. If W is an irreducible singular curve of F and Q is a point of W, we know what is

(*) Pervenuta all'Accademia il 10 ottobre 1967.

(**) This research was supported in part by a grant from the National Science Foundation.

O. Zariski *

meant by saying that F is *equisingular, at* Q, *along* W. We recall from [5] that equisingularity of F at Q, along W, implies that (1) F is equimultiple, at Q, along W, and that (2) Q is a simple point of the total singular curve S of F (S = union of the irreducible singular curves of F).

Definition: A simple point Q *of* S *such that* F *is equisingular, at* Q, *along the irreducible component of* S *passing through* Q, *is said to be a singular point* of dimensionality type 1. *All other singular points of* F *are called* exceptional *singular points.*

The set of exceptional singularities of F consists therefore of the following points of F: (1) the *isolated* singularities of F (i.e., the singular points which do not lie on singular curves); (2) all the singular points of the total singular curve S of F; (3) certain simple points of S. That the set of exceptional singularities of F is finite follows from Theorem 4.4, part (*b*), of [5], or also from Theorem 5.2, (the Jacobian criterion of equisingularity) of [5]. If F is algebroid then the center O of F is the only possible exceptional singularity of F (since the generic point of any irreducible component W of the singular curve S is not an exceptional singularity).

We have indicated in [3] (Note I, Proposition 5) and have proved in [5] (Theorem 7.4 and Corollary 7.5) that if F has no exceptional singularities, then the normalization $\bar{F}$ of F is non-singular and is obtainable from F by a finite sequence of monoidal transformations $T_i : F_{i+1} \rightarrow F_i$ ($i = 0, 1, \cdots, N$; $F_0 = F$; $F_{N+1} = \bar{F}$), such that each F_i is free from exceptional singularities and the center of each T_i is an irreducible singular curve of F_i.

For the convenience of the reader we shall state now in some detail the precise facts which underlie the cited theorem 7.4 and which have been brought out in our paper [5] for any embedded surface.

(*a*) Each irreducible component W of the total singular curve of S of F determines an *equivalence class* C (F, W) of singularities of embedded (i.e., plane) algebroid curves (in the sense of our paper [4]), with the property that if Q is *any* point of W which is not an exceptional singularity of F, and if G is *any* non-singular algebroid surface (in the affine 3-space in which F is locally embedded at Q) which is transversal to the curve W, at Q, then the section of F with G is an algebroid *curve* Γ (i.e., has no multiple components) and this curve Γ has at Q a singularity *which belongs to the equivalence class* C (F, W).

(*b*) Let C'_1 (F, W), C'_2 (F, W), $\cdots$, C'_m (F, W) be the set of equivalence classes (of algebroid plane curves) which represent the quadratic transform of C (F, W). Here m is the number of distinct tangent lines of any member Γ of the class C (F, W), so that the quadratic transform Γ′ of Γ splits therefore, into m algebroid curves $\Gamma'_1, \Gamma'_2, \cdots, \Gamma'_m$, having distinct centers $O'_1, O'_2, \cdots, O'_m$, and C'_i (F, W) is the equivalence class determined by Γ'_i. (We note that the m classes C'_i (F, W) need not be distinct). Let F′ be the transform of F by the monoidal transformation $T : F' \rightarrow F$, with center W, and let W′ be the *proper* transform of W on F′ (whence W′ is a curve, which may be reducible, and each irreducible component of W′

corresponds to W). Let Q be, as above, any point of W which is not an exceptional singularity of F. Then the following is true: (1) the total transform of Q on F′ consists exactly of m points $Q'_1, Q'_2, \cdots, Q'_m$, and these points lie on W′; (2) each point Q'_i is a simple point of W′, and if W'_i is the irreducible component of W′ which contains Q'_i then F′ is equisingular at Q'_i, along W'_i (the m curves W'_i need not be distinct); (3) for a suitable ordering of the indices we have $C(F', W'_i) = C'_i(F, W)$ $(i = 1, 2, \cdots, m)$; more explicity, if Γ is as in *(a)* and $\Gamma'_1, \Gamma'_2, \cdots, \Gamma'_m$ are as in *(b)*, then Γ'_i is a section of F′, transversal to W'_i at Q'_i.

(c) Let F* denote the surface obtained from F by deleting all the exceptional singularities of F. From the fact that $T^{-1}\{Q\}$ is a finite set for any point Q on F^* follows that if we get $F'^* = T^{-1}\{F^*\}$, then F'^* is dominated by the normalization of F* (and is the monoidal transform of F*, with center $W \cap F^*$). Furthermore, also F'^* *is free from exceptional singularities.* In view of the relations $C(F', W'_i) = C'_i(F, W)$ $(i = 1, 2, \cdots, m)$ noted in *(b)*, and since any plane algebroid curve can be desingularized by a finite number of successive quadratic transformations, it follows that by a finite number of successive monoidal transformations, centered at singular curves, we obtain the normalization $\bar{F}^*$ of F* and that $\bar{F}^*$ is non-singular.

In particular, if F has no exceptional singularities, then $F^* = F$, $\bar{F}^* = \bar{F}$. Thus, in the absence of exceptional singularities, the problem of reduction of singularities if F is essentially a *problem in dimension* 1: the reduction process consists of and runs parallel to the reduction of the singularities of the curve sections of F which are transversal to the total singular curve S of F. It is for this reason that we say that in this case all the singularities of F* are of dimensionality type 1. The situation is particularly illuminating in the case in which F is a complex-analytic surface. In this case it can be proved (see Whitney [2], § 11–12, and our forthcoming paper [6], § 7) that if Q and W are as in *(a)* and we regard F* as imbedded in affine $\mathbf{A}_3$, locally at Q, in such a way that W is a line in $\mathbf{A}_3$, then the natural vector bundle structure of $\mathbf{A}_3$, over W as base space, induces on F*, in the neighborhood of Q, the structure of fibre bundle over W, the fibre being any curve in the equivalence class C (F,W).

In the general case, F may have exceptional singularities. The reduction of singularities of F can thus be made to depend on the elimination of the exceptional singularities of the surface. The object of this paper is to exhibit an essentially canonical procedure for the elimination of the exceptional singularities of F.

§ 1. REDUCTION TO "QUASI-ORDINARY" MULTIPLE POINTS.

By a *permissible* transformation of F we shall mean a birational regular map $T: F' \to F$ of one of the following types:

(1) A locally quadratic transformation whose center is an *exceptional* singular point of F.

O. Zariski **

(2) A monoidal transformation whose center is an irreducible singular curve Γ of F, provided Γ has the following two properties: (2 *a*) if Γ is *m*–fold for F, *every* point of Γ is *m*–fold for F; (2 *b*) the only singularities of Γ are ordinary double points.

It is quite harmless to allow Γ to have ordinary double points, because it is easy to see that if O is an ordinary double point of Γ and if O is also *m*–fold for F (in accordance with condition (2 *a*)), then the monoidal transformation $T : F' \to F$, with center Γ, is *locally, at* O, the product $T_1 T_2$ of two permissible monoidal transformations

$$T_1 : F_1 \to F \quad , \quad T_2 : F' \to F_1 ,$$

centered at *regular* algebroid curves. Namely, if Γ_1 and Γ_2 are the two branches of Γ at O, then T_1 is centered at Γ_1, while T_2 is centered at the T_1^{-1}–transform $\Gamma_{2,1}$ of Γ_2 (1).

We note the following consequence: for any point P of Γ (including the point P = O) the set $T^{-1}\{P\}$ is finite, and hence F′ is dominated by the normalization of F (see [5], Proposition 7.2).

We denote by $e(F)$ the maximum of the multiplicities of the exceptional singular points of F, and we set $s = e(F)$. The object of the rest of this paper will be to show that by a finite number of permissible transformations it is possible to transform F into a surface $\bar{F}$ such that $e(\bar{F}) < e(F)$. This will achieve the object stated at the end of the introduction. The actual permissible transformations which we shall have to use in order to reduce the numerical character $e(F)$ will always have *s*–fold center ($s = e(F)$). We note that if $T : F' \to F$ is a permissible transformation and if F^* is the surface obtained from F after deleting the exceptional singularities of F, then the inverse image of F^* on F′ is dominated by the normalization of F^* (see Introduction). Therefore, the non–singular transform F_0 of F which our reduction process will ultimately lead us to will be such that the inverse image of F^* on F_0 is the normalization of F^*.

Definition 1.1. *A singular point* P *of* F is called quasi-ordinary *if there exist local transversal parameters* x_1, x_2 *of* P *on* F *such that the critical algebroid curve* $\Delta_{x_1 x_2}$ *associated with these parameters has an ordinary double point* (2).

(1) *Proof.* With F embedded in $\mathbf{A}_3$, locally at O, we may assume that Γ is defined by the equations $xy = z = 0$. Since Γ is *m*–fold, the local equation of F is of the form $\sum_{i=0}^{m} A_i(x, y, z)\, x^i y^i z^{m-i} = 0$, where the A_i are power series in x, y, z. Since O is also *m*–fold, we must have $A_0(0, 0, 0) \neq 0$. By the Weierstrass preparation theorem we may then assume that $A_0(x, y, z)$ is 1. If $\mathfrak{o}$ denotes the local ring of O on F, then it follows that z/xy is integral over $\mathfrak{o}$, and hence, locally at O, the T^{-1}–transform of F is Spec $\mathfrak{o}[z/xy]$. We set $F_1 = \operatorname{Spec} \mathfrak{o}[z/x]$, $z_1 = \dfrac{z}{x}$, $\mathfrak{o}_1$ = the local ring of the point $O_1 : x = y = z_1 = 0$ of F_1. Then Γ_1 is the branch $x = z = 0$, $\Gamma_{2,1}$ is the regular arc $y = z_1 = 0$, and $F' = \operatorname{Spec} \mathfrak{o}_1[z_1/y]$. Note that both $\Gamma_{2,1}$ and O_1 are *m*–fold for F_1.

(2) For the definition of local transversal parameters and of Δ_{x_1, x_2}, see [5], definition 2.3 and the end of section 2.

Note that a quasi-ordinary point is necessarily an exceptional singularity, in view of [5], Theorem 5.2.

The first step of our reduction process will result from the following:

Proposition 1.2. By successive locally quadratic transformations, centered at exceptional s-fold points, it is possible to obtain a transform F_1 *of* F *such that either* $e(F_1) < s$, *or* $e(F_1) = s$ *and all exceptional s-fold points of* F_1 *are quasi-ordinary.*

Proof: Let P be an exceptional singular point of F, of multiplicity s. Let $T : F' \to F$ be the locally quadratic transformation of F, with center P. Let P′ be any point of F′ which corresponds to P. We fix a system of local coordinates x_1, x_2, z at P such that x_1, x_2 are transversal parameters at P. The local equation of F at P is then of the form

(1) $$f(x_1, x_2, z) = z^s + A_1(x_1, x_2) z^{s-1} + \cdots + A_s(x_1, x_2) = 0,$$

where $A_i(x_1, x_2)$, is a power series whose initial form is of degree $\geq i$. Let $D(x_1, x_2)$ be the discriminant of f with respect to z. From (1) it follows at once that $\frac{x_1}{z}, \frac{x_2}{z}$ cannot be simultaneously zero at P′. Hence either $\frac{x_2}{x_1}, \frac{z}{x_1}$ or $\frac{x_1}{x_2}, \frac{z}{x_2}$ belong to the local ring $\mathfrak{o}'$ of P′ on F′. Let, say, $\frac{x_2}{x_1}, \frac{z}{x_1}$ belong to $\mathfrak{o}'$, and let, say, a and b be the $\mathfrak{m}'$-residues of $\frac{x_2}{x_1}, \frac{z}{x_1}$, where $\mathfrak{m}'$ is the maximal ideal of $\mathfrak{o}'$. Upon replacing x_2 and z by $x_2 - ax_1$ and $z - bx_1$ respectively, we may assume that $a = b = 0$. If we set then

$$x_1' = x_1 \quad , \quad x_2' = \frac{x_2}{x_1} \quad , \quad z' = \frac{z}{x_1},$$

then x_1', x_2', z' will be local coordinates at P′, and we will have $A_i(x_1', x_1' x_2') = x_1'^i A_i'(x_1', x_2')$, where $A_i'(x_1', x_2')$ is a power series in x_1', x_2'. We set

(2) $$f'(X_1', X_2', Z') = Z'^s + A_1'(X_1', X_2') Z'^{s-1} + \cdots + A_s'(X_1', X_2'),$$

whence

$$f(X_1', X_1' X_2', X_1' Z') = X'^s f'(X_1', X_2', Z').$$

Then

(3) $$f'(x_1', x_2', z') = 0$$

is a local equation of F′, at P′ (and necessarily, $A_s'(0, 0) = 0$, since $f'(0, 0, 0) = 0$).

Suppose now that the multiplicity of P′ for F′ is still s (by (2), it cannot be greater than s). Then two things must happen simultaneously: (1) $Z' = 0$ must be an s-fold root of the polynomial $f'(0, 0, Z')$, i.e., we must have $A_i'(0, 0) = 0$, for $i = 1, 2, \cdots, s$; (2) x_1', x_2' must be transversal local parameters at P′. The discriminant $D'(x_1', x_2')$ of F′, with respec to z', is related to the discriminant $D(x_1, x_2)$ of f, with respect to z, by the following equation:

(4) $$D(x_1', x_1' x_2') = x_1'^{s(s-1)} D'(x_1', x_2').$$

If then $\Delta'_{x'_1,x'_2}$ is the local critical curve of F' at P', associated with the parameters x'_1, x'_2, then it follows from (4) that $\Delta'_{x'_1,x'_2}$ is contained in the total transform of Δ_{x_1,x_2} under the quadratic transformation centered at the center $x_1 = x_2 = 0$ of Δ_{x_1,x_2}.

Since the number of exceptional s-fold points of F' which correspond to P is finite, and since it is known that after a finite number of locally quadratic transformations we get a total transform of Δ_{x_1,x_2} having only ordinary double points, the proposition is proved.

From now on we shall assume that all the exceptional singularities of F, of highest multiplicity s, are quasi-ordinary singularities of F.

§ 2. Analysis of quasi-ordinary multiple points.

Let P be a quasi-ordinary s-fold point of F, where $s = e(F)$, and let, then, x_1 and x_2 be local transversal parameters at P such that the local critical curve Δ_{x_1,x_2} has an ordinary double point at the origin $x_1 = x_2 = 0$. By a biholomorphic transformation of the local parameters x_1, x_2 we can arrange that Δ_{x_1,x_2} consists of the lines $x_1 = 0$ and $x_2 = 0$. It is then well known [3] (the ground field k being algebraically closed and of characteristic zero) that each of the s roots z_α of the defining equation (1) of F is a fractional power series in x_1, x_2:

$$(5) \qquad z_\alpha = \varphi_\alpha(x_1, x_2) \quad , \quad \varphi_\alpha(0, 0) = 0,$$

where by a fractional power series in x_1, x_2 we mean a power series in x_1, x_2 with rational, non-negative exponents, having bounded denominators.

By a fractional *monomial* we shall mean a monomial $x_1^{\alpha_1} x_2^{\alpha_2}$ where α_1 and α_2 are rational, non-negative numbers. Given two fractional monomials M_1, M_2, with $M_i = x_1^{\alpha_{i1}} x_2^{\alpha_{i2}}$ $(i = 1, 2)$, we say that M_1 divides M_2 if $\alpha_{1j} \leqq \alpha_{2j}$ for $j = 1, 2$, in other words: if the quotient $M_2 | M_1$ is a fractional monomial. Since the discriminant $D(x_1, x_2)$ of $f(x_1, x_2; Z)$ is, by assumption, of the form $x_1^{n_1} x_2^{n_2} \varepsilon(x_1, x_2)$, where n_1 and n_2 are positive integers and $\varepsilon(x_1, x_2)$ is a unit in the power series ring $k[[x_1, x_2]]$, it follows that we have

$$(6) \qquad z_\alpha - z_\beta = M_{\alpha\beta}\, \varepsilon_{\alpha\beta}(x_1, x_2), \qquad \alpha \neq \beta \; ; \; \alpha, \beta = 1, 2, \cdots, s,$$

where $M_{\alpha\beta}$ is a fractional monomial in x_1, x_2 and $\varepsilon_{\alpha\beta}(x_1, x_2)$ is a fractional power series such that $\varepsilon_{\alpha\beta}(0, 0) \neq 0$. Let M be the common divisor of the $s(s-1)/2$ monomials $M_{\alpha\beta}$:

$$M = x_1^{\lambda_1} x_2^{\lambda_2},$$

where λ_1 and λ_2 are therefore non-negative rational numbers.

(3) See, for instance, Abhyankar [1], Theorem 3.

Proposition 2.1. There exists an integral power series $g(x_1, x_2)$ *such that for each* $\alpha = 1, 2, \cdots, s$ *we have*

$$z_\alpha = g(x_1, x_2) + x_1^{\lambda_1} x_2^{\lambda_2} G_\alpha(x_1, x_2), \tag{7}$$

where $G_\alpha(x_1, x_2)$ *is a fractional power series and where for each index* α *there exists an index* β *such that* $G_\alpha(0, 0) \neq G_\beta(0, 0)$. *Furthermore, we have*

$$\lambda_1 + \lambda_2 \geqq 1. \tag{8}$$

Proof: From the identity

$$M_{\alpha\beta}\, \varepsilon_{\alpha\beta} + M_{\beta\gamma}\, \varepsilon_{\beta\gamma} + M_{\gamma\alpha}\, \varepsilon_{\gamma\alpha} = 0 \tag{9}$$

which holds for any three indices α, β, γ, follows that either $M_{\alpha\beta}$ divides $M_{\beta\gamma}$ or $M_{\beta\gamma}$ divides $M_{\alpha\beta}$. As a consequence, for any fixed α it is true that the $s-1$ monomial $M_{\alpha\beta}$ are completely ordered by the divisibility relation. If, then, we set, M_α = highest common divisor of the $M_{\alpha\beta}$ ($\beta = 1, 2, \cdots, s$; $\beta \neq \alpha$), then M_α is one of the monomials $M_{\alpha\beta}$, and we have, furthermore, that $x_1^{\lambda_1} x_2^{\lambda_2}$ is the highest common divisor of $M_1, M_2, \cdots, M_s$.

We assert that $M_1 = M_2 = \cdots = M_s$. For, let, say, $M_1 = M_{12}$, and let α, β be two distinct indices, different from 1; assume also that, say, $\alpha \neq 2$. We have that M_{12} divides $M_{1\alpha}$. From the identity (9) follows that if, say, $M_{\alpha\beta}$ divides $M_{\beta\gamma}$, then $M_{\alpha\beta}$ also divides $M_{\gamma\alpha}$. *Hence* M_1 $(= M_{12})$ *divides* $M_{2\alpha}$, for all $\alpha \neq 2$ (including $\alpha = 1$, since $M_{12} = M_{21}$). Hence M_1 divides M_2. Similarly, M_2 must divide M_1. Hence $M_1 = M_2$, and similarly $M_1 = M_\alpha$ for $\alpha = 2, 3, \cdots, s$.

We go back to the power series $\varphi_\alpha(x_1, x_2)$ in (5) and we denote by $g(x_1, x_2)$ the sum of those terms of $\varphi_1(x_1, x_2)$ which are not divisible by $x_1^{\lambda_1} x_2^{\lambda_2}$. Thus, we can write

$$z_1 = g(x_1, x_2) + x_1^{\lambda_1} x_2^{\lambda_2} G_1(x_1, x_2),$$

where $G_1(x_1, x_2)$ is a fractional power series. We have, for each $\alpha \neq 1$: $z_\alpha = z_1 + (z_\alpha - z_1) = g(x_1, x_2) + x_1^{\lambda_1} x_2^{\lambda_2} G_1(x_1, x_2) + M_{\alpha 1}\, \varepsilon_{\alpha,1}(x_1, x_2)$. Since $x_1^{\lambda_1} x_2^{\lambda_2}$ divides $M_{\alpha 1}$, this last expression of z_α is indeed of the form (7). For fixed α we have $z_\alpha - z_\beta = x_1^{\lambda_1} x_2^{\lambda_2}(G_\alpha(x_1, x_2) - G_\beta(x_1, x_2))$. Since we know that for some β we must have $M_{\alpha\beta} = M_\alpha = x_1^{\lambda_1} x_2^{\lambda_2}$, it follows from (6) that for that particular β (whose value depends on α) we must have $G_\alpha(x_1, x_2) - G_\beta(x_1, x_2) = \varepsilon_{\alpha\beta}(x_1, x_2)$, whence $G_\alpha(0, 0) \neq G_\beta(0, 0)$.

To prove that $g(x_1, x_2)$ is an *integral* power series, assume the contrary. Let

$$\varphi_1(x_1, x_2) = \sum_{i,j} c_{ij}\, x_1^{i/m} x_2^{j/m}, \qquad (c_{ij} \in k)$$

and let $x_1^{p/m} x_2^{q/m}$ be a monomial, *not divisible* by $x_1^{\lambda_1} x_2^{\lambda_2}$, such that $c_{pq} \neq 0$, and assume that at least one of the exponents p/m, q/m is not an integer. It ω denotes a primitive m^{th} root of unity and a, b are arbitrary integers, then the power series

$$\psi(x_1, x_2) = \sum_{i,j} c_{ij}\, \omega^{ai+bj} x_1^{i/m} x_2^{j/m}$$

represents one of the conjugates of z_1 and is therefore equal to one of the power series $\varphi_\alpha(x_1, x_2)$ $(\alpha = 1, 2, \cdots, s)$. Let d be the highest common divisor of p and q. We choose the integers a and b so as to have $ap + bq = d$. Since $d \not\equiv 0 \pmod{m}$ it follows that the coefficient $c_{pq}\,\omega^d$ of the monomial $x_1^{p/m} x_2^{q/m}$ is different from c_{pq}. Hence $\psi(x_1, x_2) = \varphi_\alpha(x_1, x_2)$, with $\alpha \neq 1$, and the monomial $x_1^{p/m} x_2^{q/m}$ actually occurs in the power series $\varphi_\alpha(x_1, x_2) - \varphi_1(x_1, x_2)$. Therefore $M_{1\alpha}$ must divide $x_1^{p/m} x_2^{q/m}$, and hence–*a fortiori*– $x_1^{\lambda_1} x_2^{\lambda_2}$ must divide $x_1^{p/m} x_2^{q/m}$ (since $x_1^{\lambda_1} x_2^{\lambda_2} = M_1 = M_2 = \cdots = M_s$). This is in contradiction with our choice of the monomial $x_1^{p/m} x_2^{q/m}$.

Finally, to prove (8), we observe that

$$f(x_1, x_2, Z) = \prod_{\alpha=1}^{s} (Z - \varphi_\alpha(x_1, x_2)) = \tag{10}$$

$$= \prod_{\alpha=1}^{s} (Z - g(x_1, x_2) - x_1^{\lambda_1} x_2^{\lambda_2} G_\alpha(x_1, x_2)).$$

Since the origin P is an s–field point of F, the terms of least degree in f must be of degree s. Now each of the s factors $Z - \varphi_\alpha(x_1, x_2)$ contains terms of degree 1 (for istance, the term Z). Hence, in each of these factors the terms of lowest degree *must be of degree* 1. Now, every term in $g(x_1, x_2)$ is of degree $\geqq 1$ (since the relation $f(0, 0, 0) = 0$ implies $g(0, 0) = 0$). Hence, for each $\alpha = 1, 2, \cdots, s$, it is true that in the fractional power series $x_1^{\lambda_1} x_2^{\lambda_2} G_\alpha(x_1, x_2)$ the terms of lowest degree must be of degree $\geqq 1$. From the fact that $G_\alpha(0, 0) \neq G_\beta(0, 0)$ for some pair of indices α, β, follows that $G_\alpha(0, 0) \neq 0$ for some index α. This implies that $\lambda_1 + \lambda_2 \geqq 1$, and completes the proof.

Proposition 2.2. *Let* P *be a quasi-ordinary s–field point of* F *and let* (1) *be the local equation of* F *at* P, *where we assume that* x_1, x_2 *are local transversal parameters and that the critical curve* Δ_{x_1,x_2} *consists of the two lines* $x_1 = 0$ *and* $x_2 = 0$. *Let* Γ_i $(i = 1, 2)$ *denote the (locally) irreducible algebroid curve through* P, *defined by the equations*

$$\Gamma_i : x_i = 0 \quad , \quad z + \frac{A_1(x_1, x_2)}{s} = 0. \tag{11}$$

Then Γ_1 *and* Γ_2 *are the only possible locally irreducible s–fold curves of* F *through* P, *and* Γ_i *is s–fold for* F *if and only if* $\lambda_i \geqq 1$. *Furthermore, if* W *is any irreducible s–fold curve of* F *passing through* P *(whence the local component of* W *at* P *is either one of the curves* Γ_1, Γ_2 *or their union) and if the monoidal transformation* $T : F' \to F$ *with center* W *has the property that the (necessarily finite) set* $T^{-1}\{P\}$ *contains a point* P′ *which is still s–fold for* F′, *then* $T^{-1}\{P\} = P'$ *and* P′, *if exceptional for* F′, *is a quasi–ordinary singularity of* F′.

Proof: Let π be the projection of F onto the (x_1, x_2)–plane defined generically by $\pi(x_1, x_2, z) = (x_1, x_2)$. Let Γ be an irreducible algebroid curve on F through P which is singular for F. Then $\pi(\Gamma)$ must be contained in Δ_{x_1,x_2}; in other words, Γ is contained in one of the planes $x_1 = 0$, $x_2 = 0$. Let, say, $x_1 = 0$ on Γ. The generic point of Γ has then coordinates $X_1 = 0$,

$X_2 = \xi$, $Z = \zeta$, where ξ is a trascendental over k and where ζ is a root of the polynomial $f(0, \xi, Z)$ in Z. If Γ is s–fold, then ζ must be an s–fold of this polynomial (which has degree s in Z). Hence, in this case, we must have $f(0, \xi, Z) = (Z - \zeta)^s$, showing that $\zeta = -\frac{A_1(0, \xi)}{s}$. In other words, Γ is necessarily the curve Γ_1 defined in (11).

Upon replacing z by $z + \frac{A_1(x_1, x_2)}{s}$ we may assume that $A_1(x_1, x_2)$ is zero and that consequently Γ_i is the line $x_i = z = 0$. This change of z affects only the power series $g(x_1, x_2)$ in (7), but not the integers λ_1, λ_2.

Assume that Γ_1, say, is indeed s–fold for F. Each term of the power series $f(x_1, x_2, Z)$ must then be of degree $\geqq s$ in x_1, Z. Using the factorization (10) of f we deduce that each of the s fractional power series $g(x_1, x_2) + x_1^{\lambda_1} x_2^{\lambda_2} G_\alpha(x_1, x_2)$ must be free from terms of degree < 1 in x_1. Since $g(x_1, x_2)$ is an integral power series and $\lambda_1 + \lambda_2 \geqq 1$, it follows that $\lambda_1 \geqq 1$. Conversely, assume that $\lambda_1 \geqq 1$. Then each factor $Z - \varphi_\alpha(x_1, x_2)$ in (10) is of degree $\geqq 1$ in $x_1, Z - g(x_1, x_2)$. Hence the curve Γ defined by $x_1 = Z - g(x_1, x_2) = 0$ is s–fold for F. This implies that $g(0, \xi) = -\frac{A_1(0, \xi)}{s}$, and thus $g(0, \xi) = 0$ (since we have assumed that $A_1(x_1, x_2)$ is zero). Hence $g(x_1, x_2)$ is divisible by x_1, and consequently the curve Γ coincides with the curve $\Gamma_1 : x_1 = z = 0$, which is thus s–fold for F.

Let now W and $T : F' \rightarrow F$ be as in the proposition. If W consists, locally at P, of both curves Γ_1 and Γ_2, T is a product of two monoidal transformations with locally irreducible s–fold curves as center. Hence, to complete the proof of the proposition, it is sufficient to consider the case in which W is locally irreducible at P, say $W = \Gamma_1 : x_1 = z = 0$. We know that in this case $g(x_1, x_2)$ is divisible by x_1 and that $\lambda_1 \geqq 1$. Hence z/x_1 is an integral function of x_1, x_2, showing that if P′ is any point of $T^{-1}\{P\}$ then z/x_1 belongs to the local ring of P′ on F′.

A set of local coordinates at P′ (i.e., a basis of the maximal ideal of the local ring of P′) is then given by x_1, x_2, z', where $z' = \frac{z}{x_1} - c$ and c is the constant term c_α of one of the fractional power series

$$g(x_1, x_2)/x_1 + x_1^{\lambda_1 - 1} x_2 G_\alpha(x_1, x_2).$$

Now, if we set $f(x_1, x_2, x_1(z' + c)) = x_1^s f'(x_1, x_2, z')$, and if we assume that P′ is an s–fold point of F′, then 0 must be an s–fold root of the polynomial $f'(0, 0, Z')$ (which has degree s in Z′). It follows that in this case all the c_α are equal to c, showing that $T^{-1}\{P\}$ consists of the single point $P' = (0, 0, 0)$. Since P′ is s–fold for F′ and since 0 is an s–fold root of $f'(0, 0, Z')$, it follows that x_1, x_2 are local transversal parameters of P′ on F′ ($f(x_1, x_2, Z') = 0$ being a local equation of F′ at P′). If $D(x_1, x_2)$ and $D'(x_1, x_2)$ are respectively the discriminants of $f(x_1, x_2, Z)$ (with respect to Z) and of $f'(x_1, x_2, Z')$ (with respect to Z′), then we have

$$D = x_1^{\frac{s(s-1)}{2}} D'.$$

This shows that the critical curve Δ'_{x_1,x_2} of F′ (at P′) is contained in Δ_{x_1,x_2}, i.e., is contained in the union of the two lines $x_1 = 0$, $x_2 = 0$. Hence, if P′ is an exceptional singularity, Δ'_{x_1,x_2} is the set of both lines $x_1 = 0$, $x_2 = 0$, and thus P′ is a quasi-ordinary multiple point of F′. This completes the proof of the proposition.

COROLLARY 2.3. *If all the exceptional singularities of* F, *of highest multiplicity* s $(= e(\mathrm{F}))$ *are quasi-ordinary, then any monoidal transformation of* F *whose center is an* s*-fold curve* W *of* F *is permissible.*

For, in the first place, no point of W can have multiplicity greater than s for F, for in the contrary case any such point would be an exceptional singularity of F, contrary to the assumption that $s = e(\mathrm{F})$. In the second place, if P is a singular point of W, then P is necessarily an exceptional singularity of F, and since P has highest multiplicity s it is quasi ordinary, and thus, by the preceding proposition, P can only be an ordinary double point of W.

Definition 2.4. *A singular point* P *of* F, *of multiplicity* s, *is said to be strictly exceptional if* P *does not lie on any* s*-fold curve of* F.

COROLLARY 2.5. *A quasi-ordinary multiple point* P *of* F *is strictly exceptional if and only if* $\lambda_1 < 1$ *and* $\lambda_2 < 1$.

This is a direct consequence of Proposition 2.2.

Proposition 2.6. *Let* P *be a strictly exceptional, quasi-ordinary multiple point of* F, *of multiplicity* s $(= e(\mathrm{F}))$, *let* $\lambda(\mathrm{P}) = \lambda_1 + \lambda_2$ *and let* $\mathrm{T} : \mathrm{F}' \to \mathrm{F}$ *be the locally quadratic transformation of* F *with center* P. *If* $\lambda(\mathrm{P}) > 1$ *then* $\mathrm{T}^{-1}\{\mathrm{P}\}$ *contains at most two points which have multiplicity* s *for* F′; *if* P′ *is one of these points then* P′ *is quasi-ordinary, strictly exceptional and* $\lambda(\mathrm{P}') < \lambda(\mathrm{P})$.

Proof: Let P′ be any point of $\mathrm{T}^{-1}\{\mathrm{P}\}$. Since x_1, x_2 are local transversal parameters at P, we may assume (see proof of Proposition 1.2) that x_2/x_1 and z/x_1 belong to the local ring of P′ on F′ and that, consequently, there exists constants a, b in k such that $x_1, x'_2 = \frac{x_2}{x_1} - a$ and $z' = \frac{z}{x_1} - b$ generate the maximal ideal of the local ring of P′. We may replace z by $z + g(x_1, x_2)$ and we may therefore assume that $g(x_1, x_2)$ is zero. Then we will have

$$f'(x_1, x'_2, \mathrm{Z}') = \prod_{\alpha=1}^{s} [\mathrm{Z}' + b - x_1^{\lambda_1+\lambda_2-1} (x'_2 + a)^{\lambda_2} \mathrm{G}_\alpha (x_1, x_1(x'_2 + a))],$$

where $f'(x_1, x'_2, \mathrm{Z}') = x_1^{s} f(x_1, x_1(x'_2 + a), x_1(\mathrm{Z}' + b))$, and $f'(x_1, x'_2, \mathrm{Z}') = 0$ is the local equation of F′ at P′.

Each of the s factors on the right hand side contains terms of degree 1 (for instance Z′), and if $\lambda_1 + \lambda_2 = 1$ and α is an index such that $\mathrm{G}_\alpha(0, 0) \neq 0$, then it is true that the associated factor contains a term of degree *less* than 1, namely the term $x^{\lambda_2} \mathrm{G}_\alpha(0, 0)$ (since, by Corollary 2.5, we have $\lambda_2 < 1$). Hence in this case, f' contains terms of degree less than s, and P′ is of multiplicity less than s.

Assume now that $\lambda_1 + \lambda_2 > 1$. For P' to be an s-fold point of F' it is necessary that $Z' = 0$ be an s-fold root of $f'(0, 0, Z')$. Now, $f'(0, 0, Z') = (Z' + b)^s$ (since $\lambda_1 + \lambda_2 > 1$). Hence we must have $b = 0$. *We assert that also* $a = 0$. For, were $a \neq 0$, then for the index α such that $G_\alpha(0, 0) \neq 0$ the associated factor $Z' - x_1^{\lambda_1+\lambda_2-1}(x_2' + a)^{\lambda_2} G_\alpha(x_1, x_1(x_2' + a))$ would contain the term $x_1^{\lambda_1+\lambda_2-1} a^{\lambda_2} G_\alpha(0, 0)$ of degree less than 1 (since $\lambda_1 < 1$ and $\lambda_2 < 1$), and P' could not be s-fold. Thus both a and b are zero. What we have shown is that there are only two points of $T^{-1}\{P\}$ which could possibly be s-fold points of F', namely the points in which either x_2/x_1 and z/x_1 or x_1/x_2 *and* z/x_2 have zero residues.

If P' is such a point which is s-fold for F', and if say, $x_2' = \frac{x_2}{x_1}$ and $z' = \frac{z}{x_1}$ have zero residues at P', then we find (as in the proof of Proposition 1.2) that the local critical curve $\Delta'_{x_1, x_2'}$ of F' at P' is contained in the union of the two lines $x_1 = 0$, $x_2' = 0$. Since P' is s-fold *but does not lie on any s-fold curve* (as P does not lie on any s-fold curve and as no irreducible component of $T^{-1}\{P\}$ is s-fold for F'), it follows that P' is a strictly exceptional, quasi-ordinary multiple point of F'. Furthermore, if we set $z'_\alpha = \frac{z_\alpha}{x_1}$ then we find that

$$z'_\alpha - z'_\beta = x_1^{\lambda_1+\lambda_2-1} x_2'^{\lambda_2} [G_\alpha(x_1, x_1 x_2') - G_\beta(x_1, x_1 x_2')] = M'_{\alpha\beta}\, \varepsilon'_{\alpha\beta}(x_1, x_2'),$$

with $\varepsilon'_{\alpha\beta}(0, 0) \neq 0$. Since for some α, β we have $G_\alpha(0) - G_\beta(0) \neq 0$, it follows that the highest common divisor of the monomial $M'_{\alpha\beta}$ is $x_1^{\lambda_1+\lambda_2-1} x_2'^{\lambda_2}$. Thus $\lambda_1' = \lambda_1 + \lambda_2 - 1$, $\lambda_2' = \lambda_2$ and $\lambda(P') = \lambda_1' + \lambda_2' = \lambda_1 + \lambda_2 + (\lambda_2 - 1) = \lambda(P) + (\lambda_2 - 1) < \lambda(P)$, since $\lambda_2 < 1$.

§ 3. ELIMINATION OF THE QUASI-ORDINARY s-FOLD POINTS ($s = e(F)$).

The proof can now be rapidly concluded. We start with a surface on which all the exceptional singularities of highest multiplicity s ($= e(F)$) are already quasi-ordinary. Our first step is to apply to F a monoidal transformation $T : F' \to F$ centered at an irreducible s-fold curve Γ, *provided* Γ carries exceptional s-fold points. By Corollary 2.3, such a transformation is permissible. *If* $e(F')$ is still equal to s, all the exceptional s-fold points of F' are still quasi-ordinary (Proposition 2.2). If F' still carries an irreducible s-fold curve Γ' which contains exceptional s-fold points, we again apply a monoidal transformation $F'' \to F'$ centered at Γ'. Since the monoidal transformations used in this step do not blow up any points (F', F'', etc. are dominated by the normalization of F), the case in which $e(F) = e(F') = \cdots = e(F^{(i)}) = \cdots$ *and* each $F^{(i)}$ contains an exceptional s-fold point which is not strictly exceptional cannot arise indefinitely. Hence, after a finite number of steps we will ultimately get a surface F_1 for which *either* $e(F_1) < s$ *or* $e(F_1) = s$ and all exceptional s-fold points of F_1 are strictly exceptional (this latter case may present itself even *before* the s–fold curves of F have been eliminated).

Assume now that already on F we have the situation in which all the exceptional s–fold points are quasi-ordinary and strictly exceptional. At this stage we begin to apply locally quadratic transformations centered at exceptional s–fold points. Proposition 2.6 shows that after a finite number of such transformations we will get a surface F_2 such that $e(F_2) < s$ (since $\lambda(P) \geqq 1$ at every quasi-ordinary s–fold point).

REFERENCES.

[1] S. ABHYANKAR, *On the ramification of algebraic functions*, « Amer. Jour. of Math. », vol. LXXVII (1955), pp. 575–592.

[2] H. WHITNEY, *Local properties of analytic varieties. Differential and Combinatorial Topology*, « A symposium in honor of Marston Morse (Princeton, 1965) », pp. 205–244.

[3] O. ZARISKI, *La risoluzione delle singolarità delle superficie algebriche immerse*, Nota I e II. « Rendiconti Accademia Nazionale dei Lincei », serie VIII, vol. XXXI, fasc. 3–4 (Settembre–Ottobre 1961) e fasc. 5 (Novembre 1961).

[4] O. ZARISKI, *Studies in equisingularity* I. *Equivalent singularities of plane algebroid curves*, « Amer. Jour. of Math. », vol. LXXXVII (1965), pp. 507–536.

[5] O. ZARISKI, *Studies in equisingularity II. Equisingularity in codimension 1 (and characteristic zero)*, « Amer. Jour. of Math. », vol. LXXXVII (1965), pp. 972–1006.

[6] O. ZARISKI, *Studies in equisingularity III. Saturation of local rings and equisingularity*, forthcoming in the « Amer. Jour. of Math. ».

ROMA, 1968 — Dott. G. Bardi, Tipografo dell'Accademia Nazionale dei Lincei

5.31.72 — 33/3
0698369